PHYSICAL PROPERTIES OF INORGANIC COMPOUNDS

SI Units

A. L. HORVATH

EDWARD ARNOLD

First published 1975
by Edward Arnold (Publishers) Limited
25 Hill Street, London W1X 8LL

ISBN: 0 7131 2435 0

Printed in Great Britain by
Fletcher & Son Ltd, Norwich

CONTENTS

INTRODUCTION

A. General Object

Although there are many books available on physical properties, there has been a shortage of material dealing with inorganic substances in working nomographs for accurate and rapid usage. Therefore, the purpose of this book is to fill this great void by presenting the selected physical properties of the inorganic compounds which are most frequently used by research workers, engineers, and designers in the chemical and gas industries. In addition to this, there is a great demand for reliable data in laboratories by scientists, technologists, and workers in cryogenics, and by postgraduate and undergraduate students at Universities and Polytechnics who need to solve problems in various fields of science.

The interest lies mainly in critically surveying and collecting the available information on physical properties, selecting the most reliable values, and generalizing the data by presenting it in a practical form. All efforts have been made to prepare graphical presentations in such a manner that the diagrams can be used directly for obtaining qualitative information, covering a wide temperature range.

The main emphasis has been directed upon the liquid and gaseous states. Unfortunately, the liquid phase has been less studied, both theoretically and experimentally than the gaseous and solid states, so that in many cases the theoretical predictions do not agree satisfactorily with experimental data. In this compilation the processing was based on the experimental data, presenting properties in graphical form, e.g. the change with temperature of such properties as vapour pressure, density, latent heat of vaporization, heat capacity, viscosity, thermal conductivity, surface tension, solubility in water, etc. The properties of solids or crystals have been given only in special cases when they are substances required most frequently by designers or scientists in the laboratory or chemical industry, and the handling is sometimes carried out in the solid state, e.g. carbon dioxide.

Reliable data on physical properties are scattered in the tens of thousands of technical journals published through the world in various languages. However, the useful values appear in not more than a hundred technical journals. Unfortunately, there are many documents which are difficult to obtain, or to identify. Designers in large companies have the advantage since they have access to a comprehensive library. However, they find it a very time-consuming business to locate the required data, while engineers in small companies lack the facilities of a large technical library which is often not conveniently located in many manufacturing centres, and therefore they are forced to estimate the needed values, often with very large errors. In this book all the essential physical properties of all relevant pure inorganic substances which are needed by design engineers have been drawn together. From the graphs covering a wide temperature range it is easy to obtain the data at the desired temperature

Undoubtedly, errors will exist in the compilation even after a careful check. It is hoped that the readers will bring them to the attention of the author. Furthermore, the author will greatly appreciate all comments and suggestions from readers concerning the contents of this book.

B. Sources and References

The books most frequently used and quoted by English-speaking engineers are: *Handbook of Chemistry and Physics* (210), *Chemical Engineers' Handbook* (144), *International Critical Tables* (209), and *Handbook of Chemistry* (110).

The *International Critical Tables of Numerical Data, Physics, Chemistry, and Technology* (209) are comprehensive multi-volume handbooks, and consist of seven volumes and Index. They have been published by McGraw-Hill Book Company, New York, for the National Academy of Sciences, Washington, between 1926 and 1933. The critically evaluated numerical data in these volumes are no more recent than 1930, so they are practically out of date. However, they contain much information which is not available elsewhere. Furthermore, the well prepared index provides for easy retrieval of the collected information.

Since the publication of the first edition of the *Chemical Engineers' Handbook* in 1934 by McGraw-Hill Book Company, New York, numerous improvements have been introduced in later editions. In the latest published fourth edition, the older tabulations were replaced by useful graphs and the list of references has been extended on several properties of substances. This handbook (144) is the most popular source of information for designers and engineers, and their private library often includes it.

The *Handbook of Chemistry and Physics* (210) is revised yearly and published by the Chemical Rubber Company, Cleveland, Ohio. No doubt, this is the most frequently used source of physical and chemical properties of substances and it is always on the shelf in the laboratories for quick access. The content is updated data and chemists often refer to it as the *Chemical Rubber Handbook* (*CRH*).

Similarly to the *CRH* the *Handbook of Chemistry* (110) is a periodically revised collection of physical and chemical properties of compounds. However, there are intervals of several years between consecutive editions. The publisher is McGraw-Hill Book Company, New York. The revised tenth edition was published in 1967. This book provides chemical and physical data used in laboratory work and manufacturing.

The three desk handbooks discussed above can provide most of the basic information on physical properties but the presented values are within very narrow temperature intervals. In addition to these handbooks, there are five comprehensive multi-volume handbooks: *Gmelins Handbuch der anorganischen Chemie* (57), *Landolt–Börnstein Zahlenwerte und Funktionen aus Physik, Chemie, Astronomie, Geophysik, und Technik* (109), *Mellor's Comprehensive Treatise on Inorganic and Theoretical Chemistry* (126), *Nouveau Traite de Chemie Minerale* (143), and *Tables de Constantes Selectionnees* (184), which were extensively studied and their recommendations weighted together with the following publications in order of merit: *Thermophysical Properties of Matter* (189), *JANAF Thermochemical Tables* (179), *NBS Circular 500* (158), and *Thermophysical Properties of Monatomic Gases and Liquids* (152).

In addition to the multi-volume and desk handbooks there are numerous review articles in various journals, periodicals, and reports, e.g. (18), (31), (67), (76), (78), (79), (86), (116), (160), (188), prepared for physical properties of various materials. Such investigations require skill and experience. Several articles have been published in Chemical Reviews, e.g. on critical properties by Kobe and Lynn (106) in 1953, Kudchadker *et al.* (107) in 1968, and Mathews (124) in 1972, and on the solubility of gases in liquids by Markham and Kobe (120) in 1941 and by Battino and Clever (9) in 1966. A new journal, *Journal of Physical and Chemical Reference Data*, provides excellent compilations of various properties of compounds, e.g. thermal conductivity of the elements (76) and surface tension of liquids (Vol. 1, No. 4, pp. 841–985, 1972). In 1963 Edwards (36) carried out a systematic analysis and interpolation of published vapour pressures at the Lawrence Radiation Laboratory, University of California. The National Engineering Laboratory, East Kilbride, Glasgow, issued several reports on various physical properties, e.g. on thermal conductivity (86). The Engineering Science Data Unit, London, provides physical data for use in engineering design, e.g. on specific heat (40) and viscosity (42, 43). The prepared data sheets (items) are monitored and guided by Technical Committees of the Institution of Chemical Engineers and of the Institution of Mechanical Engineers, London. The published items deal mainly with density, viscosity, and thermal conductivity of substances. The National Bureau of Standards, Washington, is the largest producer of review articles, reports, and bulletins on physical, thermodynamic, and transport properties of matter. Numerous citations can be found under 'References'. In 1968, the Committee on Data for Science and Technology (CODATA) of the International Council of Scientific Unions (ICSU) started to issue bulletins on key values for physical and thermodynamic data (20–26). The values in the reports are based on selections and calculations made mainly at the National Bureau of Standards, Washington, and the Institute for High Temperatures of the Academy of Science of U.S.S.R., Moscow. Naturally, the content of these bulletins are most reliable.

C. Selection and Presentation

In view of the review articles, reports, and bulletins published by large organizations and establishments, the selection of the physical properties in this compilation is based mainly on these investigations and publications. However, the Chemical Abstracts and the Citation Index have been searched for each compound covering the whole published physicochemical literature up to 1972. In this section only the general principles will be outlined, meanwhile, each substance has its own preface, describing in greater detail the method of selection and sources of references. The original sources of all work relevant to this monograph have been considered and studied in detail, but only those references are cited which are selected and the final results are based on them. High merit has been given to works carried out by the National Bureau of Standards, National Physical Laboratory, Gmelin-Institute für anorganische Chemie, Thermophysical Properties Research Centre (Data Series by Touloukian *et al.*), and Landolt-Börnstein, which present comprehensively all the important fundamental properties of the elements and inorganic compounds.

The selection of the various physical, thermodynamic, and transport properties of substances involves the consideration of several aspects regarding the reliability and accuracy of the data. During the critical examination of experimental results the most emphasis has been given to the following aspects:

(i) date of experiment,
(ii) purity of sample,
(iii) reliability of the apparatus (set-up),
(iv) applied experimental technique,
(v) competence of investigator,
(vi) consistency of data.

Obviously, there are close connections between some of the above aspects. The purity of some substances depends upon the date of preparation. For example, pure bromine was not available until the discovery that permanganate or chromic acid can remove the impurities, so that at the present bromine can be obtained about 99.98% pure. Further relations exist between the applied experimental technique, reliability of the apparatus, and the competence of the investigator. It is apparent, that a competent experimentalist will design a more reliable set-up than others, not having enough experience in the particular field of investigation. When there were discrepancies between the reported experimental results from different sources, the selection was based entirely on the above considerations, followed by correlation.

In the correlation procedures, heavy weight has been given to recent experimental work, and often the old data have been disregarded. The cited sources for a particular property mean that they have been considered during the correlation, and given various weights, depending upon the reliability of the investigations.

When no experimental data were available for a property, then the estimation or prediction values have been considered. The consideration was based on two motives: firstly, experimental measurements are not likely in the near future, e.g. for thermal conductivity of liquid I_2, HBr, HI, etc., and secondly, the estimation method used is widely applied in engineering calculations, e.g. corresponding state theory.

In the following paragraphs the general sources are discussed in order to give some guide to the principle of selection. Here the references are primary or secondary sources, but it is not necessary that the properties of the compounds have been adopted from them.

1. Nomenclature. The nomenclature of elements and inorganic compounds is simple and does not require any explanations.

2. Molecular Weight. In 1961, it was decided that the table of relative atomic weights should be based on the mass of $^{12}C = 12$ exactly. Both books: Handbook of Chemistry (110) and Handbook of Chemistry and Physics (210) present atomic and molecular weights based on the 1961 convention.

3. Critical Properties. The critical properties of substances have been critically analysed by Kobe & Lynn (106) in 1953, Kudchadker *et al.* (107) in 1968, Landolt-Börnstein (109) in 1971, and Mathews (124) in 1972. Further sources are: (5), (13), (17), (34), (35), (49), (57), (63), (94), (105), (110), (112), (123), (126), (127), (141), (143), (144), (162), (164), (172), (184), (192), (195), (209), (210), (212).

4. Triple and Transition Points. The main sources of collected and selected triple and transition points are: *JANAF Thermochemical Tables* (179), National Bureau of Standards Circular 500 (158), and Stull *et al.* (181). Further sources are: (13), (34), (35), (49), (57), (59), (63), (89), (93), (105), (109), (110), (112), (126), (127), (131), (141), (143), (144), (153), (162), (164), (172), (180), (184), (187), (192), (195), (207), (209), (210), (212).

5. Melting and Boiling Points. There is quite good agreement on the melting and boiling points of the elements and inorganic substances between various sources. In this compilation most of the data have been taken from the *Handbook of Chemistry* (110) and *Taschenbuch für Chemiker und Physiker* (112). Further sources are: (5), (13), (34), (35), (36), (49), (55), (56), (57), (59), (63), (77), (89), (93), (94), (105), (109), (123), (126), (127), (131), (139), (141), (143), (144), (152), (153), (158), (162), (163), (164), (172), (184), (187), (189), (192), (195), (196), (197), (205), (207), (209), (210), (212).

6. Heats of Transition, Fusion, Sublimation, Formation, and Entropy. Reliable sources for these values are: *JANAF Thermochemical Tables* (179), National Bureau of Standards Circular 500 (158), Glushko (56), and Zwolinski (212). However, in this compilation most data have been adopted from CODATA (21), (24), (25), (26). Further sources are: (5), (13), (17), (28), (32), (34), (35), (49), (50), (54), (55), (57), (59), (63), (69), (72), (73), (74), (81), (89), (93), (94), (99), (100), (101), (103), (109), (110), (112), (126), (127), (131), (141), (143), (144), (152), (153), (162), (163), (164), (172), (176), (180), (181), (184), (187), (195), (196), (197), (205), (207), (209), (210), (212).

7. Heat Capacity Ratios. The published literature is very diffused on the heat capacity ratios of the elements and inorganic compounds, and there is no comprehensive work on this subject. Particularly the temperature dependences of the heat capacity ratios have not been extensively investigated. The Engineering Science Data Unit, London, published some compilations (40), (41); however, most sources deal only with one substance. Further sources are: (5), (7), (13), (32), (35), (49), (55), (56), (57), (63), (73), (94), (109), (110), (112), (113), (123), (125), (126), (141), (143), (144), (164), (172), (196), (209), (210).

8. Specific Gravity and Density of Solids. The revised edition of the *Handbook of Thermophysical Properties of Solid Materials* (59) has been updated by the Thermophysical Properties Research Center and published by Macmillan in 1967 (187). Further sources are: (13), (29), (32), (35), (49), (57), (63), (81), (89), (109), (110), (112), (126), (127), (141), (143), (144), (153), (162), (164), (172), (184), (195), (209), (210).

9. Compressibilities and Thermal Expansions. Data on compressibility of gases and liquids are usually presented in pressure–volume–temperature relationships. The Cryogenic Data Centre, which is a division of the National Bureau of Standards, Washington, published compilations in journals, charts, monographs, and technical notes, particularly on cryogenic fluids. Several multi-volume handbooks present critical evaluation of these properties. Data can be obtained from the following sources: (13), (35), (49), (55), (56), (57), (59), (63), (81),

(89), (109), (110), (112), (126), (141), (143), (144), (162), (164), (187), (195), (209), (210). A recently published work (189) gives a comprehensive survey on thermal expansion values.

10. Velocity of Sound. Velocity of sound data are frequently published together with heat capacity and heat capacity ratio values. Most of the handbooks list these data: *American Institute of Physics Handbook* (63), *Handbook of Tables for Applied Engineering Science* (13), and *Handbook of Chemistry and Physics* (210). Further sources are: (10), (14), (29), (35), (49), (57), (64), (94), (109), (112), (126), (141), (143), (144), (153), (160), (162), (164), (195), (196), (209).

11. Diffusion. Thermophysical Properties Research Center *Data Series* Vol. 10 (189) provides a comprehensive collection of values. Further data can be obtained in the following publications: (13), (35), (49), (52), (57), (63), (75), (84), (90), (109), (112), (126), (143), (144), (164), (191), (209), (210).

12. Dielectric Constant and Dipole Moment. In 1969 Vaughan (201) presented a compilation which has been used together with the National Bureau of Standards publication (138) for this selection. Further sources are: (1), (13), (29), (35), (49), (57), (63), (94), (109), (110), (112), (122), (126), (133), (141), (143), (144), (152), (153), (164), (184), (195), (196), (209), (210).

13. Electrical Conductivity of Liquids. The Cryogenic Data Center has issued numerous publications on the electrical conductivity/resistivity of cryogenic fluids. A compendium by Johnson (89) covers most of the contents of these publications until 1960. Landolt-Börnstein (109) also presented a compilation which has not been updated yet. The temperature dependence of the electrical conductivity (or resistivity) of pure liquid elements (non-metallic) and inorganic substances has not been investigated extensively and there are few published works on this subject. Further sources of data are: (13), (35), (49), (57), (63), (65), (87), (88), (108), (110), (112), (126), (141), (143), (144), (162), (164), (184), (206), (209), (210).

14. Refractive Index. Most handbooks in chemistry and physics contain refractive index values, e.g. (13), (34), (35), (49), (57), (59), (63), (94), (105), (108), (109), (110), (112), (126), (127), (141), (143), (144), (152), (153), (162), (163), (164), (172), (174), (184), (192), (195), (196), (197), (206), (209), (210), (212).

15. Threshold Limit Value. The publications of the American Conference of Governmental Industrial Hygienists (3), (4), provide reliable information, and this compilation is based upon their recommendations. Further sources are: (35), (119), (174), (186).

16. Vapour Pressure. All handbooks in chemistry and physics contain vapour pressure data; however, very often there are some discrepancies between the presented values in tabulated or graphical forms. In this compilation the main sources of vapour pressure data are: Edwards (36), Honig & Hook (77). However, most values have been taken from original articles in journals, periodicals, and dissertations. Further sources are: (5), (13), (17), (29), (32), (34), (35), (49), (53), (54), (55), (56), (57), (59), (63), (72), (73), (81), (89), (94), (98), (109), (110), (112), (123), (125), (126), (127), (139), (141), (143), (144), (152), (153), (158), (162), (163), (164), (172), (176), (187), (192), (195), (196), (197), (207), (209), (210), (212).

17. Orthobaric Density. Landolt-Börnstein (109), *Handbook of Chemistry* (110), and *Taschenbuch für Chemiker und Physiker* (112) present reliable values for most of the elements and inorganic liquids and gases; however, many values have been adopted from original articles. Further sources are: (5), (13), (17), (29), (32), (34), (35), (49), (53), (54), (56), (57), (59), (63), (89), (94), (123), (125), (126), (127), (141), (143), (144), (152), (153), (162), (163), (164), (172), (187), (192), (195), (196), (197), (209), (210), (212).

18. Pressure–Volume–Temperature Relationship. No other chemical substance has such comprehensive P–V–T data as water (steam tables). The experimental measurements are difficult and expensive for wide temperature and pressure intervals and the presentation of the obtained data (tabulation or plotting) is rather laborious. Therefore, many P–V–T data have been presented in the form of equations of state, by fitting procedure on digital computers. In this compilation the data are presented in graphical form only. In addition to the collection of P–V–T data in Landolt-Börnstein (109), most of the handbooks in chemistry and physics also contain data on cryogenic fluids, such as NH_3, CO_2, SO_2, etc. Further data can be obtained from the following sources: (3), (13), (17), (32), (35), (49), (57), (63), (89), (110), (112), (123), (126), (141), (143), (144), (152), (153), (155), (164), (187), (195), (196), (197), (209), (210), (212).

19. Latent Heat of Vaporization. Thermodynamic tables on saturated vapours and liquids include latent heats of vaporization data. The handbooks in chemistry and physics mentioned above cover the values from triple to critical points for refrigerants. However, most frequently the data are given at 25°C and at the normal boiling point. National Bureau of Standards Circular 500 (158) and *Selected Values of Properties of Chemical Compounds* (212) provide reliable data; however, most values have been taken from original articles. Further sources are: (5), (13), (17), (29), (32), (34), (35), (49), (53), (54), (56), (57), (59), (63), (72), (81), (89), (94), (98), (109), (110), (112), (123), (126), (127), (141), (143), (144), (152), (153), (162), (163), (164), (172), (187), (195), (196), (197), (205), (207), (209), (210).

20. Heat Capacity. Publications dealing with thermodynamic properties usually include heat capacity data. Gases are more extensively investigated than liquids. The tabulated data for gases are presented for the ideal state at 1 atmosphere pressure (calculated from infrared or Raman spectra). *JANAF Thermochemical Tables* (179) gives the critically analysed and selected data found in the literature. A recently published work by the Thermophysical Properties Research Center (189) provides reliable heat capacity data in various states of matter. Unfortunately, there is only limited information on the heat capacity of real gases. Further sources are: (5), (13), (29), (32), (34), (35), (40), (49), (53), (54), (55), (56), (57), (59), (63), (72), (73), (81), (89), (93), (94), (98), (99), (100), (101), (102), (109), (110), (112), (113), (123), (125), (126), (127), (131), (140), (141), (143), (144), (152), (153), (158), (162), (163), (164), (172), (180), (181), (187), (188), (195), (196), (197), (205), (207), (209), (210), (212).

21. Viscosity. The Thermophysical Properties Research Center (189) also compiled viscosity data from the literature and critically analysed these values to obtain selected best values. Engineering Science Data Unit (42), (43) provides good background information for available data in the liquid state only. Landbolt-Börnstein (109) also presents reliable data collection on gas and liquid viscosities. Further sources are: (5), (13), (29), (34), (35), (49), (53), (57), (60), (61), (63), (72), (89), (94), (110), (112), (116), (123), (125), (126), (127), (140), (141), (143), (144), (152), (153), (162), (163), (164), (172), (182), (188), (192), (195), (196), (197), (209), (210).

22. Thermal Conductivity. The critically analysed work of the Thermophysical Properties Research Center (189) provides reliable data for gases, liquids, and solids in tables and graphical form. The National Bureau of Standards issued several bulletins (67), (146), (189) dealing with thermal conductivities of substances in various states. Liquid thermal conductivities have also been investigated by the National Engineering Laboratory (76), (86). Landolt-Börnstein (109) prepared a very comprehensive listing of gaseous and liquid elements and inorganic substances up to 1972. Further sources are: (5), (13), (29) (34), (35), (49), (53), (57), (59), (63), (72), (89), (94), (110), (112), (118), (123), (125), (126), (127), (128), (141), (143), (144), (152), (153), (154), (160), (162), (163), (164), (172), (182), (187), (188), (190), (192), (195), (196), (197), (198), (209), (210).

23. Surface Tension. A recently published article by J. J. Jasper, *J. Phys. Chem.* Ref. Data 1972, **1** (4), 841–1009 presented the result of a literature survey, evaluation, and compilation of surface tension data of some 2200 pure liquid compounds. Further sources are: (13), (29), (34), (35), (49), (53), (57), (63), (89), (94), (109), (110), (112), (123), (126), (127), (141), (143), (144), (152), (153), (162), (163), (164), (172), (192), (209), (210).

24. Solubility in Water. There are several compilations on solubilities including the solubilities of elements and inorganic compounds in water at various temperatures and pressures. Since the first appearance of Seidell's book in 1907, the fourth edition has been published in two volumes: Vol. I in 1958 and Vol. II in 1965 by the American Chemical Society (167), (168). Stephen and Stephen (175) collected and summarized the published solubility data for elements, inorganic and organic substances in two volumes (four parts). Landolt-Börnstein (109) also contains a comprehensive collection of solubility data up to 1962. However, a more recent publication provides information on solubility up to 1968 (91). Furthermore, there are two review articles in Chemical Reviews: an older compilation by Markham and Kobe (120) and a more recent one by Battino and Clever (9). Further sources are: (13), (34), (35), (49), (57), (63), (110), (112), (114), (123), (126), (127), (141), (143), (144), (152), (153), (164), (172), (192), (209), (210).

D. Units and Physical Constants

The rationalized system of decimal metric units, the Systeme International d'Unites, known as the SI, were introduced by the Conference Generale des Poids et Mesuros in 1960 (134). The system consists of six base-units only:

metre	m,
kilogramme	kg,
second	s,
ampere	A,
celsius	°C,
candela	cd,

from which all other required units are derived.

In this compilation all properties are presented in SI units only. To assist the readers who are not familiar with this system or desire some properties in other units, there are conversion tables in the Appendix. These tables have been compiled in such a way that the desired conversion factor can be obtained direct between various units. This type of tabulation proved to be the most practical form of conversion tables.

The fundamental constants are also presented in SI units in the Appendix.

References

1 Akhadov, Ya. Yu., *Dielectric Properties of Liquids and Solutions*, Publishing House for State Standards, Moscow, 1965.

2 Allied Chemical & Dye Corporation, *Hydrofluoric Acid, Anhydrous*, General Chemical Division, Product Information Bull., Code HC 80400, New York, 1945, 19 pp.

3 American Conference of Governmental Industrial Hygienists, *Documentation of Threshold Limit Values*, revised edition, Committee on Threshold Limit Values, Cincinnati, Ohio 45202, 1966, 203 pp.

4 American Conference of Governmental Industrial Hygienists, *Threshold Limit Values for Chemical Substances and Physical Agents in the Workroom Environment with Intended Changes for 1972*, Committee on Threshold Limit Values, Cincinnati, Ohio 45202, 1972, 94 pp.

5 American Society of Heating, Refrigerating, and Air-Conditioning Engineers, Inc., *ASHRAE Thermodynamic Properties of Refrigerants*, New York, 1969, 329 pp. (NH_3, CO_2, H_2, N_2, Air, O_2, etc.).

6 Audrieth, L. F., Ogg, B. A., *The Chemistry of Hydrazine*, Wiley, New York, 1951, 244 pp.

7 Baehr, H. D., *Der Isentropenexponent der Gase H_2, O_2, N_2, CH_4, CO_2, NH_3, und Luft für Drücke bis 300 bar*, Brennst.-Wärme-Kraft 1967, **19** (2), 65–8.

8 Baker, B. L., *Heat Capacity and Thermodynamic Properties of Saturated Deuterium Oxide*, Atomic Energy of Canada Ltd., AECL-4738, 1959.

9 Battino, R., Clever, H. L., The Solubility of Gases in Liquids, *Chem. Rev.*, 1966, 395–463.

10 Bergmann, L., *Der Ultraschall*, 5th ed., Stuttgart, 1949, 365 pp.

11 Bishop, A. A., Nelson, P. A., *Transport and Thermodynamic Properties of Saturated and Compressed Heavy Water*, Carolina's Virginia Nuclear Power Associates Inc., Report, 1961.

12 Bizjak, F., *Physical Properties of Hydrogen*, U.S. Gov. Res. Dev. Rept. AE61-0726, (AD 677550), Clearinghouse, Springfield, Va., July 26, 1961, 18 pp.

13 Bolz, R. E., Tuve, G. L., Eds., *Handbook of Tables for Applied Engineering Science*, The Chemical Rubber Co., Cleveland, Ohio, 1970.

14 Bradley, J. N., *Shock Waves in Chemistry and Physics*, Methuen & Co. Ltd., London, 1962, p. 334.

15 Brahinsky, H. S., Neel, C. A., *Tables of Equilibrium Thermodynamic Properties of Nitrogen. Vol. II. Constant Pressure*, U.S. Dept. of Commerce, Natl. Bur. Std. AEDC-TR-69-126 (AD 692712), Washington, Aug. 1969, 270 pp.

16 Burnham, C. W., Holloway, J. R., Davis, N. F., *Thermodynamic Properties of Water to 1000 °C. and 10 000 Bars*, Pennsylvania State Univ., Dept. of Geochem. & Mineralogy, College of Earth and Mineral Science, University Park, Pennsylvania, 1968.

17 Canjar, L. N., Manning, F. S., *Thermodynamic Properties and Reduced Correlations for Gases*, Gulf Publ. Co., Houston, Tex., 1967, 212 pp. (NH_3, CO_2, CO, H_2, N_2, O_2, SO_2, H_2O).

18 Chase, J. D., Searching for Physical Property Data, *Chem. Eng.*, 1966 (Sept. 12), **73**, 190–6.

19 Clark, C. C., *Hydrazine*, 1st ed., Mathieson Chemical Corp., Baltimore, Maryland, 1953, 133 pp.

20 CODATA, *International Compendium of Numerical Data Projects*, Springer-Verlag, Berlin, 1969, 295 pp.

21 CODATA, *Tentative Set of Key Values for Thermodyanamics*, Part I, Bull. 2, International Council of Scientific Unions, Committee on Data for Science and Technology, Frankfurt/Main, Oct. 1970, 5 pp.

22 CODATA, *A Catalog of Compilation and Data Evaluation Activities in Chemical Kinetics, Photochemistry and Radiation Chemistry*, Bull. 3, International Council of Scientific Unions, Committee on Data for Science and Technology, Frankfurt/Main, Dec. 1971, 28 pp.

23 CODATA, *Automated Information Handling in Data Centers*, 2nd ed., Bull. 4, International Council of Scientific Unions, Committee on Data for Science and Technology, Frankfurt/Main, Dec. 1971, 12 pp.

24 CODATA, *Final Set of Key Values for Thermodynamics*, Part I, Bull. 5, International Council of Scientific Unions, Committee on Data for Science and Technology, Frankfurt/Main, Dec. 1971, 5 pp.

25 CODATA, *Tentative Set of Key Values for Thermodynamics*, Part II, Bull. 6, International Council of Scientific Unions,

Committee on Data for Science and Technology, Frankfurt/Main, Dec. 1971, 7 pp.
26 CODATA, Report of the ICSU-CODATA Task Group on Key Values for Thermodynamics, November 1971, *J. Chem. Thermodyn.*, 1972, **4**, 331–6.
27 Colwell, J. H., *The Physical Properties of SO_3*, Univ. of Washington, Ph.D. thesis, Washington, 1961, 63 pp.
28 Coughlin, J. P., *Contributions to the Data on Theoretical Metallurgy. Heats and Free Energies of Formation of Inorganic Oxides*, Bureau of Mines, U.S. Government Printing Office, Washington, Bull. 542, 1954, 80 pp.
29 Data Book and Buyers Guide, Engineering Data Section, *Cryogenic & Ind. Gases*, 1971, **6** (3), 41–80 (O_2, N_2, H_2, F_2, etc.).
30 Dierdorff, L. H., *Hydrogen Peroxide Physical Properties Data Book*, Beccio Chemical Division, Food Machinery and Chemical Corp., Buffalo, 1955.
31 Dimiduk, P. W., *Bibliography of References for the Thermophysical Properties of He, H_2, D_2, Ne, Ar, N_2, O_2, CO, CO_2, CH_4, etc.*, U.S. Air Force Systems-Command, Res. Tech. Div., Tech. Rept. AFML 65–338, Washington, 1965, 140 pp.
32 Din, F., Ed., *Thermodynamic Functions of Gases*, Vol. 1, NH_3, CO_2, CO, (175 pp.), Vol. 2, Air, C_2H_2, C_2H_4, C_3H_8, Ar, (201 pp.), Vol. 3, CH_4, N_2, C_2H_6, (218 pp.), Butterworth, London, 1962 (reprint).
33 Dorsey, N. E., *Properties of Ordinary Water Substance*, Hafner Publ. Co., New York, 1968.
34 Dreisbach, R. R., *Physical Properties of Chemical Compounds*, Vol. III, Am. Chem. Soc., Advances in Chemistry Series 29, Washington, 1961, 489 pp. (Air, NH_3, Br_2, CO_2, CO, Cl_2, N_2H_4, H_2, HBr, HCl, HF, N_2, N_2O, O_2, SO_2, H_2O).
35 Ebert, H., *Physikalisches Taschenbuch*, 4th ed., Vieweg & Sohn, Braunschweig, 1966, 650 pp.
36 Edwards, D. G., *The Vapour Pressure of 30 Inorganic Liquids between 1 Atm and Critical Point*, Natl. Bur. Std., Washington, D.C., June 17, 1964.
37 Eisenberg, D., Kauzmann, W., *The Structure and Properties of Water*, Oxford University Press, Oxford, 1969.
38 Electrical Research Association, *1967 Steam Tables*, Edward Arnold, London, 1967.
39 Elliott, J. N., *Tables of the Thermodynamic Properties of Heavy Water*, Atomic Energy of Canada Ltd., AECL-1673, 1963.
40 Engineering Science Data Units, *Specific Heats and their Ratio as Functions of Temperature for Several Common Gases*, Aero 1.08, London, Dec. 1961, 4 pp.
41 Engineering Science Data Units, *A Simple Model of a Calorically Imperfect Diatomic Gas. Comparison with Real Gases*, Aero 1.19, London, Feb. 1962, 4 pp.
42 Engineering Science Data Units, *Approximate Data on the Viscosity of Some Common Liquids*, Item No. 66024, London, July 1966, 26 pp.
43 Engineering Science Data Units, *Introductory Memorandum on the Viscosity of Liquids and the Classification of Lubricating Oils*, Item No. 68036, London, Oct. 1968, 25 pp.
44 Esel'son, B. N., Blagoi, Yu. P., Grigor'ev, V. N., Manzhelii, V. G., Mikhailenko, S. A., Neklyudov, N. P., *Properties of Liquid and Solid Hydrogen*, Komitet Standartov, Mer i Izmeritel'nykh Priborov pri Sovete Ministrov SSSR, Moscow, 1969, 136 pp.
45 Failie, A. M., *Sulphuric Acid Manufacture*, 4th Printing, Reinhold Publ. Corp., New York, 1949, 669 pp.
46 Farkas, A., *Orthohydrogen, Parahydrogen, and Heavy Hydrogen*, Macmillan, New York, 1935.
47 Farr, T. D., *Phosphorus – Properties of the Element and Some of Its Compounds*, Tenn. Valley Authority, Chem. Eng. Rept. No. 8, 1950.
48 Fasullo, O. T., *Sulphuric Acid: Use and Handling*, McGraw-Hill, New York, 1965.
49 Forsythe, W. E., Ed., *Smithsonian Physical Tables*, 9th rev. ed., Smithsonian Institution, Washington, 1969, 827 pp.
50 Franklin, J. L., Dillard, J. G., Rosenstock, H. M., Herron, J. T., Draxl, K., Field, F. H., *Ionization Potentials, Appearance Potentials, and Heats of Formation of Gaseous Positive Ions*, Natl. Bur. Std. NSRDS-NBS-26, Washington, D.C., 1969, 285 pp.
51 Franks, F., Ed., *Water. A Comprehensive Treatise. Vol. 1. The Physics and Physical Chemistry of Water*, Plenum Press Corp., New York, 1972, 596 pp.
52 Fuller, E. N., Schettler, P. D., Giddings, J. C., A New Method for Prediction of Binary Gas-Phase Diffusion Coefficients, *Ind. Eng. Chem.*, 1966, **58** (5), 19–27.
53 Gallant, R. W., *Physical Properties of Hydrocarbons*, Vol. 2, Gulf Publ. Comp., Houston, Tex., 1970, 201 pp. (HCN, N_2H_4, H_2O, etc.).
54 Gerassimov, J. I., Krestovnikov, A. N., Shakhov, A. S., *Chemical Thermodynamics in Nonferrous Metallurgy*, Vol. 1 (1960), 231 pp., Vol. 2 (1961), 263 pp., Vol. 3 (1963), 284 pp., Vol. 4 (1966), 428 pp., Metallurgical Publishing House, Moscow.
55 Glushko, V. P., Ed., *Thermodynamic Properties of Chemical Substances*, 2 Vols. and Suppl., U.S.S.R. Academy of Sciences, Moscow, 1962–70.
56 Glushko, V. P., Ed., *Thermodynamic Constants of Substances*, 3 Parts, U.S.S.R. Academy of Sciences, All-Union Inst. of Sci. and Technology. Inform., Moscow, 1965–8.
57 *Gmelins Handbuch der Anorganischen Chemie* (Multi-volume Handbook), Publ. by Gmelin-Institut für anorganischen Chemie in the Max-Planck-Gesellschaft zur Förderund der Wissenschaften e.V., Verlag Chemie, GmbH., Weinheim/Berlin.
58 Gmitro, J. I., Vermeulen, T., *Vapour-Liquid Equilibria for Aqueous Sulphuric Acid*, Univ. of California, Lawrence Rad. Lab. UCRL-10886, Berkeley, California, June 24, 1963, 81 pp.
59 Goldsmith, A., Waterman, T. E., Hirschom, H. J., *Handbook of Thermophysical Properties of Solid Materials*, 5 Vols., Macmillan, New York, 1961.
60 Golubev, I. F., *Viscosity of Gases and Gas Mixtures*, Gosudarstvennoe Izdatel'stvo Fiziko-Matematicheskov Literatury, Moscow, 1959.
61 Golubev, I. F., Gnesdilov, N. E., *Viscosity of Gas Mixtures*, Gosudarstvennoe Izdatel'stvo Komiteta Standartov Soveta Ministrov, Moscow, 1971, 327 pp.
62 Grabau, M., Brahinsky, H. S., *Thermodynamic Properties of Nitrogen from 300 to 5000 K and from 1 to 1000 Amagats*, Natl. Bur. Std. AEDC-TR-66-69 (AD 636730), Washington, D.C., Aug. 1966.
63 Gray, D. E., Ed., *American Institute of Physics Handbook*, 2nd ed., McGraw-Hill, New York, 1963.
64 Greene, E. F., Toennies, J. P., *Chemical Reactions in Shock Waves*, Edward Arnold, London, 1964.
65 Grüttner, B., Dove, M. F. A., Clifford, A. F., *Chemistry in Anhydrous, Prototropic Inorganic Solvents. Part 1. Inorganic Chemistry in Liquid Hydrogen Cyanide and Liquid Hydrogen Fluoride*, Pergamon Press, Oxford, 1971, 315 pp.
66 Guizouarn, L., *Molecular Physical Properties of the Hydrogen Isotopes and of their Isotopic Combinations*, Commiss. Energ. At. (Fr.), Serv. Doc. Ser. Bibliogr. No. 87, Paris, 1967, 49 pp.
67 Hall, L. A., *Thermal Conductivity of Ten Cryogenic Liquids. A Bibliography*, Natl. Bur. Std. NBS-7684, Washington, D.C., April 15, 1963, 15 pp.

68 Harrison, R. H., *The Thermodynamic Properties of Ammonia*, Univ. of Texas, M.S. thesis in Chem. Eng., Austin, Texas, 1952.

69 Haywood, R. W., *Thermodynamic Tables in SI (metric) Units*, Cambridge University Press, London, 1968, 42 pp.

70 Hellwig, L. R., *Pressure–Volume–Temperature Properties of Sulphur Dioxides*, Univ. of Texas, Ph.D. dissertation in Chem. Eng., Austin, Texas, 1955.

71 Hersh, C. K., Berger, A. W., Brown, J. R. C., Physical Properties of Liquid Ozone–Oxygen Mixture: Density, Viscosity, and Surface Tension, Am. Chem. Soc., Adv. in Chemistry Ser. No. 21, *Ozone Chemistry and Technology*, Washington, D.C., 1959.

72 Hilsenrath, J., *et al.*, *Tables of Thermodynamic and Transport Properties of Air, Argon, Carbon Dioxide, Hydrogen, Nitrogen, Oxygen, Carbon Monoxide, Steam*, Pergamon Press, Oxford, 1960.

73 Hilsenrath, J., Beckett, C. W., Benedict, W. S., Fano, L., Hoge, H. J., Masi, J. F., Nuttall, R. L., Touloukian, Y. S., Woolley, H. W., *Tables of Thermal Properties of Gases*, Natl. Bur. Std. Circular 564, Washington, D. C., Nov. 1, 1955, 488 pp.

74 Hilsenrath, J., Klein, M., *Table of Thermodynamic Properties and Chemical Composition of Nitrogen in Chemical Equilibrium Including Second Virial Correlations from 1600 to 15 000 K* Natl. Bur. Std. AEDC-TR-66-65 (AD 630461), Washington, D.C., Apr. 1966.

75 Himmelblau, D. M., Diffusion of Dissolved Gases in Liquids, *Chem. Rev.*, 1964, **64**, 527–50.

76 Ho, C.Y., Powell, R. W., Liley, P. E., Thermal Conductivity of the Elements, *J. Phys. Chem. Ref. Data*, 1972, **1** (2), 279–421.

77 Honig, R. E., Hook, H. O., Vapour Pressure Data for Some Common Gases, *RCA Review*, 1960, **21**, 360–8.

78 Horvath, A. L., Reference Literature to Properties of Chemical Compounds. Inorganic Liquids and Gases, *J. Chem. Docum.*, 1970, **10** (1), 59–65.

79 Horvath, A. L., Reference Literature to Thermodynamic Diagrams, *J. Chem. Docum.*, 1971, **11** (4), 231–3.

80 Howieson, J., *Tables of Properties of Heavy Water at Saturation Temperatures*, Atomic Energy of Canada Ltd., AECL-1055, July 1966.

81 Hultgren, R. R., Orr, L., Anderson, P. D., Kelley, K. K., *Selected Values for the Thermodynamic Properties of Metals and Alloys*, Wiley, New York, 1963, 963 pp.

82 Hust, J. G., Stewart, R. B., *Thermodynamic Property Values for Gaseous and Liquid Carbon Monoxide from 70 to 300 K with Pressures to 300 Atmospheres*, Natl. Bur. Std. Techn. Note 202, Washington, D.C., Nov. 1963, 105 pp.

83 IUPAC, Atomic Weights of the Elements 1971, *Pure & Appl. Chem.*, 1972, **30** (3–4), 638–49.

84 Ibrahim, S. H., Kuloor, N. R., Diffusion in Binary Gas System, *Brit. Chem. Eng.*, 1961, **6** (12), 862–3.

85 Jacolin, O., *Proprietes Physiques et Chimiques de l'Ammoniac*, Commissariat à l'Energie Atomique, Bibliogr. No. 57, Paris, 1965, 68 pp.

86 Jamieson, D. T., Tudhope, J. S., *The Thermal Conductivity of Liquids: A Survey to 1963*, National Engineering Laboratory, Report No. 137, Glasgow, March 1964, 67 pp.

87 Jandler, J., *Die Chemie in Wasserähnlichen Lösungsmitteln*, Springer-Verlag, Berlin, 1949, 367 pp. (HF, NH_3, H_2SO_4, HCN, HNO_3, I_2, SO_2).

88 Jandler, J., Spandau, H., Addison, C. C., Eds., *Chemistry in Nonaqueous Ionizing Solvents*. Vol. I, Chemistry in Anhydrous Liquid Ammonia, Parts 1 & 2; Vol. II, Chemistry in Anhydrous, Prototropic Inorganic Solvents (HCN & HF), Parts 1 & 2; Vol. III, Chemistry in Liquid Dinitrogen Tetroxide and Sulphur Dioxide; Vol IV, Chemistry in Lower Fatty Acids and Derivatives, Interscience Publ., New York, 1960–71.

89 Johnson, V. J., Ed., *A Compendium of the Properties of Materials at Low Temperatures*, Part I, Properties of Fluids (1960), 489 pp., Part II, Properties of Solids (1960), 330 pp., Part III, Bibliography of References (1960), 161 pp., Clearinghouse for Federal Sci. & Techn. Infor., U.S. Dept. of Commerce, Springfield, Va.

90 Jost, W., *Diffusion in Solids, Liquids, and Gases*, 3rd ed., Academic Press, New York, 1960, 652 pp.

91 Kafarov, B. V., *Reference Book on Solubility*, 3 Vols., Nauka Publishing House, Moscow, 1967–8.

92 Kang, T. L., *Thermodynamic Properties of Sulphur Dioxide*, Univ. of Texas, Ph.D. dissertation in Chem. Eng., Austin, Texas, 1960.

93 Karapetyants, M. Kh., Karapetyants, M. L., *Thermodynamic Constants of Inorganic and Organic Substances*, Khemia Publishing House, Moscow, 1968.

94 Kaye, G. W. C., Laby, T. H., *Tables of Physical and Chemical Constants and Some Mathematical Functions*, 13th ed., Longmans, London, 1966, 250 pp.

95 Kazavchinskii, Ya. Z., Kessel'man, P. M., Kirillin, V. A., Rivkin, S. L., Sheindlin, A. E., Shpil'rain, E. E., Sychev, V. V., Timrot, D. L., *Heavy Water. Thermophysical Properties*, Israel Program for Scientific Translations, Jerusalem, 1971, 263 pp.

96 Keenan, J. H., Kaye, J., *Gas Tables. Thermodynamic Properties of Air Products of Combustion and Component Gases Compressible Flow Functions*, Wiley, New York, 1948, 238 pp.

97 Keenan, J. H., Keyes, F. G., Hill, P. G., Moore, J. G., *Steam Tables*, Wiley, New York, 1969.

98 Kelley, K. K., *Contributions to the Data of Theoretical Metallurgy. The Free Energies of Vaporisation and Vapour Pressures of Inorganic Substances*, Bureau of Mines, U.S. Government Printing Office, Washington, D.C., Bulletin 383, 1935, 132 pp.

99 Kelley, K. K., *Contribution to the Data of Theoretical Metallurgy. Heats of Fusion of Inorganic Substances*, Bureau of Mines, U.S. Government Printing Office, Washington, D.C., Bulletin 393, 1936, 166 pp.

100 Kelley, K. K., *Contribution to the Data of Theoretical Metallurgy. The Thermodynamic Properties of Sulphur and Its Inorganic Compounds*, Bureau of Mines, U.S. Government Printing Office, Washington, D.C., Bulletin 406, 1937, 154 pp.

101 Kelley, K. K., *Contributions to the Data on Theoretical Metallurgy. High-Temperature Heat-Content, Heat Capacity, and Entropy Data for the Elements and Inorganic Compounds, XIII*, Bureau of Mines, U.S. Government Printing Office, Washington, D.C., Bulletin 584, 1960, 232 pp.

102 Kelley, K. K., Anderson, C. T., *Contributions to the Data of Theoretical Metallurgy. Metal Carbonates, Correlation and Applications of Thermodynamic Data*, Bureau of Mines, U.S. Government Printing Office, Washington, D.C., Bulletin 384, 1935, 73 pp.

103 Kelley, K. K., King, E. G., *Contributions to the Data on Theoretical Metallurgy. Entropies of Elements and Inorganic Compounds, XIV*, Bureau of Mines, U.S. Government Printing Office, Washington, D.C., Bulletin 592, 1961, 149 pp.

104 Keyes, F. G., Brownlee, R. B., *Thermodynamic Properties of Ammonia*, Wiley, New York, 1916.

105 *Kirk-Othmer Encyclopedia of Chemical Technology*, 2nd ed., ed., Interscience Publ., New York, 1966, 22 Vols. + Suppl.

106 Kobe, K. A., Lynn, R. M., The Critical Properties of Elements and Compounds, *Chem. Rev.*, 1953, **52**, 117–236.

107 Kudchadker, A. P., Alani, G. H., Zwolinski, B. J., The Critical Constants of Organic Substances, *Chem. Rev.*, 1968, **68**, 659–735.

108 Lagowski, J. J., Ed., *The Chemistry of Non-Aqueous Solvents*, Vol. I, Principles and Techniques (1966), 403 pp, Vol. II, Acidic and Basic Solvents, HCl, HBr, HI, HF, H_2SO_4, HNO_3, NH_3 (1967), 398 pp, Vol. III, Inert, Aprotic, and Acidic Solvents, SO_2, H_2S (1970), 408 pp., Academic Press, New York.

109 *Landolt-Börnstein Zahlenwerte und Funktionen aus Physik, Chemie, Astronomie, und Technik* (Multi-volume Handbook), 6th ed., Springer-Verlag, Berlin, 1950–72.

110 Lange, N. A., *Handbook of Chemistry*, 10th ed., McGraw-Hill, New York, 1967.

111 Laporte Chemicals Limited, *Hydrogen Peroxide Data Manual*, Luton, Aug. 1960, 80 pp.

112 Lax, E., Synowietz, C., *Taschenbuch für Chemiker und Physiker*, Springer-Verlag, Berlin, Vol. 1 (1967), 1522 pp., Vol. 2 (1964), 1178 pp., Vol. 3 (1970), 670 pp.

113 Leduc, A., The Specific Heats of Gases and Vapours. A Critical Review of Methods and Results, *Chem. Rev.*, 1929, **6**, 1–16.

114 Linke, W. F., *Solubilities of Inorganic and Metal-Organic Compounds*, 4th ed., Am. Chem. Soc., Van Nostrand Co. Inc., Princeton, N.J., Vol. I (1958), 1484 pp., Vol. II (1965), 1914 pp.

115 Little, W. J., Neel, C. A., *Tables of the Thermodynamic Properties of Nitrogen from 100 to 1500 K*, Natl. Bur. Std., AEDC-TDR-62-170 (AD 283441), Washington, D.C., Sept. 1962.

116 Maitland, G. C., Smith, E. B., Critical Reassessment of Viscosities of 11 Common Gases, *J. Chem. Eng. Data*, 1972, **17** (2), 150–6.

117 Malkov, M. P., Ed., *Liquid Hydrogen. A Collection of Articles*, Mir, Moscow, 1964.

118 Mallan, G. M., *Thermal Conductivity of Liquids*, Univ. of Southern California, Ph.D. dissertation, 1968, 178 pp.

119 Manufacturing Chemists Association, *Chemical Safety Data Sheets*, Washington, D.C., NH_3, SD-8 (1960), Br_2, SD-49 (1968), Cl_2, SD-80 (1960), HCl, SD-39 (1970), HCN, SD-67 (1961), HF, SD-25 (1957), H_2O_2, SD-53 (1969), H_2S, SD-36 (1968), HNO_3, SD-5 (1961), H_3PO_4, SD-70 (1958), SO_2, SD-52 (1953), H_2SO_4, SD-20 (1963).

120 Markham, A. E., Kobe, K. A., The Solubility of Gases in Liquids, *Chem. Rev.*, 1941, **28**, 519–88.

121 Maryott, A. A., Buckley, F., *Tables of Dielectric Constants and Electric Dipole Moments of Substances in the Gaseous State*, Natl. Bur. Std. Circular 537, Washington, D.C., 1953.

122 Maryott, A. A., Smith, E. R., *Table of Dielectric Constants of Pure Liquids*, Natl. Bur. Std. Circular 514, Washington, D.C., Aug. 10, 1951, 44 pp.

123 Matheson Company, Inc., *Matheson Gas Data Books*, 4th ed., East Rutherford, N. J., 1966.

124 Mathews, J. F., The Critical Constants of Inorganic Substances, *Chem. Rev.*, 1972, **72** (1), 71–100.

125 Mayhew, Y. R., Rogers, G. F. C., *Thermodynamic and Transport Properties of Fluids. SI Units*, Blackwell, Oxford, 1972, 2nd ed., 20 pp.

126 Mellor, J. W., Ed., *A Comprehensive Treatise on Inorganic and Theoretical Chemistry* (Multi-volume Handbook), Longmans, London, 1922–67, 16 Vols. + Suppls.

127 Mishchenko, K. P., Ravdel, A. A., Eds., *Pocket Handbook of Physicochemical Constants*, Khimiya, Leningrad, 1967, 5th ed., 182 pp.

128 Missenard, A., *Thermal Conductivity of Solid, Liquids, and Gases and Their Mixtures*, Editions Eyrolles, Paris, 1965, 576 pp.

129 Monsanto Chemical Company, *Phosphoric Acid. Physical and Chemical Properties*, Techn. Bull. No. P-26, St. Lewis, Mo., 1946, 12 pp.

130 Mullins, J. C., Ziegler, W. T., Kirk, B. S., *The Thermodynamic Properties of Oxygen from 20 to 100 K* Georgia Inst. of Technol., Eng. Expt. Station, Techn. Rept. No. 2, March 1, 1962.

131 McBride, B. J., Heimel, S., Ehlers, J. G., Gordon, S., *Thermodynamic Properties to 6000 K for 210 Substances Involving the First 18 Elements*, Natl. Aeronautics and Space Administration, NASA-SP-3001, Washington, D.C., 1963, 328 pp.

132 McCarty, R. D., Weber, L. A., *Thermophysical Properties of Oxygen from the Freezing Liquid Line to 600°R for Pressure to 5000 PSIA*, Natl. Bur. Std., Note 384, Washington, D.C., July 1971, 189 pp.

133 McClellan, A. L., *Tables of Experimental Dipole Moments*, Freeman & Co., San Francisco, 1963, 713 pp.

134 McGlashan, M. L., Manual of Symbols and Terminology for Physicochemical Quantities and Units, *Pure & Appl. Chem.*, 1970, **21** (1), 1–44.

135 National Bureau of Standards, *Tables of Thermodynamic Properties of Ammonia*, 1st ed., Circular 142, Washington, D.C., 1923, 30 pp.

136 Neel, C. A., Lewis, C. H., *Interpretation of Imperfect Nitrogen Thermodynamic Data. I. At Constant Enthalpy*, Natl. Bur. Std. AEDC-TDR-64-212 (AD 449409), Washington, D.C., Oct. 1964.

137 Neel, C. A., Lewis, C. H., *Interpretation of Imperfect Nitrogen Thermodynamic Data. II. At Constant Pressure*, Natl. Bur. Std. AEDC-TDR-64-213 (AD 449381), Washington, D.C., Oct. 1964

138 Nelson, R. D., Linde, D. R., Maryott, A. A., *Selected Values of Electric Dipole Moments for Molecules in the Gas Phase*, Natl. Bur. Std. NSRDS-NBS-10, Washington, D.C., 1967, 49 pp.

139 Nesmeyanov, A. N., *Vapour Pressure of the Elements*, Publishing House of the USSR Academy of Science, Moscow, 1961, 396 pp.

140 Neugebauer, F. J., *Collection of Heat-Transfer Properties of Gases at Moderate Pressures and Rules for Rapid Estimation of Missing Data*, Natl. Bur. Std., Washington, D.C., April 30, 1962.

141 Nikol'skii, B. P., *Handbook for Chemists*, Khimiya, Moscow, 1962–8, 2nd ed., 6 Vols. + Suppl.

142 Office of Critical Tables, *Continuing Numerical Data Projects—A Survey and Analysis*, 2nd ed., National Academy of Sciences, National Research Council, Washington, D.C., 1966, 203 pp.

143 Pascal, P., Ed., *Nouveau Traite de Chimie Minerale* (Multi-volume Handbook), Masson et Cie, Éditeurs, Paris, 1956–64, 20 Vols.

144 Perry, J. H., Ed., *Chemical Engineers' Handbook*, 4th ed., McGraw-Hill, New York, 1970.

145 Phillips Petroleum Company, *Hydrofluoric Acid Alkylation*, Chemical Products Dept., Bartlesville, Oklahoma, 1946.

146 Powell, R. W., Ho, C. Y., Liley, P. E., *Thermal Conductivity of Selected Materials*, 2 Parts, NSRDS-NBS-8 and 16, Natl. Bur. Std., Washington, D.C., 1966 (168 pp.) and 1968 (146 pp.).

147 Predvoditelev, A. S., Stupochenko, E. V., Ionov, V. P., Pleshanov, A. S., Rozhdestvenskii, I. B., Samuilov, E. V., *Charts of Thermodynamic Functions of Air: 1000 to 12 000 K and 0.001 to 1000 Atm.* Publishing House of USSR, Academy of Sciences, Moscow, 1960, 56 pp.

148 Predvoditelev, A. S., Stupochenko, E. V., Pleshanov, A. S., Samuilov, E. V., Rozhdestvenskii, I. B., *Tables of Thermodynamic Functions of Air: 12 000 to 20 000 K and 0.001 to 1000 Atm.*, Publishing House of USSR, Academy of Sciences, Moscow, 1959, 230 pp.

149 Predvoditelev, A. S., Stupochenko, E. V., Pleshanov, A. S., Samuilov, E. V., Rozhdestvenskii, I. B., *Tables of Thermodynamic Functions of Air: 200 to 6000 K and 0.000 01 to 100 Atm.*, Publishing House of USSR, Academy of Sciences, Moscow, 1962, 268 pp.

150 Predvoditelev, A. S., Stupochenko, E. V., Samuilov, E. V., Stakhanov, I. P., Pleshanov, A. S., Rozhdestvenskii, I. B., *Tables of Thermodynamic Functions of Air: 6000 to 12 000 K and 0.001 to 1000 Atm.*, Publishing House of the USSR, Academy of Sciences, Moscow, 1957, 302 pp.

151 Quinn, E. L., Jones, C. L., *Carbon Dioxide*, Am. Chem. Soc. Monogr. Ser. 72, Reinhold Publ. Corp., New York, 1936.

152 Rabinovitch, V. A., *Thermophysical Properties of Monatomic Gases and Liquids*, 2 parts, Publishing House for State Standards, Moscow, 1968, 207 pp.

153 Rabinovitch, V. A., Ed., *Thermophysical Characteristics of Substances*, Publishing House for State Standards, Moscow, 1968.

154 Riedel, L., *Wärmeleitfähigkeitsmessungen an Flüssigkeiten*, No. 1, C. F. Müller Verlag, Karlsruhe, 1948, 47 pp.

155 Robertson, S. L., *P–V–T Properties of Gases to 10 000 Bars and 400°C*, Univ. of Oklahoma, Ph.D. dissertation, 1969, 135 pp.

156 Roder, H. M., Weber, L. A., Eds., *ASRDI Oxygen Technology Survey. Vol. I. Thermophysical Properties*, National Technical Information Services, Springfield, Va., 1972, 426 pp.

157 Roder, H. M., Weber, L. A., Goodwin, R. D., *Thermodynamic and Related Properties of Parahydrogen from the Triple Point to 100 K at Pressures to 340 Atmospheres*, Natl. Bur. Std. Monograph 94, Washington, D.C., Aug. 1965, 112 pp.

158 Rossini, F. D., Wagman, D. D., Evans, W. H., Levine, S., Jaffe, I., *Selected Values of Chemical Thermodynamic Properties*, Natl. Bur. Std. Circular 500, Washington, D.C., 1952, 1268 pp.

159 Sage, B. H., Lacey, W. N., *Some Properties of the Lighter Hydrocarbons*, Am. Petrol. Inst., New York, 1955 (H_2S, CO_2).

160 Sakiadis, B. C., Coates, J., *A Literature Survey of the Thermal Conductivity of Liquids. Part IV. A Literature Survey of Ultrasonic Velocities in Liquids and Solutions*, Louisiana State Univ., Eng. Expt. Station, Bulletin 46, 1954.

161 Salomon, M., *Thermodynamic Properties of Liquid H_2O and D_2O and Their Mixtures*, National Aeronautics and Space Administration, NASA-TN-D-5223, May 1969.

162 Samsonov, G. V., Ed., *Handbook of the Physicochemical Properties of the Elements*, Materials Science Institute of the Academy of Sciences of the Ukrainian S.S.R., Naukova Dumka Press, Kiev, 1965.

163 Samsonov, G. V., *Physicochemical Properties of Oxides, Handbook*, Publishing House for State Standards, Moscow, 1968.

164 Sato, K., Ed., *Physical Property Values of Substances*, Maruzen Publishing Co., Tokyo, 1967.

165 Schmidt, E., *Properties of Water and Steam in SI Units*, Springer-Verlag, Berlin, 1969.

166 Schumb, W. C., Satterfield, C. N., Wentworth, R. L., *Hydrogen Peroxide*, Am. Chem. Soc. Monogr. Ser. 128, Reinhold Publ. Corp., New York, 1955, 759 pp.

167 Seidell, A., *Solubilities of Inorganic and Metal Organic Compounds*, 4th Printing, Vol. I, Van Nostrand, New York, 1953, 1698 pp.

168 Seidell, A., Linke, W. F., *Solubilities of Inorganic and Organic Compounds*, Suppl. to 3rd ed., Van Nostrand, New York, 1952, 1254 pp.

169 Sindt, C. F., Mann, D. B., *Temperature–Entropy Diagram for Parahydrogen. Triple-Point Region*, Natl. Bur. Std. Techn. Rept. 343, Washington, June 22, 1966, 6 pp.

170 Sitting, M., *Nitrogen in Industry*, Van Nostrand Co. Inc., Princeton, 1965, 278 pp.

171 Smith, C. E., *Thermodynamic Properties of Nitrogen*, Lockheed TR-6-90-62-111, Dec. 1962.

172 Spiers, H. M., Ed., *Technical Data on Fuel*, 6th ed., The British National Committee World Power Conference, London, 1961, 360 pp.

173 Stauffer Chemical Company, *Hydrogen Chloride Anhydrous*, Techn. Service Product Rept., New York, 16 pp.

174 Steere, N. V., Ed., *Handbook of Laboratory Safety*, The Chemical Rubber Co., Cleveland, Ohio, 1967.

175 Stephen, H., Stephen, T., *Solubilities of Inorganic and Organic Compounds*, 2 Vols., Pergamon Press, Oxford, 1963.

176 Stern, K. H., Weise, E. L., *High Temperature Properties and Decomposition of Inorganic Salts. Part 1. Sulphates*, Natl. Bur. Std. NSRDS-NBS-7, Washington, D.C., 1966, 38 pp.

177 Strobridge, T. R., *The Thermodynamic Properties of Nitrogen from 64 to 300 K between 0.1 and 200 Atmospheres*, Natl. Bur. Std. Techn. Note 129 (PB-161-630), Washington, D.C., 1962, 85 pp.

178 Strobridge, T. R., *The Thermodynamic Properties of Nitrogen from 114 to 540°R. Between 1.0 and 3000 PSIA. Supplement A*, Natl. Bur. Std. Techn. Note 129A, Washington, D.C., Feb. 1963.

179 Stull, D. R., Director, *JANAF Thermochemical Tables*, The Dow Chemical Co., PB 168370 (1965), 945 pp., 1st Addendum, PB 168370-1 (1966), 197 pp., 2nd Addendum, PB 168370-2 (1967), 193 pp., 3rd Addendum, PB 168370-3 (1968), Midland, Michigan.

180 Stull, D. R., Sinke, G. C., *Thermodynamic Properties of the Elements*, Am. Chem. Soc., Advances in Chemistry Ser. 18, Washington, D.C., 1956, 234 pp.

181 Stull, D. R., Westrum, E. F., Sinke, G. C., *The Chemical Thermodynamics of Organic Compounds*, Wiley, New York, 1969.

182 Svehla, R. A., *Estimated Viscosities and Thermal Conductivities of Gases at High Temperatures*, Lewis Res. Center Techn. Rept. R-132, Cleveland, Ohio, 1962, 140 pp.

183 Szubinski, H.-W., *Die Schwefelsäurefabrikation*, Vieweg & Sohn, Braunschweig, 1961, 488 pp.

184 Tables de Constantes Sélectionnées (P. Khodadad, Director), Pergamon Press, London, 1947-up-to-date, 15 Vols.

185 Thorp, C. E., *Bibliography of Ozone Technology*, Vol. 2, Armour Res. Foundation, Illinois Inst. of Technology, Chicago, Ill., 1955.

186 *Threshold Limit Values for 1969*, Dept. of Employment and Productivity, H.M. Factory Inspectorate, Techn. Data Note 2/69, H.M. Stationery Office, London, 1969, 16 pp.

187 Touloukian, Y. S., Ed., *Thermophysical Properties of High Temperature Solid Materials*, 6 Vols., Macmillan, New York, 1967, 8500 pp.

188 Touloukian, Y. S., Garritsen, J. K., Moore, N. Y., Eds., *Thermophysical Properties Research Literature: A Retrieval Guide*, 2nd ed., 3 Vols., Plenum Press, New York, 1967, 2759 pp.

189 Touloukian, Y. S., Ho, C. Y., Eds., *Thermophysical Properties of Matter* (Multi-volume Handbook), Plenum Press, New York, 1970–3, 13 Vols.

190 Tsederberg, N. V., *Thermal Conductivity of Gases and Liquids*, M.I.T. Press, Cambridge, Mass., 1965.

191 Tyrrell, H. J. V., *Diffusion and Heat Flow in Liquids*, Butterworths, London, 1961, 329 pp.

192 *Ullmanns Encyklopädie der technischen Chemie* (Multi-volume Handbook), 3rd ed., Urban & Schwarzenberg, München, 1962, 19 Vols. + Suppl.

193 Union Carbide Corporation, *HCl Physical Properties*, Carbon Products Div. Rept. No. RD4-7.3, Cleveland, Ohio, 44101.

194 Urey, H. C., Murphy, G. M., *Physical Properties and Analysis of Heavy Water*, Div. III, Vol. 4A, McGraw-Hill, New York, 1951, 438 pp.

195 Vargaftik, N. B., *Thermophysical Properties of Substances*, Gosenergoizdat, Moscow, 1956 (Translation: F-TS-9537/v, 1959, 485 pp.).

196 Vargaftik, N. B., *Handbook of the Thermophysical Properties of Gases and Liquids*, Fizmatgiz, Moscow, 1963, 708 pp.

197 Vargaftik, N. B., *Reference Book on Thermophysical Properties of Gases and Liquids*, Nauka Publishing House, Moscow, 1964.

198 Vargaftik, N. B., Filipov, L. P., Tarzimanow, A. A., Yurchak, R. P., *Heat Conductivity of Gases and Liquids*, Publishing House for State Standards, Moscow, 1970.

199 Vassermann, A. A., Kazavchinskii, Ya. Z., Rabinovich, V. A., *Thermophysical Properties of Air and Air Components*, Izdatel'stvo Nauk, Moscow, 1966.

200 Vassermann, A. A., Rabinovich, V. A., *Thermophysical Properties of Liquid Air and Its Components*, Izdatel'stvo Komiteta Standartov, Mer i Izmeritel'nykh Priborov pri Sovete Ministrov SSSR, Moscow, 1968, 240 pp.

201 Vaughan, W. E., Tables of Dielectric Constants, Dipole Moments, and Dielectric Relaxation Times, *Digest of Literature on Dielectrics* 1969, **33**, 21–83.

202 Vedeneyev, V. I., Gurvich, L. V., Kondorat'yev, V. N., Medvedev, V. A., Frankevich, Ye. L., *Bond Energies, Ionization Potentials, and Electron Affinities*, Academy of Sciences, Moscow, 1962.

203 Vukalovich, M. P., *Thermodynamic Properties of Water and Steam*, VEB Verlag Technik, Berlin, 1958.

204 Vukalovich, M. P., Altunin, V. V., *Thermophysical Properties of Carbon Dioxide*, Collet's (Publ.) Ltd., London, 1968, 463 pp.

205 Vukalovich, M. P., Kirillin, V. A., Remizov, C. A., Siletskiy, V. S., Timofeyev, V. N., *Thermodynamic Properties of Gases*, Sudostroyeniye Publishing House, Moscow, 1953, 375 pp.

206 Waddington, T. C., Ed., *Non-Aqueous Solvent Systems*, Academic Press, London, 1965, 408 pp. (NH_3, HF, H_2SO_4, SO_2, Cl_2, Br_2, F_2, I_2).

207 Wagman, D. D., Evans, W. H., Halow, I., Parker, V. B., Bailey, S. M., Schumm, R. H., *Selected Values of Chemical Thermodynamic Properties*, 1. Tables for the First 23 Elements in the Standard Order of Arrangement (1965), 124 pp., 2. Tables for the Elements 24 through 32 in the Standard Order Arrangement (1966), 62 pp., 3. Tables for the first 34 Elements in the Standard Order of Arrangement (1968), 264 pp., Natl. Bur. Std., Washington, D.C.

208 Walker, W. A., *Thermodynamic Properties of Liquid Water to 1 kbar*, U.S. Naval Ordnance Laboratory, NOL-TR-66-217, 1967.

209 Washburn, E. W., Ed., *International Critical Tables of Numerical Data, Physics, Chemistry, and Technology* (Multi-volume Handbook), McGraw-Hill, New York, 1926–33, 7 Vols. + Index.

210 Weast, R. C., Ed., *Handbook of Chemistry and Physics*, 51st ed., The Chemical Rubber Co., Cleveland, Ohio, 1970–1.

211 Wood, R. E., Baer, F. W., Boone, W. J., *Thermal Conductivities and Prandtl Numbers of Nitrogen from 133 to 740 K between 1 and 240 Atmospheres*, U.S. Dept. of Interior, Bureau of Mines, Rept. 7541, Washington, D.C., July 1971, 34 pp.

212 Zwolinski, B. J., Dir., *Selected Values of Properties of Chemical Compounds*, Thermodynamic Research Center, Texas A & M Univ., Texas, 1968, 886 loose-leaf data sheets.

1 BROMINE

Bromine was discovered by Balard in 1826. It is the only liquid non-metallic element with a reddish-brown colour at room temperature and atmospheric pressure. Bromine is very volatile forming a red vapour with irritating effect and disagreeable odour. Liquid bromine is used in the preparation of most of the bromides, intermediate products, and in the synthesis of organic compounds. Bromine compounds are mainly used in the production of medicines and photographic preparations.

The purification of bromine was very difficult until the discovery that the removal of traces of impurities can be effected by permanganate or chromic acid oxidation, therefore the earlier measurements are not very reliable. At the present, highly pure bromine is commercially available (99.98–99.99%), far better than the purified samples used some generations ago.

Several properties of bromine, e.g., degree of dissociation (26), condensation coefficient (9), and thermodynamic properties (26), (28) are not given here, because they are outside the scope of this presentation.

Seshadri *et al.* (26) calculated the P–V–T properties, enthalpy, and entropy of bromine up to a temperature of 1000 K and a pressure of 300 bars using the Martin-Hou equation of state. A temperature-entropy diagram has also been presented. P–V–T relations for bromine vapour are also given by Washburn (29).

Michigan Chemical Corporation (22) issued a booklet (giving 506 references) on bromine in 1958, which provides collective data from the published literature and their own research results. The contents of this booklet are very useful to scientists working in the field of bromine chemistry and its handling. Further information in more detail on bromine and its properties can be found in Gmelins (10), Mellor (21), Pascal (23), and CRH (30).

The relevant physical properties of bromine (molecular) are summarised in Table 1.1 and Figs. 1.1–1.12.

References

1 American Conference of Government Industrial Hygienists, Documentation of Threshold Limit Values, revised edition, Committee on Threshold Limit Values, Cincinnati, Ohio 45202, 1966, 203 pp.

2 Braune, H., Basch, R., Wentzel, W., *Z. phys. Chem.*, 1928, **137A**, 176–92.

3 Burriel Lluna, J. A., Cragg, C. B., Rowlinson, J. S., *An. Fis. Quim.*, 1968, **64** (1), 1–8.

4 Chao, M. S., Stenger, V. A., *Talanta*, 1964, **11**, 271–81.

5 Cheesman, G. H., Scott, D. L., *Aust. J. Chem.*, 1968, **21**, 287–97.

6 Edwards, D. G., *The Vapour Pressure of 30 Inorganic Liquids between One Atmosphere and the Critical Point*, University of California, Lawrence Rad. Lab., Livermore, Calif., 1963, pp. 126–9.

7 Eng. Sci. Data Units, Item No. 66024, London, July 1966.

8 Fisher, J., Bingle, J., *J. Am. Chem. Soc.*, 1955, **77**, 6511–2.

9 Frey, M. B., Gregory, N. W., *J. Am. Chem. Soc.*, 1960, **82**, 1068–9.

10 *Gmelins Handbuch der anorganischen Chemie*, *Syst. No.* **7**, Brom, Verlag Chemie, GmbH., Berlin, 1931, pp. 58–115.

11 Hildebrand, D. L., Kramer, W. R., McDonald, R. A., Stull, D. R., *J. Am. Chem. Soc.*, 1958, **80**, 4129–32.

12 Honig, R. E., Hook, H. O., *RCA Rev.*, 1960, **21**, 360–8.

13 Ibrahim, S. H., Kuloor, N. R., *Brit. Chem. Eng.*, 1961, **6** (12), 862–3.

14 Kelley, K. K., *Contributions to the Data of Theoretical Metallurgy, III. The Free Energies of Vaporisations and Vapour Pressures of Inorganic Substances*, U.S. Dept. of the Interior, Bureau of Mines, Bulletin 383, Washington, D.C., 1935, p. 26.

15 Kobe, K. A., Lynn, R. E., *Chem. Rev.*, 1953, **52**, 117–236.

16 Landolt-Börnstein: *Zahlenwerte und Funktionen aus Physik, Chemie, Astronomie, und Technik*, 6th ed., Vol. II, Part 5b, Springer-Verlag, Berlin, 1968, p. 47.

17 Lax, E., Synowietz, C., Eds., *Taschenbuch für Chemiker und Physiker*, 3rd ed., Springer-Verlag, Berlin, 1967.

18 Liley, P. E., *The Thermal Conductivity of 46 Gases at Atmospheric Pressures*, Proc. 4th. Symp. Thermophysical Prop., Am. Soc. Mech. Eng., New York, 1968, pp. 323–49.

19 Linke, W. F., *Solubilities of Inorganic and Metal-Organic Compounds*, 4th ed., Vol. I, D. Van Nostrand Co., Inc., Princeton, N.J., 1958, p. 440.

20 Mathews, J. F., *Chem. Rev.*, 1972, **72**, 71–100.

21 Mellor, J. W., Ed., *A Comprehensive Treatise on Inorganic and Theoretical Chemistry*, Vol. II, Suppl. I, *The Halogens*, Longmans, London, 1956, pp. 689–814.

22 Michigan Chemical Corporation, *Bromine, Its Properties and Uses*, Saint Louis, Michigan, 1968, 62 pp.

23 Pascal, P., Ed., *Nouveau Traité de Chimie Minérale*,' Vol. XVI, F, Cl, Br, I, At, Mn, Tc, Re, Masson et Cie, Éditeurs, Paris, 1960, pp. 337–447.

24 Richards, T. W., *J. Am. Chem. Soc.*, 1915, **37**, 1641–56.

25 Sapper, A., Biltz, W., *Z. anorg. Chem.*, 1931, **198**, 184–90.

26 Seshadri, D. N., Viswanath, D. S., Kuloor, N. R., *Indian J. Technol.*, 1970, **8** (6), 191–8.

27 Steere, N. V., Ed., *Handbook of Laboratory Safety*, The Chemical Rubber Co., Cleveland, 1967.

28 Stull, D. R., Dir., *JANAF Thermochemical Tables*, Clearinghouse for Federal Scientific and Technical Information, PB 168 370, Springfield, Va., 1965, p. Br2.

29 Washburn, E. W., Ed., *International Critical Tables*, McGraw-Hill, New York, Vol. I (1926), pp. 102–3 and Vol. VI (1929), p. 462.

30 Weast, R. C., Ed., *Handbook of Chemistry and Physics*, 52nd edition, The Chemical Rubber Co., Cleveland, 1971–2.

31 Yosim, S. J., *J. Chem. Phys.*, 1964, **40** (10), 3069–75.

Table 1.1 Properties of Bromine

		Ref.
Chemical formula	Br_2	(17)
Molecular weight	159.81	(17)
Critical temperature	311.0°C	(20)
Critical pressure	1.0336×10^7 N m^{-2}	(20)
Critical density	1260.0 kg m^{-3}	(20)
Triple-point temperature	−7.3°C	(5)
Triple-point pressure	5.8265×10^3 N m^{-2}	(5)
Boiling point at 1.0133×10^5 N m^{-2}	58.78°C	(29)
Melting point	−7.2°C	(11)
Transition temperature	None	(9)
Density of solid at −194°C	4107.0 kg m^{-3}	(25)
Compressibility of liquid at 20°C	5.13×10^{-11} m^2 N^{-1}	(24)
Coefficient of thermal expansion of liquid at boiling point	1.21×10^{-3} °C^{-1}	(31)
Heat of formation of gas at 25°C	3.0928×10^7 J kg-mol^{-1}	(28)
Free energy of formation of gas at 25°C	2.9684×10^6 J kg-mol^{-1}	(28)
Heat of sublimation at −88.15°C	4.4715×10^7 J kg-mol^{-1}	(9)
Heat of fusion at −7.2°C	1.0580×10^7 J kg-mol^{-1}	(28)
Heat of transition	None	(9)
Entropy of gas at 25°C	2.4554×10^5 J kg-mol^{-1} °C^{-1}	(28)
Heat capacity ratio of gas at 25°C and 1.0133×10^5 N m^{-2}	1.32	(17)
Velocity of sound at 0°C and 1.0133×10^5 N m^{-2}	135.0 m s^{-1}	(29)
Diffusion in air at 0°C and 1.0133×10^5 N m^{-2}	9.0×10^{-6} m^2 s^{-1}	(13)
Dielectric constant of liquid at 20°C	3.09	(17)
Dielectric constant of gas at 180°C and 1.0133×10^5 N m^{-2}	1.0128	(22)
Refractive index of liquid at 15°C	1.661	(22)
Threshold limit value	(0.1 ppm) 0.7×10^{-6} kg m^{-3}	(1)

1.1 Vapour Pressure of Solid and Liquid Bromine (5,6,8,12)

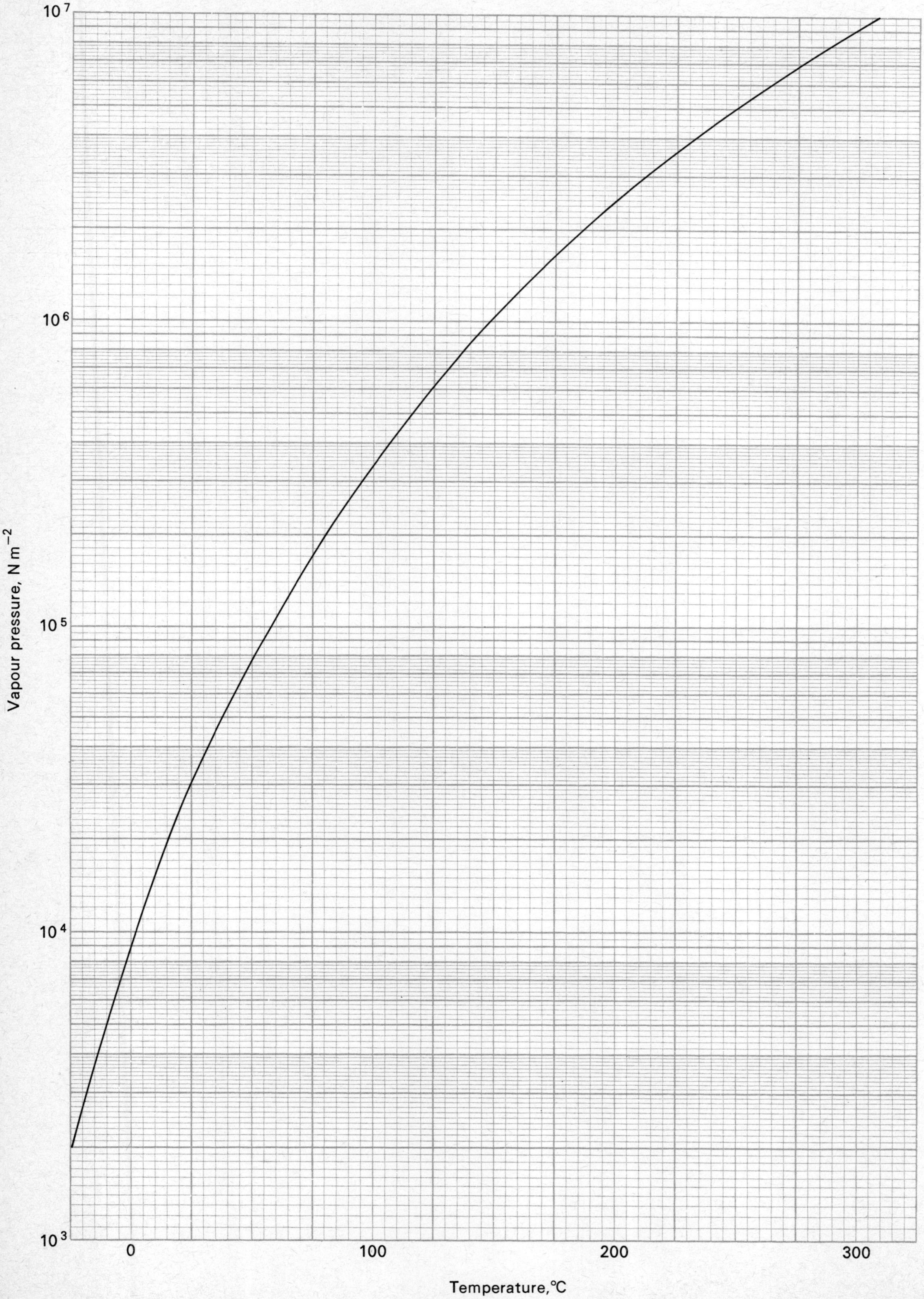

1.2 Orthobaric Density of Bromine (3,4,17)

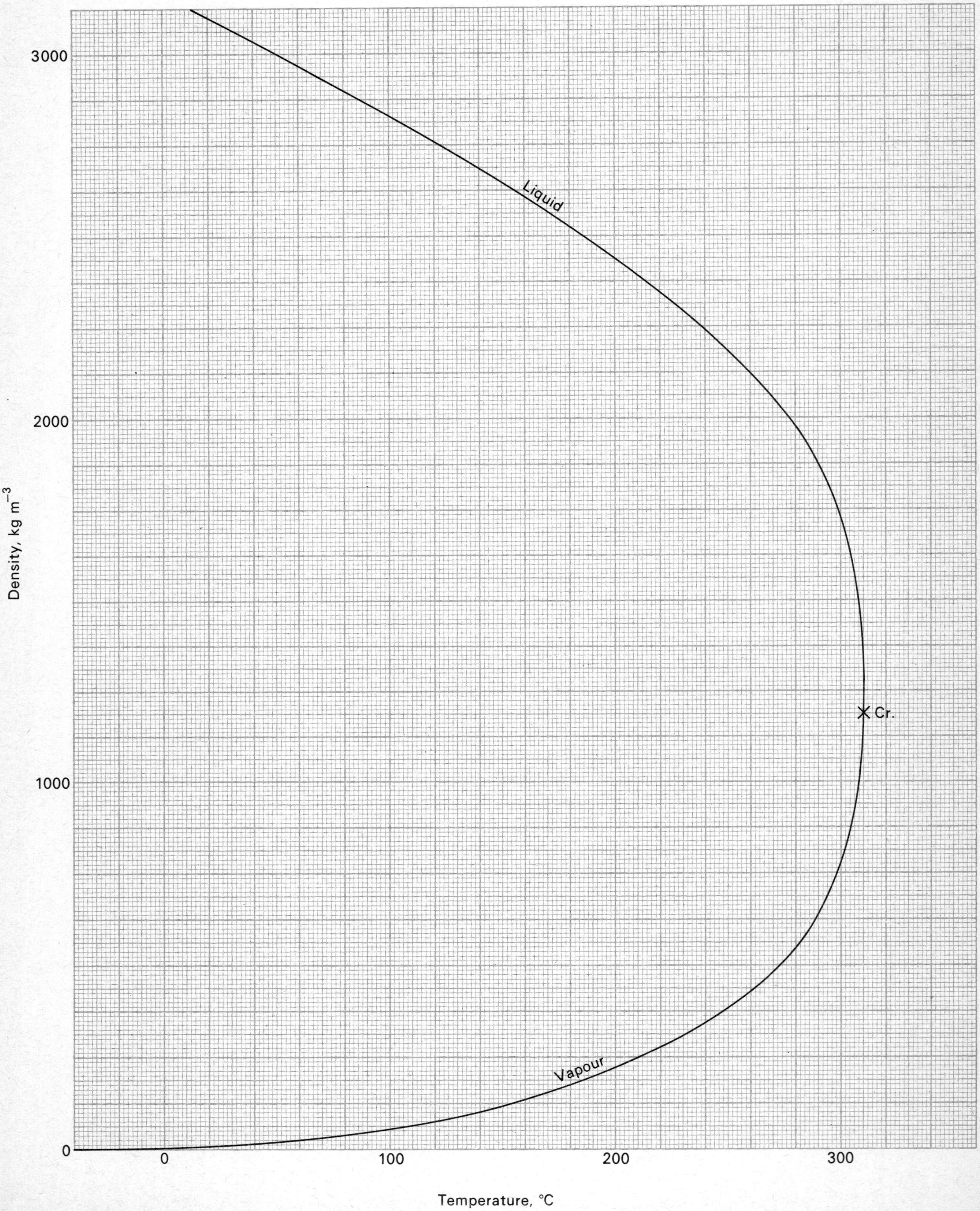

1.3 Latent Heat of Vaporisation of Bromine (11,14)

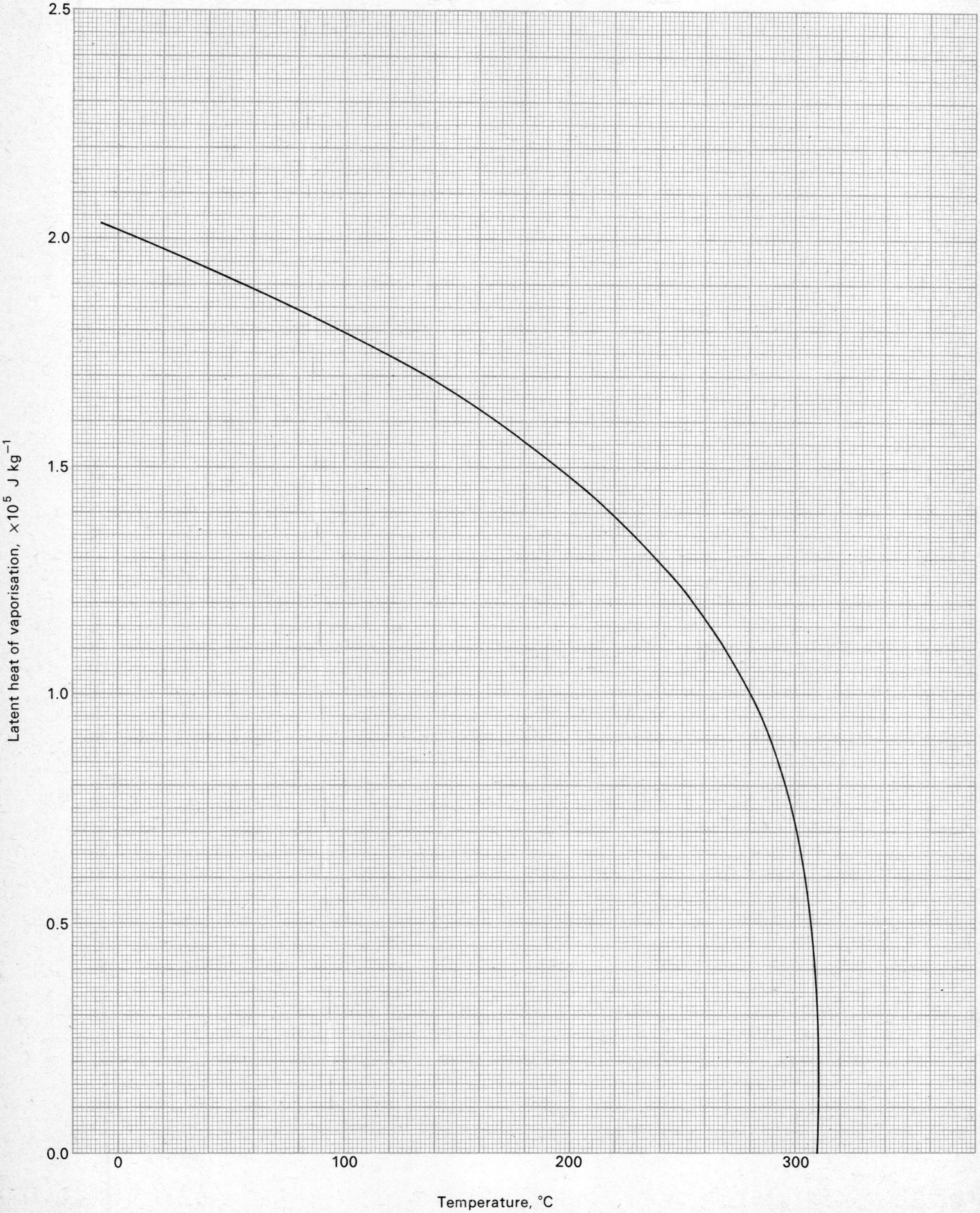

1.4 Heat Capacity of Solid and Liquid Bromine (11)

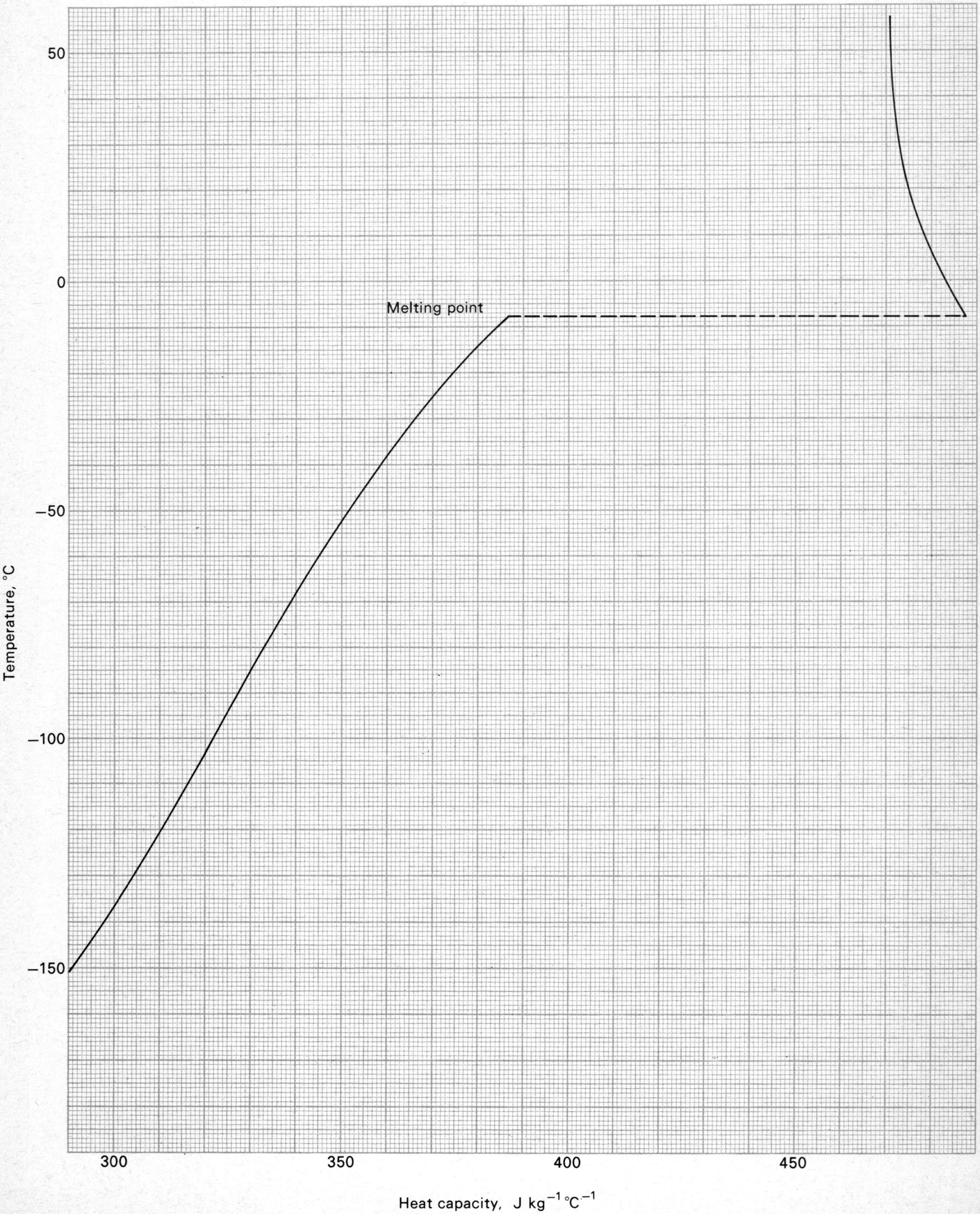

1.5 Heat Capacity of Gaseous Bromine at 1.0133×10^5 N m^{-2} (28)

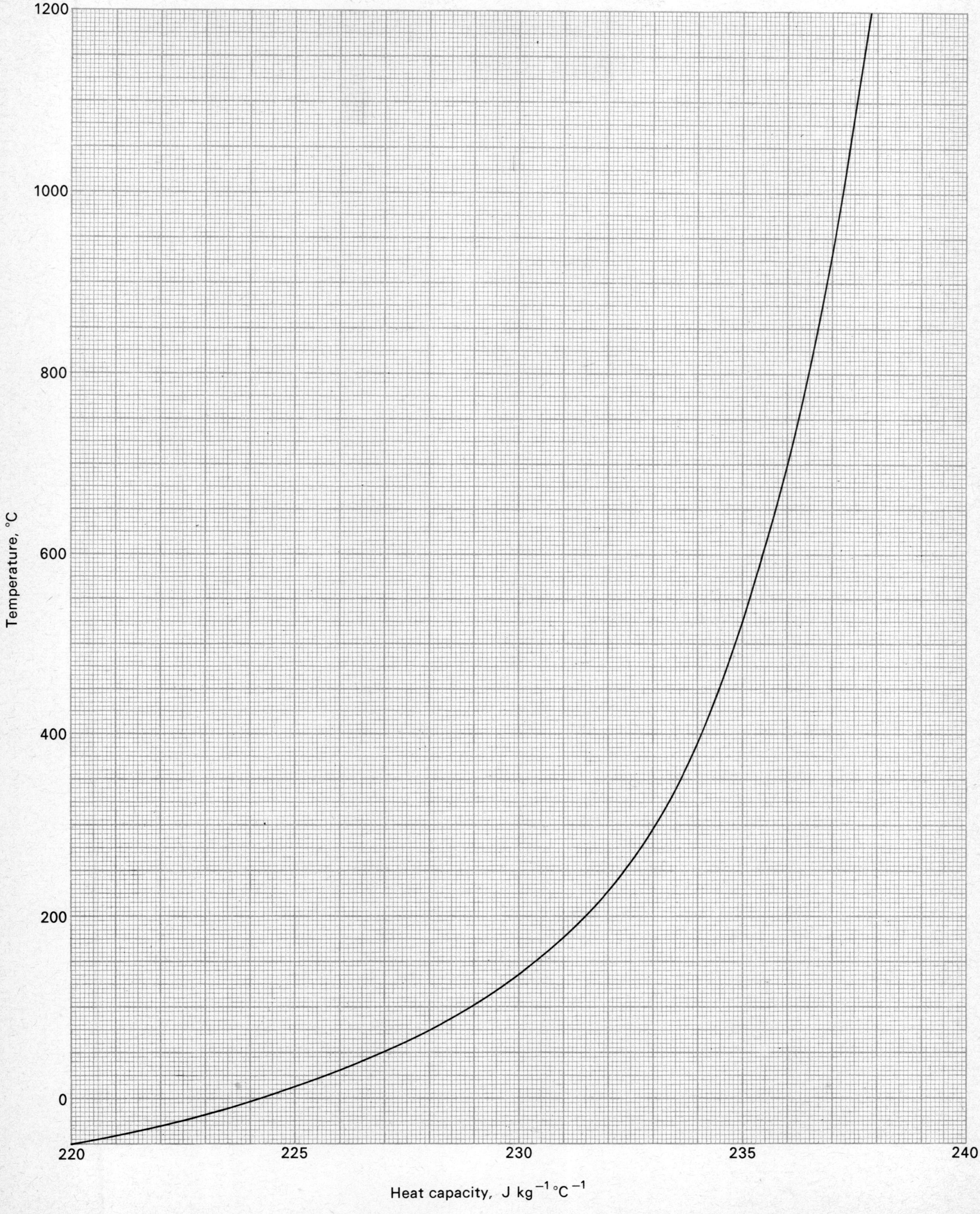

1.6 Viscosity of Liquid Bromine (7)

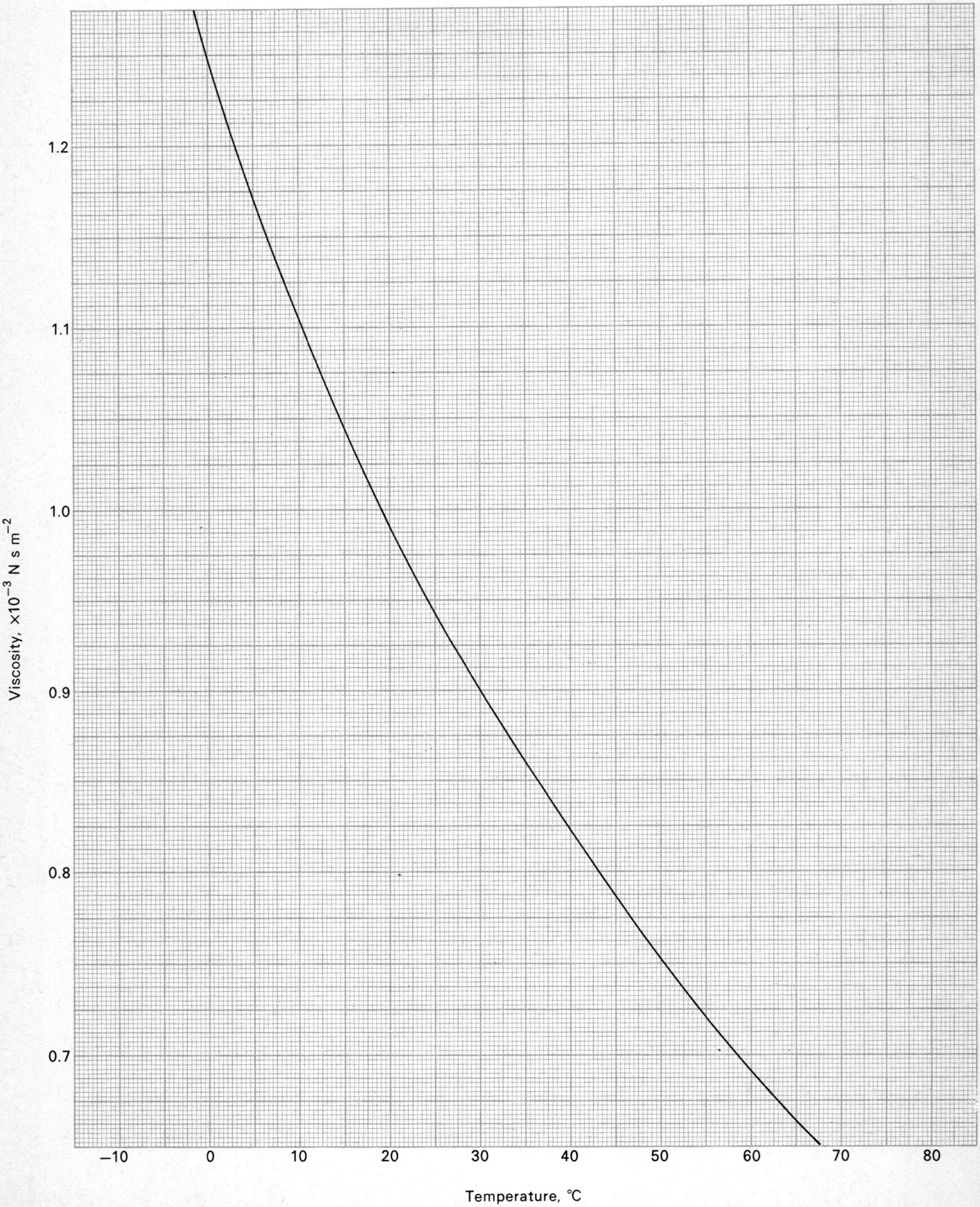

1.7 Viscosity of Gaseous Bromine at 1.0133×10^5 N m^{-2} (2)

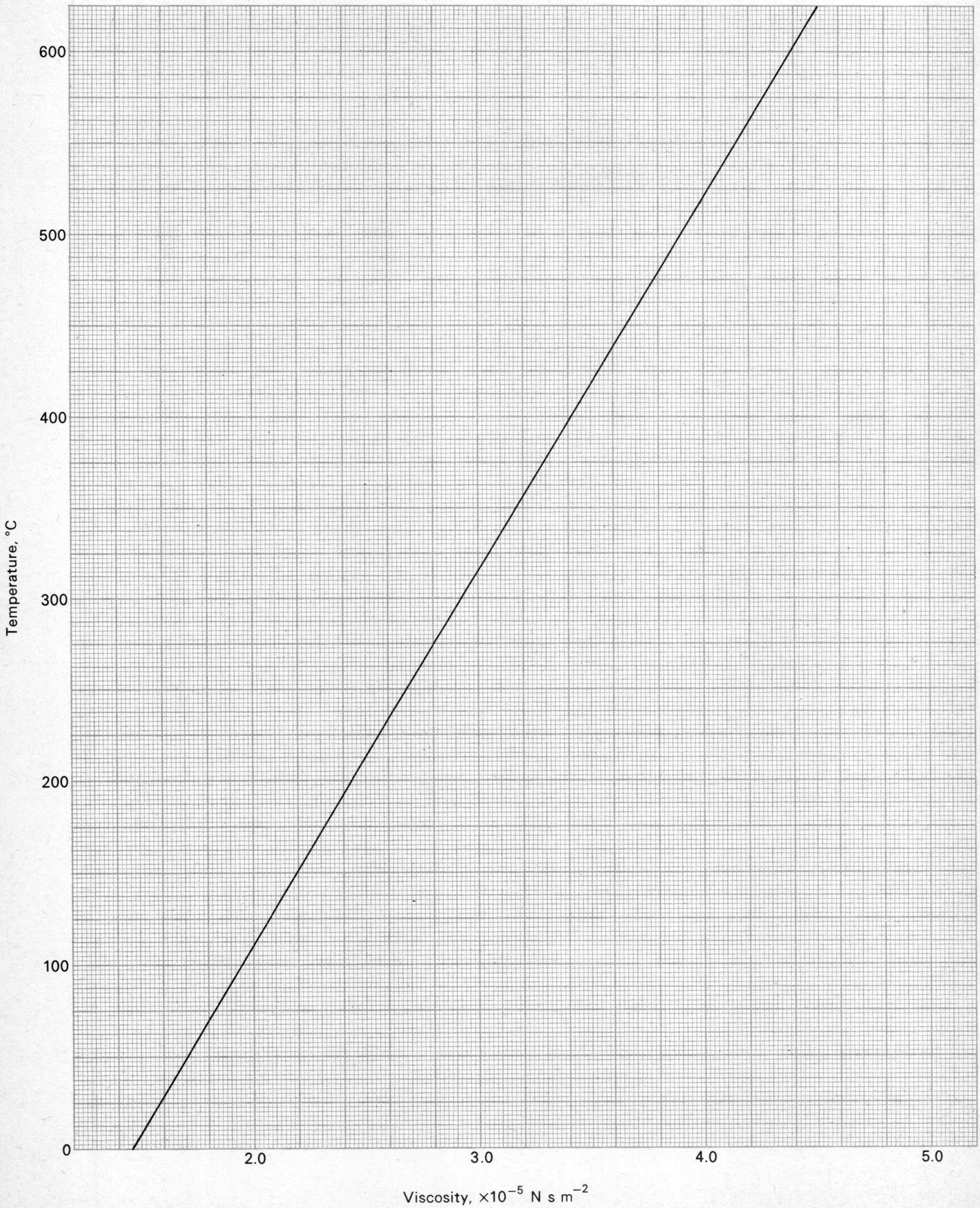

1.8 Thermal Conductivity of Liquid Bromine (**4**)

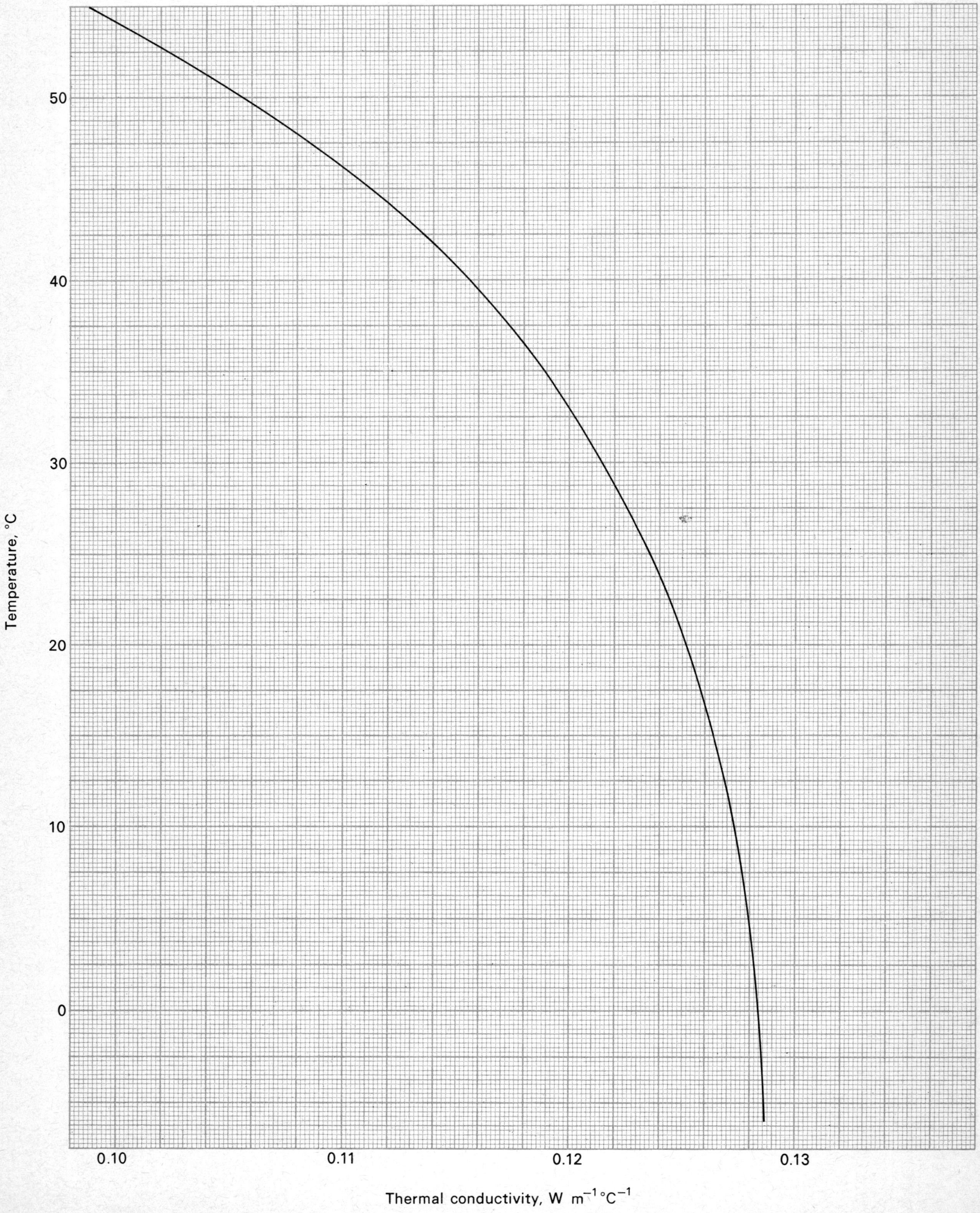

1.9 Thermal Conductivity of Gaseous Bromine at 1.0133×10^5 N m^{-2} (16,18)

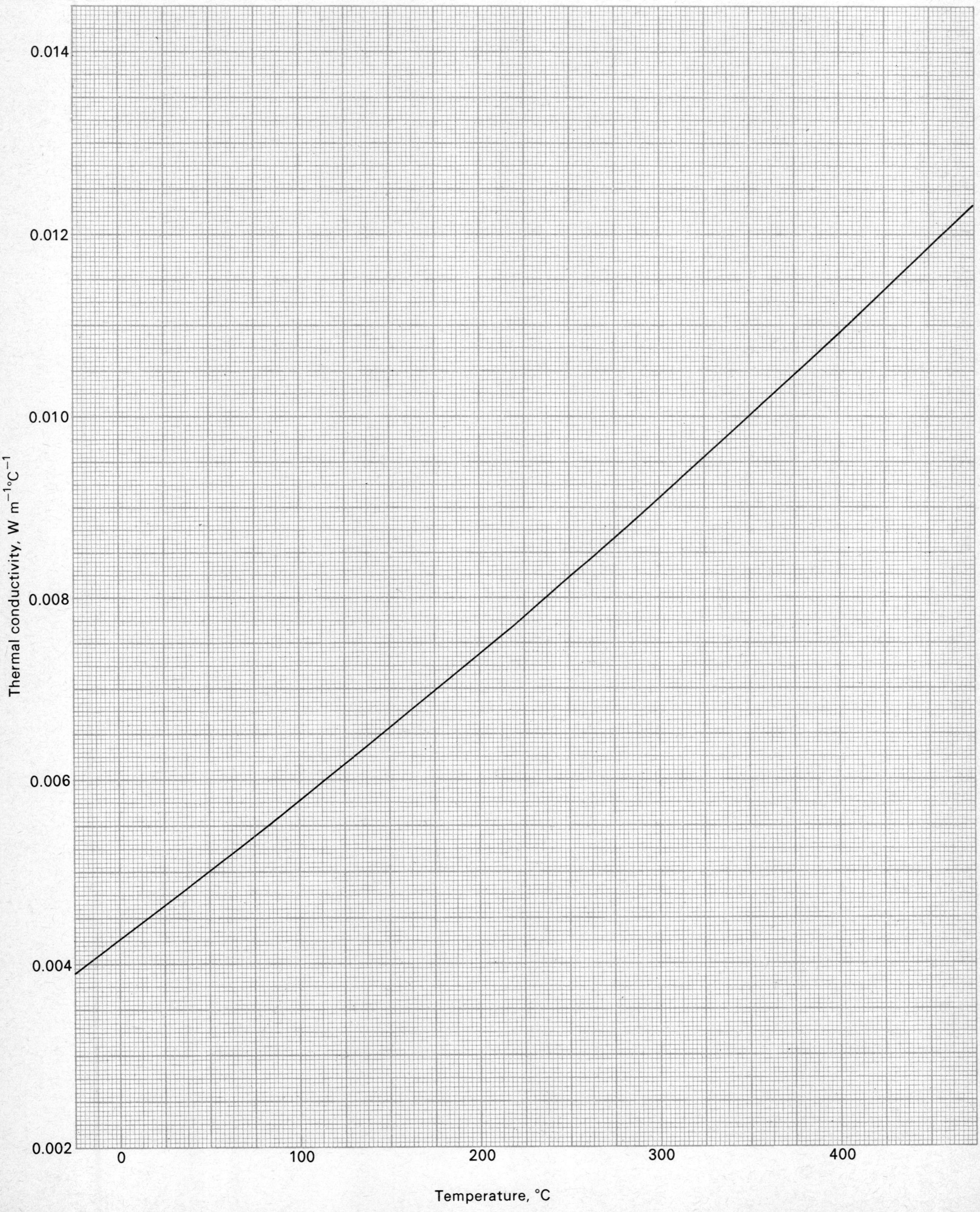

1.10 Surface Tension of Bromine (4)

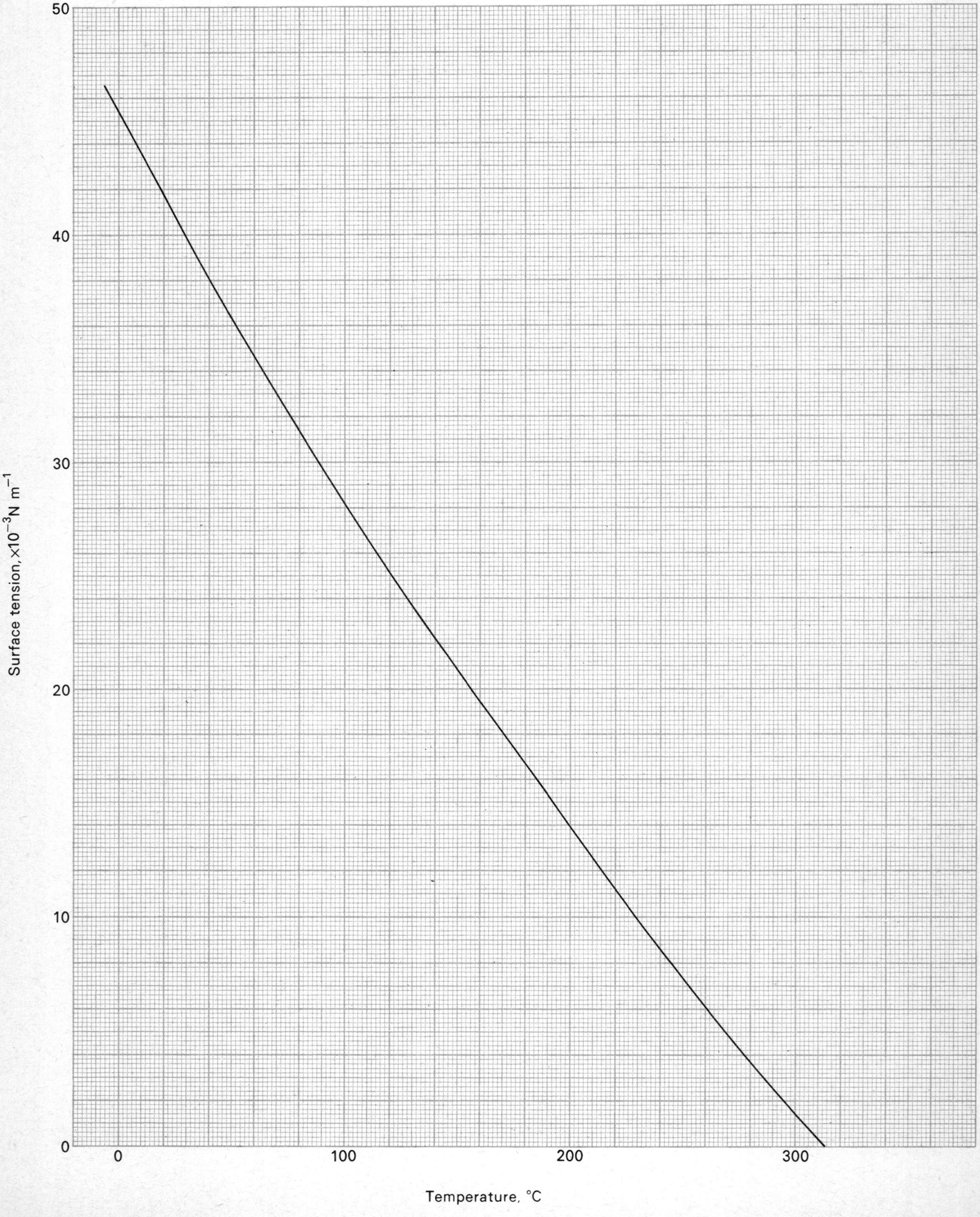

1.11 Electrical Conductivity of Solid and Liquid Bromine (4)

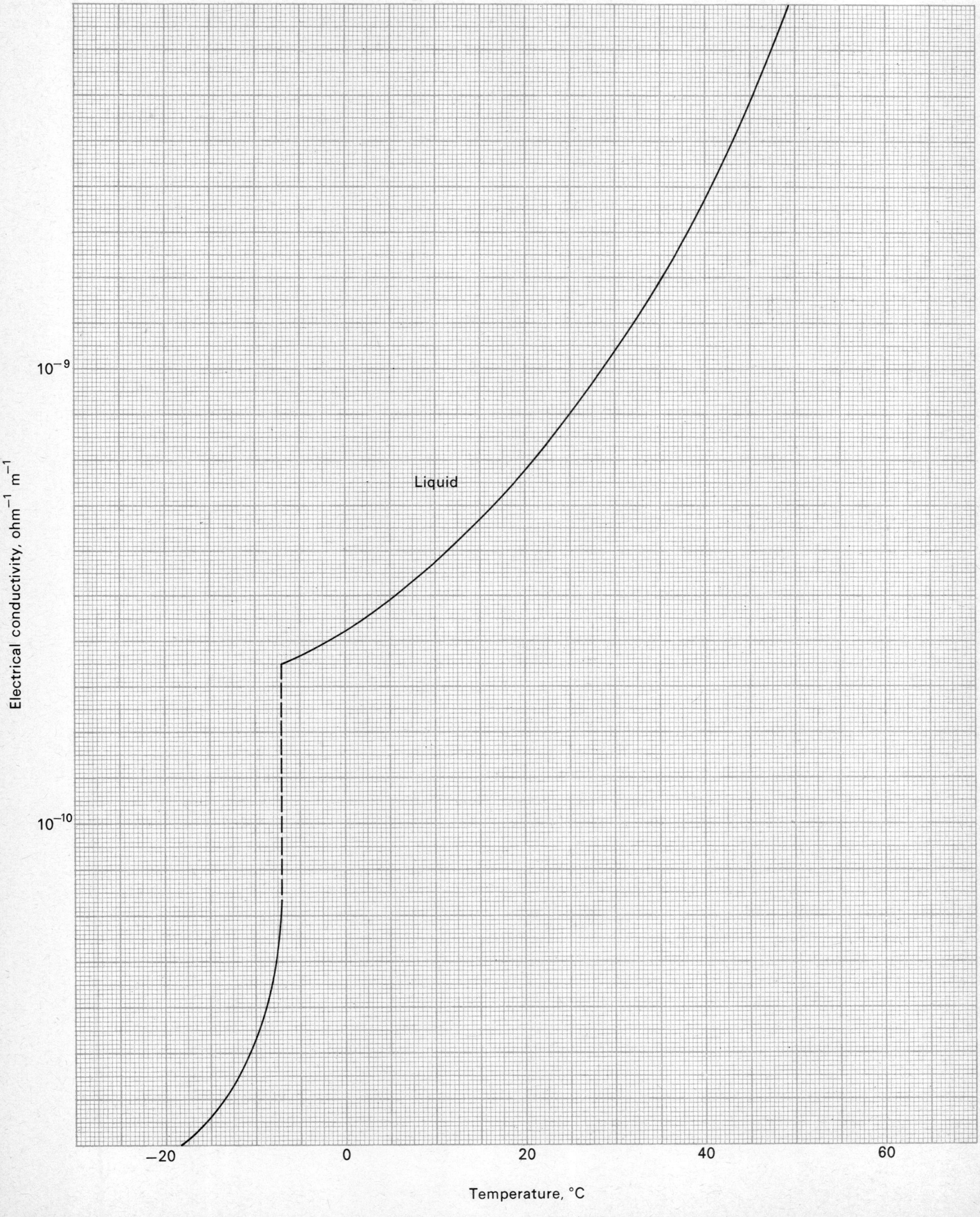

1.12 Solubility of Bromine in Water at 1.0133×10^5 N m^{-2} (19)

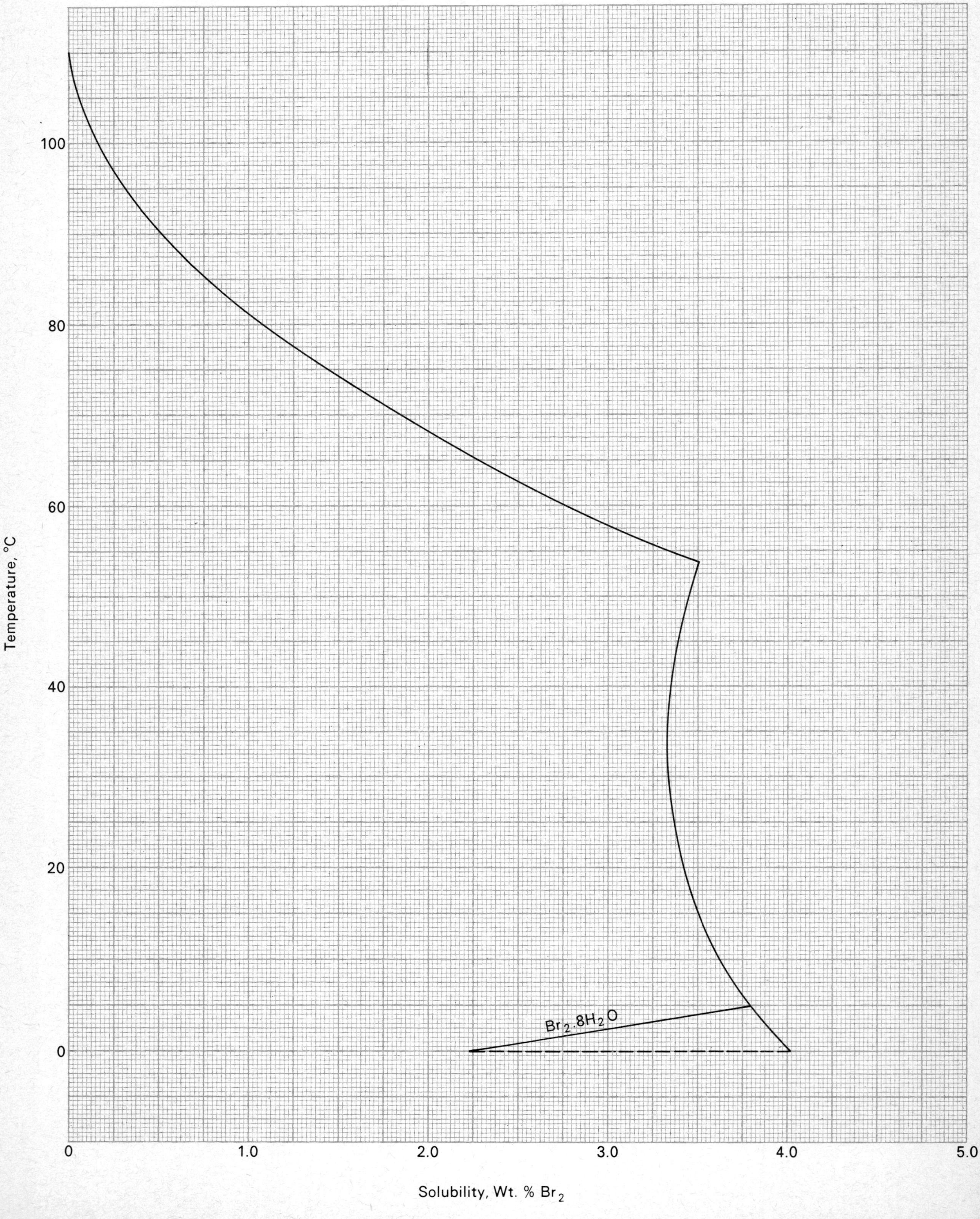

2 CHLORINE

Chlorine is a greenish-yellow gas with an irritating and suffocating odour, attacking the respiratory tract. It was discovered by Sheele in 1774 by reacting muriatic acid with manganese dioxide. Identification as an element was established by Davy in 1808, and he gave it the name of chlorine. It is a heavy gas and combines directly with nearly all elements. Chlorine has a great van der Waals attraction among its molecules imparting relatively high melting points. It has unique physical and chemical properties. The unusually high entropy of fusion and the absence of transition peaks in the heat capacity curve for solid chlorine indicates that the chlorine molecule does not rotate freely in the solid state. Also the density data on liquid chlorine suggest that rotation is inhibited in the liquid state too. Liquid chlorine has the largest compressibility of all elements at 20°C.

Since its discovery chlorine has achieved an enormously large field of applications. It is the most frequently used agent in bleaching, used as a germicide in drinking water, as a reagent in organic chemistry, etc. In spite of its wide application and daily transportation in road and rail tankers, there is no experimental data on liquid thermal conductivity. There are several empirical equations for the estimation of thermal conductivity (34); however, these are not reliable because of its unique behaviour and properties.

In general, the physical properties of chlorine reported by different investigators are in good agreement, having reasonable accuracy despite the early work. There are several major works or reviews on its physical properties: Chlorine Institute (3), Gmelins (11), Kapoor and Martin (16), Manufacturing Chemists Association (21), Matheson Company (22), Mellor (24), Pascal (27), Salmon (33), Sconce (35), Solvay (39), which provide most of the values or references to the data. Its equation of state has been investigated by Kapoor and Martin (16). The thermodynamic properties and diagrams have not been presented here, but they can be found in the following works: Dobratz (6), Feber and Herrick (8), Kapoor and Martin (16), McBride, *et al.* (26), Seshadri, *et al.* (37), Stull (41). The sound absorption, compressibility factor, virial coefficients, and heat capacity ratios at various temperatures and pressures have been reported by Sittig (38). The thermodynamic data relating to dissolution of gaseous chlorine in water at various temperatures has been presented by Kustodina (19). The phase diagram Cl_2—H_2O is discussed by Ketelaar (17).

Table 2.1 and Figs. 2.1–2.13 summarise the more relevant physical properties of molecular chlorine.

References

1 Andrew, S. P. S., *Chem. Eng. Sci.*, 1955, **4**, 269–72.

2 Braune, H., Linke, R., *Z. phys. Chem.*, 1930, **A148**, 195–215.

3 Chlorine Institute, Inc., *Chlorine Manual*, 3rd ed., New York, 1959, pp. 24–32.

4 Clark, C. H. D., *Trans. Faraday Soc.*, 1935, **31**, 585–96.

5 CODATA, Report of the ICSU-CODATA Task Group on Key Values for Thermodynamics, November, 1971, *J. Chem. Thermodyn.*, 1972, **4**, 331–6.

6 Dobratz, C. J., *Chem. Enb.*, 1958 (Feb. 10), **65**, 144–6.

7 Eng. Sci. Data Units, Item No. 66024, London, July 1966.

8 Feber, R. C., Herrick, C. C., *An Improved Calculation of the Ideal Gas Thermodynamic Functions of Selected Diatomic Molecules*, LA-3597, Los Alamos Sci. Lab., Univ. of California, Los Alamos, Sept. 1966, pp. 39–42.

9 Francis, A. W., *J. Chem. Eng. Data*, 1960, **5**, 534–9.

10 Giauque, W. F., Powell, T. M., *J. Am. Chem. Soc.*, 1939, **61**, 1970–4.

11 *Gmelins Handbuch der Anorganischen Chemie*, Syst. No. 6, Chlor, Teil A, Verlag Chemie GmbH., Berlin, 1968, pp. 188–227.

12 Hawes, L. L., Cheesman, G. H., *Acta Cryst.*, 1959, **12**, 477–8.

13 Henglein, F. A., Rosenberg, G., Muchlinski, H., *Z. Physik*, 1922, **11**, 1–12.

14 Honig, R. E., Hook, H. O., *RCA Rev.*, 1960, **21**, 360–8.

15 Huff, V. N., Gordon, S., Morrell, V. E., NACA Report No. 1037, 1951, pp. 1–57.

16 Kapoor, R., Martin, J. J., *Thermodynamic Properties of Chlorine*, Eng. Res. Inst., Univ. of Michigan, Michigan, 1957, 43 pp.

17 Ketelaar, J. A. A., *Electrochem. Technol.*, 1967, **5** (3–4), 143–7.

18 Kobe, K. A., Lynn, R. E., *Chem. Rev.*, 1953, **52**, 117–236.

19 Kustodina, V. A., *Zh. Prikl. Khim.*, 1970, **43** (9), 2096–7.

20 Leduc, A., *Chem. Rev.*, 1929, **6**, 1–16.

21 Manufacturing Chemists Association, *Chlorine*, Chemical Safety Data Sheet SD-80, Washington, D.C., 1970, 31 pp.

22 *Matheson Gas Data Book*, 4th ed., The Matheson Company, Inc., East Rutherford, N.J., 1966, pp. 101–9.

23 Mathews, J. F., *Chem. Rev.*, 1972, **72**, 71–100.

24 Mellor, J. W., Ed., *A Comprehensive Treatise on Inorganic and Theoretical Chemistry*, Vol. II, Suppl. I, *The Halogens*, Longmans, London, 1956, pp. 262–688.

25 Moszynski, J. R., Ed., *Proceedings of the 4th Symposium on Thermophysical Properties*, Am. Soc. Mech. Eng., New York, 1968, pp. 323–49.

26 McBride, B. J., Heimel, S., Ehlers, J. G., Gordon, S., *Thermodynamic Properties to 6000 K for 210 Substances Involving the First 18 Elements*, NASA SP-3001, Natl. Aeronautics and space Administration, Washington, D.C., 1963, p. 75.

27 Pascal, P., Ed., *Nouveau Traité de Chimie Minérale*, Vol. XVI, F, Cl, Br, I, At, Mn, Tc, Re, Masson et Cie, Éditeurs, Paris, 1960, pp. 151–335.

28 Pellaton, M., *J. Chim. Phys.*, 1915, **13**, 426.

29 Rabinovich, M. A., *Zh. Russ. Fiz. Khim., Obshchestva*, 1926, **58**, 229–39.

30 Ree, T. S., Ree, T., Eyring, H., *J. Chem. Phys.*, 1964, **41**, 524–30.

31 Richardson, E. G., *J. Acoust. Soc. Am.*, 1959, **31**, 152–4.

32 Rossini, F. P., Wagman, D. D., Evans, W. H., Levine, S., Jaffe, I., *Selected Values of Chemical Thermodynamic Properties*, Circular 500, Natl. Bur. Stand., Washington, D.C., 1952, p. 549.

33 Salmon, P., Magne, M., *Le Chlore*, Inst. Natl. Sec. Prev. Accid. Trav., Paris, 1960, pp. 54–9.

34 Schaefer, C. A., Thodos, G., *AIChE Journal*, 1959, **5** (3), 367–72.

35 Sconce, J. S., Ed., *Chlorine, Its Manufacture, Properties and Uses*, Reinhold Publ. Corp., New York, 1962, pp. 21–45.

36 Seitz, F., *The Modern Theory of Solids*, McGraw-Hill Book Co., New York, 1940, p. 73.

37 Seshadri, D. N., Viswanath, D. S., Kulor, N. R., *J. Indian Inst. Sci.*, 1966, **48** (1), 38–51.

38 Sittig, E., *Acustica*, 1960, **10**, 81–6.

39 Solvay & Cie, *Liquid Chlorine*, Solvay Technical and Engineering Bulletin No. 7, 3rd ed., New York, 1955.

40 Steere, N. V., Ed., *Handbook of Laboratory Safety*, The Chemical Rubber Co., Cleveland, 1967.

41 Stull, D. R., Dir., *JANAF Thermochemical Tables*, PB 168 370, Clearinghouse for Federal Scientific and Technical Information, Springfield, Va., 1965, p. Cl_2.

42 Thomson, T. R., Eyring, H., Ree, T., *Proc. Natl. Acad. Sci. U.S.*, 1960, **46**, 336–43.

43 Trauts, M., Ruf, F., *Ann. Phys.*, 1934, **20**, 127–34.

44 Wagenbreth, H., *PTB-Mitteilungen*, 1968, **2**, 91–4.

45 Wagman, D. D., Evans, W. H., Halow, I., Parker, V. B.,

Bailey, S. M., Schumm, R. H., *Selected Values of Chemical Thermodynamic Properties*, 3 Parts, Natl. Bur. Std., Washington, D.C., 1965–8.

46 Weast, R. C., Ed., *Handbook of Chemistry and Physics*, 52nd ed., The Chemical Rubber Co., Cleveland, 1971–2.

47 Ziegler, L., *Chem.-Ing.-Technik*, 1950, **22** (11), 229–48.

Table 2.1 Properties of Chlorine

		Ref.
Chemical formula	Cl_2	(41)
Molecular weight	70.906	(41)
Critical temperature	144.0°C	(23)
Critical pressure	7.7011×10^6 N m^{-2}	(23)
Critical density	573.0 kg m^{-3}	(23)
Triple-point temperature	−100.98°C	(10)
Triple-point pressure	1391.86 N m^{-2}	(10)
Boiling point at 1.0133×10^5 N m^{-2}	−34.6°C	(28)
Melting point	−101.0°C	(32)
Transition temperature	None	(32)
Heat of formation of gas at 25°C	None	(5)
Free energy of formation of gas at 25°C	None	(41)
Heat of sublimation at triple-point temperature	2.512×10^7 J kg-mol^{-1}	(36)
Heat of fusion	6.410×10^6 J kg-mol^{-1}	(10)
Heat of transition	None	(32)
Entropy of gas at 25°C	2.22965×10^5 J kg-mol^{-1} °C^{-1}	(5)
Heat capacity ratio of gas at 15°C and 1.0133×10^5 N m^{-2}	1.355	(20)
Compressibility of liquid at −33.1°C	87.7×10^{-11} m^2 N^{-1}	(42)
Coefficient of thermal expansion of liquid at boiling point	1.84×10^{-3} °C^{-1}	(47)
Velocity of sound at 19.7°C and 1.0133×10^5 N m^{-2}	219 m s^{-1}	(31)
Diffusivity in air at 20°C and 1.0133×10^5 N m^{-2}	1.24×10^{-5} m^2 s^{-1}	(1)
Dielectric constant of liquid at −50°C	2.10	(11)
Dielectric constant of gas at 25°C and 1.0133×10^5 N m^{-2}	1.001 52	(4)
Electrical conductivity of liquid near melting point	10^{-14} ohm^{-1} m^{-1}	(29)
Refractive index of liquid at 25°C	1.3788	(9)
Threshold limit value	(1 ppm) 3×10^{-6} kg m^{-3}	(33)

2.1 Vapour Pressure of Solid and Liquid Chlorine (14,33)

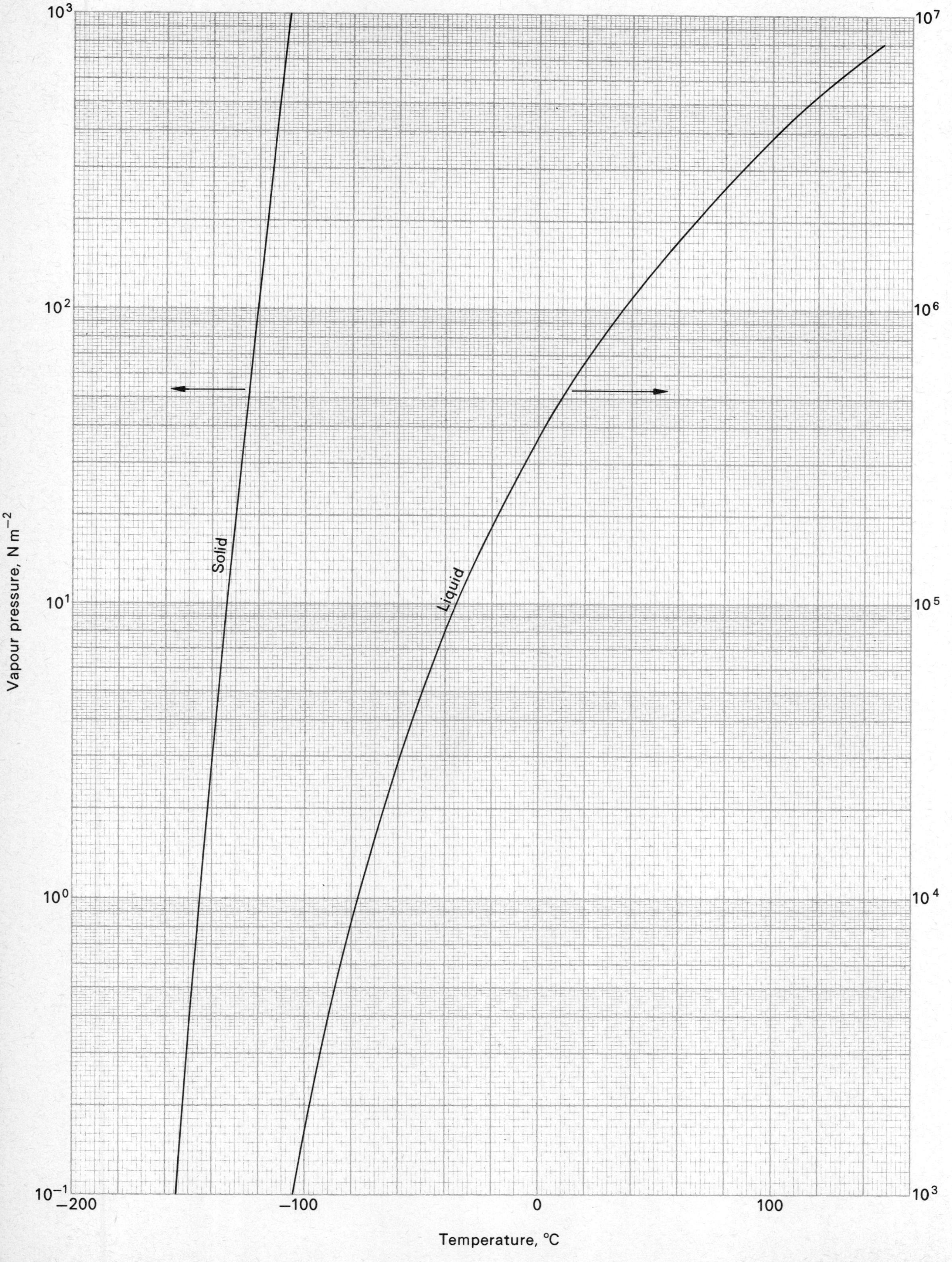

2.2 Density of Solid Chlorine (12)

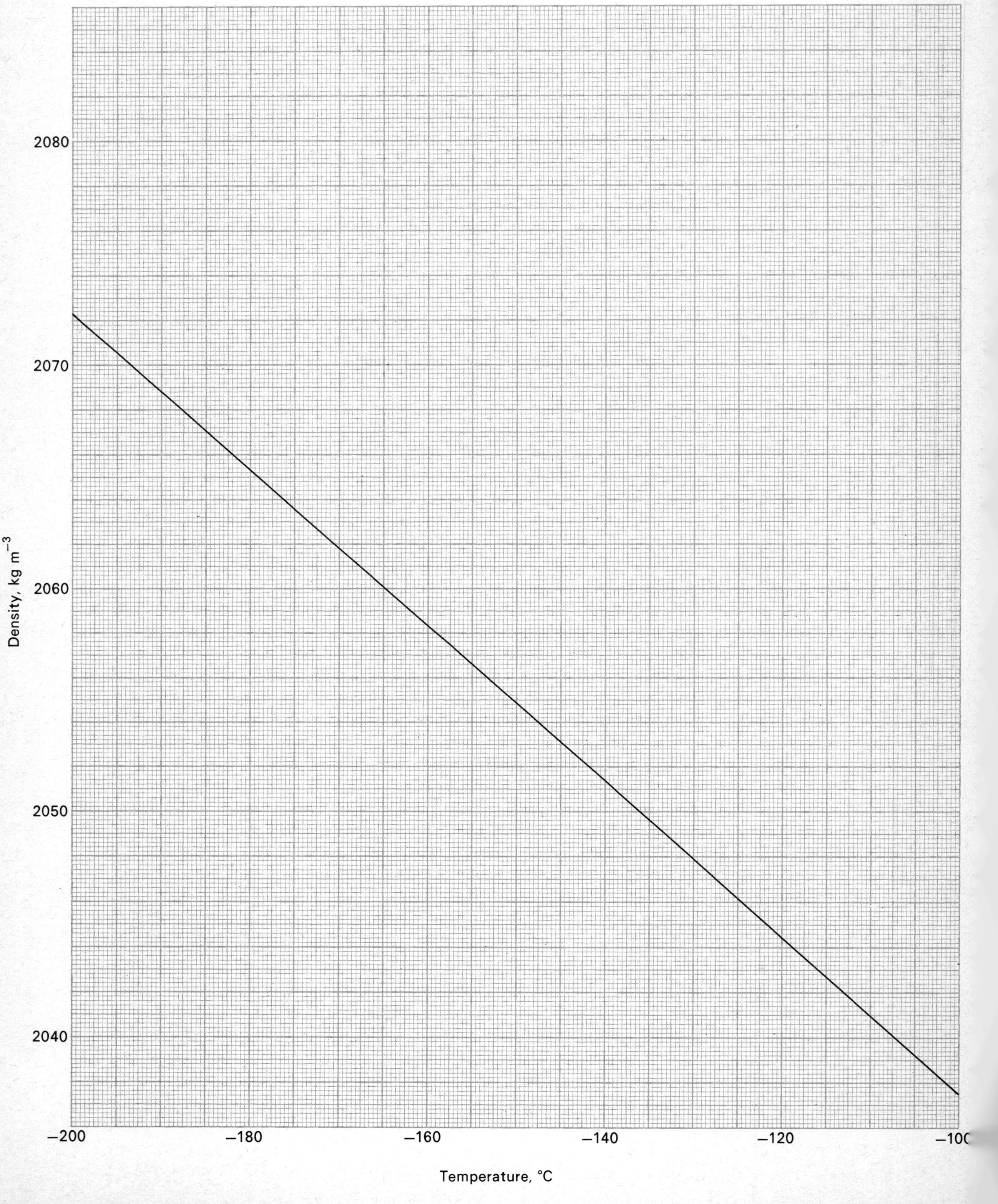

2.3 Orthobaric Density of Chlorine (33)

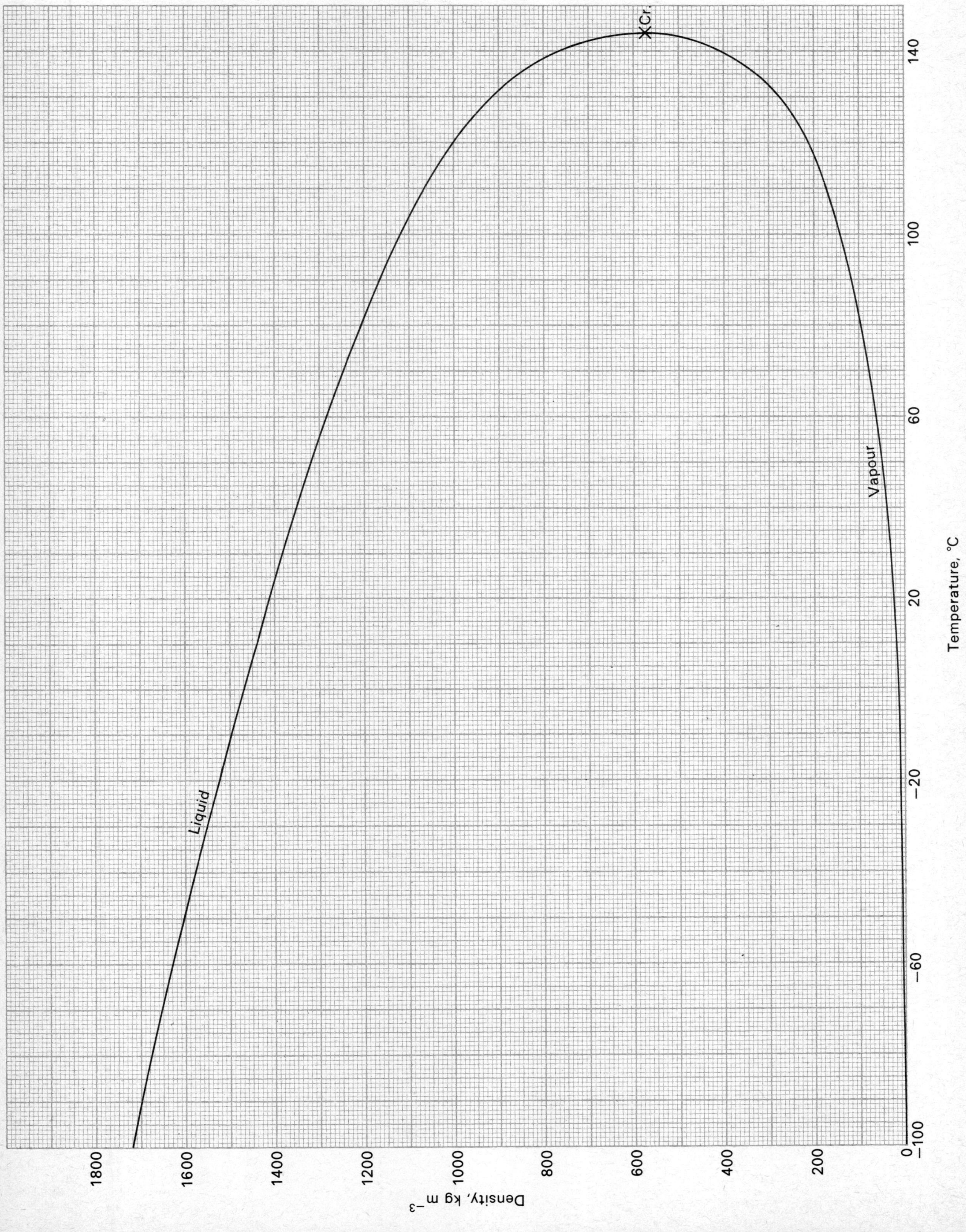

2.4 Supercritical Density of Chlorine (44)

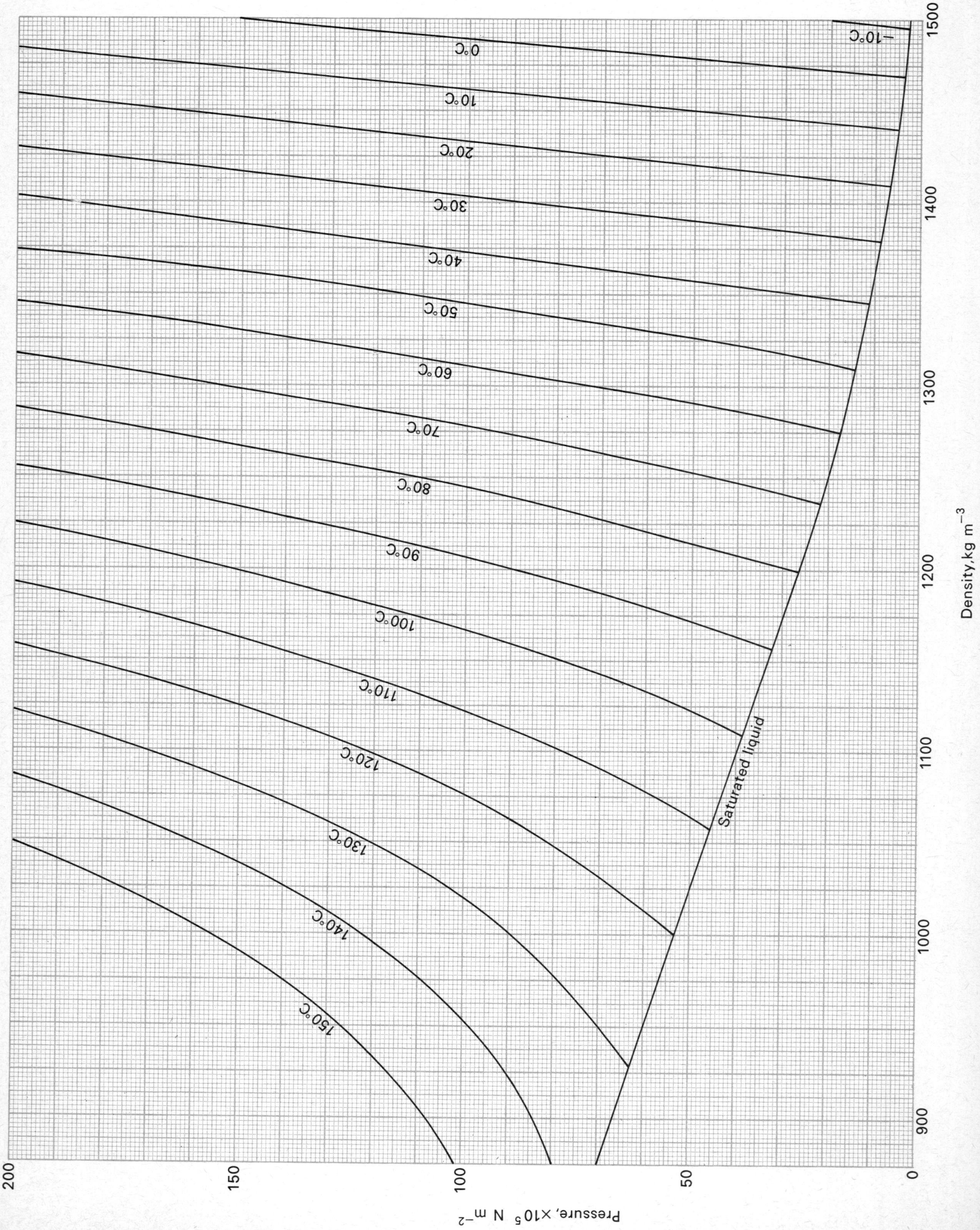

2.5 Latent Heat of Vaporisation of Chlorine (47)

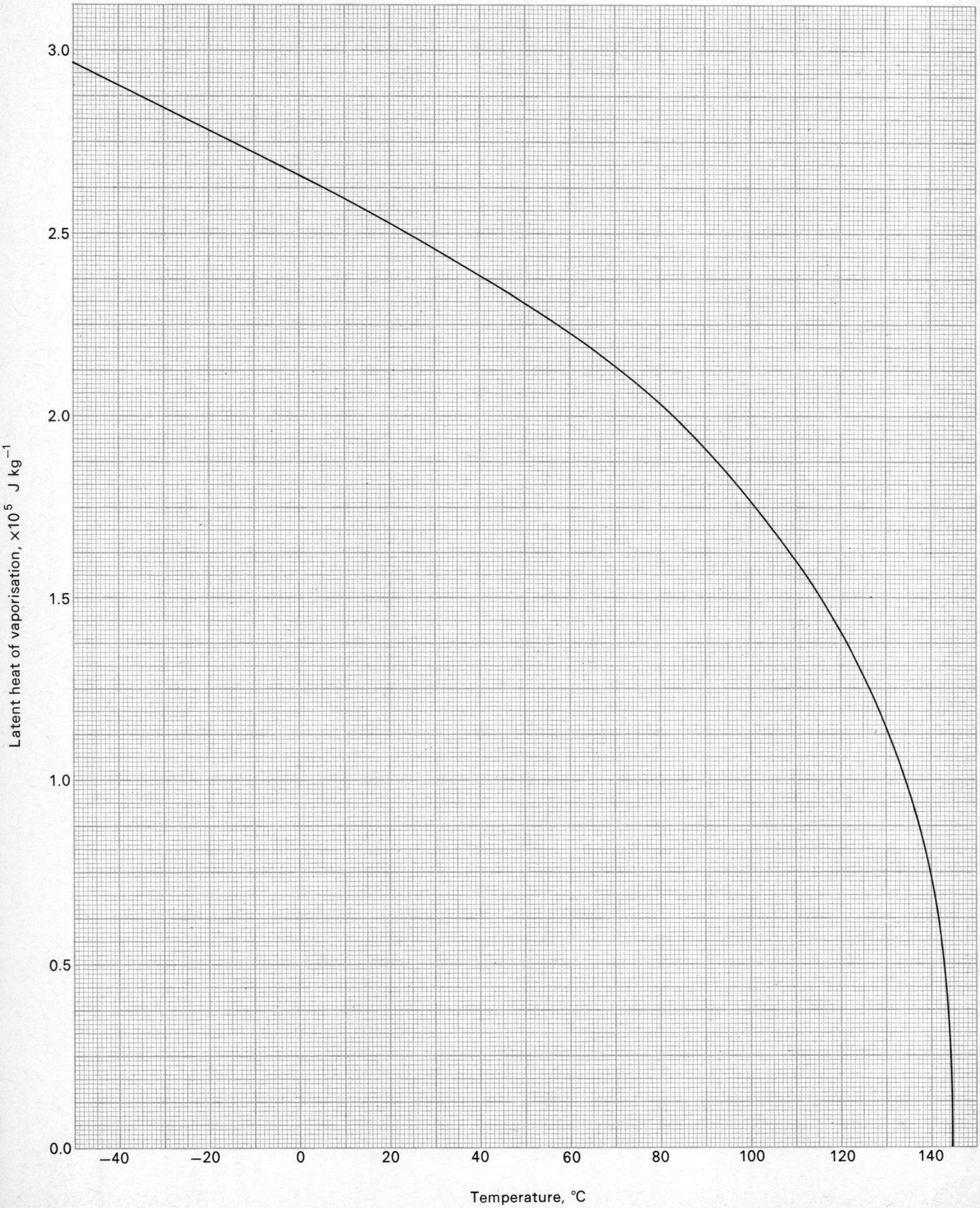

2.6 Heat Capacity of Solid and Liquid Chlorine (10)

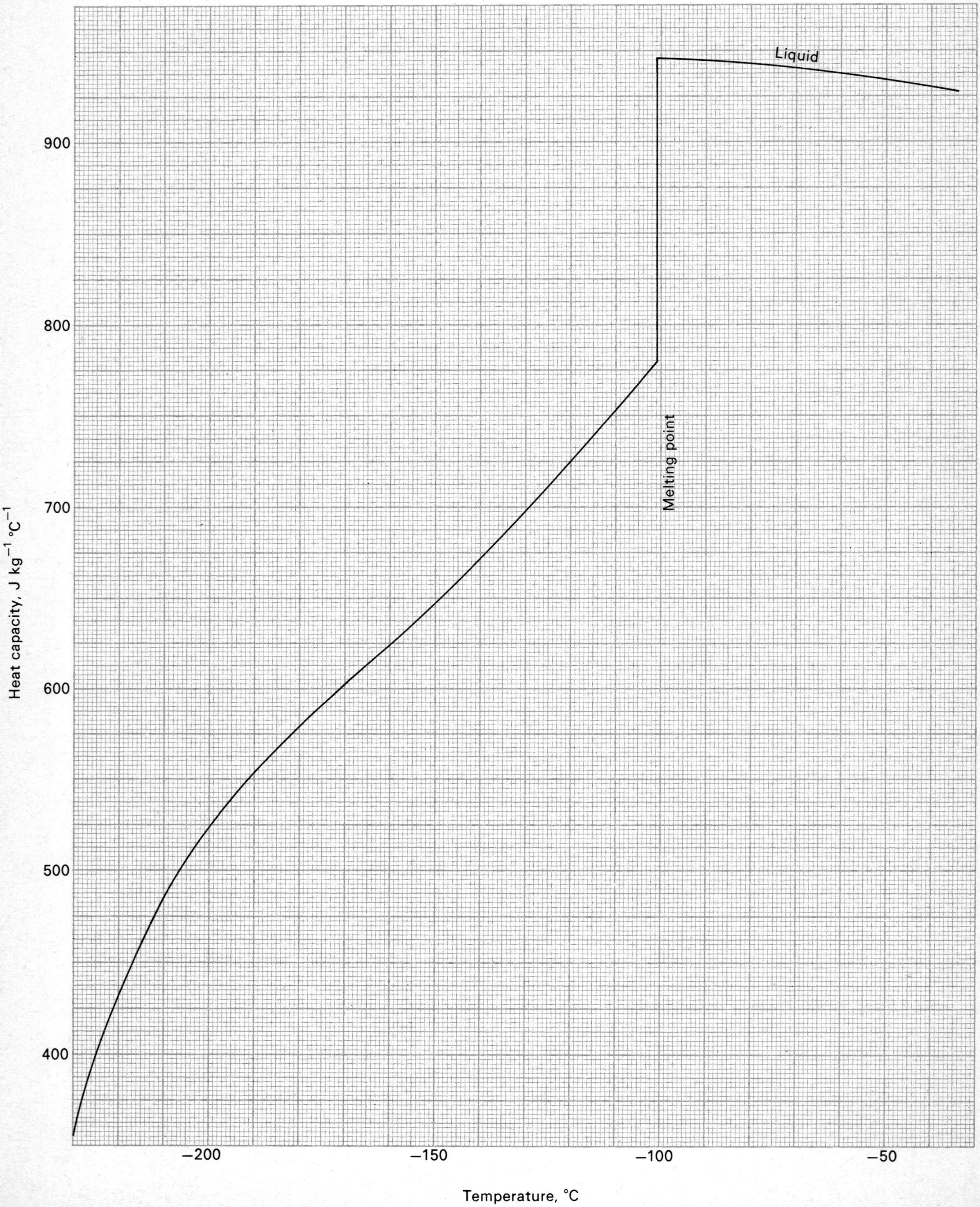

2.7 Heat Capacity of Gaseous Chlorine at 1.0133×10^5 N m^{-2} (15)

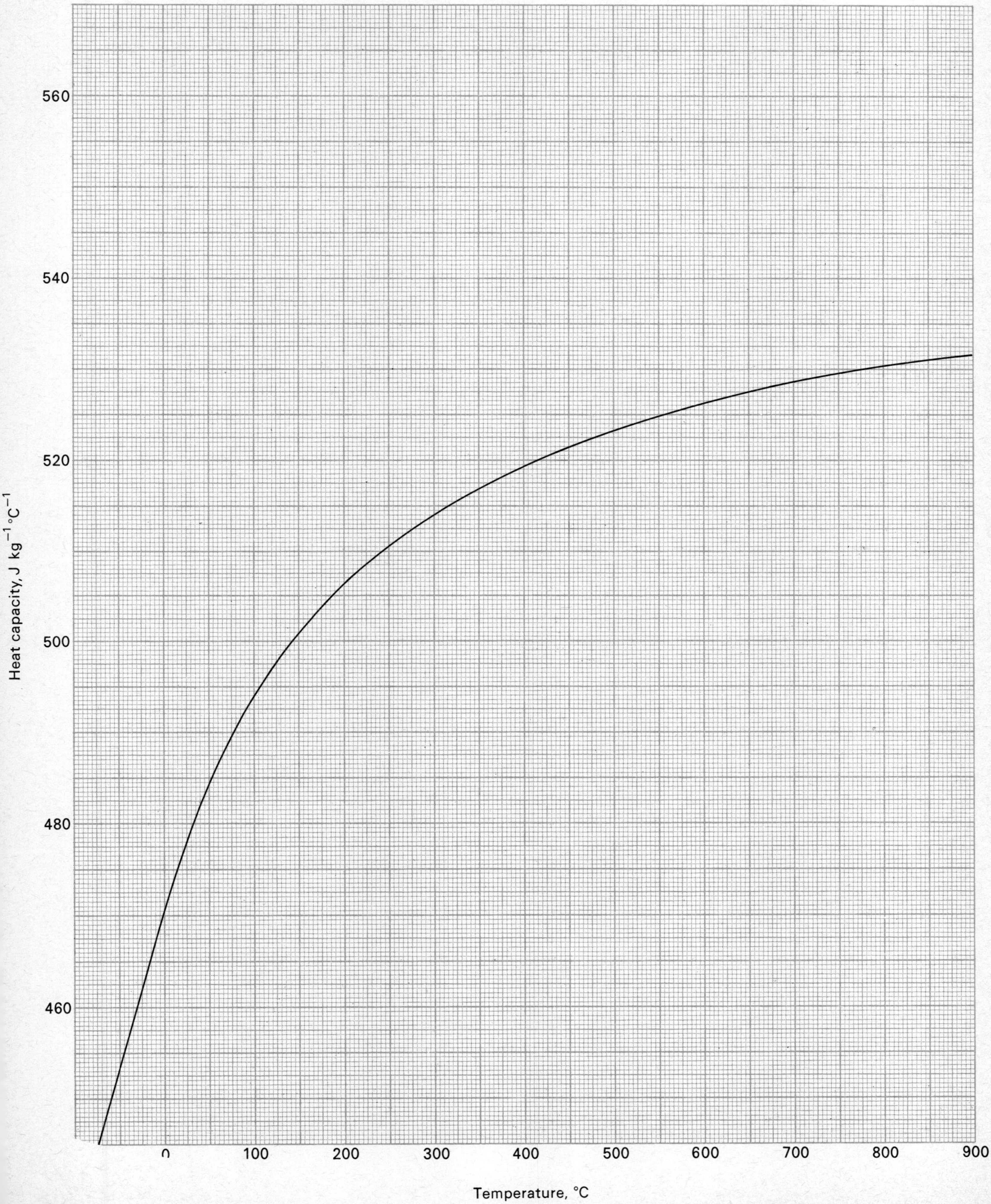

2.8 Viscosity of Liquid Chlorine (7)

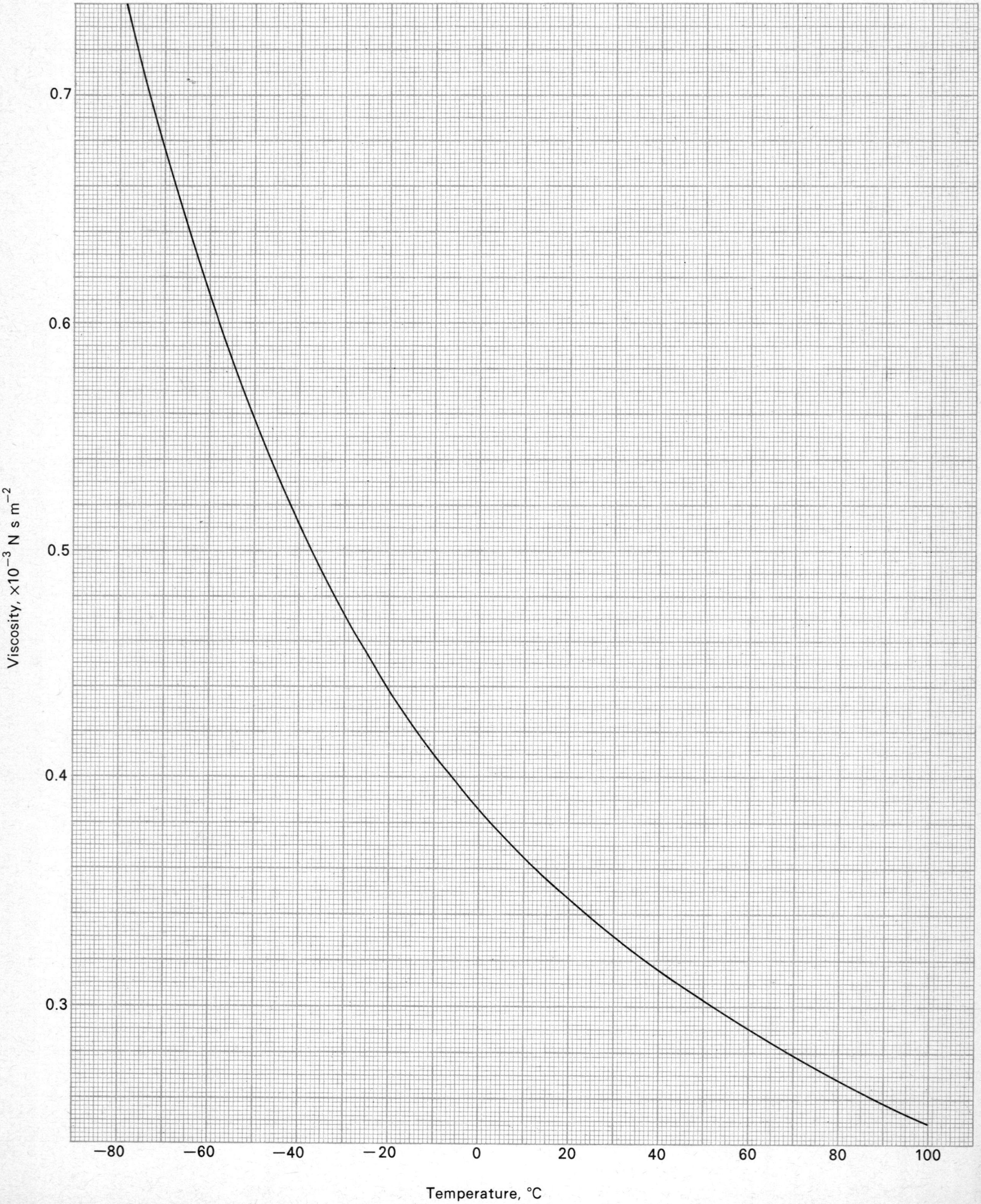

2.9 Viscosity of Gaseous Chlorine at 1.0133×10^5 N m^{-2} (2,43)

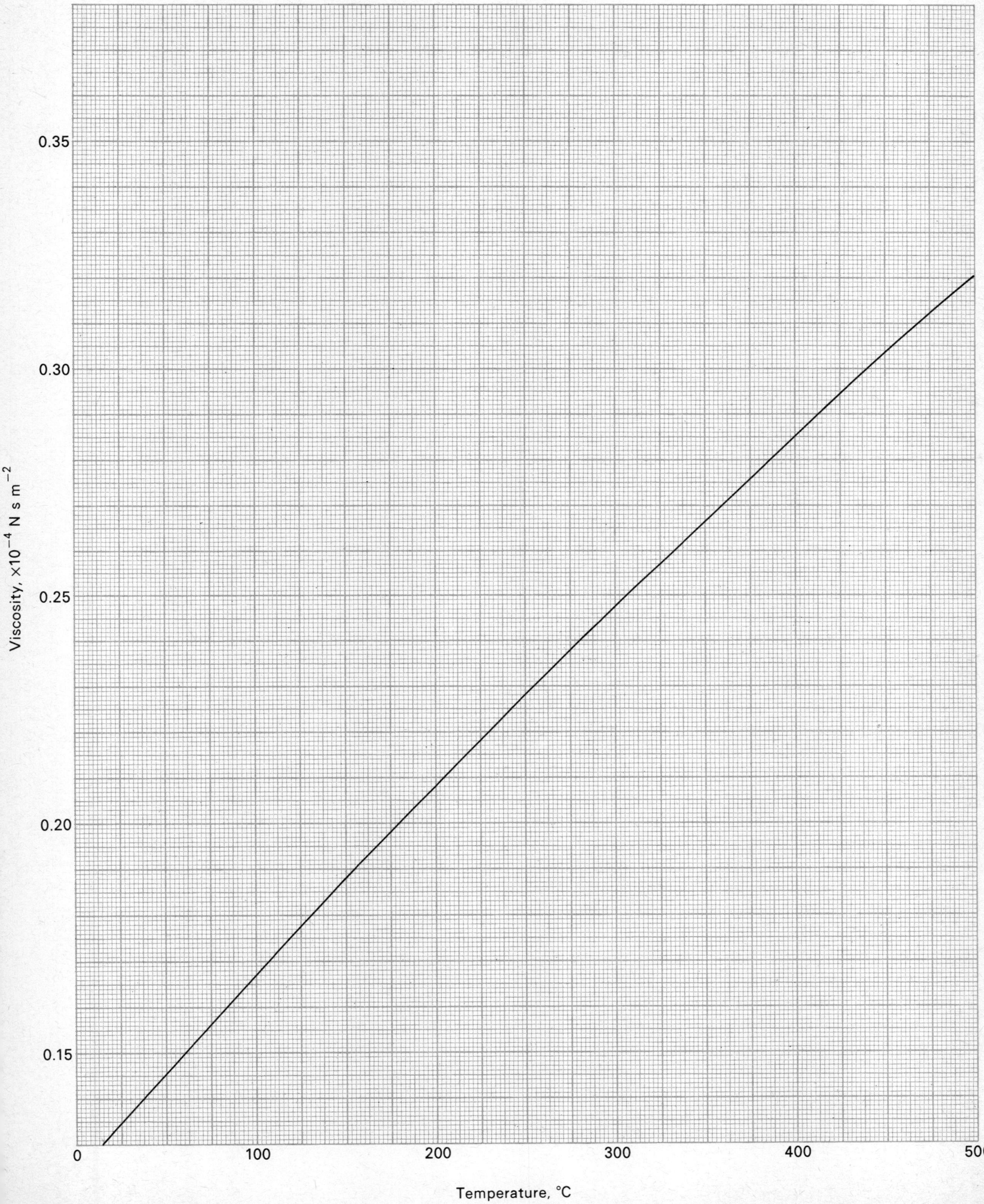

2.10 Thermal Conductivity of Liquid Chlorine (34)

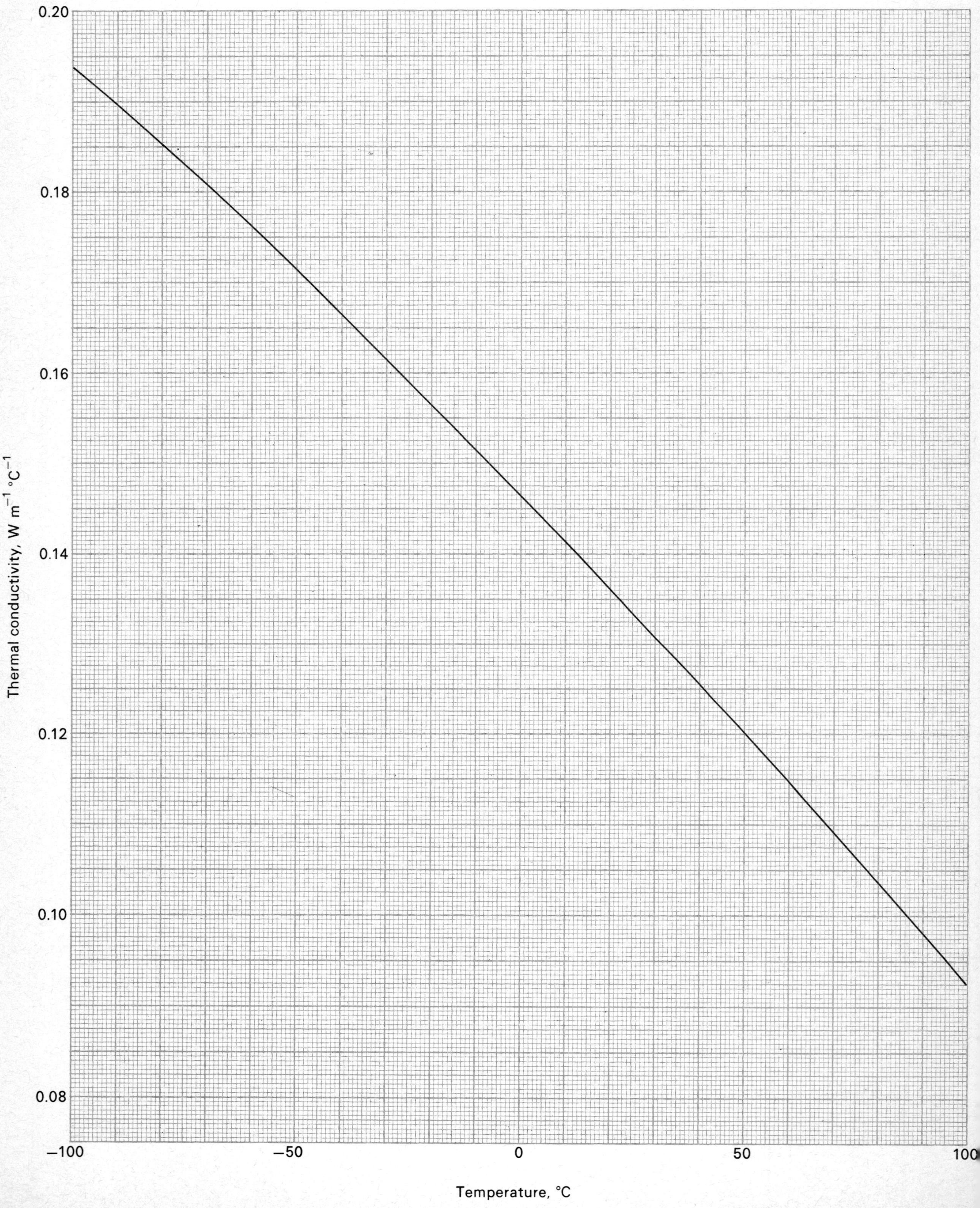

2.11 Thermal Conductivity of Gaseous Chlorine at 1.0133×10^5 N m^{-2} (25)

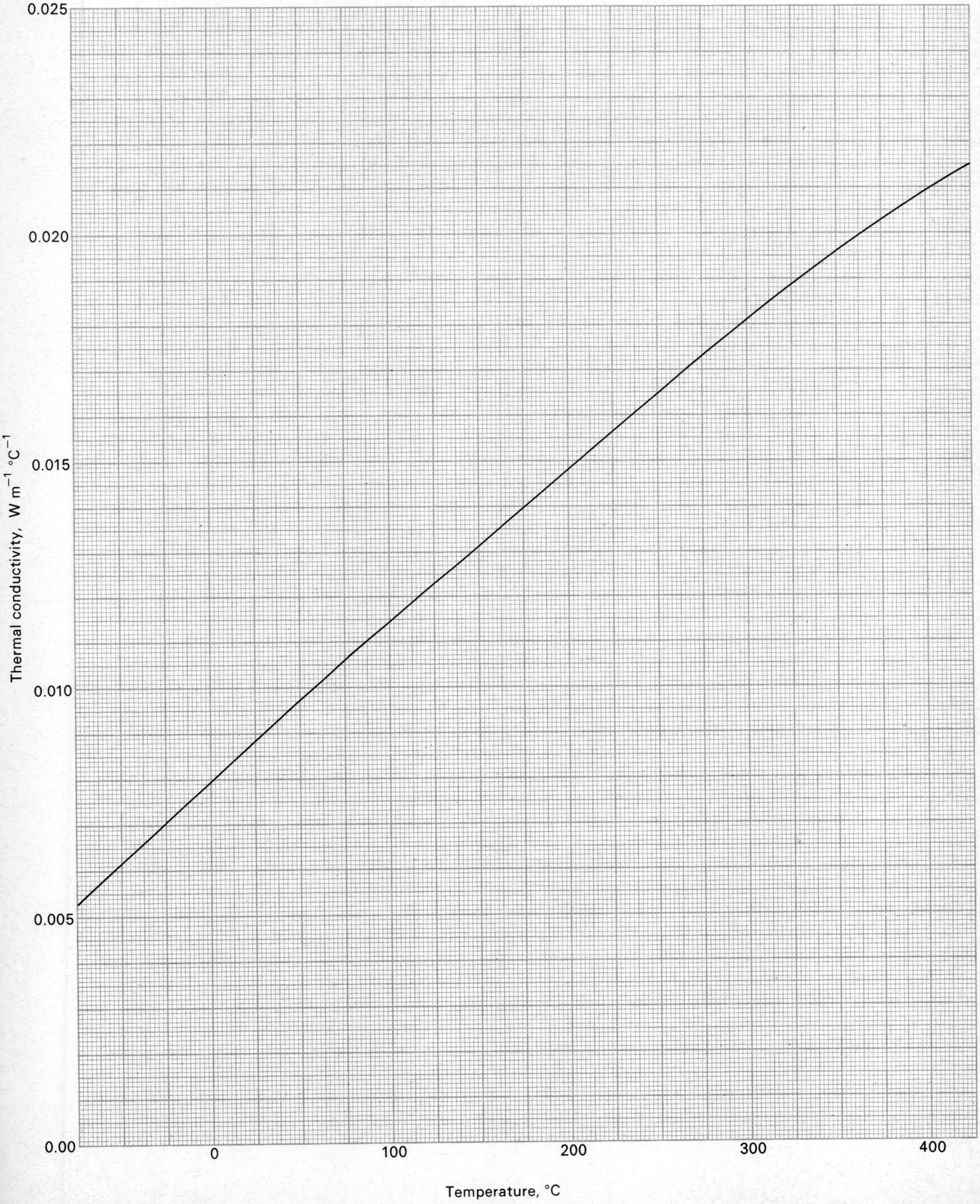

2.12 Surface Tension of Chlorine (30)

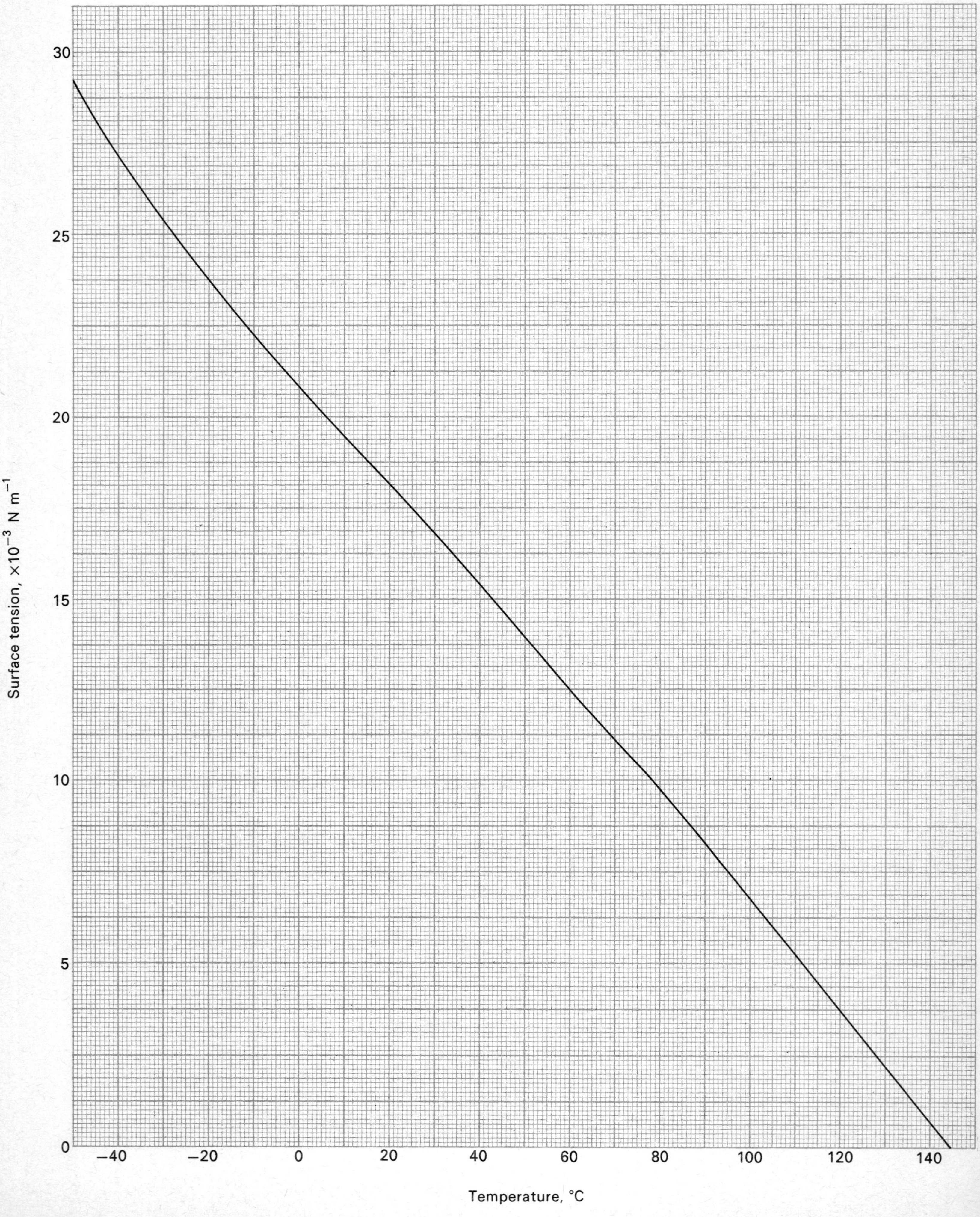

2.13 Solubility of Chlorine in Water at 1.0133×10^5 N m^{-2} Total Pressure (17,33)

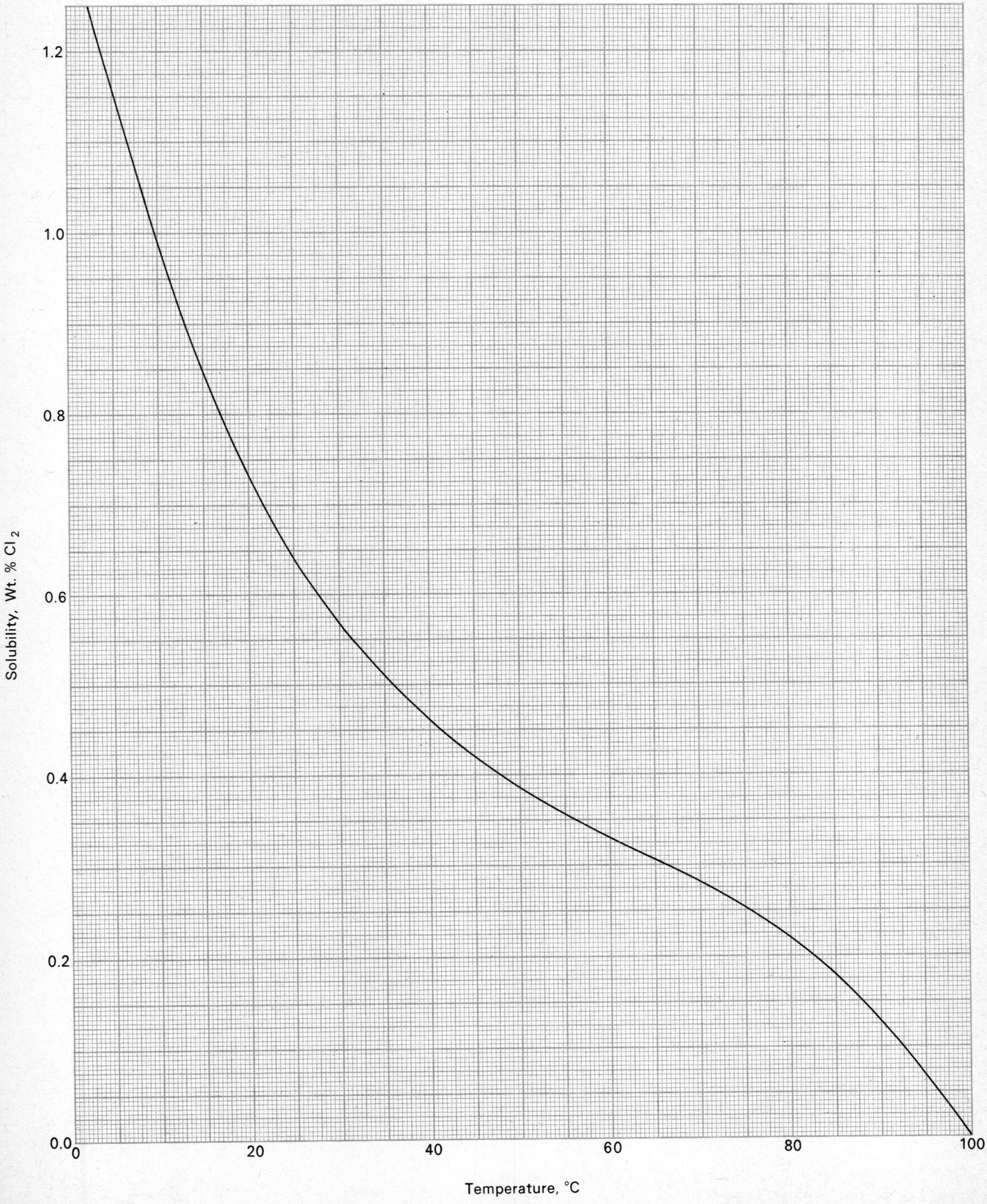

3 FLUORINE

Molecular fluorine is a pale yellow gas at ambient temperature and pressure. It condenses to a yellow liquid at −188.2°C, and this solidifies to a yellow solid at −219.62°C. At −253.0°C the colour of the solid changes to white. Fluorine is diamagnetic. It reacts directly with all elements except chlorine, nitrogen, oxygen, and the noble gases. Fluorine reacts vigorously with liquid water, liberating not only oxygen but ozone.

Both gaseous and liquid fluorine are generally used in processes leading to inorganic and organic fluorine compounds. In particular industries manufacturing refrigerants, solvents, propellants, etc. require large quantities of fluorine. Liquid fluorine is a very energetic oxidising agent and therefore can be used as a propellant oxidiser in rocket propulsion systems.

Recently the National Bureau of Standards carried out measurements on the physical and thermodynamic properties of fluorine under the sponsorship of the United States Air Force. These publications (6), (14), (15), (16), (20), (21), are the basic sources of selection. However, there are several review articles (1), (5), (7), (8), (11), which have been examined in connection with the final selection of values for the physical properties of fluorine.

Some properties of fluorine are not presented here, e.g. intermolecular force constants (24), melting point curve (20), heat capacity ratio of solid (10), etc., which are outside the scope of this work. Furthermore, there are no reliable data on liquid thermal and electrical conductivity of fluorine. Schaefer and Thodos (17) have developed the corresponding state theory of thermal conductivity for diatomic molecules; however, since 1959, it has been proved that this theory gives for example about 50% error for the thermal conductivity of liquid hydrogen chloride (26).

The selected physical properties of fluorine are presented in Table 3.1 and Figs. 3.1–3.11.

References

1 Allied Chemical Corporation, *Fluorine*, Product Data Bull. PD-TA-85413, Gen. Chem. Div., New York, Revised Aug. 15, 1958.

2 Data Book and Buyers Guide, Engineering Data Section, *Cryogenic & Ind. Gases* 1971, **6** (3), 41–80.

3 Elverum, G. W., Doescher, R. N., *J. Chem. Phys.*, 1952, **20** (12), 1834–6.

4 Franck, E. U., Stöber, W., *Z. Naturfors.*, 1952, **7a**, 822–9.

5 Gmelins Handbuch der anorganischen Chemie, Syst. No. 5, Fluor, Verlag-Chemie, GmbH, Weinheim, 1959, pp. 81–98.

6 Goodwin, R. D., Prydz, R., *J. Res. Natl. Bur. Std.*, 1970, **74A** (4), 499–505.

7 Hall, L. A., McCarty, R. D., *A Bibliography of Thermophysical Properties for Fluorine from 0 to 300 K*, Natl. Bur. Stand., Report 7676, April 1, 1963.

8 Horvath, A. L., *J. Chem. Doc.*, 1970, **10** (1), 59–65.

9 Hu, J.-H., White, D., Johnston, H. L., *J. Am. Chem. Soc.*, 1953, **75**, 5642–8.

10 Kanda, E., *Bull. Chem. Soc. Japan*, 1937, **12**, 511–20.

11 Landolt-Börnstein: *Zahlenwerte und Funktionen aus Physik, Chemie, Astronomie, Geophysik, und Technik*, Springer-Verlag, Berlin, Vols. I–IV.

12 Lange, N. A., Ed., *Handbook of Chemistry*, 10th ed., McGraw-Hill, New York, 1967, pp. 112–13.

13 Liley, P. E., *The Thermal Conductivity of 46 Gases at Atmospheric Pressure*, Proc. 4th Symp. Thermophys. Prop., The Am. Soc. Mech. Eng., New York, 1968, pp. 323–49.

14 Prydz, R., Goodwin, R. C., *J. Res. Natl. Bur. Stand.*, 1970, **74A** (5), 661–5.

15 Prydz, R., Straty, G. C., *J. Res. Natl. Bur. Stand.*, 1970, **74A** (6), 747–60.

16 Prydz, R., Straty, G. C., Timmerhaus, K. D., *J. Chem. Phys.*, 1970, **53** (6), 2359–63.

17 Schaefer, C. A., Thodos, G., *AIChE. Journal*, 1959, **5** (3), 367–72.

18 Slesser, C., Ed., *Preparation and Technology of Fluorine and Organic Fluoro Compounds*, McGraw-Hill, New York, 1951, p. 137.

19 Steere, N. V., Ed., Handbook of Laboratory Safety, The Chemical Rubber Co., Cleveland, 1967.

20 Straty, G. C., Prydz, R., *Phys. Lett.*, 1970, **31A** (6), 301–2.

21 Straty, G. C., Prydz, R., *The Vapor Pressure of Liquid Fluorine*, Advances in Cryogenic Engineering, Vol. 16, Plenum Publ. Co., London, 1970, 36–41.

22 Stull, D. R., *et al.*, *JANAF Thermodynamic Tables*, Dow Chemical Co., Midland, Mich., 1965–6.

23 Weast, R. C., Ed., *Handbook of Chemistry and Physics*, 51st ed., ed., The Chemical Rubber Co., Cleveland, 1970.

24 White, D., Hu, J.-H., Johnston, H. L., *J. Chem. Phys.*, 1953, **21** (7), 1149–52.

25 Wicke, E., Franck, E. U., *Angew. Chem.*, 1954, **66** (22), 701–10.

26 Ziebland, H., Needham, D. P., *Recent Measurements of the Thermal Conductivity of Liquid Anhydrous Hydrogen Chloride*, Proc. 4th Symp. Thermophys. Prop., The Am. Soc. Mech. Eng., New York, 1968, pp. 296–303.

Table 3.1 Properties of Fluorine

		Ref.
Chemical formula	F_2	(22)
Molecular weight	37.9968	(22)
Critical temperature	−128.84°C	(16)
Critical pressure	5.215×10^6 N m^{-2}	(16)
Critical density	573.7 kg m^{-3}	(16)
Triple-point temperature	−219.669°C	(20)
Triple-point pressure	252.0 N m^{-2}	(16)
Boiling point at 1.0133×10^5 N m^{-2}	−188.20°C	(16)
Melting point	−219.62°C	(25)
Transition point temperature	−227.60°C	(9)
Density of solid at −219.67°C	1907.0 kg m^{-3}	(5, 11)
Heat of formation of gas at 25°C	None	(22)
Heat of sublimation at boiling point temperature	8.248×10^6 J kg-mol^{-1}	(5)
Heat of fusion	5.107×10^5 J kg-mol^{-1}	(9)
Heat of transition	7.281×10^5 J kg-mol^{-1}	(9)
Entropy of gas at 25°C	2.0305×10^5 J kg-mol^{-1} °C^{-1}	(22)
Heat capacity ratio of gas at 25°C and 1.0133×10^5 N m^{-2}	1.360	(18)
Specific gravity of solid at −223°C	1300 kg m^{-3}	(18)
Coefficient of thermal expansion of liquid at −200°C	3×10^{-3} °C^{-1}	(12)
Velocity of sound at 25°C and 1.0133×10^5 N m^{-2}	284.0 m s^{-1}	(5, 11)
Self-diffusion of gas at 0°C and 1.0133×10^5 N m^{-2}	1.70×10^{-5} m^2 s^{-1}	(25)
Dielectric constant of liquid at boiling point	1.43	(25)
Refractive index of gas at 0°C and 1.0133×10^5 N m^{-2}	1.000 195	(25)
Threshold limit value	(0.1 ppm) 0.2×10^{-6} kg m^{-3}	(1)

3.1 Vapour Pressure of Fluorine (21)

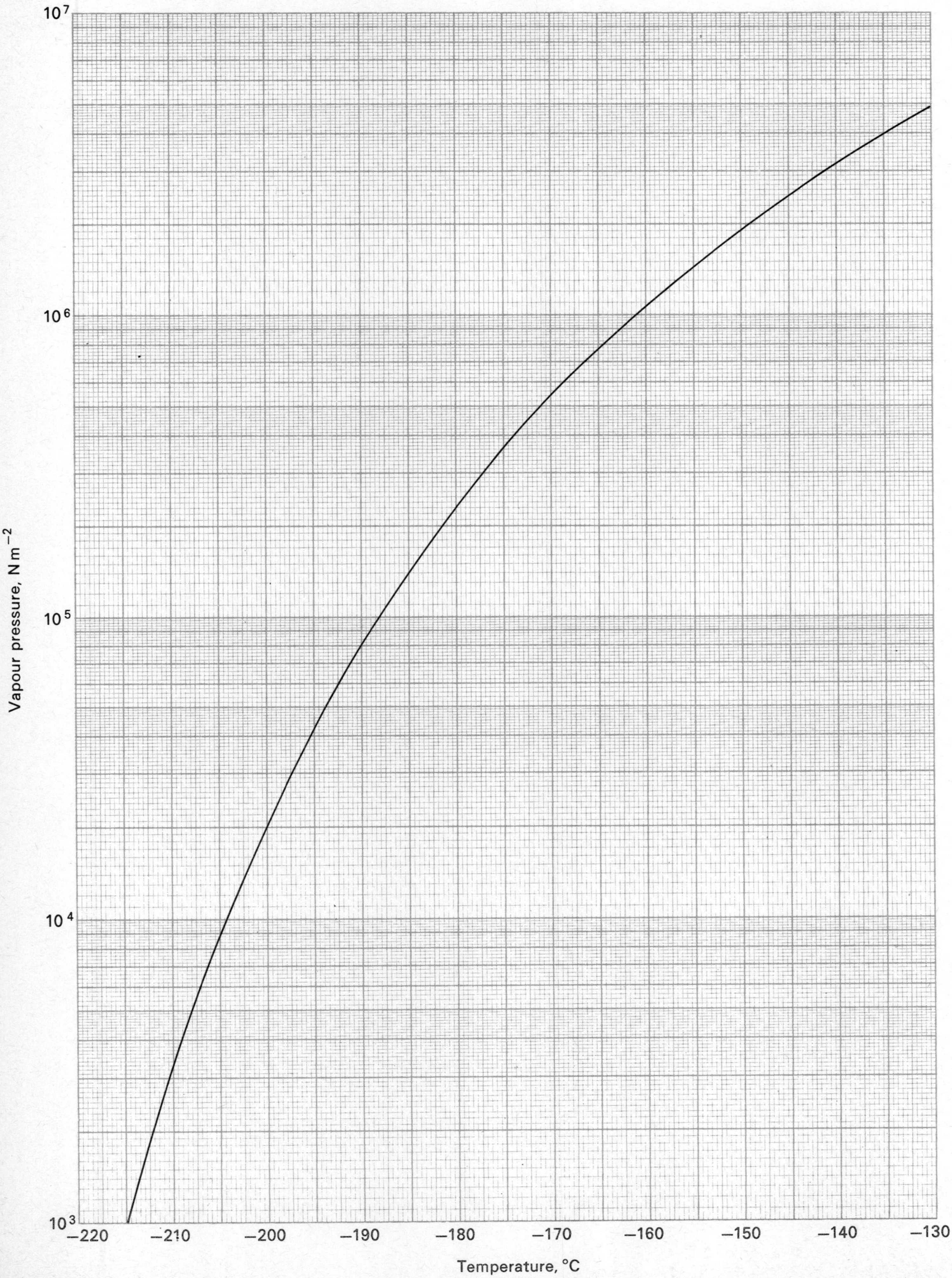

3.2 Orthobaric Density of Fluorine (16)

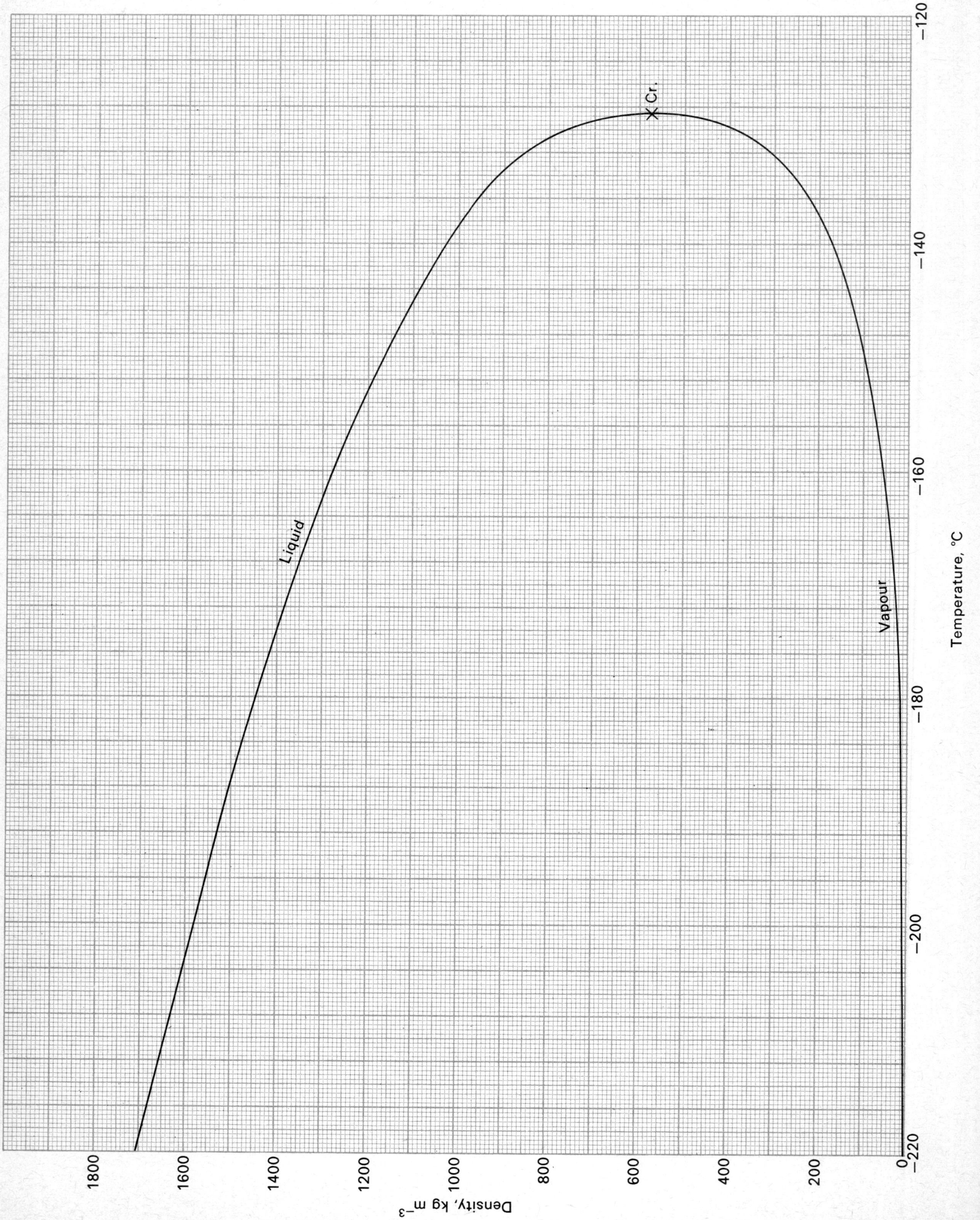

3.3 Supercritical Density of Fluorine (15)

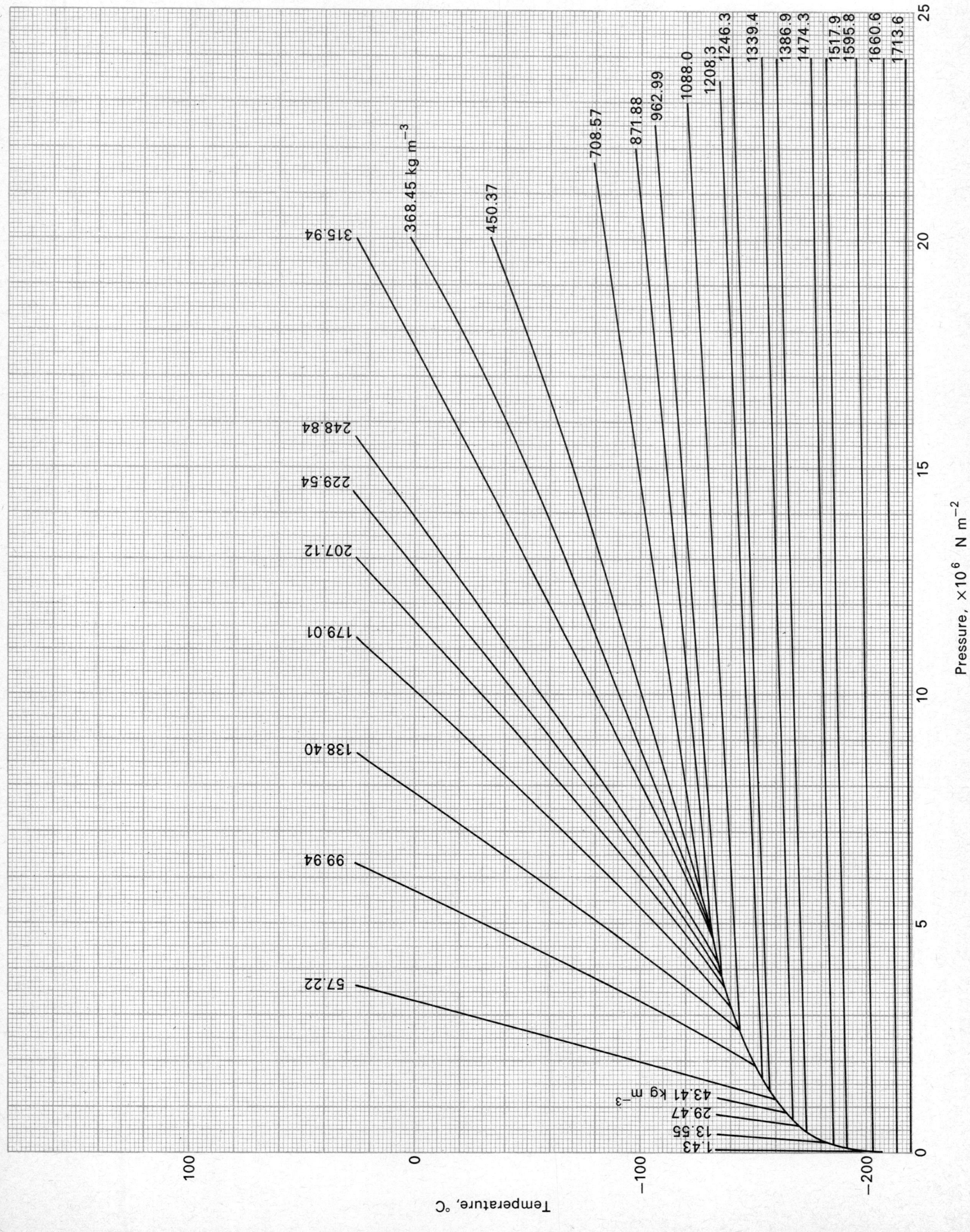

3.4 Latent Heat of Vaporisation of Fluorine (16)

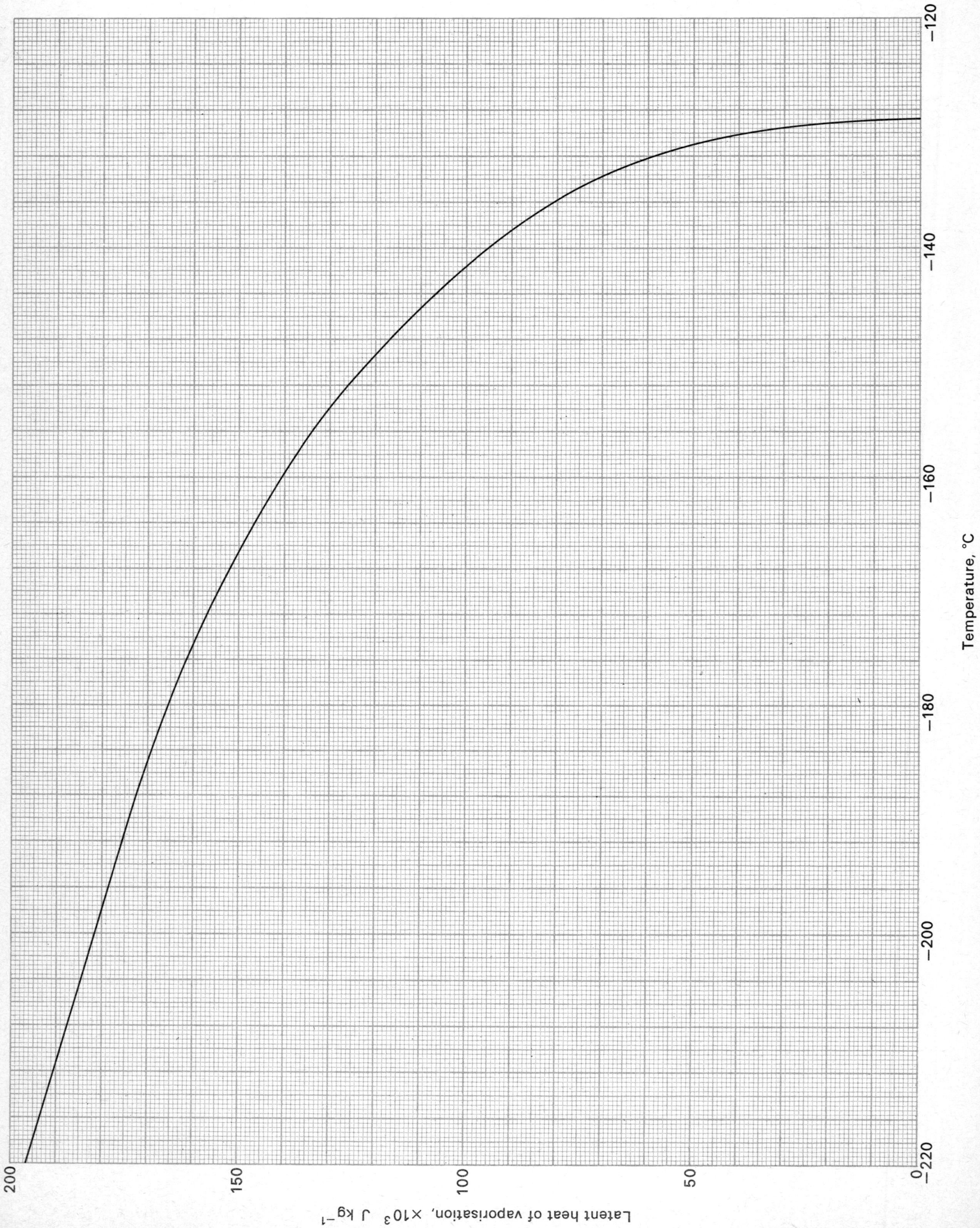

3.5 Heat Capacity of Solid (9) and Liquid Fluorine (6)

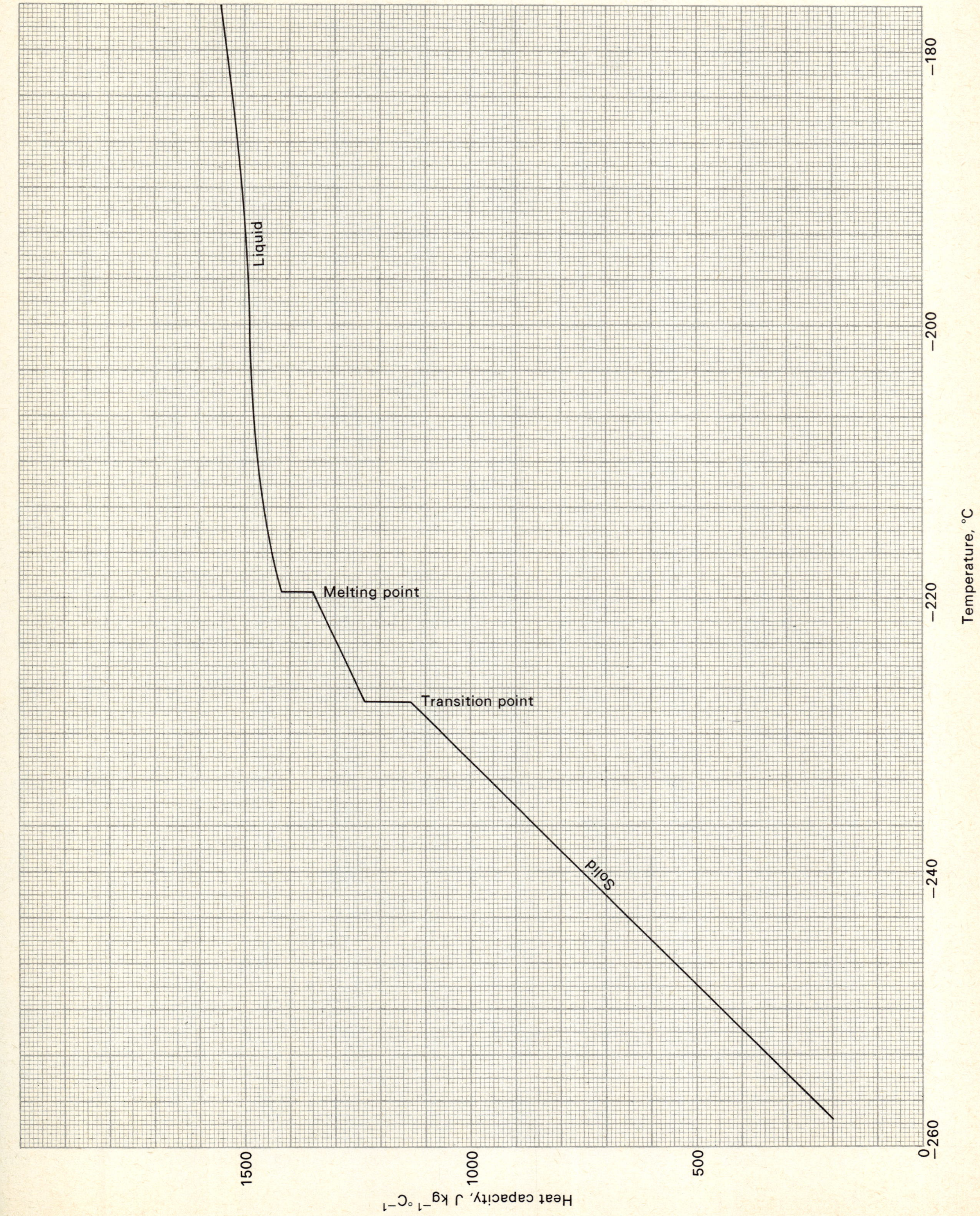

3.6 Heat Capacity of Compressed Liquid and Gaseous Fluorine (14)

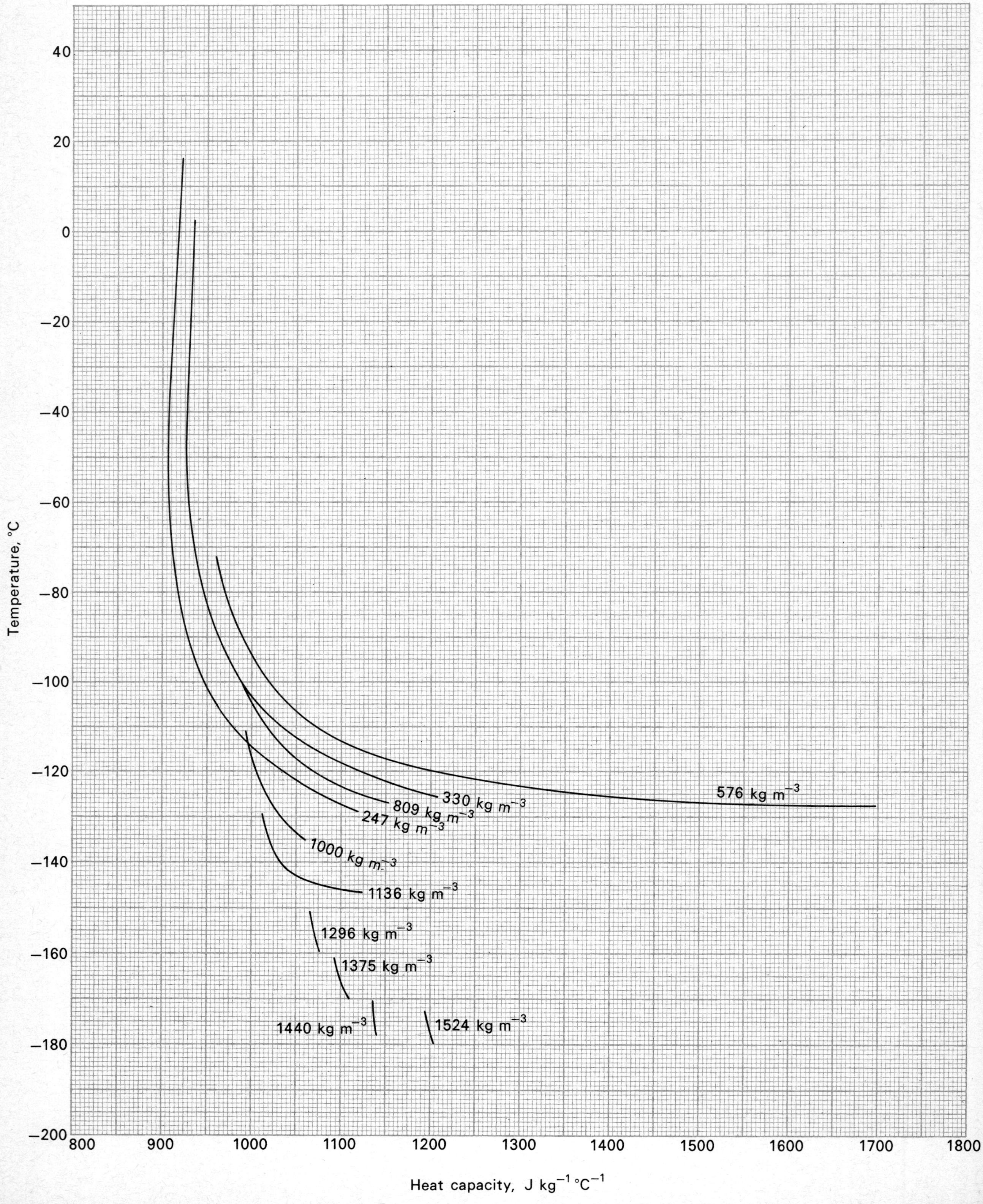

3.7 Viscosity of Liquid Fluorine (3)

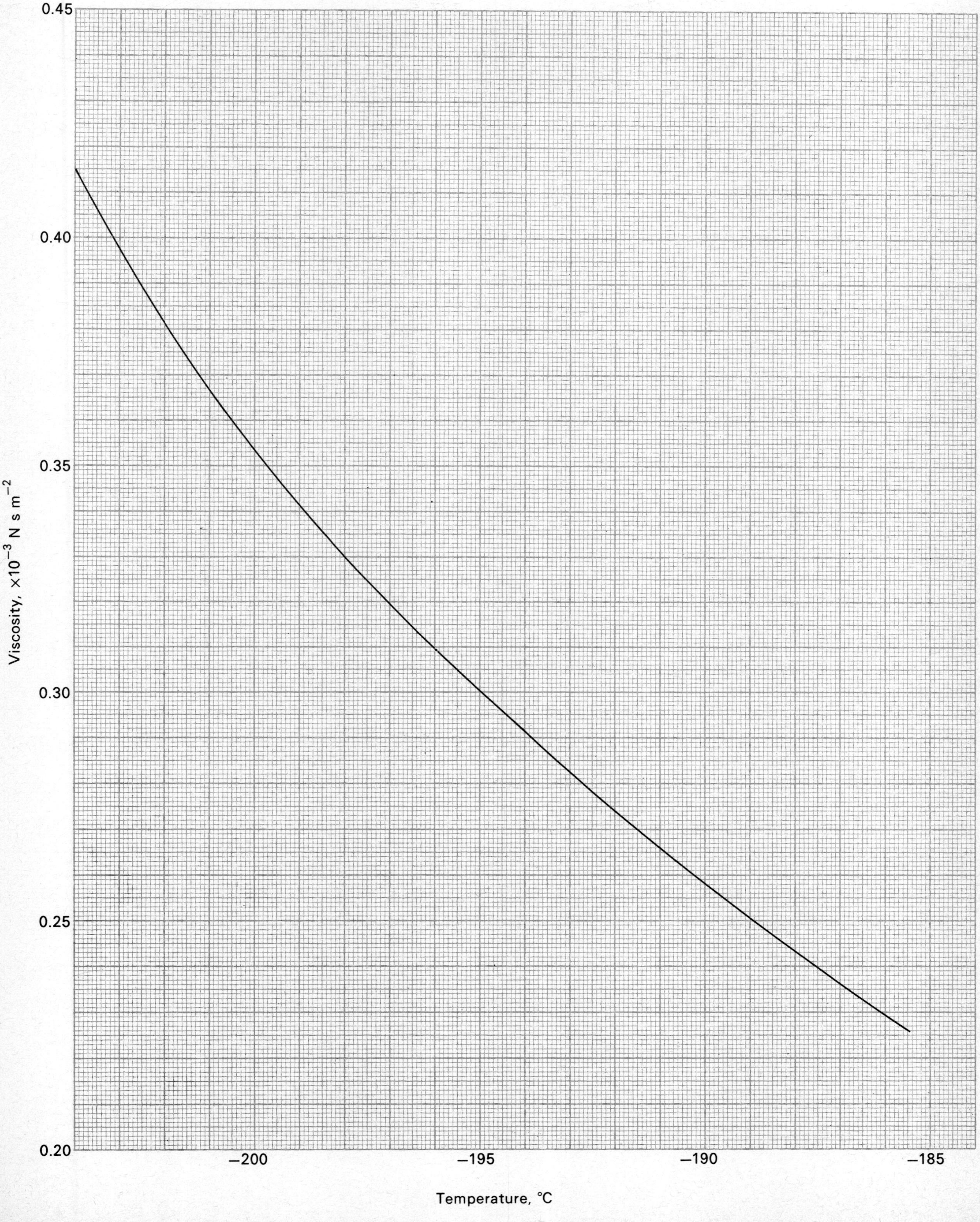

3.8 Viscosity of Gaseous Fluorine at 1.0133×10^5 N m^{-2} (4)

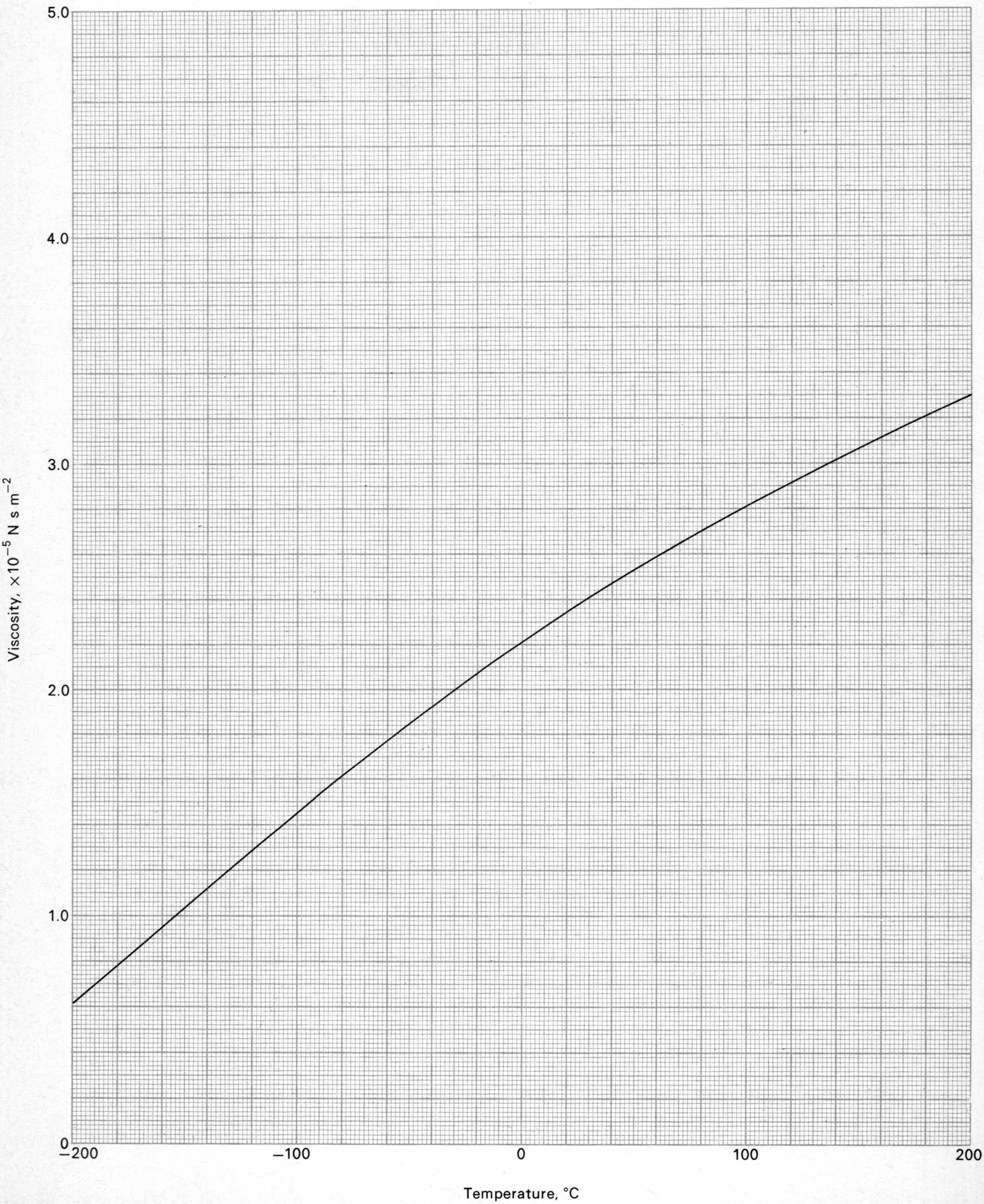

3.9 Thermal Conductivity of Liquid Fluorine (2)

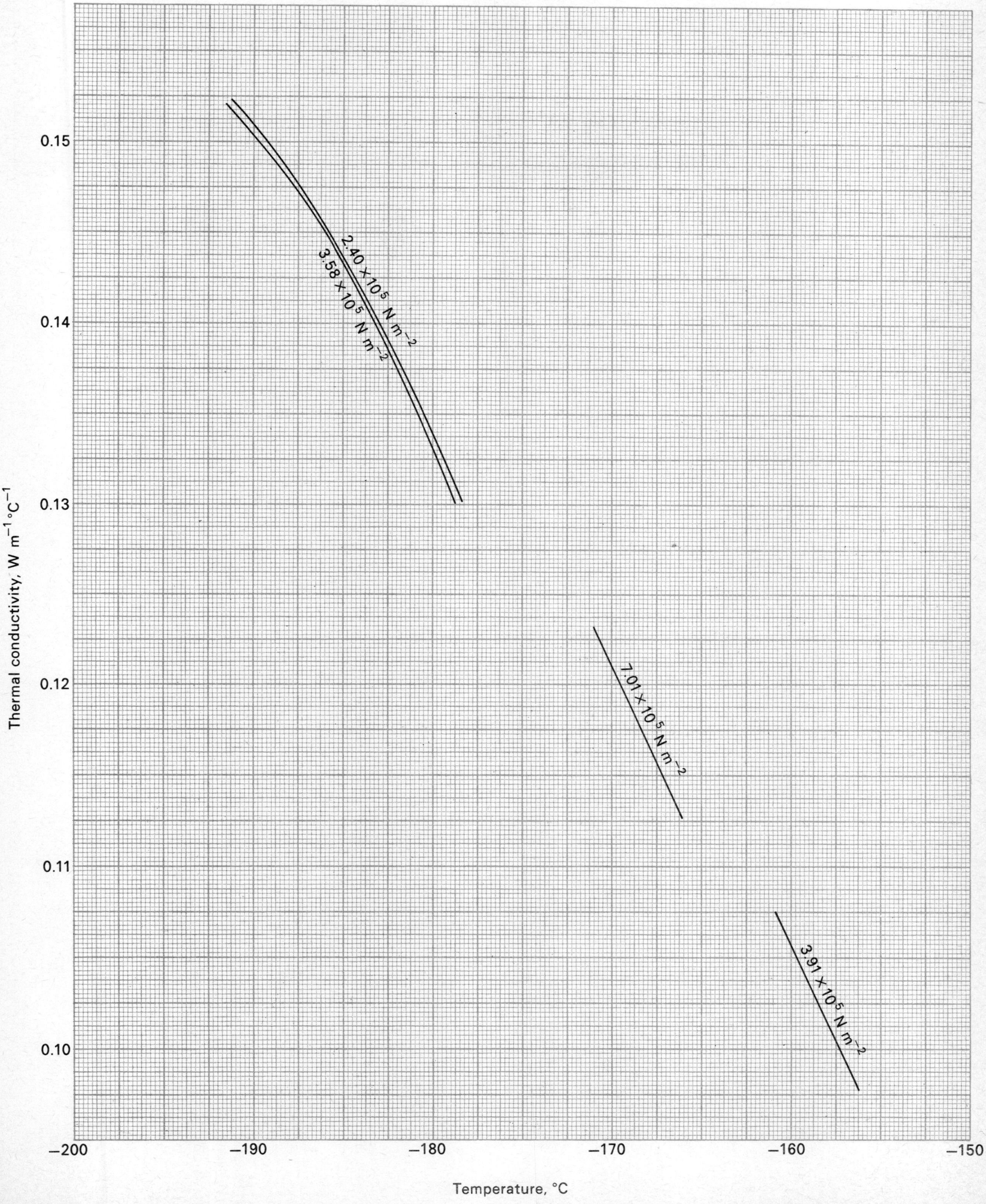

3.10 Thermal Conductivity of Gaseous Fluorine at 1.0133×10^5 N m^{-2} (13)

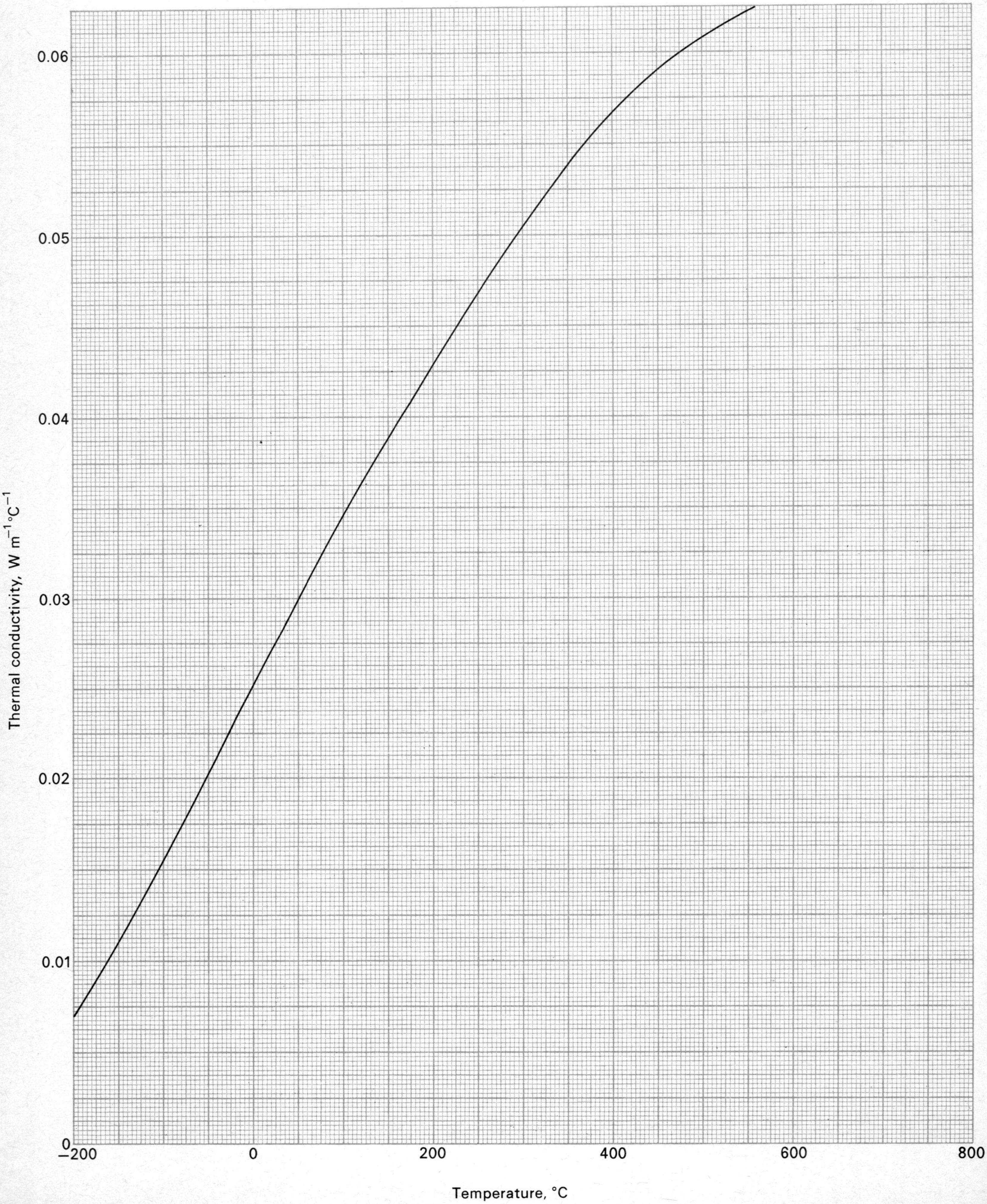

3.11 Surface Tension of Fluorine (3)

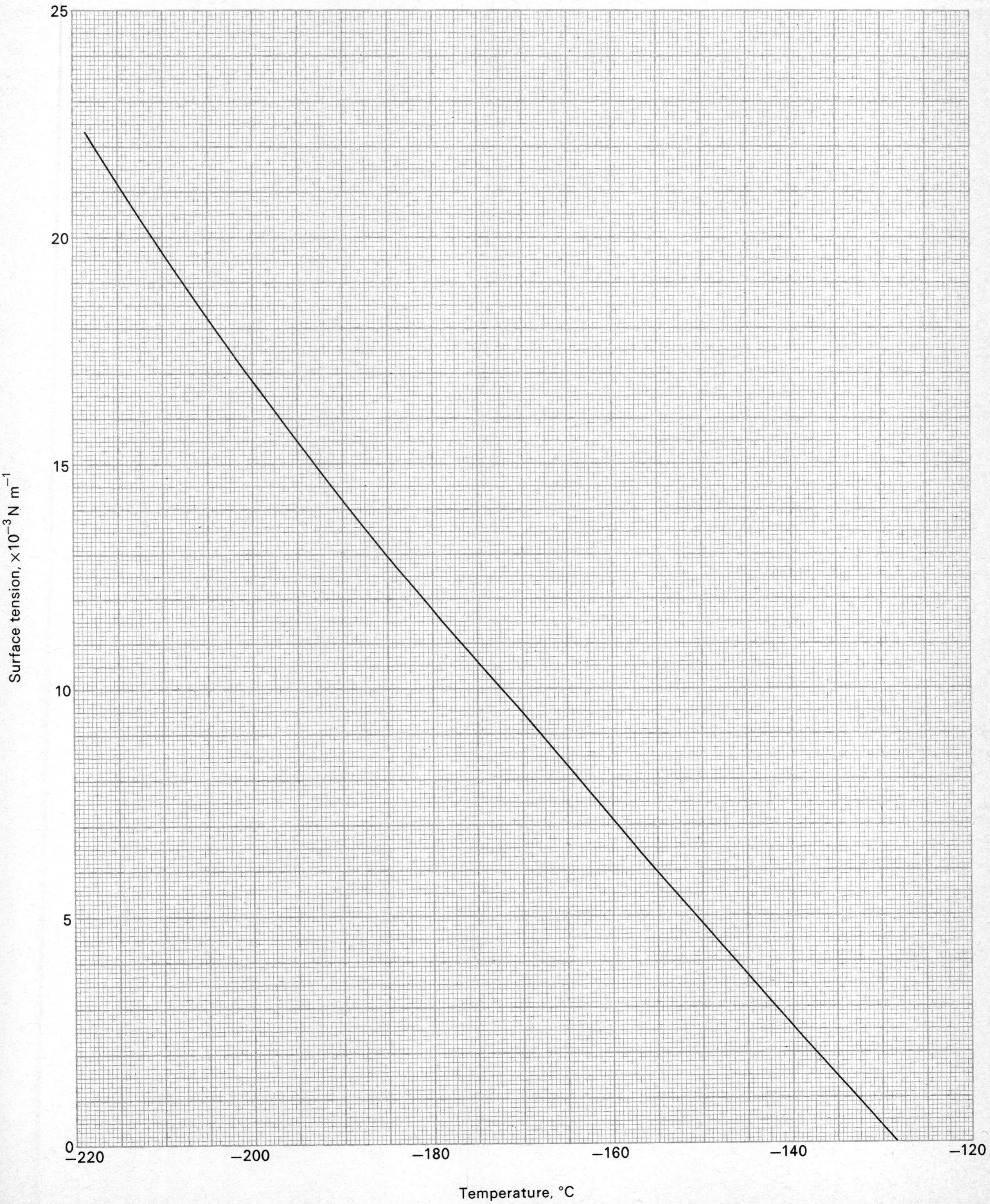

4 IODINE

Iodine is black or grey in the solid state, practically black in the liquid, and a fine violet in the vapour. Iodine crystallizes in the orthorhombic system and has a molecular lattice of I_2 molecules. Its unit cell consists of 8 atoms arranged as a symmetrical bipyramid. There is no evidence of a monoclinic variety of iodine under ordinary conditions, although Brigman (3) has observed transition to other modifications at very high pressures. Its molecular weight in neutral (non-ionised) solution is always that of I_2.

Gaseous iodine dissociates on heating more readily than any of the other halogens; its heat of dissociation is 1.5114×10^5 J/I_2 (18). Under 1.0133×10^5 N m^{-2} pressure the dissociation begins to be perceptible at 600°C, is 36% at 800°C, 80% at 900°C, and practically complete at 1000°C.

The physical, thermodynamic, and transport properties of iodine are less known than any of the other halogens. There is no reliable information on the P–V–T properties in the supercritical region or on the liquid thermal conductivity. Schaefer and Thodos (23) applied the corresponding states theory to estimate gas and liquid thermal conductivities; however, this approach was not successful enough for liquid hydrogen chloride and therefore it can only be accepted with reservation.

The presentation of additional properties; entropy of iodine (24, 5), further data on heat of fusion (8), liquid electrical conductivity (22), melting curves (1), compressibility of solid iodine (9), condensation coefficient (28), enthalpy of sublimation (17), and evidence for molecular dimers, $(I_2)_2$, are outside the scope of this work. The physical properties of iodine are summarised in Table 4.1 and Figs. 4.1–4.11.

References

1 Babb, S. E., *J. Chem. Phys.*, 1969, **50** (12), 5271–4.
2 Braune, H., Linke, R., *Z. phys. Chem.*, 1930, **A148**, 195–214.
3 Brigman, P. W., *Phys. Rev.*, 1935, **48**, 893–906.
4 Burriel Lluna, J. A., Cragg, C. B., Rowlinson, J. S., *An. Fis. Quim.*, 1968, **64** (1), 1–8.
5 Chihara, H., Nakamura, N., *Bull. Chem. Soc. Japan*, 1968, **41**, 1787–92.
6 CODATA Bulletin No. 2, P2425, Internat. Council of Sci. Union, Frankfurt/Main, Germany, Nov. 1970.
7 Franck, E. U., Wicke, E., Z. *Elektrochem.*, 1951, **55** (7), 643–7.
8 Frederick, K. J., Hildebrand, J. H., *J. Am. Chem. Soc.*, 1938, **60**, 1436–9.
9 Grover, R., Kusubov, A. S., Stromberg, H. D., *J. Chem. Phys.*, 1967, **47** (11), 4398–9.
10 Honig, R. E., Hook, H. O., *RCA Rev.*, 1960, **21**, 360–8.
11 Ibrahim, S. H., Kuloor, N. R., *Brit. Chem. Eng.*, 1961, **6** (12), 862–3.
12 Kim, S., Chang, S., *Daehan Hwahak Hwoejee*, 1965, **9** (2), 110–12.
13 Kirk-Othmer: *Encyclopedia of Chemical Technology*, 2nd ed., Vol. 11, Interscience Publ., New York, 1966, pp. 850–3.
14 Landolt-Börnstein: *Zahlenwerte und Funktionen aus Physik, Chemie, Astronomie und Technik*, 6th ed., Vol. 4, Kalorische Zustandsgrössen, Springer-Verlag, Berlin, 1961, p. 403.
15 Landolt-Börnstein: *Zahlenwerte und Funktionen aus Physik, Chemie, Astronomie und Technik*, 6th ed., Vol. 2, Part b, Lösungsgleichgewichte I, Springer-Verlag, Berlin, 1962, pp. 3–9.
16 Lange, N. A., Ed., *Handbook of Chemistry*, 10th ed., McGraw-Hill, New York, 1967.
17 Lukashenko, E. E., Pogodaev, A. M., *Russ. J. Phys. Chem.*, 1968, **42** (1), 122–4.
18 Perlman, M.L., Rollefson, C. K., *J. Chem. Phys.*, 1941, **9**, 362.
19 Rabinovich, M., *Z. phys. Chem.*, 1926, **119**, 82.
20 Richards, T. W., *J. Am. Chem. Soc.*, 1915, **37** (7), 1643–56.
21 Rossini, F. P., Wagman, D. D., Evans, W. H., Levine, S., Jaffe, I., *Selected Values of Chemical Thermodynamic Properties*, Natl. Bur. Stand., Washington, Circular 500, 1952.
22 Saito, K., Ichikawa, K., Shimoji, M., *Bull. Chem. Soc. Japan*, 1968, **41**, 1104–9.
23 Schaefer, C. A., Thodos, G., *AIChE Journal*, 1955, **5** (3), 367–72.
24 Shirley, D. A., Giauque, W. F., *J. Am. Chem. Soc.*, 1959, **81**, 4778–9.
25 Smith, N. O., *J. Chem. Educ.*, 1965, **42** (12), 654–5.
26 Steacie, E. W. R., Johnson, F. M., *J. Am. Chem. Soc.*, 1925, **47**, 754–62.
27 Steere, N. V., Ed., *Handbook of Laboratory Safety*, The Chemical Rubber Co., Cleveland, 1967.
28 Stern, J. H., Gregory, N. W., *J. Phys. Chem.*, 1957, **61**, 1226–32.
29 Stull, D. R., *et al.*, *JANAF Thermodynamic Tables*, Dow Chem. Co., Midland, Mich., 1965–6.
30 van Laar, J. J., *Z. anorg. Chem.*, 1917, **104**, 107.
31 Washburn, E. W., Ed., *International Critical Tables*, 7 Vols. + Index, McGraw-Hill, New York, 1926–33.
32 Weast, R. C., Ed., *Handbook of Chemistry and Physics*, 51st ed., The Chemical Rubber Co., Cleveland, 1970–71.

Table 4.1 Properties of Iodine

		Ref.
Chemical formula	I_2	(29)
Molecular weight	253.820	(29)
Critical temperature	546.0°C	(4)
Critical pressure	116.54×10^5 N m^{-2}	(13, 16)
Critical density	1640.0 kg m^{-3}	(4)
Triple-point temperature	113.7°C	(30)
Triple-point density	3882.0 kg m^{-3}	(30)
Boiling point at 1.0133×10^5 N m^{-2}	184.35°C	(13, 16)
Melting point	113.5°C	(8)
Transition temperature	None	(24)
Heat of formation of gas at 25°C	6.2421×10^7 J kg-mol^{-1}	(6)
Heat of sublimation at 25°C	6.2299×10^4 J kg-mol^{-1}	(21)
Heat of fusion	$1.528\,08 \times 10^7$ J kg-mol^{-1}	(8)
Heat of transition	None	(24)
Entropy of gas at 25°C	2.6057×10^5 J kg-mol^{-1} °C^{-1}	(6)
Heat capacity of liquid between 113.6°–184.0°C	318.0 J kg^{-1} °C^{-1}	(29)
Heat capacity ratio of gas at 185°C and 1.0133×10^5 N m^{-2}	1.30	(16)
Thermal conductivity of solid at 24.4°C	0.4212 W m^{-1} °C^{-1}	(13)
Specific gravity of solid at 20°C	4940.0 kg m^{-3}	(20)
Compressibility of solid at 20°C	12.8×10^{-11} m^2 N^{-1}	(20)
Coefficient of thermal expansion of solid at 20°C	2.64×10^{-4} deg^{-1}	(25)
Velocity of sound at 0°C and 1.0133×10^5 N m^{-2}	107.7 m s^{-1}	(31)
Diffusion in air at 0°C and 1.0133×10^5 N m^{-2}	0.0834×10^{-4} m^2 s^{-1}	(11)
Dielectric constant of liquid at 140°C	11.7	(16)
Electrical conductivity of liquid at 138.2°C	4.48×10^{-3} ohm^{-1} m^{-1}	(19)
Refractive index of solid at 20°C	3.34	(16)
Threshold limit value	(0.1 ppm) 10^{-6} kg m^{-3}	(16)

4.1 Vapour Pressure of Solid and Liquid Iodine (10,32)

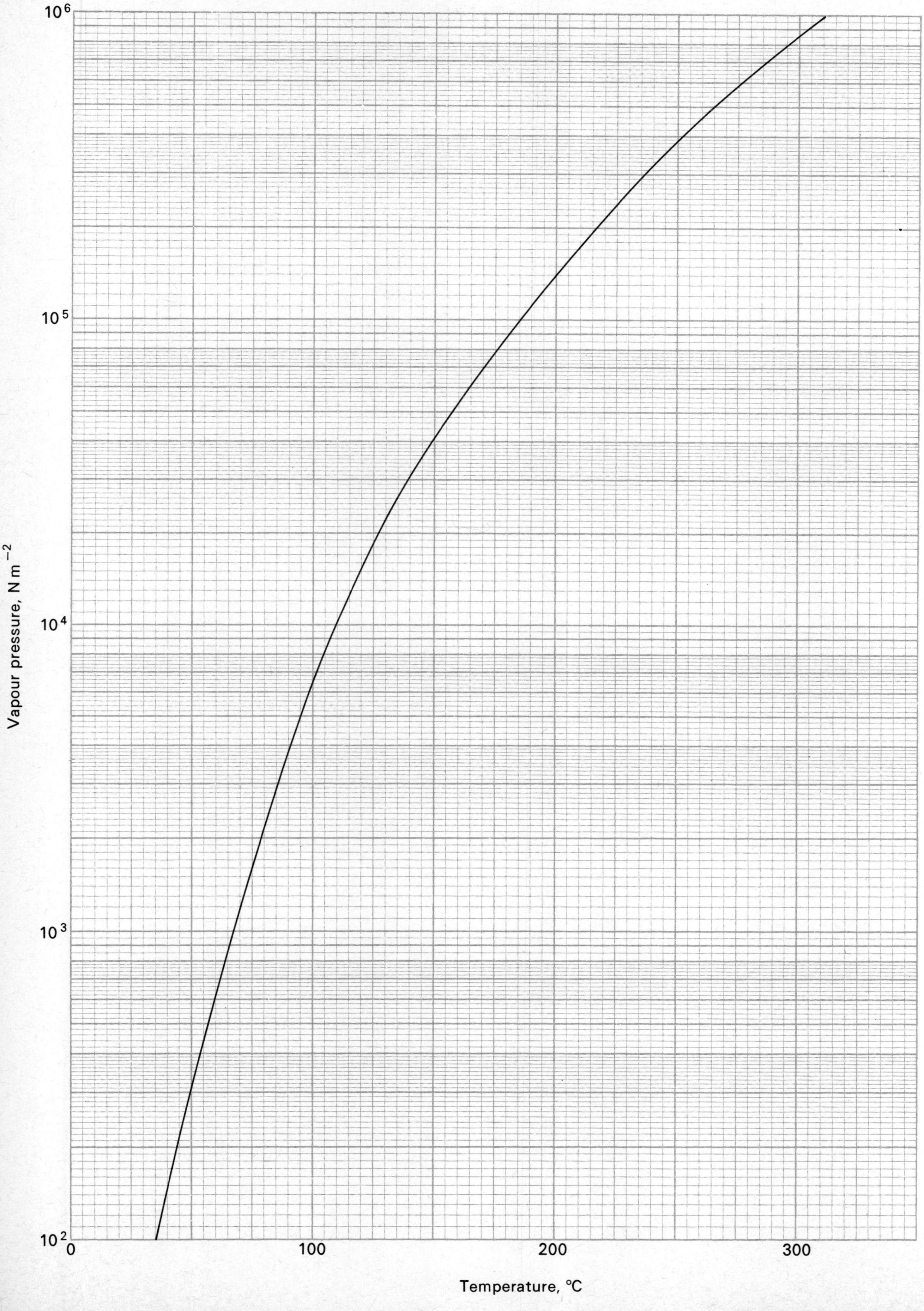

4.2 Orthobaric Density of Iodine (4)

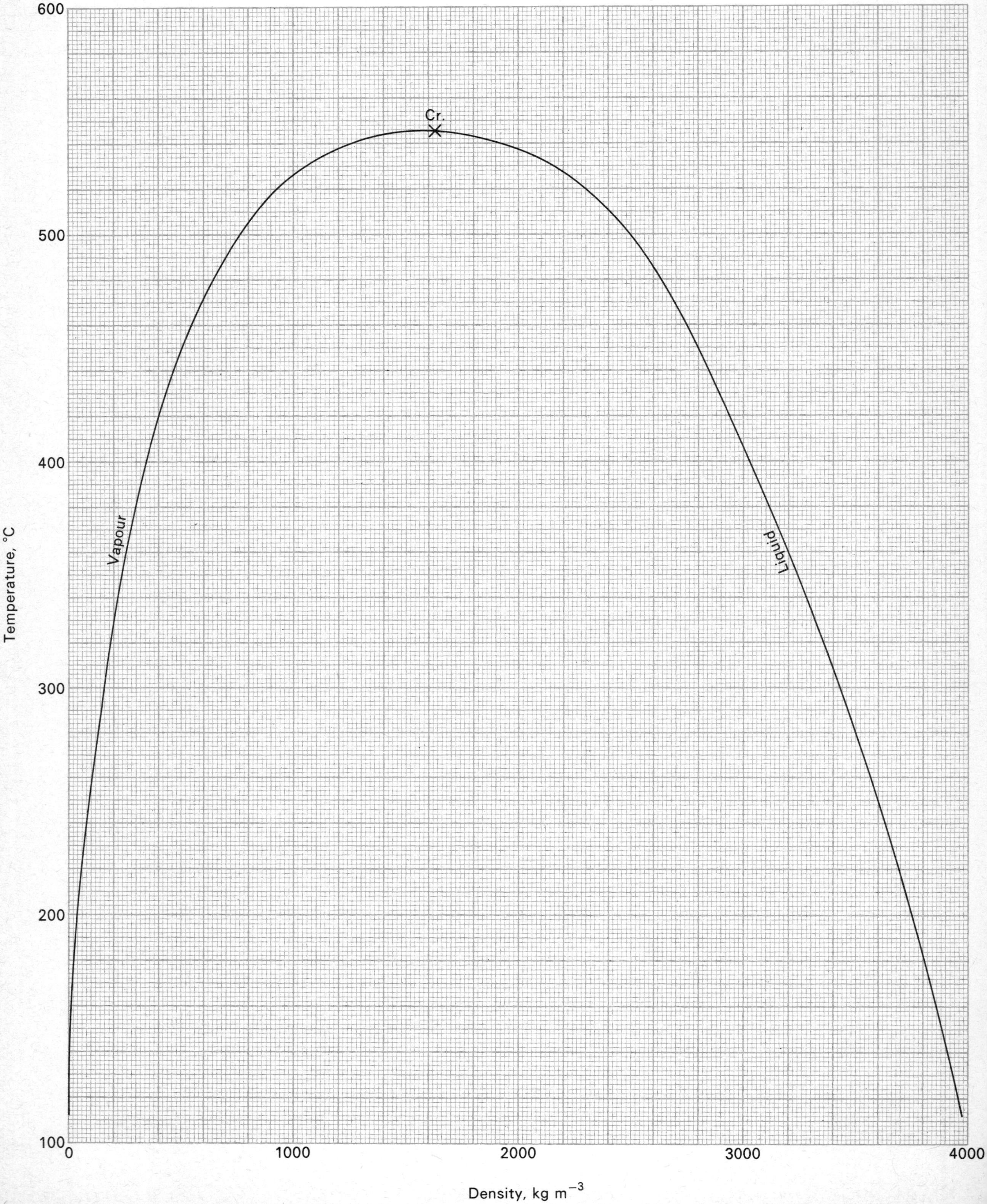

4.3 Latent Heat of Vaporisation of Iodine (29)

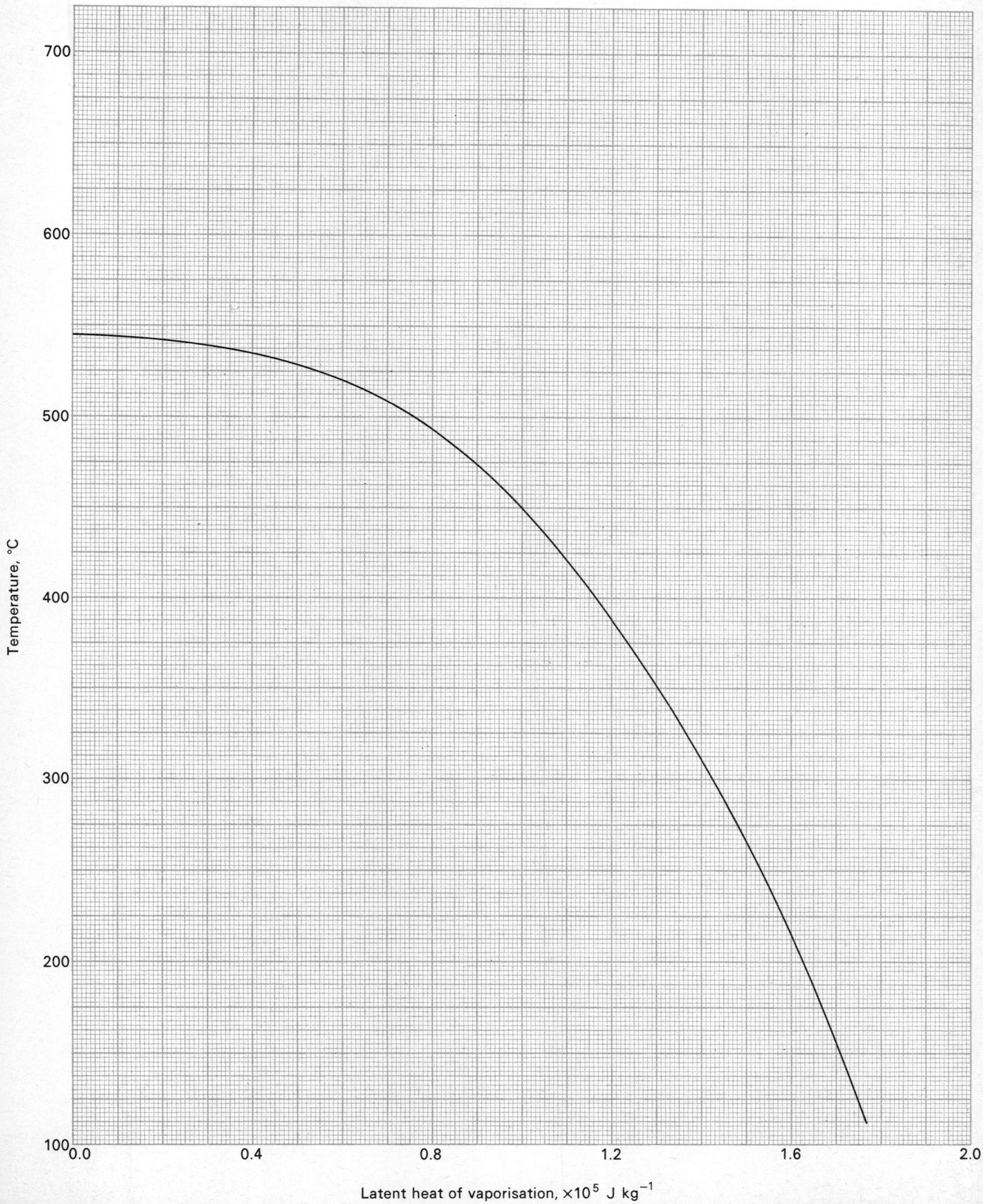

4.4 Heat Capacity of Solid Iodine (14,29)

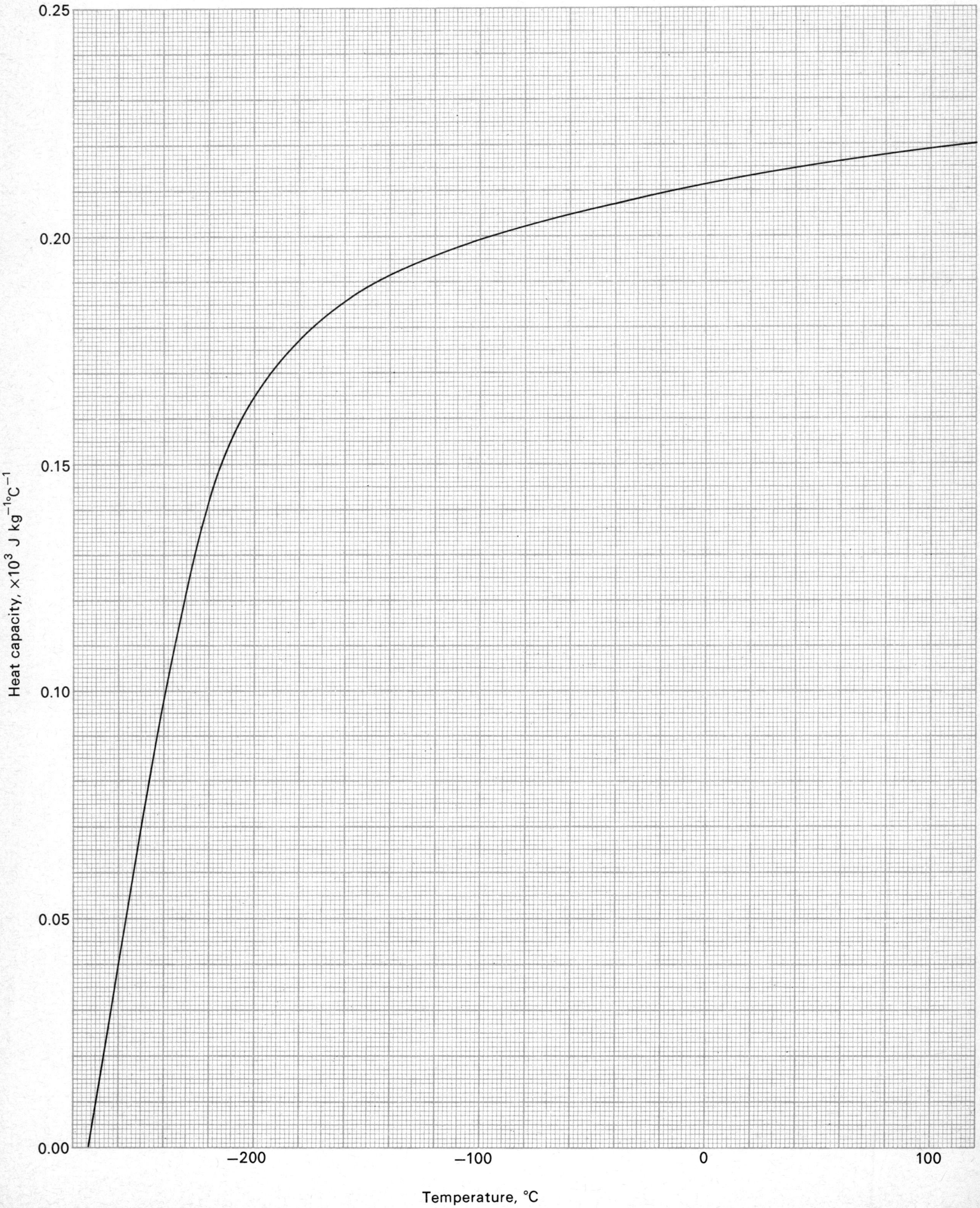

4.5 Heat Capacity of Gaseous Iodine at 1.0133×10^5 N m^{-2} (29)

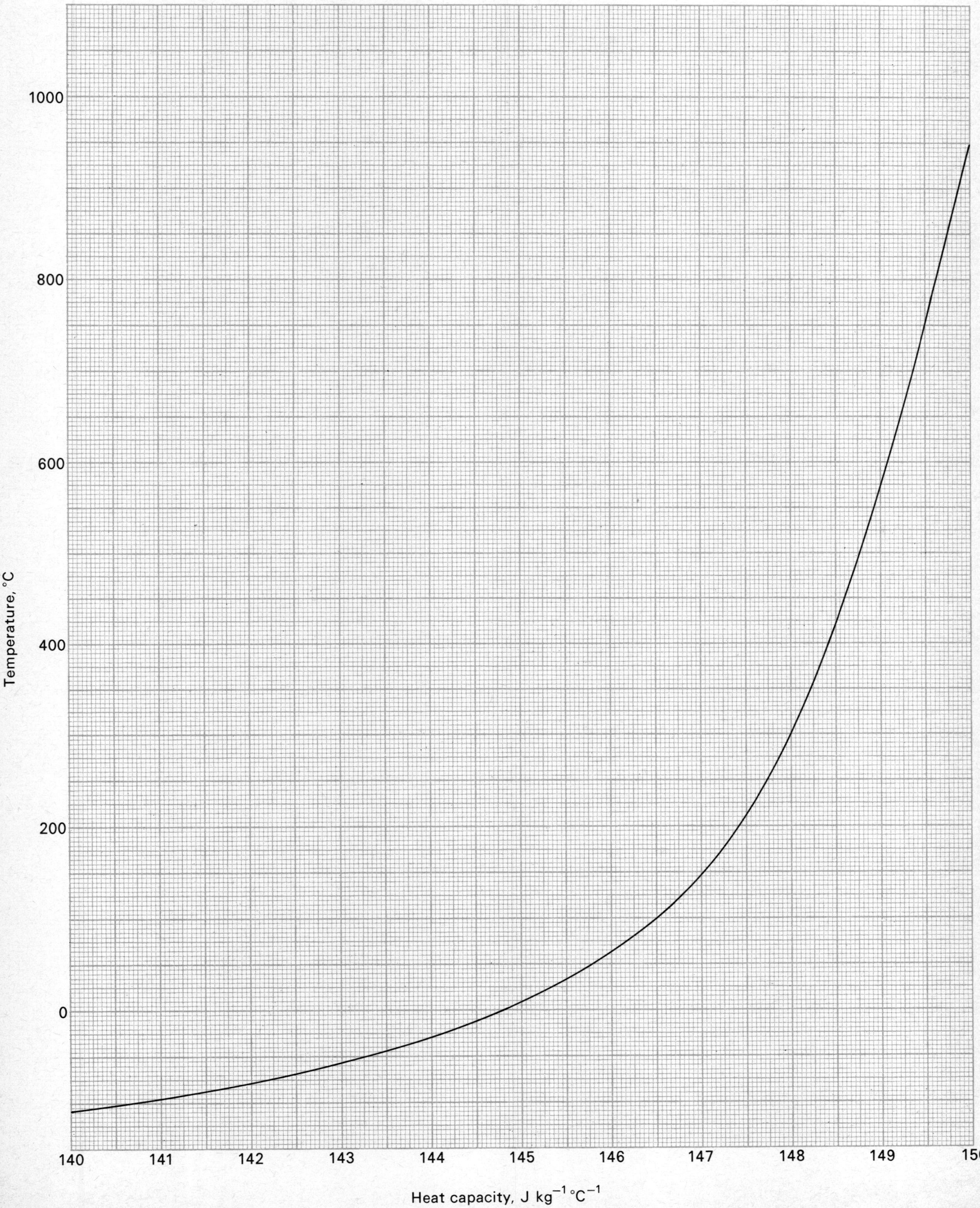

4.6 Viscosity of Liquid Iodine (26)

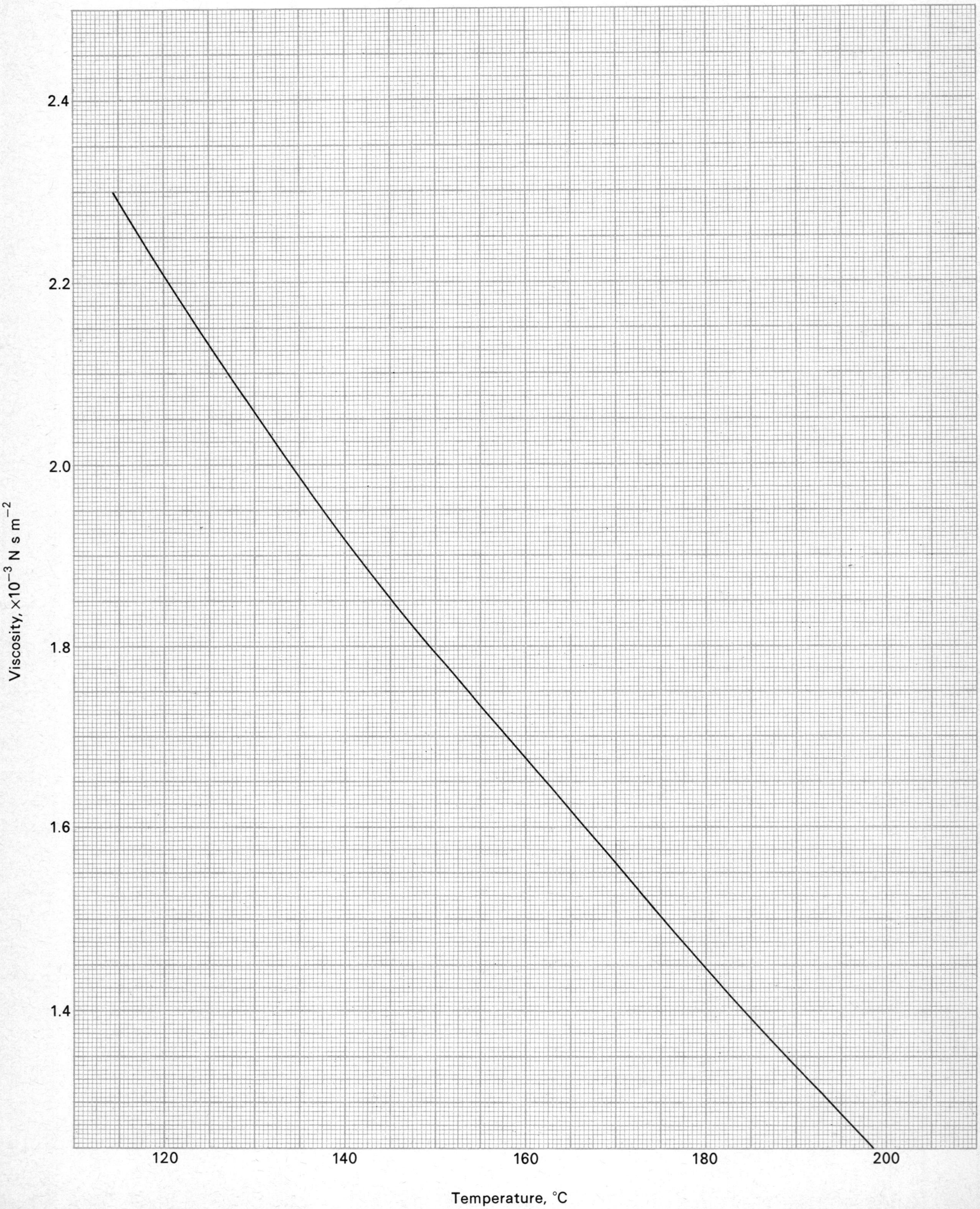

4.7 Viscosity of Gaseous Iodine at 1.0133×10^5 N m^{-2} (2)

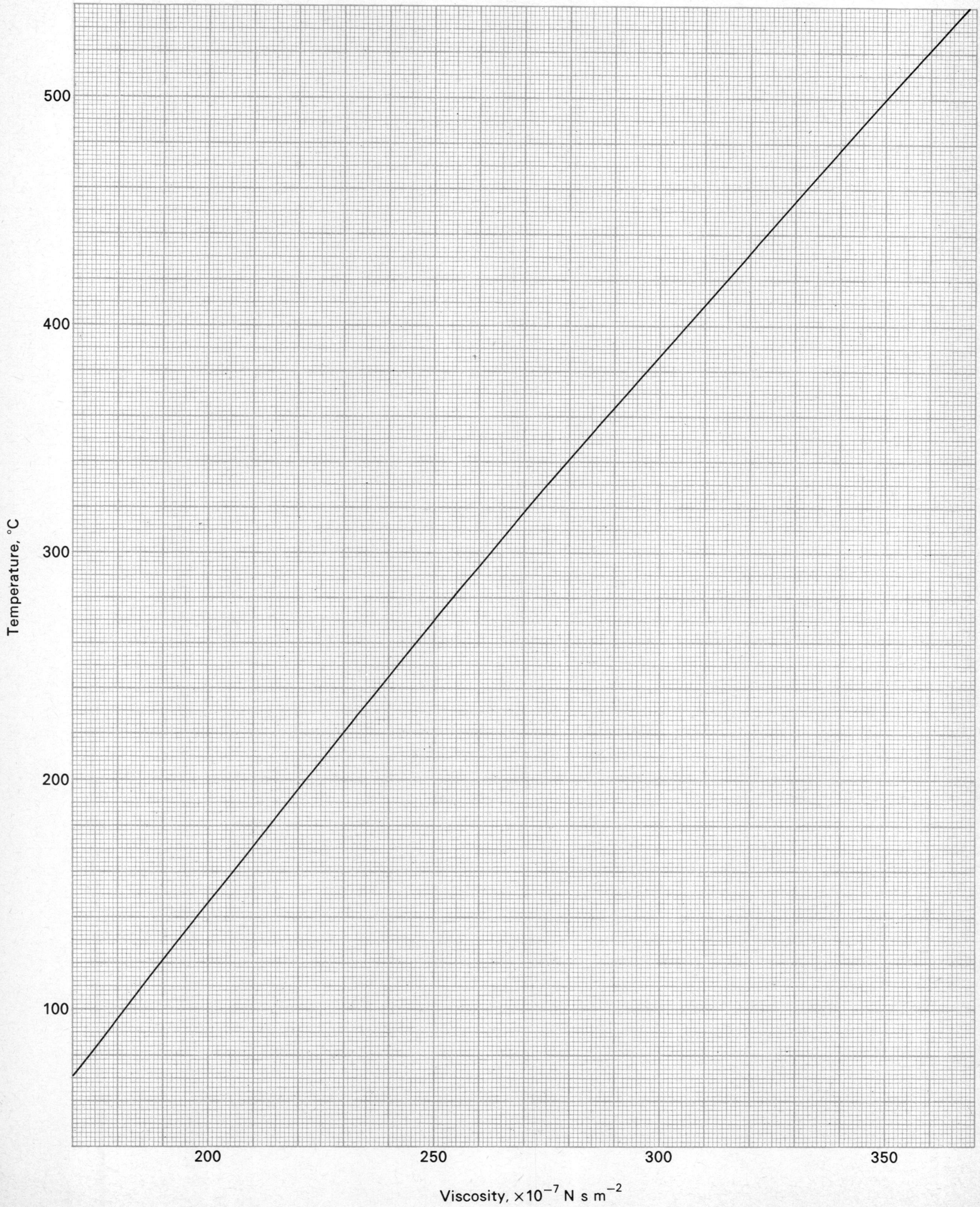

4.8 Thermal Conductivity of Liquid Iodine (23)

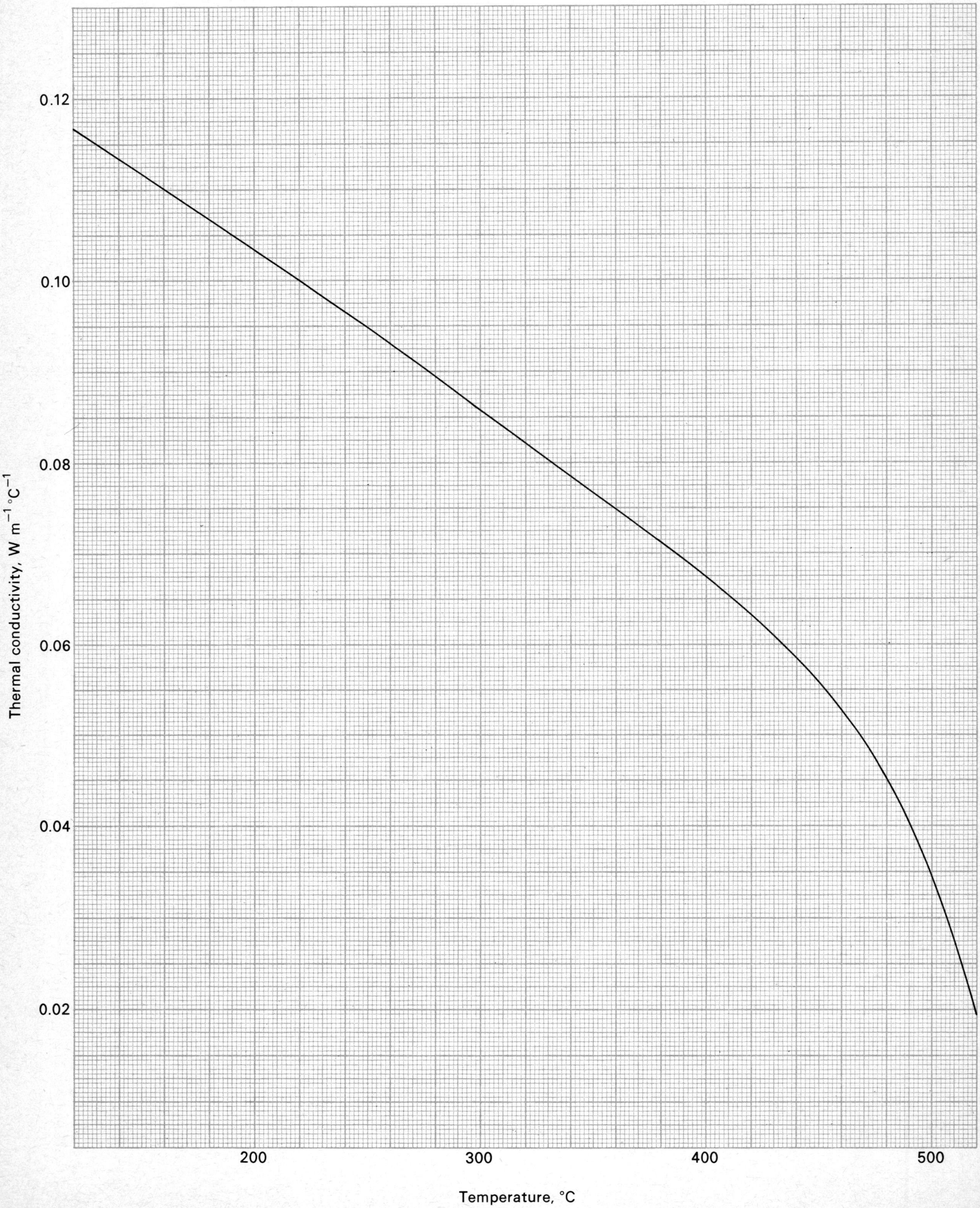

4.9 Thermal Conductivity of Gaseous Iodine (7)

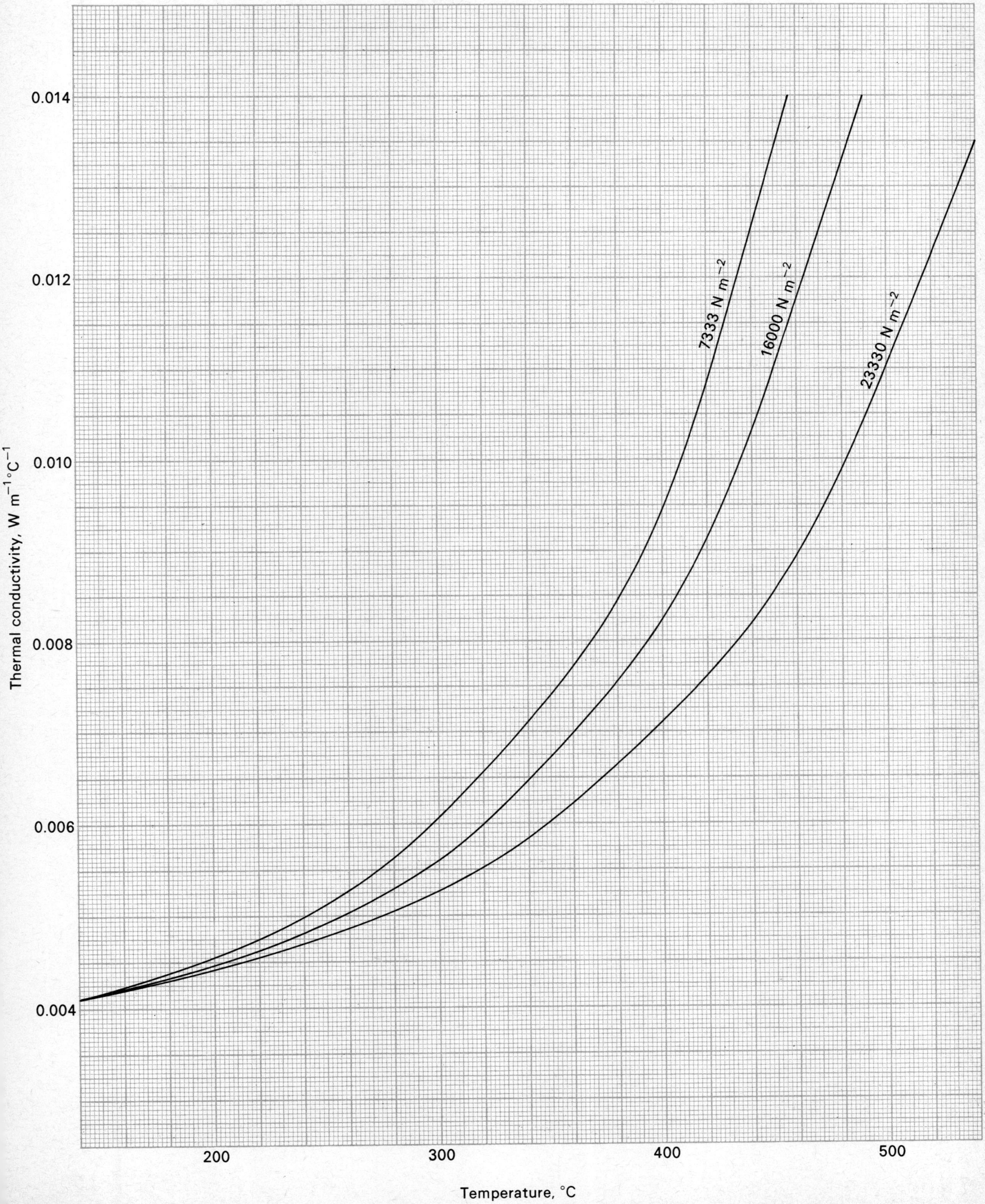

4.10 Surface Tension of Iodine (12)

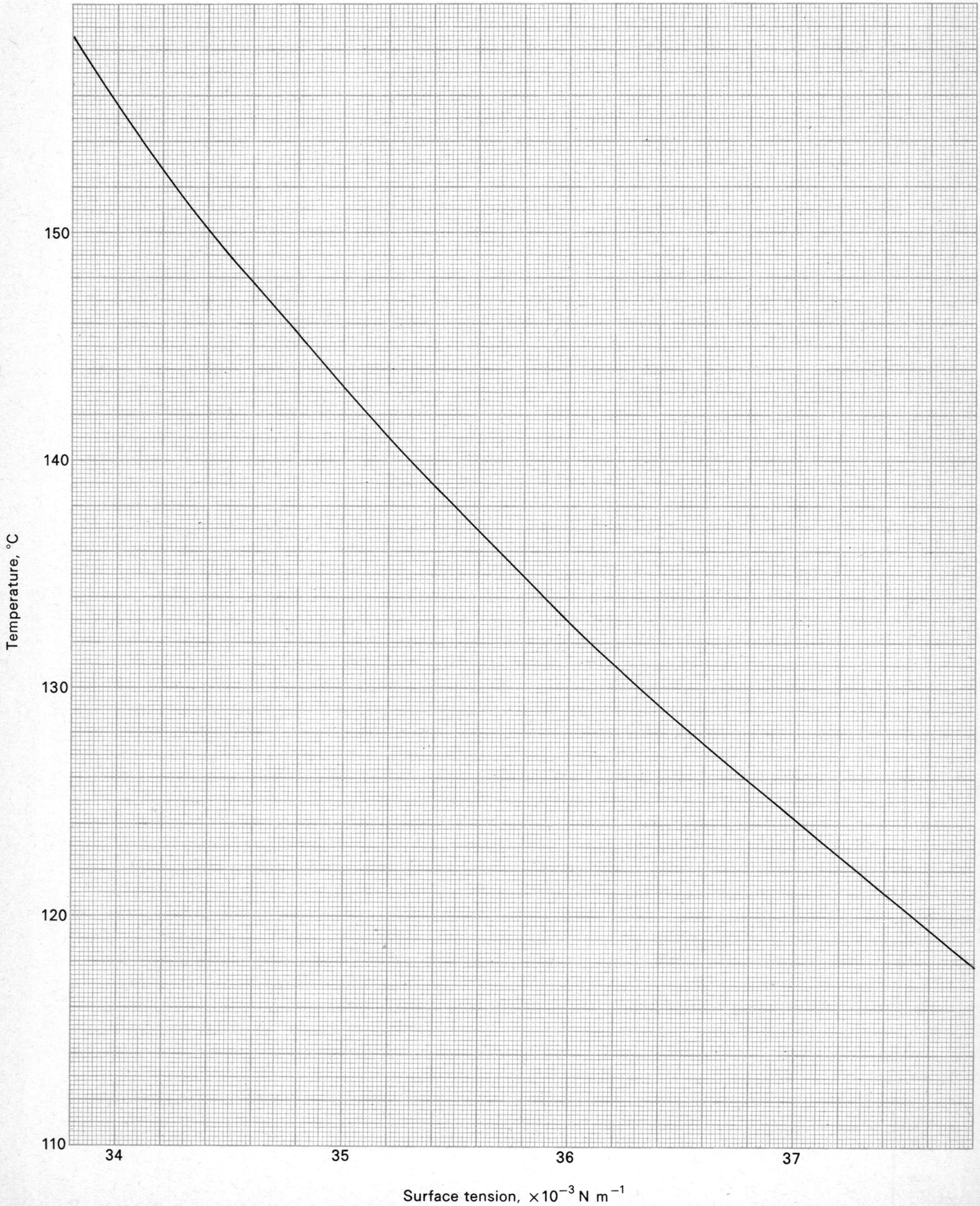

4.11 Solubility of Iodine (solid) in Water at 1.0133×10^5 N m^{-2} Total Pressure (15)

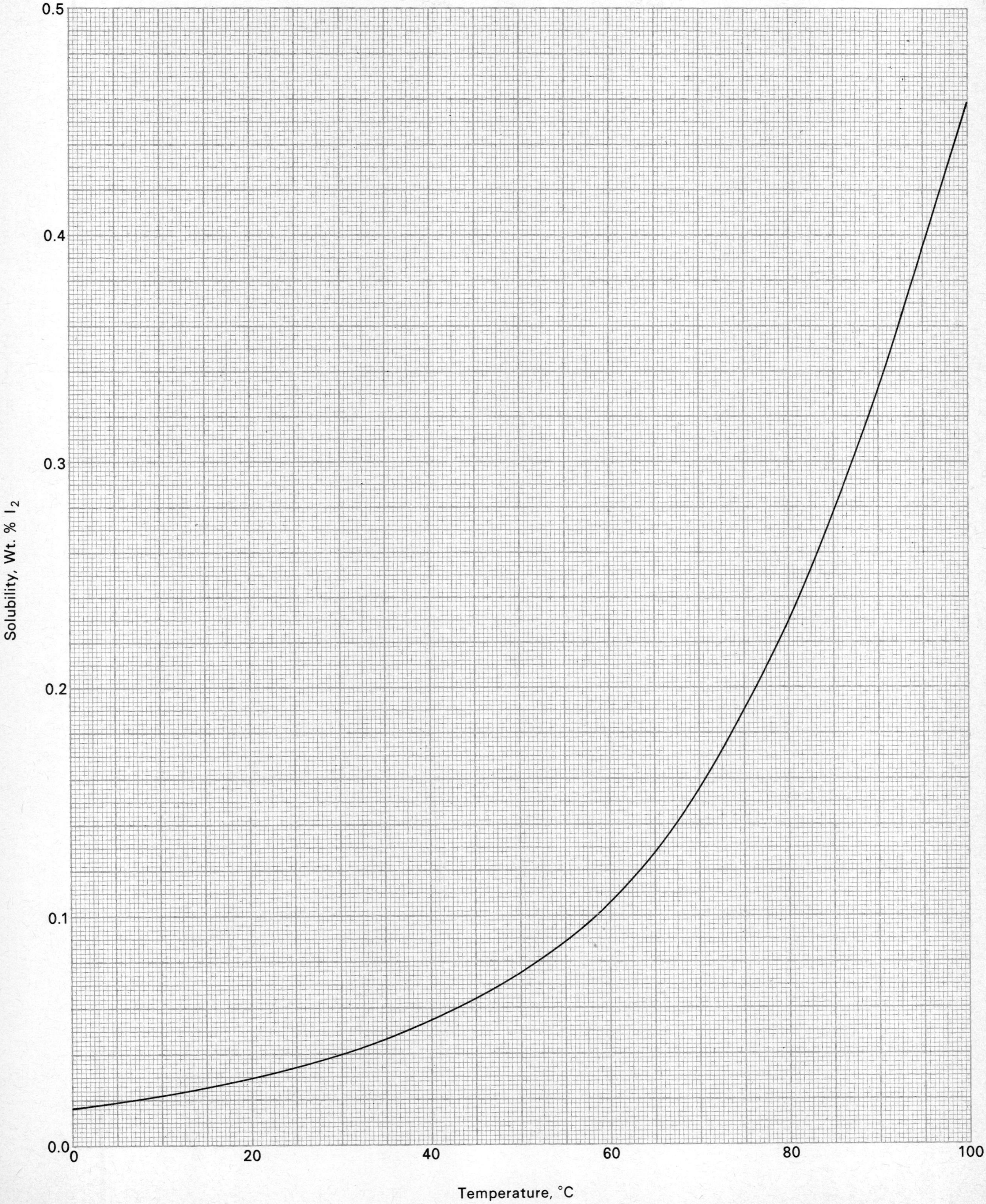

5 HYDROGEN BROMIDE

Hydrogen bromide is a colourless gas at ambient temperature and atmospheric pressure. In moist air it is a very irritating gas and develops toxic fumes. The vapour is soluble in water, giving an acidic solution, in non-ionizing solvents it is a covalent compound just as in the pure state. On heating it dissociates into hydrogen and bromine.

The solid compound is trimorphous. The heat capacity data indicate that there are three transitions of internal molecular nature. Liquid hydrogen bromide does not show any association properties. The electrical conductivity is low in the pure liquid state; about the same magnitude as that of pure water.

The main application of hydrogen bromide is in the preparation of various inorganic and organic compounds, and as an alkylation catalyst in the petroleum industry.

There are only a few compilations on the physical, thermodynamic, and transport properties of solid, liquid, and gaseous hydrogen bromide, such as: Gmelin (11), Mellor (21), and Pascal (24). Most of the experimental determinations were carried out several decades ago. However, there is some degree of confidence in the results. The P–V–T properties have not been determined and the only available information on gas density is given by Baxter (2). No experimental values exist on the liquid thermal conductivity, and the presented graph (Fig. 5.8) has been derived from an estimation method (29). Caution is recommended in the usage of these data. Gopala Rao and Swamy (13) used the significant structure theory to evaluate some thermodynamic properties of liquid hydrogen bromide, meanwhile Campbell & Fristrom (5) surveyed some of its properties for flame studies.

References

1 American Conference of Government Industrial Hygienists, *Documentation of Threshold Limit Values*, revised edition, Committee on Threshold Limit Values, Cincinnati, Ohio, 45202, 1966, 203 pp.

2 Baxter, G. P., *J. Am. Chem. Soc.*, 1922, **44**, 595–600.

3 Bleekrode, L., *Proc. Roy. Soc.*, 1884, **37**, 352.

4 Bradley, J. N., *Shock Waves in Chemistry and Physics*, Methuen & Co. Ltd., London, 1962, p. 334.

5 Campbell, E. S., Fristrom, R. M., *Chem. Rev.*, 1958, **58**, 173–234.

6 Drozdowski, E., Pietrzak, J., *Bull. Acad. Crac.*, 1913, **A**, 234–9.

7 Edwards, D. G., *The Vapour Pressure of 30 Inorganic Liquids between One Atmosphere and the Critical Point*, Univ. of California, Lawrence Rad. Lab. UCRL 7167, Livermore, California, June 13, 1963, 148 pp.

8 Eucken, A., Karwat, E., *Z. phys. Chem.* 1924, **112**, 473.

9 Faull, J. H., *J. Am. Chem. Soc.*, 1934, **56**, 522–6.

10 Giauque, W. F., Wiebe, R., *J. Am. Chem. Soc.*, 1928, **50**, 2193–2202.

11 Gmelins Handbuch der anorganischen Chemie, Syst. No. 7, Brom, Verlag Chemie GmbH, Berlin, 1931, pp. 168–216.

12 Golubev, I. F., *Viscosity of Gases and Gas Mixtures*, Gosudarstvennoe Izdatel'stvo Fiziko-Matematicheskoi Literatury, Moskva, 1959, Table 4.

13 Gopala Rao, R. V., Swamy, K. N., *Indian J. Pure & Appl. Phys.*, 1970, **8** (10), 609–11.

14 Henglein, F. A., Roth, R., Andres, P., *Z. Phys.*, 1923, **18**, 69.

15 Honig, R. E., Hook, H. O., *Vapour Pressure Data for some Common Gases, RCA Rev.*, 1960, **21**, 360–8.

16 IUPAC, *Pure & Appl. Chem.*, 1972, **30** (3–4), 638–49.

17 Karapet'yants, M. Kh., *Zh. Fiz. Khim.*, 1958, **32**, 556–68.

18 Lange, N. A., Ed., *Handbook of Chemistry*, 10th ed., McGraw-Hill, New York, 1967.

19 Lax, E., Synowietz, C., Eds., *Taschenbuch für Chemiker und Physiker*, 3rd ed., Springer-Verlag, Berlin, 1967.

20 Maryott, A. A., Smith, E. R., *Table of Dielectric Constants of Pure Liquids*, Natl. Bur. Std. Circular 514, Washington, Aug. 10, 1951, 44 pp.

21 Mellor's Comprehensive Treatise on Inorganic and Theoretical Chemistry, *The Halogens*, Vol. II, Suppl. 1, Longmans, London, 1956, pp. 724–41.

22 McIntosh, D., Steele, B. D., *Proc. Roy. Soc.*, 1904, **73**, 453.

23 McIntosh, D., Steele, B. D., *Z. phys. Chem.*, 1906, **55**, 142–50.

24 Pascal, P., Ed., *Nouveau Traité de Chimie Minérale*, Vol. XVI, F, Cl, Br, I, At, Mn, Tc, Re, Masson et Cie, Editeurs, Paris, 1960, pp. 381–94.

25 Pearson, T. G., Robinson, P. L., *J. Chem. Soc.*, 1934, 880.

26 Rossini, F. D., Wagman, D. D., Evans, W. H., Levine, S., Jaffe, I., *Selected Values of Chemical Thermodynamic Properties*, Natl. Bur. Std. Circular 500, Washington, Feb. 1, 1952.

27 Roozeboom, H. W. B., *Rec. trav. chim. Pays-Bas*, 1885, **4**, 103.

28 Roozeboom, H. W. B., *Z. phys. Chem.*, 1888, **2**, 454.

29 Schaefer, C. A., Thodos, G., *AIChE Journal*, 1959, **5** (3), 367–72.

30 Stewart, J. W., *J. Chem. Phys.*, 1962, **36** (2), 400–5.

31 Stull, D. R., Dir., *JANAF Thermochemical Tables*, Dow Chemical Co., Midland, Mich., 1965–8.

32 Strunk, W. G., Wingate, W. H., *J. Am. Chem. Soc.*, 1954, **76**, 1025–7

33 Syrkin, J. K., *Z. phys. Chem.*, 1929, **5B**, 159.

34 Walden, P., *Z. phys. Chem.*, 1909, **65**, 158.

Table 5.1 Properties of Hydrogen Bromide

		Ref.
Chemical formula	HBr	(19)
Molecular weight	80.9119	(16)
Critical temperature	90.0°C	(6)
Critical pressure	85.52×10^5 N m^{-2}	(6)
Critical density	807.kg m^{-3}	(33)
Triple-point temperature	−88.0°C	(14)
Triple-point pressure	0.2989×10^5 N m^{-2}	(14)
Boiling point at 1.0133×10^5 N m^{-2}	−67.1°C	(14)
Melting point	−86.1°C	(8)
Transition temperatures	−183.45°C −159.55°C −156.25°C	(30) (30) (30)
Density of solid at −123.15°C	2550.0 kg m^{-3}	(30)
Coefficient of thermal expansion of liquid at boiling point	1.23×10^{-3} °C^{-1}	(34)
Heat of formation of gas at 25°C	-3.6467×10^7 J kg-mol^{-1}	(31)
Free energy of formation of gas at 25°C	-5.3298×10^7 J kg-mol^{-1}	(31)
Heat of sublimation at triple-point temperature	2.1185×10^7 J kg-mol^{-1}	(14)
Heat of fusion	2.4074×10^6 J kg-mol^{-1}	(26)
Heats of transition	2.6544×10^5 J kg-mol^{-1} 3.2992×10^5 J kg-mol^{-1} 0.3588×10^5 J kg-mol^{-1}	(15) (15) (15)
Entropy of gas at 25°C	1.9873×10^5 J kg-mol^{-1} °C^{-1}	(31)
Heat capacity ratio of gas at 30°C and 1.0133×10^5 N m^{-2}	1.42	(18)
Velocity of sound at 0°C and 1.0133×10^5 N m^{-2}	199.8 m s^{-1}	(4)
Dielectric constant of liquid at 25°C	3.8	(20)
Electrical conductivity of liquid at −80°C	8×10^{-7} ohm^{-1} m^{-1}	(18)
Refractive index of liquid at 10°C	1.325	(3)
Threshold limit value	(3 ppm) 1×10^{-5} kg m^{-3}	(1)

5.1 Vapour Pressure of Hydrogen Bromide (6,7)

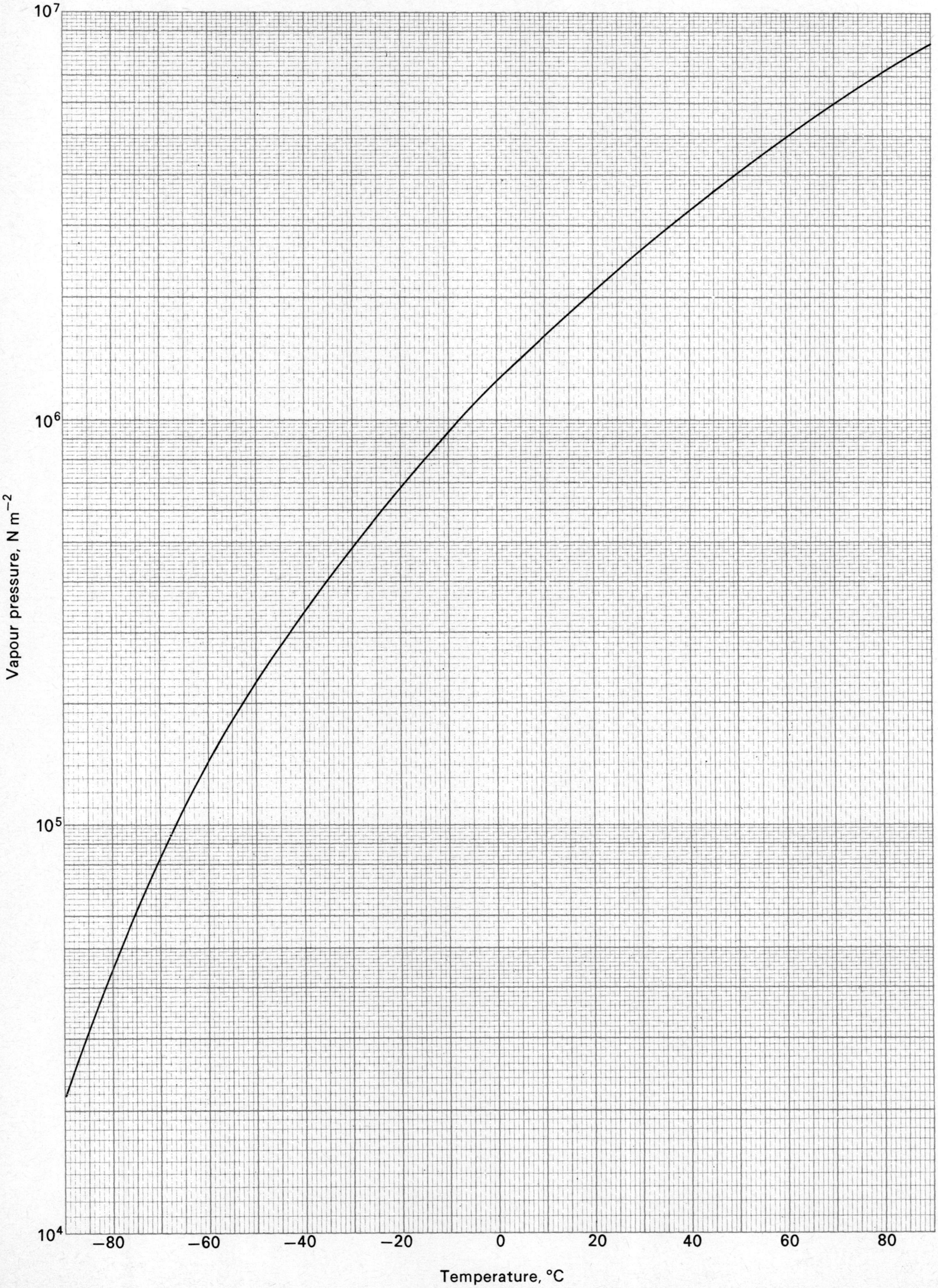

5.2 Orthobaric Density of Hydrogen Bromide (22,23,32)

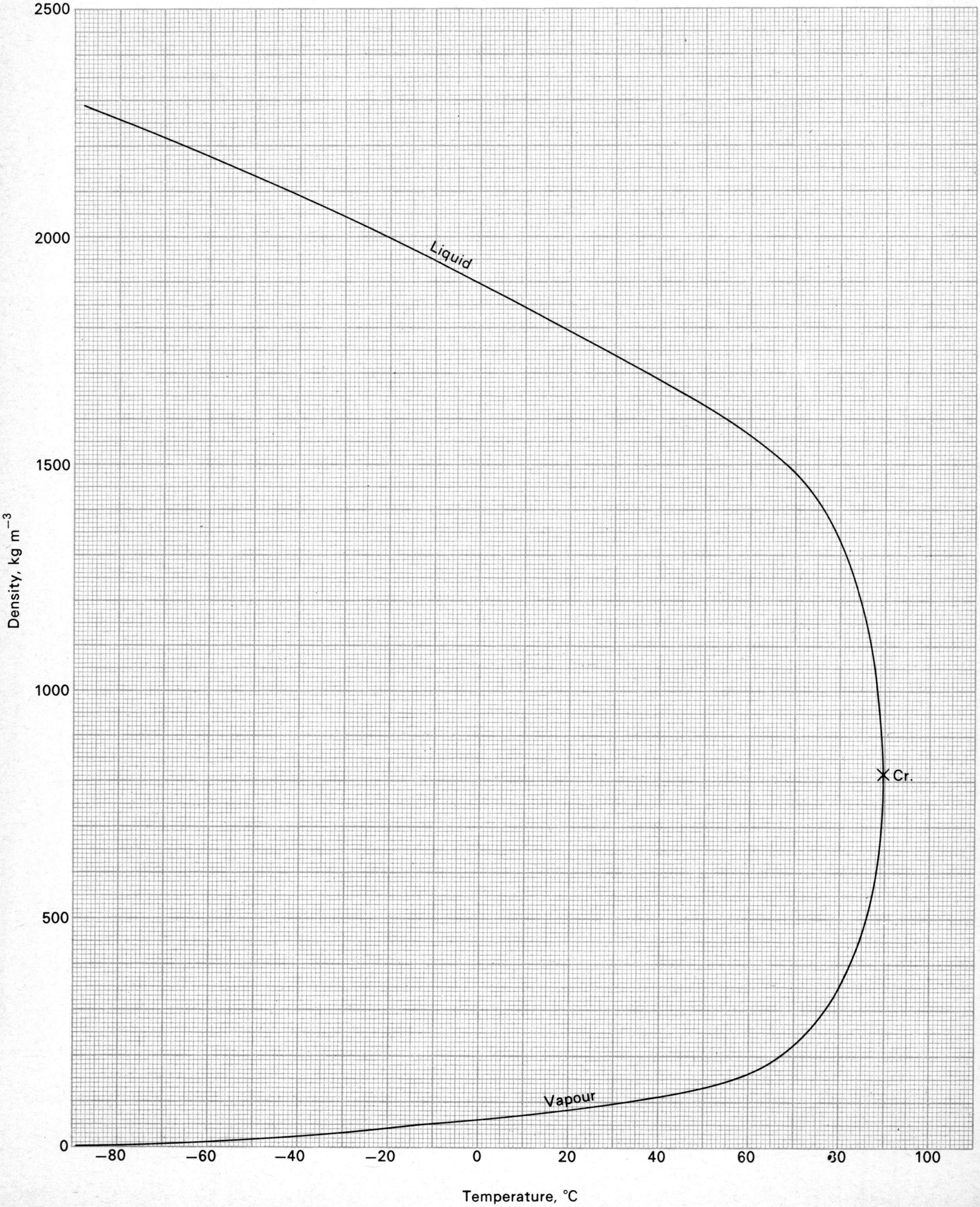

5.3 Latent Heat of Vaporisation of Hydrogen Bromide (17)

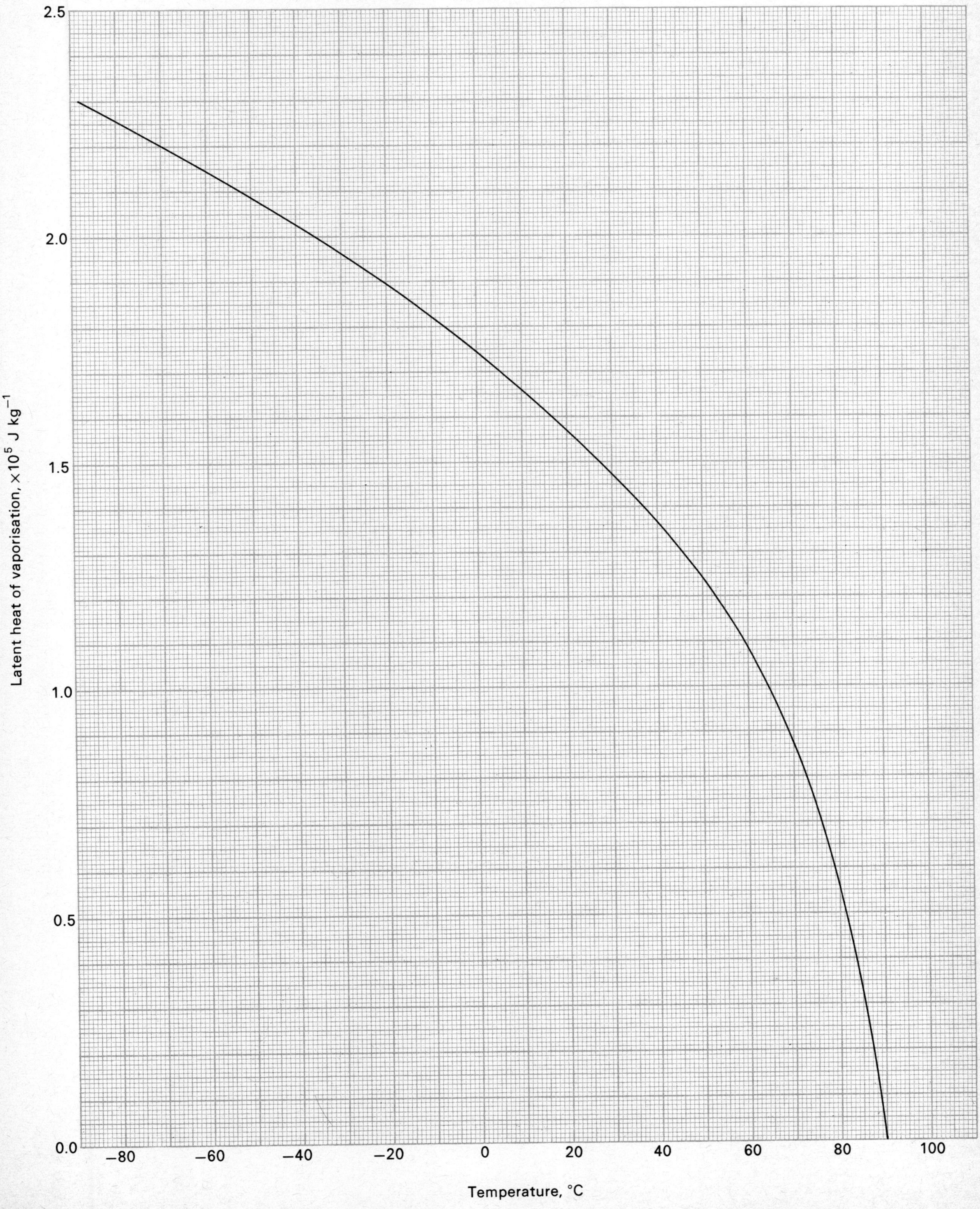

5.4 Heat Capacity of Solid and Liquid Hydrogen Bromide (10)

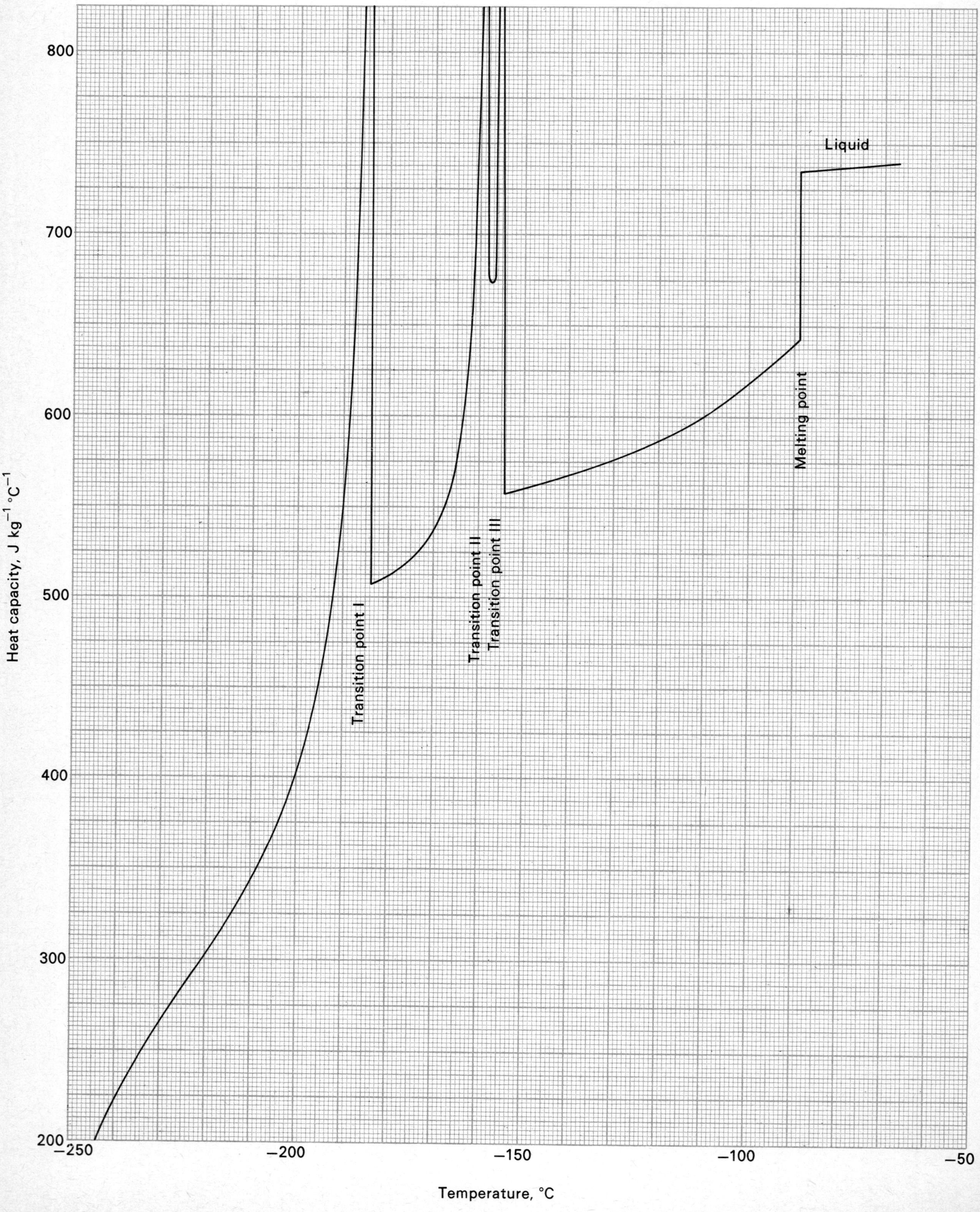

5.5 Heat Capacity of Gaseous Hydrogen Bromide at 1.0133×10^5 N m^{-2} (31)

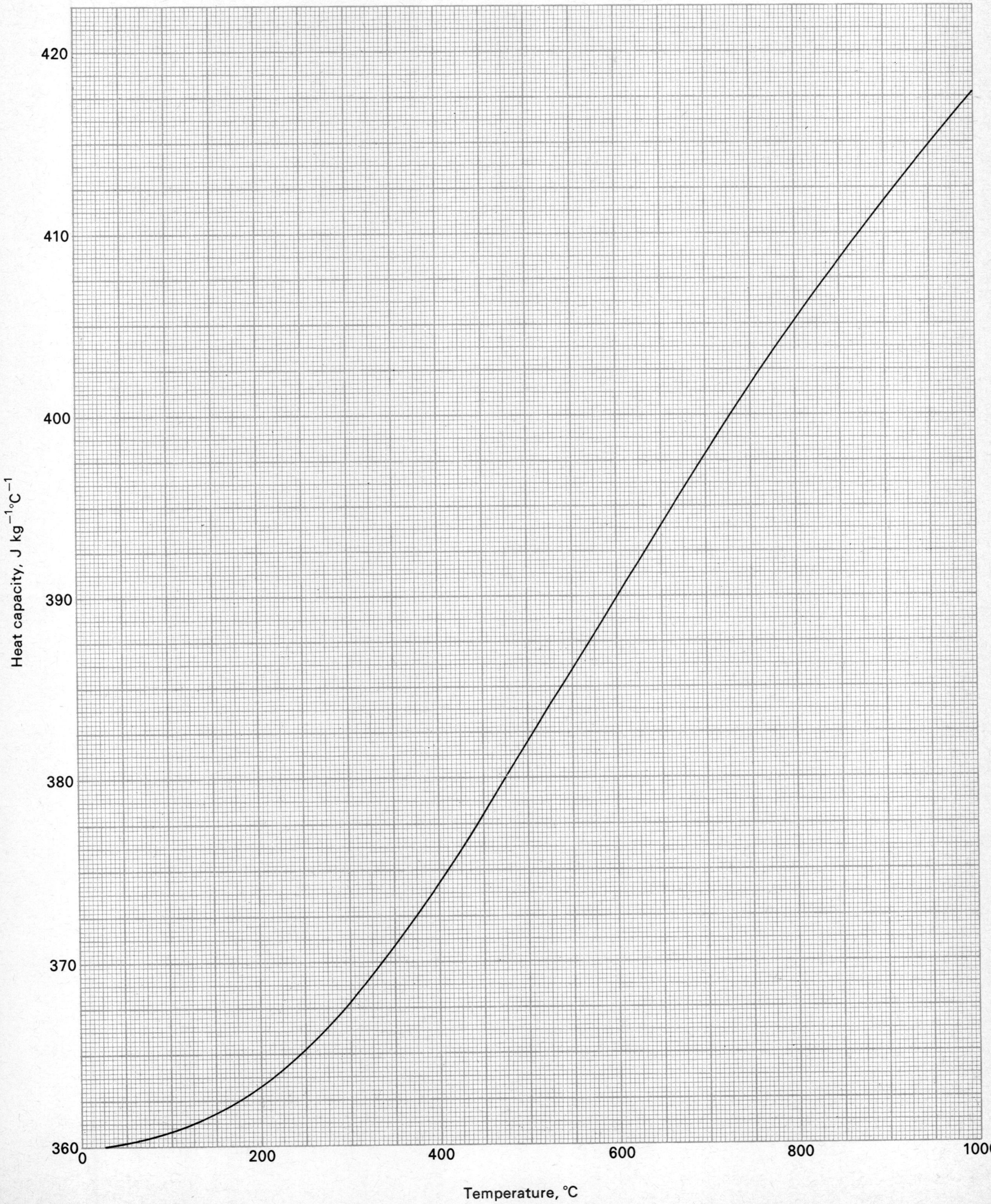

5.6 Viscosity of Liquid Hydrogen Bromide (22,23)

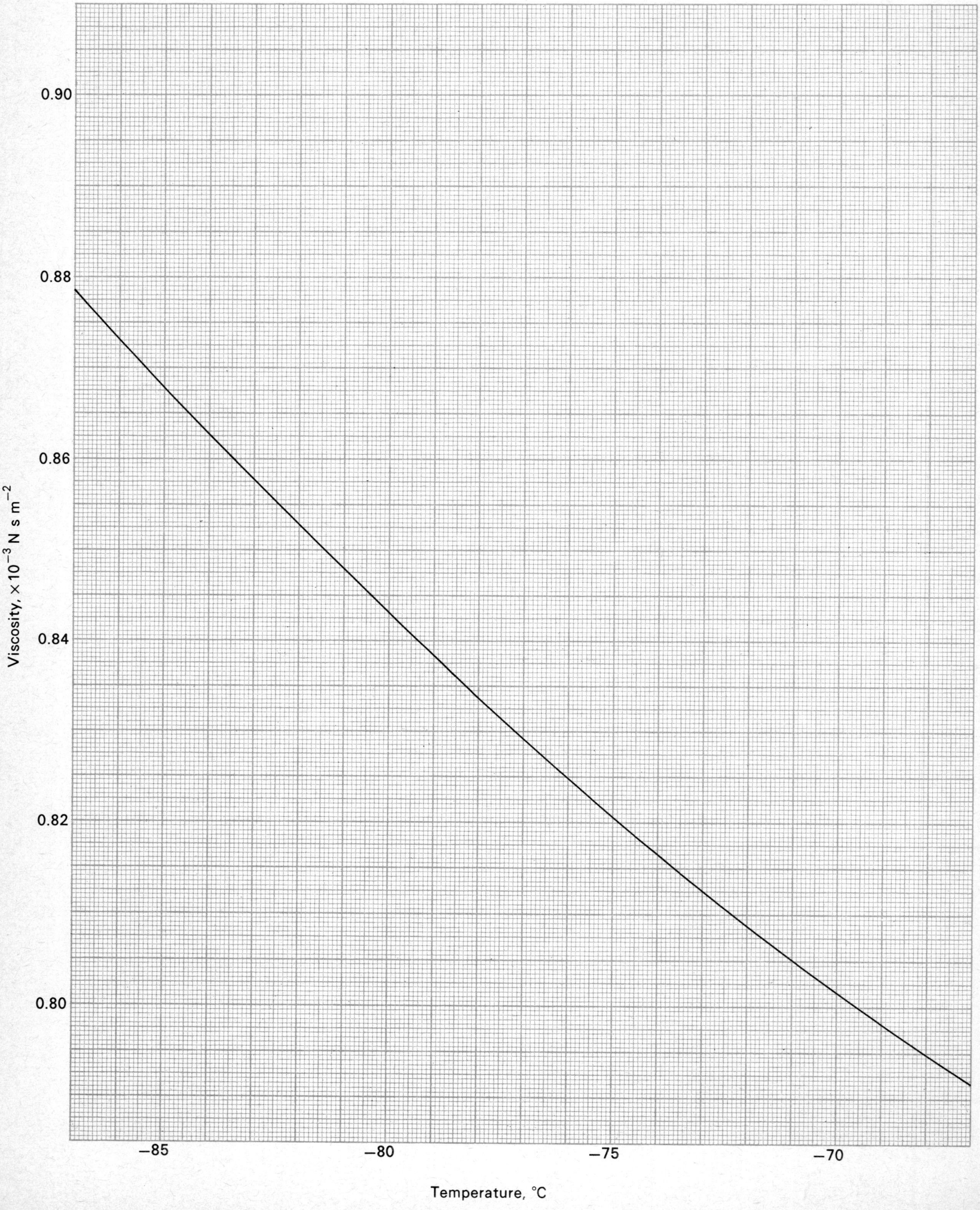

5.7 Viscosity of Gaseous Hydrogen Bromide at 1.0133×10^5 N m^{-2} (12)

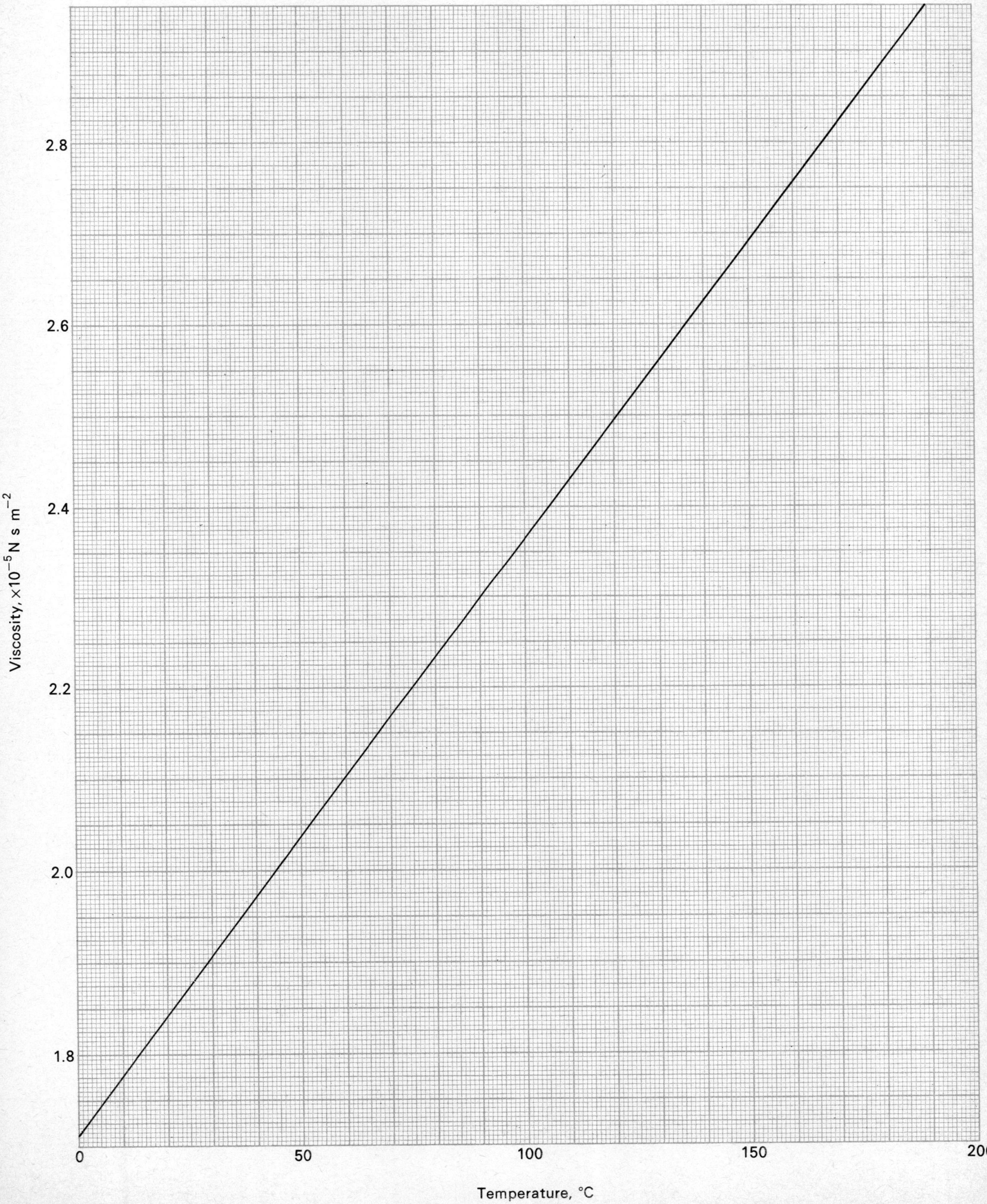

5.8 Thermal Conductivity of Liquid Hydrogen Bromide (29)

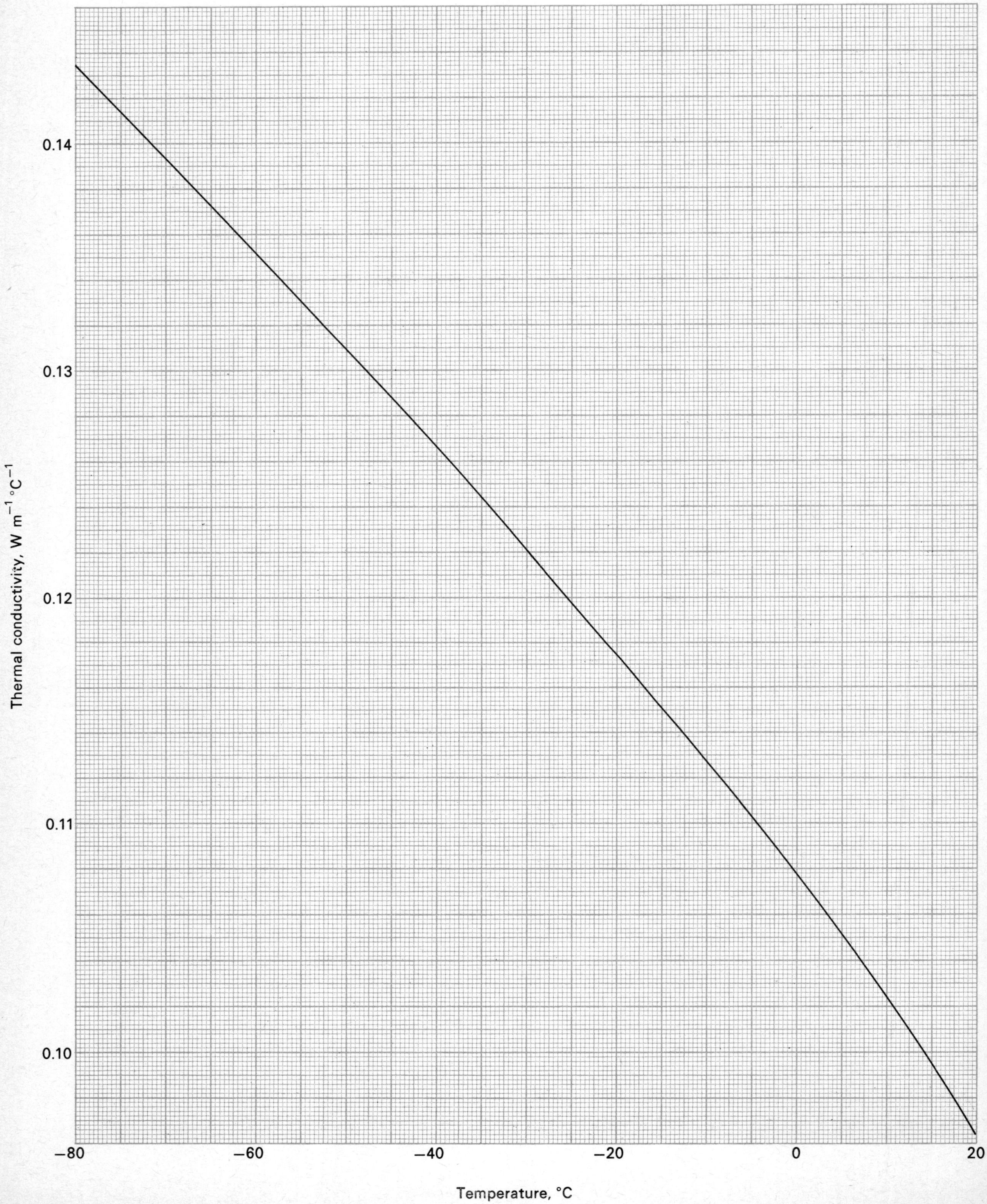

5.9 Thermal Conductivity of Gaseous Hydrogen Bromide at 1.0133×10^5 N m^{-2} (19)

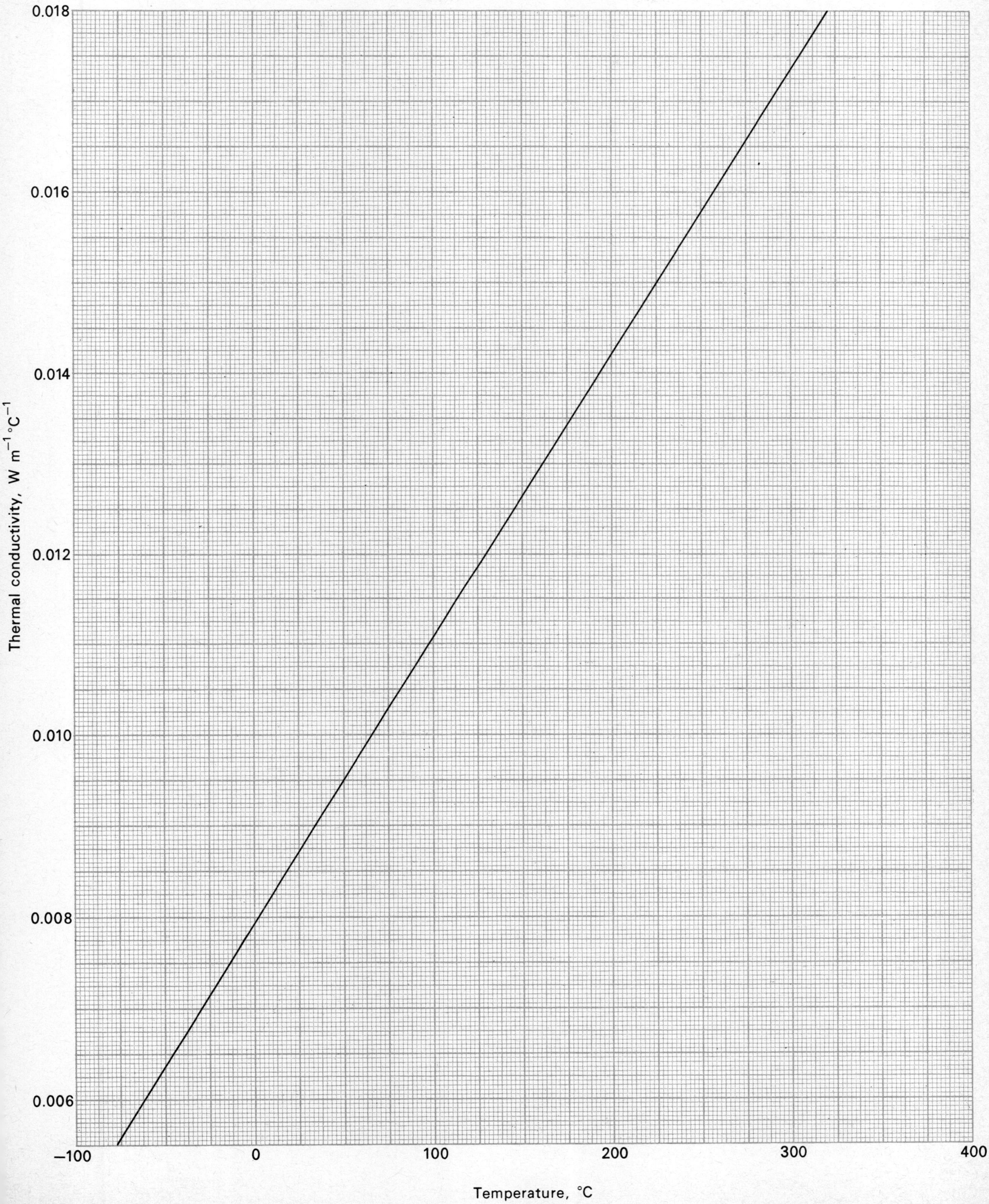

5.10 Surface Tension of Hydrogen Bromide (9,22,25)

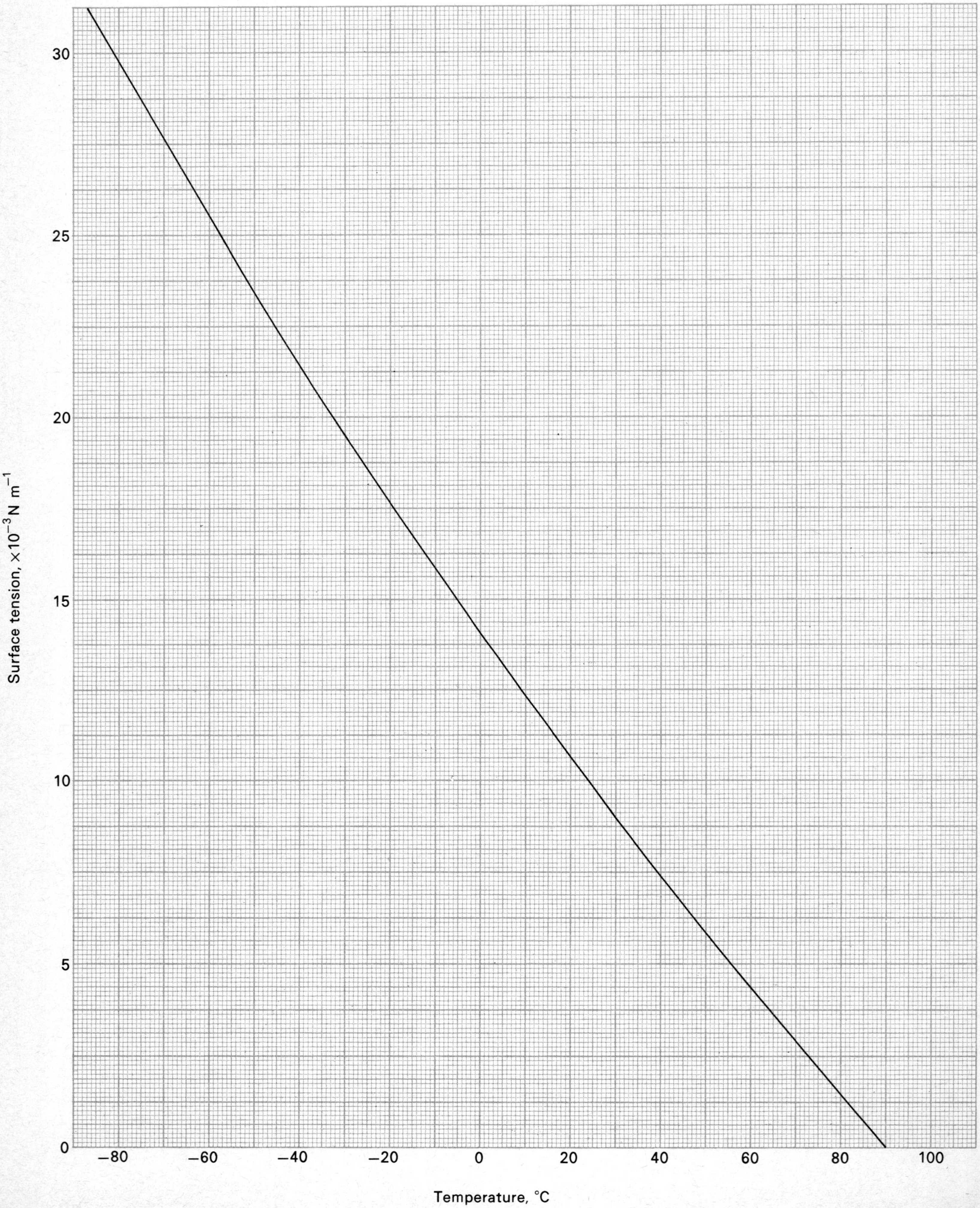

5.11 Solubility of Hydrogen Bromide in Water at 1.0133×10^5 N m^{-2} Total Pressure (19,27,28)

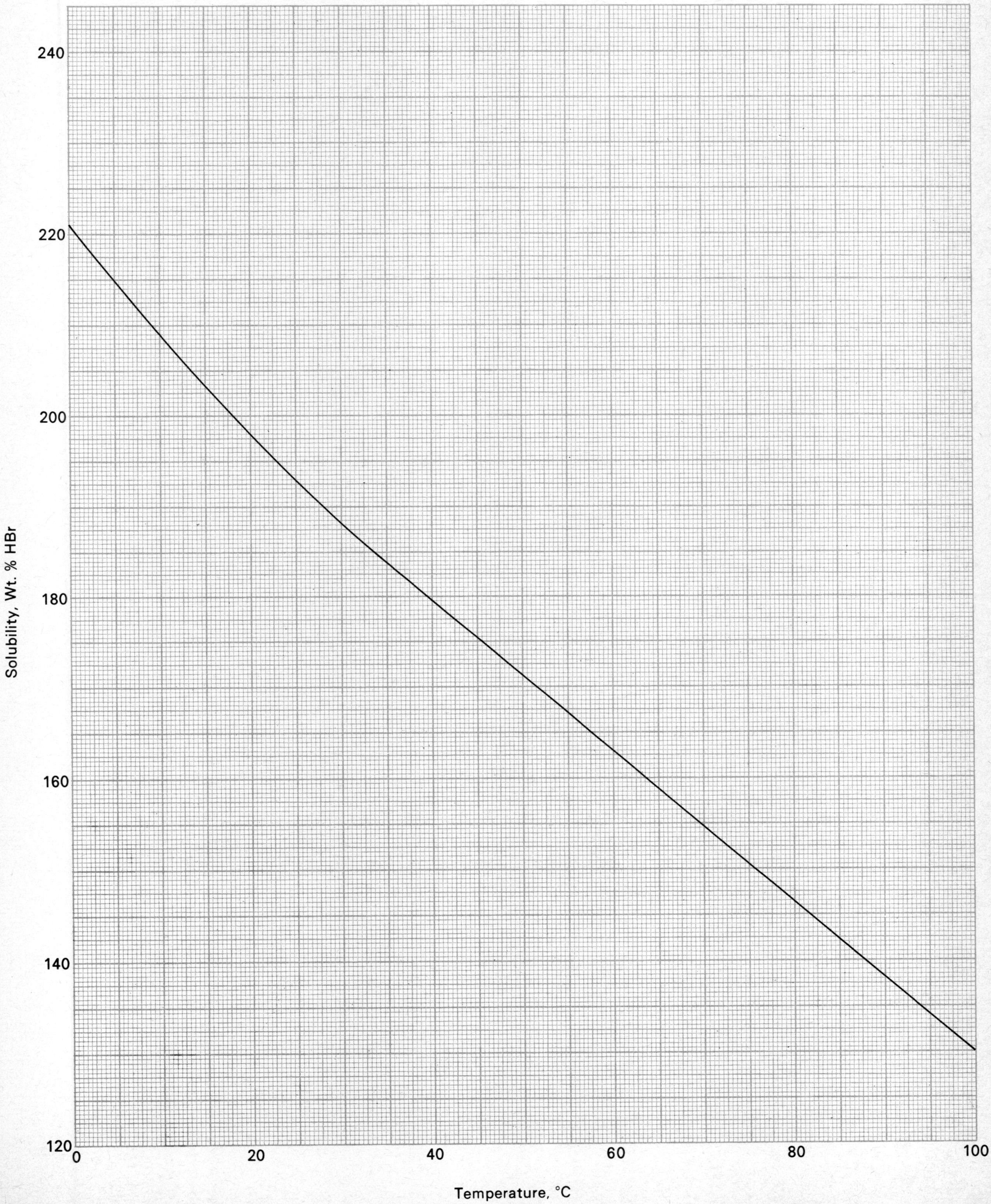

6 HYDROGEN CHLORIDE

Hydrogen chloride is a colourless pungent gas at room temperature and atmospheric pressure. It is a very stable and heat resistant compound. On heating it does not dissociate readily below 1500°C. It behaves as a covalent substance in the pure form and in non-ionizing solvents. Gaseous hydrogen chloride is toxic, but not flammable, and soluble in water. In cylinders, it is under pressure in the form of gas over liquid HCl. Gaseous HCl easily condenses to a colourless liquid, which fumes strongly in a moist atmosphere.

Solid hydrogen chloride forms white crystals in rhombic and cubic modifications. It has one transition point, where its density shows a radical change. The solid compound has a molecular lattice.

Because hydrogen chloride is one of the most frequently used substances in the chemical industry, several manufacturers issued bulletins, e.g. (32) and (36), dealing with its properties and uses. There are review articles and compilations in books on the physical, thermodynamic, and transport properties of solid, liquid, and gaseous hydrogen chloride, e.g. Gmelin (10), MCA (25), Mellor (27), and Pascal (28).

The various properties of hydrogen chloride have been investigated very intensively. Most of the experimental measurements have been carried out during the last decade, and the results are very reliable. However, there are no acceptable data on the surface tension (13). Some of the relevant sources for further study are: gas density (3), equation of state (37), and bond effect, geometry, and physical properties (29).

References

1 American Conference of Government Industrial Hygienists, *Documentation of Threshold Limit Values*, revised edition, Committee on Threshold Limit Values, Cincinnati, Ohio 45202, 1966, 203 pp.

2 Baker, C. E., Brokaw, R. S., *J. Chem. Phys.*, 1965, **43**, 3519–28.

3 Beaume, R., Coulon, *R. Compt. rend.*, 1968, **266B**, 1024–6.

4 Brose, M., Univ. of Göttingen, *Diplomarbeit*, Göttingen, 1958.

5 Clusius, K., *Z. phys. Chem.*, 1941, **49B**, 1–12.

6 Edwards, D. G., *The Vapour Pressure of 30 Inorganic Liquids between One Atmosphere and the Critical Point*, Univ. of California, Lawrence Rad. Lab. UCRL 7167, Livermore, California, June 13, 1963, 148 pp.

7 Franck, E. U., *Z. Elektrochem.*, 1951, **55**, 636–43.

8 Franck, E. U., Brose, M., Mangold, K., *Supercritical Hydrogen Chloride. Specific Heat up to 300°C and 300 atm. PVT–Data up to 400°C and 2000 atm*, Progr. Internat. Res. Thermodyn. Transport Properties, 2nd Symp. Thermophys. Properties, (Masi, J. F., Tsai, D. H., Eds.,) *Am. Soc. Mech. Eng.*, Academic Press, New York, 1962, pp. 159–65.

9 Giauque, W. F., Wiebe, R., *J. Am. Chem. Soc.*, 1928, **50**, 101–22.

10 *Gmelins Handbuch der anorganischen Chemie*, Syst. No. 6, Chlor, Teil B, Lief. 1, Verlag Chemie GmbH, Weinheim, 1968, pp. 90–121.

11 Golubev, I. F., *Viscosity of Gases and Gas Mixtures*, Gosudarstvennoe Izdatel'stvo Fiziko-Matematicheskoi Literatury, Moskva, 1959, Table 4.

12 Henvis, B. W., *Electronics*, 1947, **20** (3), 134–6.

13 Horvath, A. L., *Proc. Techn. Internat.*, 1973, **18** (1), 67–9.

14 IUPAC, Atomic Weights of the Elements 1971, *Pure & Appl. Chem.*, 1972, **30** (3–4), 638–49.

15 Iacocca, D., Iacocca, A. L., Cerceau, O., *Acta Cient. Venezolana*, 1964, **15** (5), 189–92.

16 Karapet'yants, M. Kh., *Zh. Fiz. Khim.*, 1958, **32**, 556–68.

17 Kordes, E., *Z. Elektrochem.*, 1954, **58**, 424–31.

18 Krynicki, K., Hennel, J. W., *Inst. Nucl. Phys. Cracow Rept. No. 258*, Cracow, 1963.

19 Krynicki, K., Hennel, J. W., *Acta Phys. Polon.*, 1963, **24**, 269–71.

20 Lange, N. A., Ed., *Handbook of Chemistry*, 10th ed., McGraw-Hill, New York, 1967.

21 Lax, E., Synowietz, C., Eds., *Taschenbuch für Chemiker und Physiker*, 3rd ed., Springer-Verlag, Berlin, 1967.

22 Leduc, A., *Chem. Rev.*, 1929, **6**, 1–16.

23 Lewis, G. N., MacDonald, R. T., Schutz, P. W., *J. Am. Chem. Soc.*, 1934, **56**, 494–5.

24 Mangold, K., Franck, E. U., *Z. Elektrochem.*, 1962, **66** (3), 260–6.

25 Manufacturing Chemists Association, *Hydrochloric Acid, Aqueous and Hydrogen Chloride, Anhydrous*, Chemical Safety Data Sheet SD-39, Revised, Washington, 1970, 28 pp.

26 Maryott, A. A., Smith, E. R., *Table of Dielectric Constants of Pure Liquids*, Natl. Bur. Std. Circular 514, Washington, Aug. 10, 1951, 44 pp.

27 Mellor's Comprehensive Treatise on Inorganic and Theoretical Chemistry, *The Halogens*, Vol. II, Suppl. 1, Longmans, London, 1956, pp. 413–51.

28 Pascal, P., Ed., *Nouveau Traité de Chimie Minerale*, Vol. XVI, F, Cl, Br, I, At, Mn, Tc, Re, Masson et Cie, Éditeurs, Paris, 1960, pp. 201–15.

29 Petke, J. D., Whitten, J. L., *J. Chem. Phys.*, 1972, **56** (2), 830–8.

30 Rossini, F. D., Wagman, D. D., Evans, W. H., Levine, S., Jaffe, I., *Selected Values of Chemical Thermodynamic Properties*, Natl. Bur. Std. Circular 500, Washington, Feb. 1, 1952.

31 Spangenberg, C. D., *Chem. Eng.*, 1951, **58** (9), 170–1.

32 Stauffer Chemical Company, *Hydrogen Chloride Anhydrous*, Techn. Service Product Rept., New York, 16 pp.

33 Stewart, J. W., *J. Chem. Phys.*, 1962, **36** (2), 400–5.

34 Stull, D. R., Dir., *JANAF Thermochemical Tables*, Dow Chemical Co., Midland, Mich., 1965–8.

35 Thomas, W., *Volumetric Behaviour of Hydrogen Chloride*, Progr. Internat. Res. Thermodyn. Transport Properties, 2nd Symp. Thermophys. Properties (Masi, J. F., Tsai, D. H., Eds.), Am. Soc. Mech. Eng., Academic Press, New York, 1962, pp. 166–71.

36 Union Carbide Corporation, *HCl Physical Properties*, Carbon Products Division, Rept. No. RD4-7.3, Cleveland, Ohio 44101.

37 Voloshin, A. P., *Equation of State for Hydrogen Chloride*, Teplofiz. Svoistva Gazov, Mater. Vses Teplofiz. Konf. Svoistvam Veshchestv Vys. Temp. 3rd (Vargaftik, N. B., Ed.,) 1968, (Publ. 1970), pp. 92–5.

38 Yosim, S. J., *J. Chem. Phys.*, 1964, **40** (10), 3069–75.

39 Ziebland, H., Needham, D. P., *Recent Measurements of the Thermal Conductivity of Liquid Anhydrous Hydrogen Chloride*, Ministry of Technology, Explosives Research and Development Establishment, Report No. 27/R/67, Waltham Abbey, 1967, 16 pp.

40 Ziebland, H., Needham, D. P., *Recent Measurements of the Thermal Conductivity of Liquid Anhydrous Hydrogen Chloride*, Proc. 4th Symp. Thermophys. Properties, (Maszynski, J. R., Ed.,) Univ. of Maryland, Am. Soc. Mech. Eng., New York, 1968, pp. 296–303.

Table 6.1 Properties of Hydrogen Chloride

		Ref.
Chemical formula	HCl	(21)
Molecular weight	36.4609	(14)
Critical temperature	51.5°C	(8, 35)
Critical pressure	83.09×10^5 N m^{-2}	(8, 35)
Critical density	450.0 kg m^{-3}	(8, 35)
Triple-point temperature	−114.24°C	(5, 9)
Triple-point pressure	0.1383×10^5 N m^{-2}	(9)
Boiling point at 1.0133×10^5 N m^{-2}	−85.1°C	(30)
Melting point	−114.22°C	(30)
Transition temperature at 1.0133×10^5 N m^{-2}	−174.79°C	(9)
Density of solid at −166.15°C	1469.0 kg m^{-3}	(33)
Compressibility of liquid at −114.2°C	104.6×10^{-11} m^2 N^{-1}	(15)
Coefficient of thermal expansion of liquid at boiling point	2.40×10^{-3} C^{-1}	(38)
Heat of formation of gas at 25°C	-9.2373×10^7 J kg-mol^{-1}	(34)
Free energy of formation of gas at 25°C	-9.5333×10^7 J kg-mol^{-1}	(34)
Heat of sublimation at triple-point temperature	1.9079×10^7 J kg-mol^{-1}	(23)
Heat of fusion	1.9929×10^6 J kg-mol^{-1}	(30)
Heat of transition	1.1891×10^6 J kg-mol^{-1}	(9)
Entropy of gas at 25°C	1.8692×10^5 J kg-mol^{-1} C^{-1}	(34)
Heat capacity ratio of gas at 15°C and 1.0133×10^5 N m^{-2}	1.41	(22)
Velocity of sound at 0°C and 1.0133×10^5 N m^{-2}	296.0 m s^{-1}	(12)
Diffusion in air at 80°C and 1.0133×10^5 N m^{-2}	0.26×10^{-4} m^2 s^{-1}	(31)
Dielectric constant of liquid at 28°C	4.6	(26)
Electrical conductivity of liquid at −96°C	1×10^{-6} ohm^{-1} m^{-1}	(20)
Refractive index of liquid at 18°C	1.3287	(21)
Threshold limit value	(5 ppm) 7×10^{-6} kg m^{-3}	(1)

6.1 Vapour Pressure of Hydrogen Chloride (6,9,35)

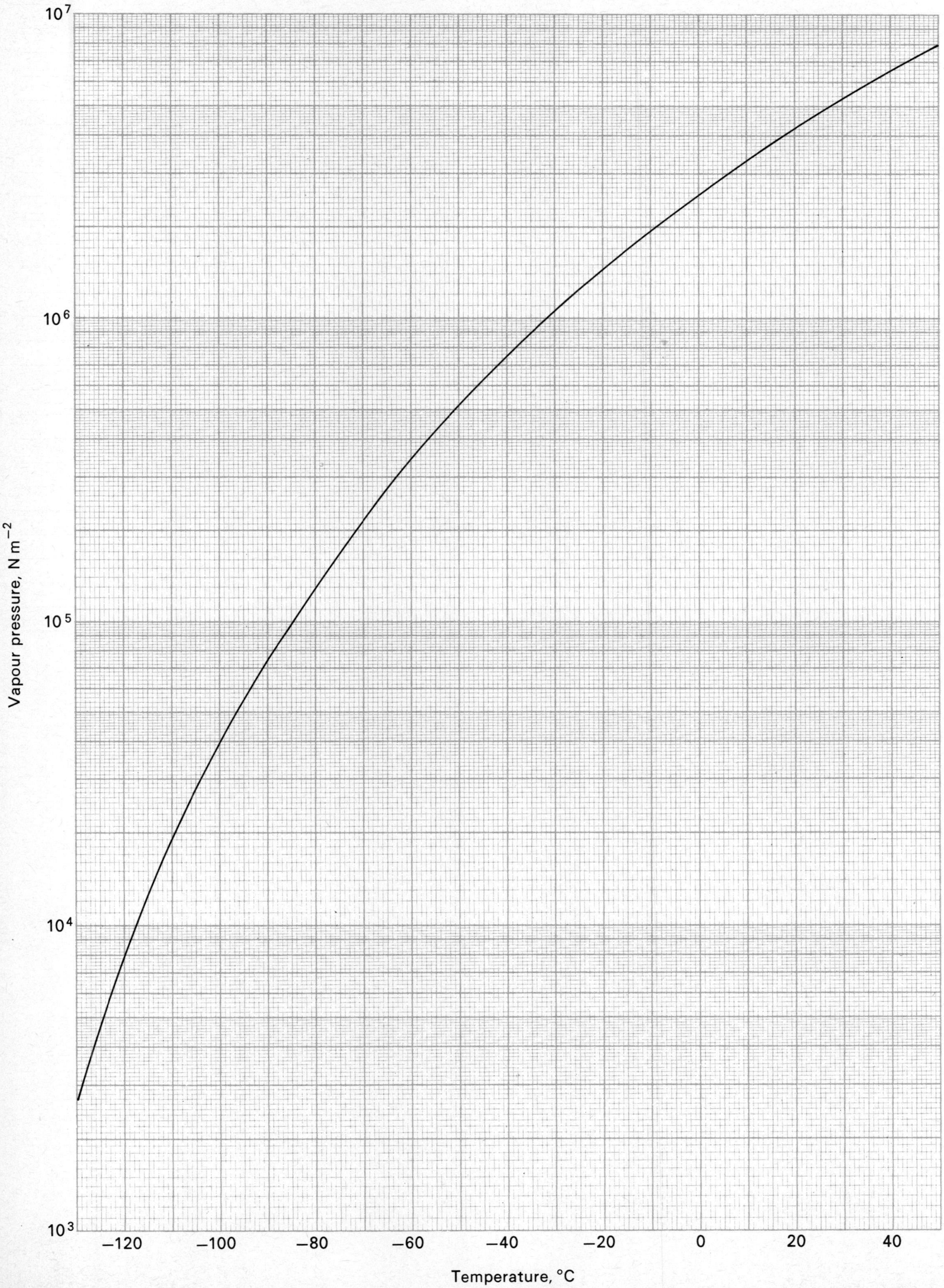

6.2 Orthobaric Density of Hydrogen Chloride (20,21,35)

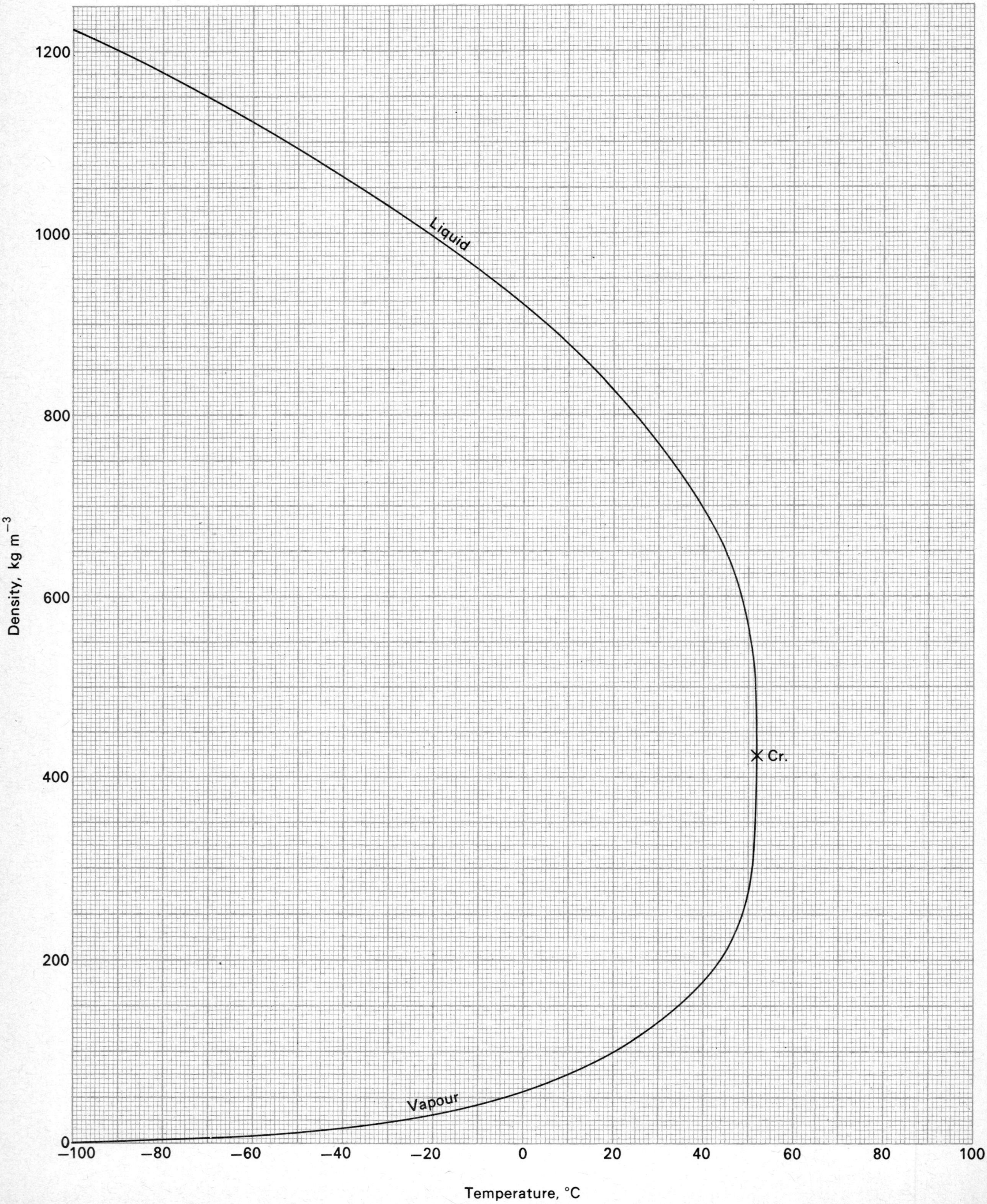

6.3 P—V—T Data for Hydrogen Chloride (4,24,35)

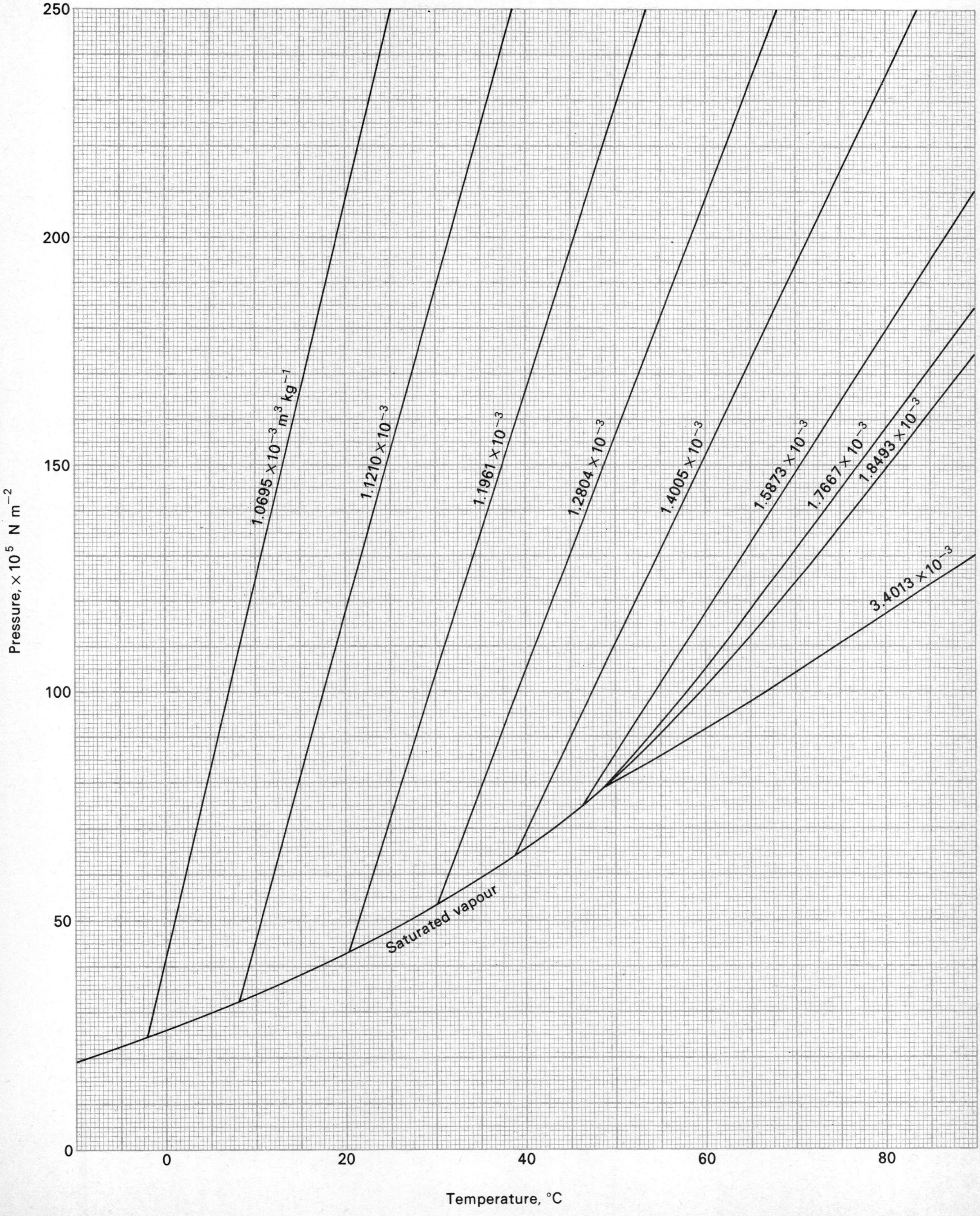

6.4 P—V—T Data for Hydrogen Chloride (Supercritical) (4,24,35)

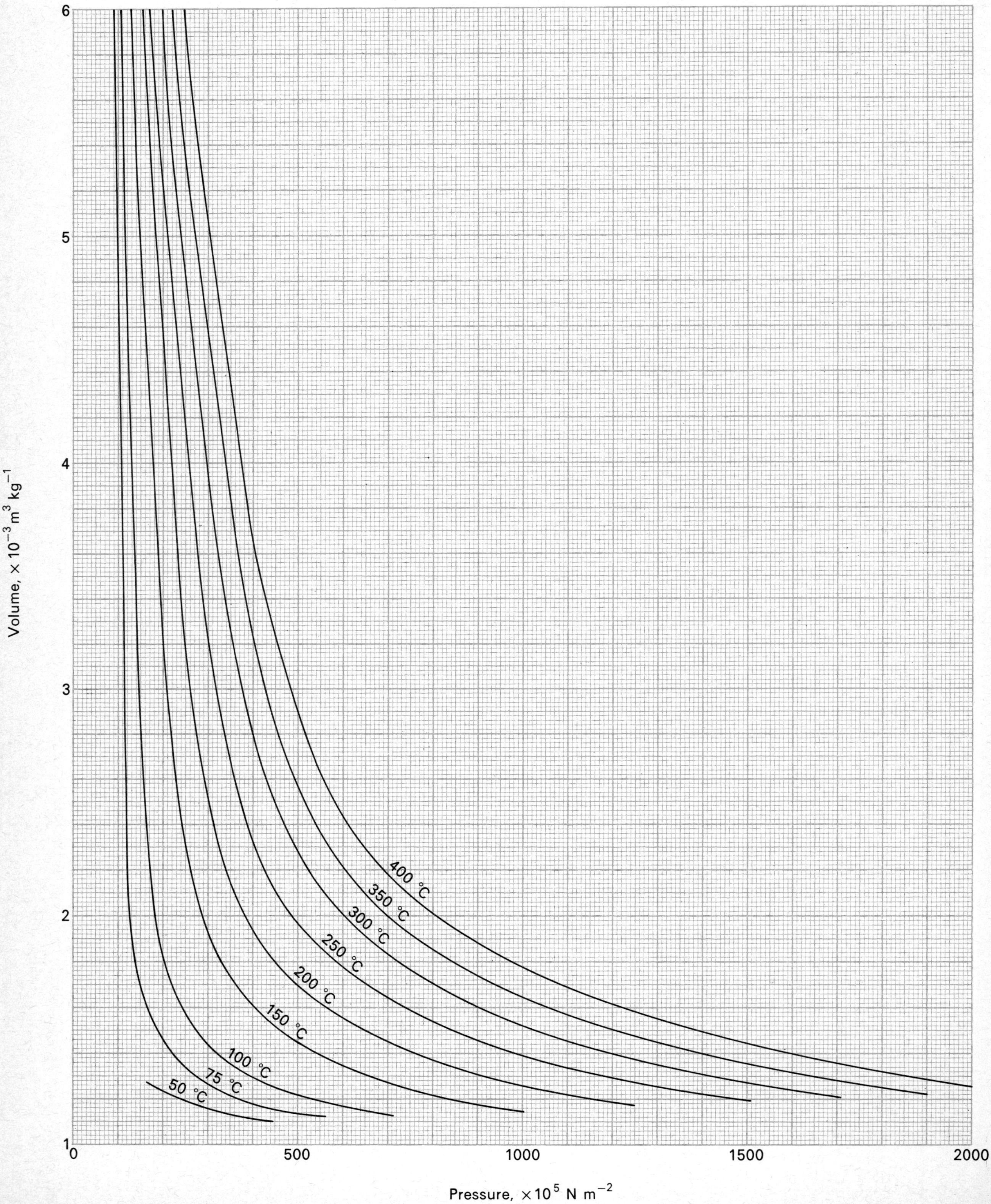

6.5 Latent Heat of Vaporisation of Hydrogen Chloride (10,16,17)

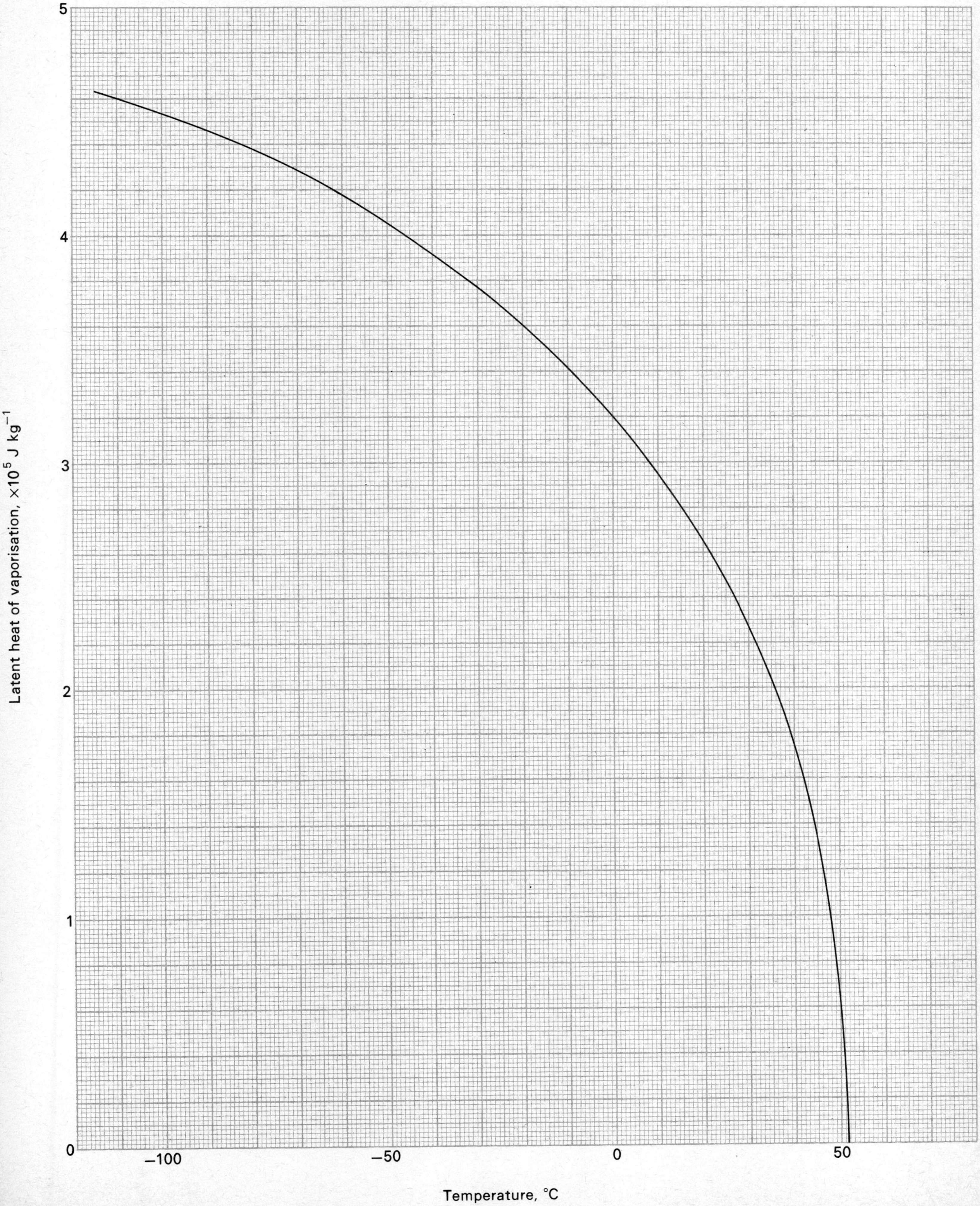

6.6 Heat Capacity of Solid and Liquid Hydrogen Chloride (9)

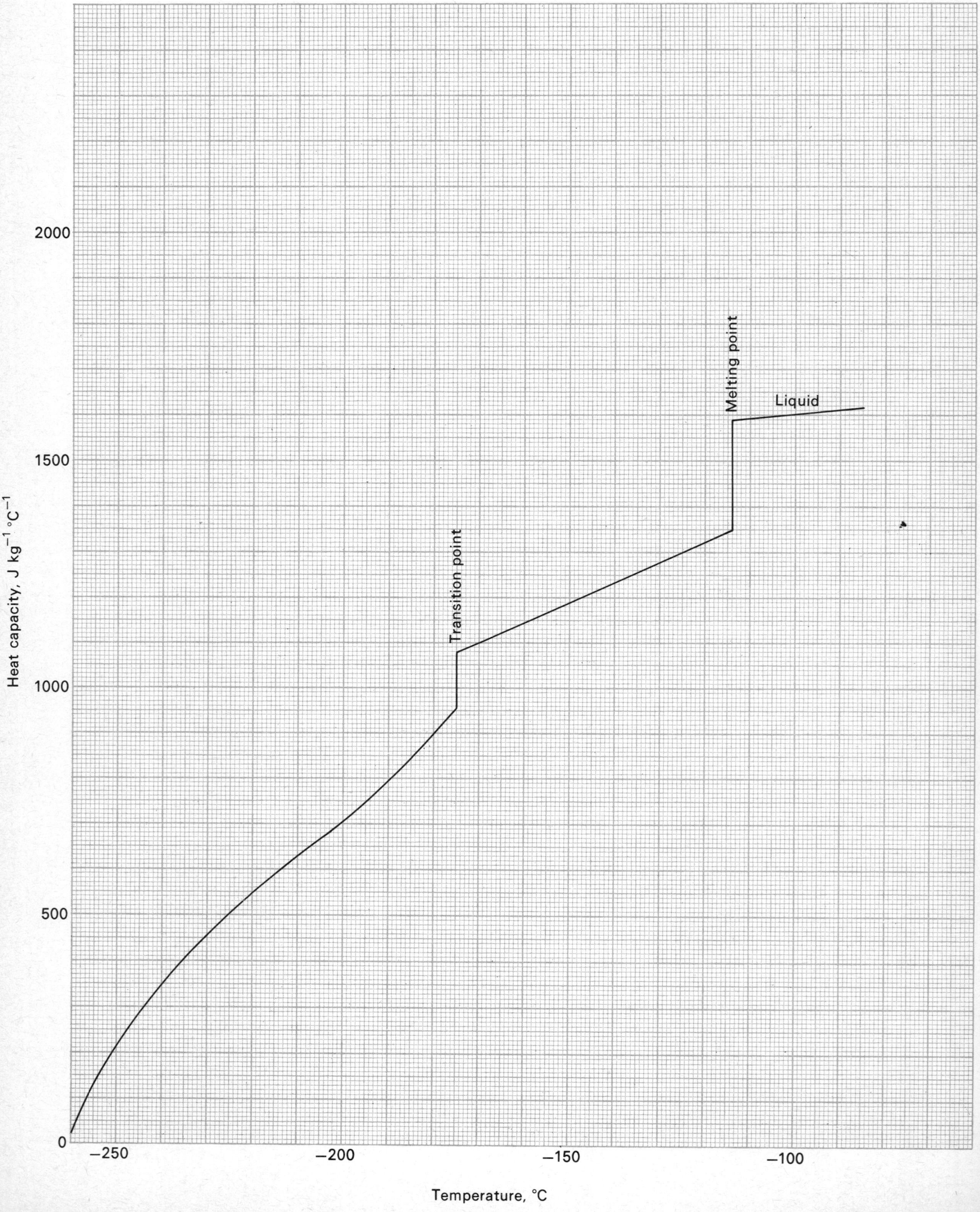

6.7 Heat Capacity of Gaseous Hydrogen Chloride at Constant Volume (8)

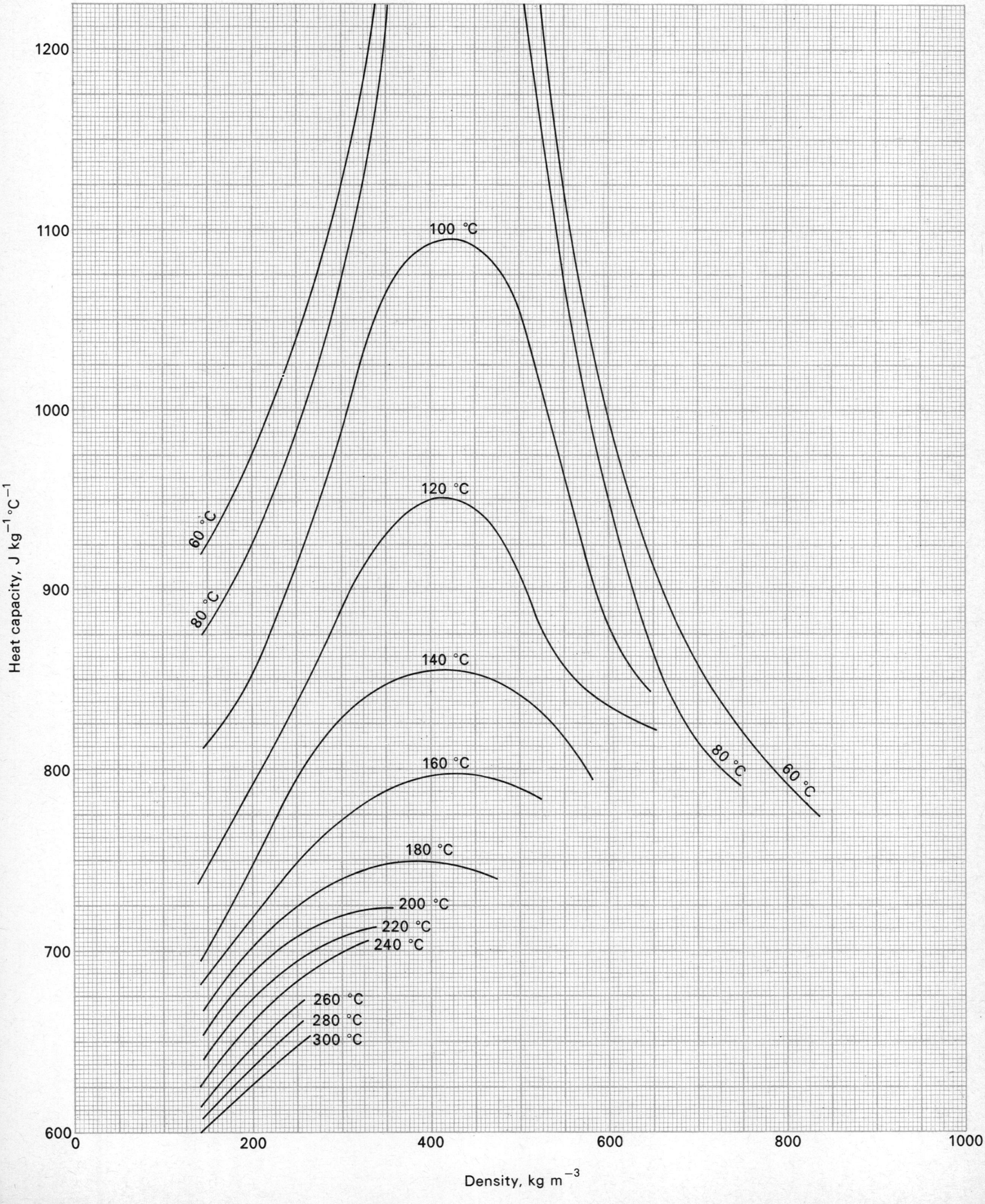

6.8 Viscosity of Liquid Hydrogen Chloride (10,18,19,20,21)

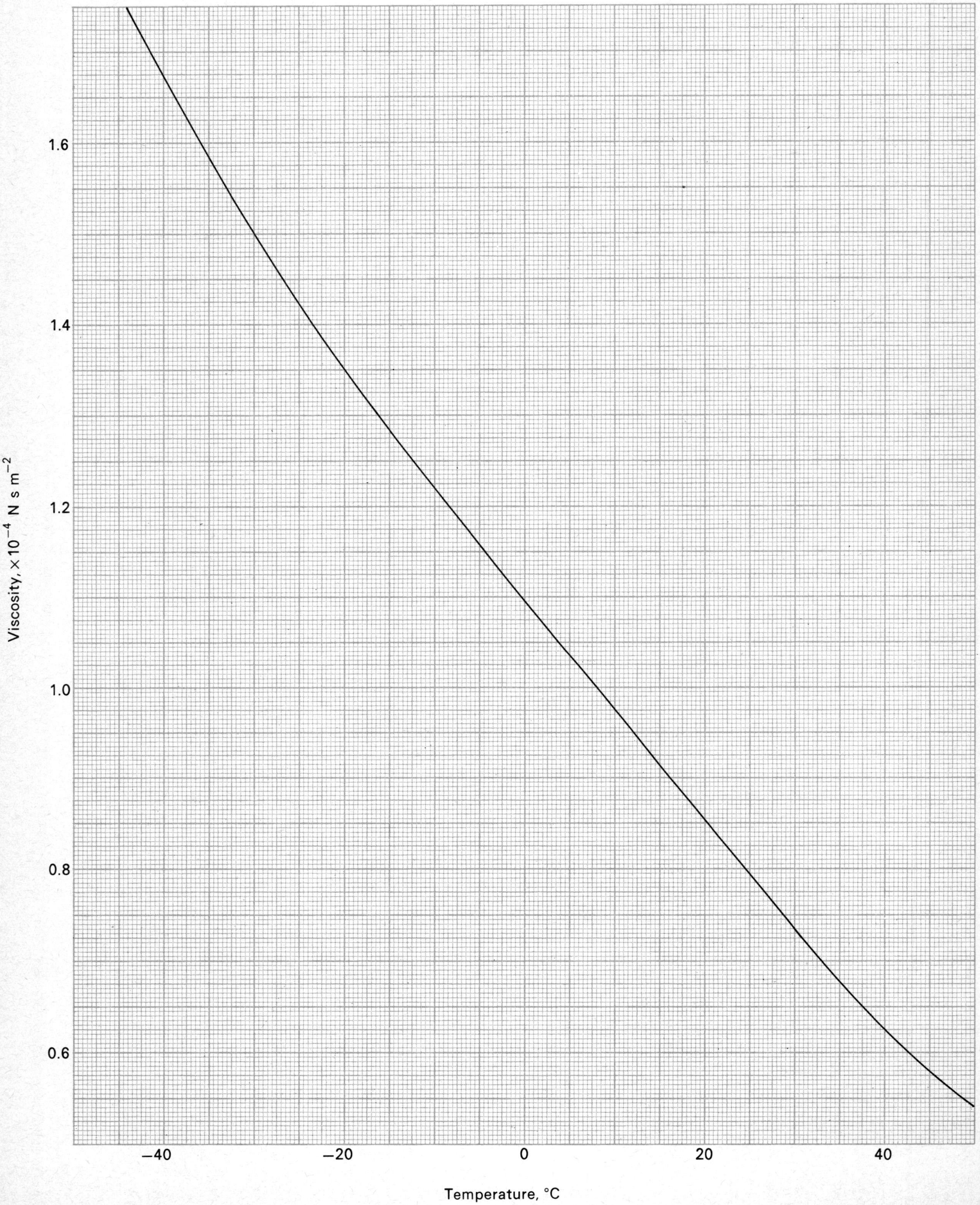

6.9 Viscosity of Gaseous Hydrogen Chloride at 1.0133×10^5 N m^{-2} (10,11)

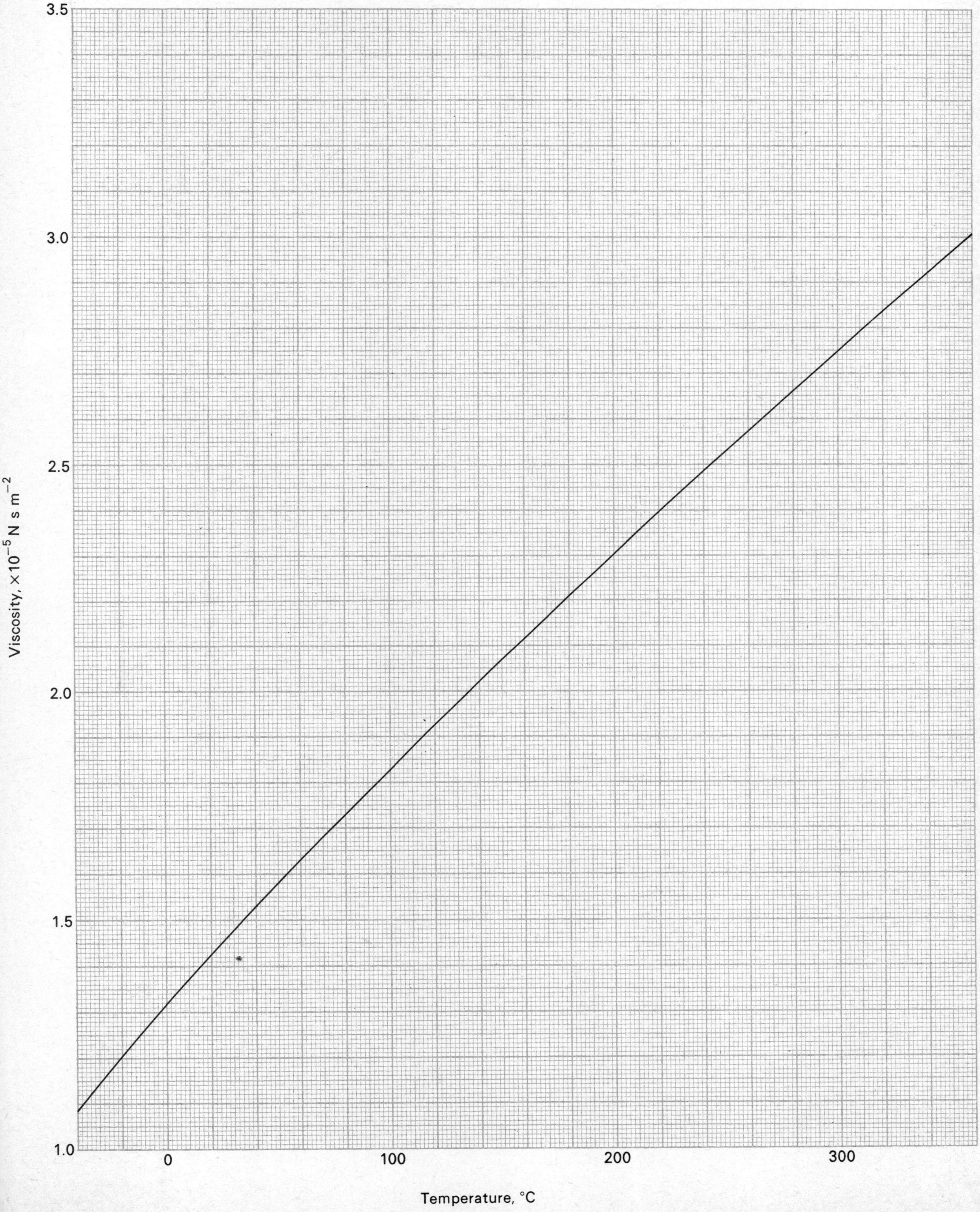

6.10 Thermal Conductivity of Liquid Hydrogen Chloride (39,40)

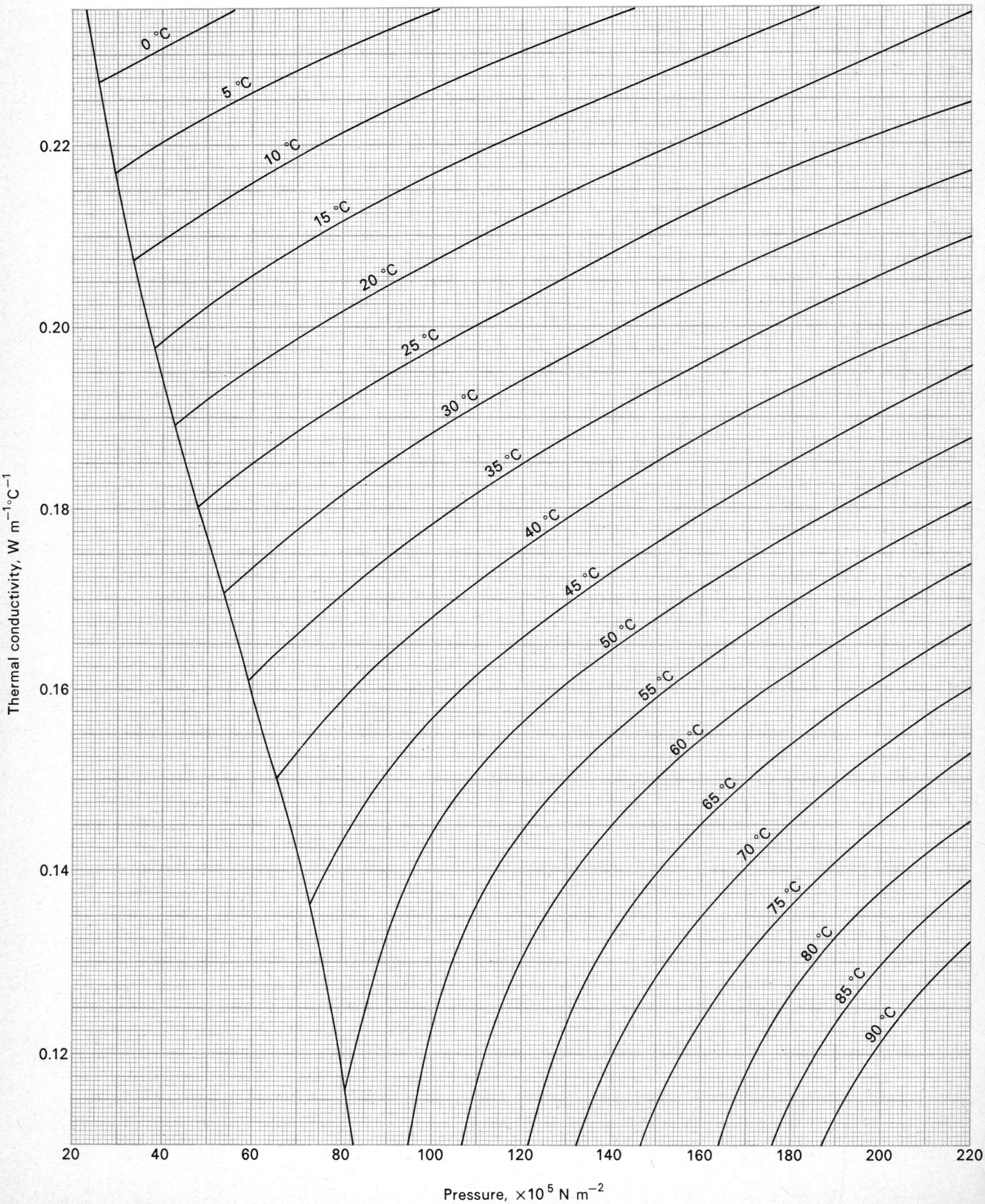

6.11 Thermal Conductivity of Gaseous Hydrogen Chloride at 1.0133×10^5 N m^{-2} (2,7)

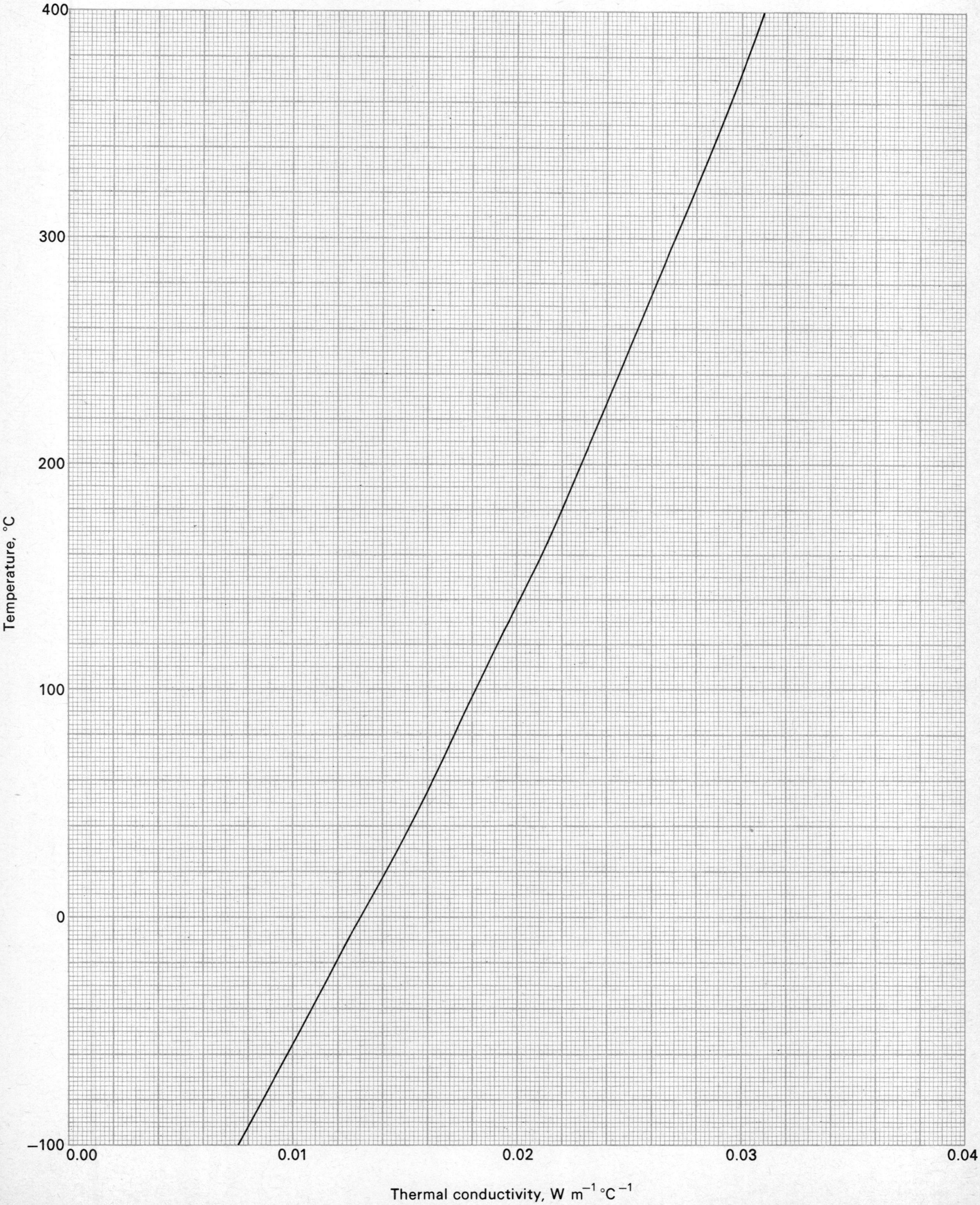

6.12 Solubility of Hydrogen Chloride in Water at 1.0133×10^5 N m^{-2} Total Pressure (10,20,21)

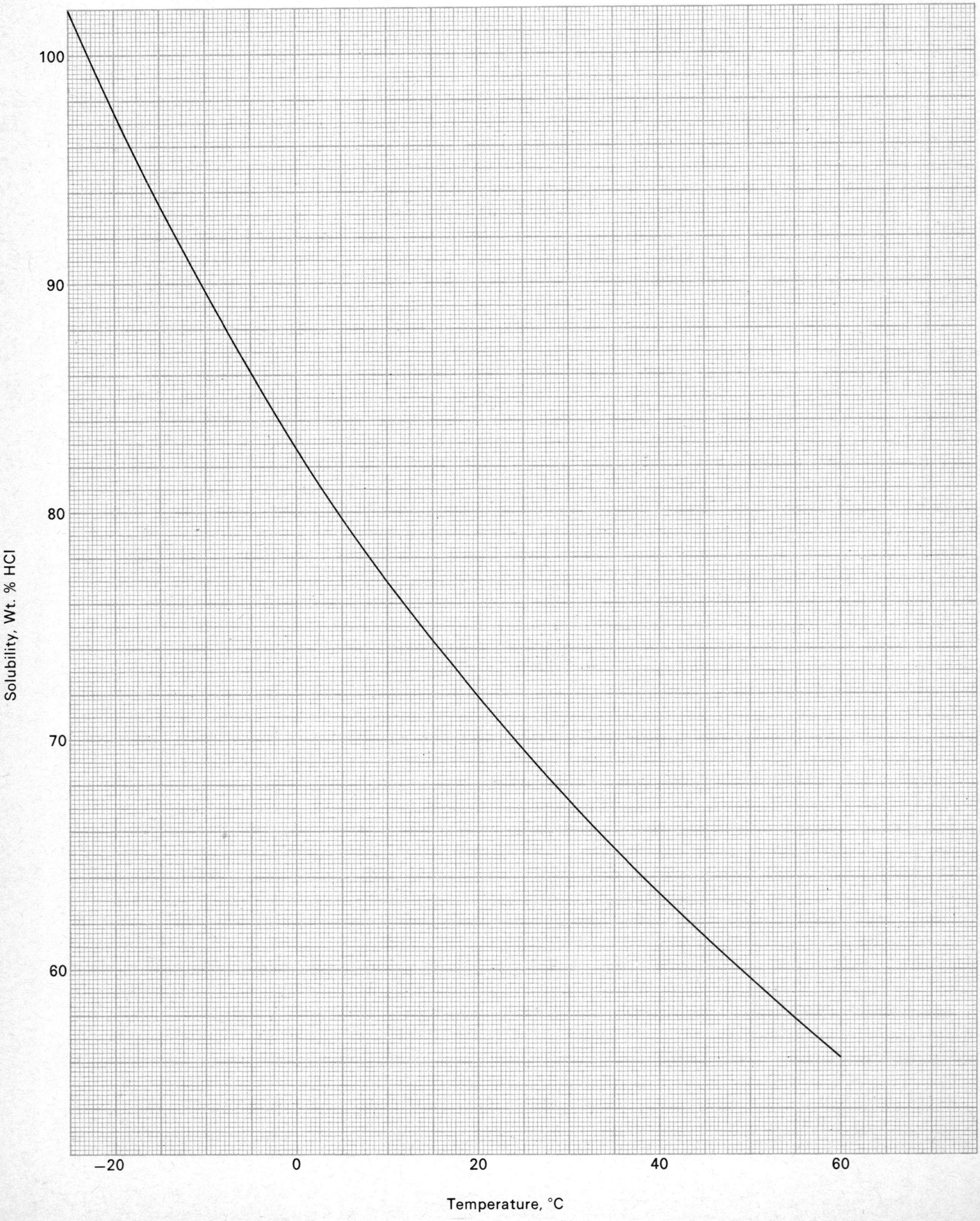

7 HYDROGEN FLUORIDE

Hydrogen fluoride is a colourless gas at room temperature and atmospheric pressure. It possesses a disagreeable pungent odour and fumes strongly in moist air.

The properties of hydrogen fluoride are different from the other halides, HBr, HCl, and HI. It shows an abnormally high boiling point and internal pressure. The molecular weight is much higher than one would expect from the molecular formula. This is due to polymerization or association in the liquid and gaseous states. Because of the high degree of association (polymerization) the properties of the gas show large deviations from ideal gas behaviour.

In the liquid state $(HF)_6$ has been found and in the gaseous state there is an equilibrium between polymerized and normal (HF) molecules. The molecular weight of the polymerized molecule has been determined from the density at various temperatures:

Temperature, °C	Molecular weight
26.4	51.2
36.3	32.2
47.3	23.7
69.4	21.0
88.1	20.6

Hydrogen fluoride is miscible with water in all proportions, with evolution of heat. The hydrofluoric acid formed is a much weaker acid than the other hydrogen halides.

The compound exhibits strong hydrogen bonding and short-range interactions. In the solid state it does not show any phase transition, in contrast to the other halides.

Anhydrous HF has been available commercially for about 40 years. The distribution is carried out in steel cylinders, tank cars, and pipelines. The main application of HF is as a commercial catalytic agent in alkylation processes in the petroleum industry. It can alkylate alkyl halides, olefins, alcohols, esters, ethers, etc.

Physical, thermodynamic, and transport properties of HF have been investigated in various states. However, some of the reported data show discrepancies. The disagreements between different investigators are due to the abnormal behaviour of the molecule. A typical discrepancy, e.g. liquid heat capacity, has been analysed (17) using a comparative test. The major investigation on various properties has been carried out by Franck and co-workers (9, 10, 11, 36); meanwhile a recent article by Vanderzee and Rodenburg (40) critically examined the thermodynamic properties of HF. The thermodynamic diagrams have been constructed by Yabroff *et al.* (41), on the basis of the critical evaluation of published data. Various properties of HF have been surveyed in articles (25, 32, 33), industrial bulletins (1, 29), and books (14, 16, 24, 26, 28).

References

1 Allied Chemical & Dye Corporation, *Hydrofluoric Acid, Anhydrous*, General Chemical Division, Product Information Bull. Code HC 8040, New York, 1945, 19 pp.

2 American Conference of Government Industrial Hygienists, *Documentation of Threshold Limit Values*, revised edition, Committee on Threshold Limit Values, Cincinnati, Ohio 45202, 1966, 203 pp.

3 Armstrong, G. T., *Hydrogen Fluoride and the Thermochemistry of Fluorine*, Natl. Bur. Std., Techn. Note 513, Washington, Feb. 1970, 21 pp.

4 Atoji, M., Lipscomb, W. N., *Acta Cryst.*, 1954, **7**, 173–5.

5 Axmann, A., Biem, W., Borsch, P., Hossfeld, F., Stiller, H., *Disc. Faraday Soc.*, 1969, **48**, 69–77.

6 Baker, C. E., *J. Chem. Phys.*, 1967, **46**, 2846–8.

7 Cady, G. H., Hildebrand, J. H., *J. Am. Chem. Soc.*, 1930, **52**, 3843–6.

8 Dahmlos, J., Jung, G., *Z. phys. Chem.*, 1933, **21B**, 317–22.

9 Franck, E. U., Meyer, F., *Z. Elektrochem.*, 1959, **63** (5), 571–82.

10 Franck, E. U., Spalthoff, W., *Naturwissensch.*, 1953, **40** (22), 580.

11 Franck, E. U., Spalthoff, W., *Z. Elektrochem.*, 1957, **61** (3), 348–57.

12 Fredenhagen, K., Cadenbach, G., *Z. phys. Chem.*, 1933, **164A**, 201–8.

13 Gerhard, R., Univ. of Göttingen, *Thesis*, Göttingen, 1958.

14 *Gmelins Handbuch der anorganischen Chemie*, Syst. No. 5, Fluor, Verlag Chemie GmbH, Weinheim, 1959, pp. 160–81.

15 Greene, E. F., Toennies, J. P., *Chemical Reactions in Shock Waves*, Edward Arnold, London, 1964, pp. 333–4.

16 Grüttner, B., Dove, M. F. A., Clifford, A. F., *Chemistry in Anhydrous, Prototropic Inorganic Solvents. Part 1. Inorganic Chemistry in Liquid Hydrogen Cyanide and Liquid Hydrogen Fluoride*, Pergamon Press, Oxford, 1971, pp. 128–48.

17 Horvath, A. L., *Z. phys. Chem. Neue Folge*, 1972, **78** (3/4), 209–10.

18 Hu, J., White, D., Johnston, H. L., *J. Am. Chem. Soc.*, 1953, **75**, 1232–6.

19 IUPAC, Atomic Weights of the Elements 1971, *Pure & Appl. Chem.*, 1972, **30** (3–4), 638–49.

20 Jarry, R. L., Davis, W., *J. Phys. Chem.*, 1953, **57**, 600–4.

21 Lagemann, R. T., Knowles, C. H., *J. Chem. Phys.*, 1960, **32** (2), 561–4.

22 Lax, E., Synowietz, C., Eds., *Taschenbuch für Chemiker und Physiker*, 3rd ed., Springer-Verlag, Berlin, 1967.

23 Maryott, A. A., Smith, E. R., *Table of Dielectric Constants of Pure Liquids*, Natl. Bur. Std. Circular 514, Washington, Aug. 10, 1951, 44 pp.

24 Mellor's Comprehensive Treatise on Inorganic and Theoretical Chemistry, *The Halogens*, Vol. II, Suppl. 1, Longmans, London, 1956, pp. 85–128.

25 National Petroleum News, *Anhydrous Hydrofluoric Acid. Properties as an Alkylation Catalyst*, 1942 (Oct. 28), **34** (43), R-366–8.

26 Pascal, P., Ed., *Nouveau Traité de Chimie Minerale*, Vol. XVI, F, Cl, Br, I, At, Mn, Tc, Re, Masson et Cie, Éditeurs, Paris, 1960, pp. 105–26.

27 Perkins, A. J., *J. Phys. Chem.*, 1964, **68**, 654–5.

28 Phillips Petroleum Company, *Hydrofluoric Acid Alkylation*, Chemical Products Dept., Bartlesville, Oklahoma, 1946, pp. 249–66.

29 Posey, J. C., Carbide & Carbon Chemical Co., Rept. K-1063, K-25 Plant, Sept. 1953.

30 Rossini, F. D., Wagman, D. D., Evans, W. H., Levine, S., Jaffe, I., *Selected Values of Chemical Thermodynamic Properties*, Natl. Bur. Std. Circular 500, Washington, Feb. 1, 1952.

31 Schaefer, C. A., Thodos, G., *AIChE Journal*, 1959, **5** (3), 367–72.

32 Simons, J. H., *Chem. Rev.*, 1931, **8**, 213–35.

33 Simons, J. H., *Petrol. Ref.*, 1943, **22** (5), 83–9.

34 Simons, J. H., Bouknight, J. W., *J. Am. Chem. Soc.*, 1932, **54**, 129–35.

35 Simons, J. H., Dresdner, R. D., *J. Am. Chem. Soc.*, 1944, **66**, 1070–2.

36 Spalthoff, W., Franck, E. U., *Z. Elektrochem.*, 1957, **61** (8), 993–1000.

37 Stull, D. R., Dir., *JANAF Thermochemical Tables*, Dow Chemical Co., Midland, Mich., 1965–8.

38 Svehla, R. A., *Estimated Viscosities and Thermal Conductivities of Gases at High Temperatures*, Lewis Res. Center, Techn. Rept. R-132, Cleveland, Ohio, 1962, 140 pp.
39 Tyuzyo, K., *Bull. Chem. Soc. Japan*, 1957, **30** (7), 782–9.
40 Vanderzee, C. E., Rodenburg, W. Wm., *J. Chem. Thermodyn.*, 1970, **2**, 461–78.
41 Yabroff, R. M., Smith, J. C., Lightcap, E. H., *J. Chem. Eng. Data*, 1964, **9** (2), 178–82.

Table 7.1 Properties of Hydrogen Fluoride

		Ref.
Chemical formula	HF	(22)
Molecular weight	20.0063	(19)
Critical temperature	188.0°C	(11)
Critical pressure	64.85×10^5 N m^{-2}	(11)
Critical density	290.0 kg m^{-3}	(11)
Triple-point temperature	None	(18)
Triple-point pressure	None	(18)
Boiling point at 1.0133×10^5 N m^{-2}	19.54°C	(12)
Melting point	−83.01°C	(8)
Transition temperature	None	(5, 18)
Density of solid at −125°C	1663 kg m^{-3}	(4)
Compressibility (isothermal) of liquid at 15°C	56.23×10^{-12} m^2 N^{-1}	(21)
Coefficient of thermal expansion of liquid at 15°C	2.239×10^{-3} °C^{-1}	(21)
Heat of formation of gas at 25°C	-2.7130×10^8 J kg-mol^{-1}	(3, 37)
Free energy of formation of gas at 25°C	-2.7340×10^8 J kg-mol^{-1}	(37)
Heat of fusion	4.5804×10^6 J kg-mol^{-1}	(30)
Heat of transition	None	(5, 18)
Entropy of gas at 25°C	1.7379×10^5 J kg-mol^{-1} C^{-1}	(37)
Heat capacity ratio of gas at 27°C and 1.0133×10^5 N m^{-2}	1.40	(15)
Velocity of sound at 15°C and 1.0133×10^5 N m^{-2}	452.1 m s^{-1}	(21)
Dielectric constant of liquid at 0°C	84.0	(23)
Electrical conductivity of liquid at 25°C	2×10^{-3} ohm^{-1} m^{-1}	(27)
Refractive index of liquid at 25°C and 5892.6 Å	1.1574	(27)
Threshold limit value	(3 ppm) 2×10^{-6} kg m^{-3}	(2)

7.1 Vapour Pressure of Hydrogen Fluoride (11,20)

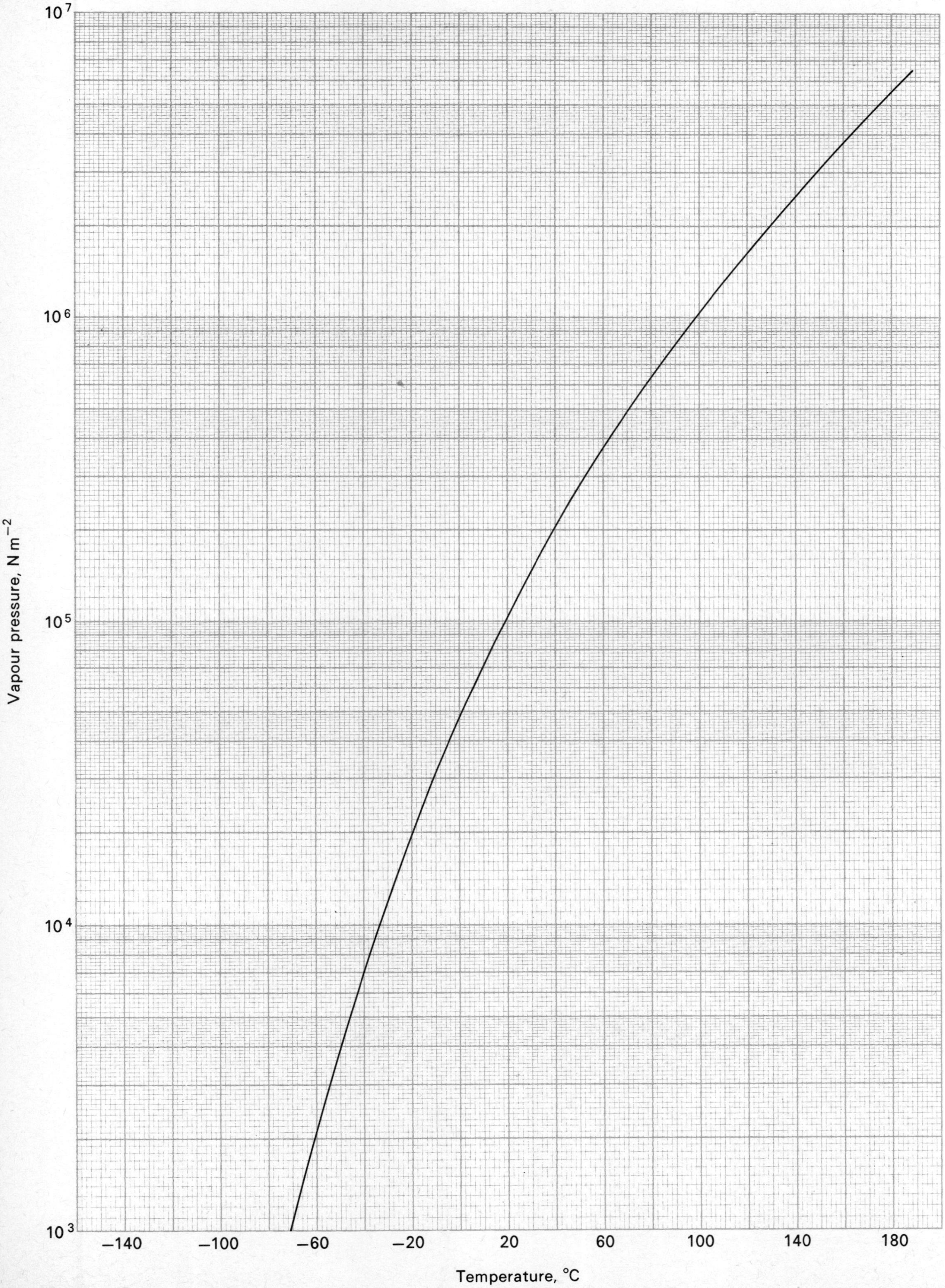

7.2 Orthobaric Density of Hydrogen Fluoride (11,20,33,34)

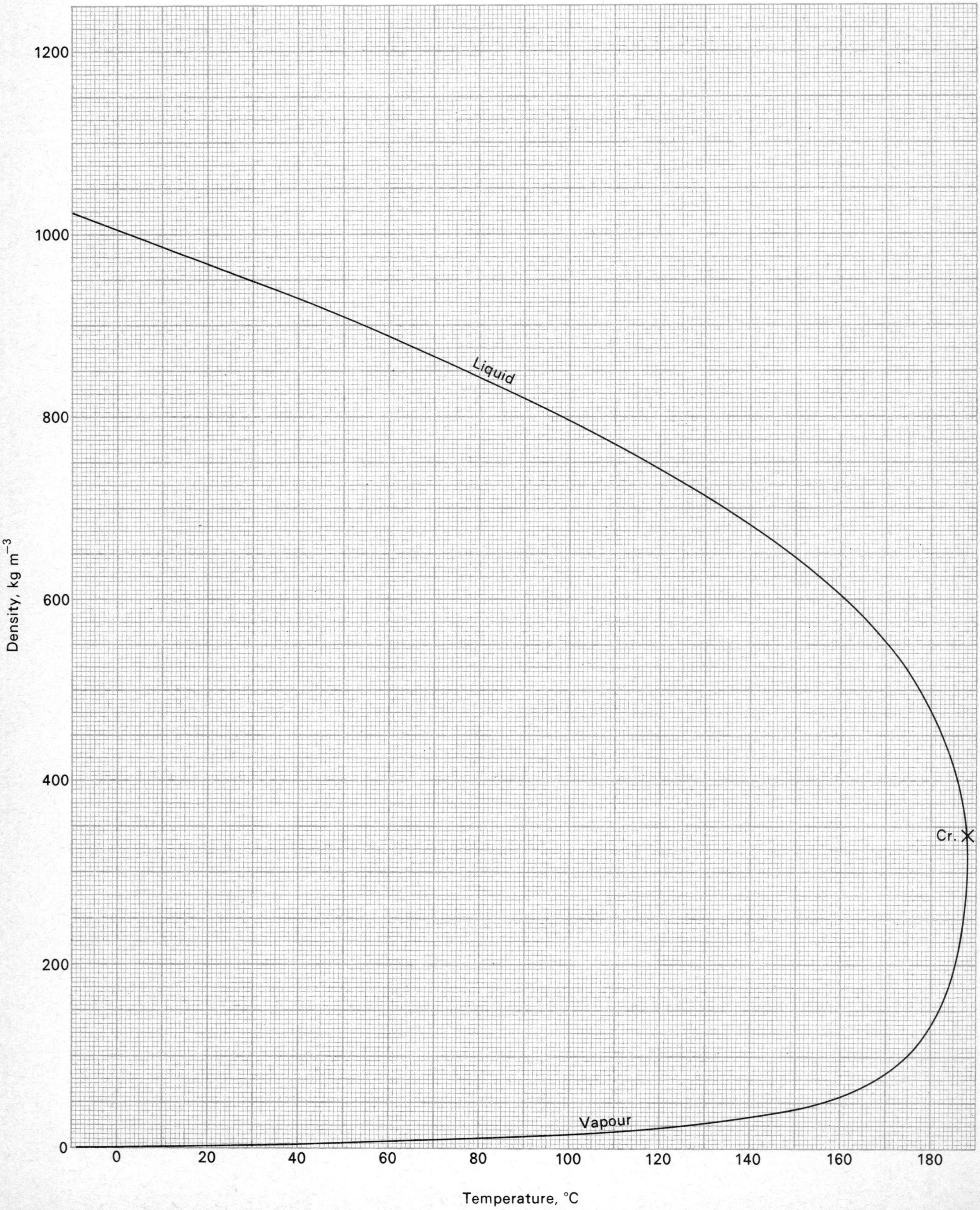

7.3 P–V–T Data for Hydrogen Fluoride (11,13)

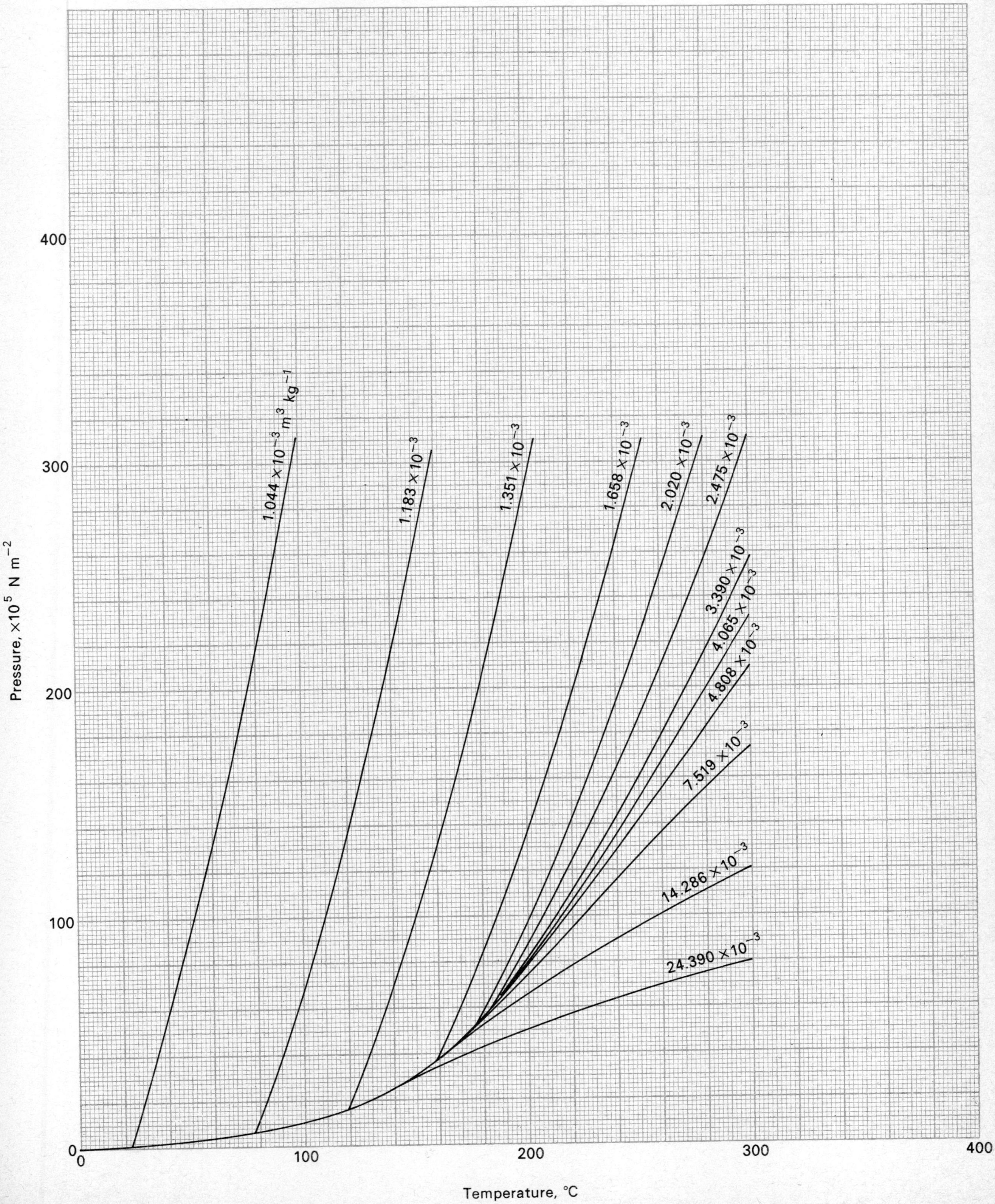

7.4 Latent Heat of Vaporisation of Hydrogen Fluoride (11,20)

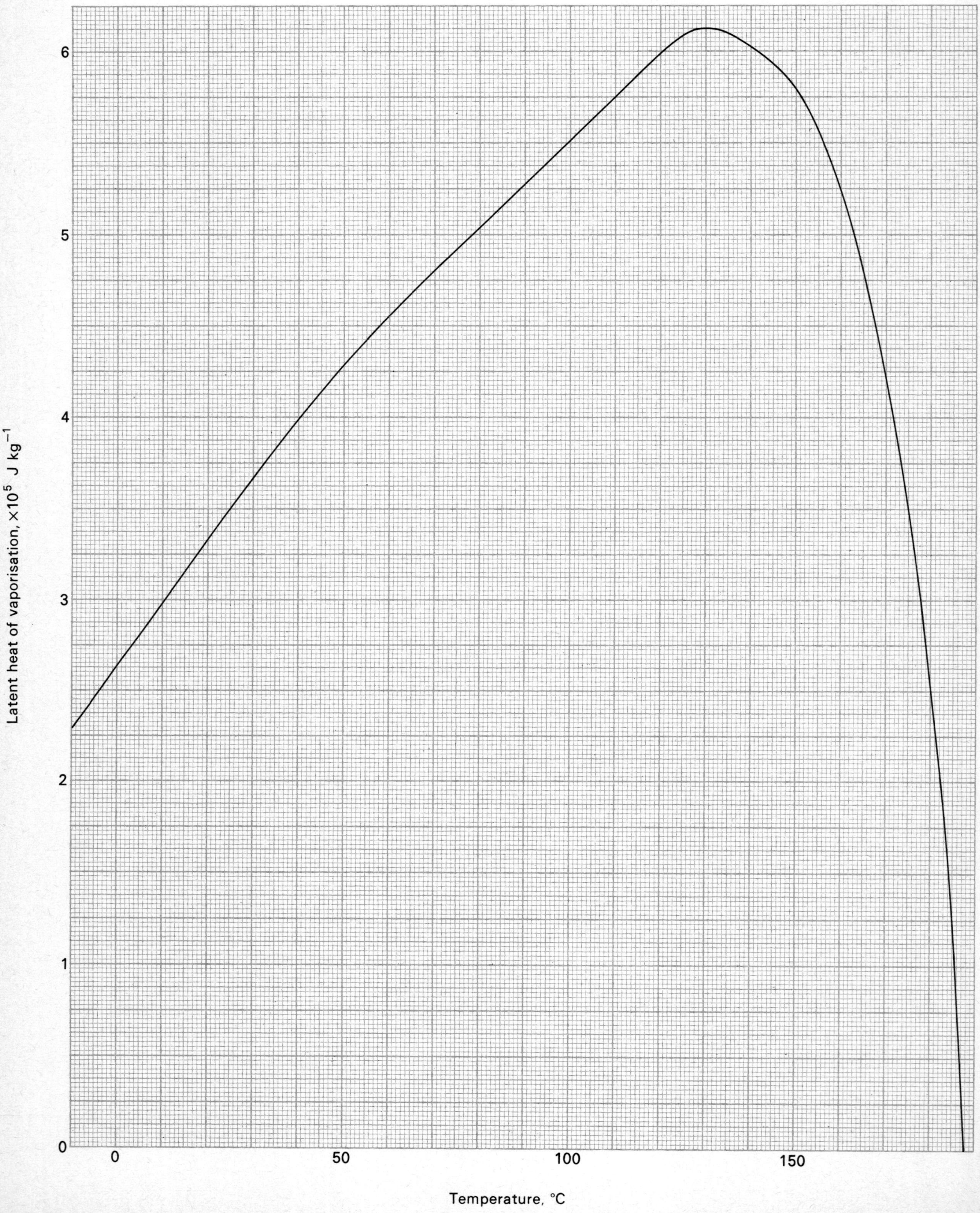

7.5 Heat Capacity of Solid and Liquid Hydrogen Fluoride (18)

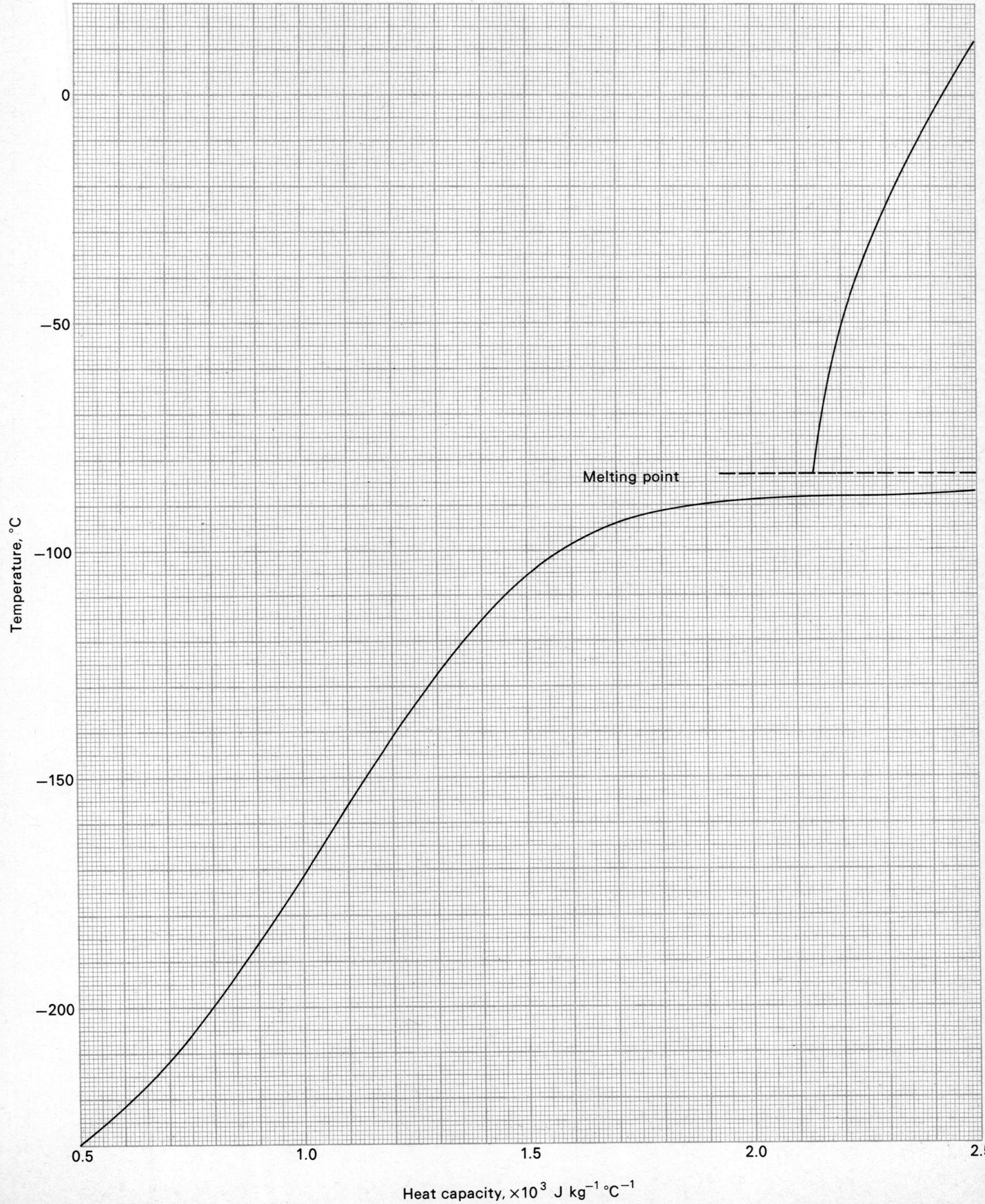

7.6 Heat Capacity of Gaseous Hydrogen Fluoride at Constant Pressure (9,40)

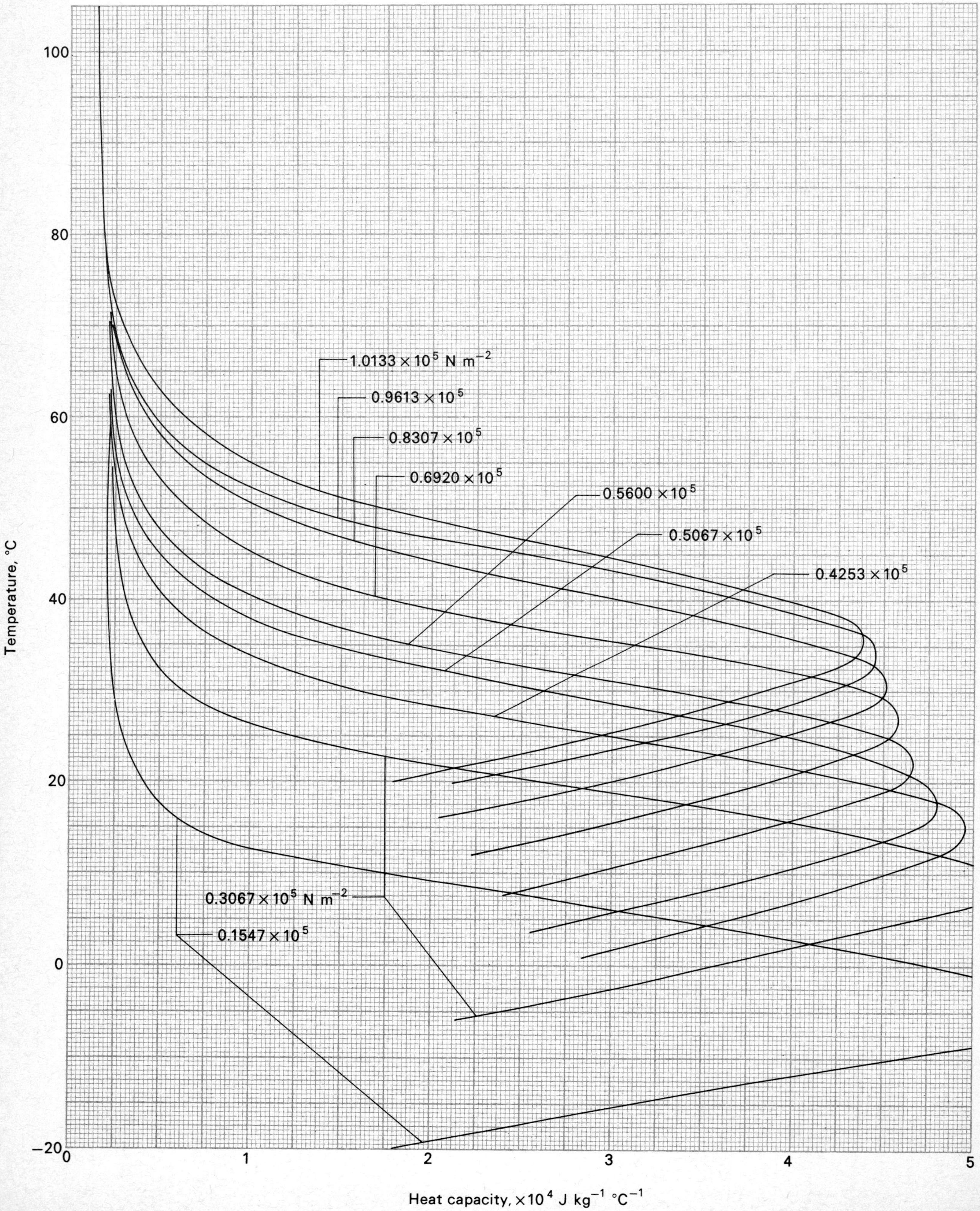

7.7 Heat Capacity of Gaseous Hydrogen Fluoride at Constant Volume (40)

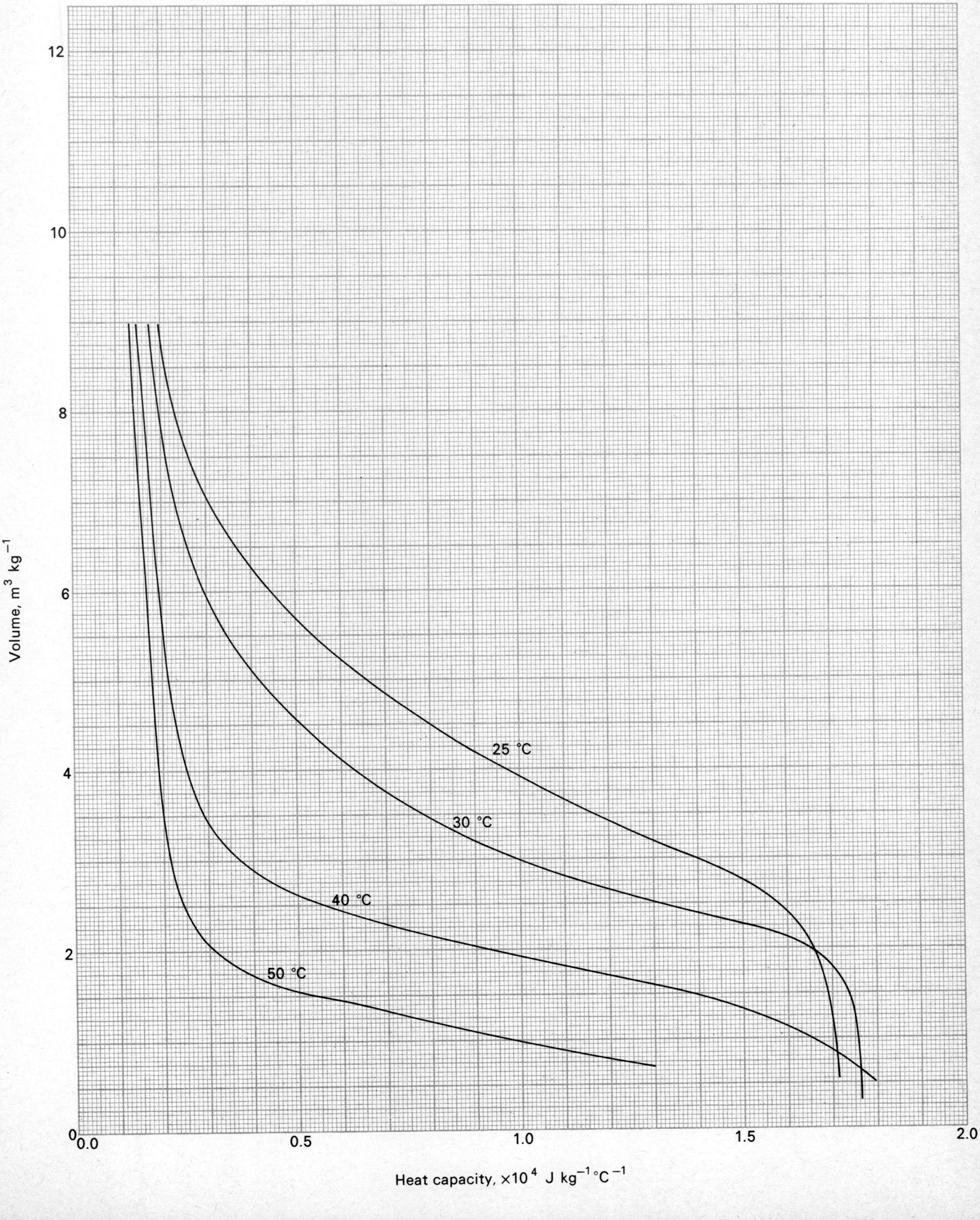

7.8 Viscosity of Liquid Hydrogen Fluoride (35,39)

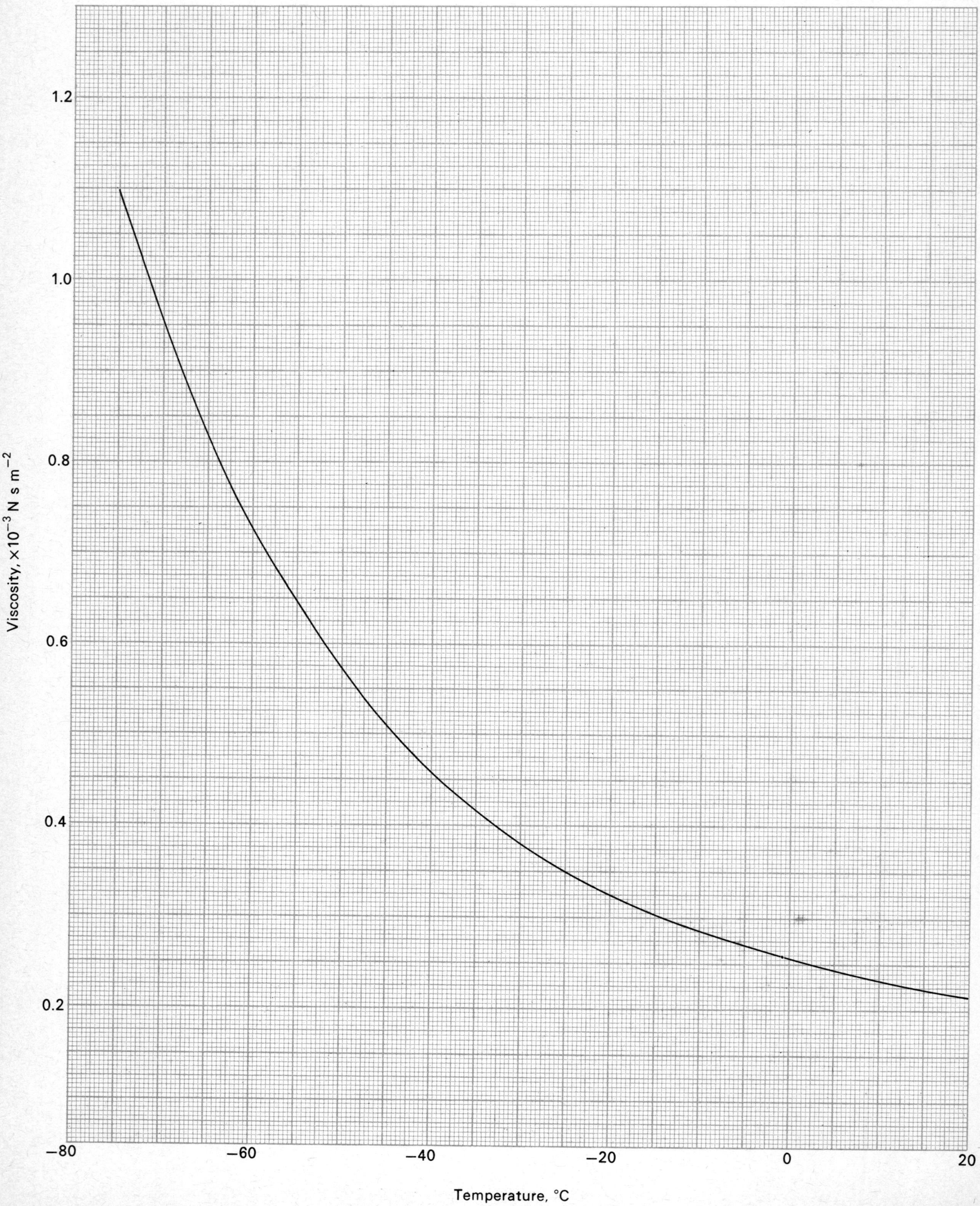

7.9 Viscosity of Gaseous Hydrogen Fluoride at 1.0133×10^5 N m^{-2} (6,29,38)

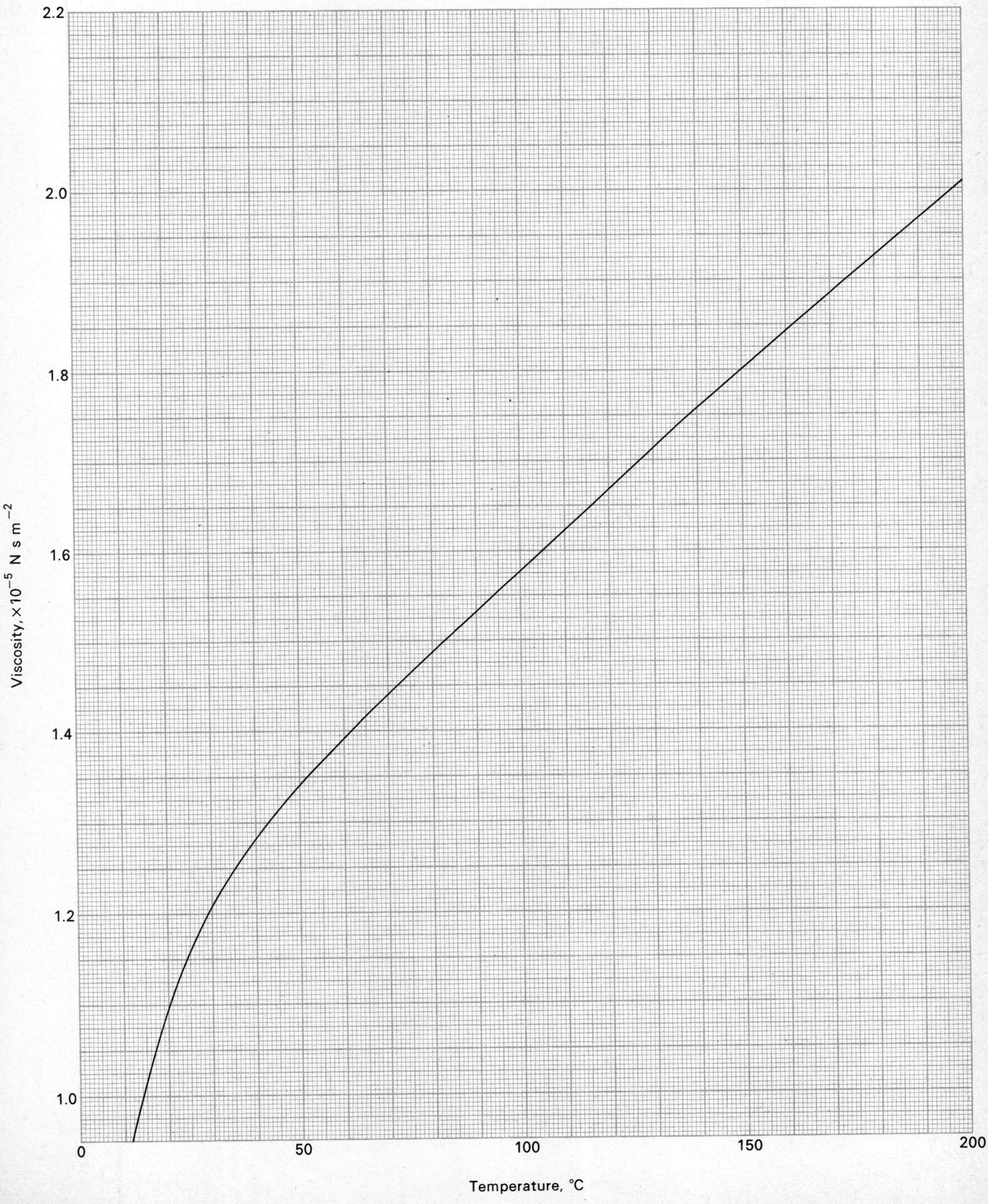

7.10 Thermal Conductivity of Liquid Hydrogen Fluoride (31)

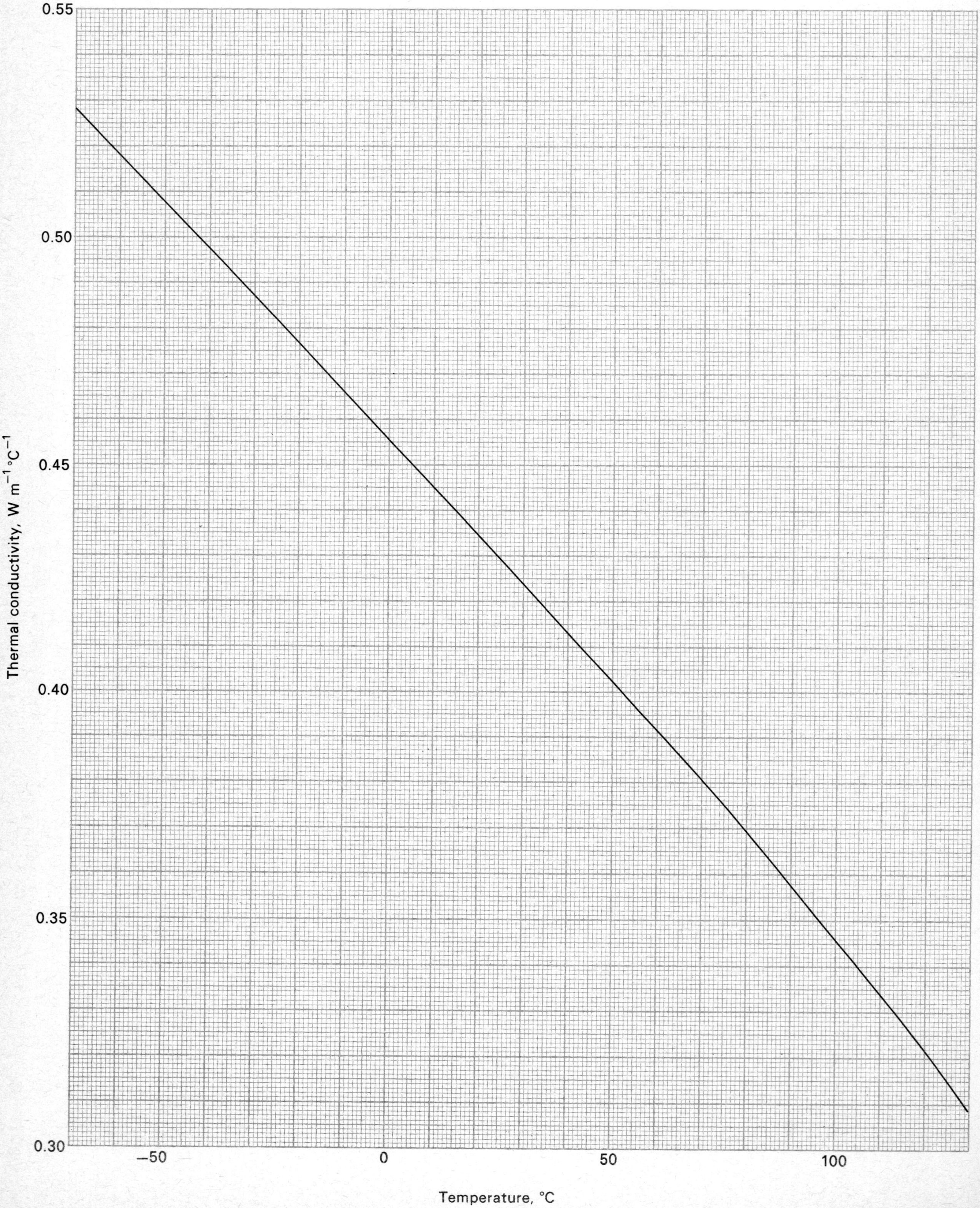

7.11 Thermal Conductivity of Gaseous Hydrogen Fluoride (6,10)

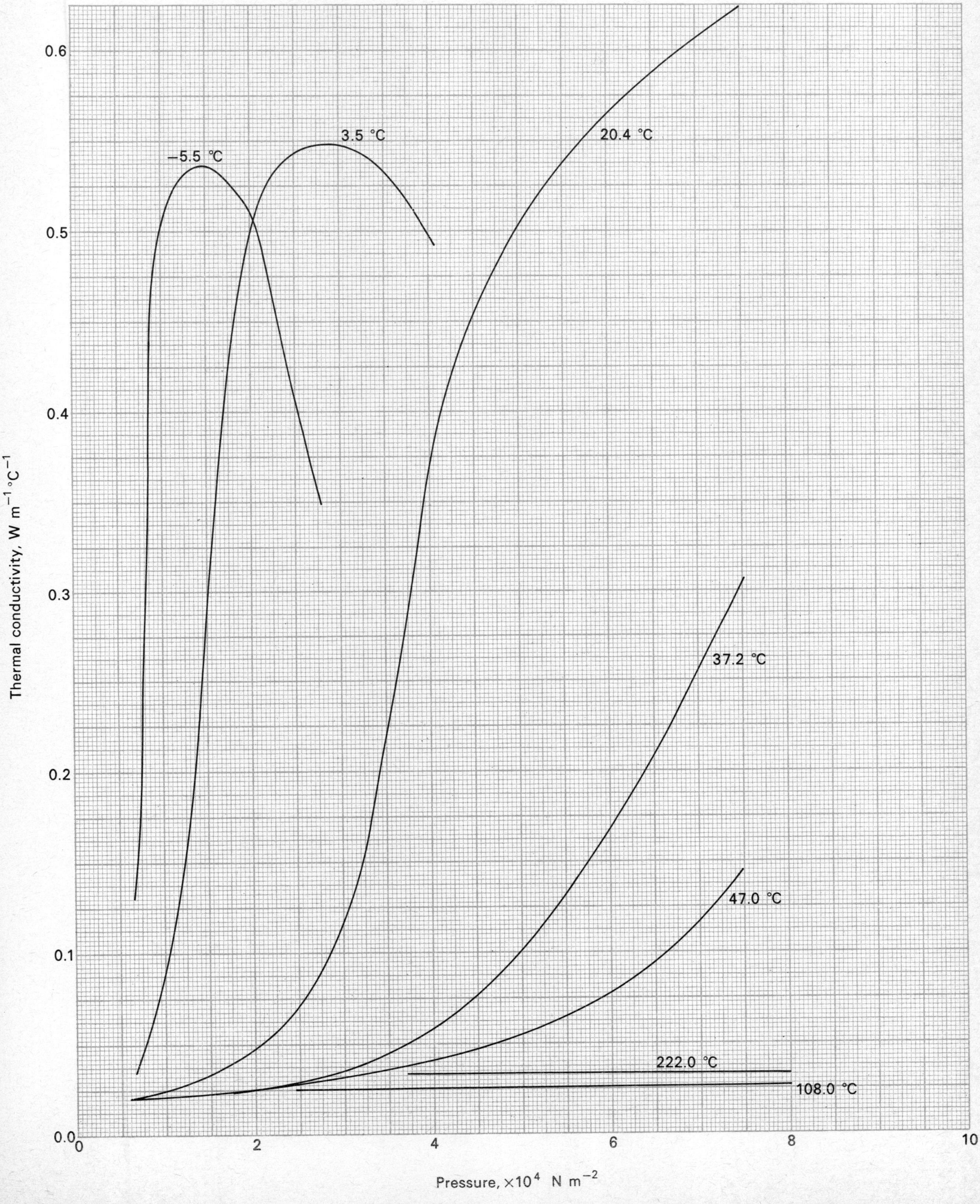

7.12 Surface Tension of Hydrogen Fluoride (34)

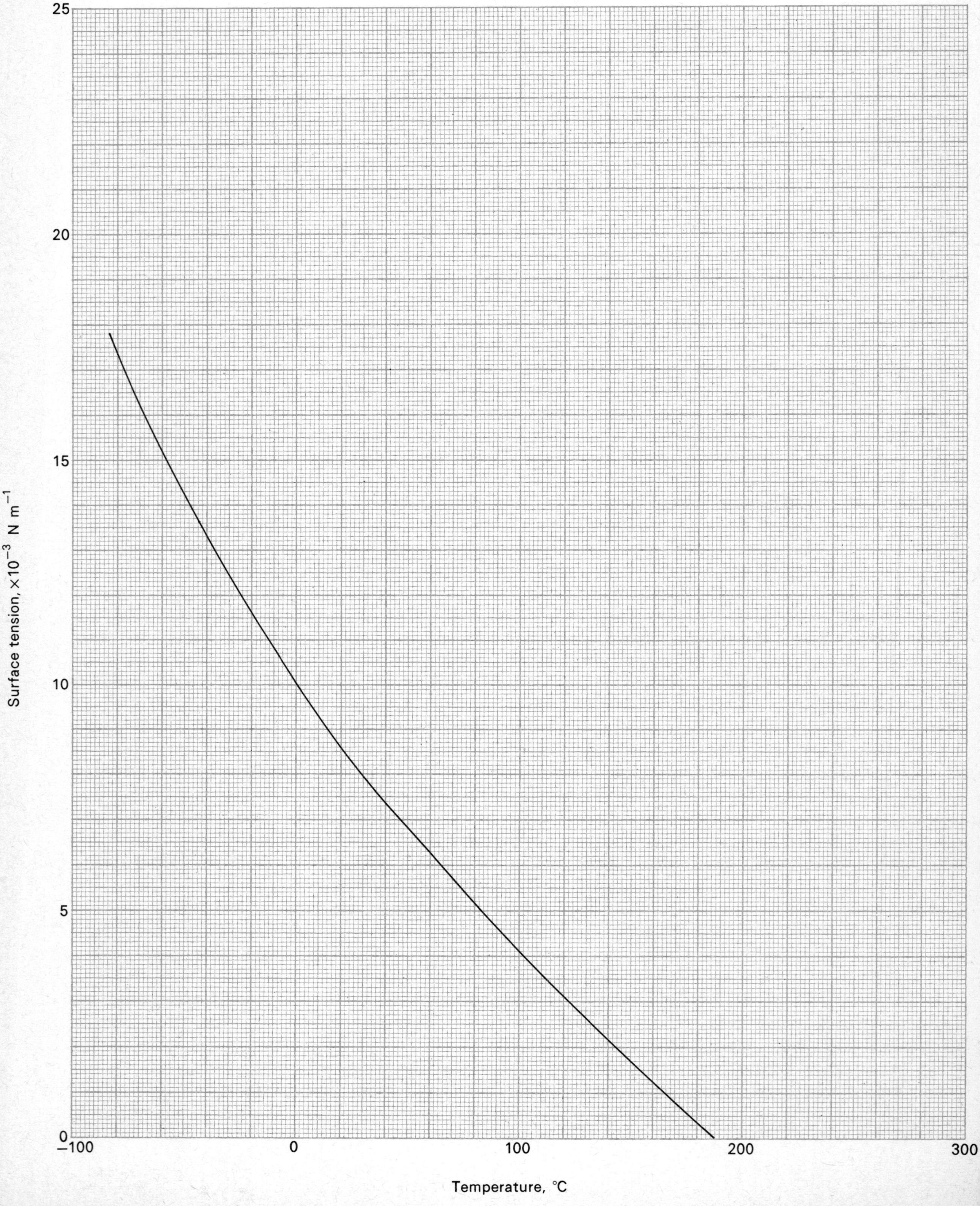

7.13 Phase Diagram for HF—H_2O System (7)

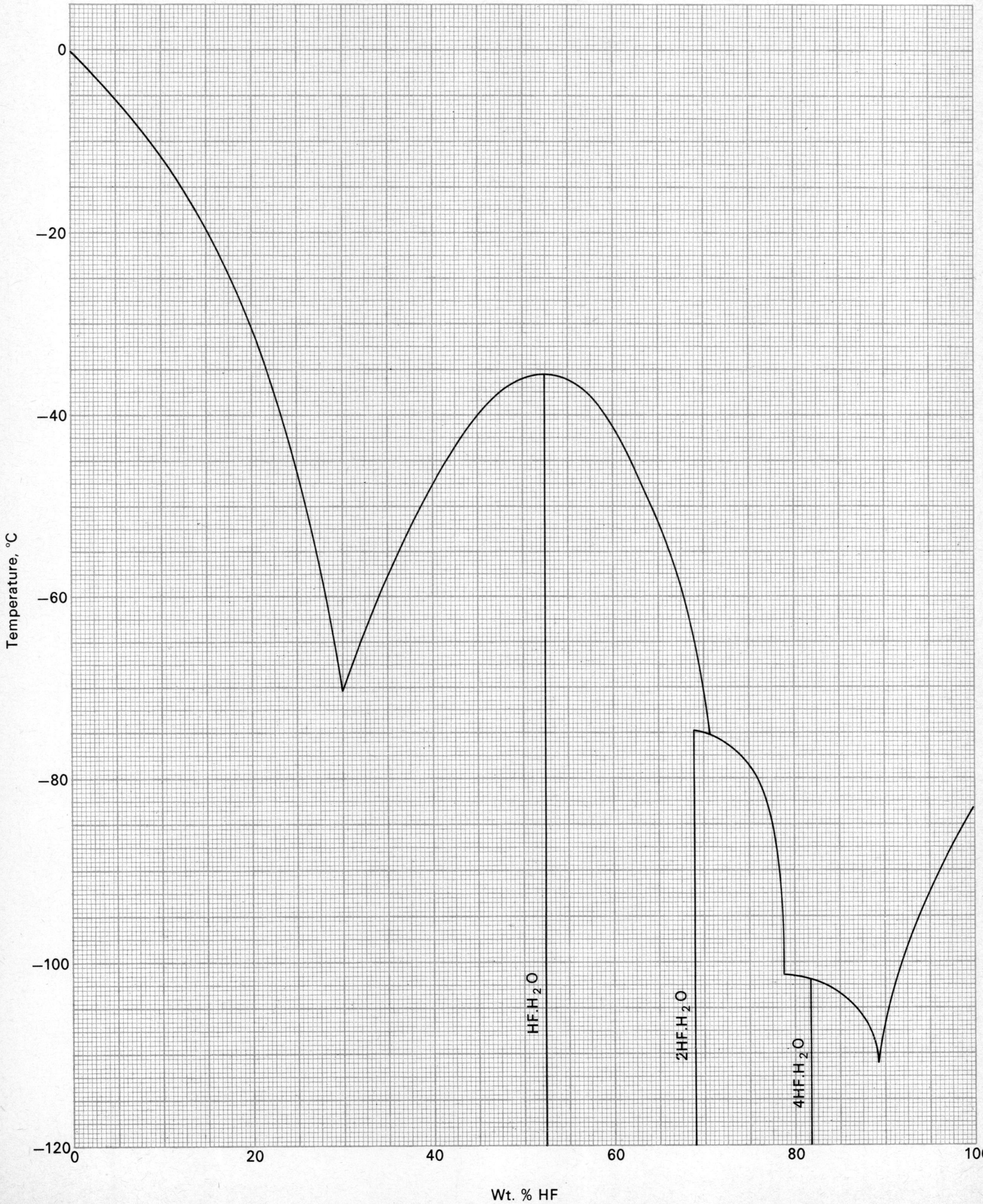

8 HYDROGEN IODIDE

Hydrogen iodide is a colourless gas at room temperature and atmospheric pressure with a strong sour taste. It dissociates into hydrogen and iodine above 100°C. It is a stronger oxidizing agent than hydrogen bromide. Gaseous hydrogen iodide is very soluble in water, forming three hydrates. In the gaseous state it forms H_2I_2 molecules; therefore, there is considerable deviation from ideal gas behaviour.

Solid HI is colourless and transparent, having a crystal structure in the tetragonal face-centred lattice, with four HI molecules in each unit cell. The colourless solid has two transition points.

Hydrogen iodide is also colourless in the liquid state. In a moist atmosphere, it fumes strongly. The evolved fumes are toxic, but not flammable.

Because hydrogen iodide is the least used chemical among the hydrogen halides, its physical, thermodynamic, and transport properties have not been investigated extensively. Most of the experimental measurements were carried out several decades ago. The published literature has been surveyed only in multi-volume handbooks, such as: Gmelin (9), Mellor (21), and Pascal (27). There is no information on the P–V–T relation of hydrogen iodide in the published literature. Some of the properties in the gaseous phase have been estimated (32), because of the lack of experimental data. The thermal conductivity of liquid hydrogen iodide has also been estimated (30), in the same way as the other halides and hydrogen halides (diatomic molecules).

References

1 Alpert, N. L., *Phys. Rev.*, 1949, **75**, 398–410.
2 Berthelot, M., *Bull. Soc. Chim. France*, 1873, **19**, 351.
3 Bleekrode, L., *Proc. Roy. Soc.*, 1884, **37**, 352.
4 Drozdowski, E., Pietrzak, J., *Bull. Acad. Crac.*, 1913, **A**, 234–9.
5 Edwards, D. G., *The Vapour Pressure of 30 Inorganic Liquids between One Atmosphere and the Critical Point*, Univ. of California, Lawrence Rad. Lab., UCRL 7167, Livermore, California, June 13, 1963, 148 pp.
6 Estreicher, Th., Schnerr, A. A., *Z. kompr. fl. Gase*, 1913, **15**, 161.
7 Eucken, A., Karwat, E., *Z. phys. Chem.*, 1924, **112**, 473–8.
8 Giauque, W. F., Wiebe, R., *J. Am. Chem. Soc.*, 1929, **51**, 1441–9.
9 *Gmelins Handbuch der anorganischen Chemie*, Syst. No. 8, Jod, Verlag Chemie GmbH, Weinheim, 1955 (2nd printing), pp. 264–84.
10 Golubev, I. F., *Viscosity of Gases and Gas Mixtures*, Gosudarstvennoe Izdatel'stvo Fiziko-Matematicheskoi Literatury, Moskva, 1959, Table 4.
11 Greene, E. F., Toennies, J. P., *Chemical Reactions in Shock Waves*, Edward Arnold, London, 1964, pp. 333–4.
12 Henglein, F. A., Roth, R., Andres, P., *Z. Phys.*, 1923, **18**, 68–9.
13 IUPAC, Atomic Weights of the Elements 1971, *Pure & Appl. Chem.*, 1972, **30** (3–4), 638–49.
14 Karapet'yants, M. Kh., *Zh. Fiz. Khim.*, 1958, **32**, 556–68.
15 Kobe, K. A., Long, E. G., *Petrol. Ref.*, 1950, **29**, (2), 124–8.
16 Lange, N. A., Ed., *Handbook of Chemistry*, 10th ed., McGraw-Hill, New York, 1967.
17 Lax, E., Synowietz, C., Eds., *Taschenbuch für Chemiker und Physiker*, 3rd ed., Springer-Verlag, Berlin, 1967.
18 Lielmezs, J., *Thermochim. Acta*, 1971, **2**, 101–5.
19 Lutugina, N. V., Kokovkina, L. I., *J. Appl. Chem. USSR*, 1965, **38** (7), 1460–5.
20 Maryott, A. A., Smith, E. R., *Table of Dielectric Constants of Pure Liquids*, Natl. Bur. Std. Circular 514, Washington, Aug. 10, 1951, 44 pp.
21 Mellor's Comprehensive Treatise on Inorganic and Theoretical Chemistry, *The Halogens*, Vol. II, Suppl. 1, Longmans, London, 1956, pp. 859–64.
22 Miravalles, R., Moles, E., *An. Espan.*, 1925, **23**, 518.
23 McIntosh, D., Steele, B. D., *Proc. Roy. Soc.*, 1904, **73**, 451–3.
24 McIntosh, D., Steele, B. D., *Z. phys. Chem.*, 1906, **55**, 142–50.
25 McIntosh, D., Steele, B. D., *Phil. Trans.*, 1906, **205A**, 114–19.
26 Noninski, Khr. Iv., *Khim. Ind. (Sofia)*, 1961, **33**, 179–82.
27 Pascal, P., Ed., *Nouveau Traité de Chimie Minérale*, Vol. XVI, F, Cl, Br, I, At, Mn, Tc, Re, Masson et Cie, Éditeurs, Paris, 1960, pp. 517–26.
28 Rossini, F. D., Wagman, D. D., Evans, W. H., Levine, S., Jaffe, I., *Selected Values of Chemical Thermodynamic Properties*, Natl. Bur. Std. Circular 500, Washington, Feb. 1, 1952.
29 Ruhemann, B., Simon, F., *Z. phys. Chem.*, 1931/32, **15B**, 406.
30 Schaefer, C. A., Thodos, G., *AIChE Journal*, 1959, **5** (3), 367–72.
31 Stull, D. R., Dir., *JANAF Thermochemical Tables*, Dow Chemical Co., Midland, Mich., 1965–8.
32 Svehla, R. A., *Estimated Viscosities and Thermal Conductivities of Gases at High Temperatures*, Lewis Res. Center, Techn. Rept. R-132, Cleveland, Ohio, 1962, 140 pp.
33 Syrkin, J. K., *Z. phys. Chem.*, 1929, **5B**, 159.
34 Trautz, M., Ruf, F., *Ann. Phys.*, 1934, **20** (5), 127–34.
35 Vowles, C., Massachusetts Inst. of Technology, M.S. thesis in Chem. Eng., 1951.
36 Yosim, S. J., *J. Chem. Phys.*, 1964, **40** (10), 3069–75.

Table 8.1 Properties of Hydrogen Iodide

		Ref.
Chemical formula	HI	(17)
Molecular weight	127.9124	(13)
Critical temperature	150.8°C	(6)
Critical pressure	8.309×10^6 N m^{-2}	(35)
Critical density	1049.0 kg m^{-3}	(33)
Triple-point temperature	−50.7°C	(22)
Triple-point pressure	0.5244×10^5 N m^{-2}	(22)
Boiling point at 1.0133×10^5 N m^{-2}	−35.35°C	(8)
Melting point	−50.8°C	(8)
Transition temperatures	−203.15°C −148.15°C	(1) (1)
Density of solid at −82°C	3425.0 kg m^{-3}	(29)
Coefficient of thermal expansion of liquid at boiling point	1.44×10^{-3} °C^{-1}	(36)
Heat of formation of gas at 25°C	2.6377×10^7 J kg-mol^{-1}	(31)
Free energy of formation of gas at 25°C	1.5910×10^6 J kg-mol^{-1}	(31)
Heat of sublimation at −50.7°C	2.2064×10^7 J kg-mol^{-1}	(12)
Heat of fusion	2.8721×10^6 J kg-mol^{-1}	(28)
Heats of transition	2.9517×10^5 J kg-mol^{-1} 5.2670×10^5 J kg-mol^{-1}	(7) (7)
Entropy of gas at 25°C	2.0662×10^5 J kg-mol^{-1} °C^{-1}	(31)
Heat capacity ratio of gas at 20°C and 1.0133×10^5 N m^{-2}	1.40	(16)
Velocity of sound at 27.0°C and 1.0133×10^5 N m^{-2}	165.3 m s^{-1}	(11)
Dielectric constant of liquid at 22°C	2.90	(20)
Electrical conductivity of liquid at −35°C	2×10^{-5} ohm^{-1} m^{-1}	(16)
Refractive index of liquid at 16.5°C	1.466	(3)

8.1 Vapour Pressure of Hydrogen Iodide (4,5)

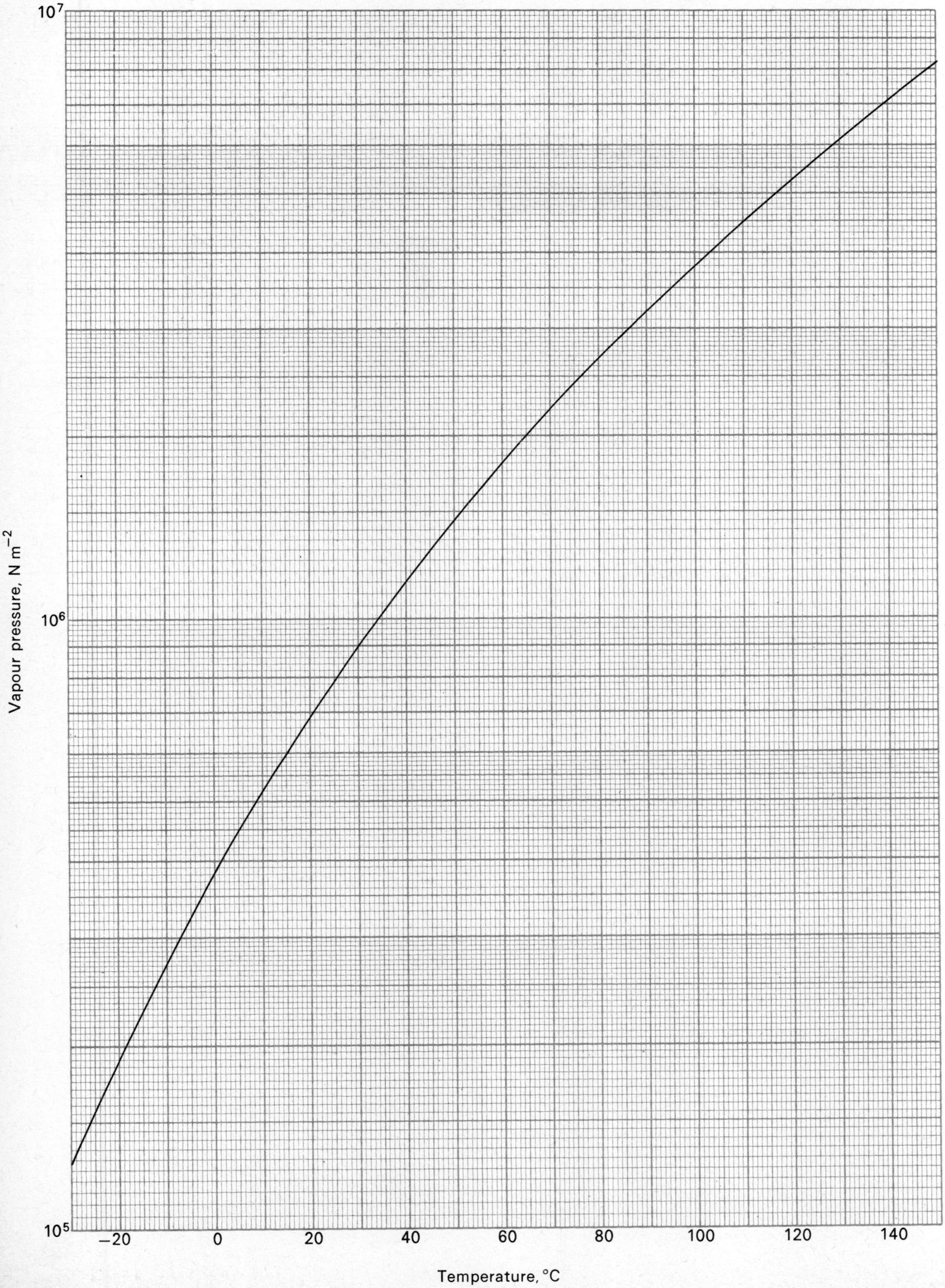

8.2 Orthobaric Density of Hydrogen Iodide (23,24,25)

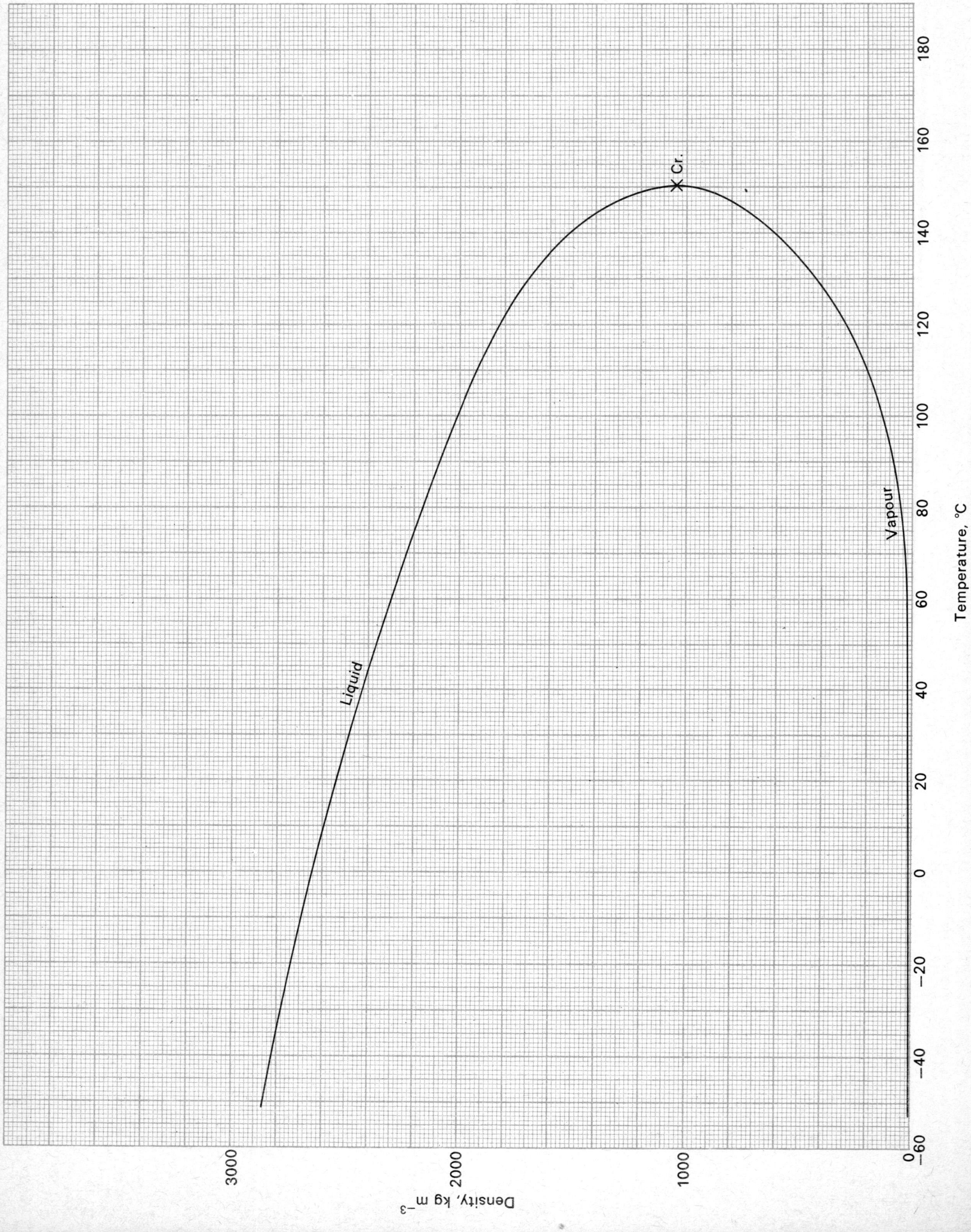

8.3 Latent Heat of Vaporisation of Hydrogen Iodide (14,18,26)

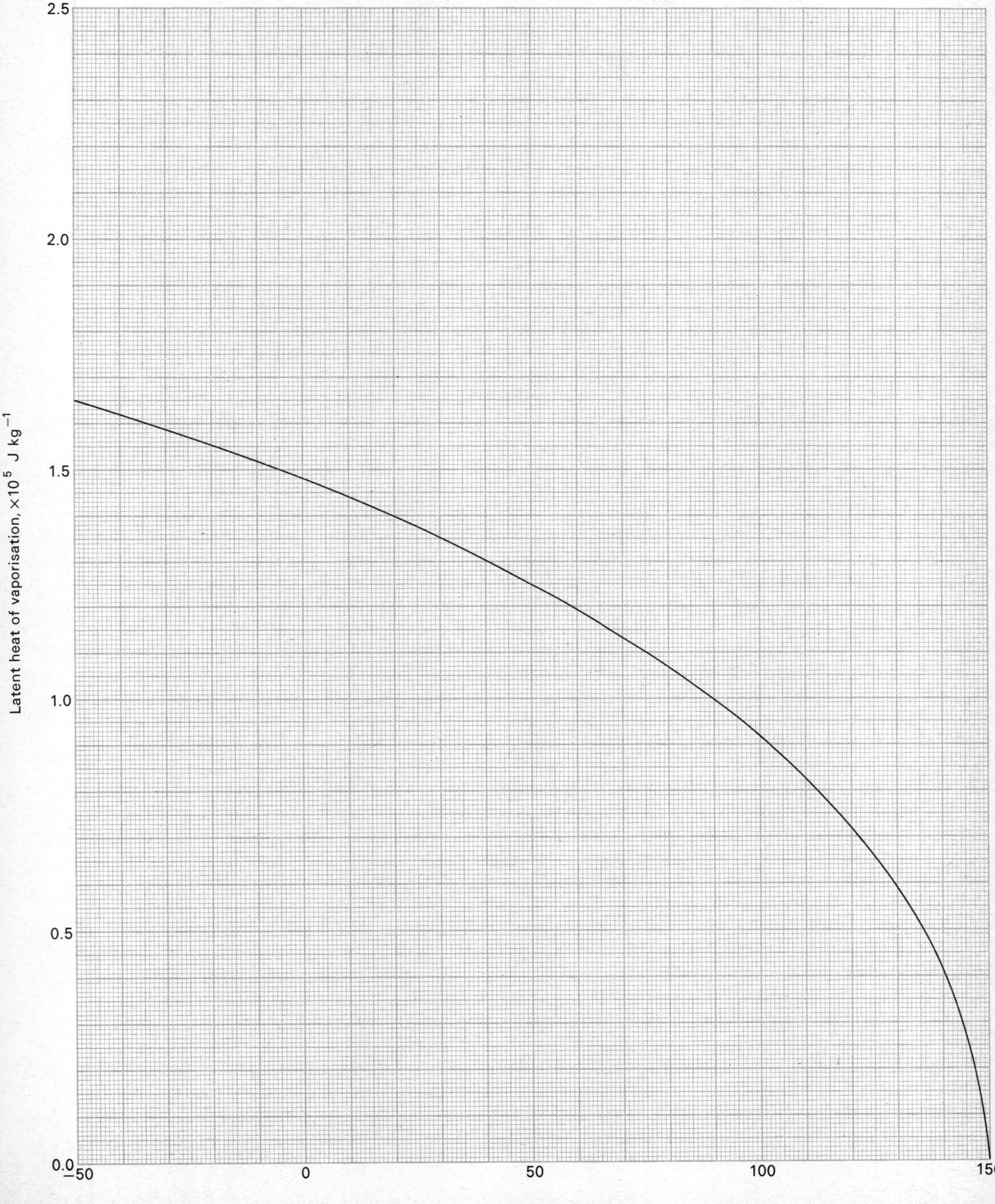

8.4 Heat Capacity of Solid and Liquid Hydrogen Iodide (8)

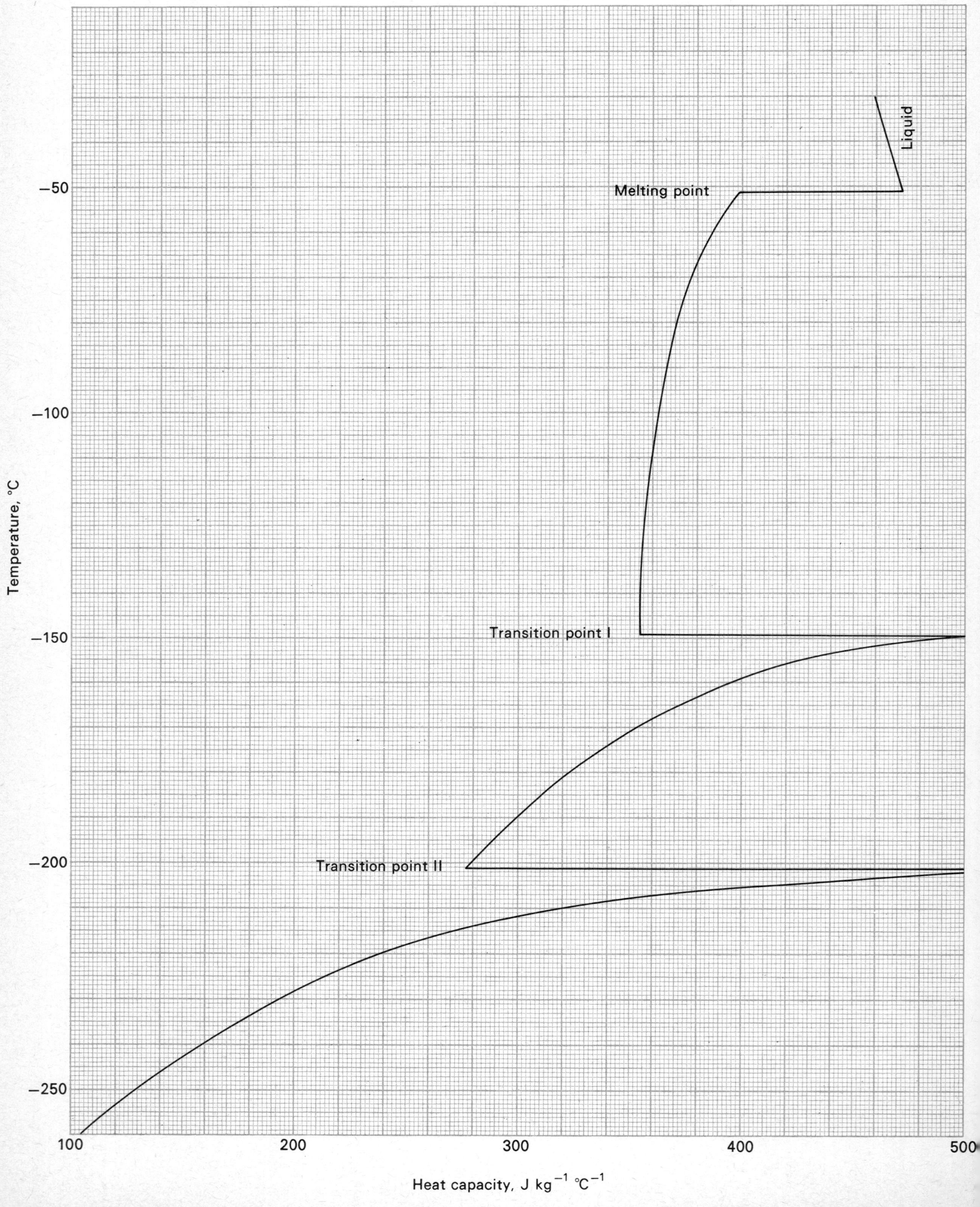

8.5 Heat Capacity of Gaseous Hydrogen Iodide at 1.0133×10^5 N m^{-2} (15,31)

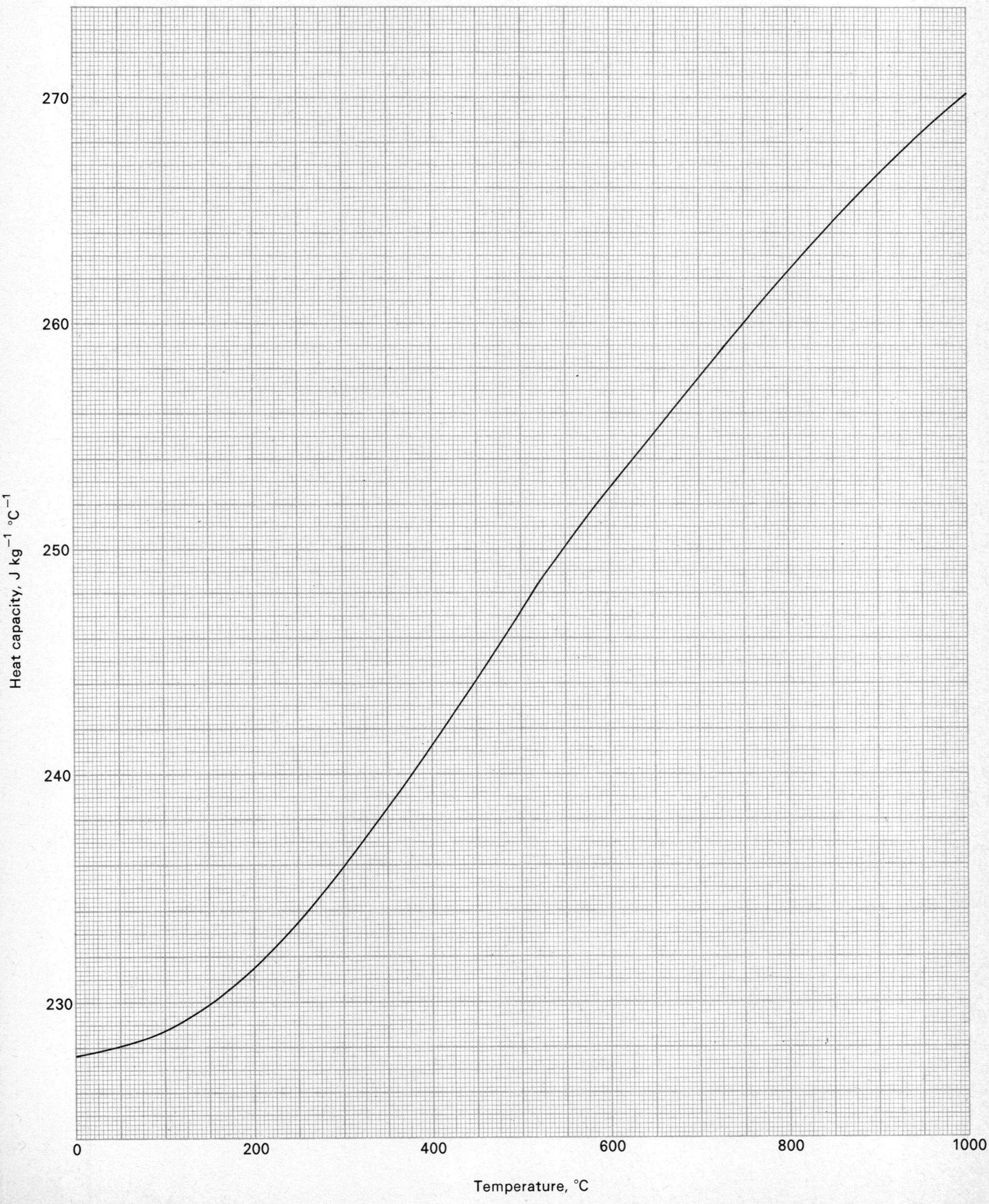

8.6 Viscosity of Liquid Hydrogen Iodide (24,25)

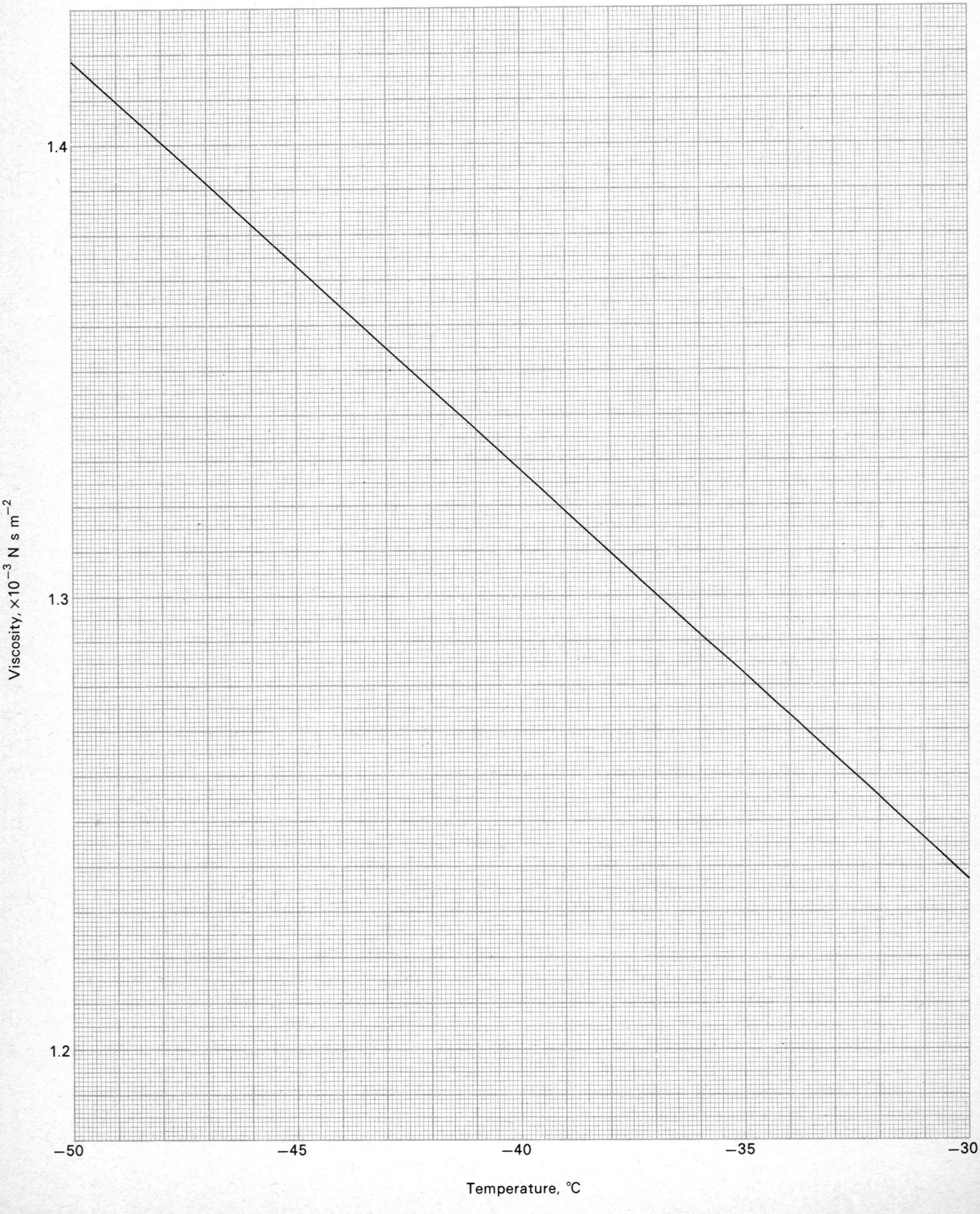

8.7 Viscosity of Gaseous Hydrogen Iodide at 1.0133×10^5 N m^{-2} (10,34)

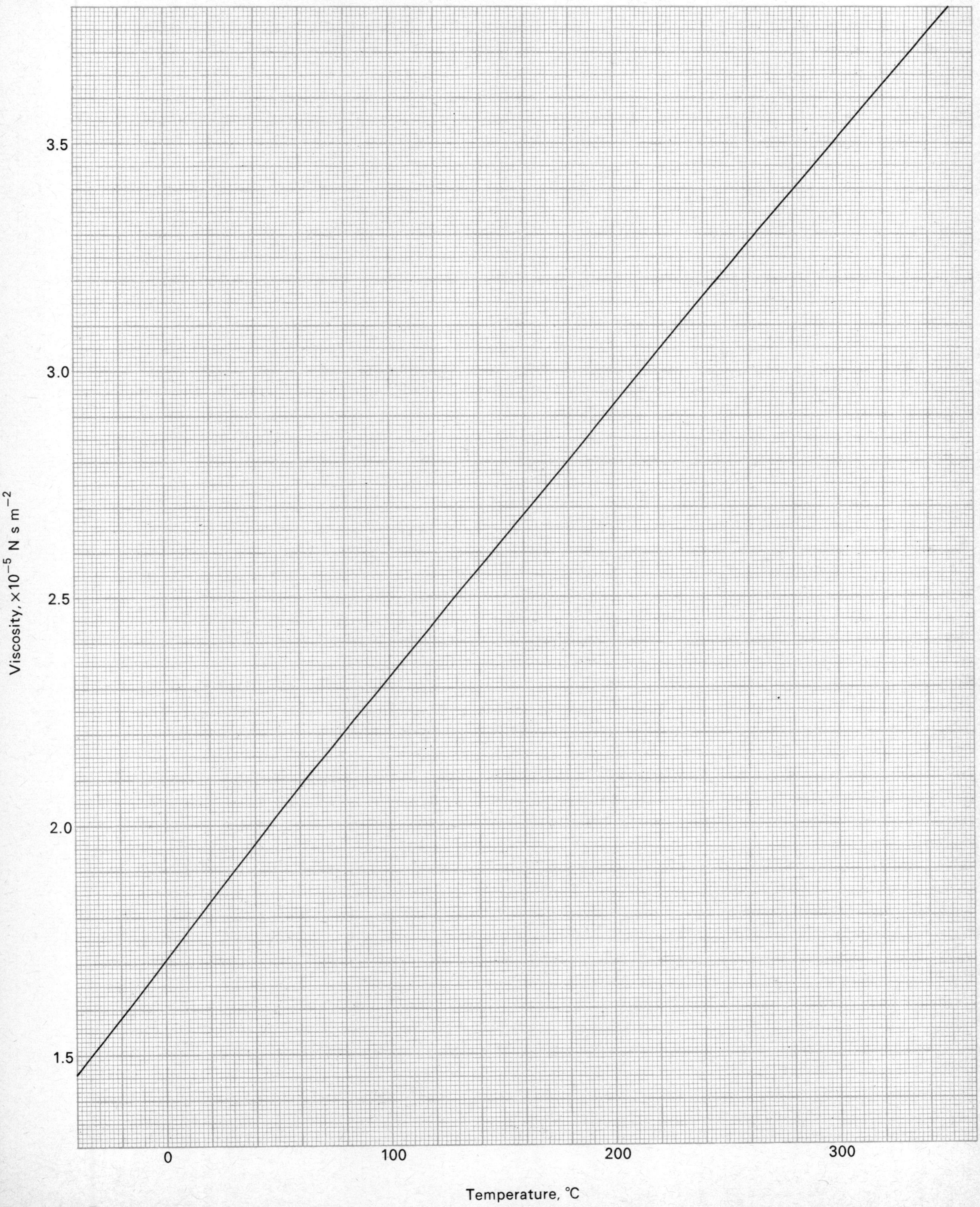

8.8 Thermal Conductivity of Liquid Hydrogen Iodide (30)

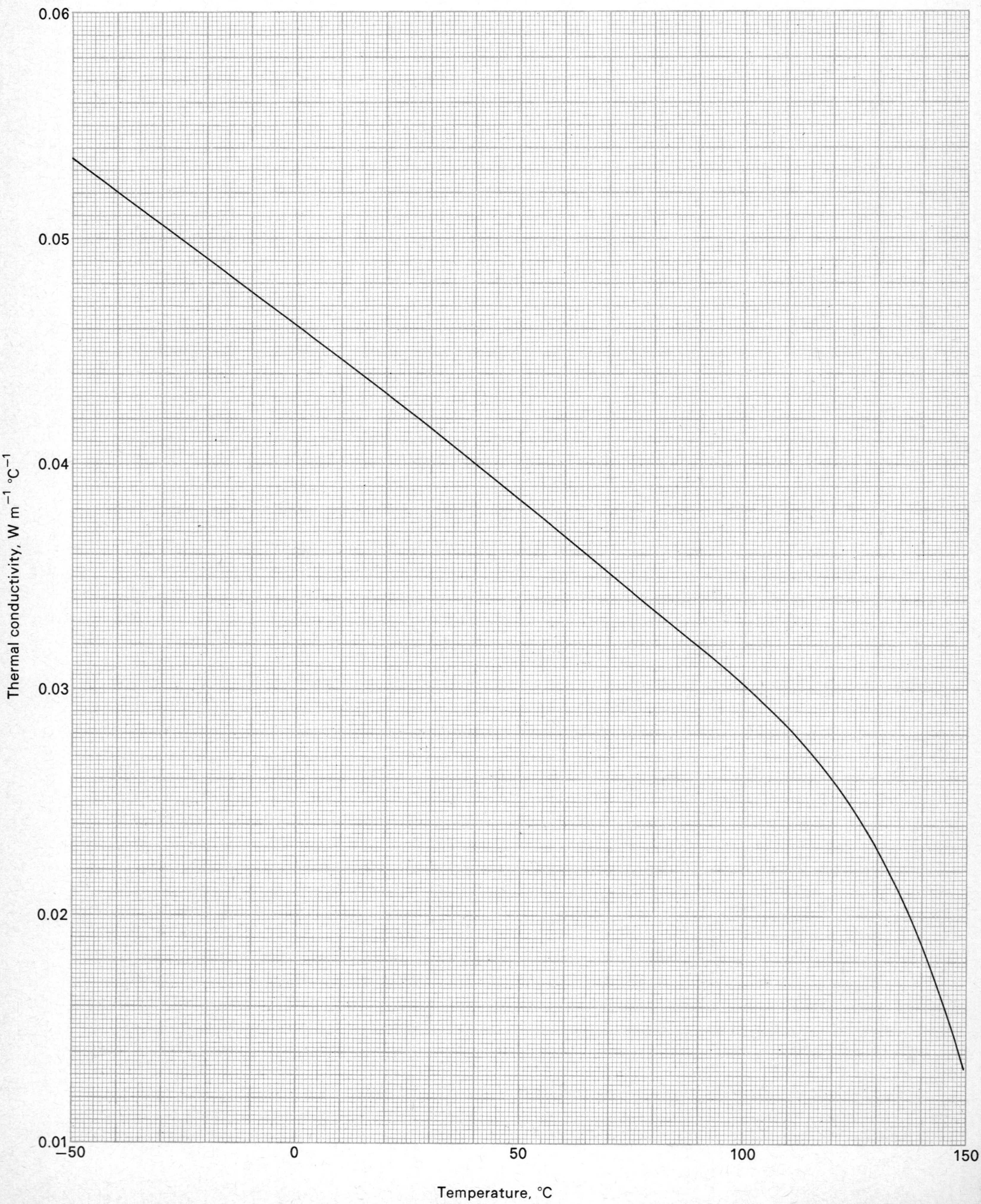

8.9 Thermal Conductivity of Gaseous Hydrogen Iodide at 1.0133×10^5 N m^{-2} (32)

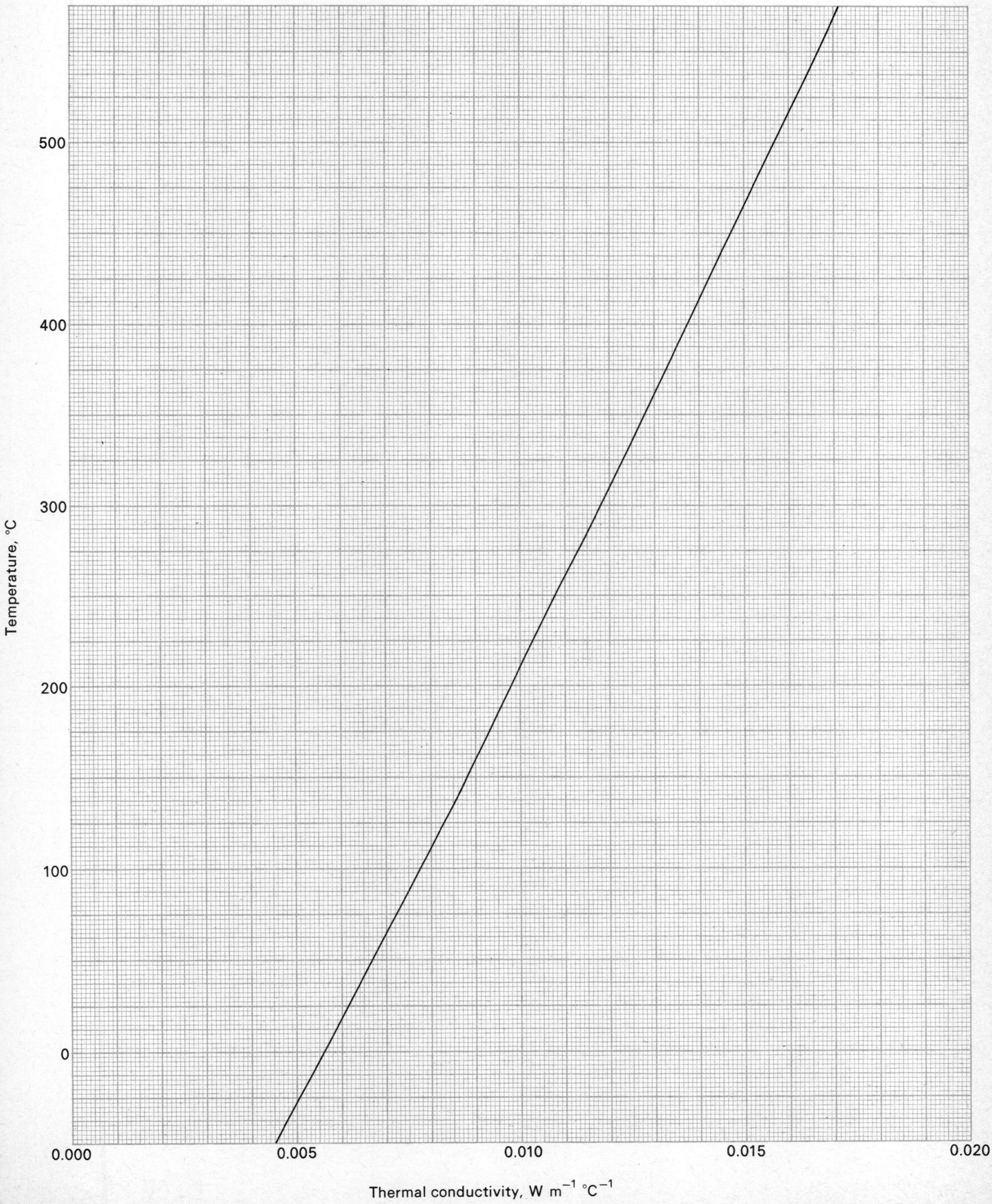

8.10 Surface Tension of Hydrogen Iodide (17)

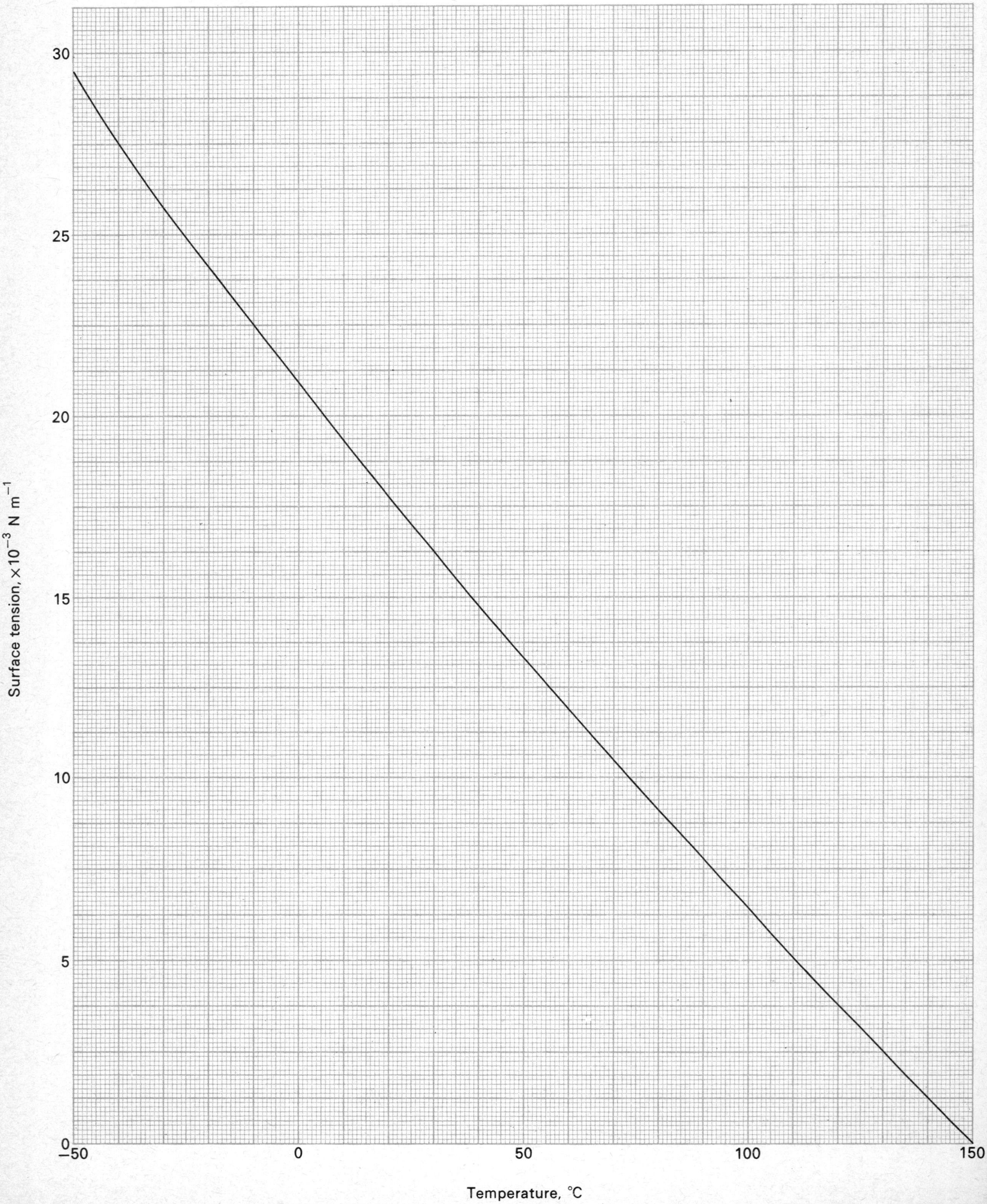

8.11 Solubility of Hydrogen Iodide in Water at 1.0133×10^5 N m^{-2} Total Pressure (2,19)

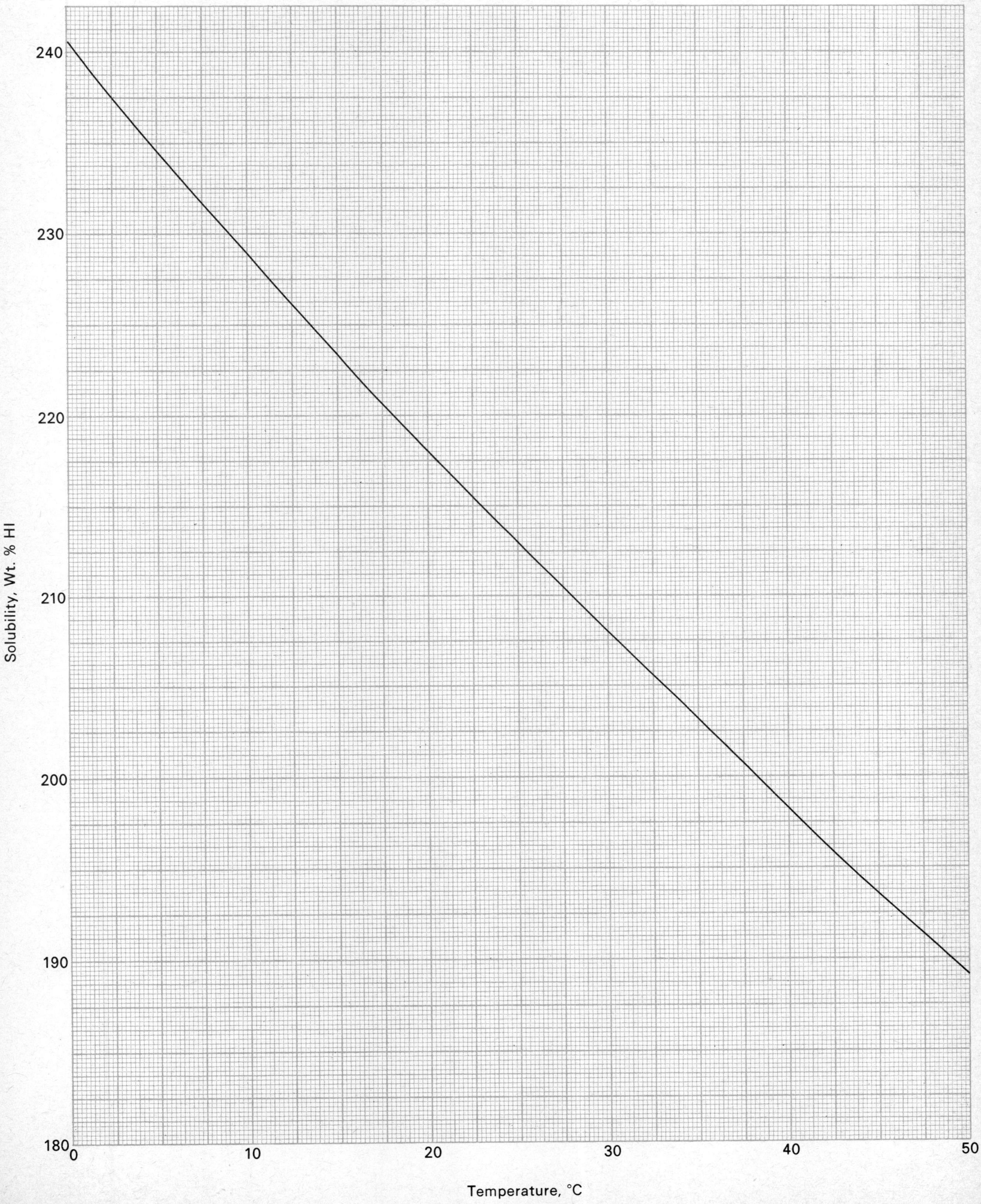

9 SULPHUR HEXAFLUORIDE

Sulphur hexafluoride was discovered by Moissan and Lebeau (6) in 1900. On heating, it is a very stable compound with inert behaviour similar to nitrogen. It is a colourless, odourless, and tasteless gas, which is not flammable, non-toxic, and very slightly soluble in water. The vapour pressure of sulphur hexafluoride is above 1.0133×10^5 N m^{-2} at the triple-point temperature, so that it has no normal boiling point. Its sublimation temperature is −63.7°C. The presented normal boiling point in Table 9.1 is an extrapolated value (15).

Solid sulphur hexafluoride is dimorphic, having only one transition point. Sulphur hexafluoride forms symmetrical octahedral molecules, and is very slowly decomposed by a powerful electrical discharge. The molecule has no dipole moment. Because of its thermal stability, chemical inertness, low condensation temperature, low toxicity, heat transfer ability, and good electrical characteristics, the electrical industry is the major user of SF_6. Its applications in various electrical equipment as an electrical insulating gas, include: transformers, capacitors, microwave components, circuit breakers, etc. The electrical conductivity has been determined by Belo *et al.* (3) at high temperatures and high pressures.

The gas has been produced commercially since 1947 and several manufacturers have issued bulletins on the electrical characteristics and properties. However, comprehensive collections of the published articles on the properties of SF_6 were compiled only in multi-volume handbooks, such as: Gmelin (13), Mellor (21), and Pascal (23).

The liquid viscosity data have been provided by Matheson Gas Products, Division of Will Ross, Inc., East Rutherford, N.J. (private communication), in addition to the single value found in the *Matheson Gas Data Book* (19).

Most of the presented data in Table 9.1 and Figs. 9.1–9.12 are based on selected experimental measurements; however, some properties have been correlated from several sources.

References

1 American Conference of Government Industrial Hygienists, *Documentation of Threshold Limit Values*, revised edition, Committee on Threshold Limit Values, Cincinnati, Ohio 45202, 1966, 203 pp.

2 Ashton, J. T., Dawe, R. A., Miller, K. W., Smith, E. B., Stickings, B. J., *J. Chem. Soc. (A)*, 1968, 1793–6.

3 Belov, V. A., Semenov, A. M., *High Temperature*, 1971, **9** (4), 654–7.

4 Borisoglebskii, V. P., Rozov, A. N., Smirnov, O. A., *Russ. J. Phys. Chem.*, 1972, **46** (2), 308–9.

5 Bradley, J. N., *Shock Waves in Chemistry and Physics*, Methuen & Co. Ltd., London, 1962, p. 334.

6 Brown, J. A., *The Use of Sulphur Hexafluoride as a Gaseous Dielectric*, Symp. on Electric Insul. Gases, Am. Soc. Testing Mater. Spec. Publ. No. 346, Philadelphia, 1964, pp. 24–34.

7 Chang, K.-C., *Thermal Conductivities of Sulphur Dioxide and Sulphuryl Fluoride and Binary Mixtures Containing These Gases*, St. Louis Univ., Ph. D. Dissertation, 1969, 139 pp.

8 Dawe, R. A., Maitland, G. C., Rigby, M., Smith, E. B., *Trans. Faraday Soc.*, 1970, **66** (572), 1955–65.

9 Edwards, D. G., *The Vapour Pressure of 30 Inorganic Liquids between One Atmosphere and the Critical Point*, Univ. of California, Lawrence Rad. Lab., UCRL 7167, Livermore, California, June 13, 1963, 148 pp.

10 Eucken, A., Bertran, A., *Z. phys. Chem.*, 1936, **31B**, 361–81.

11 Eucken A., Schröder, E., *Z. phys. Chem.*, 1938, **41B**, 307–19.

12 Friedman, H. L., *J. Am. Chem. Soc.*, 1954, **76**, 3294–7.

13 *Gmelins Handbuch der anorganischen Chemie*, Syst. No. 9, Schwefel, Verlag Chemie GmbH, Berlin, 1963, pp. 1714–21.

14 IUPAC, Atomic Weights of the Elements 1971, *Pure & Appl. Chem.*, 1972, **30** (3–4), 638–49.

15 Klemm, W., Henkel, P., *Z. anorg. Chem.*, 1932, **207**, 73–86.

16 Lax, E., Synowietz, C., Eds., *Taschenbuch für Chemiker und Physiker*, 3rd ed., Springer-Verlag, Berlin, 1967.

17 Lis, J., Kellard, P. O., *Brit. J. Appl. Phys.*, 1965, **16**, 1099–1104.

18 Mannheimer, M., *Chem. Analyst*, 1956, **45** (1), 8–10.

19 *Matheson Gas Data Book*, 4th ed., The Matheson Co. Inc., East Rutherford, N.J., 1966.

20 Mears, W. H., Rosenthal, E., Sinka, J. V., *J. Phys. Chem.*, 1969, **73** (7), 2254–61.

21 *Mellor's Comprehensive Treatise on Inorganic and Theoretical Chemistry*, Vol. X, Longmans, London, 1930, pp. 630–1.

22 Morrison, T. J., Johnstone, N. B., *J. Chem. Soc.*, 1955, 3655–9.

23 Pascal, P., Ed., *Nouveau Traité de Chimie Minérale*, Vol. XIII, Part II, Masson et Cie, Éditeurs, Paris, 1960, pp. 1132–4.

24 Pearson, T. G., Robinson, P. L., *J. Chem. Soc.*, 1933, 1427–8.

25 Perel'stein, J. J., *Scientific Communications of the Institute for Cryogenic Research, Moskva*, Gostorgisdat, Moscow, 1961.

26 Rossini, F. D., Wagman, D. D., Evans, W. H., Levine, S., Jaffe, I., *Selected Values of Chemical Thermodynamic Properties*, Natl. Bur. Std. Circular 500, Washington, Feb. 1, 1952.

27 Schumb, W. C., *Ind. Eng. Chem.*, 1947, **39** (3), 421–3.

28 Stewart, J. W., *J. Chem. Phys.*, 1962, **36** (2), 400–5.

29 Stull, D. R., Dir., *JANAF Thermochemical Tables*, Dow Chemical Co., Midland, Mich., 1965–8.

30 Tison, J. K., Hunt, E. R., *J. Chem. Phys.*, 1971, **54** (4), 1526–31.

31 Ulybin, S. A., Zherdev, E. P., *Dokl. Akad. Nauk, USSR*, 1970, **191** (3), 572–3.

32 Ulybin, S. A., Zherdev, E. P., *Teploenergetika*, 1971, **18** (11), 76–8.

33 Walden, P., *Z. phys. Chem.*, 1909, **65**, 129–225.

34 Watson, H. E., Ramswamy, K. L., *Proc. Roy. Soc.*, 1936, **156A**, 144–57.

35 Watson, H. E., Rao, G. G., Ramswamy, K. L., *Proc. Roy. Soc.*, 1934, **143A**, 558–88.

Table 9.1 Properties of Sulphur Hexafluoride

		Ref.
Chemical formula	SF_6	(16)
Molecular weight	146.0504	(14)
Critical temperature	45.54 °C	(30)
Critical pressure	37.600×10^5 N m^{-2}	(30)
Critical density	736.0 kg m^{-3}	(20)
Triple-point temperature	−50.0 °C	(15)
Triple-point pressure	2.3267×10^5 N m^{-2}	(15)
Boiling point at 1.0133×10^5 N m^{-2}	−68.0 °C	(15)
Melting point	−50.7 °C	(16)
Transition temperature	−179.15 °C	(11)
Density of solid at −91.15 °C	2550.0 kg m^{-3}	(28)
Coefficient of thermal expansion of liquid at melting point	2.90×10^{-3} C^{-1}	(15)
Heat of formation of gas at 25 °C	-1.2217×10^9 J kg-mol^{-1}	(29)
Free energy of formation of gas at 25 °C	-1.1177×10^9 J kg-mol^{-1}	(29)
Heat of sublimation at −63.7 °C	2.2860×10^7 J kg-mol^{-1}	(26)
Heat of fusion	5.0242×10^6 J kg-mol^{-1}	(26)
Heat of transition	1.6077×10^6 J kg-mol^{-1}	(11)
Entropy of gas at 25 °C	2.9187×10^5 J kg-mol^{-1} C^{-1}	(29)
Heat capacity ratio of gas at 25 °C and 1.0133×10^5 N m^{-2}	1.29	(10, 29)
Velocity of sound at 0 °C and 1.0133×10^5 N m^{-2}	131.2 m s^{-1}	(5)
Dielectric constant of gas at 25 °C and 1.0133×10^5 N m^{-2}	1.002049	(35)
Refractive index of gas at 25 °C and 1.0133×10^5 N m^{-2}	1.000766	(34)
Threshold limit value	(1000 ppm) 6×10^{-3} kg m^{-3}	(1)

9.1 Vapour Pressure of Sulphur Hexafluoride (9,15,20)

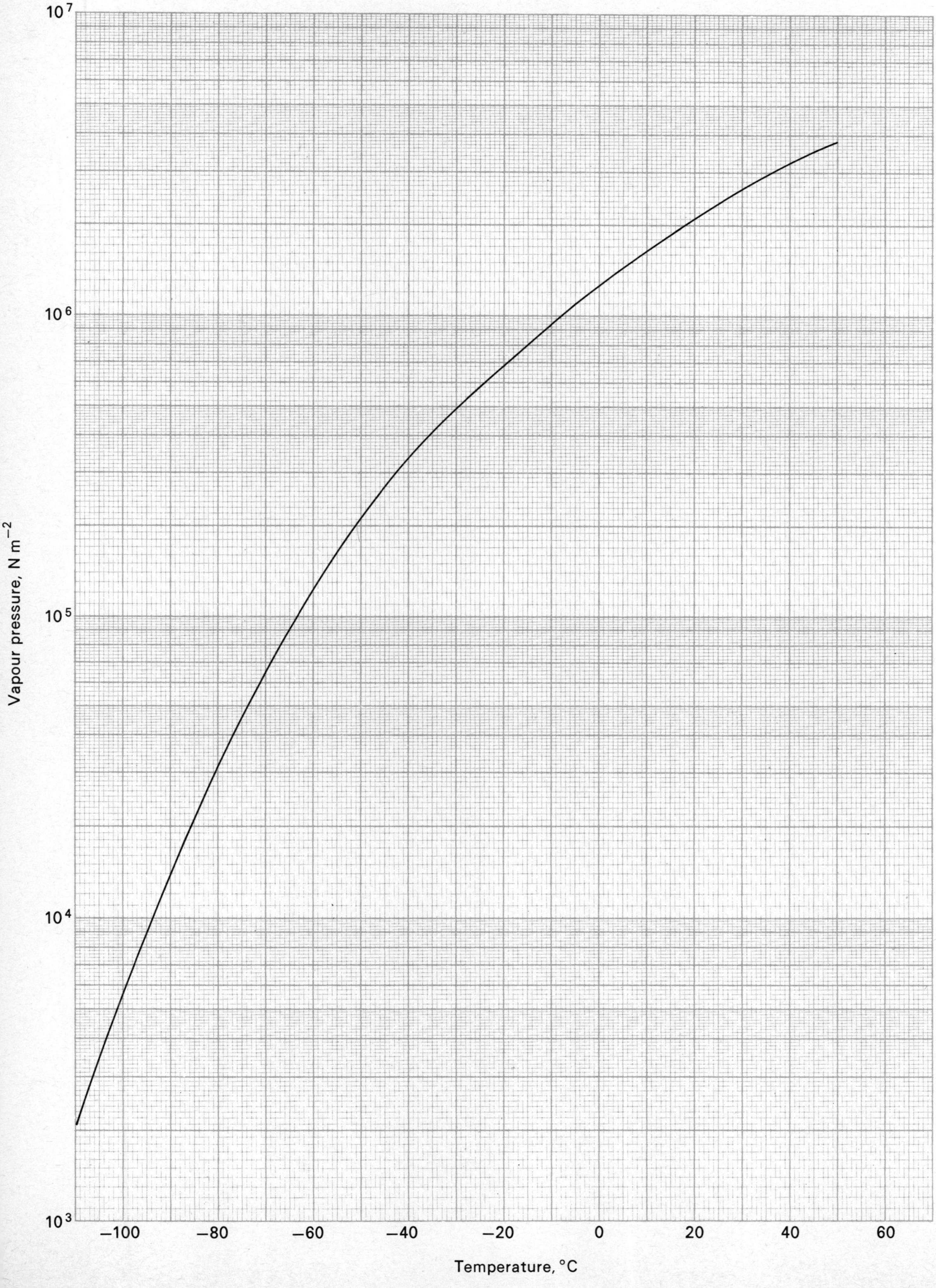

9.2 Orthobaric Density of Sulphur Hexafluoride (20,30,31)

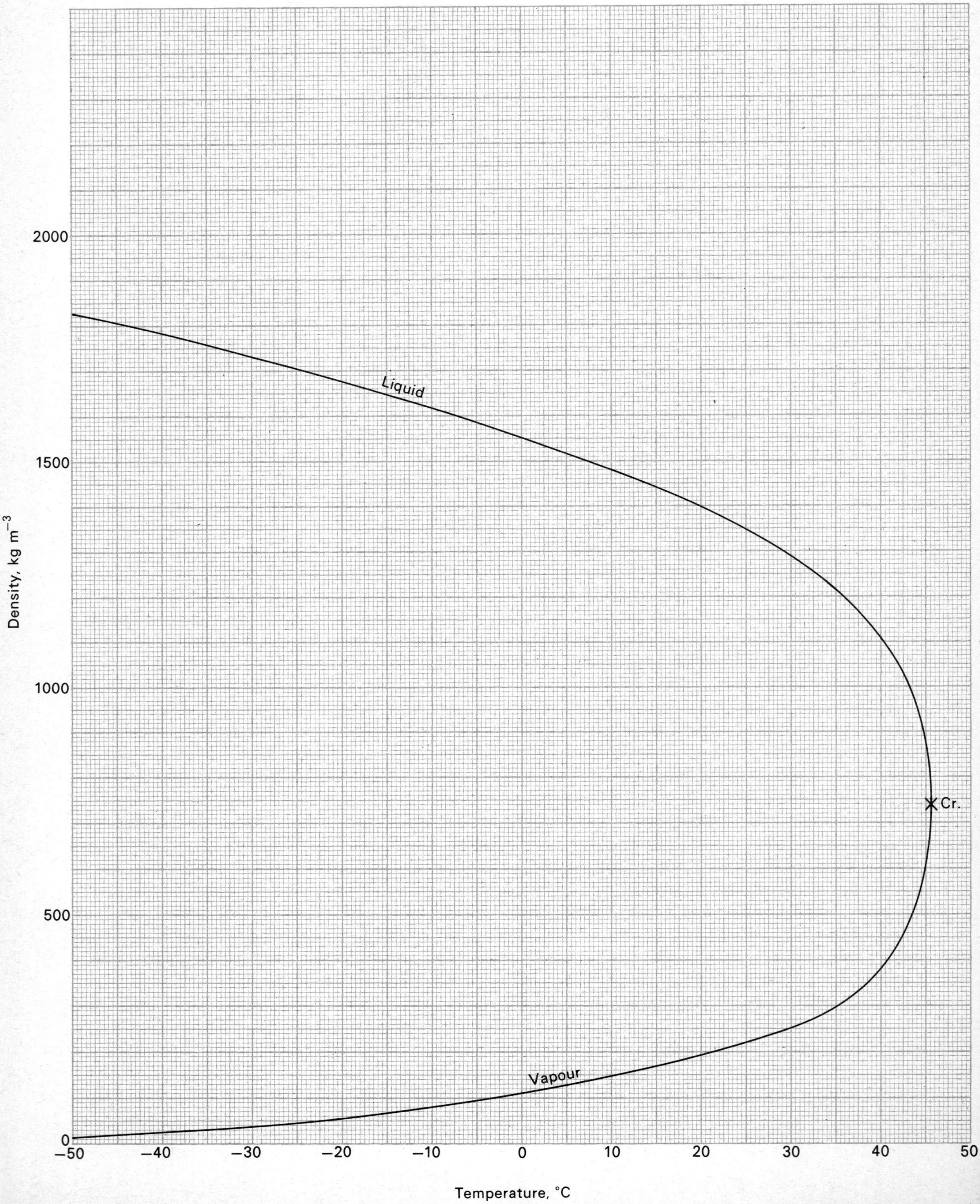

9.3 P-V-T Data for Sulphur Hexafluoride (6,20,31,32)

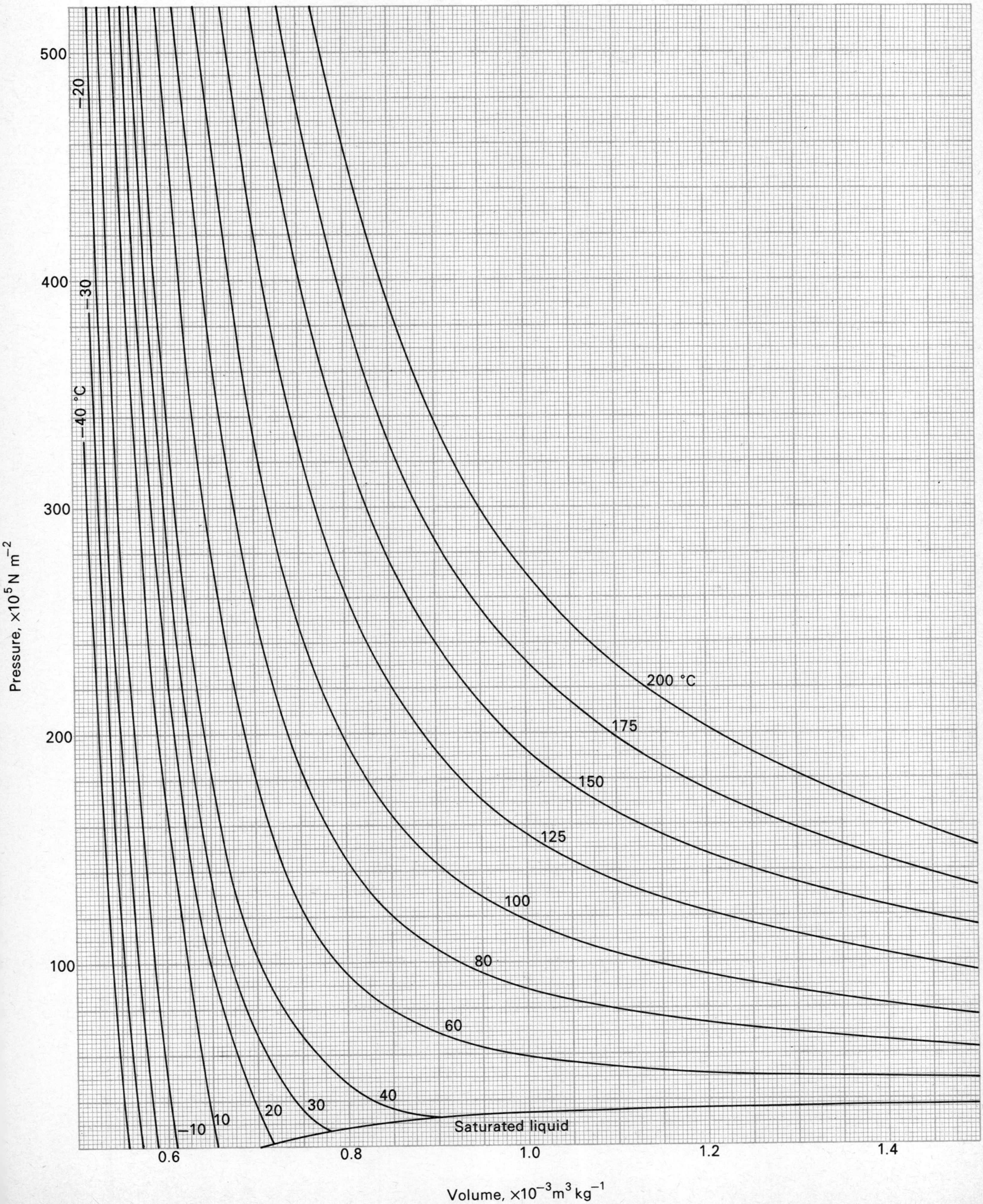

9.4 Latent Heat of Vaporisation of Sulphur Hexafluoride (15,25)

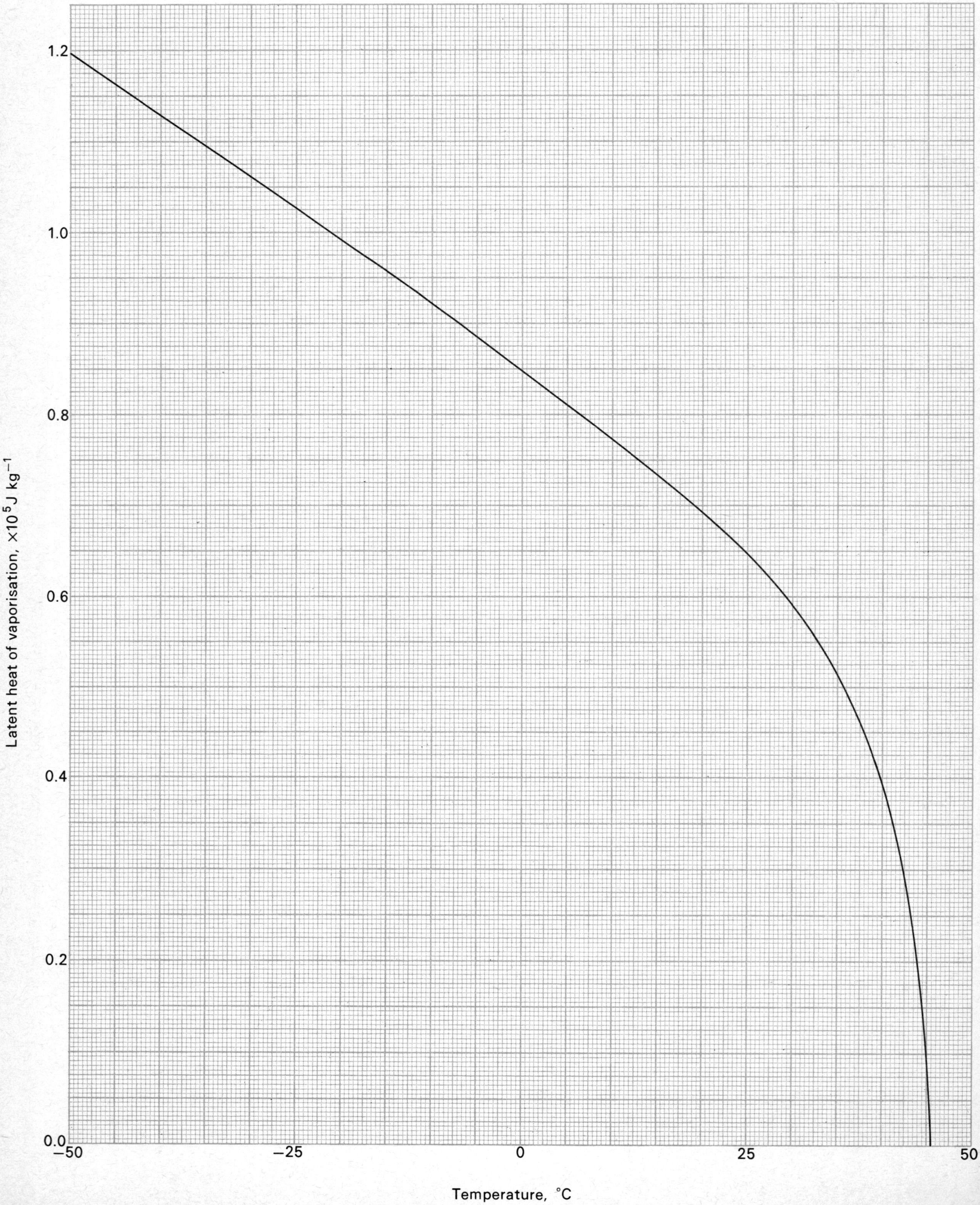

9.5 Heat Capacity of Solid and Liquid Sulphur Hexafluoride (11,27)

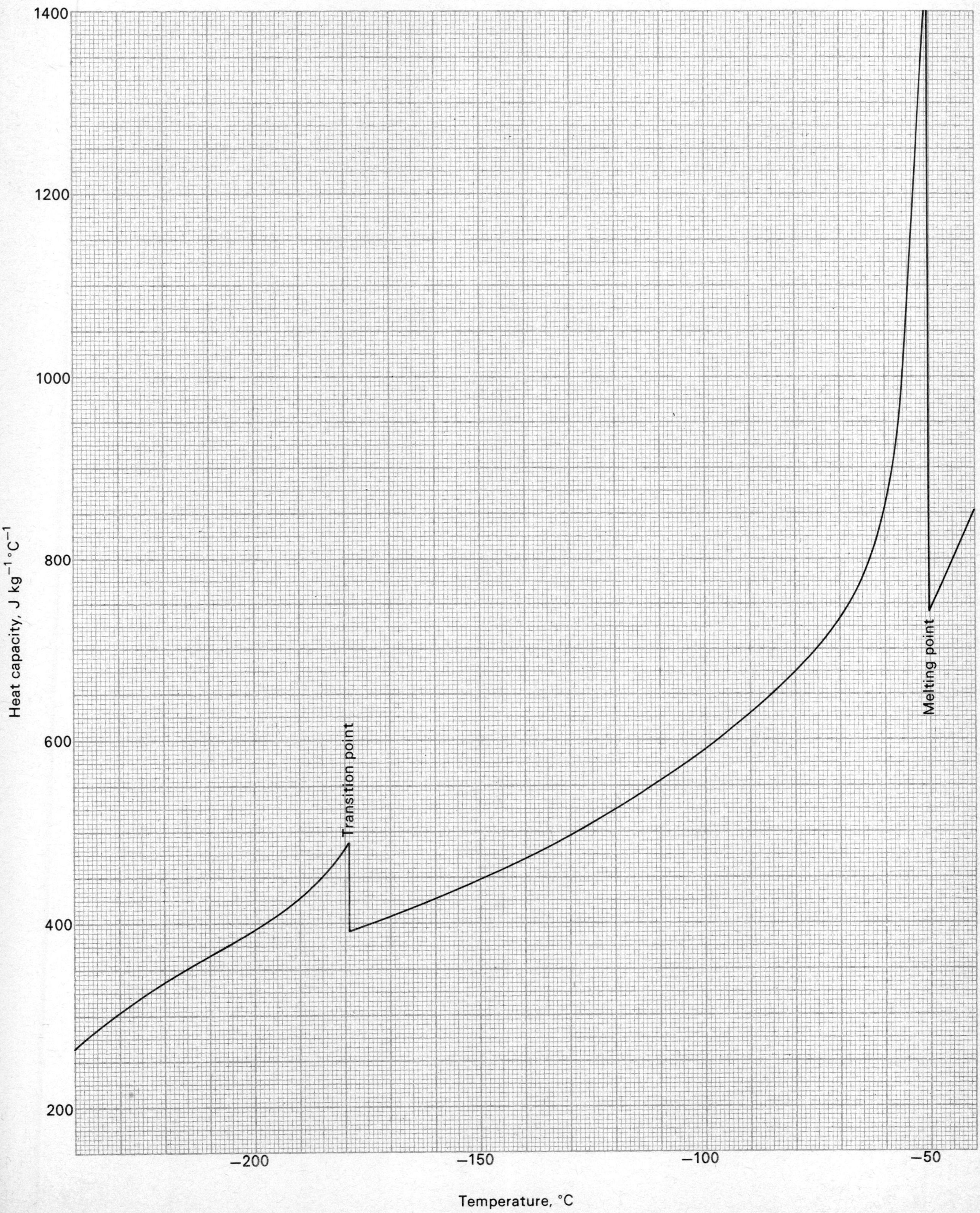

9.6 Heat Capacity of Gaseous Sulphur Hexafluoride at 1.0133×10^5 N m^{-2}(29)

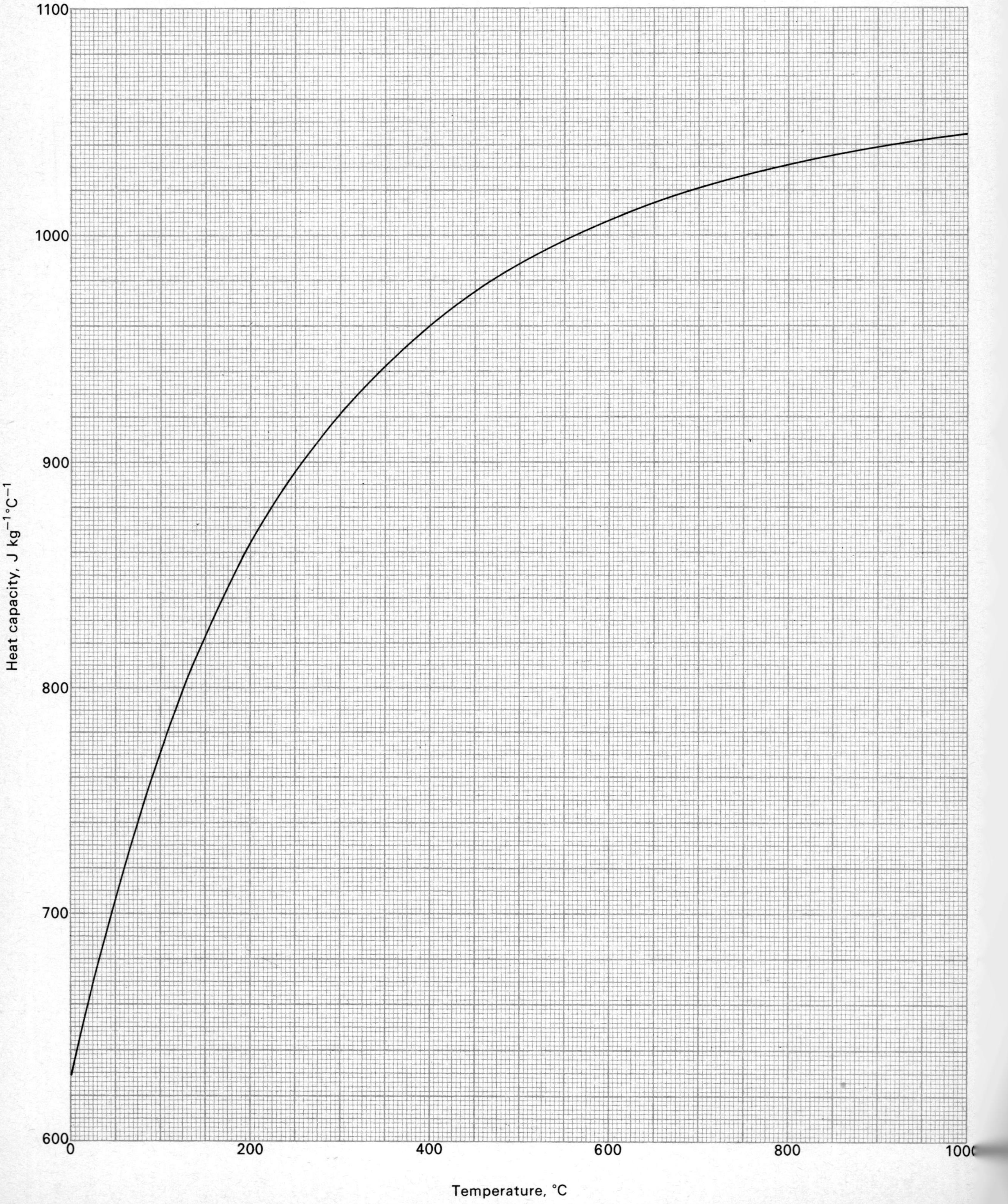

9.7 Viscosity of Liquid Sulphur Hexafluoride (19)

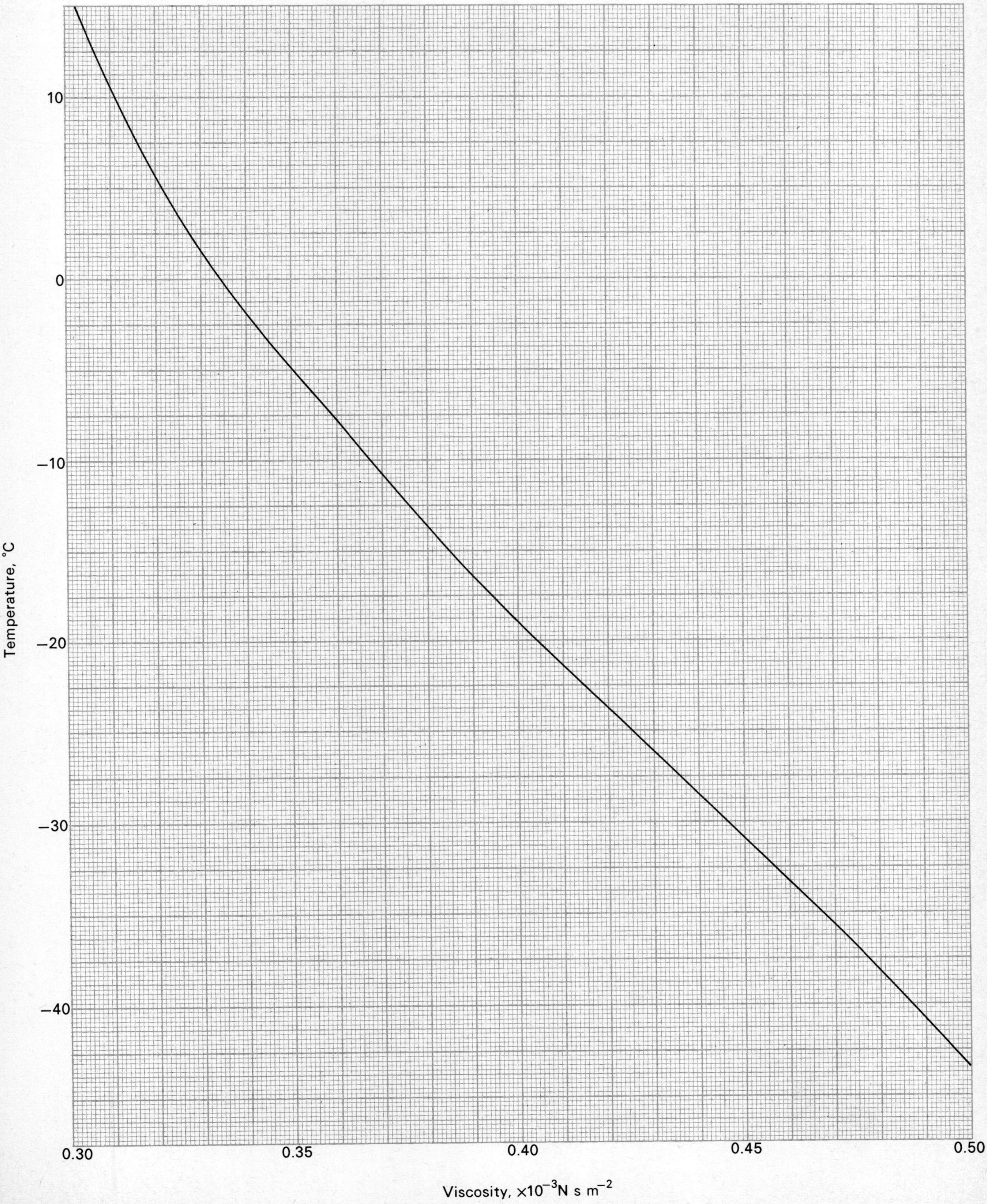

9.8 Viscosity of Gaseous Sulphur Hexafluoride at 1.0133×10^5 N m^{-2}(3,4,8)

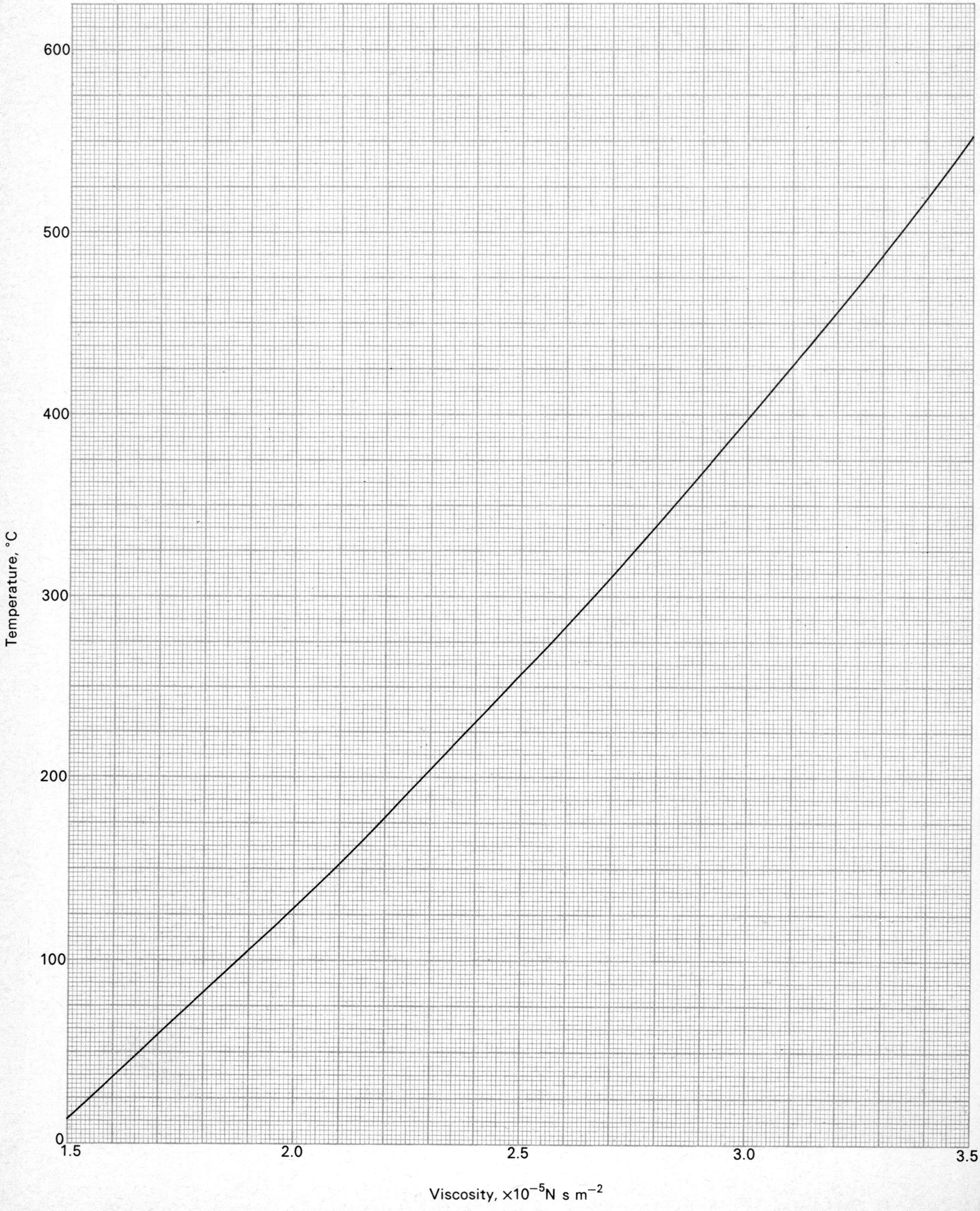

9.9 Thermal Conductivity of Liquid and Gaseous Sulphur Hexafluoride (17)

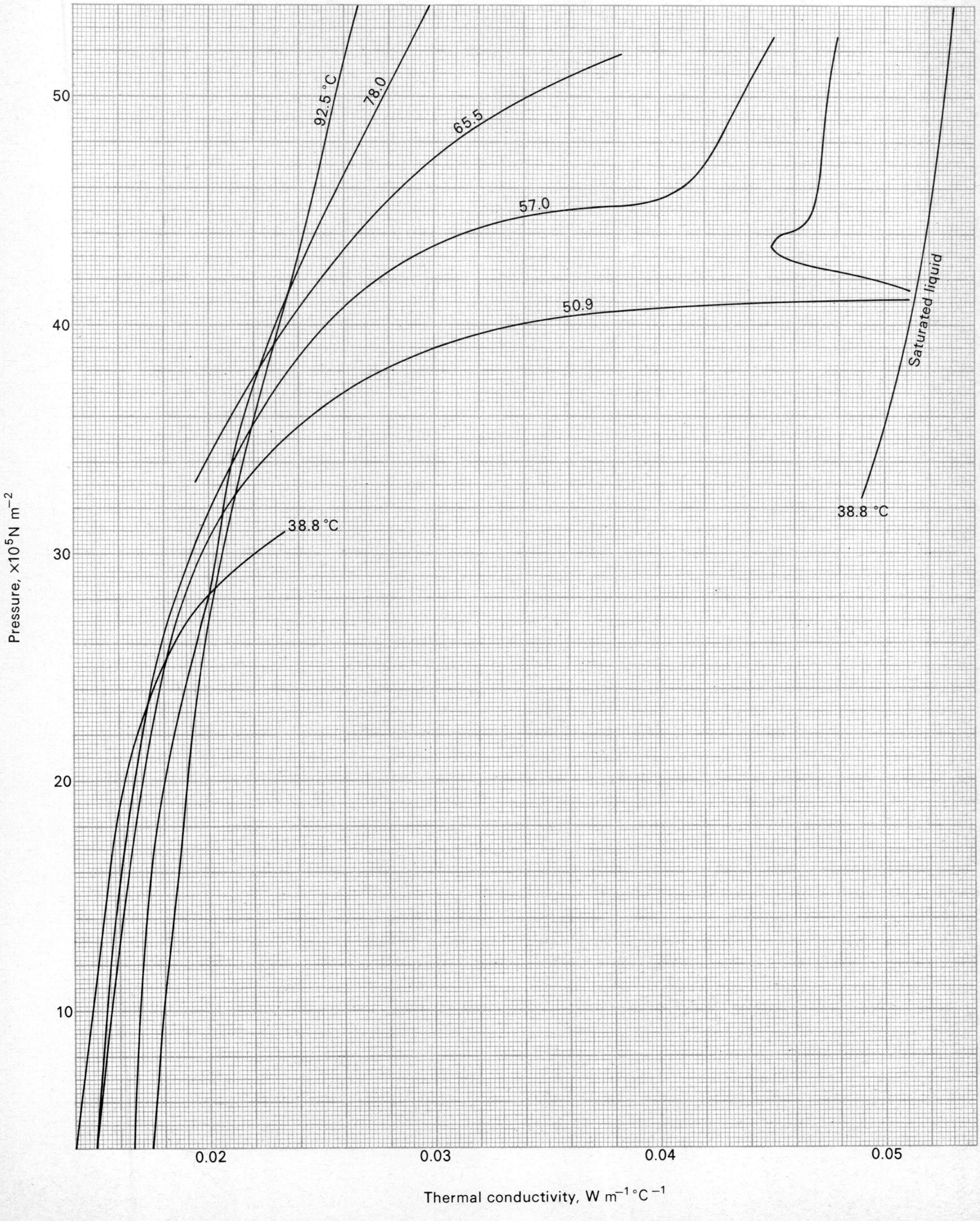

9.10 Thermal Conductivity of Gaseous Sulphur Hexafluoride at $1.0133 \times 10^5 N\ m^{-2}$(7)

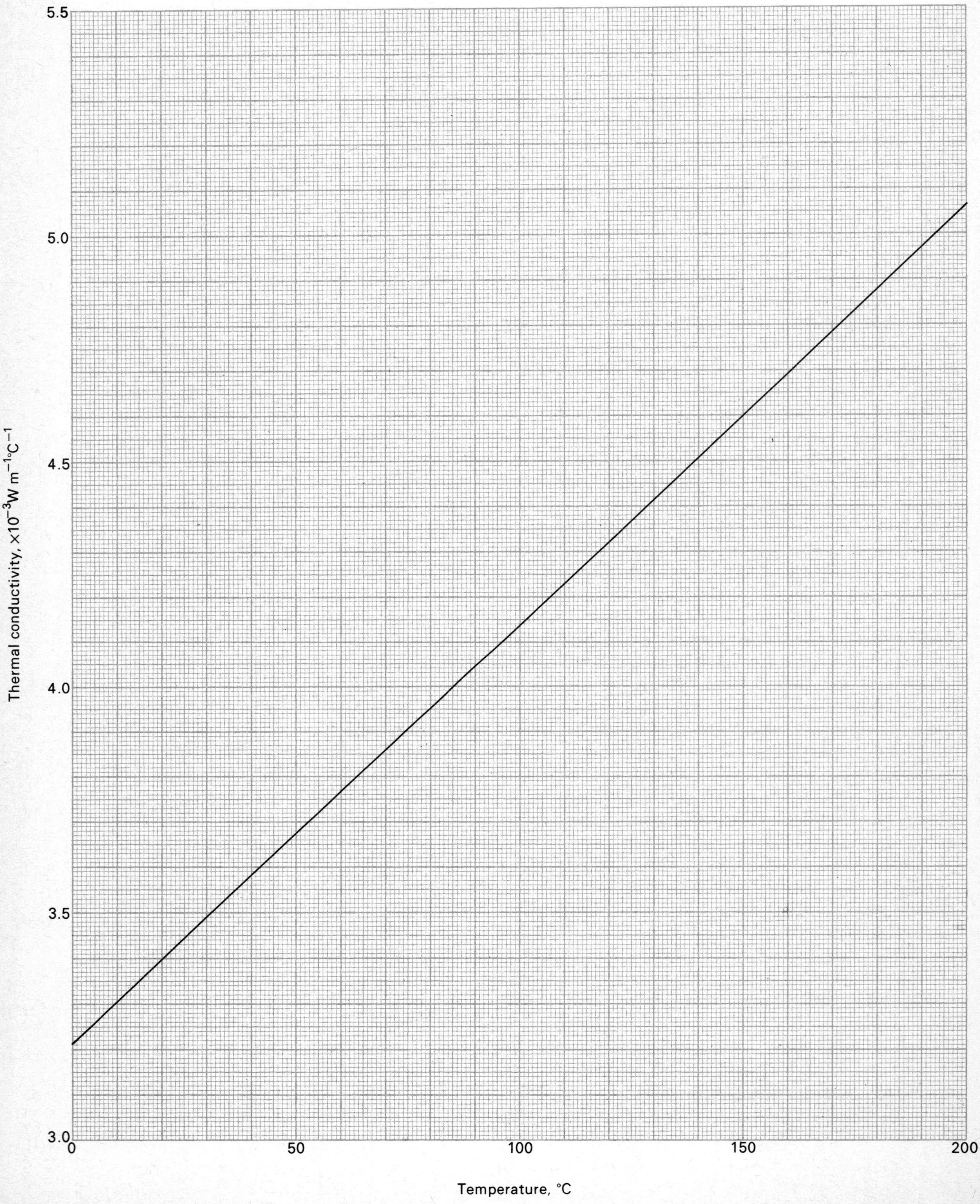

9.11 Surface Tension of Sulphur Hexafluoride (24,33)

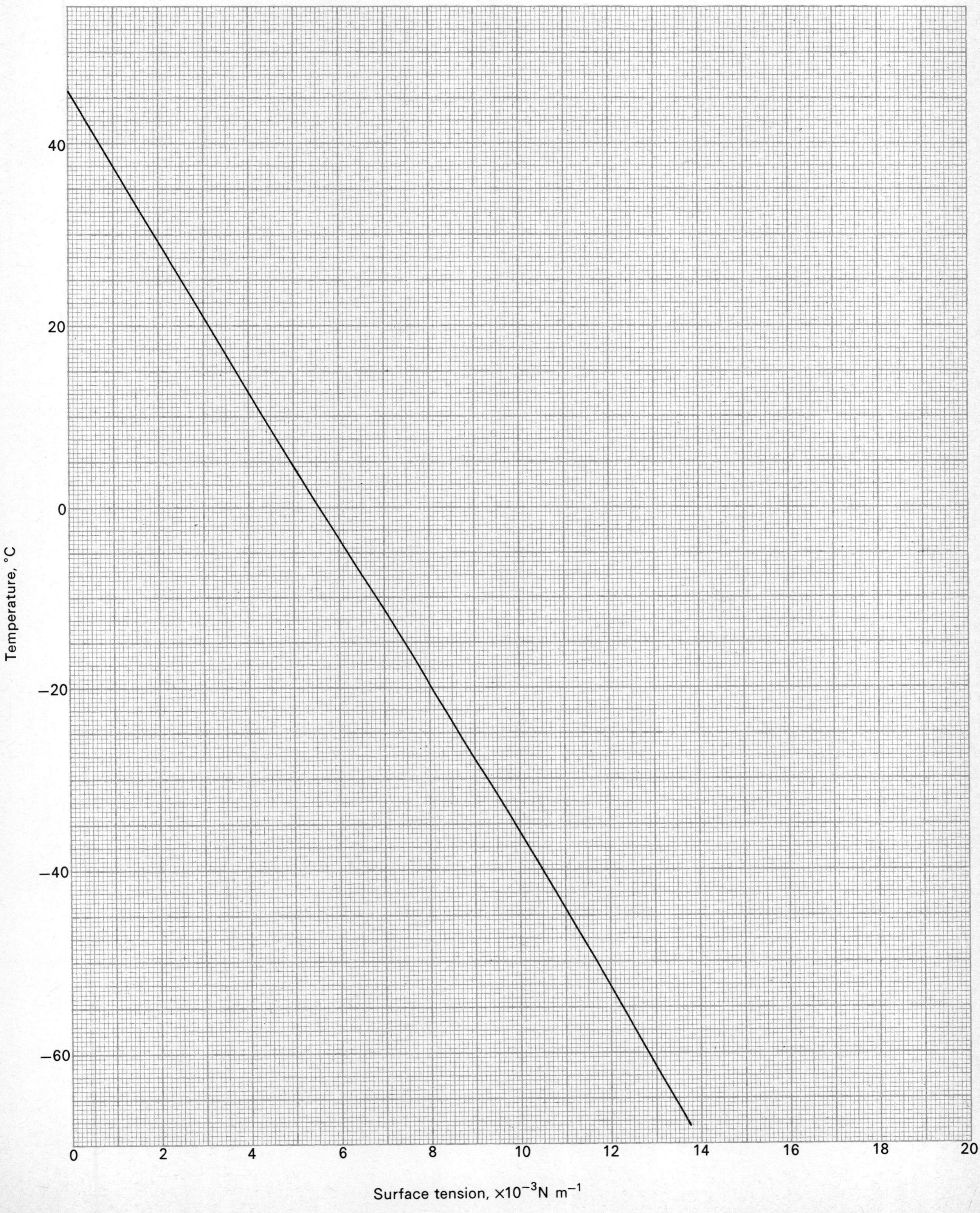

9.12 Solubility of Sulphur Hexafluoride in Water at $1.0133 \times 10^5 N\ m^{-2}$ Partial Pressure (2,12,18,22)

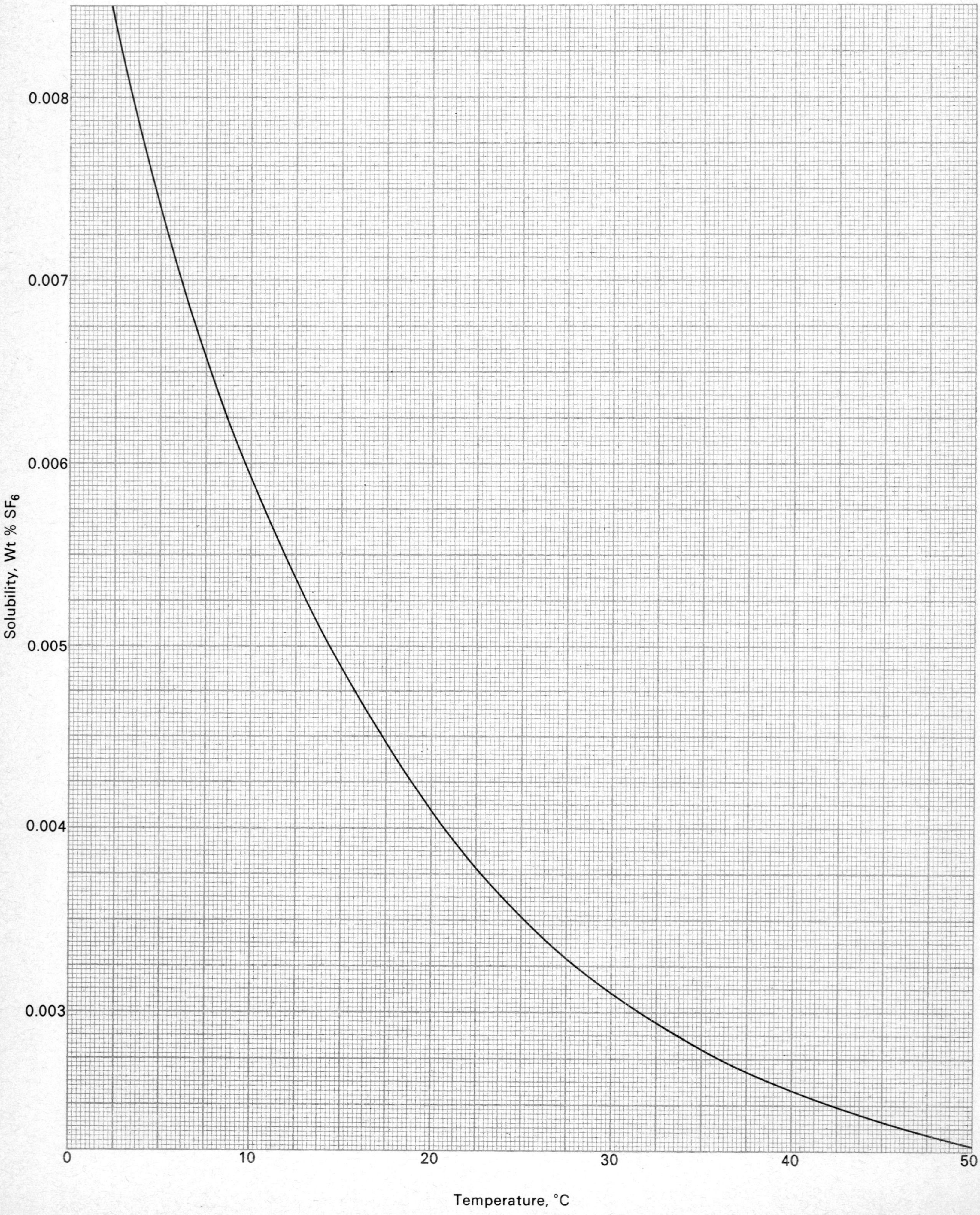

10 AIR

The atmosphere has about one millionth the mass of the earth. Dry air is mainly a mixture of nitrogen, oxygen, and argon, however, it also contains other gases in trace quantities. The following table shows the composition of air at ground level.

Gas	*Mol. wt.*	*Vol. %*
Nitrogen, N_2	28.0134	78.09
Oxygen, O_2	31.9988	20.95
Argon, A	39.948	0.93
Carbon dioxide, CO_2	44.01	0.02–0.04
Neon, Ne	20.183	18.0×10^{-4}
Helium, He	4.0026	5.3×10^{-4}
Krypton, Kr	83.80	1.1×10^{-4}
Hydrogen, H_2	2.0158	0.5×10^{-4}
Xenon, Xe	131.30	0.08×10^{-4}
Ozone, O_3	47.9982	0.02×10^{-4}
Radon, Rn	222.0	7×10^{-18}
Water, H_2O	18.0148	0.2–4.0

Since air is a mixture of gases, (principally consisting of nitrogen and oxygen, the other constituents being present in negligible amounts) there are two vapour pressure curves, i.e. the dew (saturated liquid) and the bubble point (saturated vapour) curves. The determined properties of air at saturation do not imply that nitrogen and oxygen are in equilibrium, but that each is present in amounts corresponding to the composition of air. The latent heat of vaporisation presented in Fig. 10.4 is defined as the heat required to evaporate unit quantity liquid from the bubble point to the dew point temperature at constant pressure. Therefore, the temperature is not constant during the evaporation. The only available information on liquid surface tension is given at −190.3°C and at various oxygen concentrations (16).

Various properties of air have been determined with high accuracy and there are numerous excellent compilations of the selected properties in solid, liquid, and gaseous states. National Bureau of Standards issued a bibliography of thermophysical properties of air from 0 to 300 K (L. A. Hall, NBS Tech. Note 383, Oct. 1969), giving a total of 610 references. Furthermore, there are excellent presentations of the critically selected physical, thermodynamic and transport properties of air by Vasserman *et al.* (37, 38, 39). Additional compilations in journals and multi-volume handbooks are as follows: Din (9), Gmelin (14), Mellor (29), Pascal (30), Romberg (32), Tans (35), Ullmann (36), and Williams (41).

References

1 American Conference of Government Industrial Hygienists, *Documentation of Threshold Limit Values*, revised edition, Committee on Threshold Limit Values, Cincinnati, Ohio 45202, 1966, 203 pp.
2 Arnold, J. H., *Ind. Eng. Chem.*, 1930, **22**, 1091–5.
3 Baehr, H. D., *Brännst.-Wärme-Kraft*, 1967, **19** (2), 65–8.
4 Bergmann, L., *Der Ultraschall*, 5th ed., Stuttgart, 1949, p. 365.
5 Braune, H., Basch, R., Wentzel, W., *Z. phys. Chem.*, 1928, **137A**, 176–92.
6 Carroll, D. L., Lo, H. Y., Stiel, L. I., *J. Chem. Eng. Data*, 1968, **13** (1), 53–7.
7 Dana, L. I., *Proc. Am. Acad. Arts Sci.*, 1925, **60**, 241–67.
8 Dewar, A., *Proc. Roy. Soc.*, 1896/98, **15**, 133–46.
9 Din, F., Ed., *Thermodynamic Functions of Gases*, Vol. 2, Butterworths, London, 1956.
10 Dommett, R. L., RAE-TN-GW 429, (AD 115386), 1956, 1–39.
11 Engineering Science Data Units, *Approximate Data on the Viscosity of some Common Liquids*, Item No. 66024, London, July 1966, 26 pp.
12 Eucken, A., *Phys. Zeitschr.*, 1911, **12**, 1101–7.
13 Geier, H., Schafer, K., *Allgem. Wärmetech.*, 1961, **10**, 70–5.
14 *Gmelins Handbuch der anorganischen Chemie*, Syst. No. 3, Sauerstoff, Verlag Chemie GmbH, Weinheim, 1960, pp. 819–976.
15 Goring, G. E., Eagan, D. P., *Can. J. Chem. Eng.*, 1971, **49**, 25–31.
16 Grunmach, L., *Ann. Phys.*, 1901, **6** (4), 559–64.
17 Hilsenrath, J., Beckett, C. W., Benedict, W. S., Fano, L., Hoge, H. J., Masi, J. F., Nuttal, R. L., Touloukian, Y. S., Woolley, H. W., Natl. Bur. Std. Circular 564, Washington, 1955, 488 pp.
18 IUPAC, Atomic Weights of the Elements 1971, *Pure & Appl. Chem.*, 1972, **30** (3–4), 638–49.
19 Kallman, H. K., RM 442, (AD 103216), 1950, 1–44.
20 Karpenko, A. S., Markevich, A. M., Ryabinin, Ya. N., *Zh. Eksper. Theor. Fiz.*, 1952, **23**, 468–76.
21 Keesom, W. H., Tuyn, W., *Comm. Leiden. Suppl.*, 1932, **72**, 26–39.
22 Kestin, J., Leidenfrost, W., *The Effect of Moderate Pressures on the Viscosity of Five Gases*, Progr. Internat. Res. Thermodyn. Transport Properties, 2nd Symp. Thermophys. Properties (Masi, J. F., Tsai, D. H., Eds.), Am. Soc. Mech. Eng., Academic Press, New York, 1962, pp. 321–38.
23 Keyes, F. G., Techn. Rept. 37, Project Squid, (ATI 16173), 1952, 1–33.
24 Kobeko, P. P., Kuvshinskii, E. V., Shishkin, N. I., *Zh. Tech. Fiz.*, 1944, **14**, 3–9.
25 Lange, N. A., Ed., *Handbook of Chemistry*, 10th ed., McGraw-Hill, New York, 1967.
26 Lax, E., Synowietz, C., Eds., *Taschenbuch für Chemiker und Physiker*, 3rd ed., Springer-Verlag, Berlin, 1967.
27 Lien, W. H., Phillips, N. E., *J. Chem. Phys.*, 1961, **34**, 1073–4.
28 Lo, H. Y., Carroll, D. L., Stiel, L. I., *J. Chem. Eng. Data*, 1966, **11** (4), 540–4.
29 *Mellor's Comprehensive Treatise on Inorganic and Theoretical Chemistry*, Vol. 1, Longmans, London, 1922, pp. 147–50.
30 Pascal, P., Ed., *Nouveau Traite de Chimie Minerale*, Vol. I, Généralités, Air, Eau, Hydrogéne, Deutérium, Tritium, Helium et Gas Inertes, Masson et Cie, Éditeurs, Paris, 1956, pp. 386–412.
31 Roebuck, J. R., *Proc. Am. Acad.*, 1928/30, **64**, 287–334.
32 Romberg, H., *Kältetechnik*, 1970, **22** (12), 398–403.
33 Rudenko, N. S., *Zh. Expt. Theor. Fiz.*, 1939, **9**, 1078–80.
34 Scheel, K., *Verh. phys. Ges.*, 1907, **9** (2), 24–36.
35 Tans, A. M. P., *Brit. Chem. Eng.*, 1962, **7** (10), 792.
36 *Ullmans Encyklopädie der technischen Chemie*, Vol. 12, Luftzerlegung, Urban & Schwarzenberg, München, 1960, pp. 54–74.
37 Vasserman, A. A., Kazavchinskii, Ya. Z., Rabinovich, V. A., *Thermophysical Properties of Air and Air Components*, Izdatel'stvo Nauk, Moskva, 1966, 383 pp.
38 Vasserman, A. A., Rabinovich, V. A., *Thermophysical Properties of Liquid Air and Its Components*, Izdatel'stvo Komiteta Standartov, Mer i Izmeritel'nykh Priborov pri Sovete Ministrov SSSR, Moskva, 1968, 235 pp.
39 Vasserman, A. A., Rabinovich, V. A., *Teplo Massoperenos*, 1968, 7, 47–57.
40 Verain, L., *Ann. Physique*, 1914, **1** (9), 255–96.
41 Williams, V. C., *Trans. Am. Inst. Chem. Eng.*, 1943, **39**, 93–111.
42 Winkler, L. W., *Ber.*, 1901, **34**, 1409.

Table 10.1 Properties of Air

		Ref.
Chemical formula	$0.79N_2 + 0.21\ O_2$	(26)
Molecular weight	28.9752	(18)
Critical temperature	−140.7 °C	(37)
Critical pressure	37.74×10^5 N m^{-2}	(37)
Critical density	313.0 kg m^{-3}	(37)
Boiling point at 1.0133×10^5 N m^{-2}	−191.4 °C	(21)
Melting point at 1.0133×10^5 N m^{-2}	−214.0 °C	(8)
Coefficient of thermal expansion of gas at 25 °C and 1.0133×10^5 N m^{-2}	336×10^{-5} °C^{-1}	(31)
Density of gas at 0 °C and 1.0133×10^5 N m^{-2}	1.2928 kg m^{-3}	(26)
Vapour pressure of solid at −216 °C	1866.67 N m^{-2}	(8)
Heat capacity of solid at −269.15 °C	28.1 J kg^{-1} °C^{-1}	(27)
Heat capacity ratio of gas at 0 °C and 1.0133×10^5 N m^{-2}	1.403	(25)
Enthalpy of gas at 27 °C and 1.0133×10^5 N m^{-2}	8.7032×10^6 J kg-mol^{-1}	(26)
Entropy of gas at 27 °C and 1.0133×10^5 N m^{-2}	1.9887×10^5 J kg-mol^{-1} C^{-1}	(26)
Velocity of sound at 0 °C and 1.0133×10^5 N m^{-2}	331.45 m s^{-1}	(4)
Diffusion in steam at 42 °C and 1.0133×10^5 N m^{-2}	0.238×10^{-4} m^2 s^{-1}	(2)
Dielectric constant of gas at 22 °C and 1.0133×10^5 N m^{-2}	1.000 539	(40)
Electrical conductivity of gas at 2000 °C and 9×10^7 N m^{-2}	7.697×10^{-7} ohm^{-1} m^{-1}	(20)
Refractive index of liquid at −191.16 °C and 5460 Å	1.001 021	(34)

10.1 Vapour Pressure of Air (37,38)

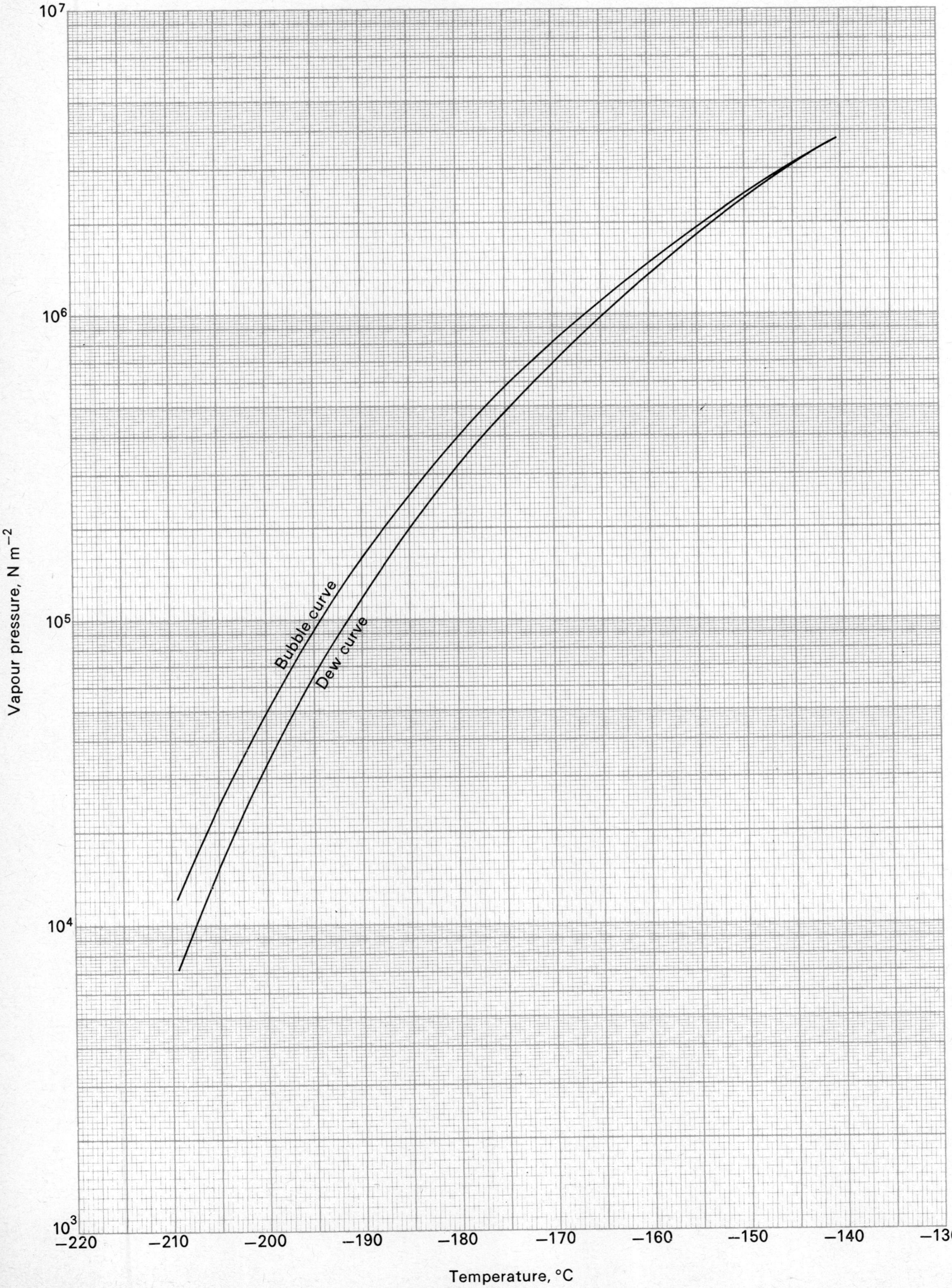

10.2 Orthobaric Density of Air (38)

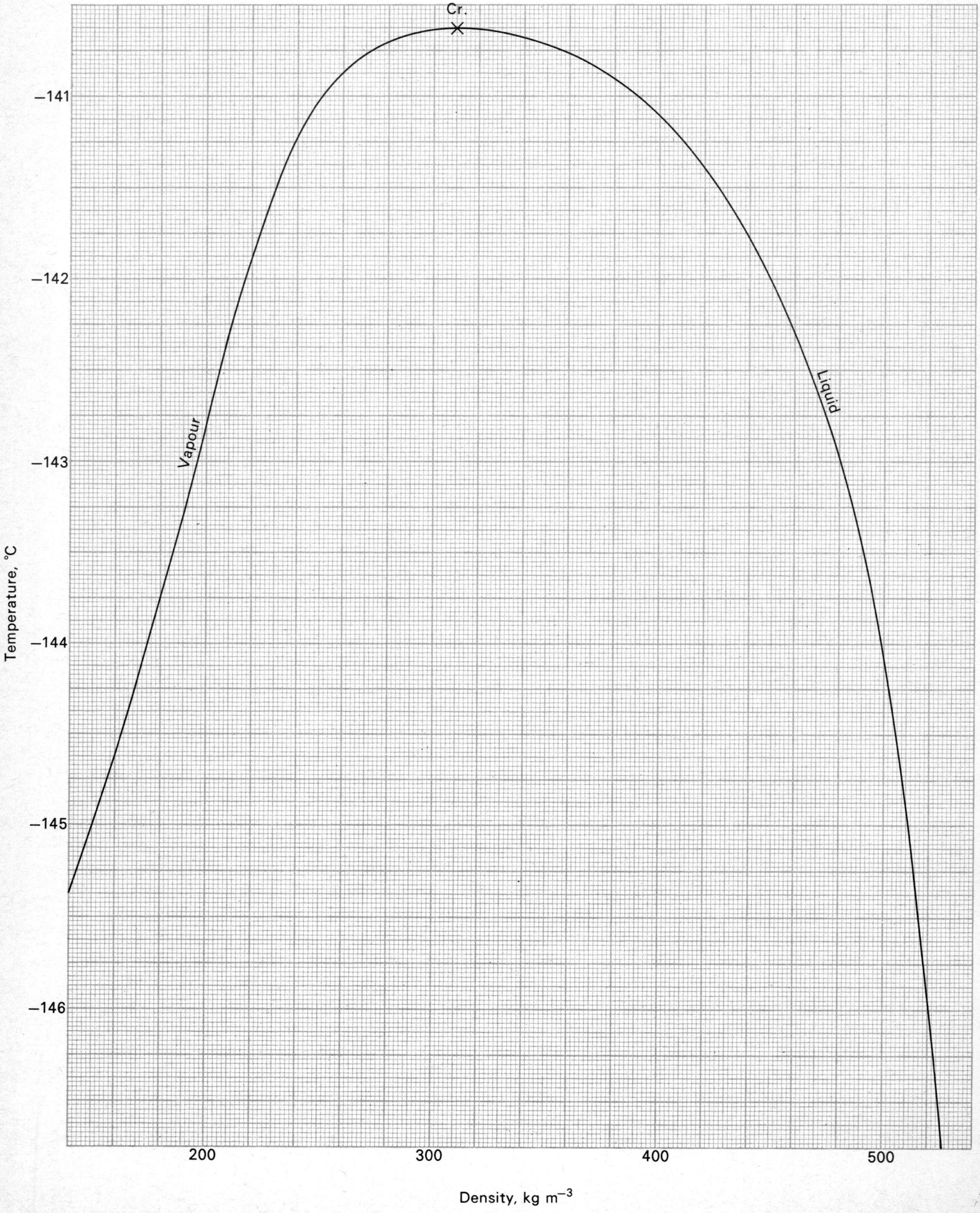

10.3 P-V-T Data for Air (38)

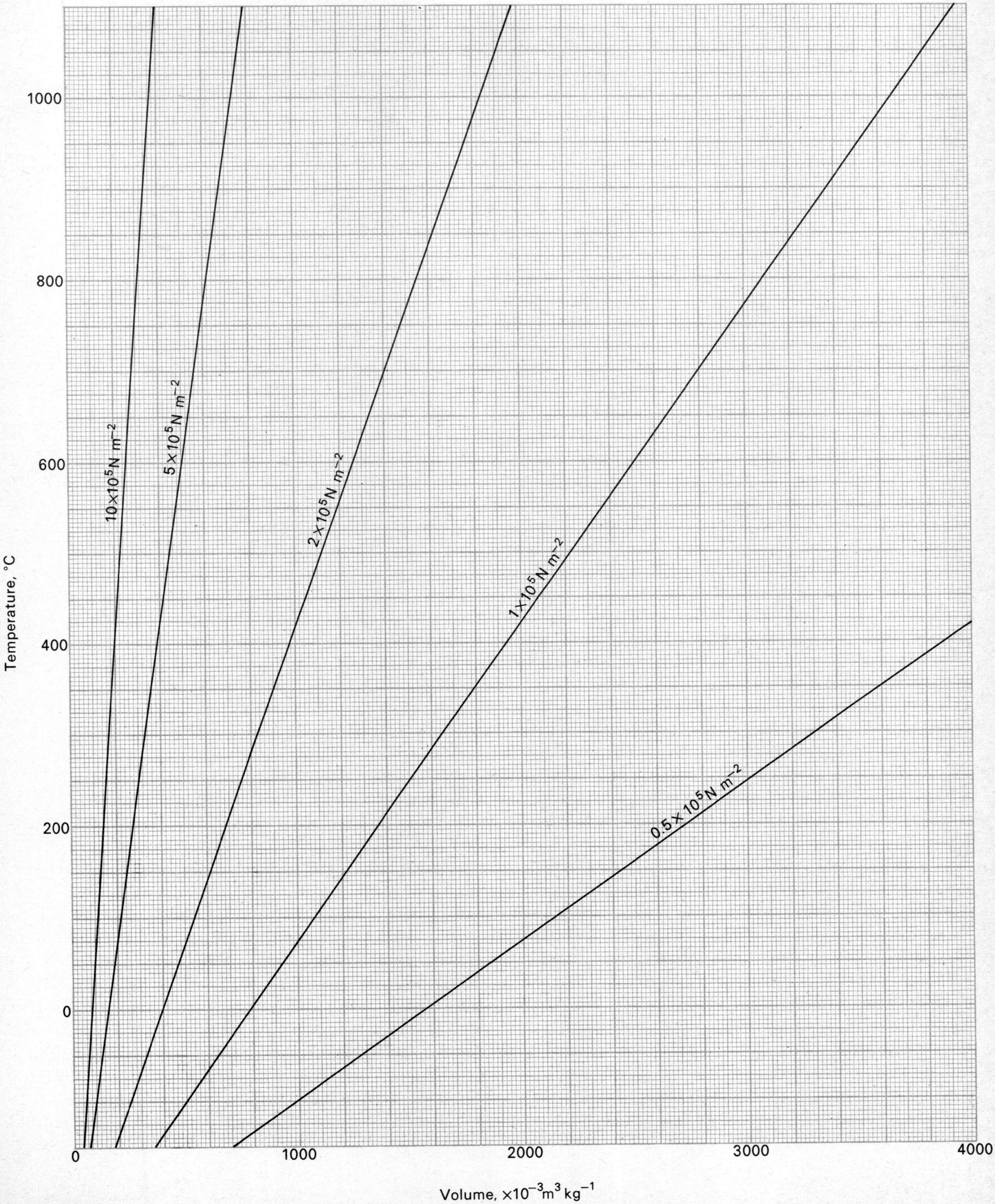

10.4 Latent Heat of Vaporisation of Air (7,38)

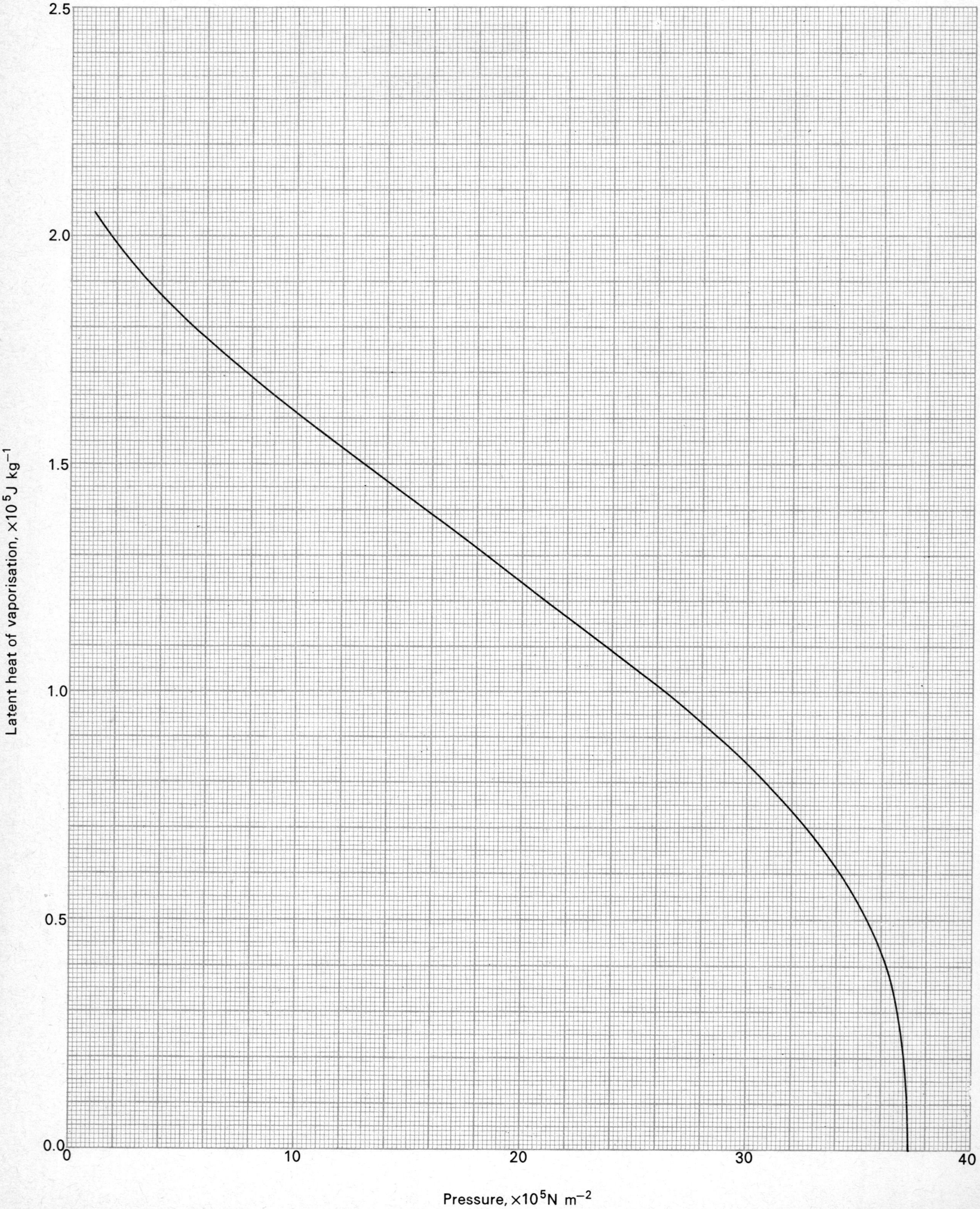

10.5 Heat Capacity of Liquid Air (38)

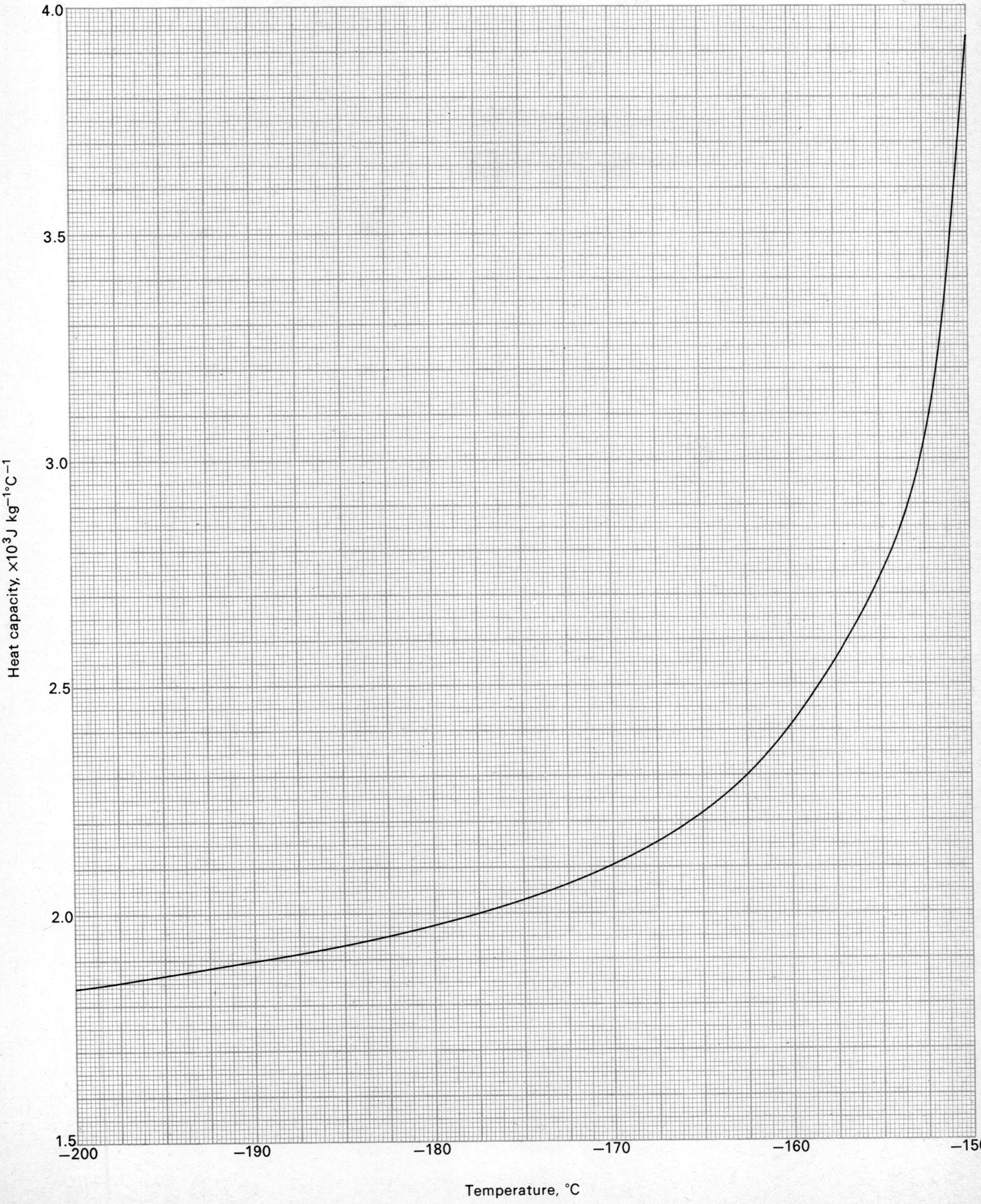

10.6 Heat Capacity of Gaseous Air at $1.0133 \times 10^5 N\ m^{-2}$(10,17,19)

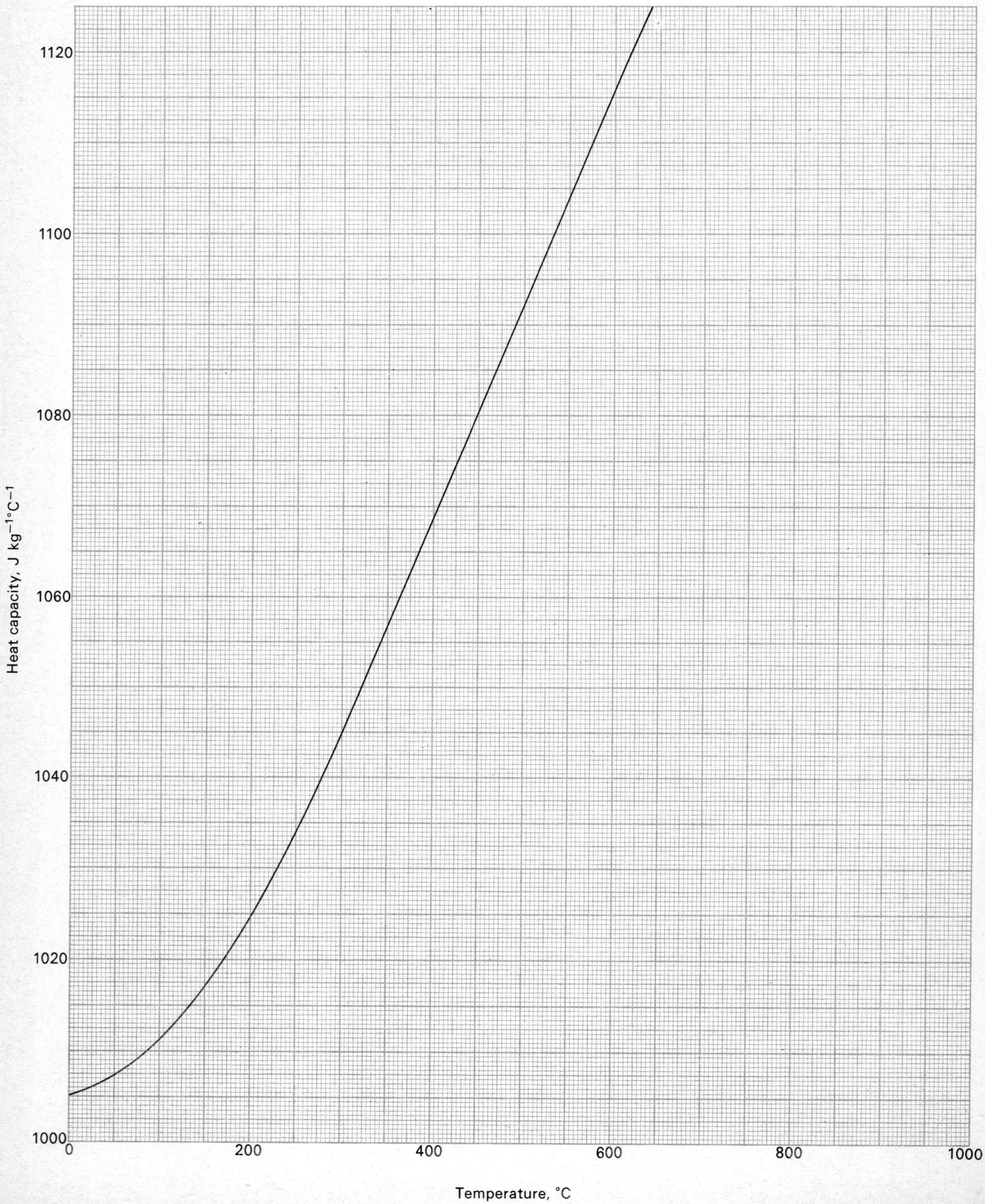

10.7 Heat Capacity Ratio (C_p/C_v) of Gaseous Air (3,26)

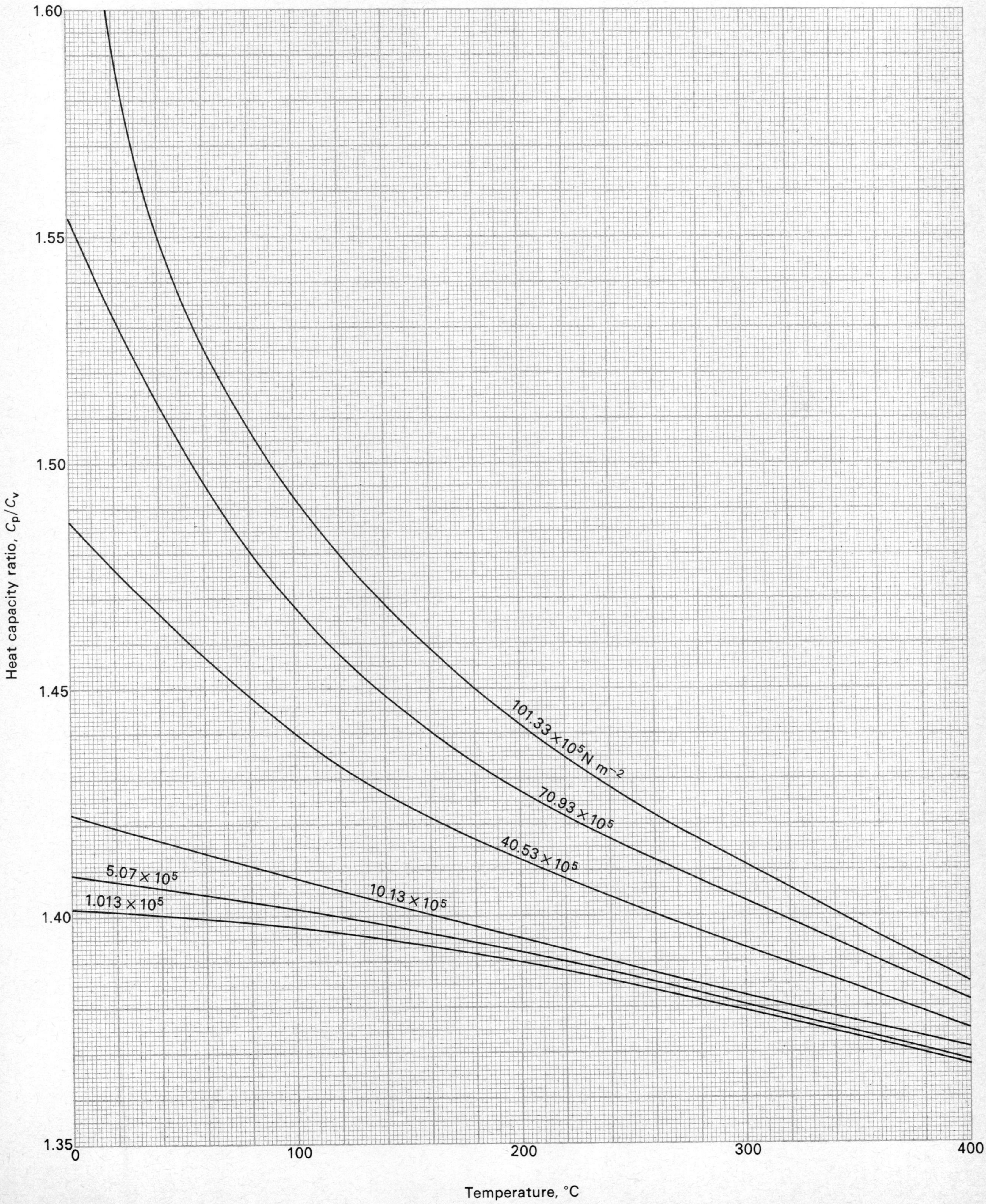

10.8 Viscosity of Liquid Air (11,24,33)

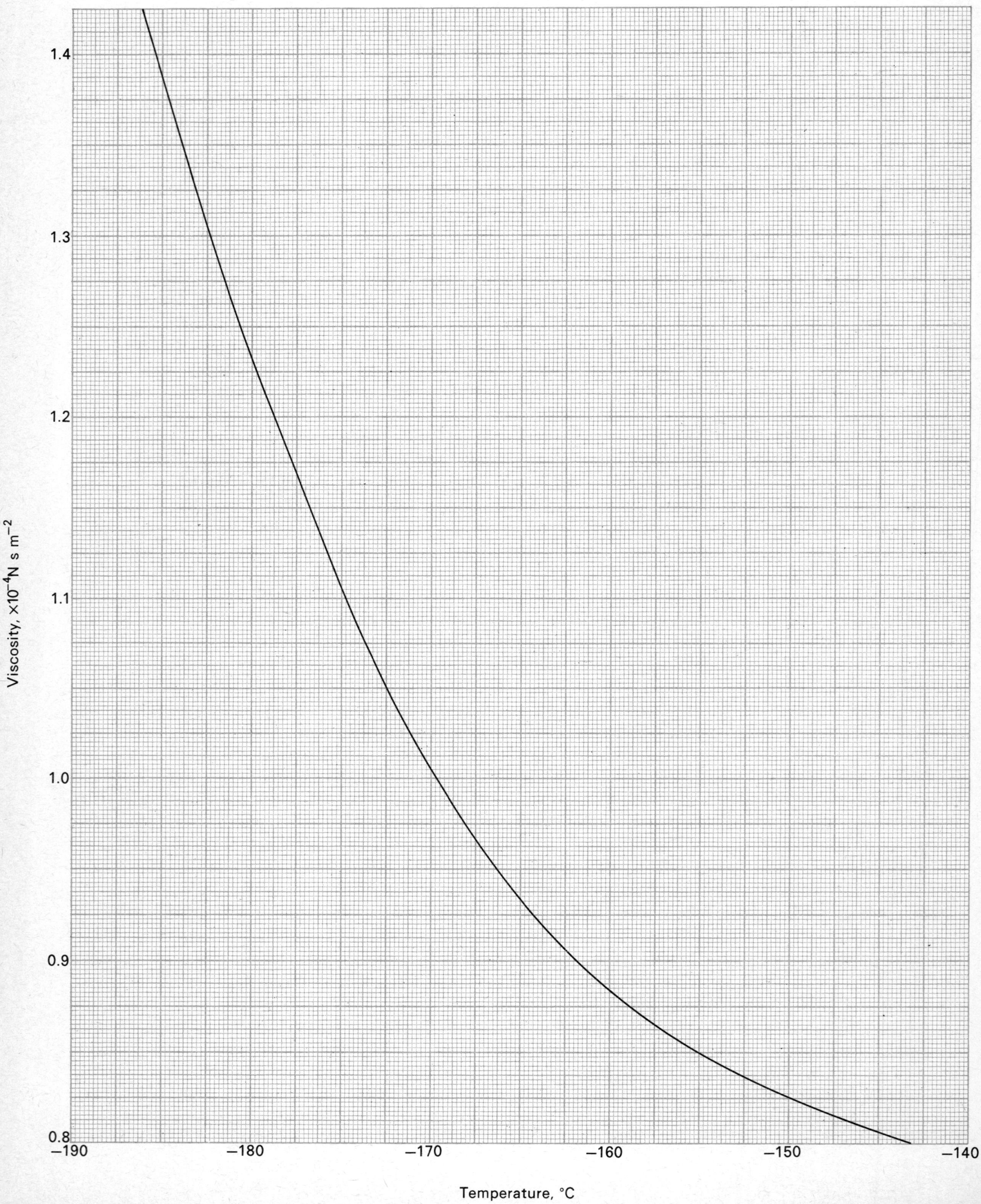

10.9 Vicosity of Gaseous Air at $1.0133 \times 10^5 N\ m^{-2}$(5,15,22,28)

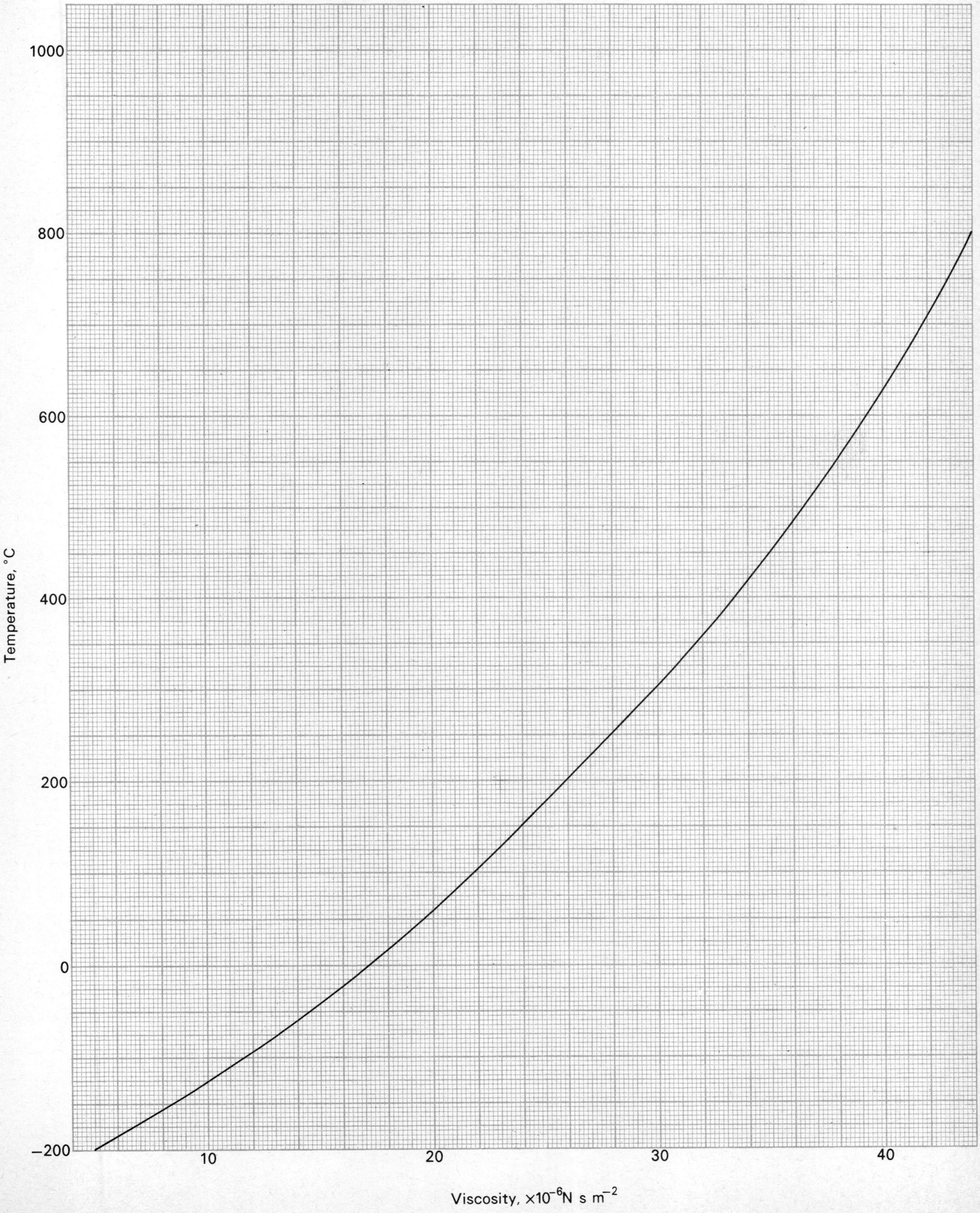

10.10 Viscosity of Gaseous Air (38)

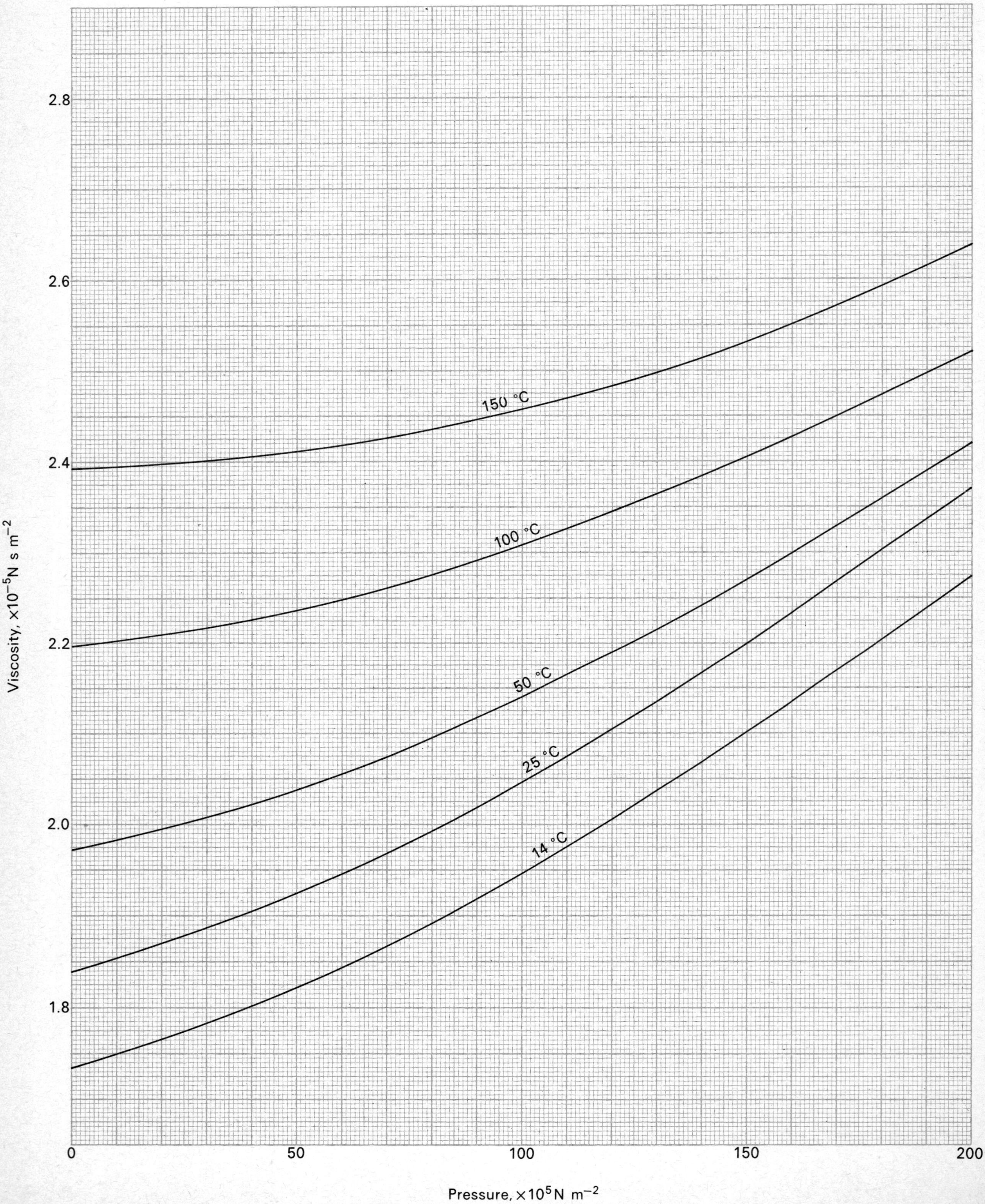

10.11 Thermal Conductivity of Liquid Air (12,38)

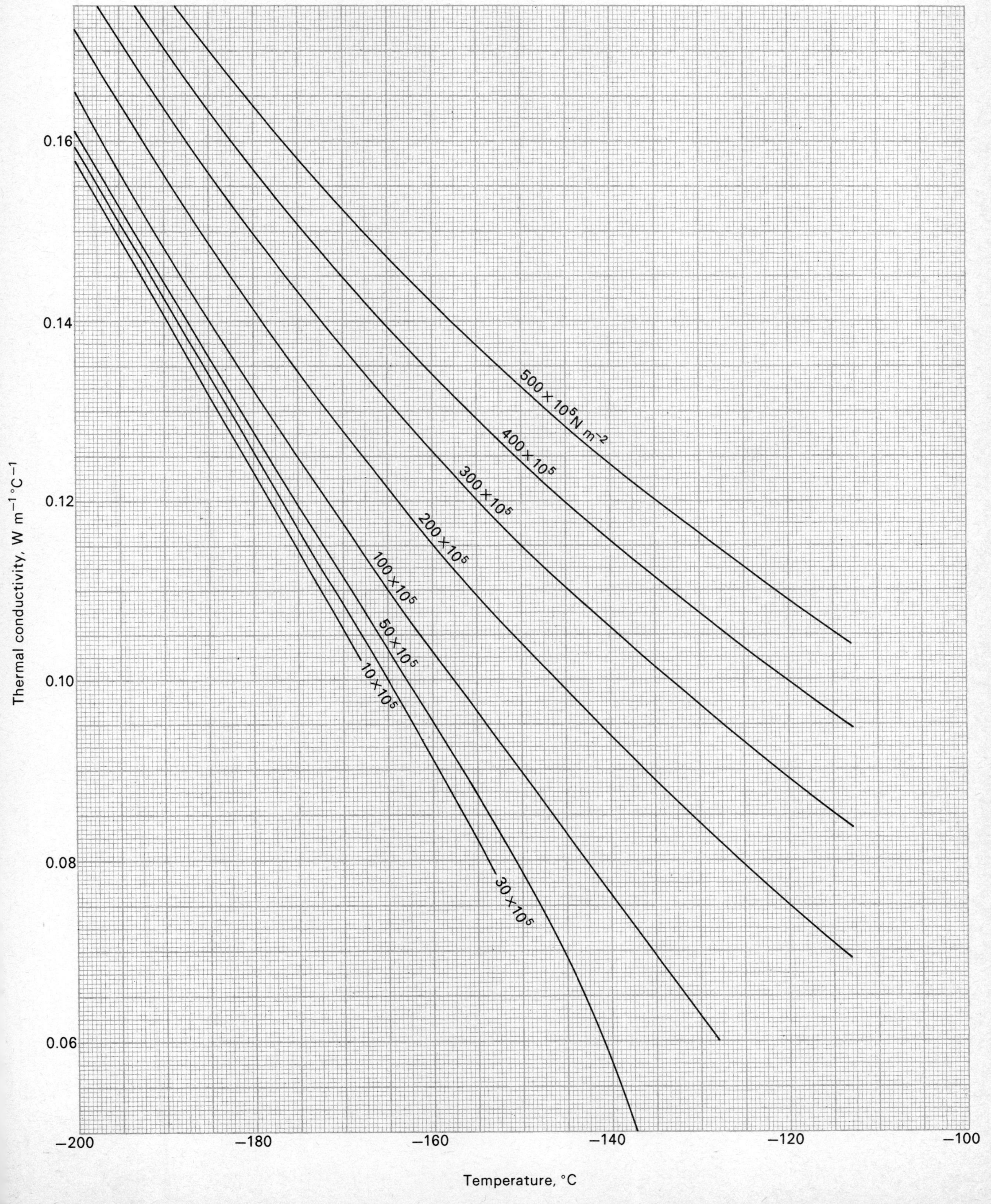

10.12 Thermal Conductivity of Gaseous Air at 1.0133×10^5 N m^{-2} (12,13,23,26)

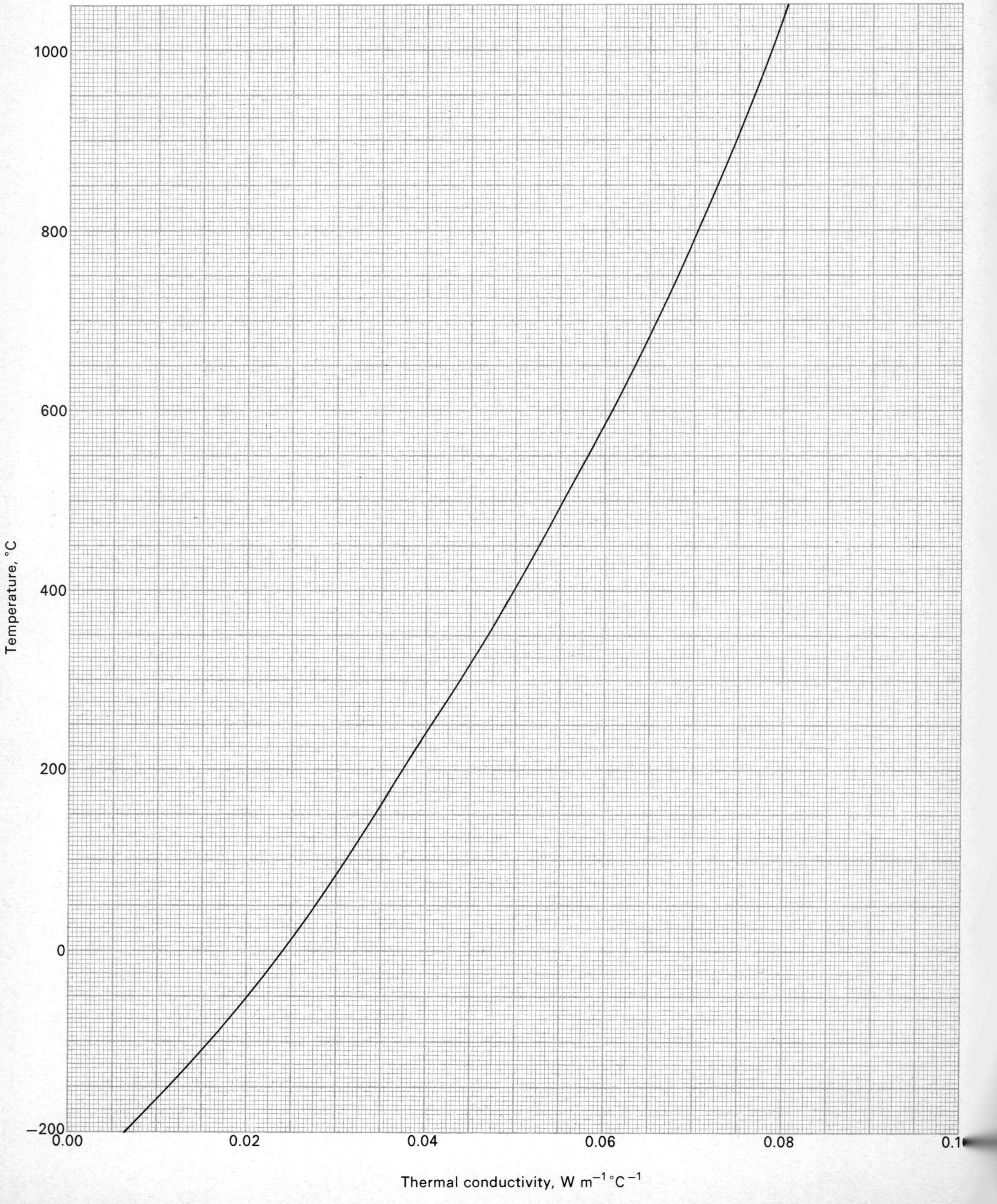

10.13 Thermal Conductivity of Gaseous Air (6,37,38)

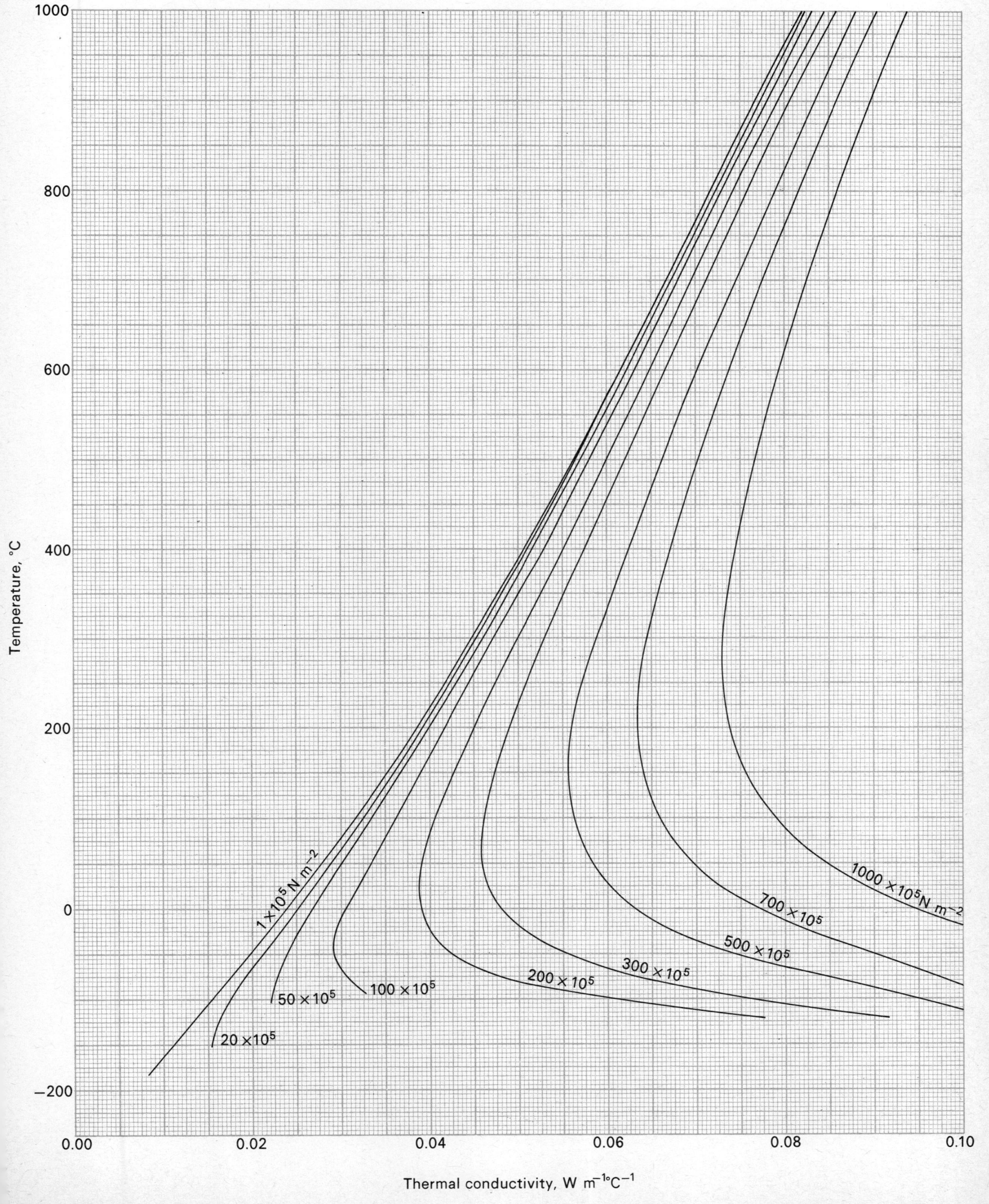

10.14 Surface Tension of Air (16)

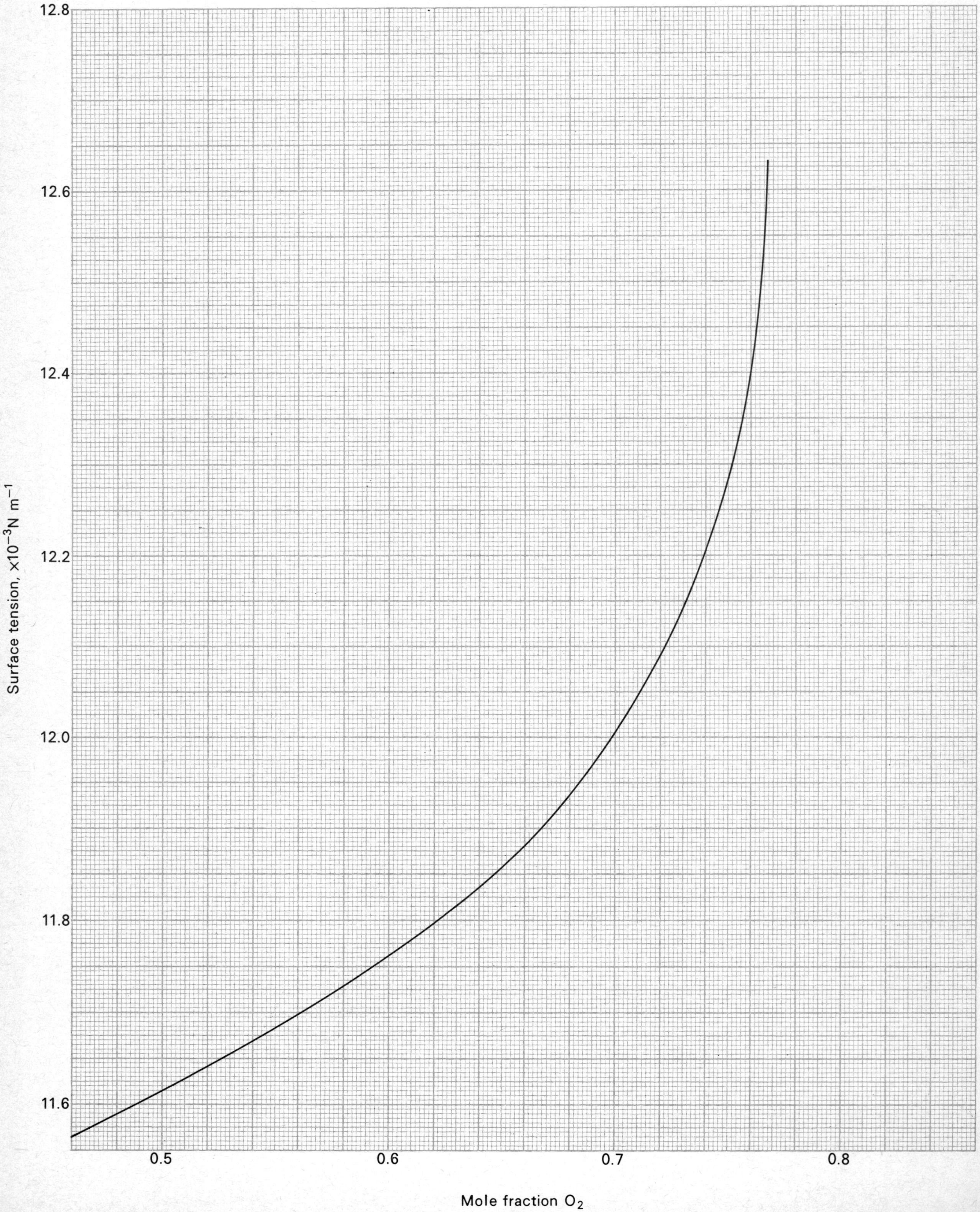

10.15 Solubility of Air in Water at $1.0133 \times 10^5 N\ m^{-2}$ Total Pressure (26,42)

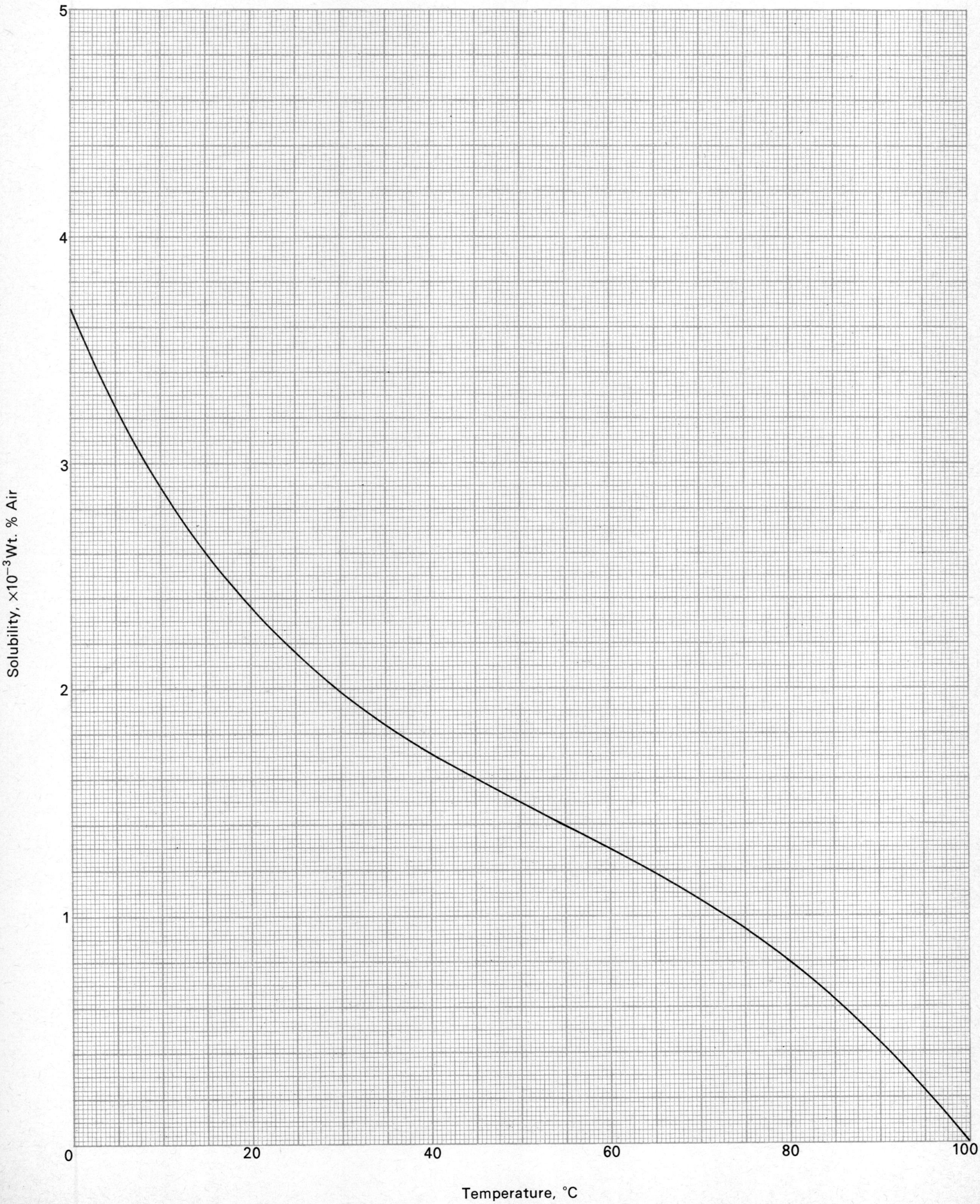

11 CARBON DIOXIDE

Carbon dioxide is a colourless and odourless gas at ambient temperature and atmospheric pressure. It occurs in the atmosphere with a concentration of between 0.02–0.04 vol. %. The gas is not poisonous and is non-flammable.

In the solid state it consists of large cubic crystals as clear as water. This solid gasifies without melting first.

Liquid carbon dioxide is obtained readily by compression of the gas below the critical temperature. Carbon dioxide has zero dipole moment, because the molecule is linear. Chemically it is not a very active compound.

Commercial carbon dioxide is distributed as a liquid in steel cylinders. The main user of carbon dioxide is the beverage industry. However, there are several atomic power stations using carbon dioxide as a coolant. Refrigerating cycles use it in the liquid state, while the solid is used in cold storage (dricold), instead of ice. Liquid carbon dioxide is also used under pressure as a fire extinguisher medium.

The physical, thermodynamic, and transport properties of solid, liquid, and gaseous carbon dioxide have been compiled by Vucalovich and Altunin (38). Most of the presented data were critically examined on the basis of the experimental techniques and other factors which might affect the final results. The methods of selection are discussed in full detail, and the reliability of the data is graded. Because of the high standards of this selection, Table 11.1 and Figs. 11.1–11.18 are based mainly on this excellent compilation. Further comprehensive works of interest on the physicochemical properties of carbon dioxide are: Canjar and Manning (4), Chen (5), Din (8), Gmelin (13), Hilsenrath *et al.* (17, 18), Kalafati and Rumshiskii (22), Landolt-Börnstein (25), Pascal (31), and Quinn and Jones (33).

References

1 American Conference of Government Industrial Hygienists, *Documentation of Threshold Limit Values*, revised edition, Committee on Threshold Limit Values, Cincinnati, Ohio 45202, 1966, 203 pp.

2 Angelino, G., Macchi, E., *Trans. Am. Soc. Mech. Eng., J. Eng. Power*, 1970 (July), pp. 301–9.

3 Borovik, E., *Zh. Eksp. Teor. Fiz.*, 1949, **19**, 561–4.

4 Canjar, N. J., Manning, F. S., *Thermodynamic Properties and Reduced Correlations of Gases*, Gulf Publ. Co., Houston, Texas, 1967.

5 Chen, L. H., *Thermodynamic and Transport Properties of Gaseous Carbon Dioxide*, 1st Symp. Thermophys. Properties, Am. Soc. Mech. Eng., New York, 1959, pp. 358–69.

6 Cheung, H., U.S. Atomic Energy Comm. Report UCRL-8230, Washington, 1958, 146 pp.

7 Cheung, H., LeBromley, R. A., Wilke, C. R., U.S. Atomic Energy Comm. Report UCRL-8230, Washington, 1959, 64 pp.

8 Din, F., Ed., *Thermodynamic Functions of Gases*, Vol. 1, Ammonia, Carbon Dioxide, Carbon Monoxide, Butterworths, London, 1956, 175 pp.

9 Eucken, A., Hauck, F., *Z. phys. Chem.*, 1928, **134**, 161–77.

10 Eucken, A., Sarstedt, B., *Z. phys. Chem.*, 1941, **50B**, 143–70.

11 Geier, H., Schafer, K., *Allgem. Wärmetech.*, 1961, **10**, 70–5.

12 Gibbs, R. E., Van Ness, H. C., *I&EC. Fundam.*, 1971, **10** (2), 312–15.

13 *Gmelins Handbuch der anorganischen Chemie*, Syst. No. 14, Kohlenstoff, Teil C, Lief. 1, Verlag Chemie GmbH, Weinheim, 1970, pp. 368–504.

14 Gupta, G. P., Saxena, S. C., *Molecular Phys.*, 1970, **19** (6), 871–880.

15 Heck, R. C. H., *Mech. Eng.*, 1940, **62**, 9–12.

16 Heck, R. C. H., *Mech. Eng.*, 1941, **63**, 126–35.

17 Hilsenrath, J., *et al.*, *Tables of Thermodynamic and Transport Properties of Air, Argon, Carbon Dioxide, Hydrogen, Nitrogen, Oxygen, Carbon Monoxide, Steam*, Pergamon Press, Oxford, 1960.

18 Hilsenrath, J., Beckett, C. W., Benedict, W. S., Fano, L., Hoge, H. J., Masi, J. F., Nuttal, R. L., Touloukian, Y. S., Woolley, H. W., *Tables of Thermal Properties of Gases*, Natl. Bur. Std. Circular 564, Washington, 1955, 488 pp.

19 Hilsenrath, J., Touloukian, Y. S., *Trans. Am. Soc. Mech. Eng.*, 1954, **76**, 967–85.

20 IUPAC, Atomic Weights of the Elements 1971, *Pure & Appl. Chem.*, 1972, **30** (3–4), 638–49.

21 Ibrahim, S. H., Kuloor, N. R., *Brit. Chem. Eng.*, 1961, **6** (12), 862–3.

22 Kalafati, D. D., Rumshiskii, L. Z., *Thermodynamic Properties of Gaseous Carbon Dioxide*, Izd. Mosk. Energ. Inst., Moskva, 1957.

23 Kestin, J., Ro, S. T., Wakeham, W. A., *J. Chem. Phys.*, 1972, **56** (8), 4114–18.

24 Koch, B., *Grundlagen des Wärmeaustausches*, Verlagsanstalt H. Beucke & Sohne, Berlin, 1950.

25 *Landolt-Börnstein Zahlenwerte und Funktionen aus Physik, Chemei, Astronomie, und Technik*, 6th ed., 4 Teil, Bandteil a, Technik, Springer-Verlag, Berlin, 1967, pp. 659–70.

26 Lax, E., Synowietz, C., Eds., *Taschenbuch für Chemiker und Physiker*, 3rd ed., Springer-Verlag, Berlin, 1967.

27 Leduc, A., *Chem. Rev.*, 1929, **6**, 1–16.

28 Li, Y-H., Tsui, T-F., *J. Geophys. Res.*, 1971, **76** (18), 4203–7.

29 Maitland, G. C., Smith, E. B., *J. Chem. Eng. Data*, 1972, **17** (2), 150–6.

30 Maryott, A. A., Smith, E. R., *Table of Dielectric Constants of Pure Liquids*, Natl. Bur. Std. Circular 514, Washington, Aug. 10, 1951, 44 pp.

31 Pascal, P., Ed., *Nouveau Traite de Chimie Minérale*, Vol. VIII, Carbone, Masson et Cie, Éditeurs, Paris, 1968, pp. 663–97.

32 Quinn, E. L., *J. Am. Chem. Soc.*, 1927, **49**, 2704–11.

33 Quinn, E. L., Jones, C. L., *Carbon Dioxide*, Am. Chem. Soc. Monogr. Ser. 72, Reinhold Publ. Corp., New York, 1936.

34 Rossini, F. D., Wagman, D. D., Evans, W. H., Levine, S., Jaffe, I., *Selected Values of Chemical Thermodynamic Properties*, Natl. Bur. Std. Circular 500, Washington, Feb. 1, 1952.

35 Rothman, A. J., Bromley, L. A., *Ind. Eng. Chem.*, 1955, **47**, 899–906.

36 Sengers, J. V., Keyes, P. H., *Phys. Rev. Letters*, 1971, **26** (2), 70–3.

37 Stull, D. R., Dir., *JANAF Thermochemical Tables*, Dow Chemical Co., Midland, Mich., 1965–8.

38 Vukalovich, M. P., Altunin, V. V., *Thermophysical Properties of Carbon Dioxide*, Collets (Publ.) Ltd., London, 1968, 463 pp.

Table 11.1 Properties of Carbon Dioxide

		Ref.
Chemical formula	CO_2	(26)
Molecular weight	44.0090	(20)
Critical temperature	31.0°C	(38)
Critical pressure	73.77×10^5 N m^{-2}	(38)
Critical density	468.0 kg m^{-3}	(38)
Triple-point temperature	−56.59°C	(38)
Triple-point pressure	5.1810×10^5 N m^{-2}	(38)
Sublimation temperature at 1.0133×10^5 N m^{-2}	−78.50°C	(38)
Melting point at 5.3705×10^5 N m^{-2}	−56.20°C	(34)
Transition temperature	None	(34)
Density of solid at −90°C	1530.0 kg m^{-3}	(33)
Compressibility of liquid at 0°C	4.24×10^{-9} N^{-1} m^2	(33)
Coefficient of thermal expansion of liquid at −90°C	1.28×10^{-3} C^{-1}	(26)
Heat of formation of gas at 25°C	-3.9378×10^8 J kg-mol^{-1}	(37)
Free energy of formation of gas at 25°C	-3.9465×10^8 J kg-mol^{-1}	(37)
Heat of sublimation at −78.48°C	2.5250×10^7 J kg-mol^{-1}	(34)
Heat of fusion	8.3317×10^6 J kg-mol^{-1}	(34)
Heat of transition	None	(38)
Entropy of gas at 25°C	2.1382×10^5 J kg-mol^{-1} C^{-1}	(37)
Heat capacity ratio of gas at 15°C and 1.0133×10^5 N m^{-2}	1.304	(27)
Velocity of sound at 0°C and 1.0133×10^5 N m^{-2}	258.0 m s^{-1}	(5)
Diffusion in air at 0°C and 1.0133×10^5 N m^{-2}	1.36×10^{-5} m^2 s^{-1}	(14, 21)
Dielectric constant of liquid at 20°C	1.60	(30)
Refractive index of gas at 0°C and 1.0133×10^5 N m^{-2}	1.0004505	(26)
Threshold limit value	(5000 ppm) 9×10^{-3} kg m^{-3}	(1)

11.1 Vapour Pressure of Carbon Dioxide (38)

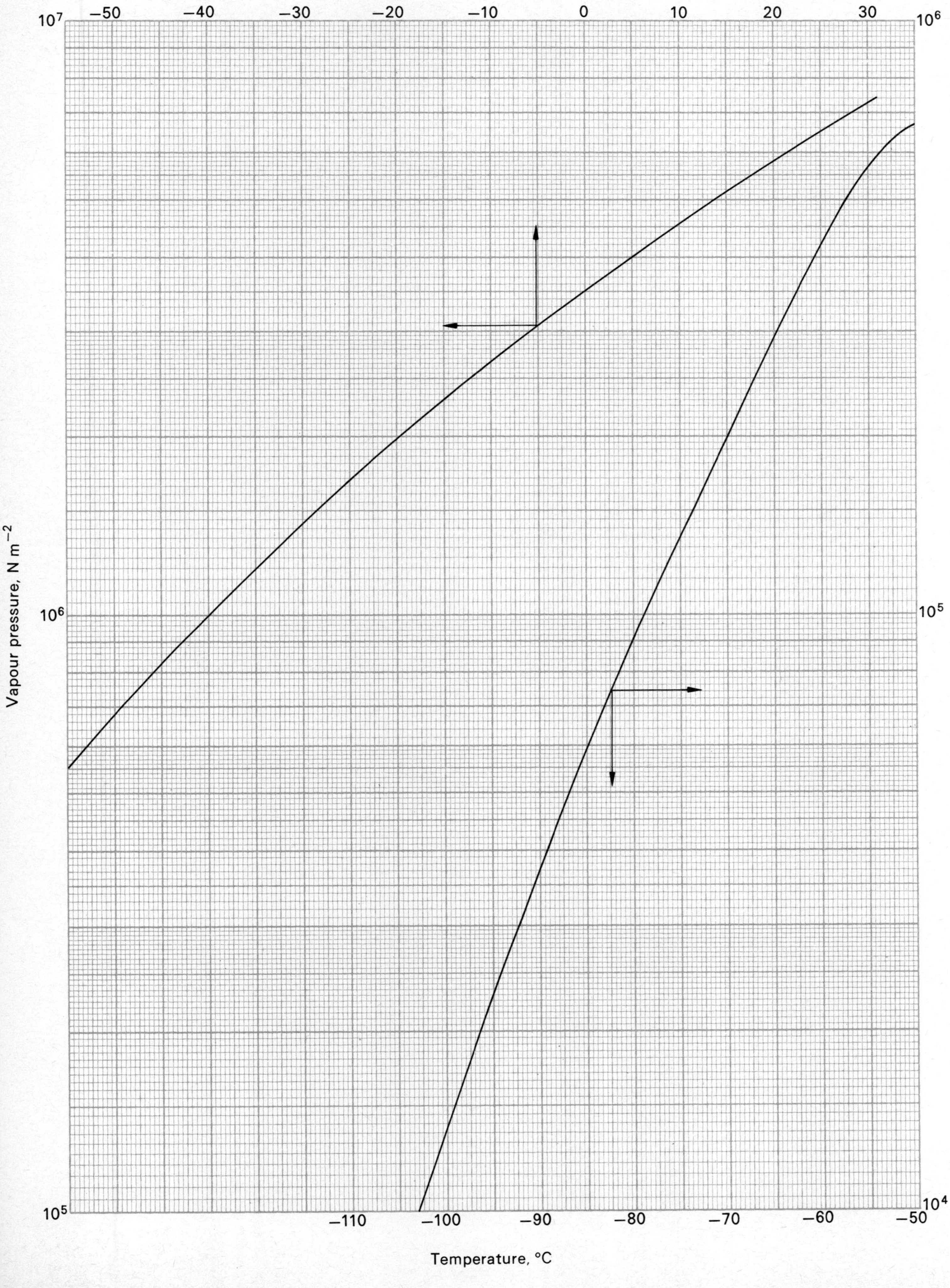

11.2 Orthobaric Density of Carbon Dioxide (2,26,38)

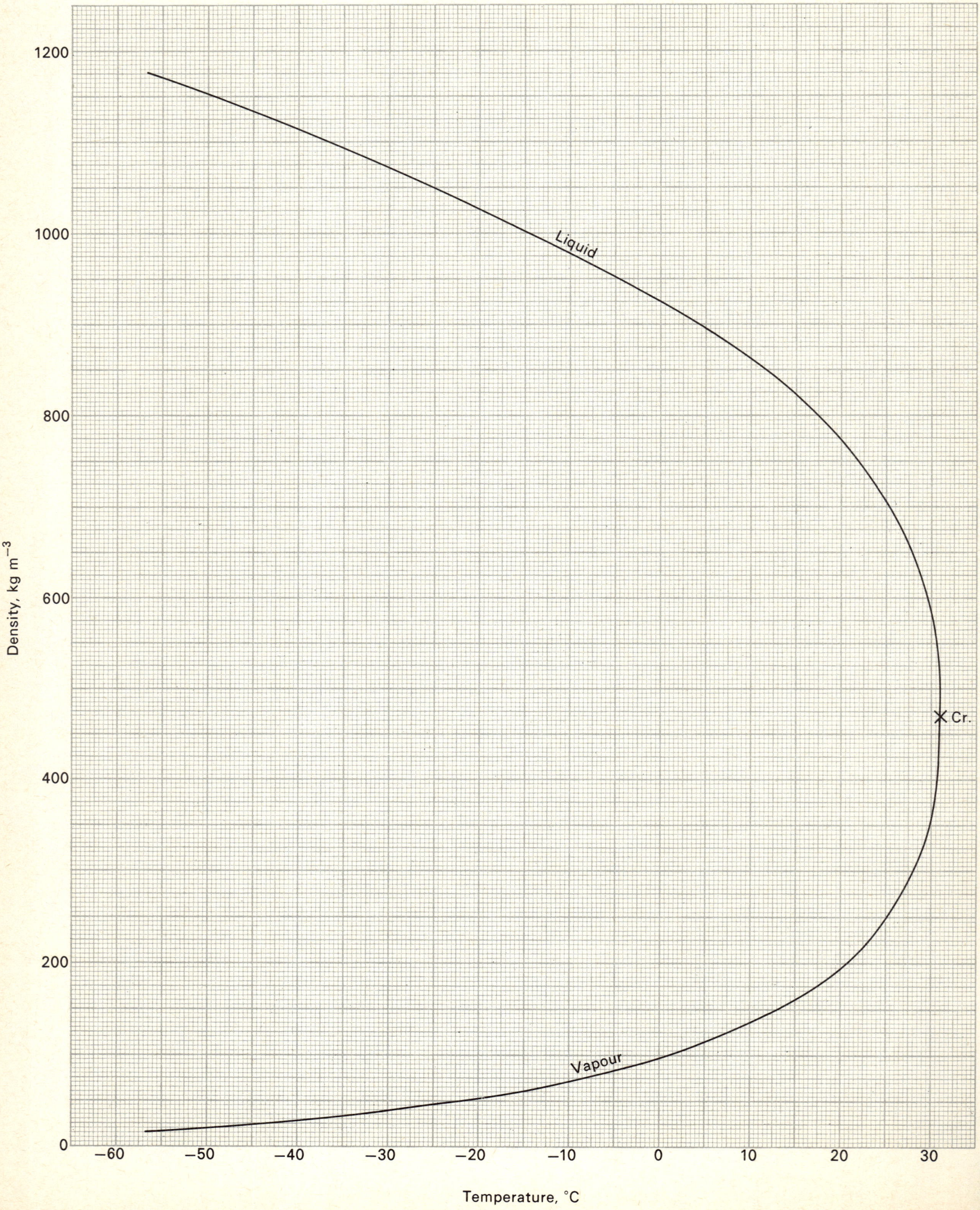

11.3 Density of Liquid Carbon Dioxide (38)

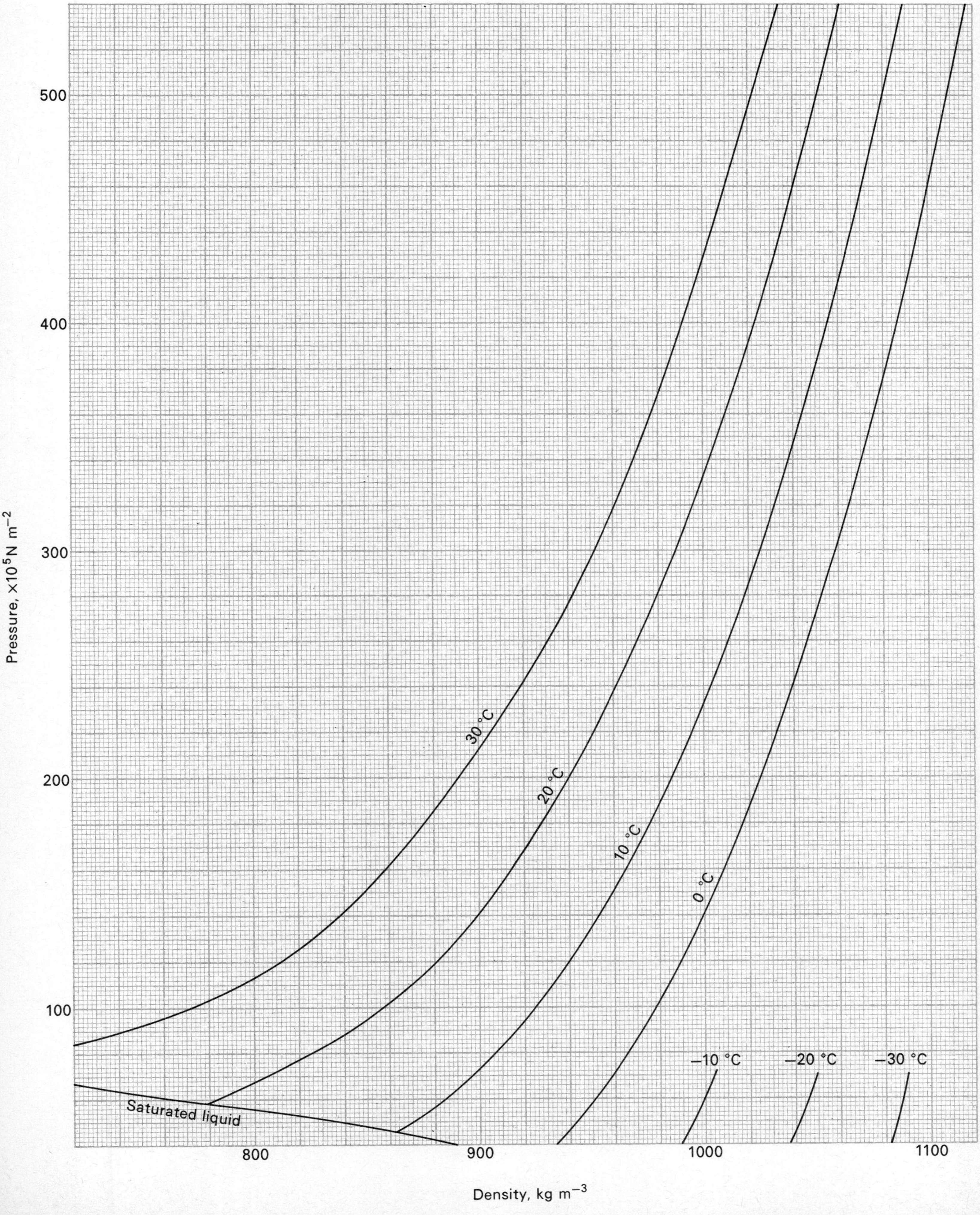

11.4 Supercritical Density of Carbon Dioxide (2,5,38)

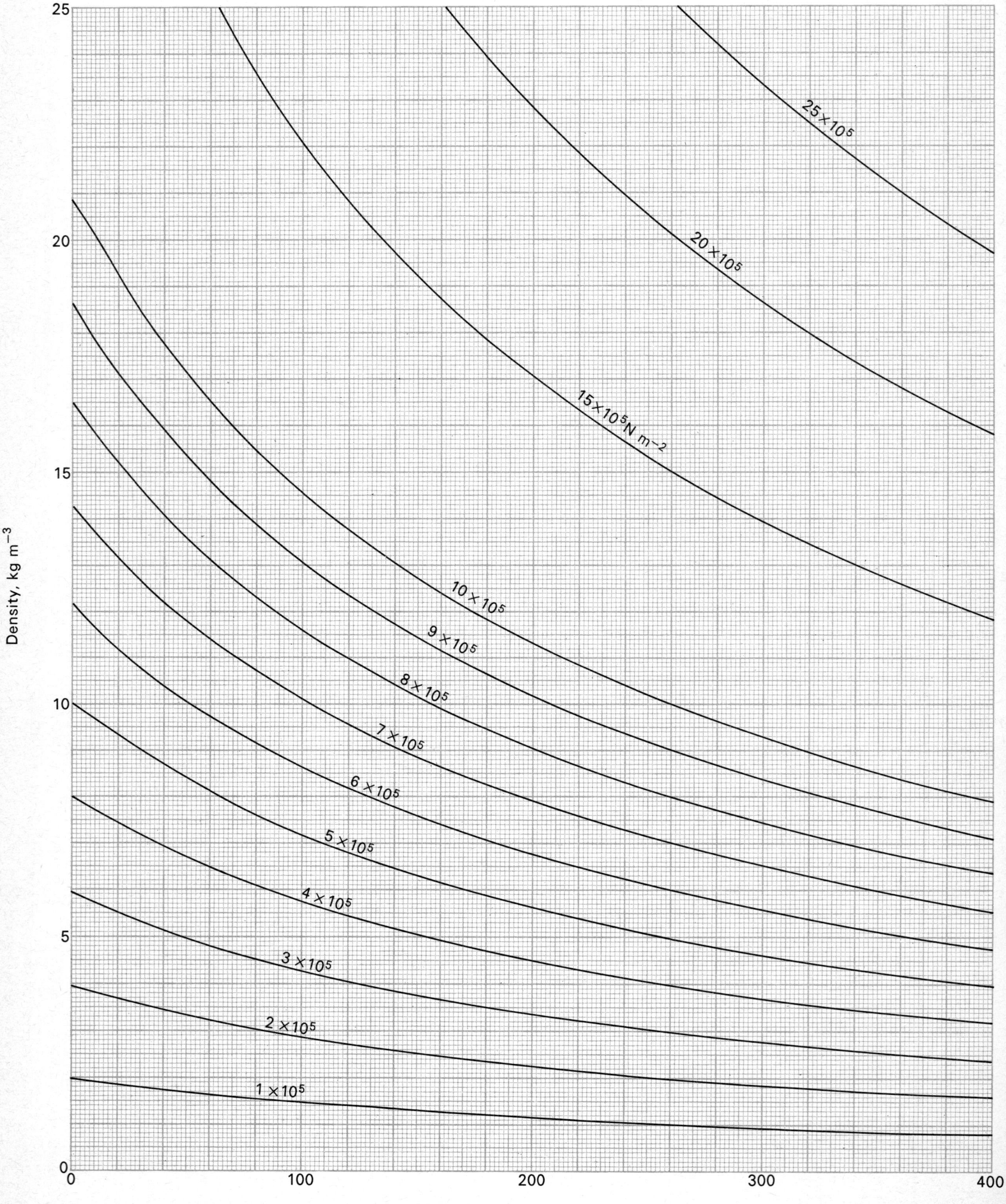

11.5 Latent Heat of Vaporisation of Carbon Dioxide (38)

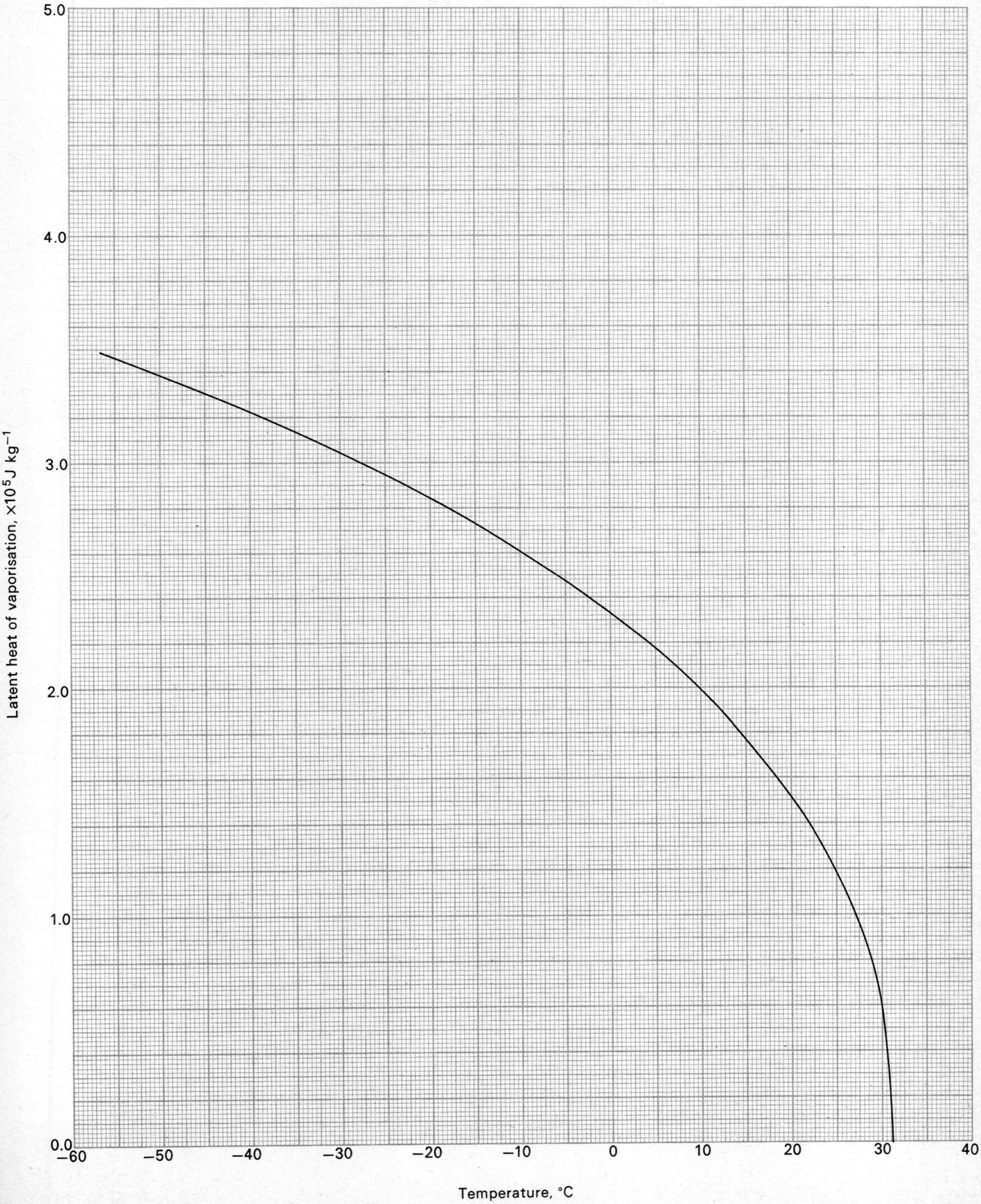

11.6 Heat Capacity of Liquid Carbon Dioxide (9,24,38)

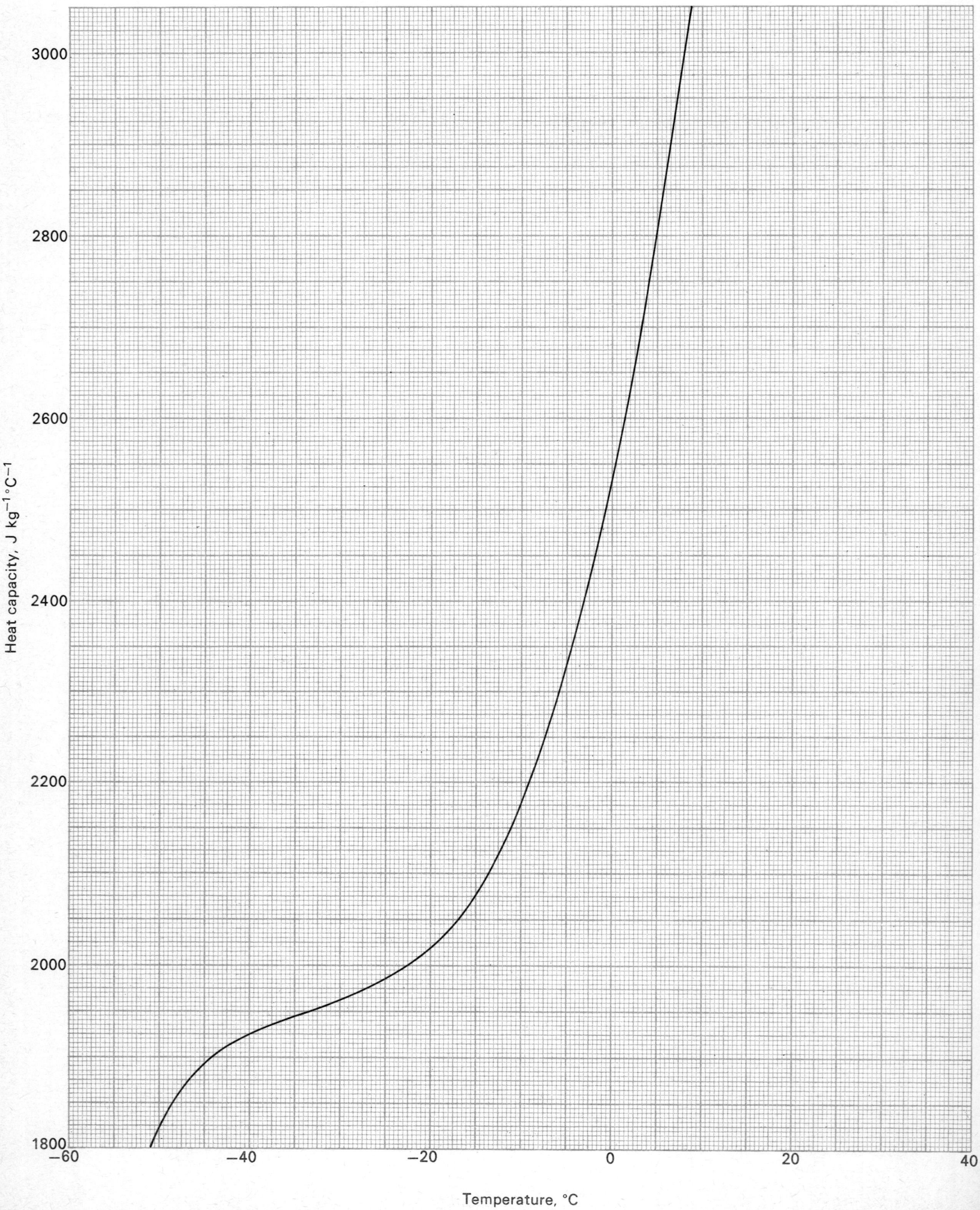

11.7 Heat Capacity of Liquid Carbon Dioxide (38)

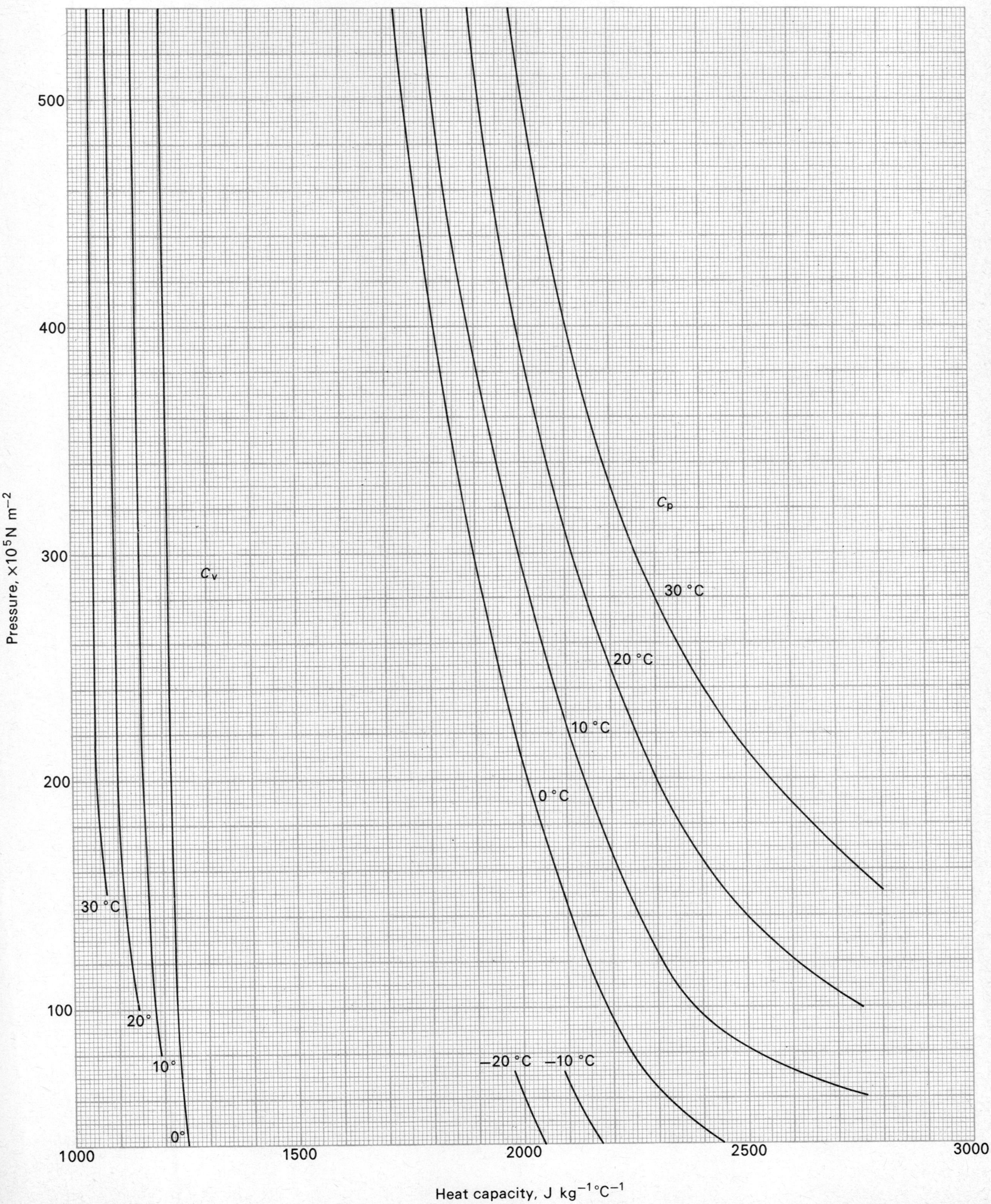

11.8 Heat Capacity of Gaseous Carbon Dioxide at $1.0133 \times 10^5 N\ m^{-2}$(10,15,16,18,37)

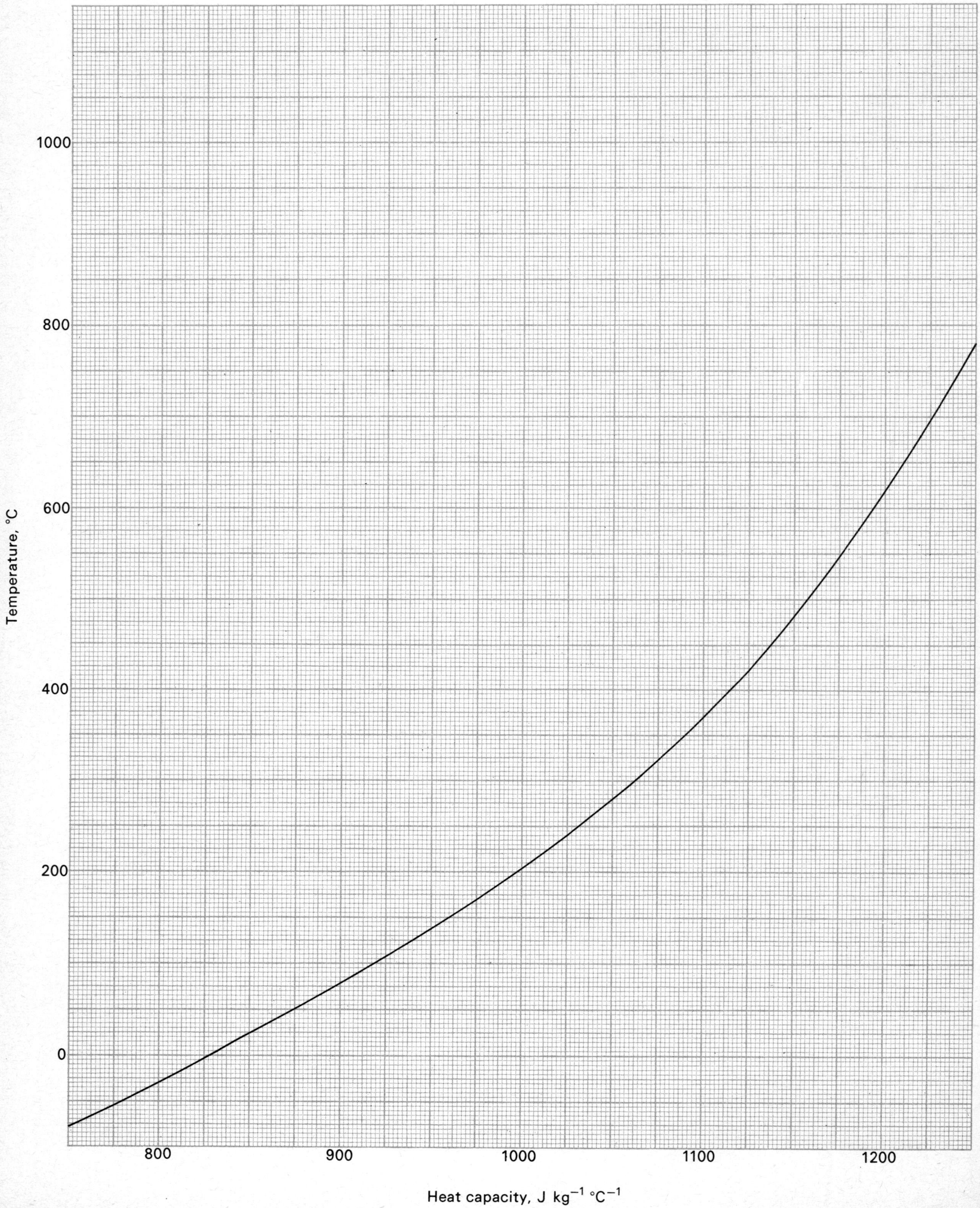

11.9 Heat Capacity of Gaseous Carbon Dioxide (38)

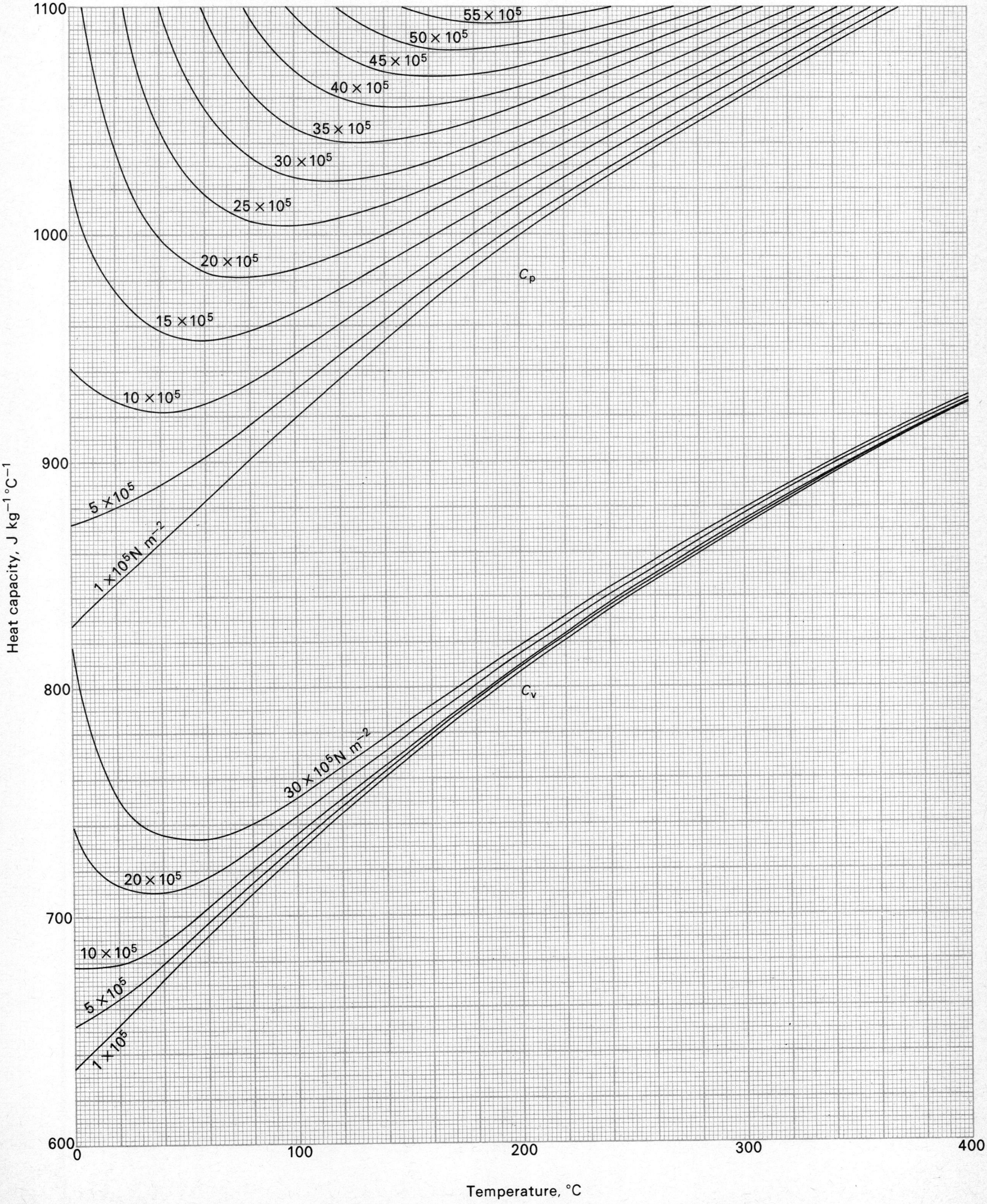

11.10 Heat Capacity Ratio (C_p/C_v) of Gaseous Carbon Dioxide at 1.0133×10^5 N m^{-2} (5,38)

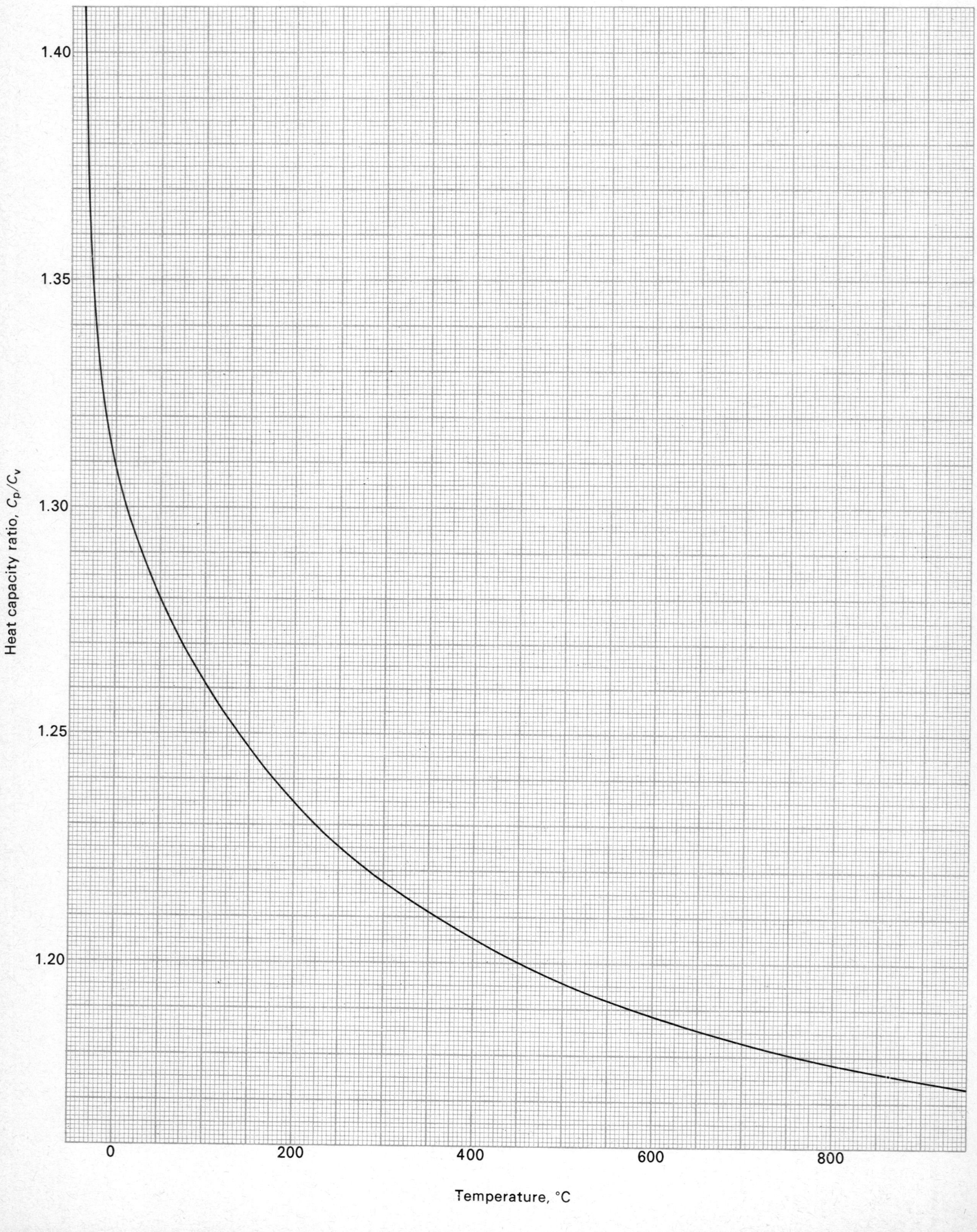

11.11 Viscosity of Saturated Carbon Dioxide (38)

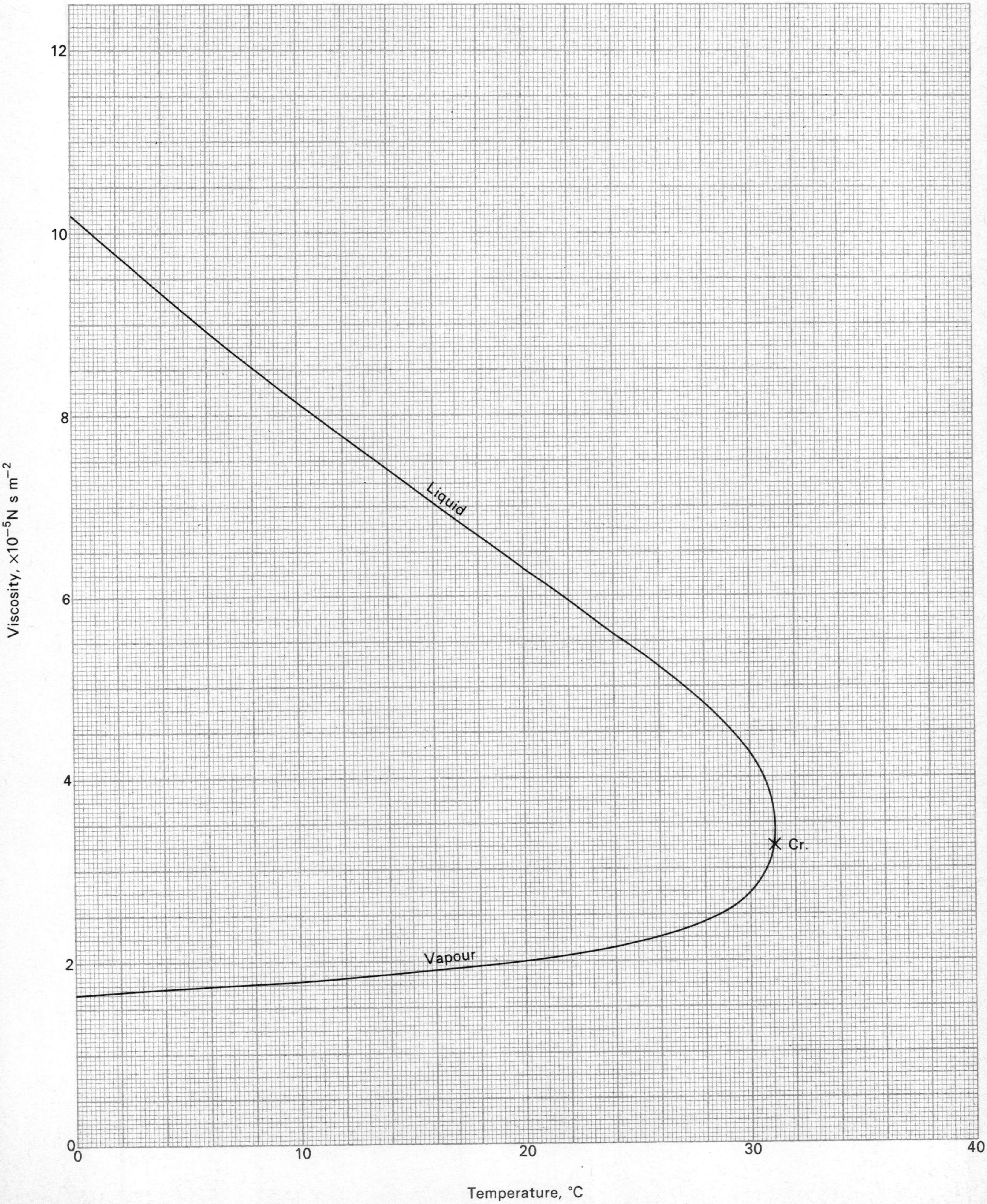

11.12 Viscosity of Gaseous Carbon Dioxide at $1.0133 \times 10^5 N\ m^{-2}$(19,23,29)

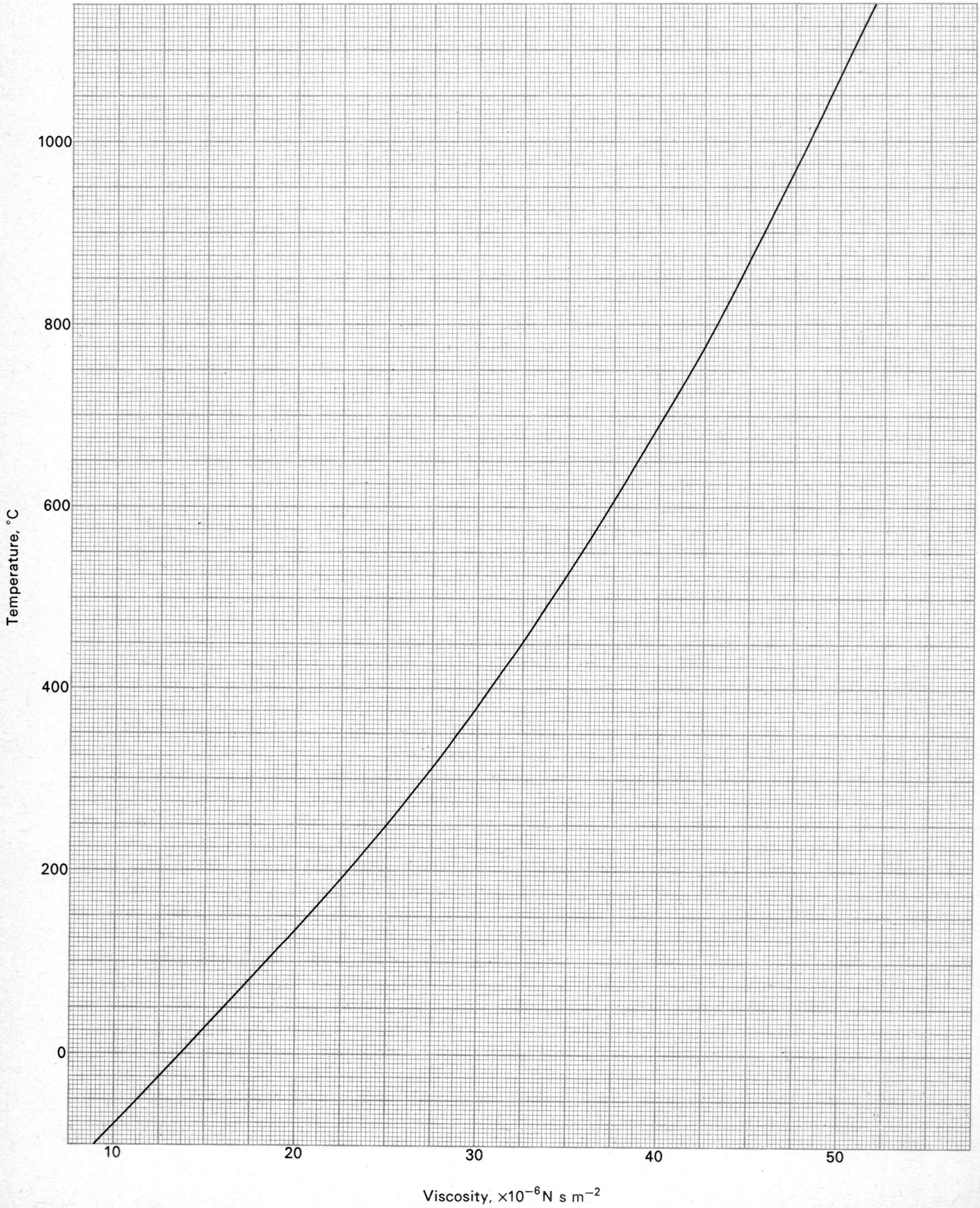

11.13 Viscosity of Gaseous Carbon Dioxide (5,38)

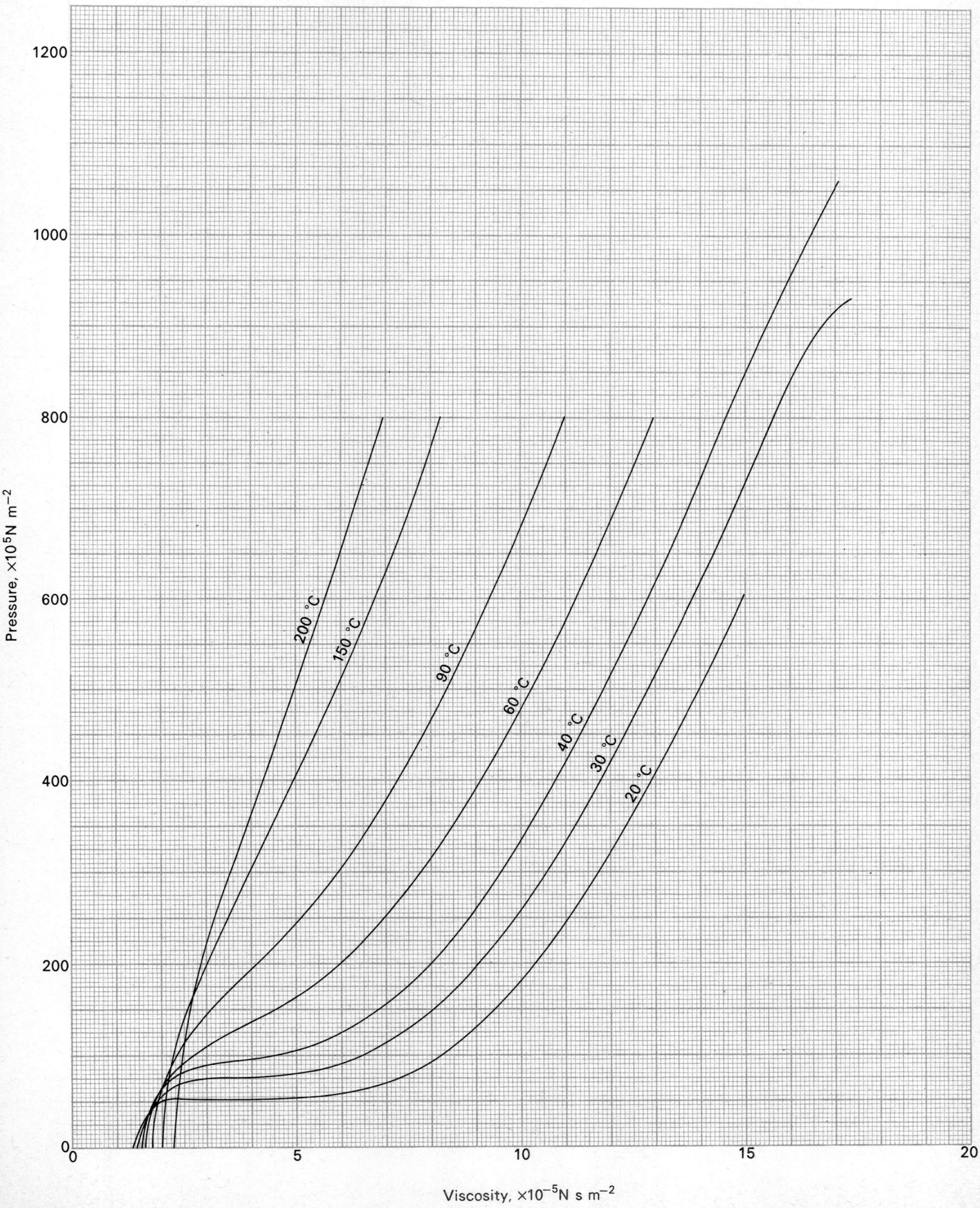

11.14 Thermal Conductivity of Liquid Carbon Dioxide (3)

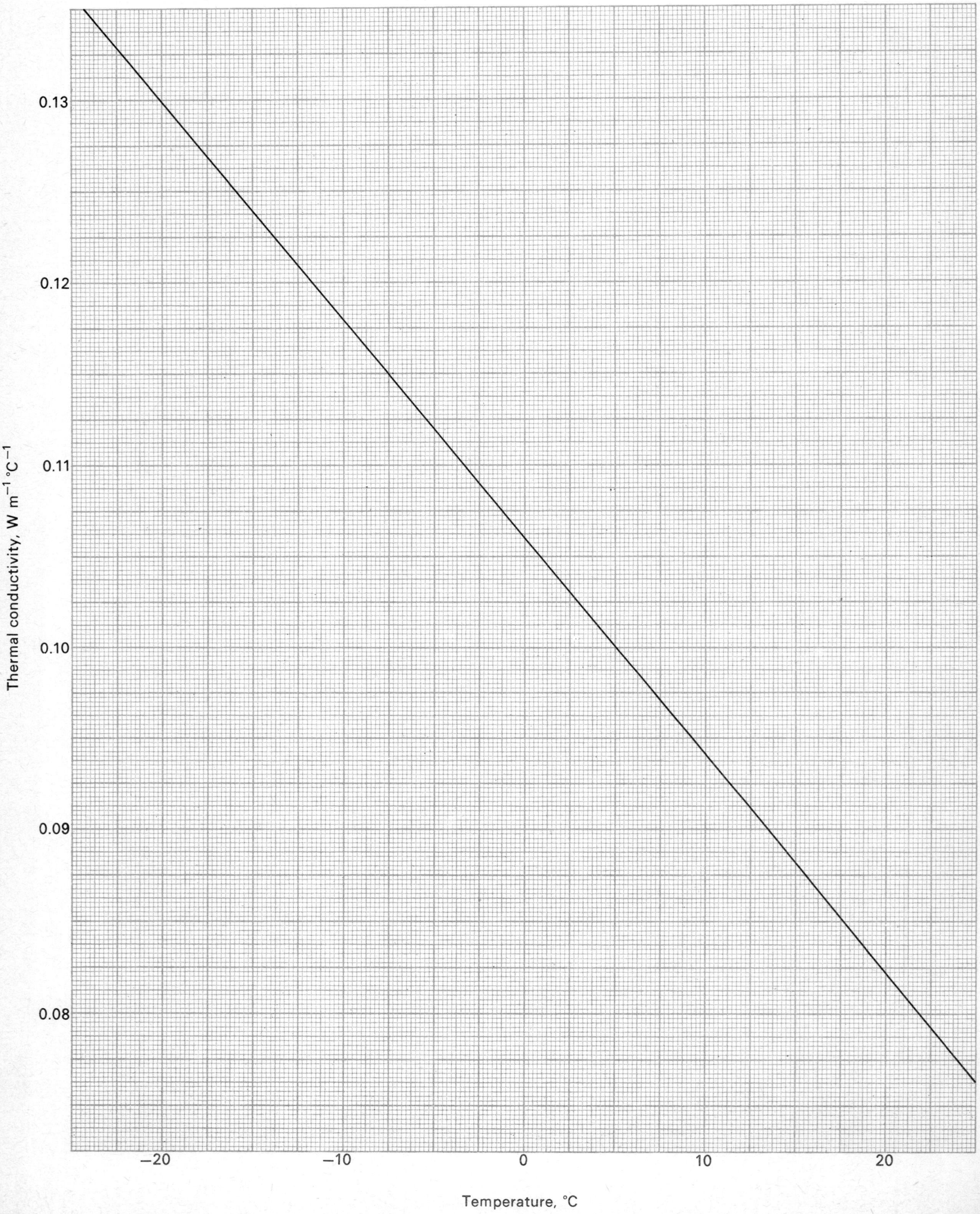

11.15 Thermal Conductivity of Gaseous Carbon Dioxide at $1.0133 \times 10^5 N\ m^{-2}$ (6,7,11,19,35)

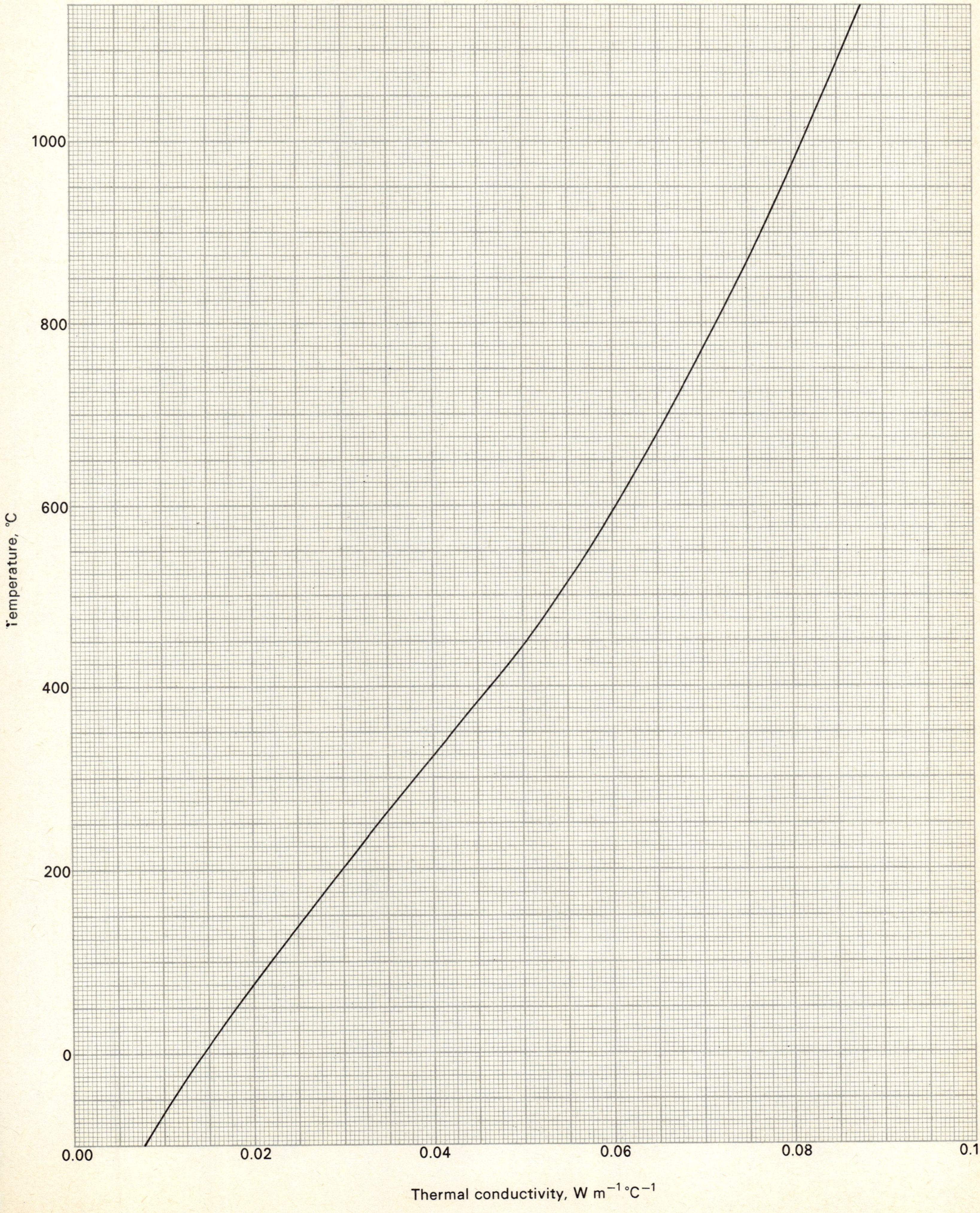

11.16 Thermal Conductivity of Gaseous Carbon Dioxide (5,14,36,38)

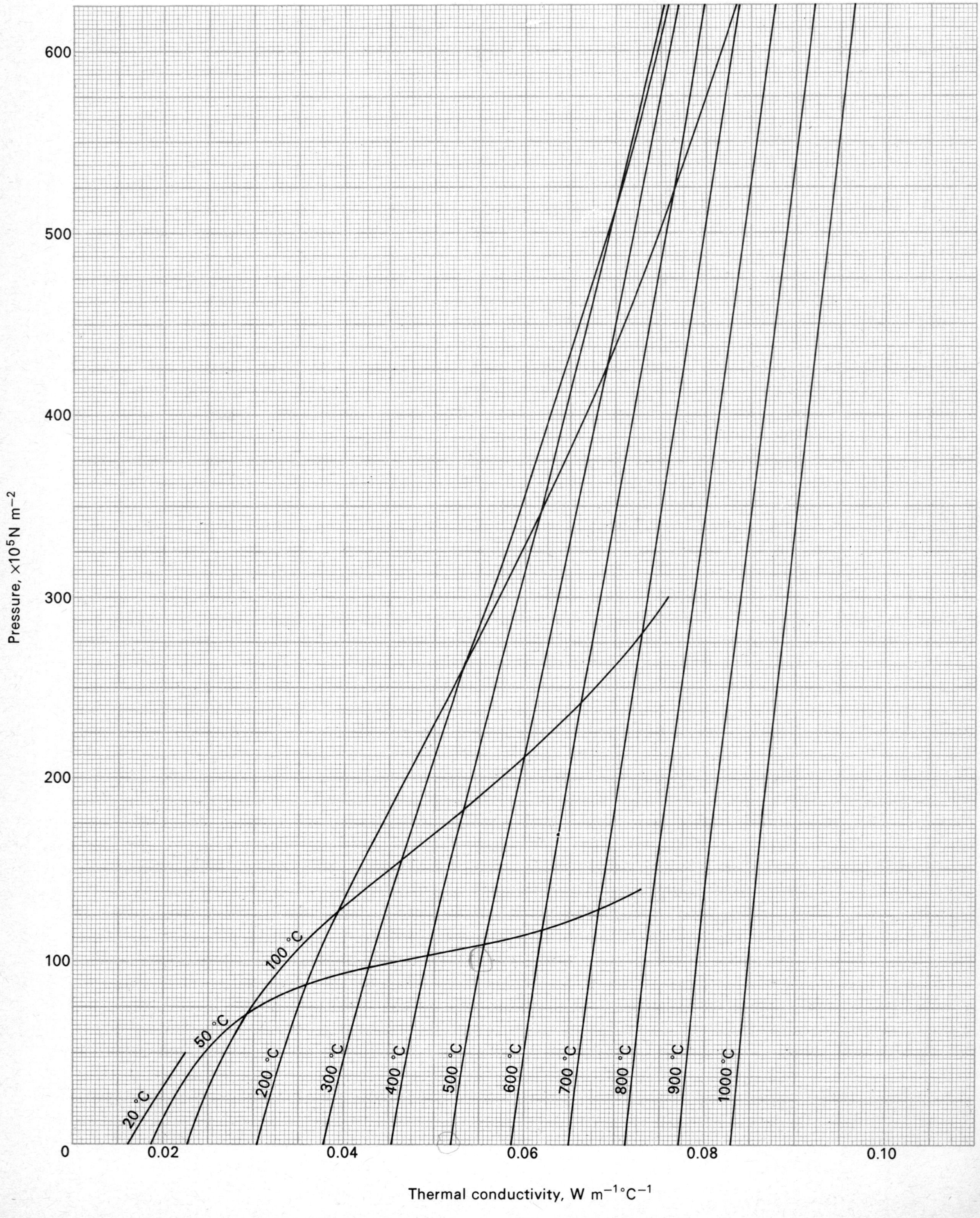

11.17 Surface Tension of Carbon Dioxide (32,33)

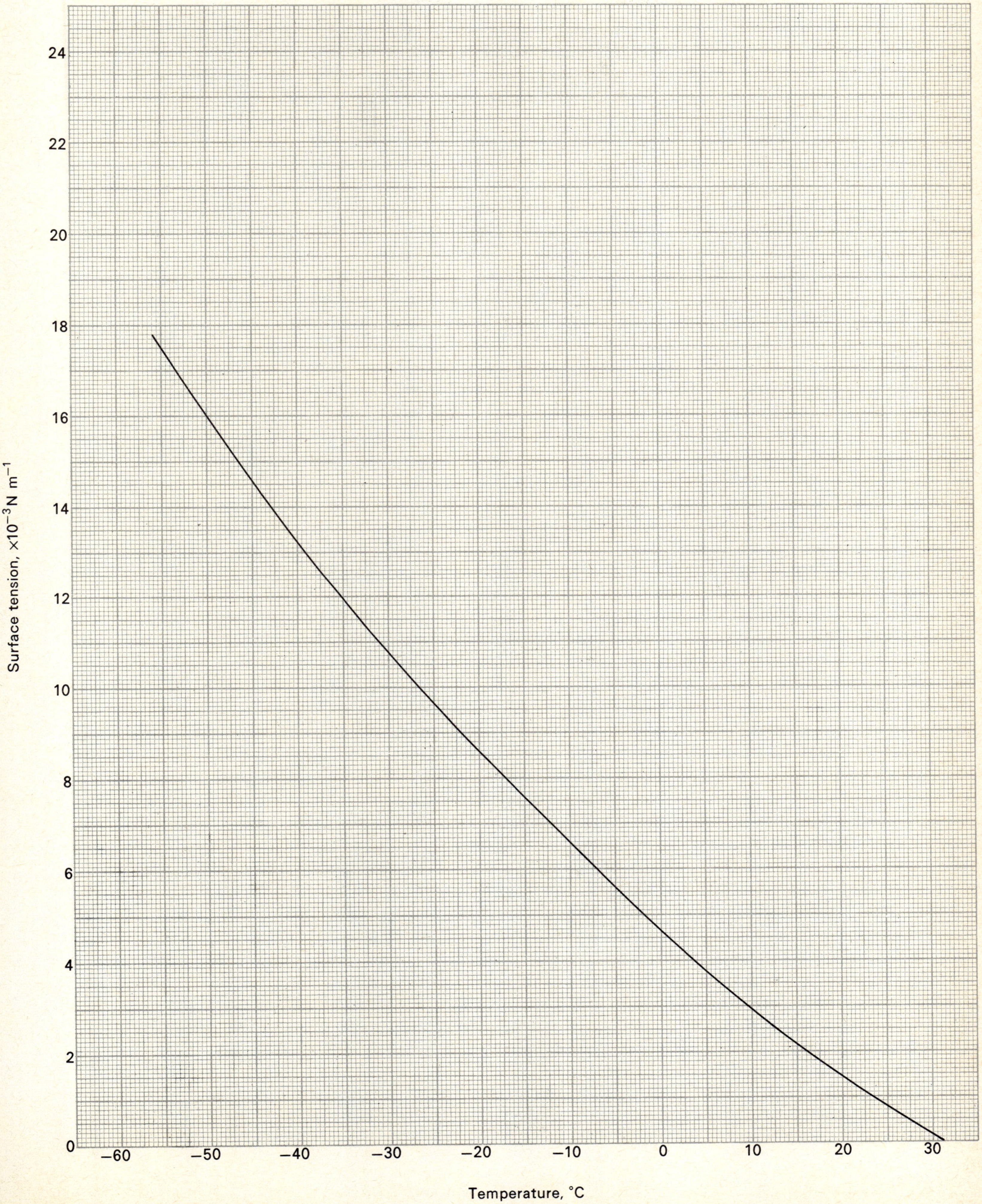

11.18 Solubility of Carbon Dioxide in Water at $1.0133 \times 10^5 N\ m^{-2}$ Total Pressure (12,26,28)

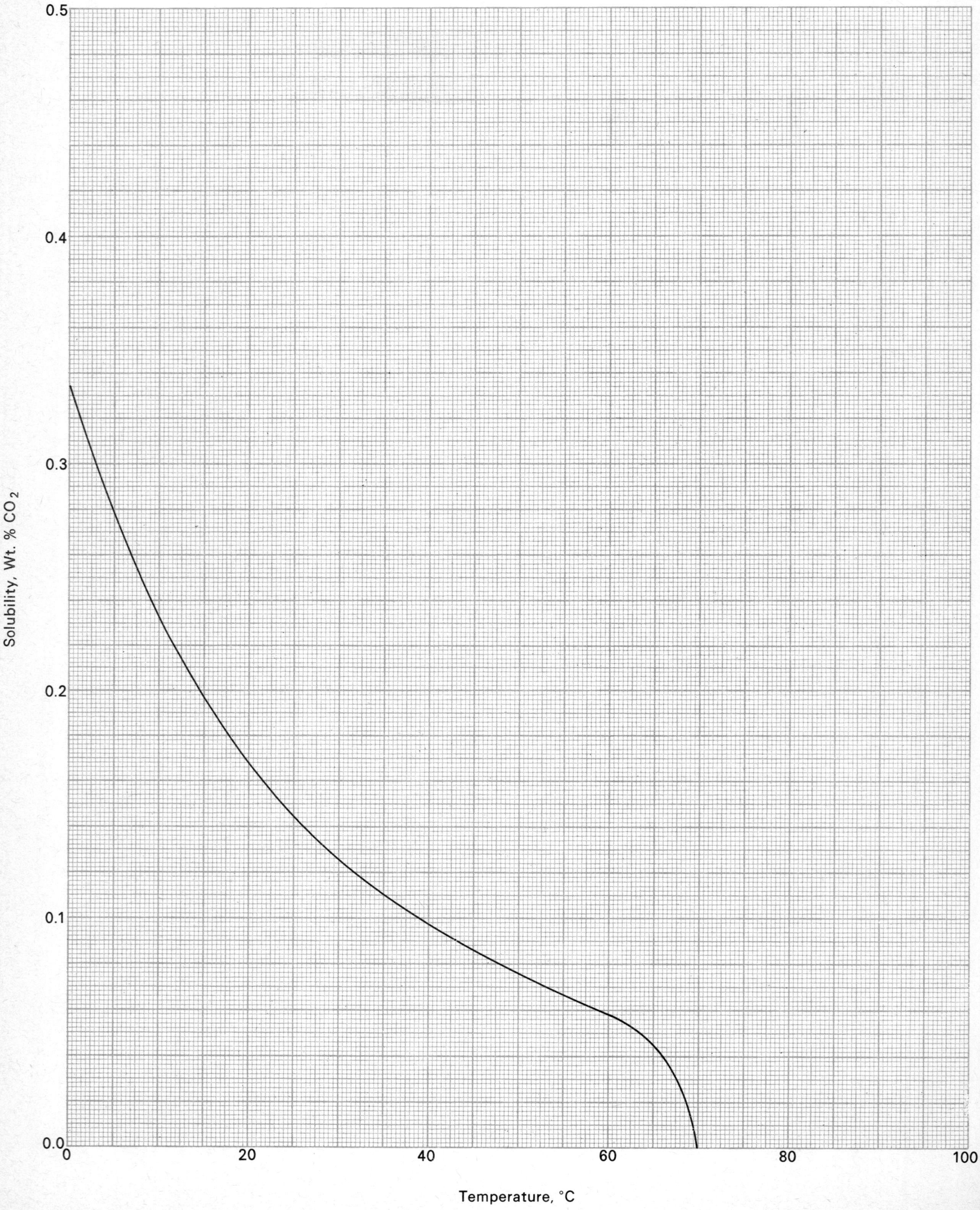

12 CARBON MONOXIDE

Carbon monoxide is a colourless, odourless, and very poisonous gas. It is quite stable in the cold, and used as a reducing agent at moderate temperatures. It reacts with many metals, forming carbonyls.

Solid carbon monoxide has one transition point. Below this temperature the solid forms cubic crystals (alpha–carbon monoxide), while above the transition temperature the hexagonal modification forms (beta–carbon monoxide), whose crystalline structure shows a more dense packing.

In the liquid state, it is also colourless. The gas is only slightly soluble in water. The molecule shows a slight dipole moment and the observed C—C bond distance is 1.13 Å.

The physical, thermodynamic, and transport properties of carbon monoxide in the solid, liquid, and gaseous phases have been extensively investigated. Several surveys provide a good selection of the critically examined data. The thermodynamic values for gaseous and liquid carbon monoxide from 70 to 300 K with pressures to 300 atmospheres have been reported by Hurst and Stewart (20). Further compilations can be found in: Gmelin (15), Hilsenrath *et al.* (18), Pascal (31), and Din (see under CO_2, ref. 8).

The presented data in Table 12.1 and Figs. 12.1–12.13 are based on selected experimental measurements; however, some properties have been correlated from several sources.

References

1 American Conference of Government Industrial Hygienists, *Documentation of Threshold Limit Values*, revised edition, Committee on Threshold Limit Values, Cincinnati, Ohio 45202, 1966, 203 pp.

2 Barua, A. K., Afzal, M., Flynn, G. P., Ross, T., *J. Chem. Phys.*, 1964, **41** (2), 374–8.

3 Boon, J. P., Legros, J. C., Thomaes, G., *Physica*, 1967, **33**, 547–57.

4 Borovik, E., Matveev, A., Panina, E., *J. Tech. Phys. USSR*, 1940, **10**, 988–98.

5 Clayton, J. O., Giauque, W. F., *J. Am. Chem. Soc.*, 1932, **54**, 2610–26.

6 Clusius, K., *Z. phys. Chem. (Leipzig)*, 1929, **3B**, 41–79.

7 Cuthbertson, C., Cuthbertson, M., *Proc. Roy. Soc.*, 1920, **97A**, 152–9.

8 Davis, B. W., Rice, O. K., *J. Chem. Phys.*, 1967, **47**, 5043–53.

9 Engineering Science Data Units, *Specific Heats and Their Ratio as Functions of Temperature for Several Common Gases*, Royal Aeronautical Society Data Sheet 00.01.08, London, 1965, 4 pp.

10 Engineering Science Data Units, *Approximate Data on the Viscosity of Some Common Liquids*, Item No. 66024, London, July 1966, 26 pp.

11 Eucken A., *Phys. Zeitschr.*, 1913, **14**, 324–32.

12 Gambil, W. R., *Chem. Eng.*, 1959, **66** (14), 157–60.

13 Geier, H., Schafer, K., *Allgem. Wärmetech.*, 1961, **10**, 70–5.

14 Gerf, S. F., Galkov, G. I., *Zh. Tech. Fiz.*, 1941, **11**, 801–8.

15 *Gmelins Handbuch der anorganischen Chemie*, Syst. No. 14, Kohlenstoff, Teil C, Lief. 1, Verlag Chemie GmbH, Weinheim, 1970, pp. 162–260.

16 Golubev, I. F., *Viscosity of Gases and Gas Mixtures*, Gosudarstvennoe Izdatel'stvo Fiziko-Matematicheskoi Literatury, Moskva, 1959, Table 4.

17 Greene, E. F., Toennies, J. P., *Chemical Reactions in Shock Waves*, Edward Arnold, London, 1964, pp. 333–4.

18 Hilsenrath, J., Beckett, C. W., Benedict, W. S., Fano, L., Hoge, H. J., Masi, J. F., Nuttal, R. L., Touloukian, Y. S., Woolley, H. W., Natl. Bur. Std. Circular 564, 1955, 488 pp.

19 Honig, R. E., Hook, H. O., Vapour Pressure Data for some Common Gases, *RCA Rev.*, 1960, **21**, 360–8.

20 Hust, J. G., Stewart, R. B., *Thermodynamic Property Values for Gaseous and Liquid Carbon Monoxide from 70 to 300 K with Pressures to 300 Atmospheres*, Natl. Bur. Std., Techn. Note 202, Washington, Nov. 30, 1963, 109 pp.

21 IUPAC, Atomic Weights of the Elements 1971, *Pure & Appl. Chem.*, 1972, **30** (3–4), 638–49.

22 Johnston, H. L., Grilly, E. R., *J. Chem. Phys.*, 1946, **14**, 233–8.

23 Kordes, E., *Z. Elektrochem.*, 1954, **58**, 424–31.

24 Kudashev, V. I., Serdyuk, L. S., *High Temperature*, 1969, **7** (5), 841–7.

25 Lax, E., Synowietz, C., Eds., *Taschenbuch für Chemiker und Physiker*, 3rd ed., Springer-Verlag, Berlin, 1967.

26 Leduc, A., *Chem. Rev.*, 1929, **6**, 1–16.

27 Masi, J. F., *Trans. Am. Soc. Mech. Eng.*, 1954, **76**, 1067–74.

28 Michels, A., Lupton, J. M., Wassenaar, T., de Graaff, W., *Physica*, 1952, **18**, 121.

29 Mullins, J. C., Kirk, B. S., Ziegler, W. T., NP-13862, 1963, **3**, 70–3.

30 O'Brien, H. R., Parker, W. L., *J. Biol. Chem.*, 1922, **50**, 289–300.

31 Pascal, P., Ed., *Nouveau Traité de Chimie Minérale*, Vol. VIII, Carbone, Masson et Cie, Éditeurs, Paris, 1968, pp. 550–66.

32 Ree, T. S., Ree, T., Eyring, H., *J. Chem. Phys.*, 1964, **41**, 524–30.

33 Riedel, L., *Kältetechnik*, 1957, **9**, 127–34.

34 Rossini, F. D., Wagman, D. D., Evans, W. H., Levine, S., Jaffe, I., *Selected Values of Chemical Thermodynamic Properties*, Natl. Bur. Std. Circular 500, Washington, Feb. 1, 1952.

35 Spangenberg, C. D., *Chem. Eng.*, 1951, **58** (9), 170–1.

36 Sprow, F. B., Prausnitz, J. M., *Trans. Faraday Soc.*, 1966, **62**, 1097–1104.

37 Stull, D. R., Dir., *JANAF Thermochemical Tables*, Dow Chemical Co., Midland, Mich., 1965–8.

38 Vegard, L., *Z. Physik*, 1934, **88**, 235–41.

39 Watson, H. E., Rao, G. G., Ramaswamy, K. L., *Proc. Roy. Soc.*, 1920, **143A**, 558–88.

40 Willis, W. L., *Cryogenics*, 1966, **6** (5), 279–84.

41 Winkler, L. W., *Ber.*, 1901, **34**, 1409–21.

42 Yosim, S. J., *J. Chem. Phys.*, 1964, **40** (10), 3069–75.

Table 12.1 Properties of Carbon Monoxide

		Ref.
Chemical formula	CO	(25)
Molecular weight	28.0100	(21)
Critical temperature	−140.23 °C	(8, 12, 33)
Critical pressure	3.499×10^6 N m^{-2}	(8, 12, 33)
Critical density	301.0 kg m^{-3}	(8, 12, 33)
Triple-point temperature	−205.07 °C	(25)
Triple-point pressure	0.1537×10^5 N m^{-2}	(25)
Boiling point at 1.0133×10^5 N m^{-2}	−191.50 °C	(34)
Melting point	−199.00 °C	(34)
Transition temperature	−211.63 °C	(34)
Density of solid (Hexagonal) at −208.15 °C (Cubic) at −253.15 °C	929.0 kg m^{-3} 1028.8 kg m^{-3}	(38) (38)
Coefficient of thermal expansion of liquid at boiling point	5.27×10^{-3} $°C^{-1}$	(42)
Heat of formation of gas at 25 °C	-1.1062×10^8 J kg-mol^{-1}	(37)
Free energy of formation of gas at 25 °C	-1.3737×10^8 J kg-mol^{-1}	(37)
Heat of sublimation at triple-point temperature (Cubic)	7.366×10^6 J kg-mol^{-1}	(29)
Heat of fusion	8.3740×10^5 J kg-mol^{-1}	(34)
Heat of transition	6.3211×10^5 J kg-mol^{-1}	(19)
Entropy of gas at 25 °C	1.9766×10^5 J kg-mol^{-1} $°C^{-1}$	(37)
Heat capacity ratio of gas at 25 °C and 1.0133×10^5 N m^{-2}	1.404	(26)
Velocity of sound at 15 °C and 1.0133×10^5 N m^{-2}	353.6 m s^{-1}	(17)
Diffusion in air at 80 °C and 1.0133×10^5 N m^{-2}	0.20×10^{-4} m^2 s^{-1}	(35)
Dielectric constant of gas at 25 °C and 1.0133×10^5 N m^{-2}	1.000634	(39)
Electrical conductivity of liquid at −188.1 °C	9.43×10^{-19} ohm^{-1} m^{-1}	(40)
Refractive index of gas at 0 °C and 5461 Å	1.0003364	(7)
Threshold limit value	(50 ppm) 5.5×10^{-5} kg m^{-3}	(1)

12.1 Vapour Pressure of Carbon Monoxide (20)

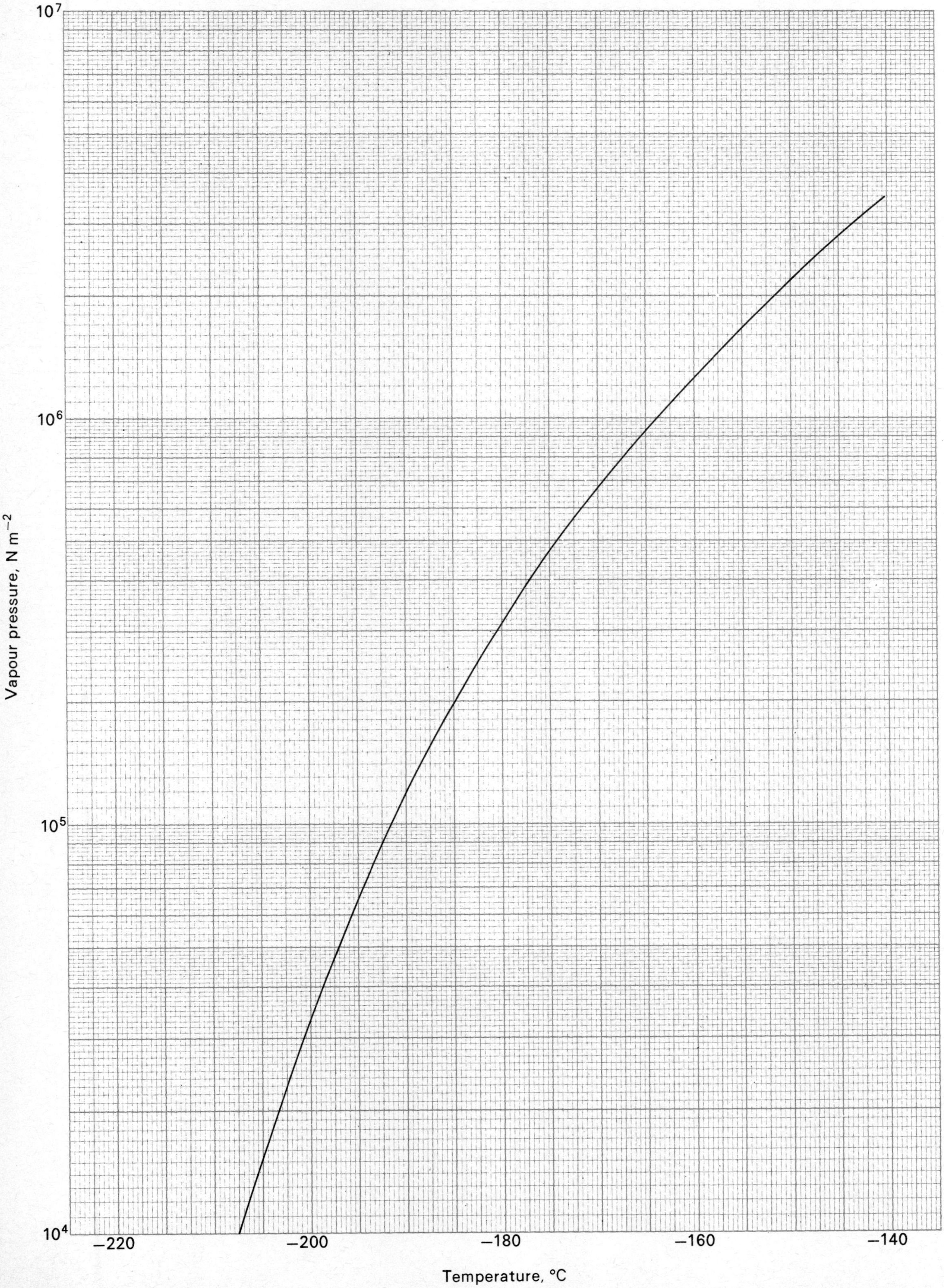

12.2 Orthobaric Density of Carbon Monoxide (20,25)

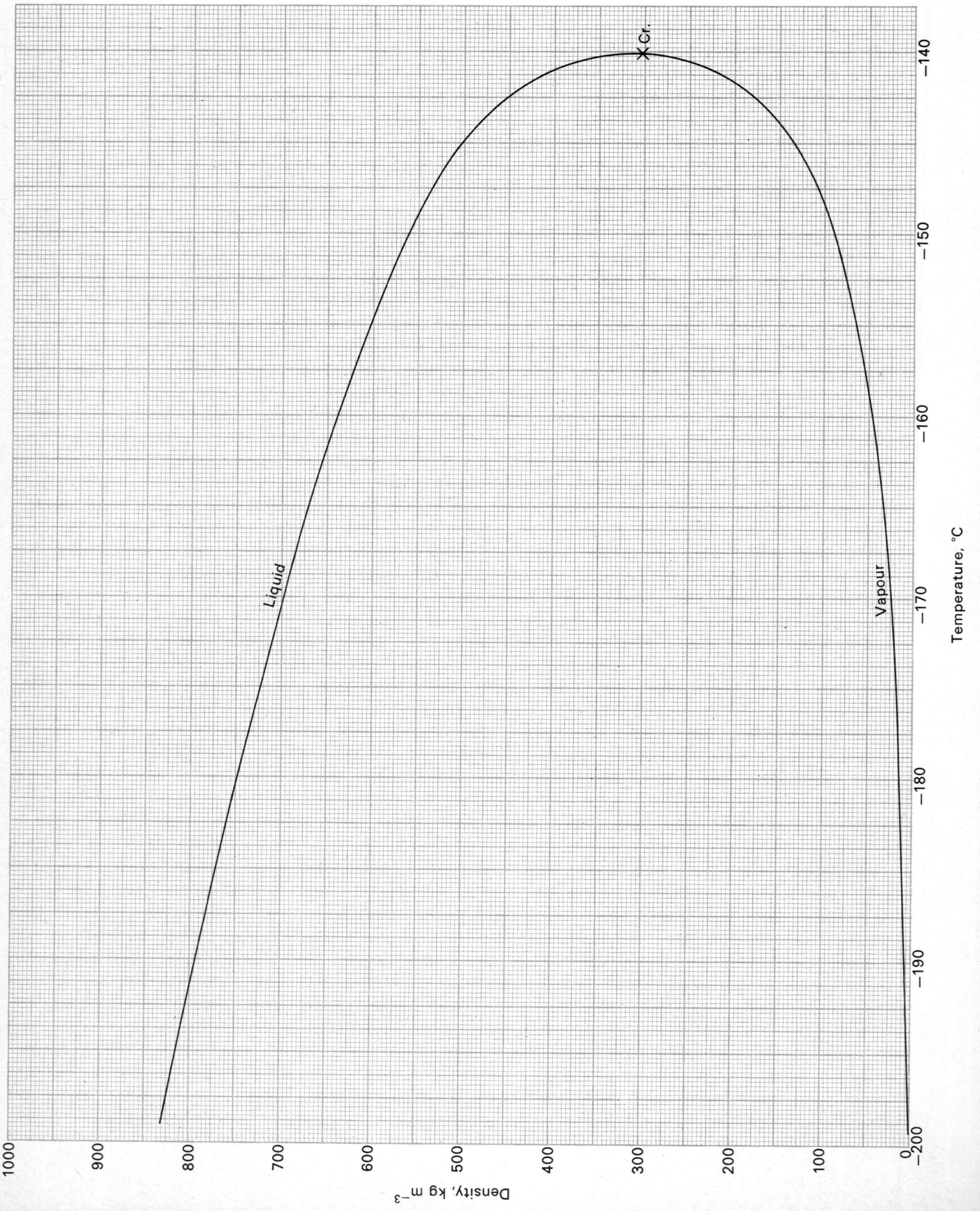

12.3 P-V-T Data for Carbon Monoxide (20,24,28)

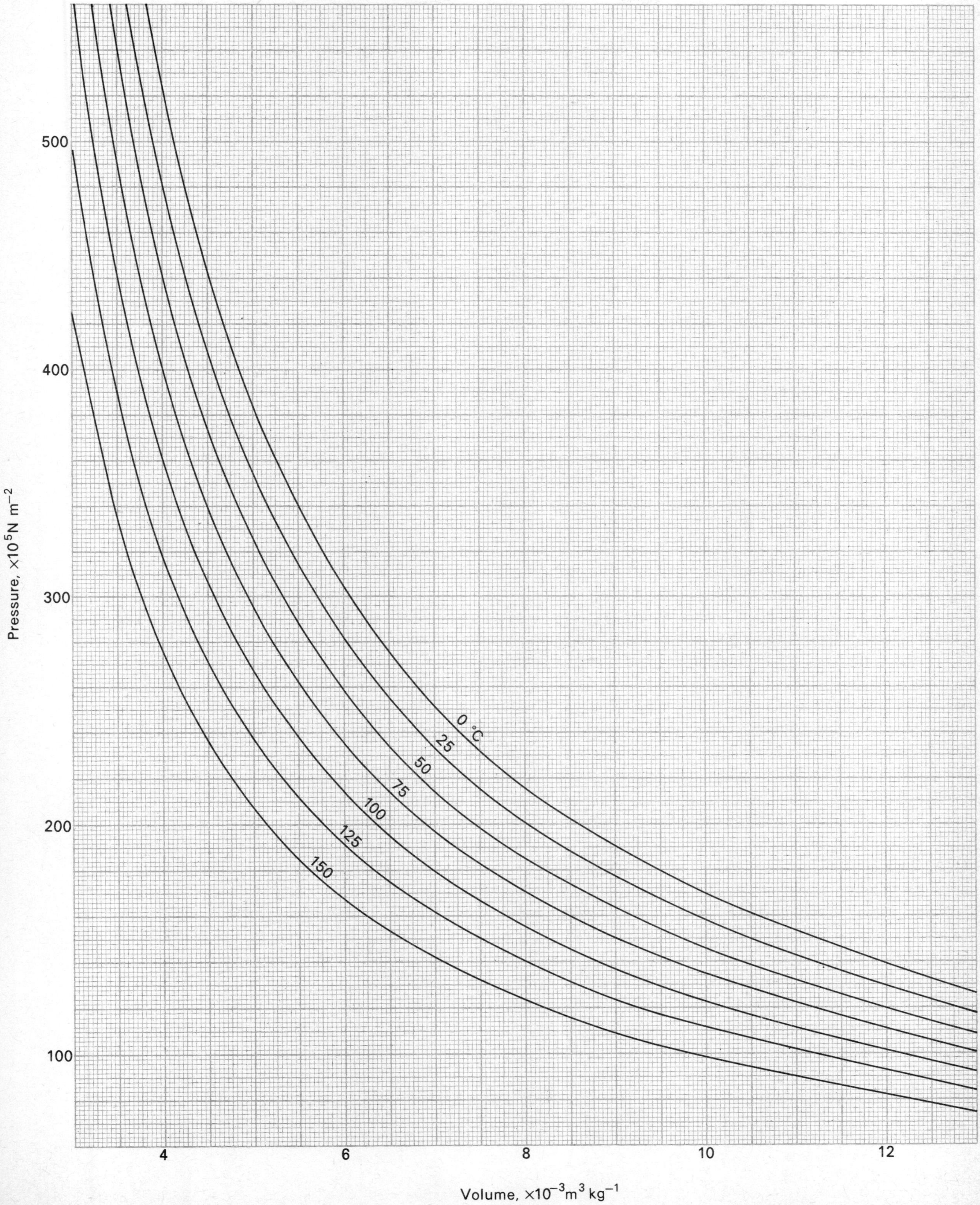

12.4 Latent Heat of Vaporisation of Carbon Monoxide (20,23)

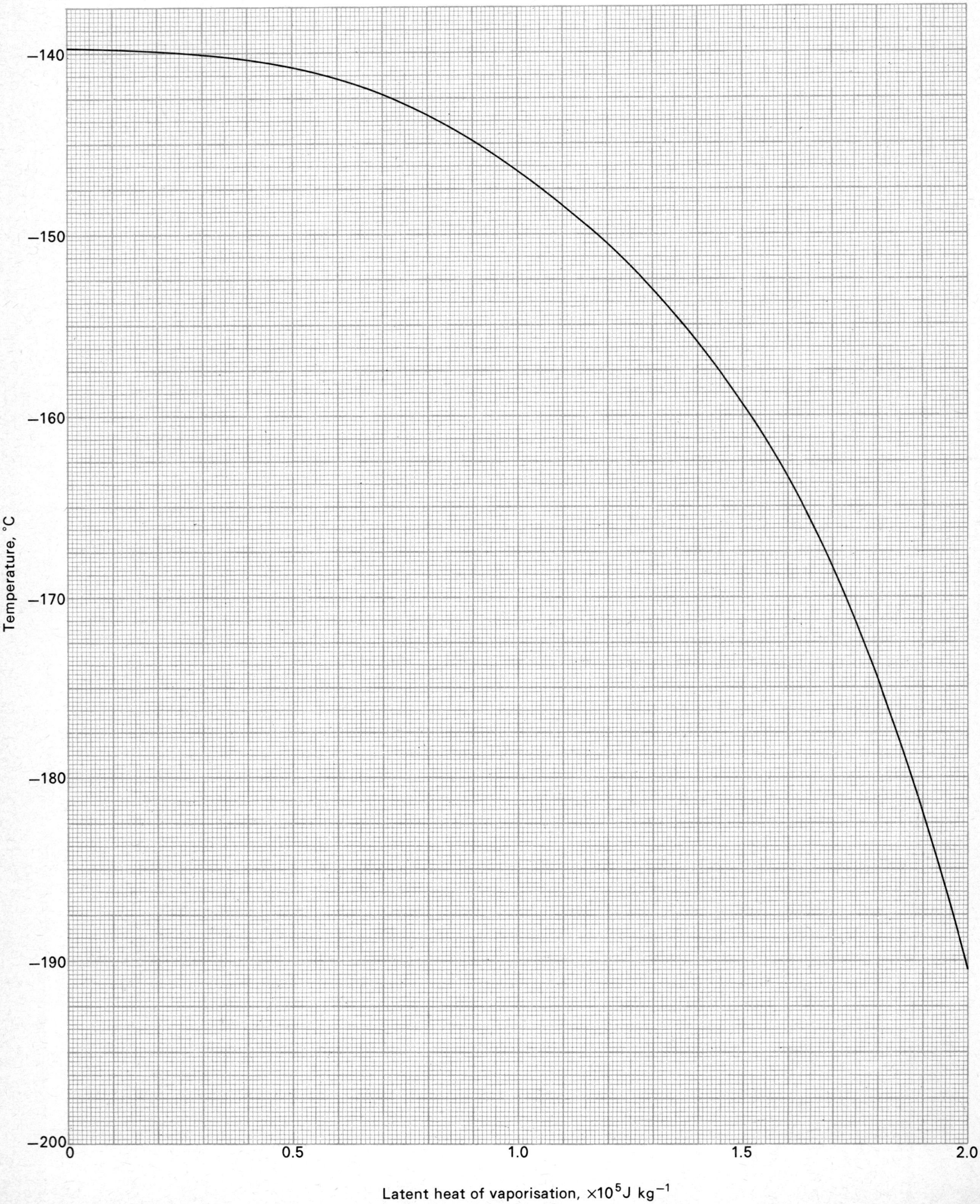

12.5 Heat Capacity of Liquid Carbon Monoxide (5,6)

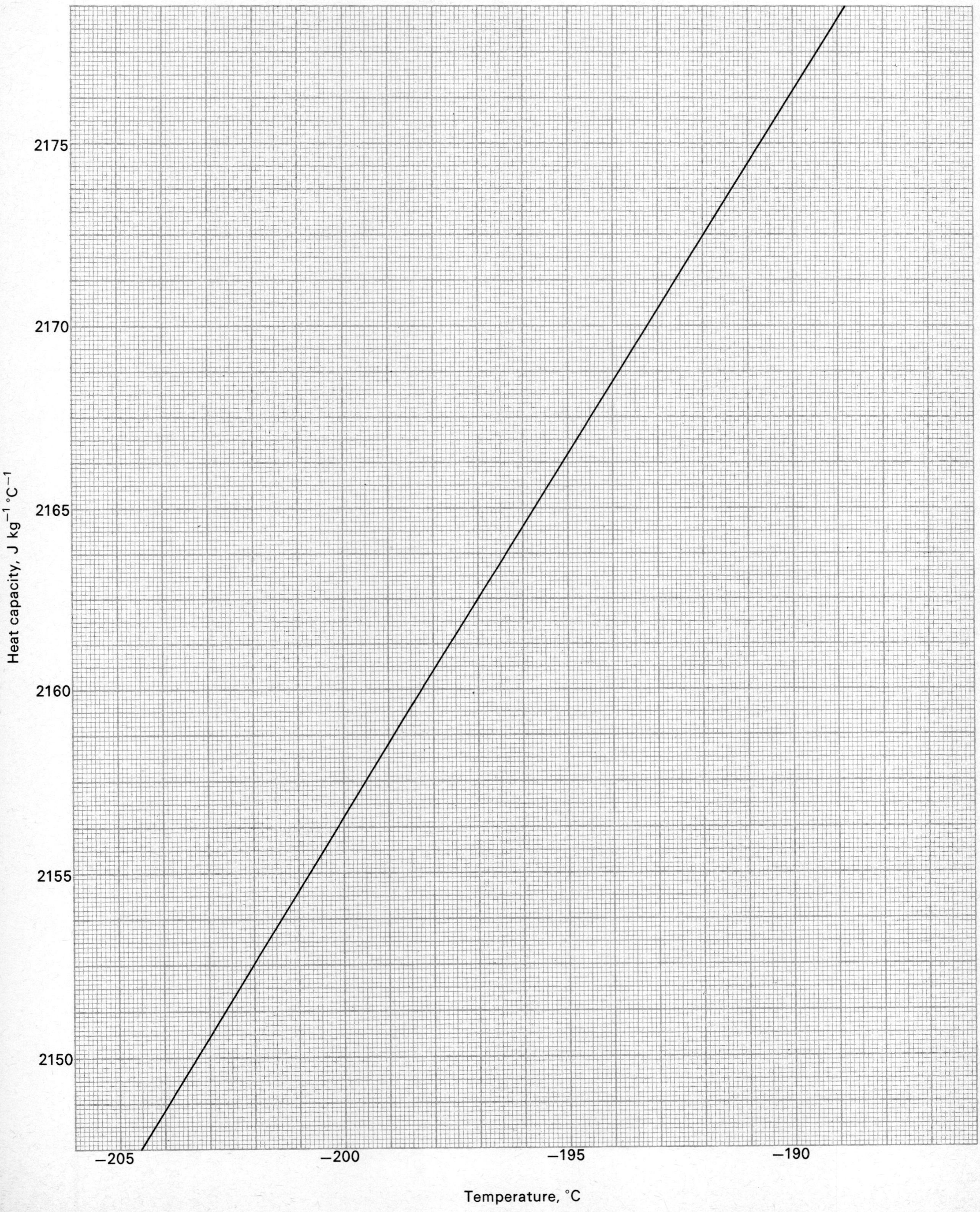

12.6 Heat Capacity of Gaseous Carbon Monoxide at $1.0133 \times 10^5 N\ m^{-2}$ (18,27,37)

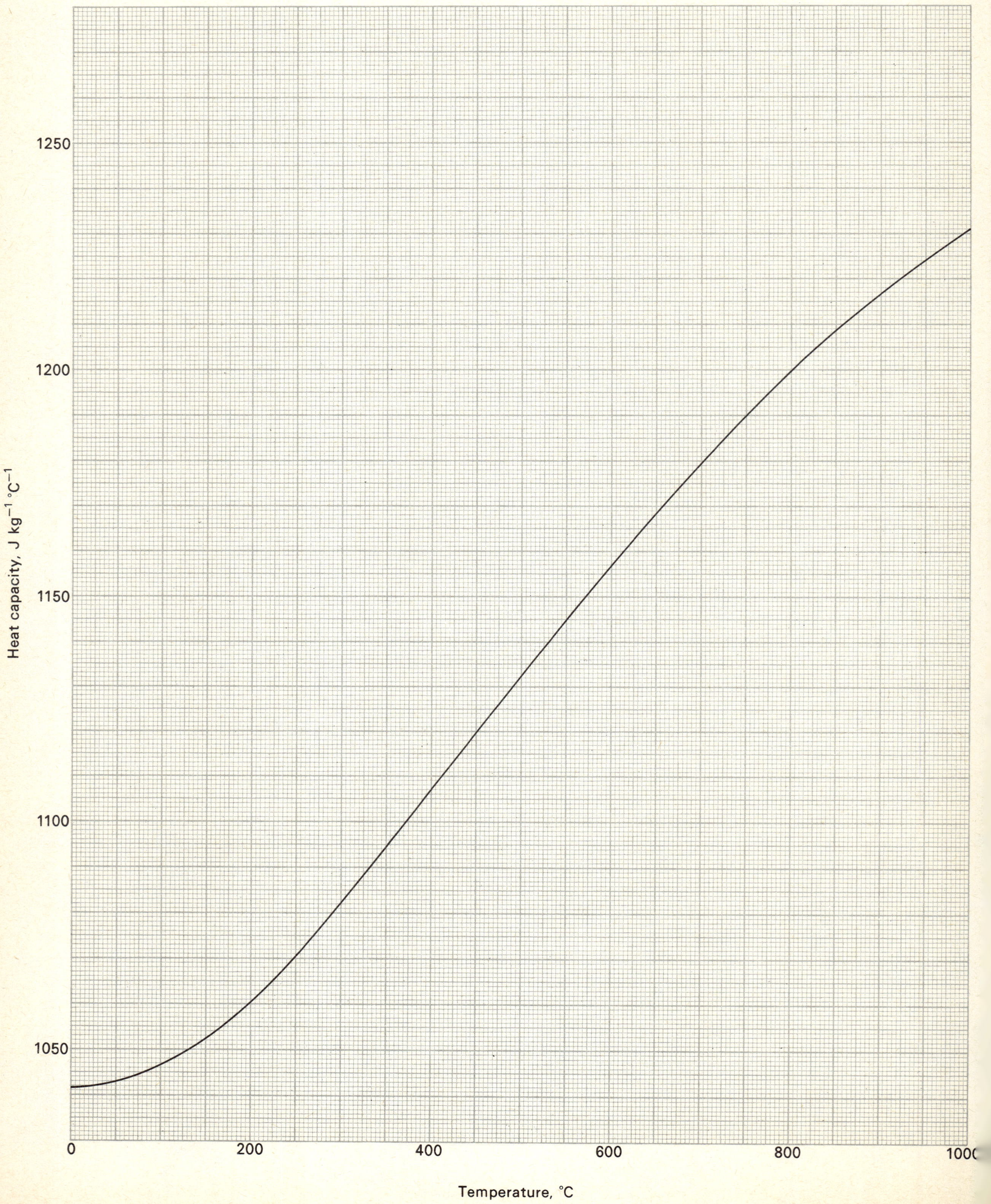

12.7 Heat Capacity Ratio (C_p/C_v) of Gaseous Carbon Monoxide at 1.0133×10^5N m^{-2}(9)

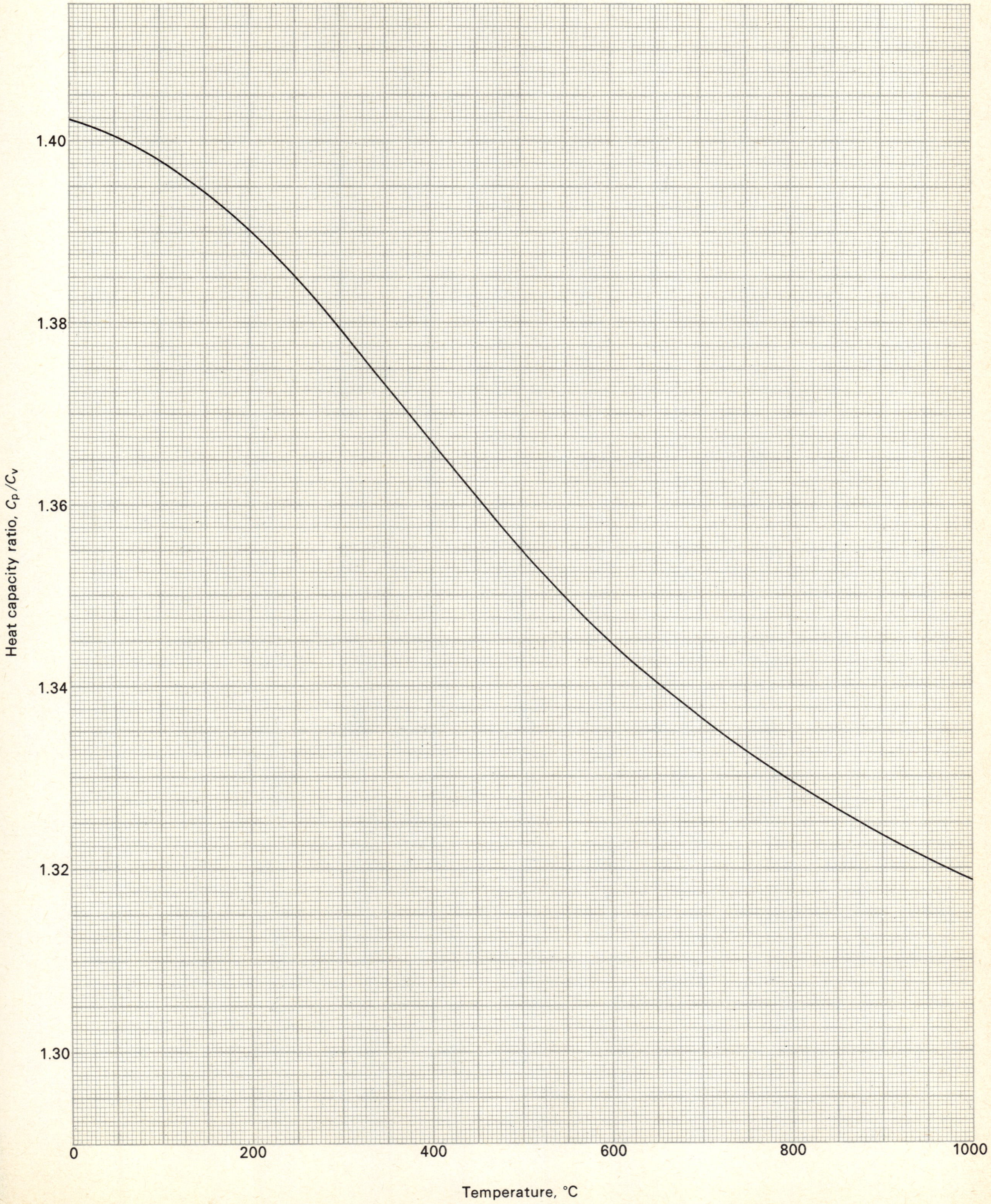

12.8 Viscosity of Liquid Carbon Monoxide (3,10,14)

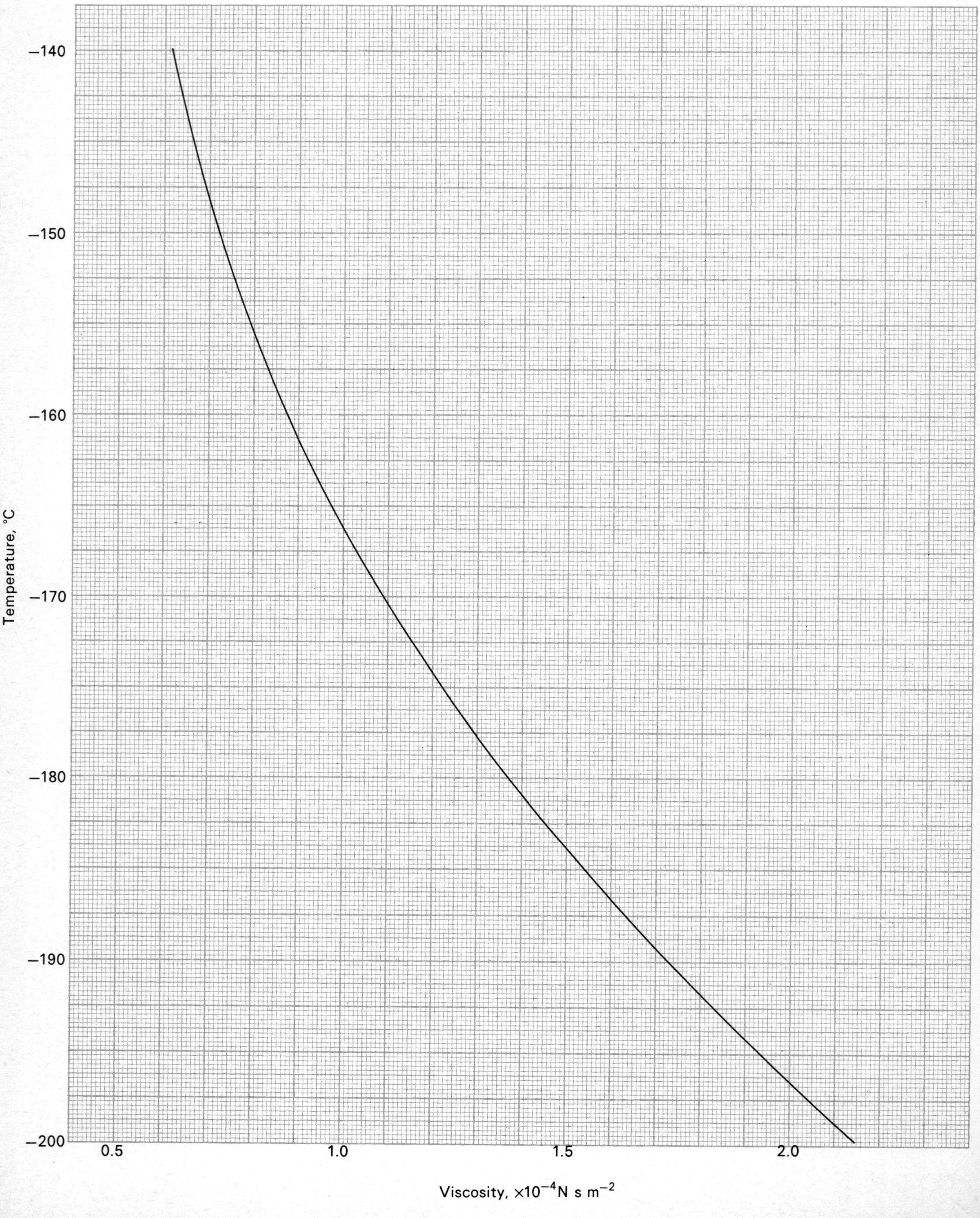

12.9 Viscosity of Gaseous Carbon Monoxide at $1.0133 \times 10^5 N\ m^{-2}$(2,16)

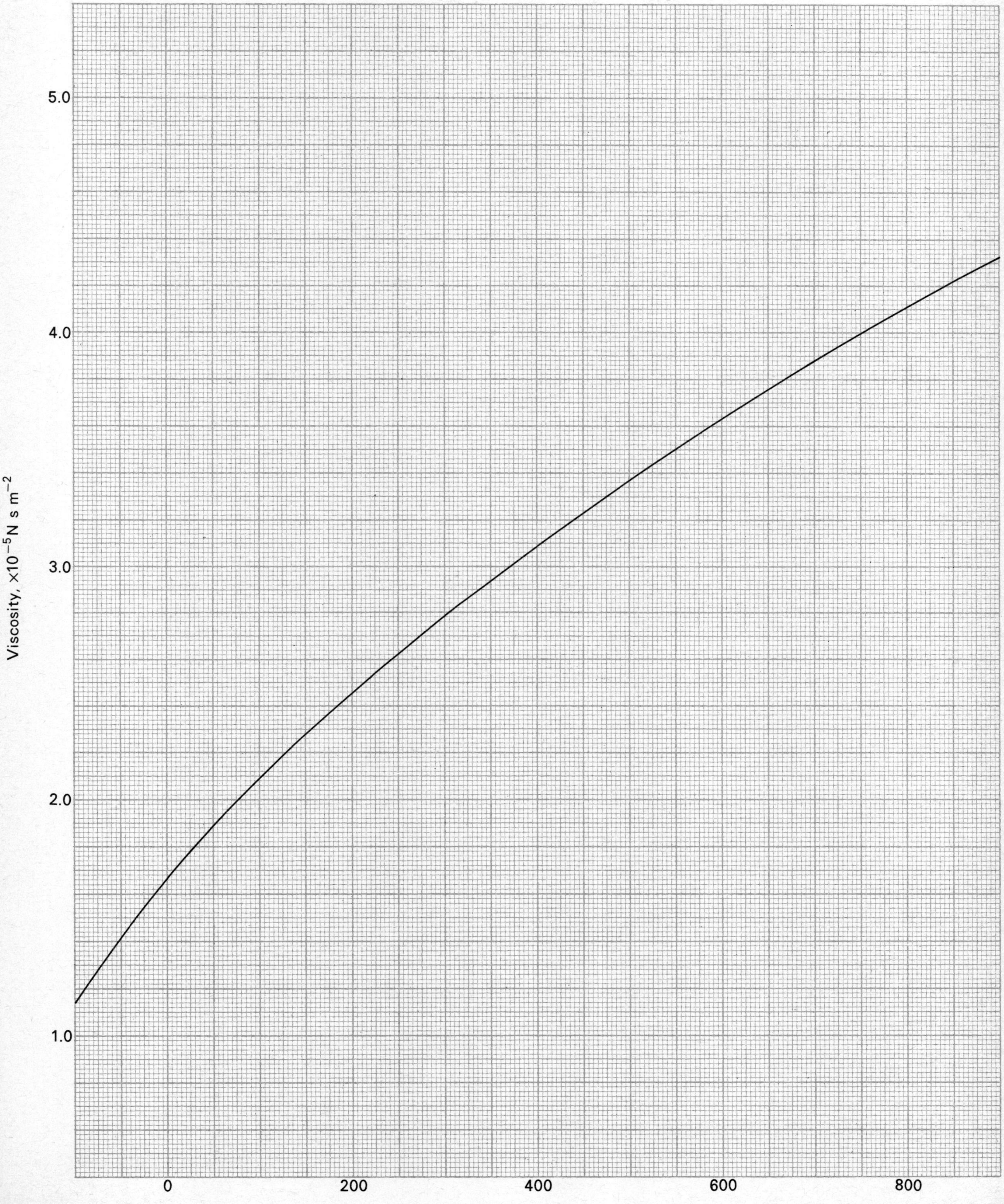

12.10 Thermal Conductivity of Liquid Carbon Monoxide (4)

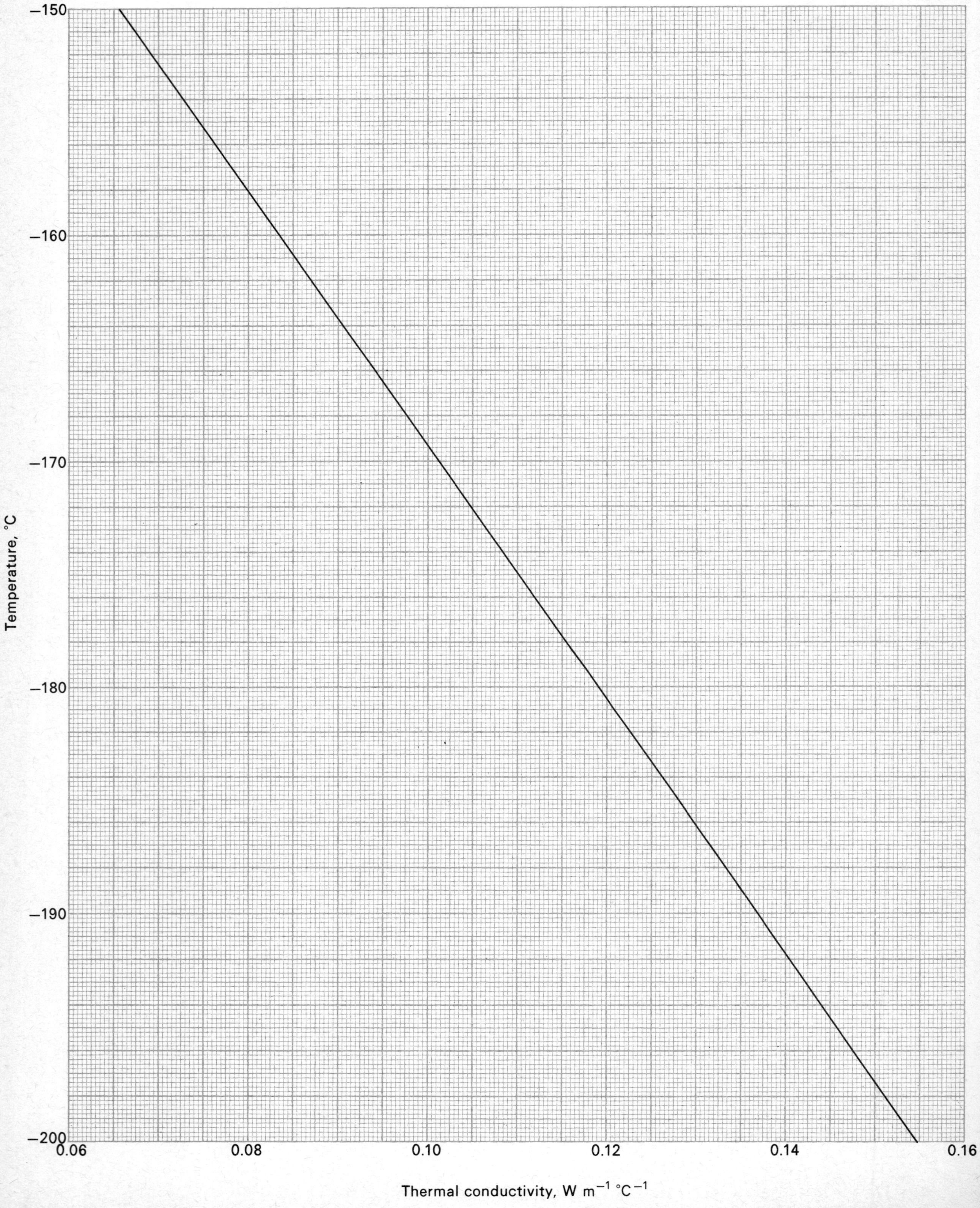

12.11 Thermal Conductivity of Gaseous Carbon Monoxide at $1.0133 \times 10^5 N\ m^{-2}$ (11,13,22)

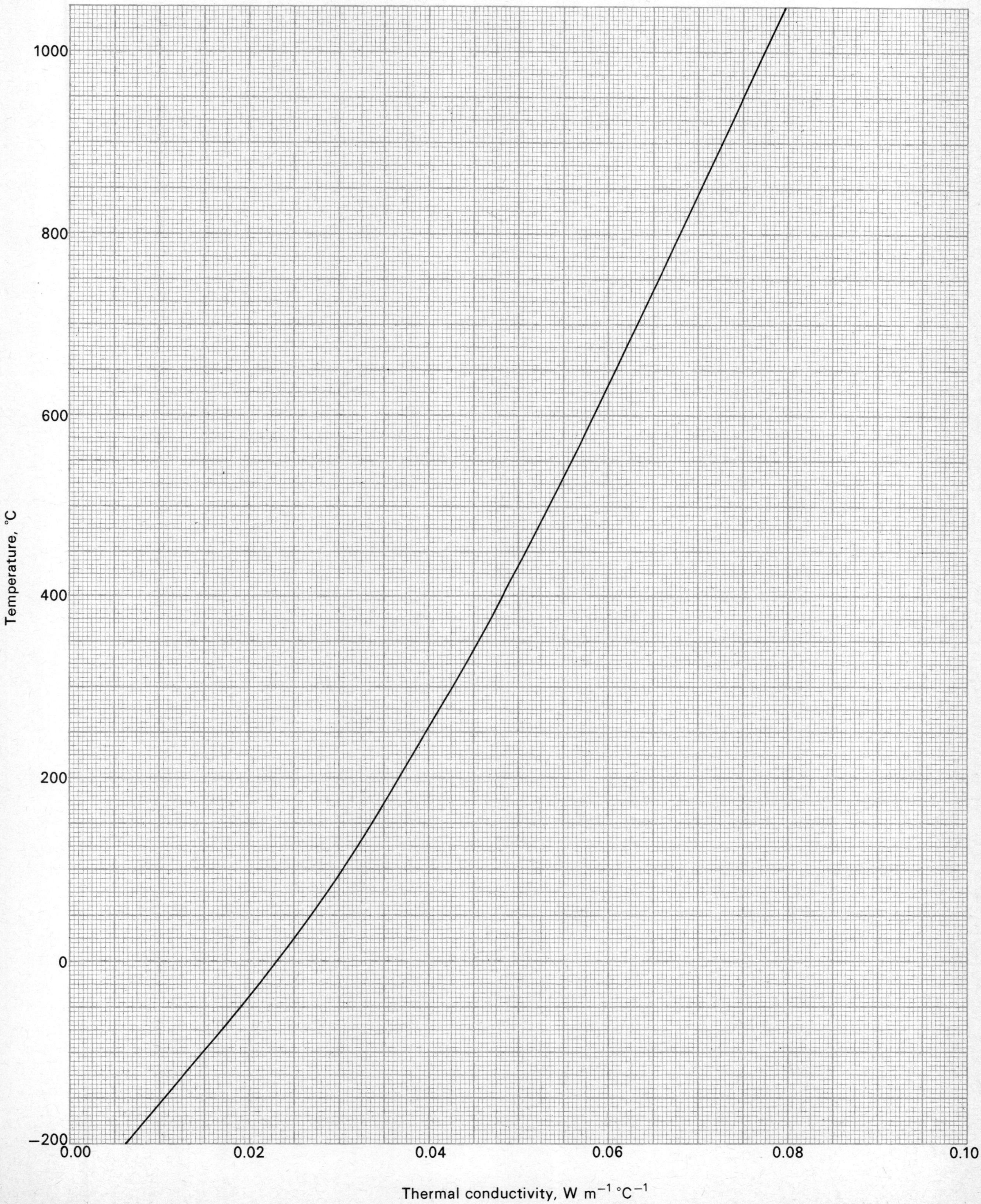

12.12 Surface Tension of Carbon Monoxide (32,36)

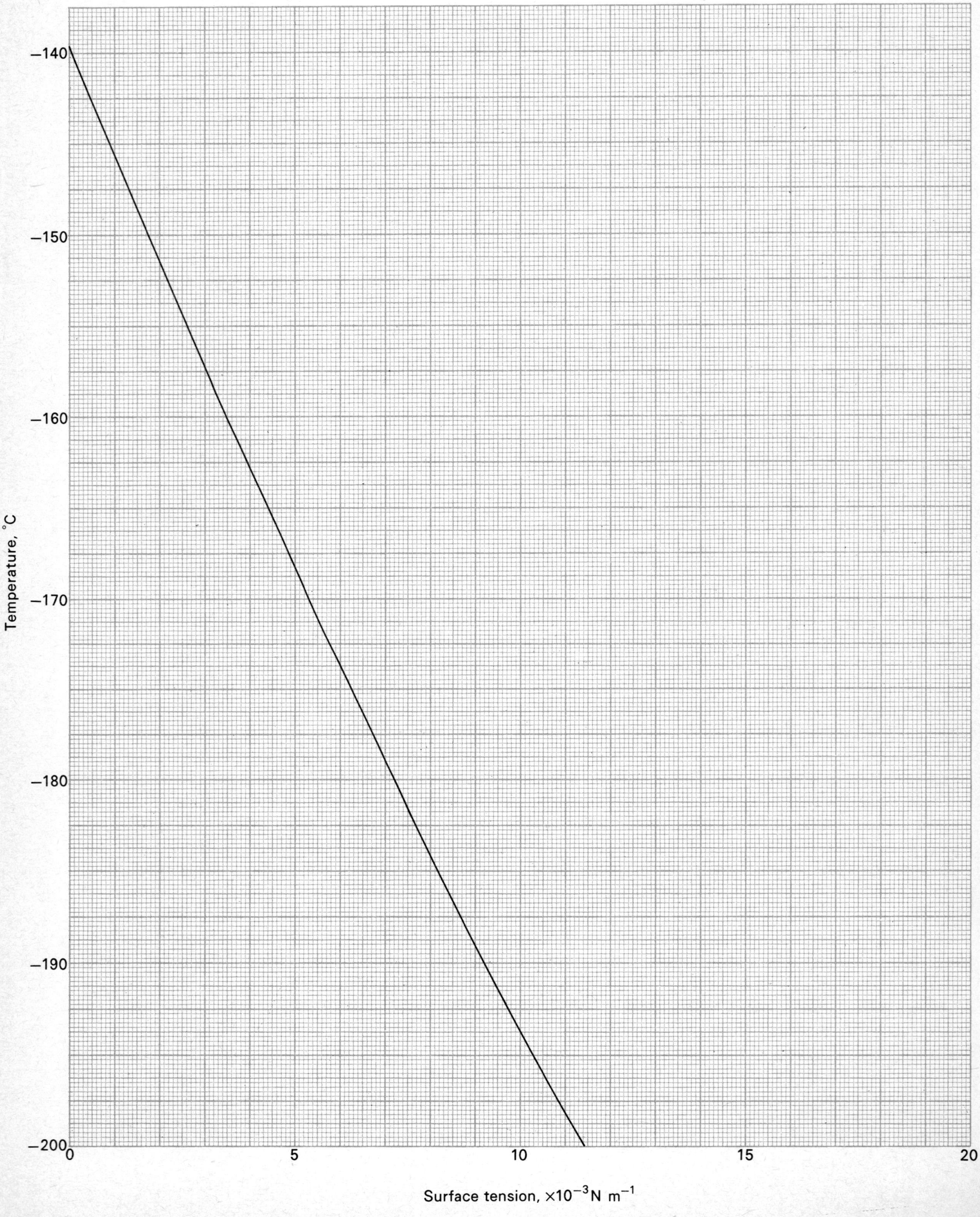

12.13 Solubility of Carbon Monoxide in Water at $1.0133 \times 10^5 N\ m^{-2}$ Total Pressure (25,30,41)

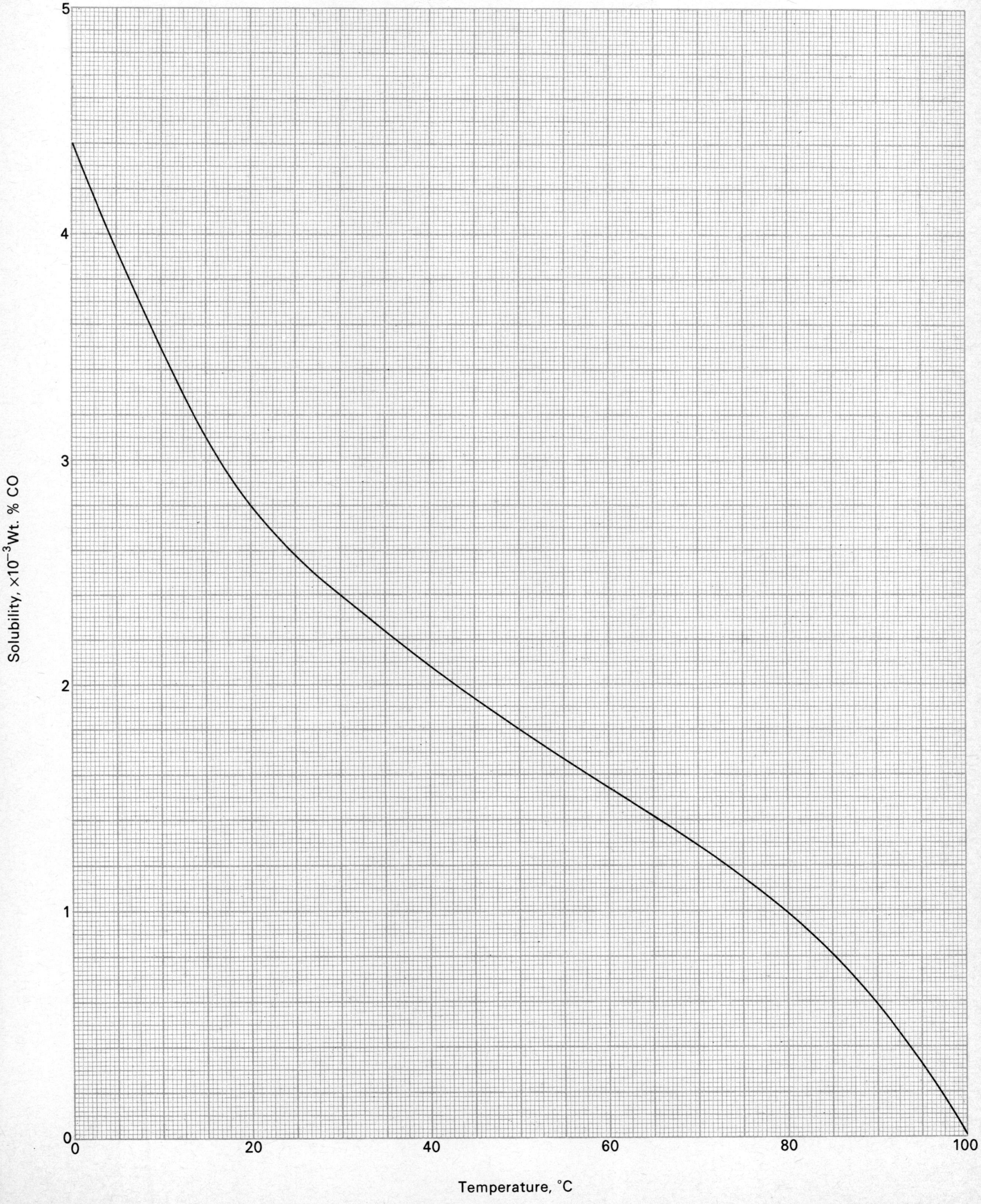

13 HEAVY WATER

There are two effects which contribute to the differences in physicochemical properties between heavy and ordinary water:

(i) the nuclear mass of deuterium is approximately twice that of hydrogen and
(ii) heavy water is more associated than ordinary water.

The liquidus and solidus of the solid solutions of heavy water and ordinary water are never separated by more than 0.02°C. The maximum densities of both H_2O and D_2O occur just a few degrees celsius above their melting points.

There is no significant difference in crystal structure between ice and solid heavy water. All chemical substances which are soluble in ordinary water, are somewhat less soluble in heavy water at the same temperatures and pressures.

Heavy water is widely used in nuclear reactors and atomic power plants as a moderator and coolant.

The most comprehensive compilation of the thermophysical properties of heavy water at various temperatures and pressures has been reported by Kazavchinskii *et al.* (22). Further surveys and reviews are available as follows: Bishop and Nelson (5), Elliott (12), Howieson (17), Landolt-Börnstein (27), Pascal (37), Salomon (44), Urey and Murphy (48), and Whalley (53).

Most of the experimental measurements on the properties of heavy water have been carried out fairly recently in connection with its large scale production and application, so that the reported data are reliable as far as the purity and experimental technique are concerned. The presented values in Table 13.1 and Figs. 13.1–13.12 have mainly been adopted from Kazavchinskii *et al.* (22).

References

1 Arakawa, K., Sasaki, K., Endo, Y., *Bull. Chem. Soc. Japan*, 1969, **42**, 2079–81.
2 Baker, B. L., *Heat Capacity and Thermodynamic Properties of Saturated Deuterium Oxide*, Atomic Energy of Canada Ltd., AECL-4738, 1959.
3 Baker, C. E., Brokaw, R. S., *J. Chem. Phys.*, 1964, **40** (6), 1523–8.
4 Bartholeme, E., Clusius, K., *Z. phys. Chem.*, 1935, **28B**, 167–77.
5 Bishop, A. A., Nelson, P. A., *Transport and Thermodynamic Properties of Saturated and Compressed Heavy Water*, Carolina's Virginia Nuclear Power Assoc., Inc., Report, 1961.
6 Bonilla, C. F., Wang, S. J., *Thermal Conductivity of Deuterium Oxide*, U.S. At. Energy Commis. USAEC DP-12, Columbia Univ., May 1953.
7 Bonilla, C. F., Wang, S. J., Weiner, H., *Trans. Am. Soc. Mech. Eng.*, 1956, **78** (6), 1285–9.
8 Brown, R. S., Barnes, W. H., Maass, O., *Can. J. Res.*, 1935, **13**, 167–9.
9 Bryant, F. J., Atomic Energy Res. Establishment C/R 1927, 1956, 15 pp.
10 Cockett, A. H., Ferguson, A., *Phil. Mag.*, 1939, **28**, 685–93.
11 Cockett, A. H., Ferguson, A., *Phil. Mag.*, 1940, **29**, 185–99.
12 Elliott, J. N., *Tables of the Thermodynamic Properties of Heavy Water*, Atomic Energy of Canada Ltd., AECL-1673, 1963.
13 Engineering Science Data Units *Approximate Data on the Viscosity of some Common Liquids*, Item No. 66024, London, July 1966, 26 pp.
14 Eucken, A., Eigen, M., *Z. Elektrochem.*, 1951, **55**, 343–54.
15 Flood, H., Tronstad, L., *Z. phys. Chem.*, 1936, **175A**, 347–52.
16 Friedmann, A. S., Haar, L., *J. Chem. Phys.*, 1954, **22**, 2051–8.
17 Howieson, J., *Tables of Properties of Heavy Water at Saturation Temperatures*, Atomic Energy of Canada Ltd., AECL-1055, July 1960.
18 IUPAC, Atomic Weights of the Elements 1971, *Pure & Appl. Chem.*, 1972, **30** (3–4), 638–49.
19 Jones, G., Ray, W. A., *J. Chem. Phys.*, 1937, **5**, 505–8.
20 Jones, W. M., *Vapour Pressures of Titanium Oxide and Deuterium Oxide; Interpretation of the Isotope Effect*, Los Alamos Scientific Lab., Los Alamos, New Mexico, LADC-5905, 1964.
21 Jones, W. M., *J. Chem. Phys.*, 1968, **48**, 207.
22 Kazavchinskii, Ya. Z., Kessel'man, P. M., Kirillin, V. A., Rivkin, S. L., Sheindlin, A. E., Shpil'rain, E. E., Sychev, V. V., Timrot, D. L., *Heavy Water. Thermophysical Properties*, Gosudarstvennoe Energeticheskoe Izdatel'stvo, Moskva, 1963.
23 Kirillin, V. A., Ulibin, S. A., Experimental Determination of the Specific Volume of Heavy Water over a Wide Range of Temperatures and Pressures, *Proc. Unit. Nat. Internat. Conf. Peaceful Uses At. Energy*, 2nd, Geneva, 1958, **28**, 159–65.
24 Kudish, A. I., Wolf, D., Steckel, F., *J. Chem. Soc. Faraday Trans. I*, 1972, **68** (11), 2041–6.
25 La Mer, V. K., Baker, W. N., *J. Am. Chem. Soc.*, 1934, **56**, 2641–3.
26 La Mer, V. K., Eichelberger, W. C., Urey, H. C., *J. Am. Chem. Soc.*, 1934, **56**, 248–9.
27 *Landolt-Börnstein, Zahlenwerte und Funktionen aus Physik, Chemie, Astronomie, Geophysik, und Technik*, II Band, 4 Teil, Kalorische Zustandgrössen, Springer-Verlag, Berlin, 1961, p. 483.
28 Lax, E., Synowietz, C., Eds., *Taschenbuch für Chemiker und Physiker*, 3rd ed., Springer-Verlag, Berlin, 1967.
29 Lewis, G. N., Macdonald, R. T., *J. Am. Chem. Soc.*, 1933, **55**, 3057–9.
30 Liu, C.-T., Lindsay, W. T., *J. Chem. Eng. Data*, 1970, **15** (4), 510–13.
31 Long, E. A., Kemp, J. D., *J. Am. Chem. Soc.*, 1936, **58**, 1829–34.
32 Lonsdale, D. K., *Proc. Roy. Soc.*, 1958, **247A**, 424–34.
33 Maryott, A. A., Smith, E. R., *Table of Dielectric Constants of Pure Liquids*, Natl. Bur. Std. Circular 514, Washington, Aug. 10, 1951, 44 pp.
34 Mathews, J. F., *Chem. Rev.*, 1972, **72** (1), 71–100.
35 Meyer, F., Eigen, M., *Z. Naturfors.*, 1953, **8a**, 500.
36 Nowak, E. S., Chan, J., *Trans. Am. Soc. Mech. Eng.*, 1971, **93** (4), 422–6.
37 Pascal, P., Ed., *Nouveau Traité de Chimie Minérale*, Vol. I, Generalites, Air, Eau, Hydrogéne, Deuterium, Tritium, Helium et Gas Inertes, Masson et Cie, Éditeurs, Paris, 1956, pp. 813–25.
38 Plank, R., *Brennstoff-Wärme-Kraft*, 1961, **13**, 257–8.
39 Rivkin, S. L., Akhundov, T. S., *Teploenergetika*, 1962, **9** (5), 62.
40 Rivkin, S. L., Egorov, B. N., *Atomnaya Energiya*, 1959, **7** (5), 462–5.
41 Rivkin, S. L., Egorov, B. N., *Atomnaya Energiya*, 1963, **14** (4), 416–18.
42 Rivkin, S. L., Egorov, B. N., *Experimental Investigation of the Isobaric Specific Heat of Heavy Water*, Proc. 7th Internat. Conf. Properties of Steam, Tokyo, 1968, paper E.7.
43 Rossini, F. D., Wagman, D. D., Evans, W. H., Levine, S., Jaffe, I., *Selected Values of Chemical Thermodynamic Properties*, Natl. Bur. Std. Circular 500, Washington, Feb. 1, 1952.
44 Salomon, M., *Thermodynamic Properties of Liquid H_2O and D_2O and their Mixtures*, National Aeronautics and Space Administration, NASA TN D-5223, Washington, May 1969.
45 Timmermans, J., Bodson, H., *Compt. Rend.*, 1937, **204**, 1804–7.

46 Timrot, D. L., Shuiskaya, K. F., *Atomnaya Energiya*, 1959, **7** (5), 459–62.
47 Trakhtengerts, M. S., *Teploenergetika*, 1970, **17** (5), 70–3.
48 Urey, H. C., Murphy, G. M., *Physical Properties and Analysis of Heavy Water*, Div. III, Vol. 4A, McGraw-Hill, New York, 1951, 438 pp.
49 Vargaftik, N. B., Oleshchuk, O. N., *Teploenergetika*, 1962, **5** (12), 64.
50 Vargaftik, N. B., Oleshchuk, O. N., Belyakov, P. E., *Atomnaya Energiya*, 1959, **7** (5), 465–8.
51 Vargaftik, N. B., Oleshchuk, O. N., Beljakova, P. E., *Kernenergie*, 1960, **3**, 585.
52 Venart, J. E. S., Mani, N., *J. Mech. Eng. Sci.*, 1971, **13** (3), 205–16.
53 Whalley, E., *The Thermodynamic and Transport Properties of Heavy Water*, Proc. Communications of the Conf. on Thermodyn. Transport Properties of Fluids, Inst. of Mech. Eng., London, 1957, pp. 15–26.
54 Wilson, W. D., *J. Acoust. Soc. Am.*, 1961, **33**, 374.
55 Yosioka, K., *Sci. Papers Inst. Phys. Chem. Res.*, Tokyo, 1938, **34**, 843.
56 Ziebland, H., *Allg. Wärmetechn.*, 1959, **9**, 37.
57 Ziebland, H., Burton, J. T. A., *Intern. J. Heat Mass Transfer*, 1960, **1** (213), 242–54.

Table 13.1 Properties of Heavy Water

		Ref.
Chemical formula	D_2O	(28)
Molecular weight	20.0310	(18)
Critical temperature	370.8°C	(34)
Critical pressure	216.6×10^5 N m^{-2}	(34)
Critical density	360.0 kg m^{-3}	(34)
Triple-point temperature	3.82°C	(29)
Triple-point pressure	674.5 N m^{-2}	(29)
Boiling point at 1.0133×10^5 N m^{-2}	101.42°C	(22)
Melting point	3.80°C	(22)
Density of solid (hexagonal) at −30.0°C	1020.8 kg m^{-3}	(32)
Compressibility (adiabatic) of liquid at 25.0°C	46.48×10^{-11} m^2 N^{-1}	(55)
Coefficient of thermal expansion of liquid at 20.0°C	12.8×10^{-5} °C^{-1}	(14)
Heat of formation of gas at 25°C	-2.4938×10^8 J kg-mol^{-1}	(43)
Free energy of formation of gas at 25°C	-2.3475×10^8 J kg-mol^{-1}	(43)
Heat of sublimation at triple-point temperature	5.2883×10^7 J kg-mol^{-1}	(4)
Heat of fusion	6.2844×10^6 J kg-mol^{-1}	(43)
Entropy of gas at 25°C	1.9837×10^5 J kg-mol^{-1} °C^{-1}	(43)
Heat capacity ratio of solid at −23.15°C	1.178	(31)
Velocity of sound at 20°C and 1.0133×10^5 N m^{-2}	1384.2 m s^{-1}	(54)
Diffusion in air at 24.8°C and 1.0133×10^5 N m^{-2}	0.247×10^{-4} m^2 s^{-1}	(28)
Dielectric constant of liquid at 25°C	78.25	(33)
Electrical conductivity of liquid at 25°C	2.5×10^{-4} ohm^{-1} m^{-1}	(9)
Refractive index of liquid at 20°C	1.33844	(28)

13.1 Vapour Pressure of Heavy Water (20,22,23,29,30,39)

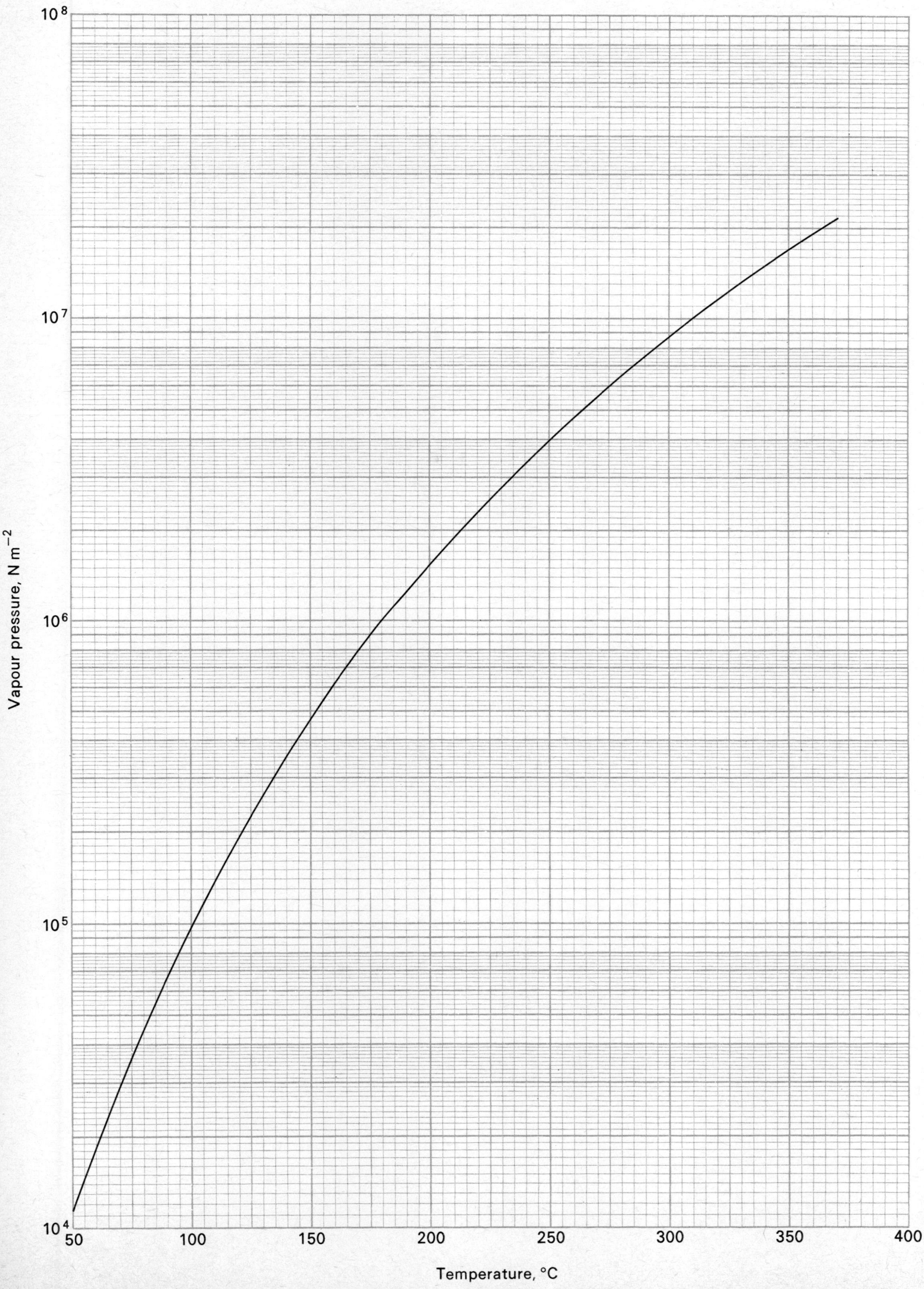

13.2 Orthobaric Density of Heavy Water (17,22,23,47)

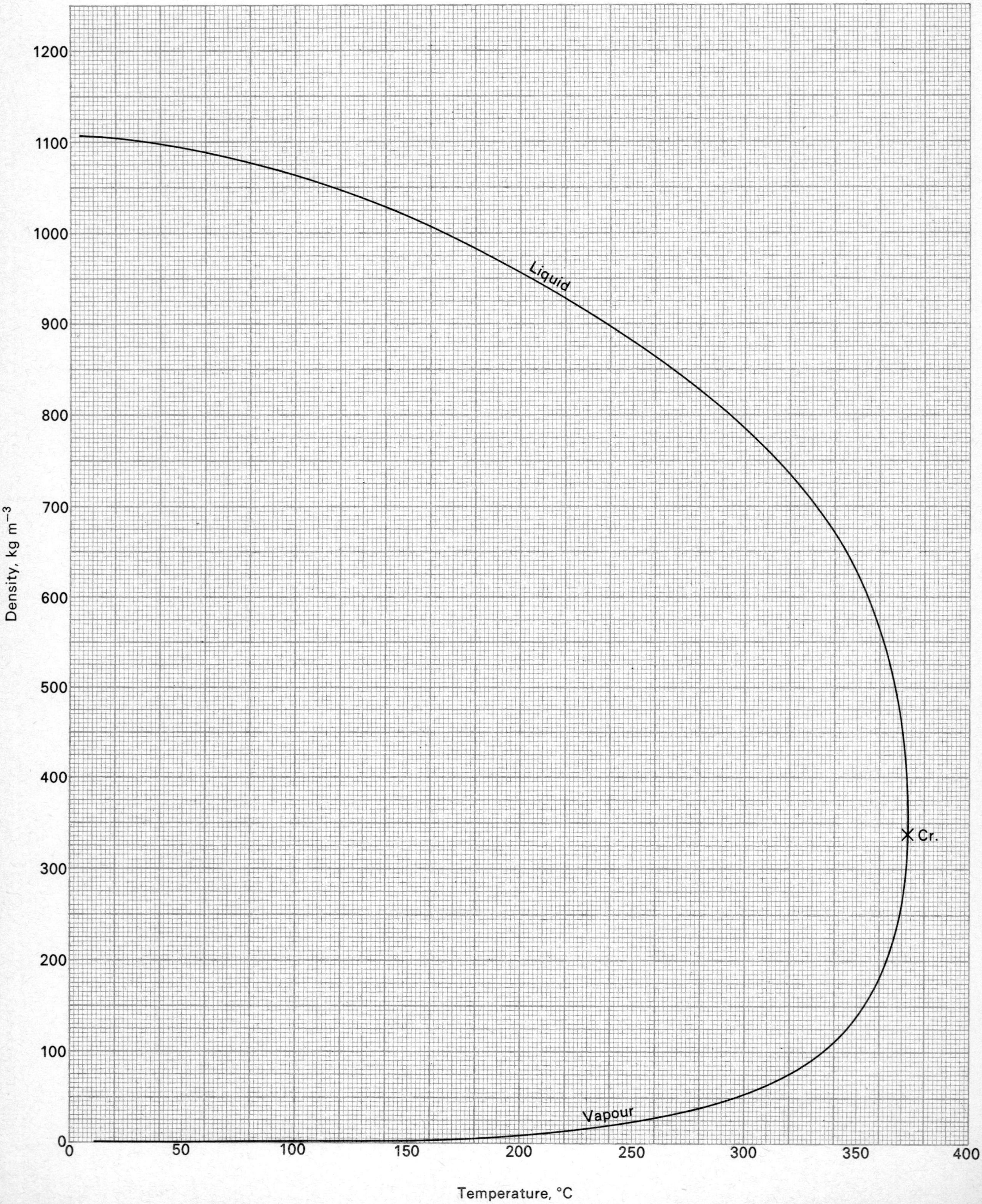

13.3 P-V-T Data for Heavy Water (22,23,38,47)

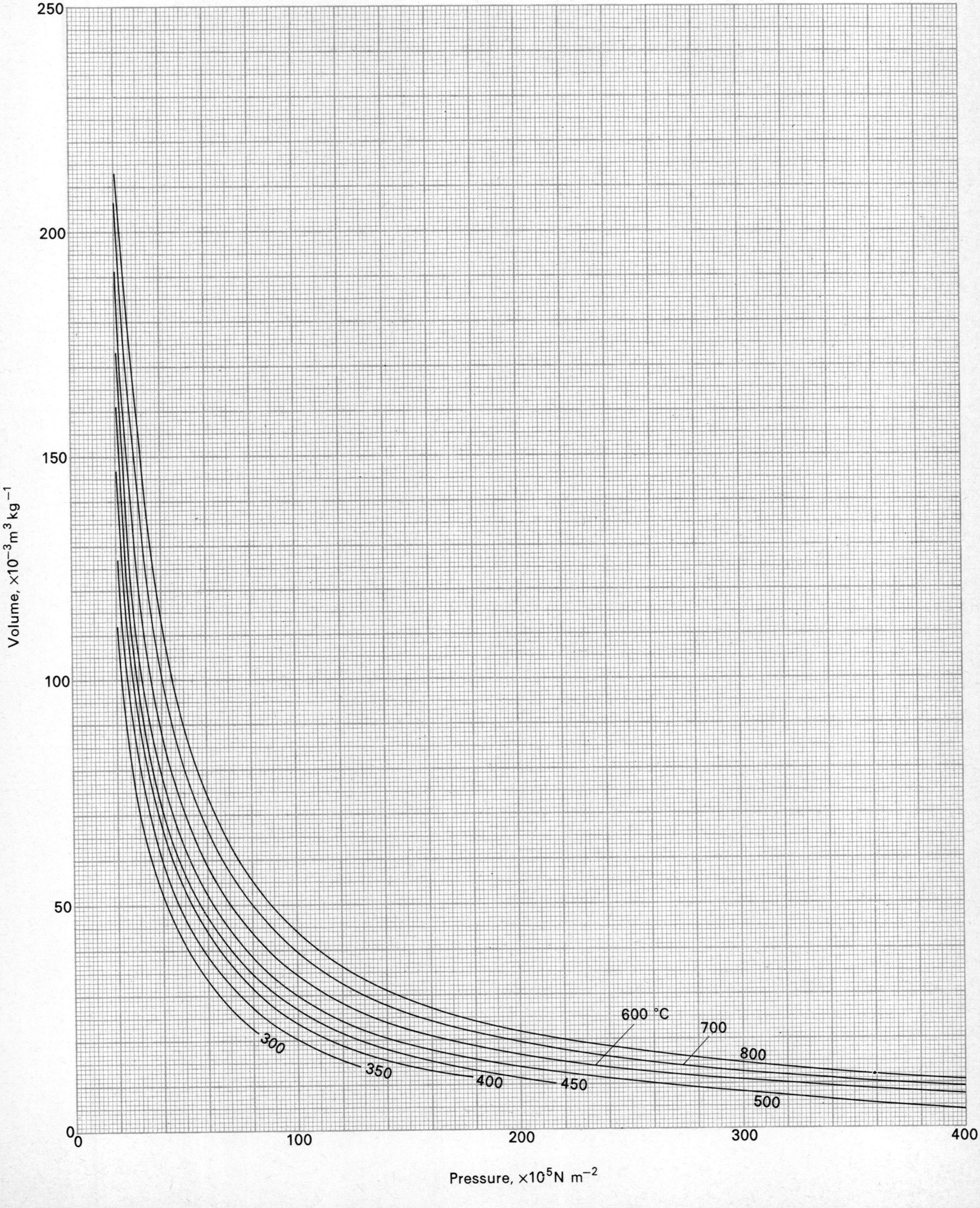

13.4 Latent Heat of Vaporisation of Heavy Water (5,12,17,22,44,47)

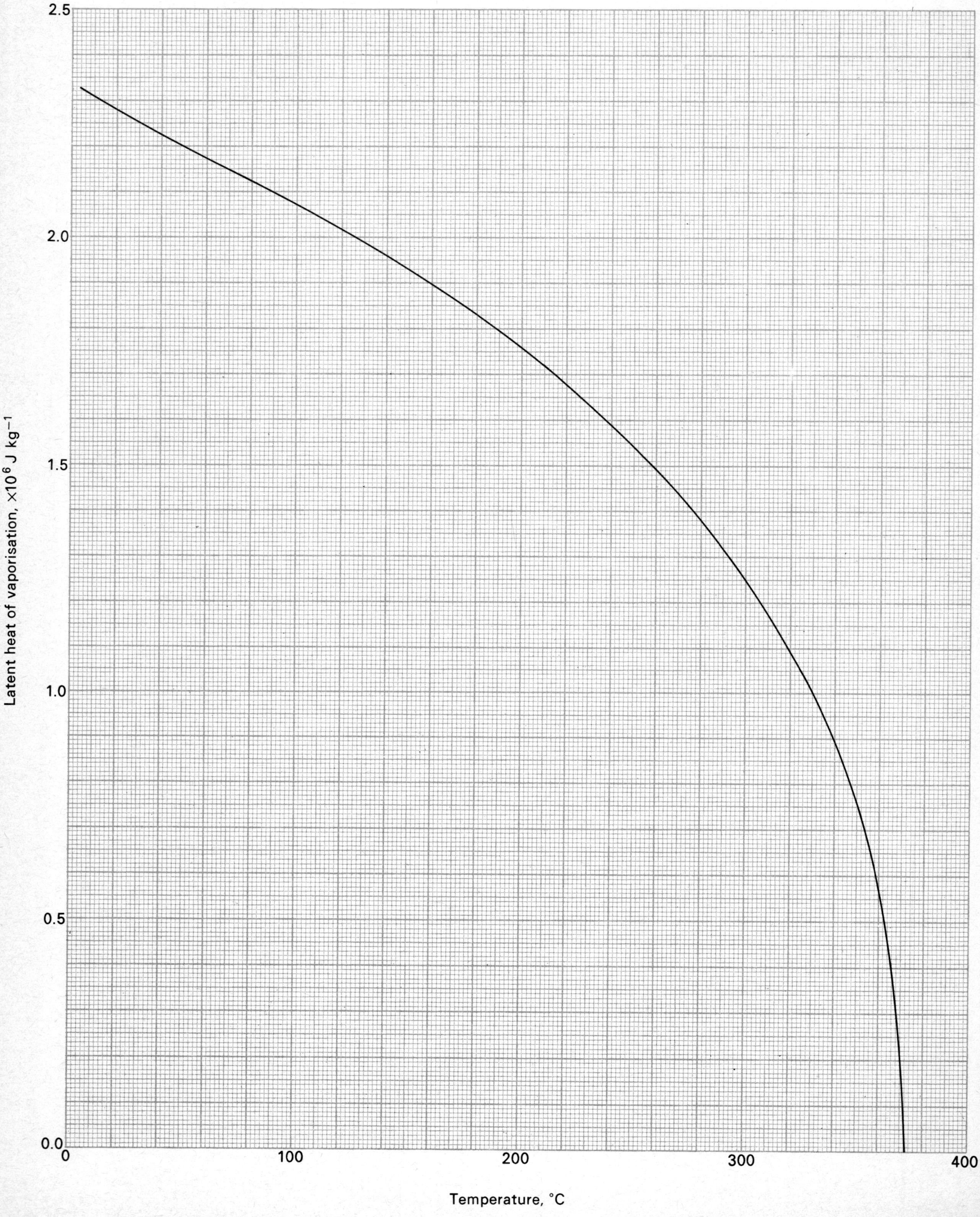

13.5 Heat Capacity of Solid and Liquid Heavy Water (1,2,8,11,16,22,27,31,40,41)

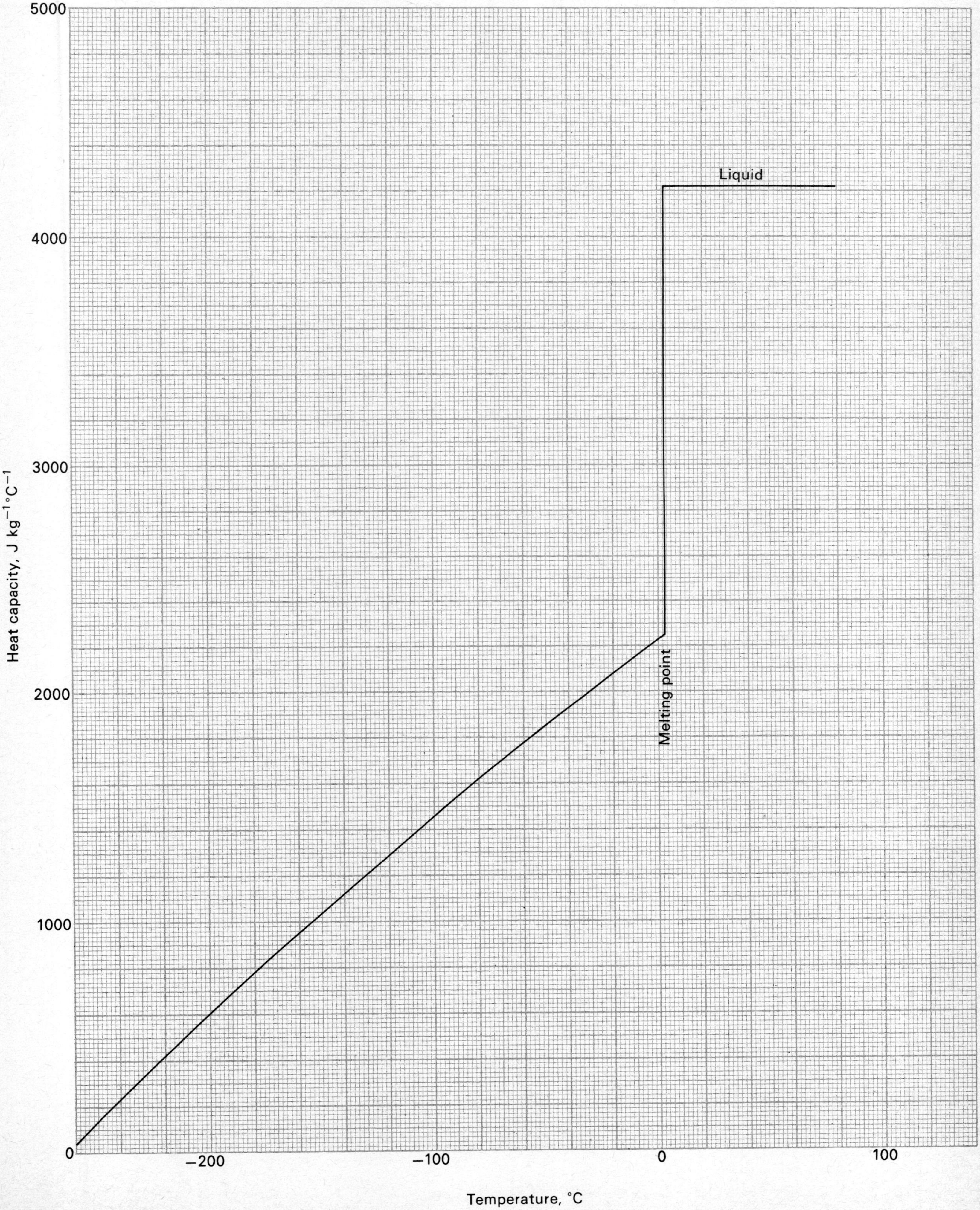

13.6 Heat Capacity of Gaseous Heavy Water at $1.0133 \times 10^5 N\ m^{-2}$(11,12,22,23,27,36,42,44,47)

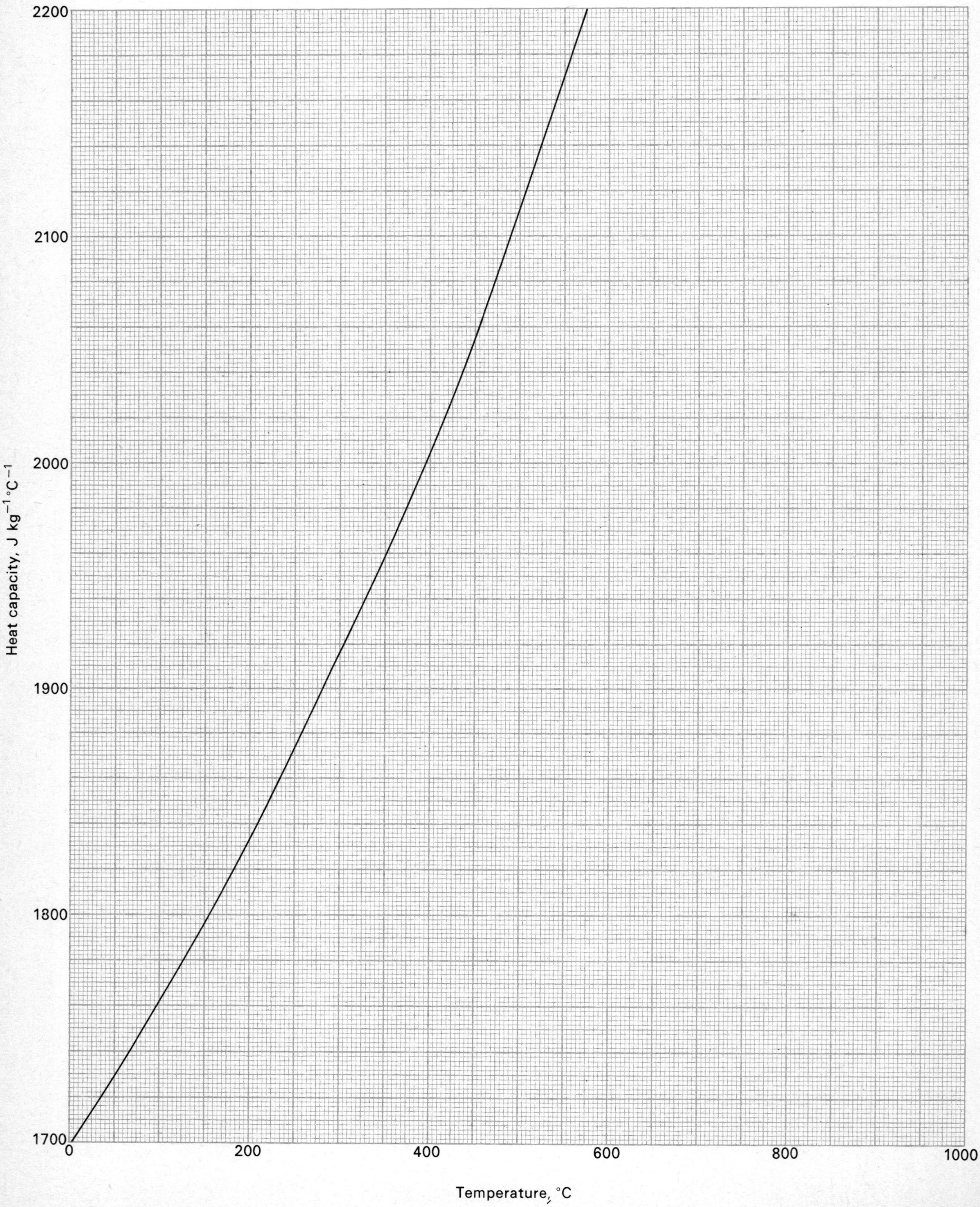

13.7 Viscosity of Liquid Heavy Water (13,22,24,46)

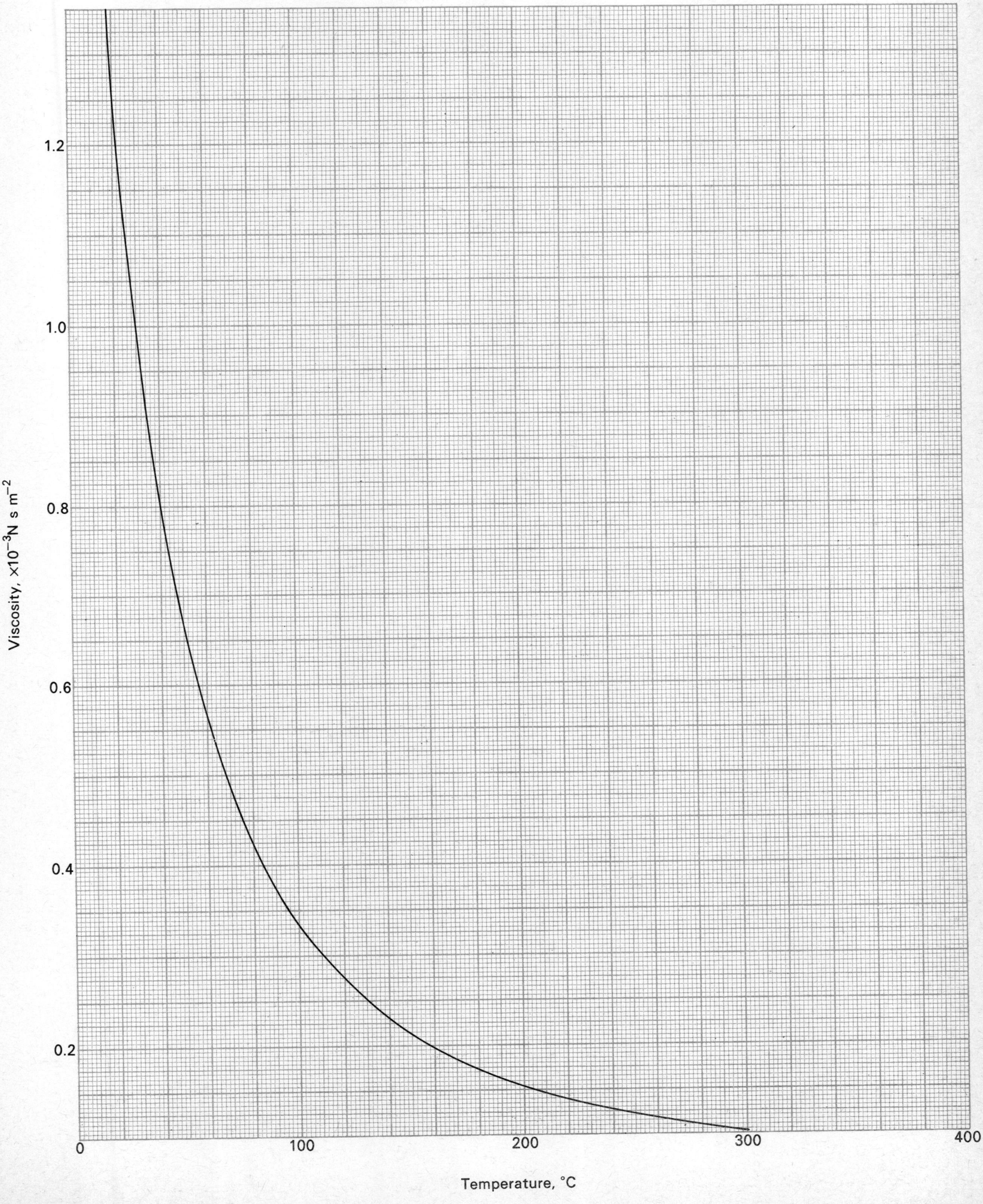

13.8 Viscosity of Gaseous Heavy Water at 1.0133×10^5N m^{-2}(6,7,22,48,53)

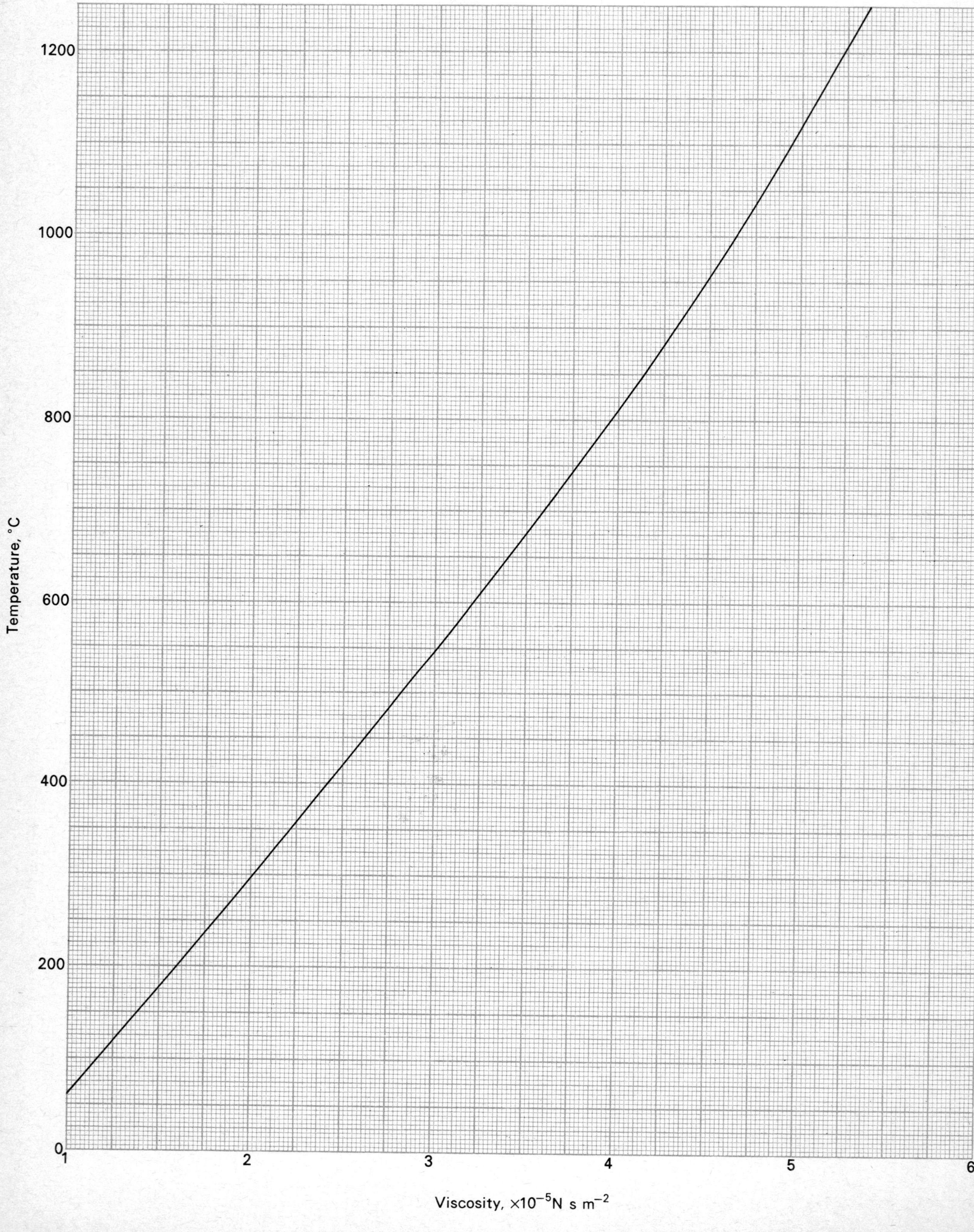

13.9 Thermal Conductivity of Liquid Heavy Water (22,35,50,51,52,57)

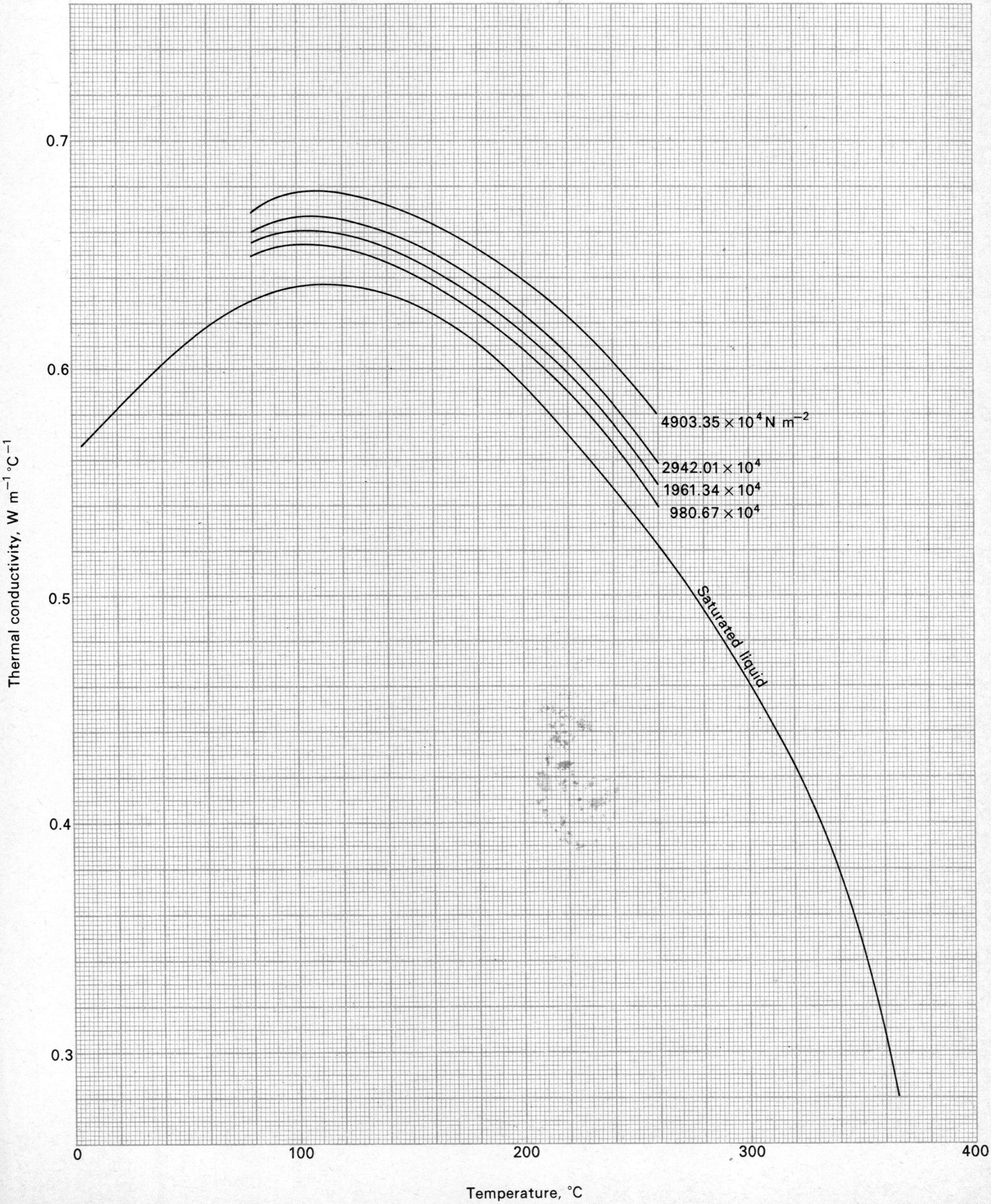

13.10 Thermal Conductivity of Gaseous Heavy Water (3,6,22,49,52,56)

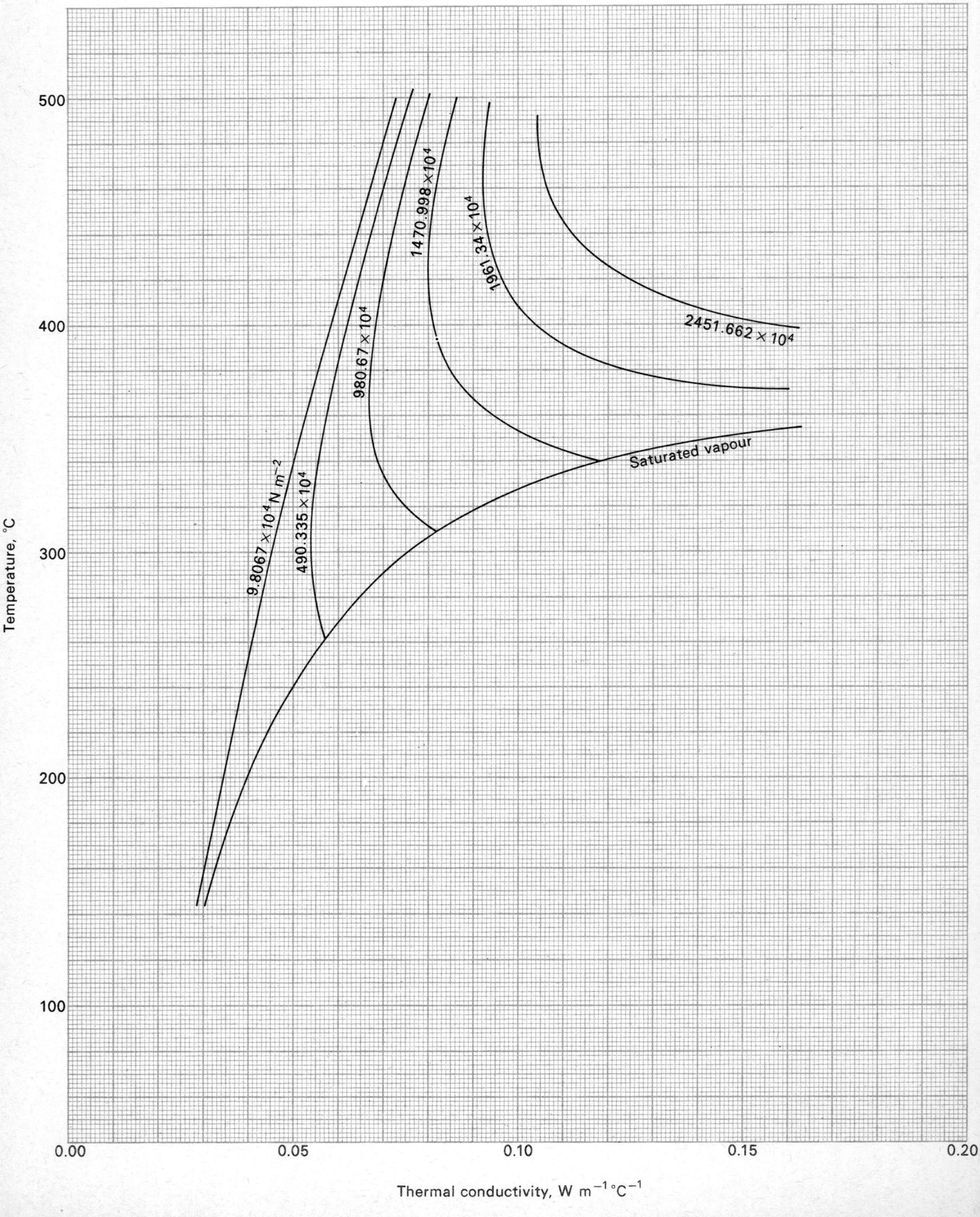

13.11 Surface Tension of Heavy Water (10,15,19,45)

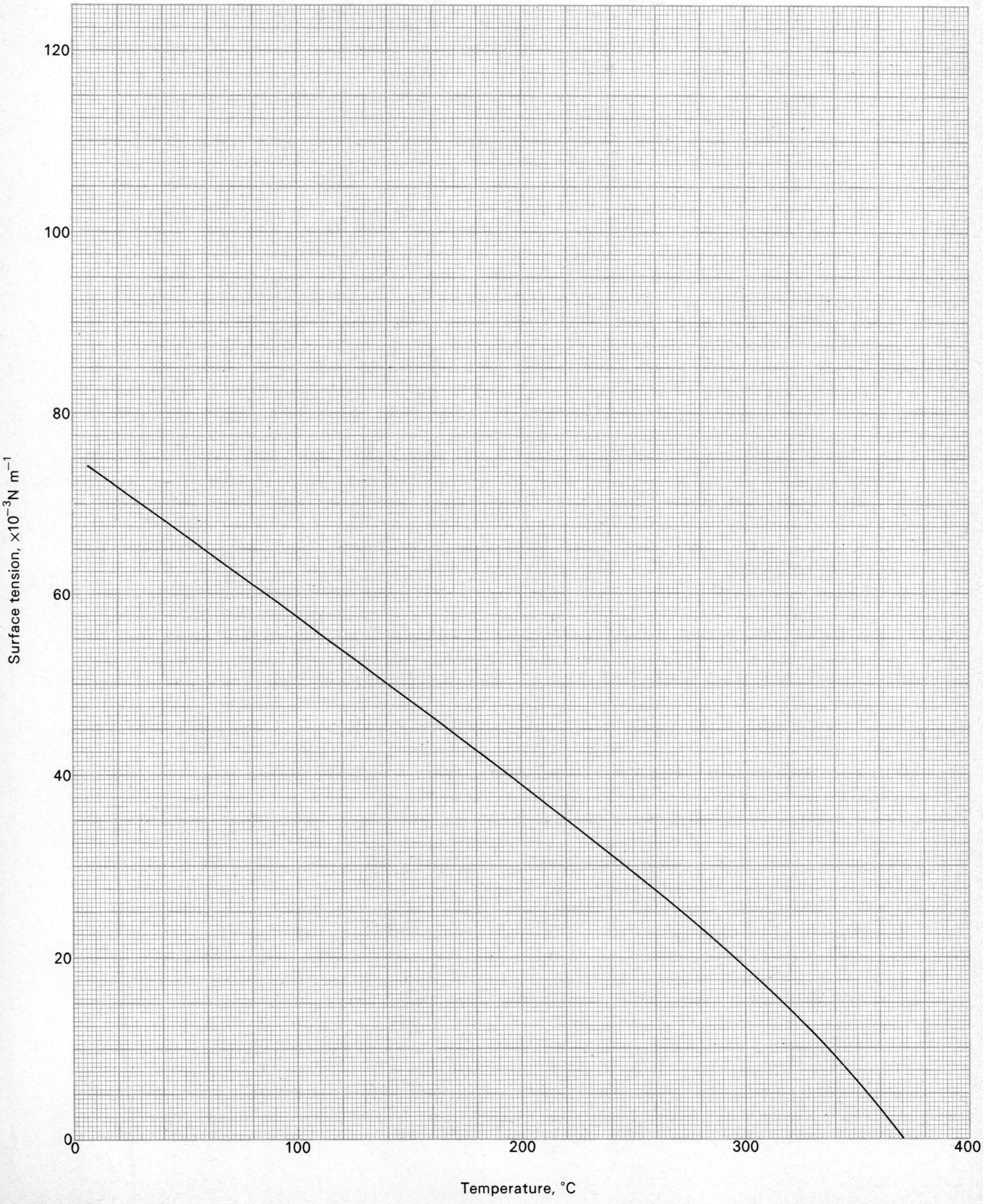

13.12 Phase Diagram for D_2O-H_2O System (25,26)

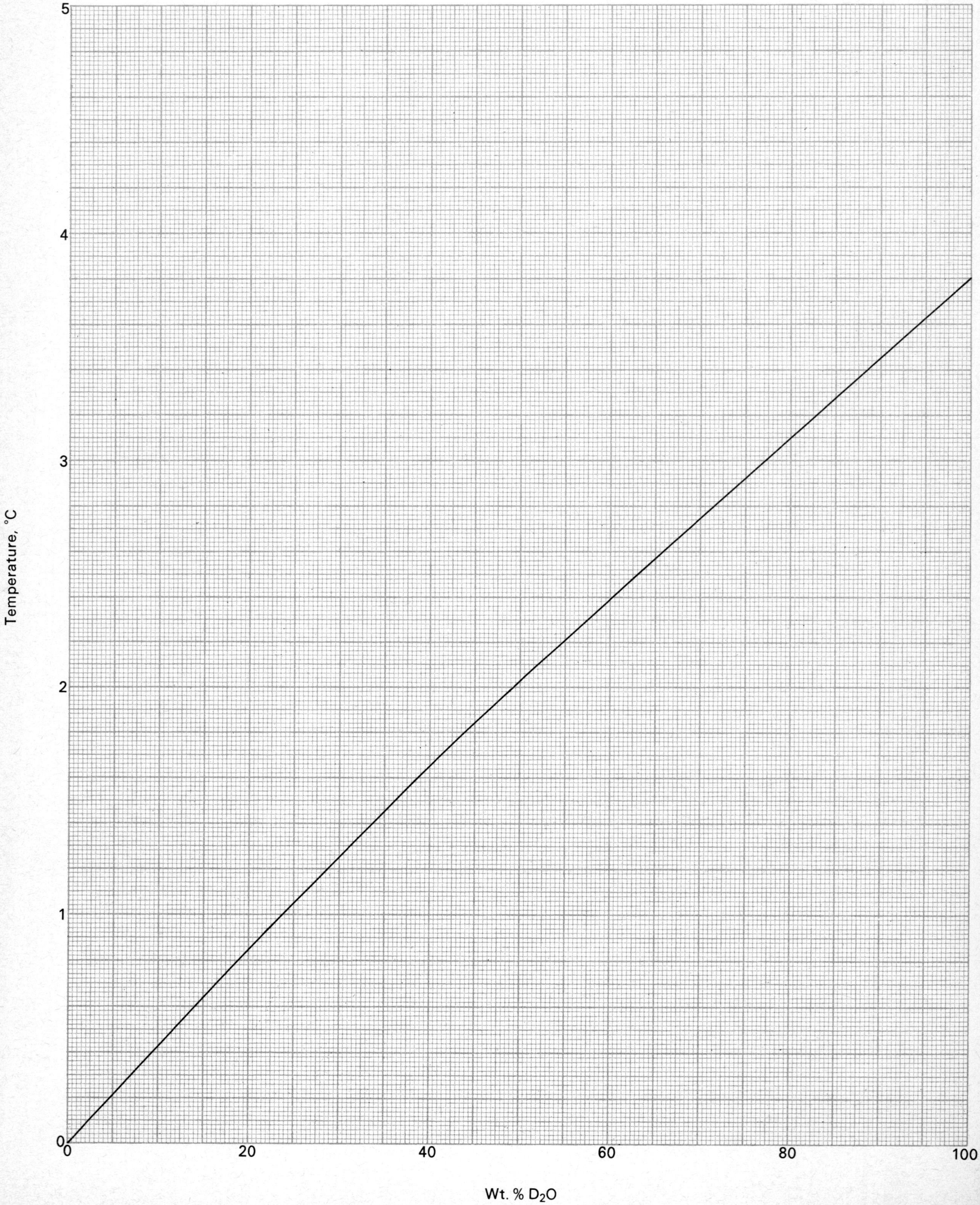

14 HYDROGEN PEROXIDE

Hydrogen peroxdie was discovered by Louis-Jacques Thenard in 1818. He concluded from his observations that it is a rather unstable compound. At room temperature and atmospheric pressure it is a water-white viscous liquid, miscible with water in all proportions. It decomposes into water and oxygen if heated much above 100°C. A rapid dissociation of H_2O_2 vapour occurs only at temperatures above 500°C and moderate pressures (5). The vapour of H_2O_2 possesses a pungent acid odour.

It is a powerful oxidizing agent in both acid and alkaline solutions, the oxygen being reduced from an oxidation state of −1 to −2. In the liquid phase (97–100% H_2O_2), it is possible to obtain high order propagating detonations. Even the vapour phase of the 40 wt. % H_2O_2—H_2O system is explosive at atmospheric pressure. Hydrogen peroxide is used as a powerful fuel oxidant in rockets and missiles, as an antiseptic, and as a bleaching agent for hair, silk, etc.

Solid anhydrous hydrogen peroxide forms in a tetragonal crystal structure, each cell containing four H_2O_2 molecules.

A comprehensive discussion of physical and thermodynamic properties has been given by Schumb *et al.* (30). The collected properties of H_2O_2 have been tabulated and presented in several books: Dierdoff (4), Gmelin (11), Kirk-Othmer (13), Mellor (20), Pascal (23), Tsykalo and Tabachnikov (36), and Ullmann (37). Further review works of interest are: Laporte (14) and Matheson and Maass (19).

References

1 Abrahams, S. C., Collins, R. L., Lipscomb, W. N., *Acta Cryst.*, 1951, **4**, 15–20.

2 American Conference of Government Industrial Hygienists, *Documentation of Threshold Limit Values*, revised edition, Committee on Threshold Limit Values, Cincinnati, Ohio 45202, 1966, 203 pp.

3 Chao, J., Wilhoit, R. C., Zwolinski, B. J., *J. Chem. Thermodyn.*, 1971, **3**, 497–505.

4 Dierdorff, L. H., *Hydrogen Peroxide Physical Properties Data Book*, Beccio Chemical Division, Food Machinery and Chemical Corp., Buffalo, 1955.

5 Elder, L. W., Rideal, E. K., *Trans. Faraday Soc.*, 1929, **23**, 545–52.

6 Engineering Science Data Units, *Approximate Data on the Viscosity of some Common Liquids*, Item No. 66024, London, July 1966, 26 pp.

7 Foley, W. T., Giguere, P. A., *Can. J. Chem.*, 1951, **29**, 895–903.

8 Giguere, P. A., Geoffrion, P., *Can. J. Chem.*, 1949, **27B**, 168–73.

9 Giguere, P. A., Geoffrion, P., *Can. J. Chem.*, 1950, **28B**, 599–607.

10 Giguere, P. A., Liu, I. D., Dugdale, J. S., Morrison, J. A., *Can. J. Chem.*, 1954, **32**, 117–28.

11 *Gmelins Handbuch der anorganischen Chemie*, Syst. No. 3, Sauerstoff, Lief. 7, Verlag Chemie, GmbH, Weinheim, 1966, pp. 2162–84.

12 IUPAC, Atomic Weights of the Elements 1971, *Pure & Appl. Chem.*, 1972, **30** (3–4), 638–49.

13 *Kirk-Othmer Encyclopedia of Chemical Technology*, 2nd ed., Vol. 11, Interscience Publ., New York, 1966, pp. 391–417.

14 Laporte Chemicals Limited, *Hydrogen Peroxide Data Manual*, Luton, Aug. 1960, 80 pp.

15 Lax, E., Synowietz, C., Eds., *Taschenbuch für Chemiker und Physiker*, 3rd ed., Springer-Verlag, Berlin, 1967.

16 Maass, O., Hatcher, W. H., *J. Am. Chem. Soc.*, 1920, **42**, 2548–69.

17 Maass, O., Herzberg, O. W., *J. Am. Chem. Soc.*, 1920, **42**, 2569–70.

18 Maryott, A. A., Smith, E. R., *Table of Dielectric Constants of Pure Liquids*, Natl. Bur. Std. Circular 514, Washington, Aug. 10, 1951, 44 pp.

19 Matheson, G. L., Maass, O., *J. Am. Chem. Soc.*, 1929, **51**, 674–87.

20 *Mellor's Comprehensive Treatise on Inorganic and Theoretical Chemistry*, Vol. 1, H. O. Longmans, London, 1922, pp. 929–36.

21 McMurtrie, R. L., Keyes, F. G., *J. Am. Chem. Soc.*, 1948, **70**, 3755–8.

22 Papousek, D., *Coll. Czech. Chem. Comm.*, 1958, **23**, 1767–73.

23 Pascal, P., Ed., *Nouveau Traité de Chimie Minéralé*, Vol. XIII, Part 1, Masson et Cie, Éditeurs, Paris, 1960, pp. 469–501.

24 Pauling, L., *The Nature of the Chemical Bond*, 2nd ed., Cornell Univ. Press, New York, 1948.

25 Phibbs, M. K., Giguere, P. A., *Can. J. Chem.*, 1951, **29**, 173–81.

26 Rossini, F. D., Wagman, D. D., Evans, W. H., Levine, S., Jaffe, I., *Selected Values of Chemical Thermodynamic Properties*, Natl., Bur. Std. Circular 500, Washington, Feb. 1, 1952.

27 Roth, E. M., Shanley, E. S., *Ind. Eng. Chem.*, 1953, **45**, 2343–9.

28 Rudenko, H. C., Konareva, B. G., *Zh. Fiz. Khim.*, 1963, **37** (12), 2761–2.

29 Scatchard, G., Kavanagh, G. M., Ticknor, L. B., *J. Am. Chem. Soc.*, 1952, **74**, 3715–20.

30 Schumb, W. C., Satterfield, C. N., Wentworth, R. L., *Hydrogen Peroxide*, Am. Chem. Soc. Monograph Ser. No. 128, Reinhold Publ. Corp., New York, 1955, 759 pp.

31 Simkin, D. J., Hurd, C. O., *Chem. Eng.*, 1958 (Jan. 13), **65**, 155–6.

32 Spar, C., *Mech. Eng.*, 1966, **88** (2), 49–54.

33 Stiel, L. I., Thodos, G., *AIChE Journal*, 1964, **10** (1), 26–30.

34 Stull, D. R., Dir., *JANAF Thermochemical Tables*, Dow Chemical Co., Midland, Mich., 1965–8.

35 Svehla, R. A., *Estimated Viscosities and Thermal Conductivities of Gases at High Temperatures*, Lewis Res. Center Techn. Rept. R-132, Cleveland, Ohio, 1962, 140 pp.

36 Tsykalo, A. L., Tabachnikov, A. G., Thermodynamic Properties of Dissociating Hydrogen Peroxide, in *Thermophysical Properties of Gases and Liquids*, No. 1, (Rabinovich, V. A., Ed), Izdatel'stvo Komiteta Standartov, Mer i Izmeritel'nykh Priborov pri Sovete Ministrov SSSR, Moskva, 1968.

37 *Ullmanns Encyklopädie der technischen Chemie*, 3rd ed., Vol. 13, Urban & Schwarzenberg, München, 1962, pp. 211–6.

Table 14.1 Properties of Hydrogen Peroxide

		Ref.
Chemical formula	H_2O_2	(15)
Molecular weight	34.0138	(12)
Critical temperature	435.35 °C	(36)
Critical pressure	155.3×10^5 N m^{-2}	(36)
Critical density	352.48 kg m^{-3}	(36)
Triple-point temperature	−0.42 °C	(29)
Triple-point pressure	34.66 N m^{-2}	(29)
Boiling point at 1.0133×10^5 N m^{-2}	150.2 °C	(15)
Melting point	−0.41 °C	(15)
Transition temperature	None	(10)
Density of solid at −20.40 °C	1710.0 kg m^{-3}	(1)
Compressibility (adiabatic) of liquid at 20 °C	2.1492×10^{-10} m^2 N^{-1}	(22)
Coefficient of thermal expansion of liquid at 0 °C	7.9×10^{-4} °C^{-1}	(9)
Heat of formation of gas at 25 °C	-1.3620×10^8 J kg-mol^{-1}	(34)
Free energy of formation of gas at 25 °C	-1.0136×10^8 J kg-mol^{-1}	(34)
Heat of sublimation at triple-point temperature	6.5230×10^7 J kg-mol^{-1}	(24)
Heat of fusion	1.0551×10^7 J kg-mol^{-1}	(26)
Heat of transition	None	(10)
Entropy of gas at 25 °C	2.3304×10^5 J kg-mol^{-1} °C^{-1}	(34)
Heat capacity ratio of liquid at 20 °C	1.233	(22)
Velocity of sound at 20 °C and 1.0133×10^5 N m^{-2}	1792.5 m s^{-1}	(22)
Diffusion in air at 60 °C and 1.0133×10^5 N m^{-2}	1.88×10^{-5} m^2 s^{-1}	(21)
Dielectric constant of liquid at 0 °C and 1.0133×10^5 N m^{-2}	84.2	(18)
Electrical conductivity of liquid at 25 °C	3.9×10^{-5} ohm^{-1} m^{-1}	(27)
Refractive index of liquid at 25 °C	1.4067	(8)
Threshold limit value	(1 ppm) 1.4×10^{-6} kg m^{-3}	(2)

14.1 Vapour Pressure of Hydrogen Peroxide (36)

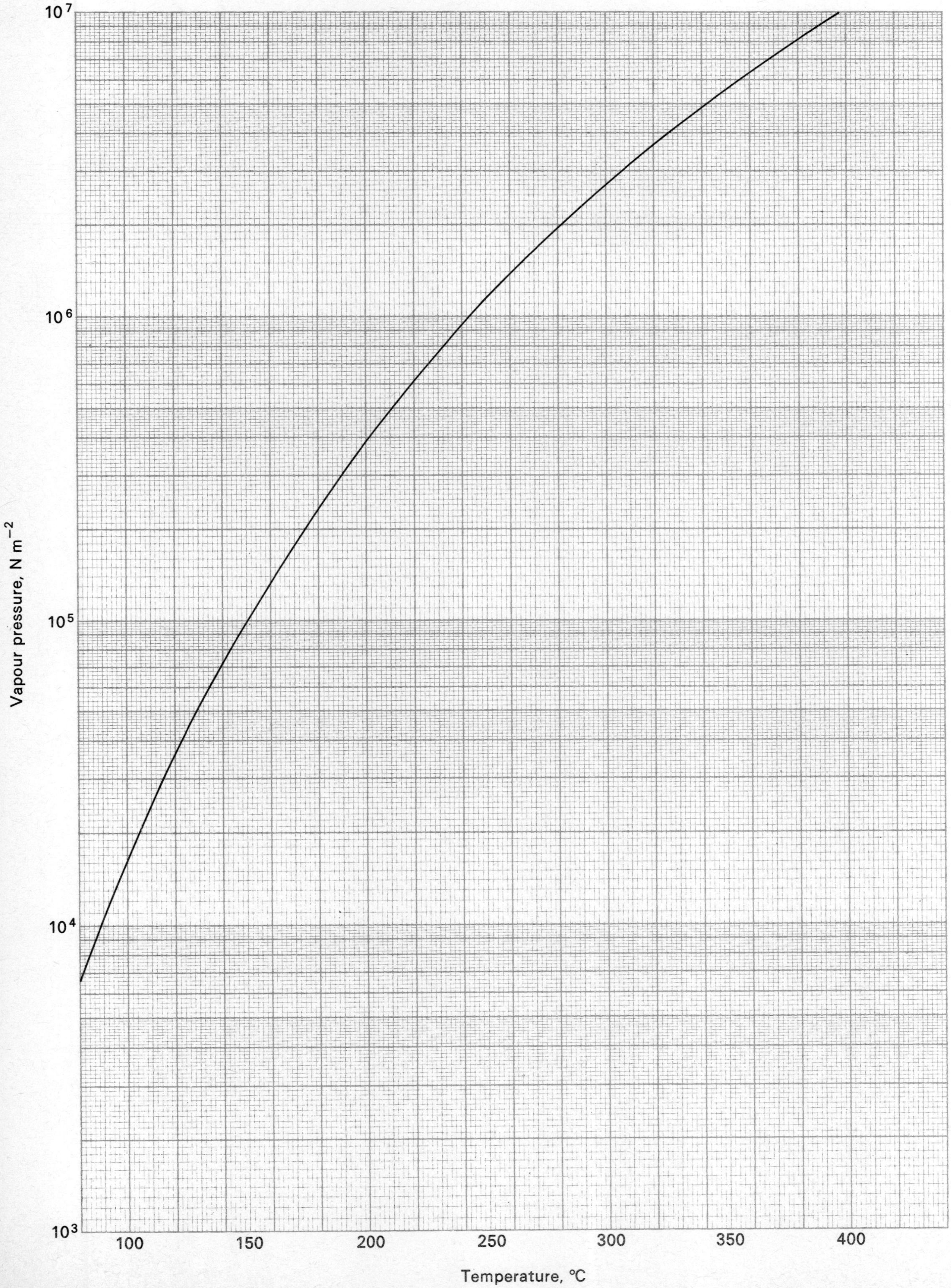

14.2 Orthobaric Density of Hydrogen Peroxide (36)

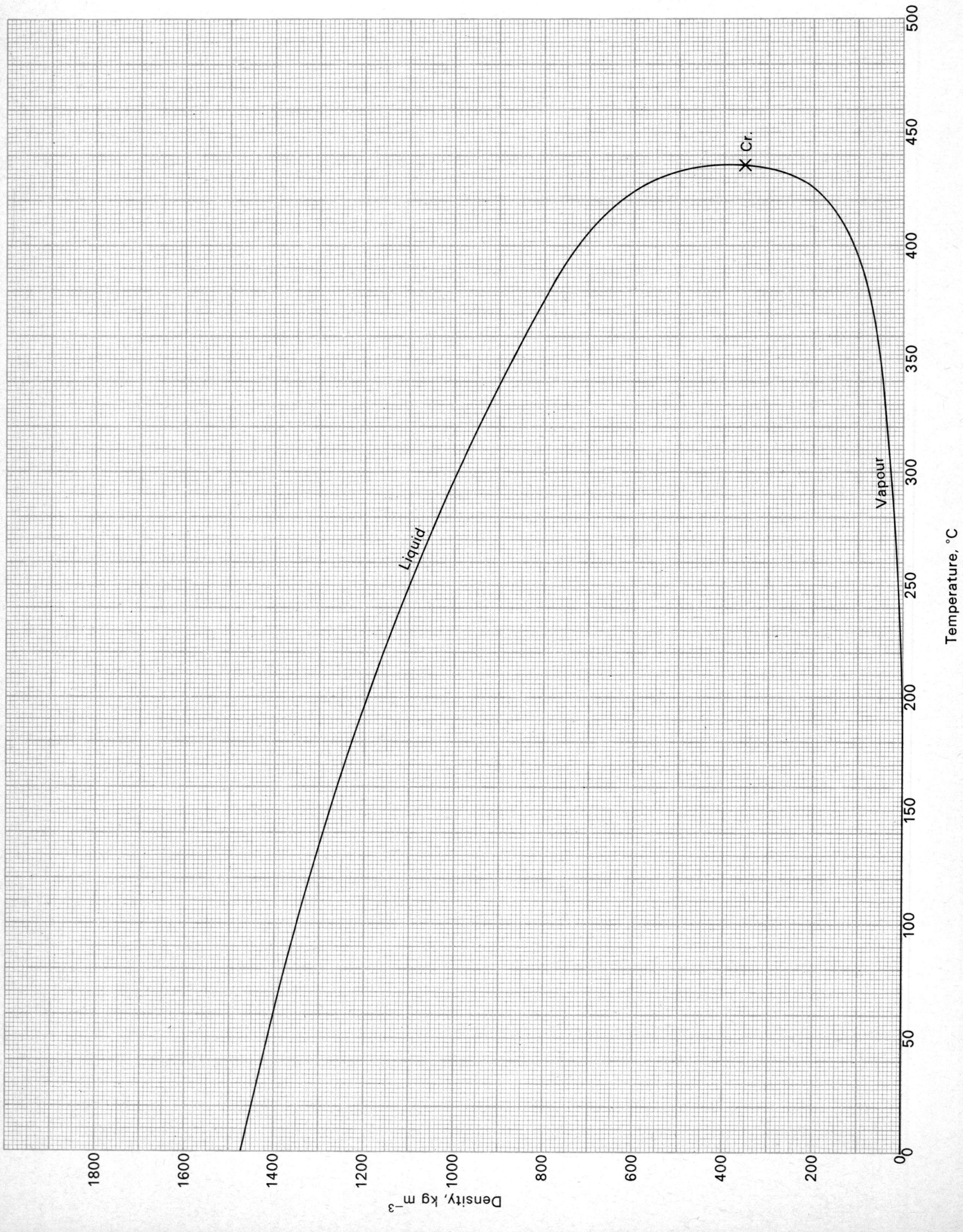

14.3 P-V-T Data for Gaseous Hydrogen Peroxide (36)

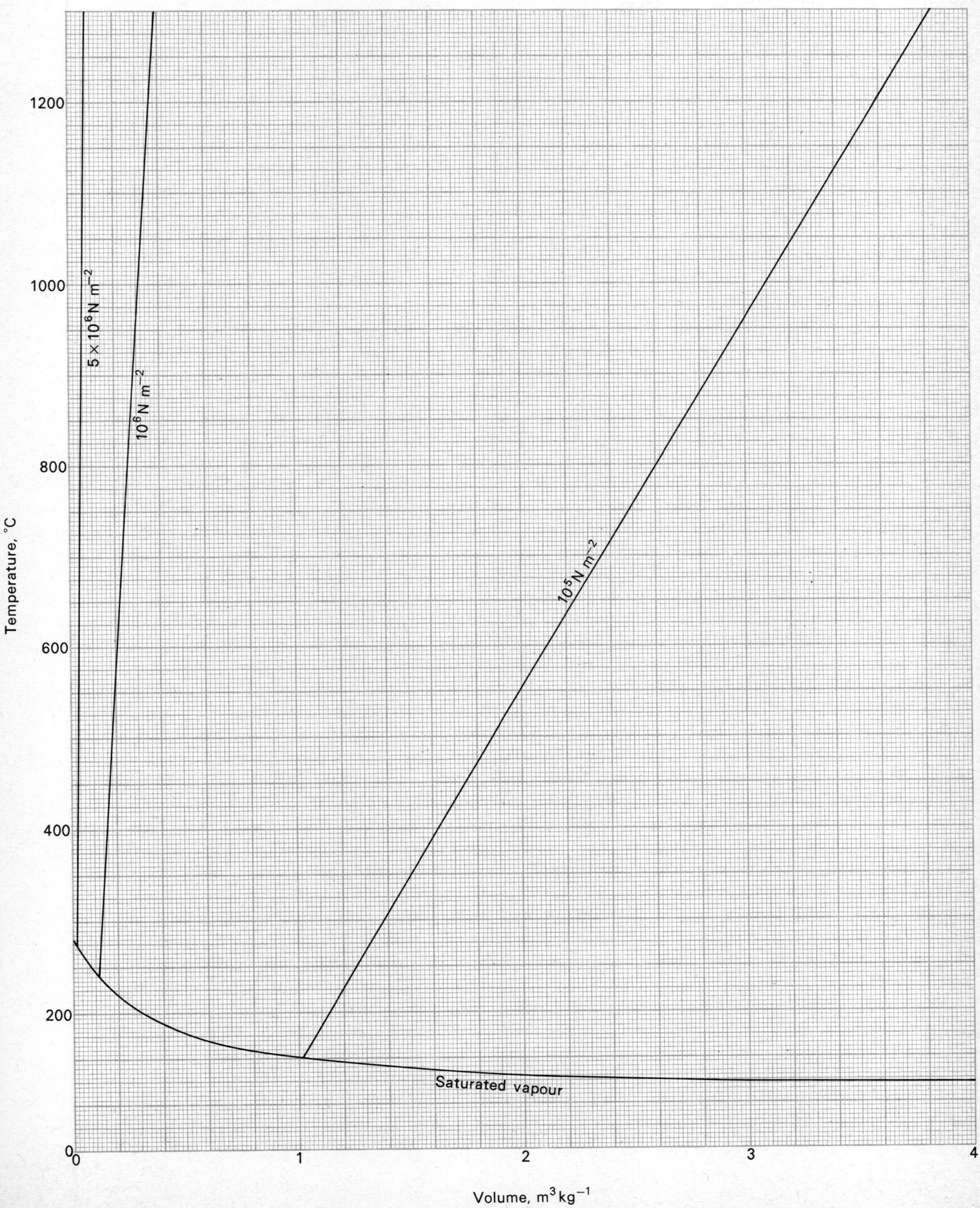

14.4 Latent Heat of Vaporisation of Hydrogen Peroxide (36)

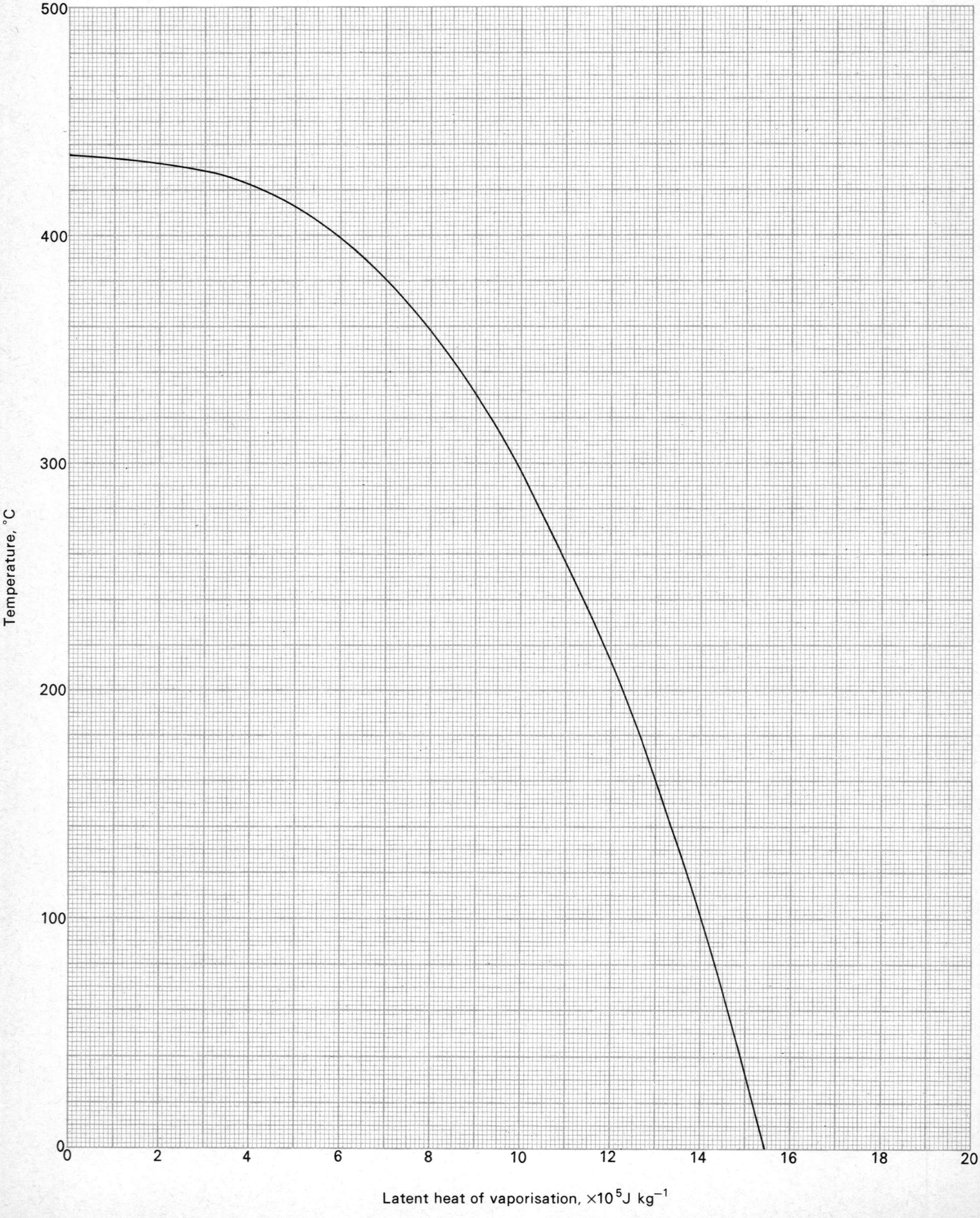

14.5 Heat Capacity of Solid Hydrogen Peroxide (10)

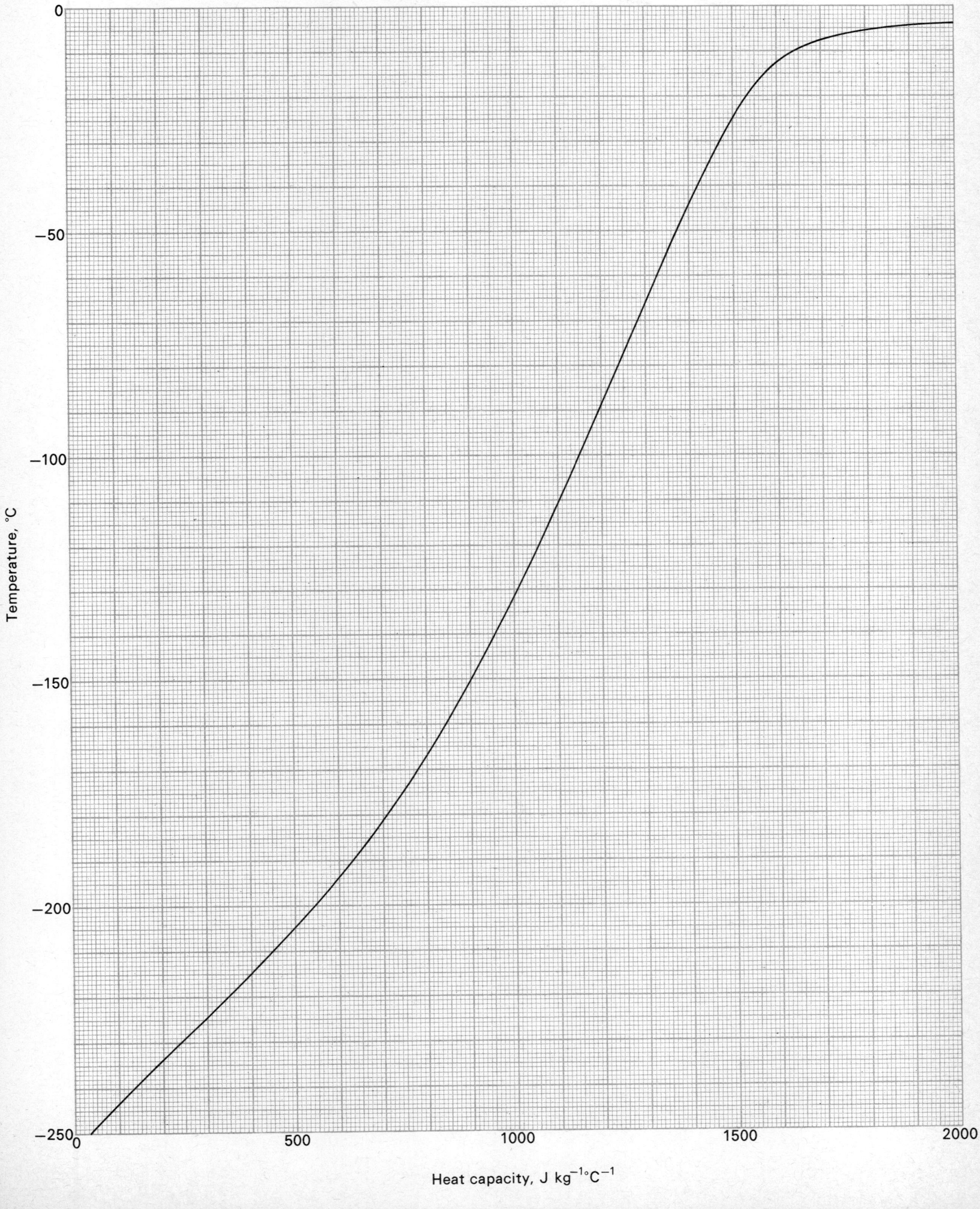

14.6 Heat Capacity of Liquid Hydrogen Peroxide (7,10,31,)

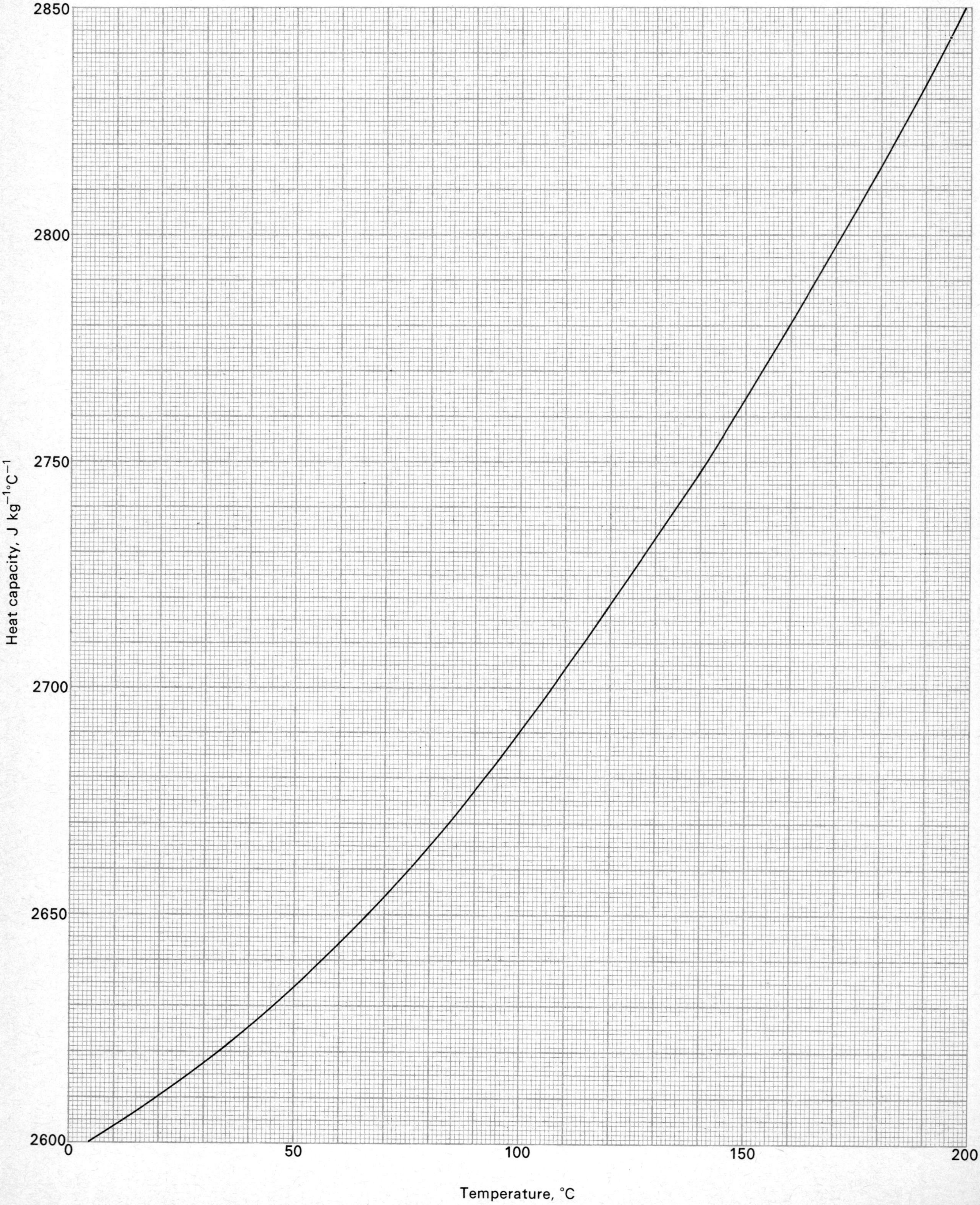

14.7 Heat Capacity of Gaseous Hydrogen Peroxide at $1.0133 \times 10^5 N\ m^{-2}$(3,34)

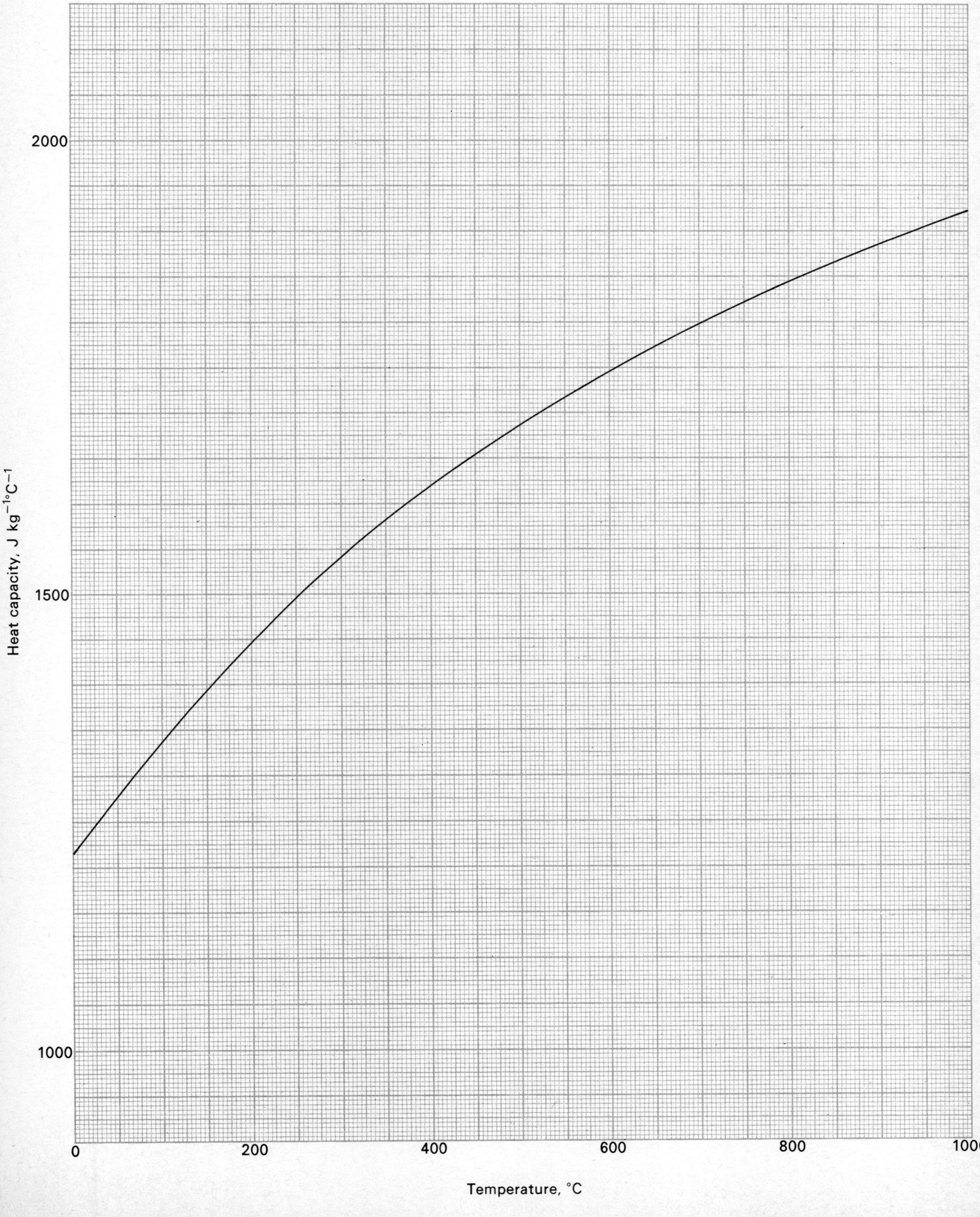

14.8 Viscosity of Liquid Hydrogen Peroxide (6,28,32)

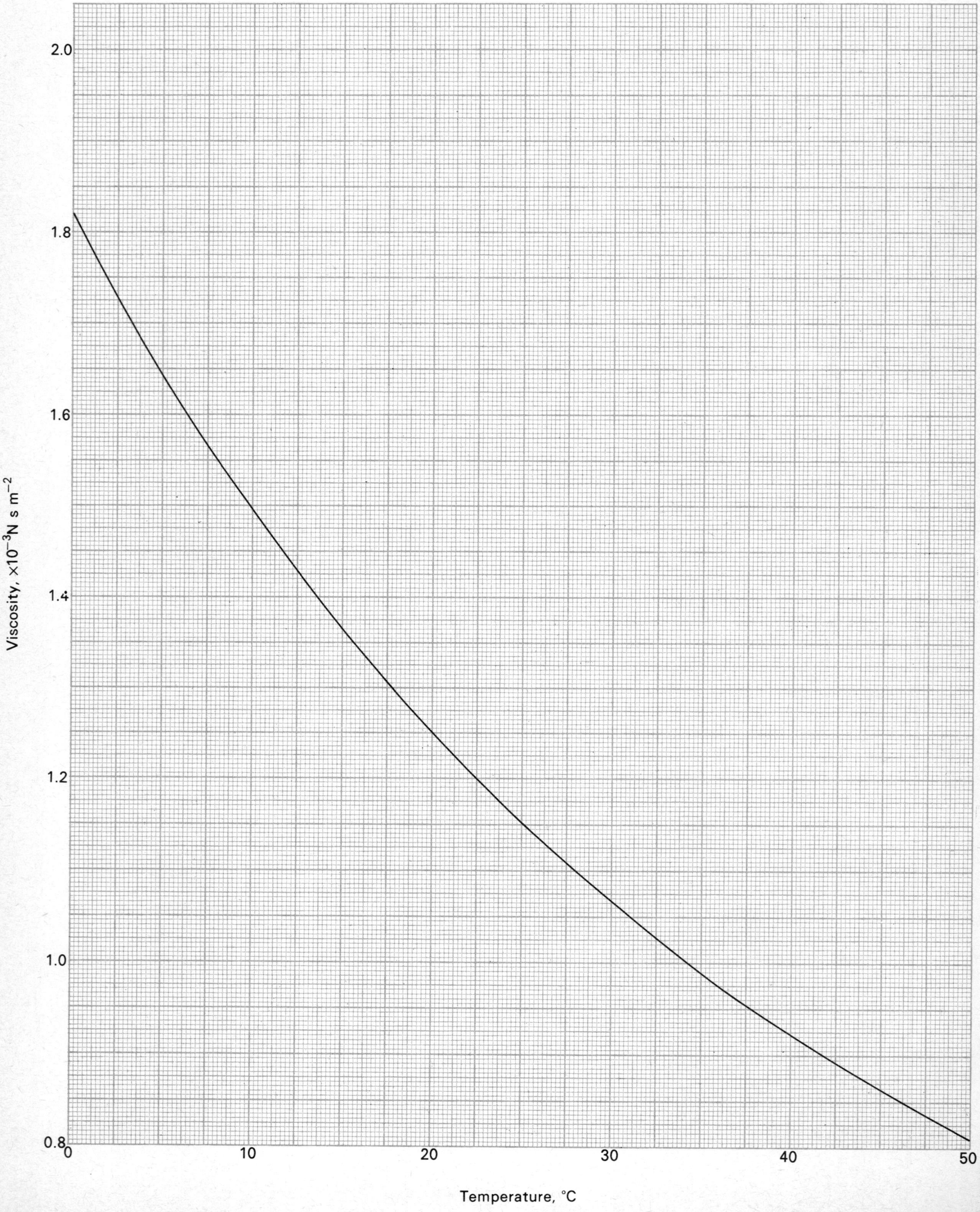

14.9 Viscosity of Gaseous Hydrogen Peroxide at $1.0133 \times 10^5 N\ m^{-2}$(35)

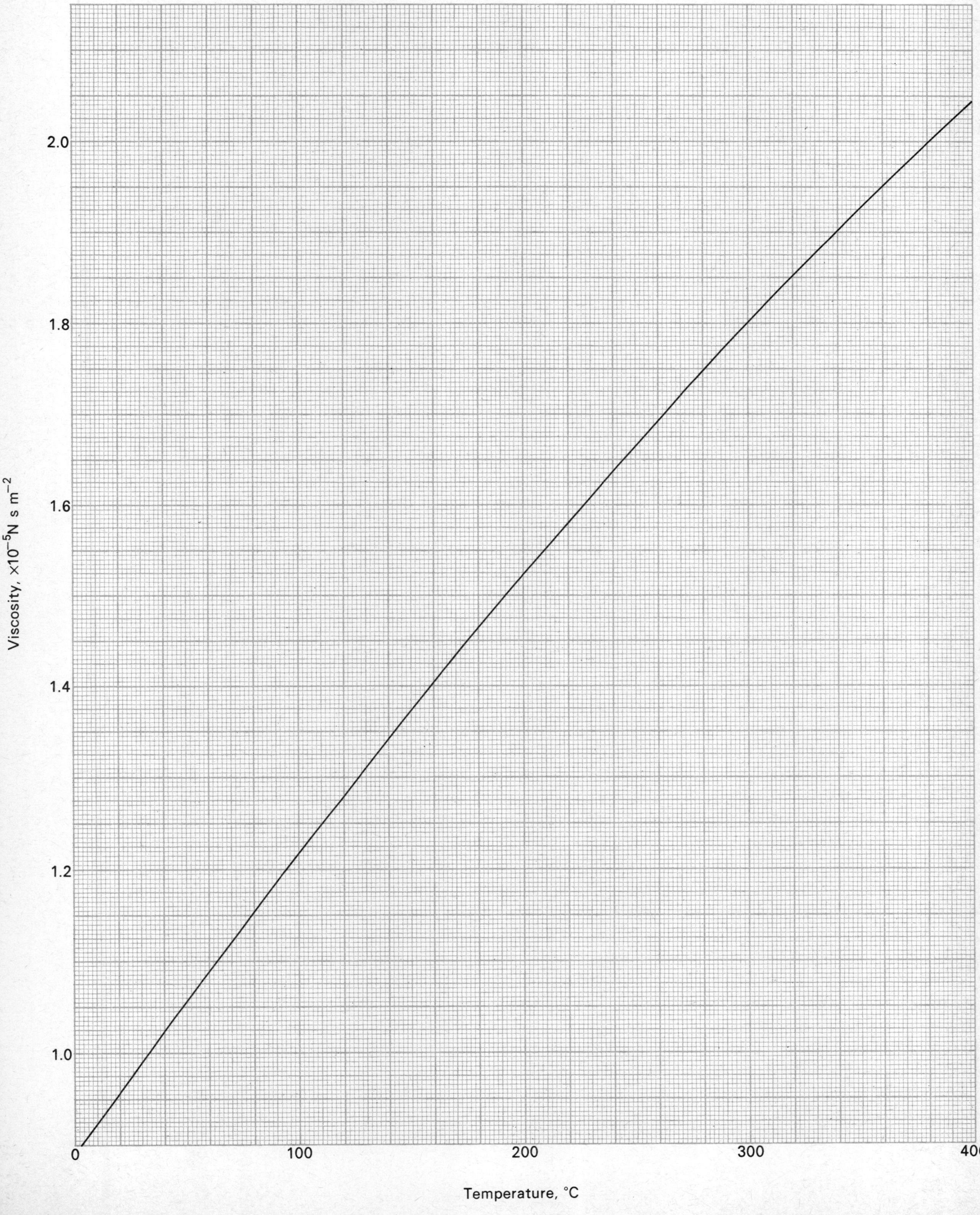

14.10 Thermal Conductivity of Liquid Hydrogen Peroxide (33)

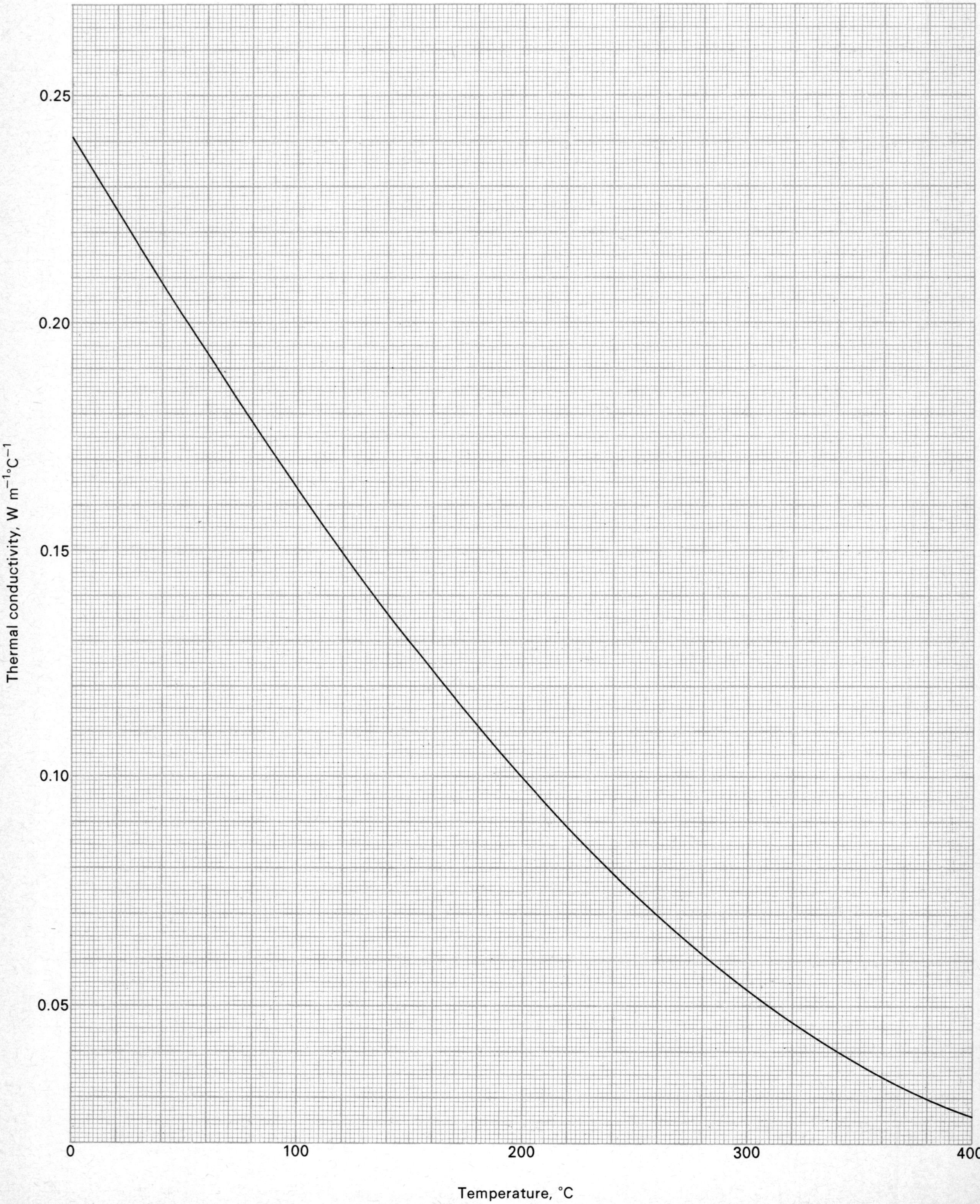

14.11 Thermal Conductivity of Gaseous Hydrogen Peroxide at $1.0133 \times 10^5 N\ m^{-2}$(35)

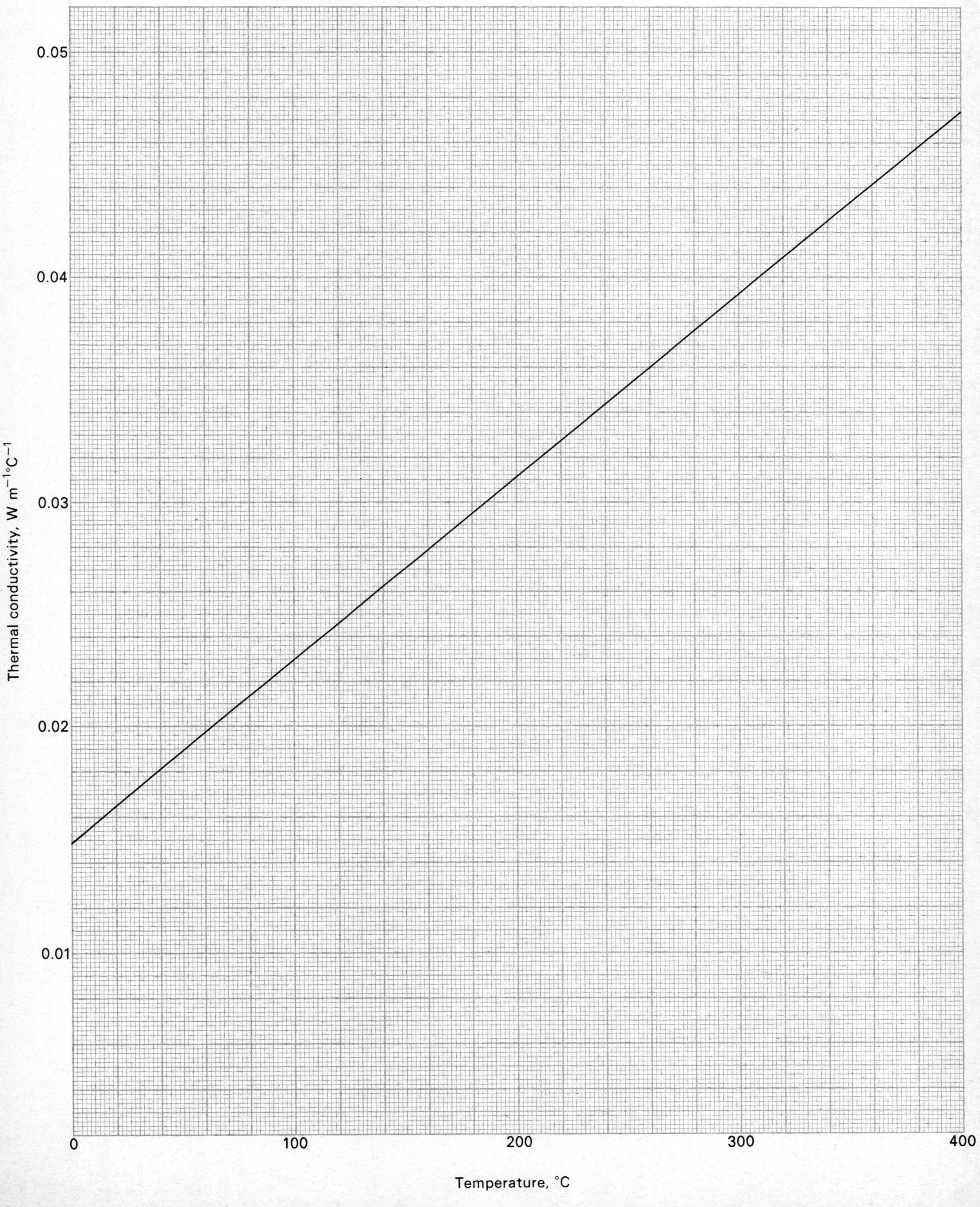

14.12 Surface Tension of Hydrogen Peroxide (16,25)

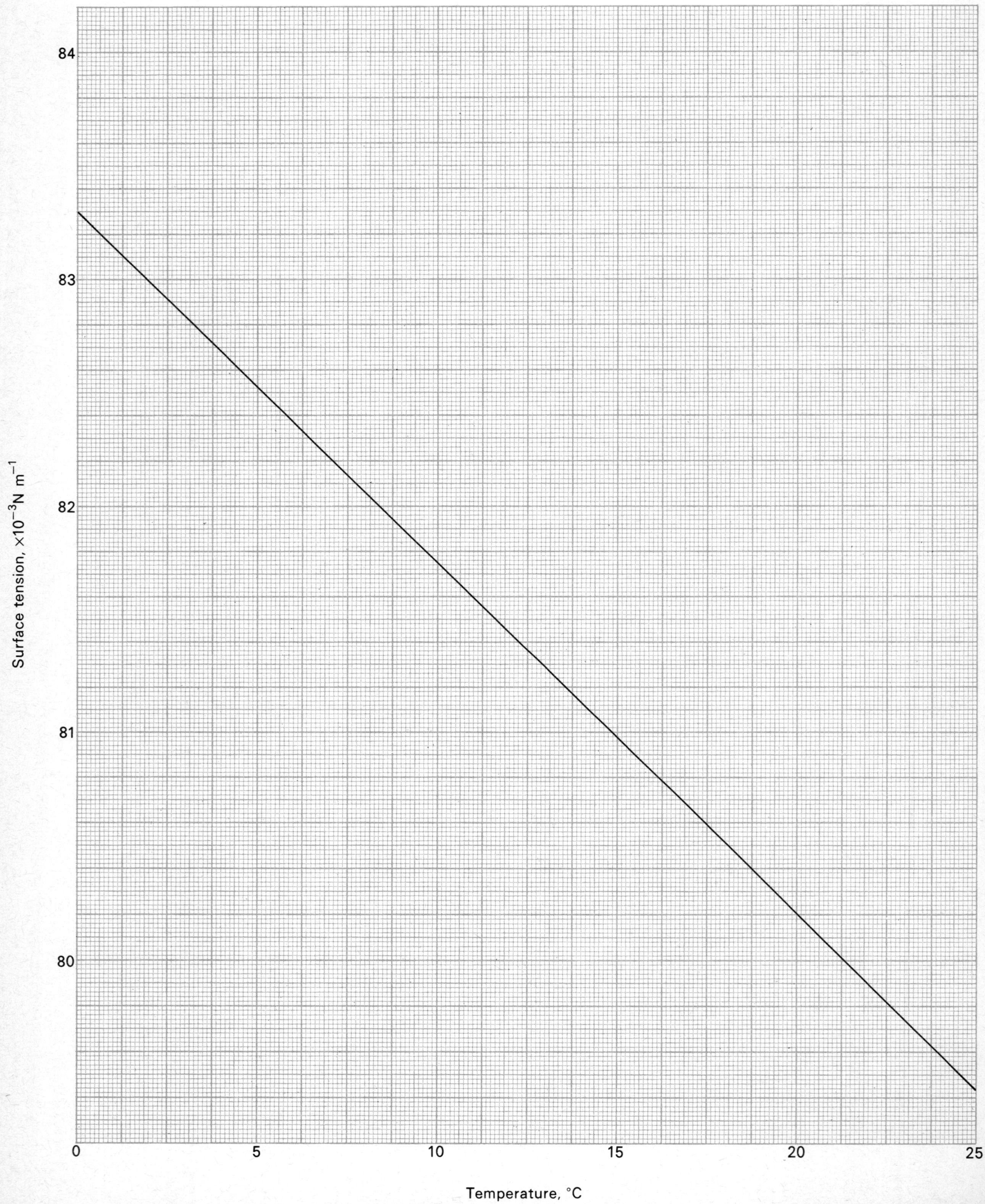

14.13 Phase Diagram for H_2O_2-H_2O System (14,17)

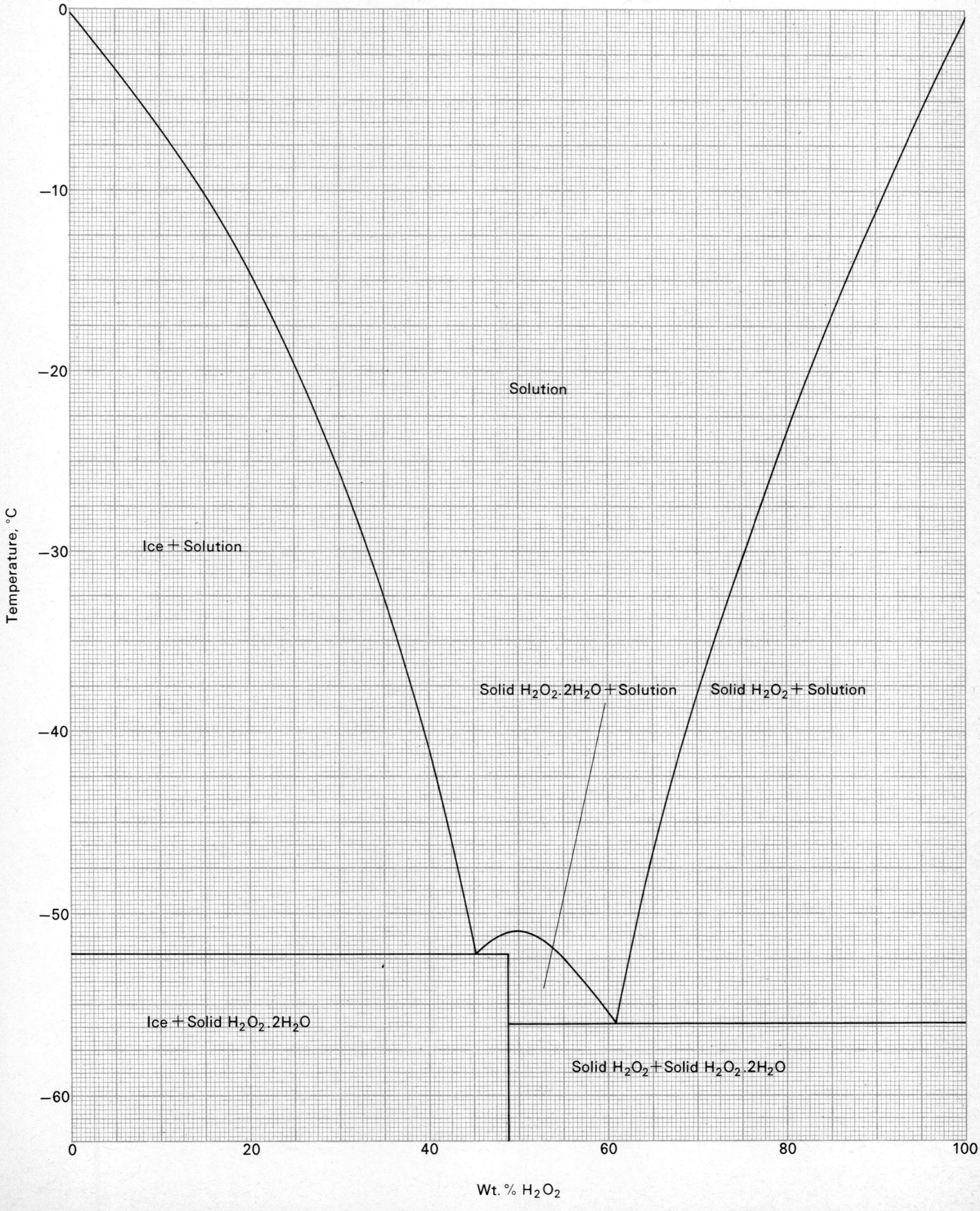

15 NITRIC ACID

Nitric acid is a colourless, fuming liquid at ambient temperature and atmospheric pressure. Above its freezing point the pure acid undergoes thermal dissociation according to the overall equation: $2\ HNO_3 \rightleftharpoons 2\,NO_2 + H_2O + \frac{1}{2}\,O_2$. During the decomposition the originally colourless liquid takes on a yellow tinge, indicating the presence of nitrogen oxides. In the pure liquid state it is made up of more than one molecular species. The HNO_3 molecule is a proton donor and proton acceptor. The acid is a strong oxidizing agent. Liquid HNO_3 is miscible with water in all proportions, forming two solid hydrates; $HNO_3.3H_2O$ (m.p. = −18.5°C) and $HNO_3.H_2O$ (m.p. = −37.7°C).

Solid nitric acid forms white orthorhombic crystals. Each unit cell contains 16 molecules of HNO_3. The crystal has an intermolecular hydrogen bond, which explains the large decrease in N—OH bond length during freezing.

The electron diffraction pattern of the vapour of HNO_3 indicates that the molecule is planar with the following structure (15):

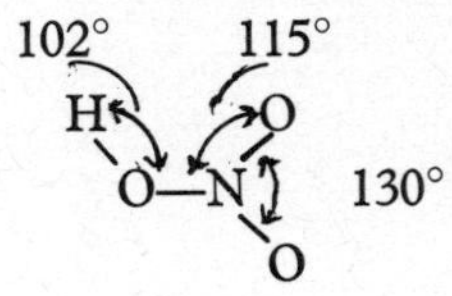

Because of dissociation of the HNO_3 molecule above its melting point, some of the physical properties have been obtained by extrapolation of data for aqueous solutions, or by thermodynamic calculations, assuming that the vapour pressure is equal to the fugacity, particularly at higher temperatures.

Compilations of various thermophysical properties of solid, liquid, and gaseous nitric acid are available in Gmelin (9), Mellor (13), Pascal (16), and Stern *et al.* (21).

The saturated vapour density, gas viscosity, and gas thermal conductivity data are not included in this presentation, because some of these are hypothetical and the predicted values are not reliable. The available estimation methods are not accurate enough for dissociating inorganic compounds. The volumetric and phase behaviour of red and white fuming nitric acids have been investigated by Reamer *et al.* (18, 19), and they might be of interest in order to derive the orthobaric vapour density of pure nitric acid, in combination with the corresponding state theory.

References

1 American Conference of Government Industrial Hygienists, *Documentation of Threshold Limit Values*, revised edition, Committee on Threshold Limit Values, Cincinnati, Ohio 45202, 1966, 203 pp.

2 Atipenko, G. L., Beletskaya, E. S., Krylova, A. G., *Zh. Prikl. Khim.*, 1958, **31**, 859.

3 Biltz, W., Hülsmann, O., Eickholz, W., *Nach. Götting. Ges.*, 1935, **1** (2), 99.

4 Bruylants, P., Mund, W., *Bull. Acad. Belg.*, 1919, p. 117.

5 Bump, T. R., Sibbitt, W. L., *Ind. Eng. Chem.*, 1955, **47** (8), 1665–70.

6 Duisman, J. A., Stern, S. A., *J. Chem. Eng. Data*, 1969, **14** (4), 457–9.

7 Forsythe, W. R., Giauque, W. F., *J. Am. Chem. Soc.*, 1942, **64**, 48.

8 Friend, J. N., Hargreaves, W. D., *Phil. Mag.*, 1943, **34**, 810–16.

9 *Gmelins Handbuch der anorganischen Chemie*, Syst. No. 4, Stickstoff, Verlag Chemie GmbH, Berlin, 1936, pp. 962–91.

10 IUPAC, Atomic Weights of the Elements 1971, *Pure & Appl. Chem.*, 1972, **30** (3–4), 638–49.

11 Küster, F. W., Kremann, R., *Z. anorg. Chem.*, 1904, **41**, 19.

12 Lax, E., Synowietz, C., Eds., *Taschenbuch für Chemiker und Physiker*, 3rd ed., Springer-verlag, Berlin, 1967.

13 *Mellor's Comprehensive Treatise on Inorganic and Theoretical Chemistry*, Vol. VIII, Suppl. 1, Part II, Nitrogen, Longmans, London, 1967, pp. 296–314.

14 Miskidzh'yan, S. P., Trifonov, N. A., *Zh. Obsh. Khim.*, 1947, **17**, 2216.

15 McGraw, G. E., Bernitt, D. L., Hisatsune, I. C., *J. Chem. Phys.*, 1965, **42** (1), 237–44.

16 Pascal, P., Ed., *Nouveau Traité de Chimie Minérale*, Vol. X, Azote, Phosphore, Masson et Cie, Éditeurs, Paris, 1956, pp. 460–98.

17 Pickering, S. U., *J. Chem. Soc.*, 1893, **63**, 440.

18 Reamer, H. H., Corcoran, W. H., Sage, B. H., *Ind. Eng. Chem.*, 1953, **45** (12), 2699–704.

19 Reamer, H. H., Mason, D. M., Sage, B. H., *Ind. Eng. Chem.*, 1953, **45** (5), 1094–7.

20 Rossini, F. D., Wagman, D. D., Evans, W. H., Levine, S., Jaffe, I., *Selected Values of Chemical Thermodynamic Properties*, Natl. Bur. Std. Circular 500, Washington, Feb. 1, 1952.

21 Stern, S. A., Mullhaupt, J. T., Kay, W. B., *Chem. Rev.*, 1960, **60**, 185–207.

22 Stull, D. R., Dir., *JANAF Thermochemical Tables*, Dow Chemical Co., Midland, Mich., 1965–8.

Table 15.1 Properties of Nitric Acid

		Ref.
Chemical formula	HNO_3	(12)
Molecular weight	63.0116	(10)
Boiling point at 1.0133×10^5 N m^{-2}	83.90°C	(6)
Melting point	−41.59°C	(6)
Transition temperature	None	(21)
Density of solid at −96.0°C	1841.0 kg m^{-3}	(21)
Compressibility (adiabatic) of liquid at 25°C	42.61×10^{-11} m^2 N^{-1}	(13)
Coefficient of thermal expansion of liquid at 18°C	1.24×10^{-3} °C^{-1}	(12)
Heat of formation of gas at 25°C	-1.3440×10^8 J kg-mol^{-1}	(22)
Free energy of formation of gas at 25°C	-8.2145×10^7 J kg-mol^{-1}	(22)
Heat of fusion	1.0480×10^7 J kg-mol^{-1}	(20)
Heat of transition	None	(21)
Entropy of gas at 25°C	2.6654×10^5 J kg-mol^{-1} C^{-1}	(22)
Heat capacity ratio of gas at 25°C and 1.0133×10^5 N m^{-2}	1.26	(13)
Velocity of sound at 25°C and 1.0133×10^5 N m^{-2}	1239.7 m s^{-1}	(13)
Electrical conductivity of liquid at 0°C	3.72 ohm^{-1} m^{-1}	(21)
Refractive index of liquid at 24°C	1.3970	(21)
Threshold limit value	(2 ppm) 5×10^{-6} kg m^{-3}	(1)

15.1 Vapour Pressure of Nitric Acid (6)

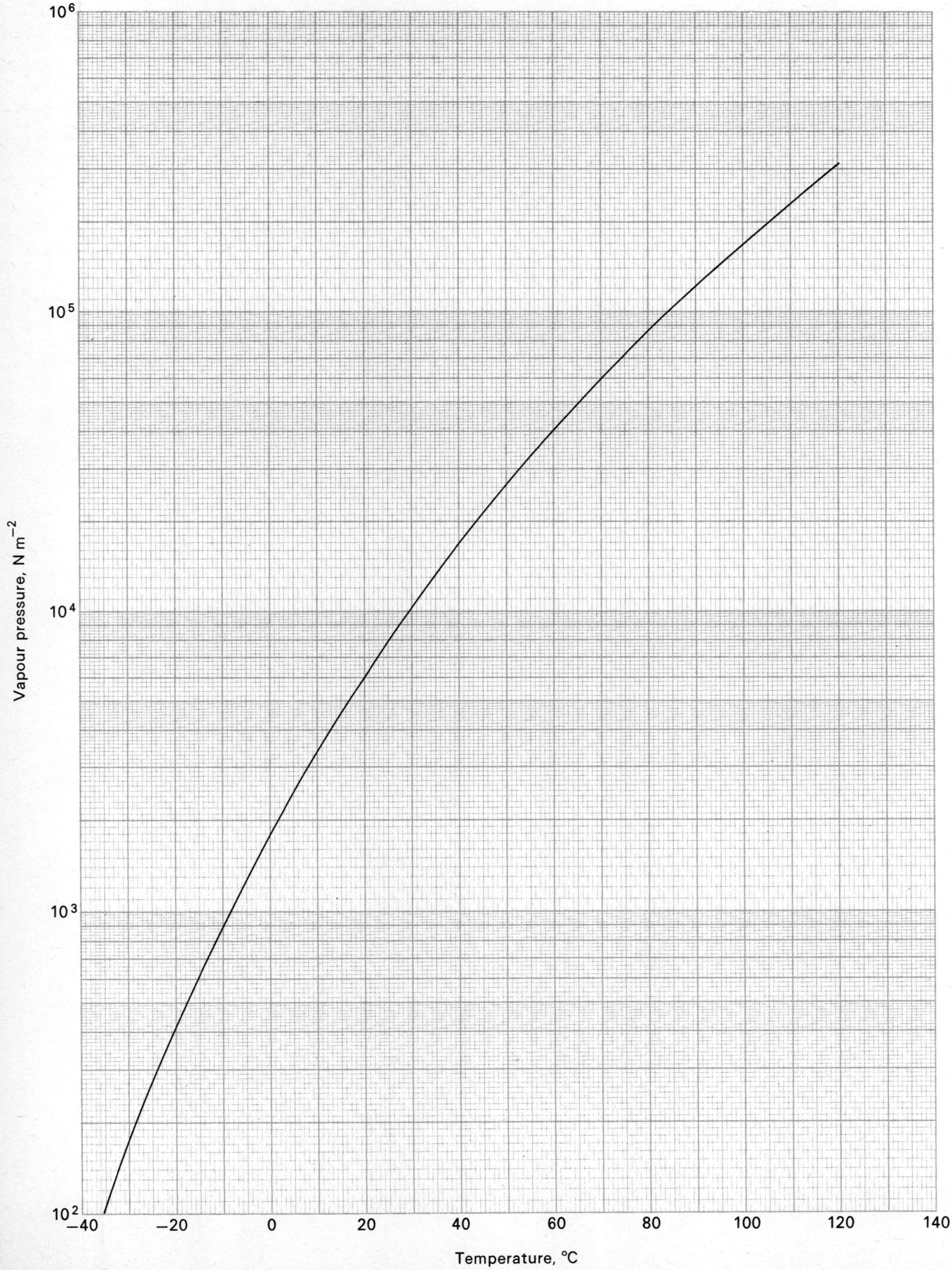

15.2 Liquid Density of Nitric Acid (2,12)

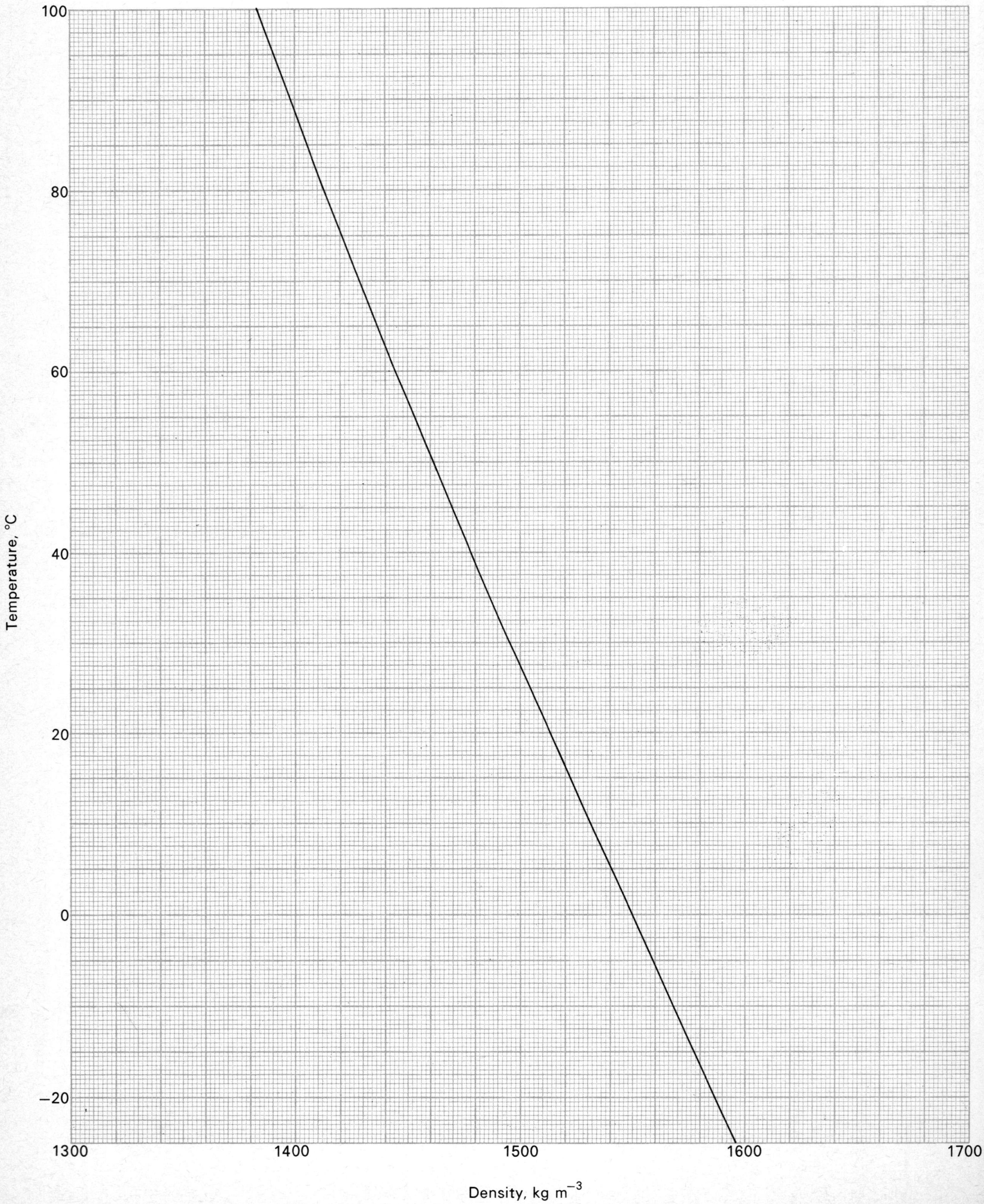

15.3 P-V-T Data for Nitric Acid (16)

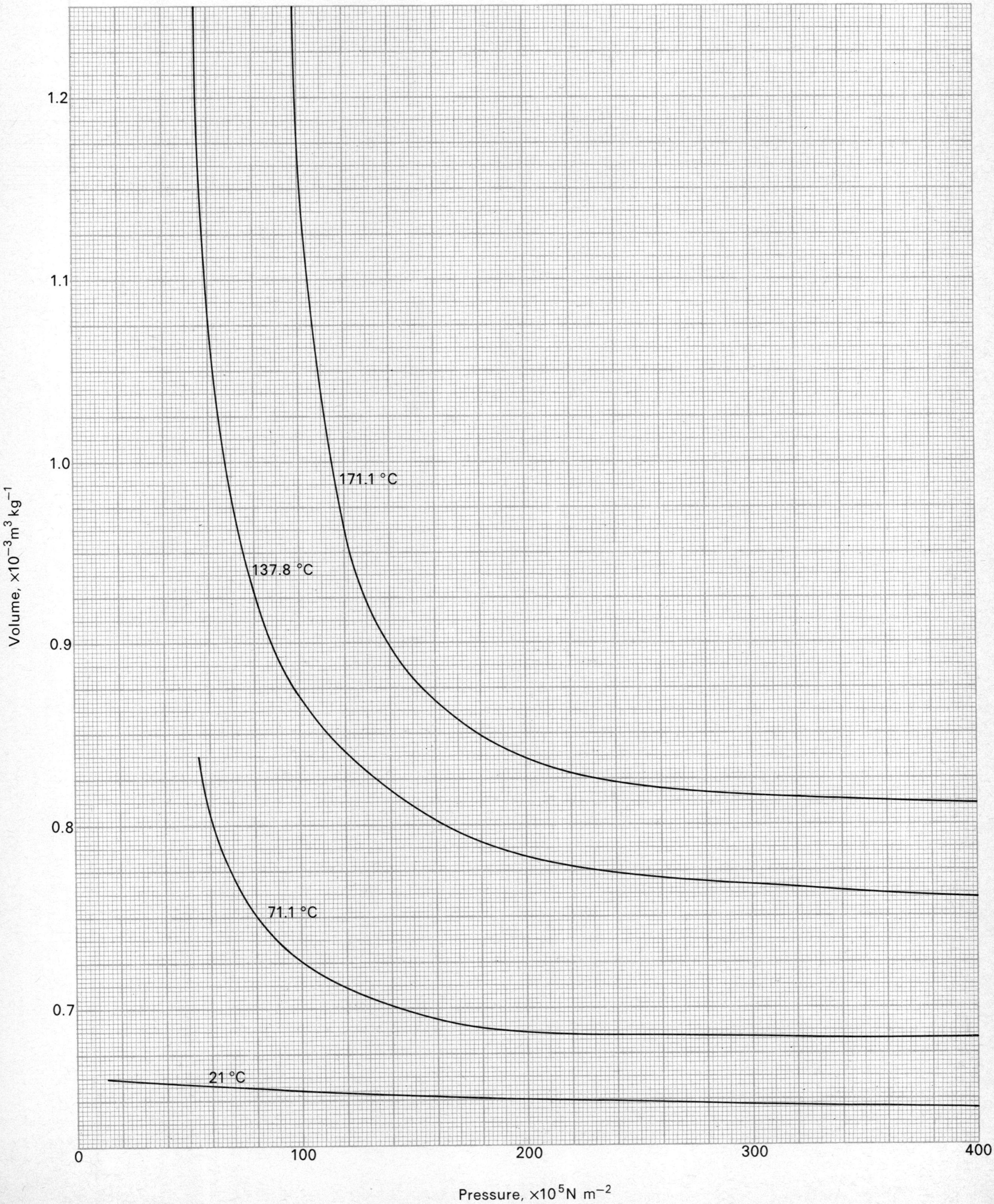

15.4 Latent Heat of Vaporisation of Nitric Acid (6,21)

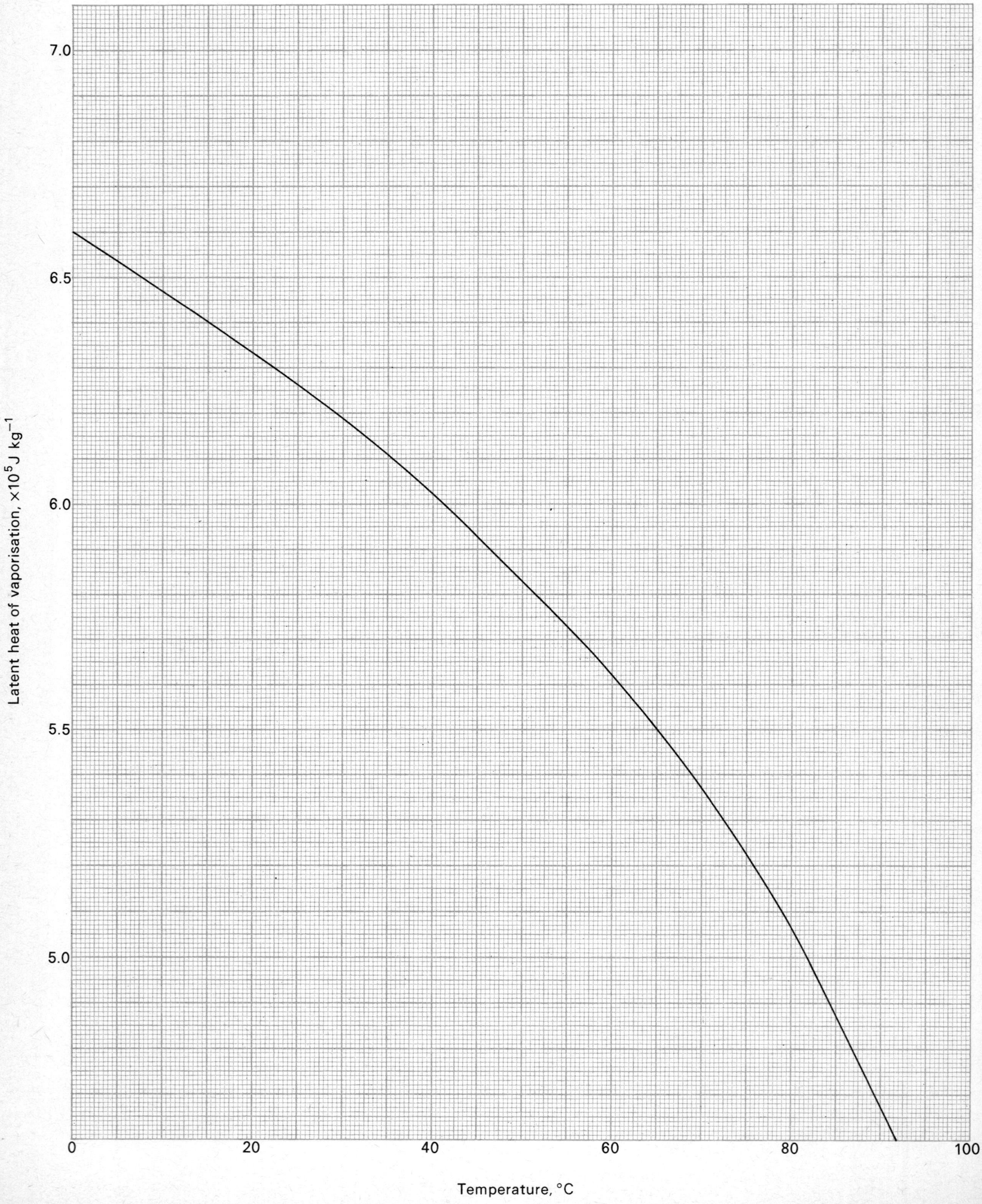

15.5 Heat Capacity of Liquid Nitric Acid (5,7)

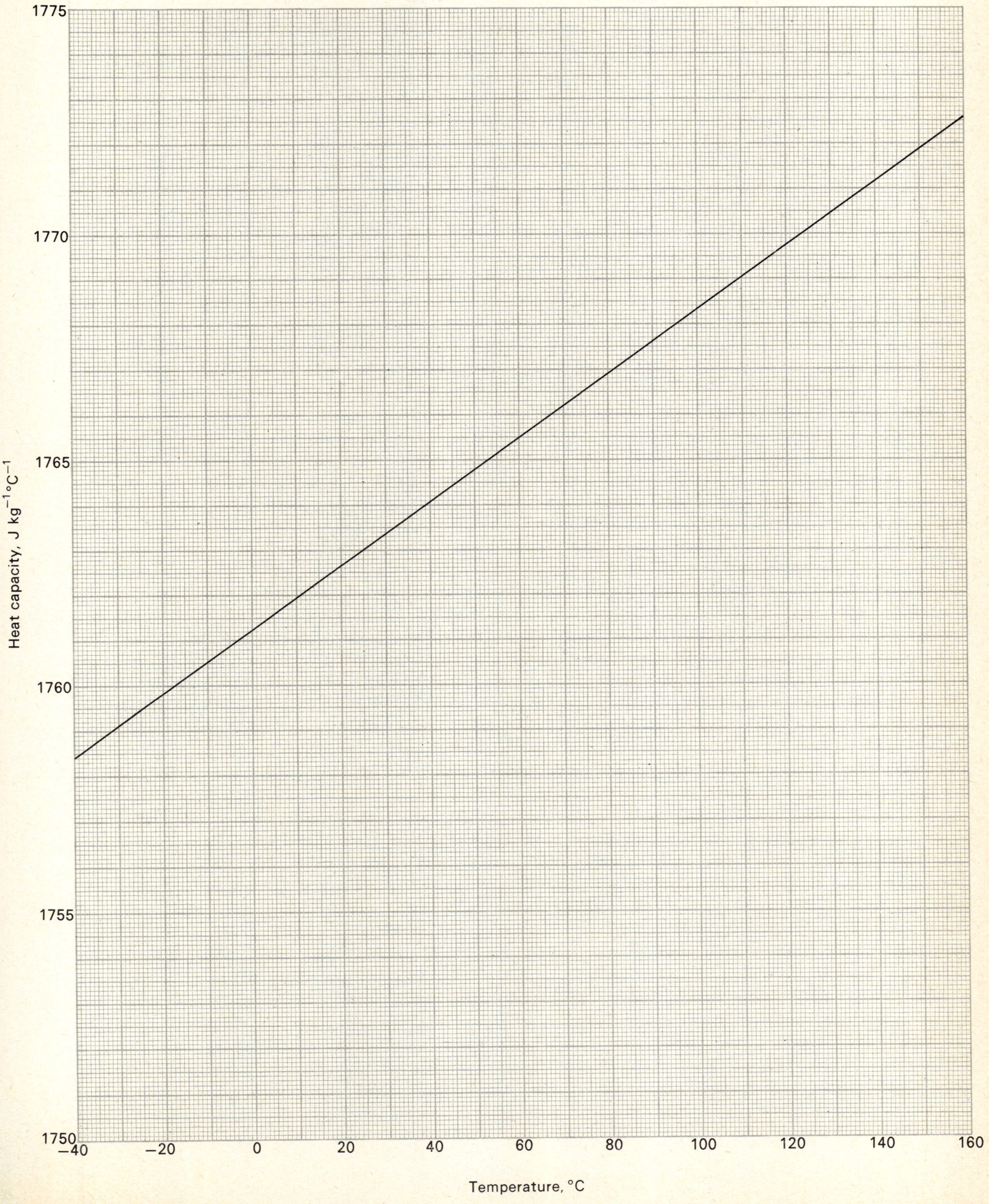

15.6 Heat Capacity of Gaseous Nitric Acid at 1.0133×10^5 N m^{-2} (6,22)

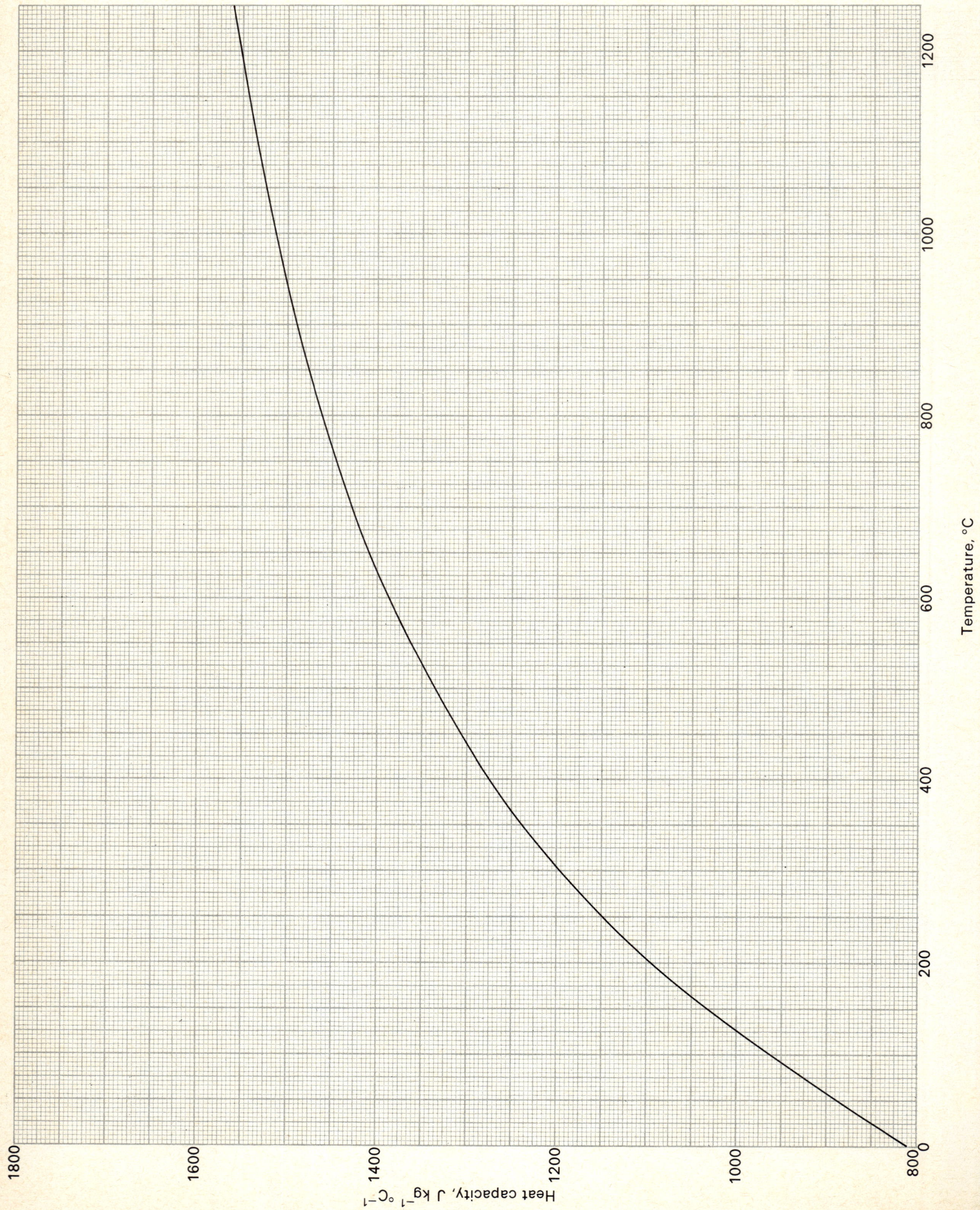

15.7 Viscosity of Liquid Nitric Acid (8,21)

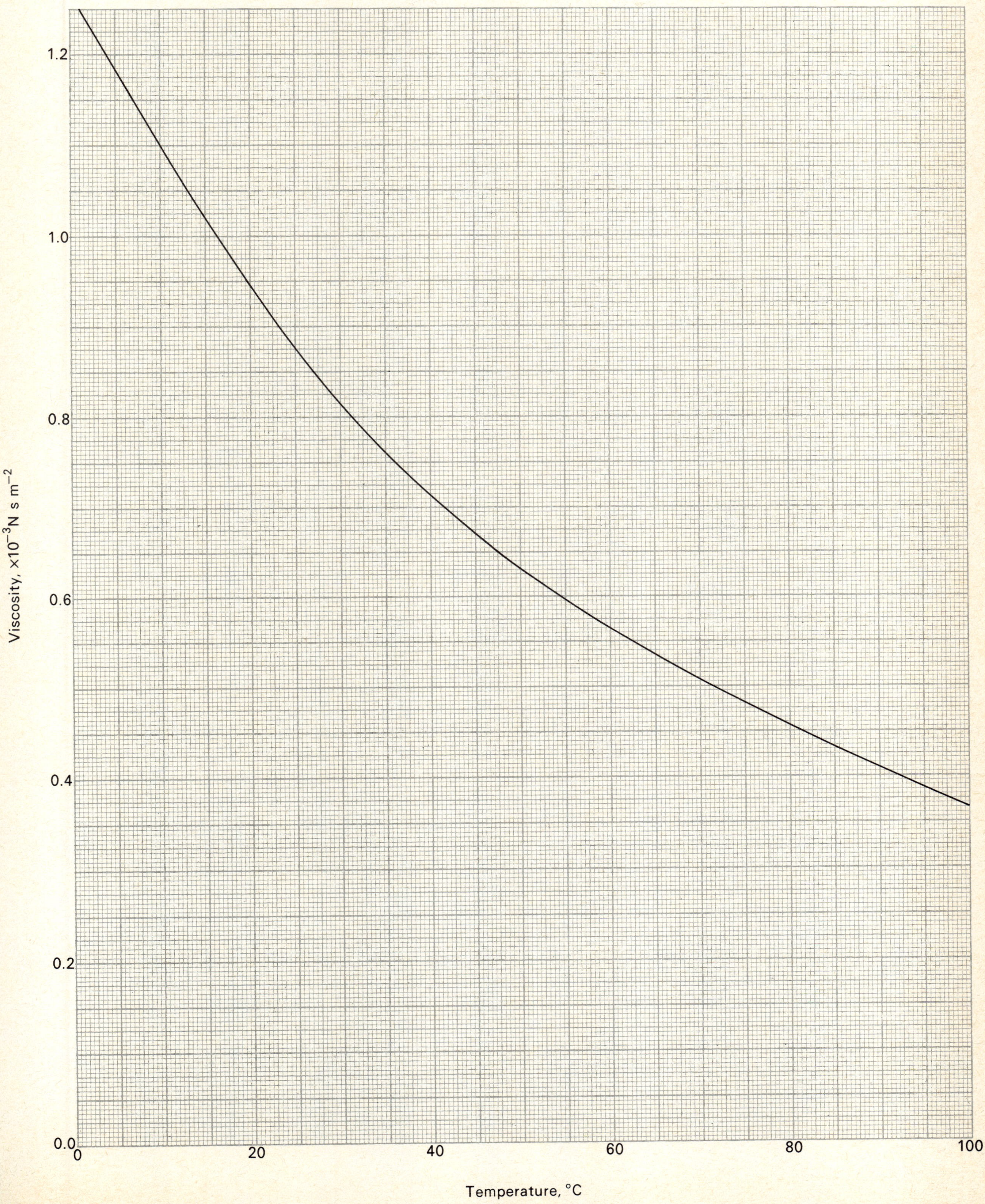

15.8 Thermal Conductivity of Liquid Nitric Acid (5)

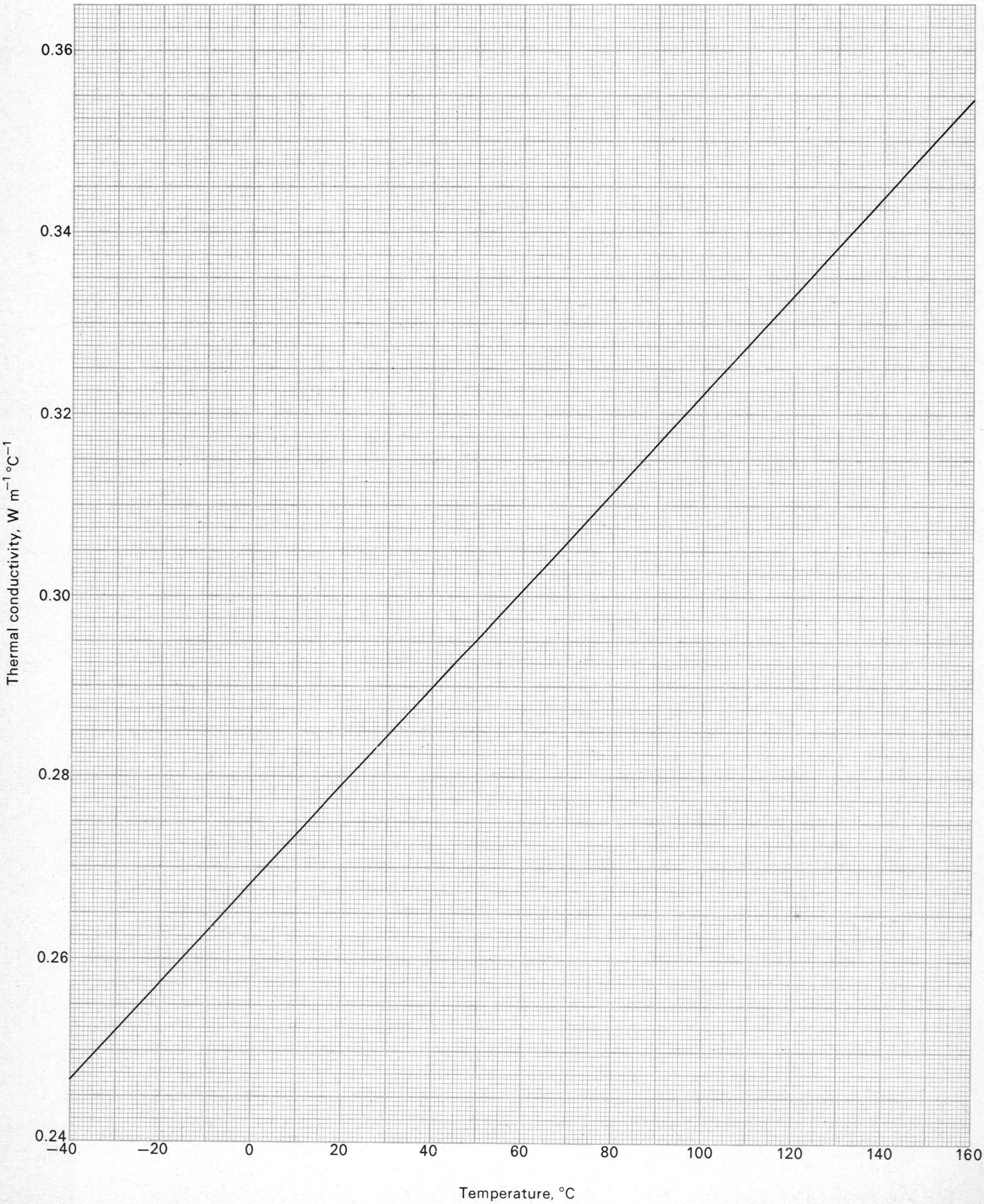

15.9 Surface Tension of Nitric Acid (14)

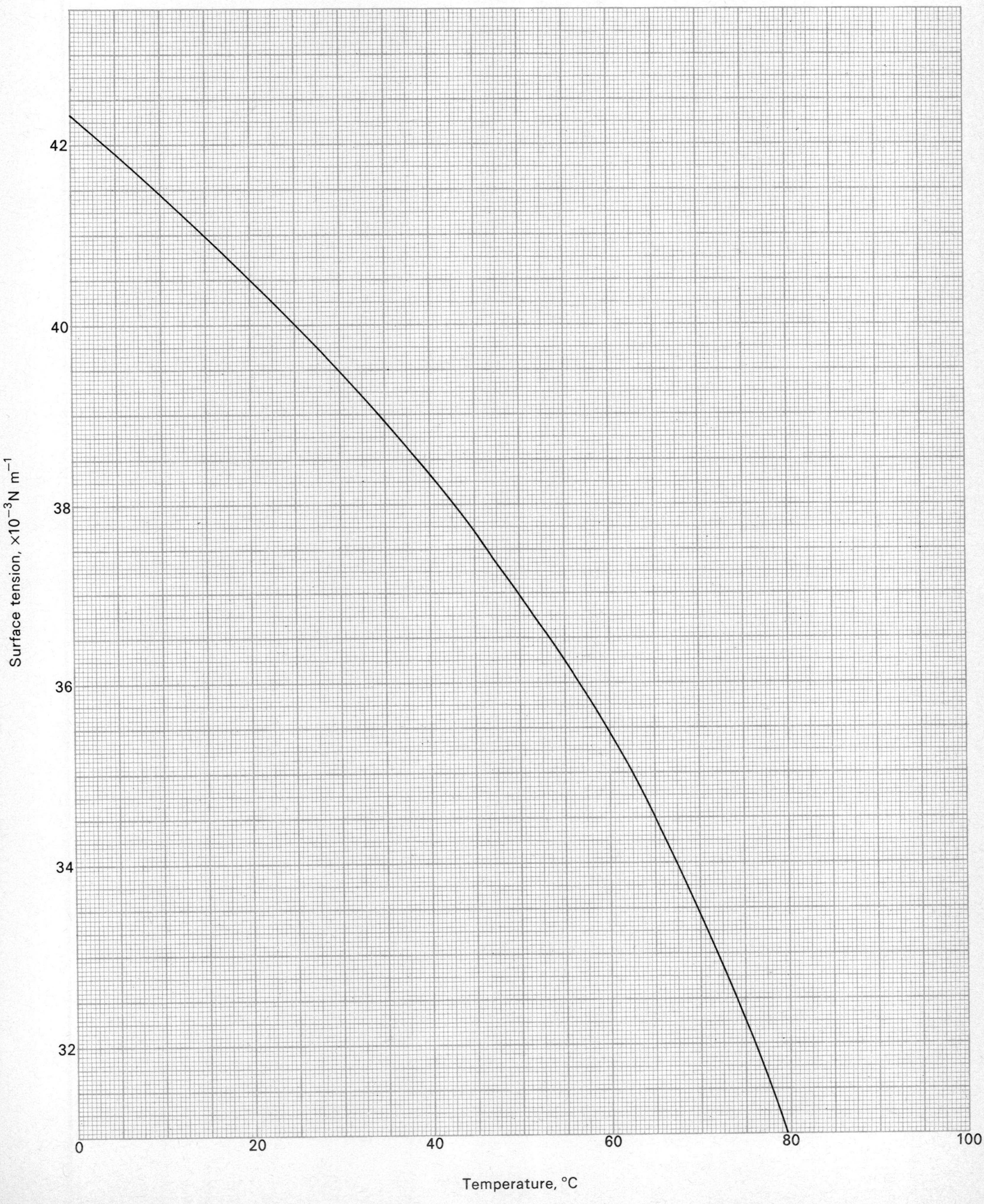

15.10 Phase Diagram for Nitric Acid–H_2O System (3,4,11,17)

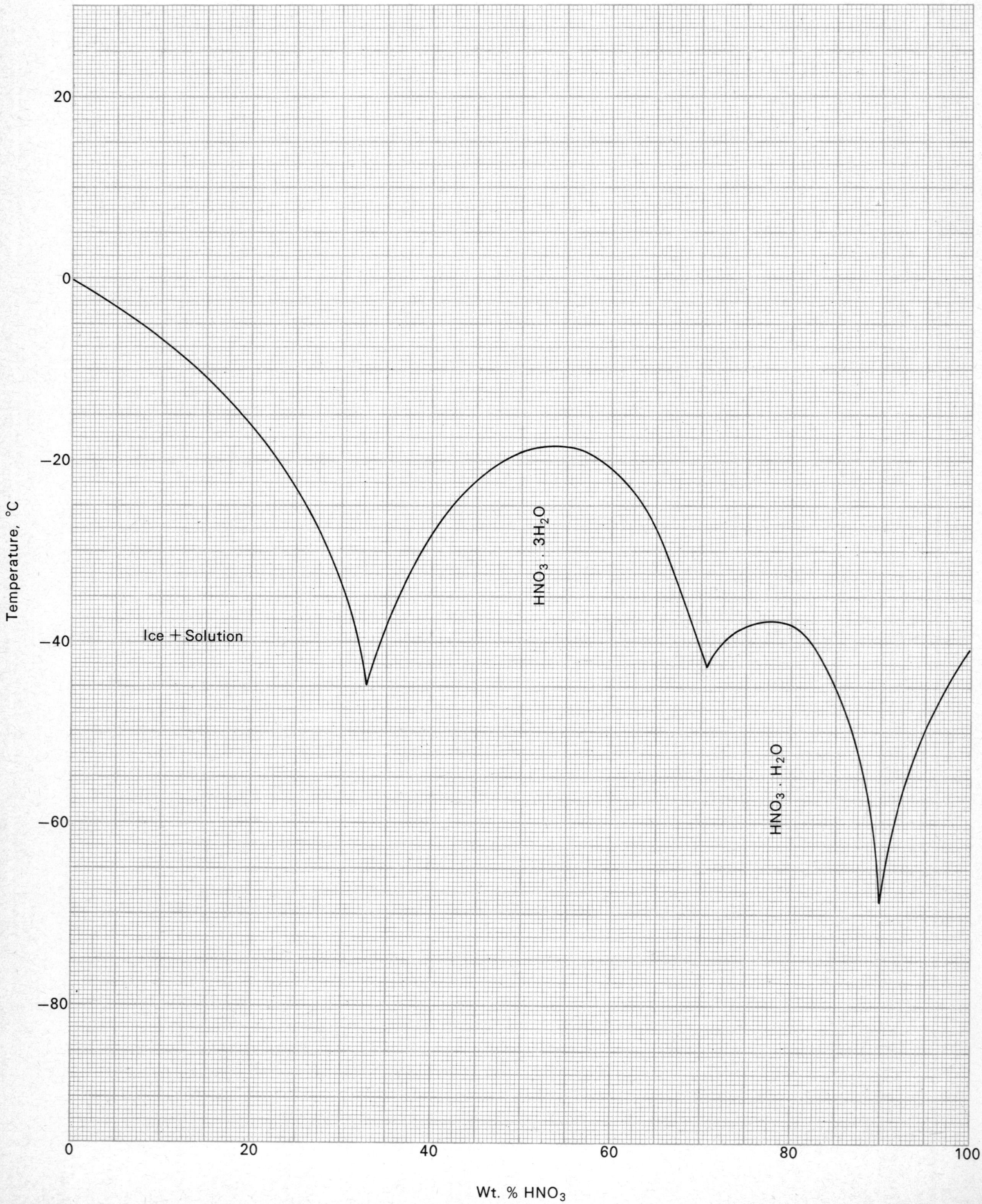

16 NITRIC OXIDE

Nitric oxide is a colourless and paramagnetic gas at room temperature and atmospheric pressure, showing a slight degree of association only at high pressures. The paramagnetism of the monomeric gas indicates that there is one unpaired electron in the molecule. The gas is somewhat soluble in water.

Solid nitric oxide forms monoclinic crystals, each unit cell consisting of dimer $(NO)_2$ molecules. Its dipole moment is 0.17 debye. The infra-red and Raman spectra show the presence of dimer molecules, which is also evident from the residual entropy, the magnetic susceptibility, and the heat of sublimation. The crystal is ionic.

Liquid nitric oxide is blue and the molecules are associated into dimers. In the liquid phase the dimer molecule is not linear.

Nitric oxide is highly endothermic. Chemically, it is very reactive. At room temperature it reacts with air, forming brown nitrogen dioxide. At high temperatures it is the most stable gas among the nitrogen oxides.

The compilations of physical, thermodynamic, and transport properties of solid, liquid, and gaseous nitric oxide can be found in the following multi-volume handbooks: Gmelin (12), Landolt-Börnstein (23), Mellor (28), and Pascal (31).

There is no reliable information on the viscosity of liquid nitric oxide in the published literature.

References

1 Adwentowsky, K., *Bull. Acad. Crac.*, 1909, 742.

2 American Conference of Government Industrial Hygienists, *Documentation of Threshold Limit Values*, revised edition, Committee on Threshold Limit Values, Cincinnati, Ohio 45202, 1966, 203 pp.

3 Becker, H. G., *Ind. Eng. Chem.*, 1924, **16**, 1220–4.

4 Biltz, W., Fischer, W., Wünnenberg, E., *Z. anorg. Chem.*, 1930, **193**, 321.

5 Byrne, R., Thodos, G., *AIChE Journal*, 1961, **7**, 185–9.

6 Cheesman, G. H., *J. Chem. Soc.*, 1932, 889–90.

7 Choy, P., Raw, C. J. G., *J. Chem. Phys.*, 1966, **45**, 1413.

8 Coppock, J. B. M., Whytlaw-Gray, R., *Proc. Roy. Soc.*, 1934, **143**, 500.

9 Edwards, D. G., *The Vapour Pressure of 30 Inorganic Liquids between One Atmosphere and the Critical Point*, Univ. of California, Lawrence Rad., Lab. UCRL 7167, Livermore, California, June 13, 1963, 148 pp.

10 Eucken, A., Karwat, E., *Z. phys. Chem.*, 1924, **112**, 467–85.

11 Gardiner, W. C., Schafer, K., *Z. Elektrochem.*, 1956, **60**, 588–94.

12 *Gmelins Handbuch der anorganischen Chemie*, Syst. No. 4, Stickstoff, Verlag Chemie GmbH, Berlin, 1936, pp. 687–718.

13 Golubev, I. F., *Viscosity of Gases and Gas Mixtures*, Gosudarstvennoe Izdatel'stvo Fiziko-Mathematicheskoi Literatury, Moskva, 1959, Table 4.

14 Greene, E. F., Toennies, J. P., *Chemical Reactions in Shock Waves*, Edward Arnold, London, 1964, pp. 333–4.

15 Honig, R. E., Hook, H. O., Vapour Pressure Data for some Common Gases, *RCA Rev.*, 1960, **21**, 360–8.

16 IUPAC, Atomic Weights of the Elements 1971, *Pure & Appl. Chem.*, 1972, **30** (3–4), 638–49.

17 Johnston, H. L., Giauque, W. F., *J. Am. Chem. Soc.*, 1929, **51**, 3194–3214.

18 Johnston, H. L., Grilly, E. R., *J. Chem. Phys.*, 1946, **14**, 233–8.

19 Johnston, H. L., McCloskey, K. E., *J. Phys. Chem.*, 1940, **44**, 1038–58.

20 Johnston, H. L., Weimer, H. R., *J. Am. Chem. Soc.*, 1934, **56**, 625–30.

21 Karapet'yants, M. Kh., *Zh. Fiz. Khim.*, 1958, **32**, 556–68.

22 Kerr, E. C., Univ. Microfilm Publ. 21480, 1957, 80 pp.

23 *Landolt-Börnstein, Zahlenwerte und Funktionen aus Physik, Chemie, Astronomie, Geophysik, und Technik*, Technik, 4 Teil, Bandteil a, Springer-Verlag, Berlin, 1967, pp. 278–80.

24 Lange, N. A., Ed., *Handbook of Chemistry*, 10th ed., McGraw-Hill, New York, 1967.

25 Lax, E., Synowietz, C., Eds., *Taschenbuch für Chemiker und Physiker*, 3rd ed., Springer-Verlag, Berlin, 1967.

26 Leduc, A., *Chem. Rev.*, 1929, **6**, 1–16.

27 Liveing, G. D., Dewar, J., *Phil. Mag.*, 1892, **34** (2), 205.

28 *Mellor's Comprehensive Treatise on Inorganic and Theoretical Chemistry*, Vol. VIII, Suppl. 1, Part II, Nitrogen, Longmans, London, 1967, pp. 219–29.

29 Mott-Smith, L. M., Daily, C. R., *Phys. Rev.*, 1926, **28** (2), 976–9.

30 McKown, P., Techn. Data Rept. 02.984.040, (AD 142607), 1957, 428 pp.

31 Pascal, P., Ed., *Nouveau Traité de Chimie Minérale*, Vol. X, Azote, Phosphore, Masson et Cie, Éditeurs, Paris, 1956, pp. 301–38.

32 Schaefer, C. A., Thodos, G., *AIChE Journal*, 1959, **5** (3), 367–72.

33 Spangenberg, C. D., *Chem. Eng.*, 1951, **58** (9), 170–1.

34 Stull, D. R., Dir., *JANAF Thermochemical Tables*, Dow Chemical Co., Midland, Mich., 1965–8.

35 Tabachnikov, A. G., Serdyuk, L. S., *Inzh.-Fiz. Zh. Akad. Nauk. Belorussk. SSR.*, 1965, **9** (3), 332–6.

36 Tammann, G., *Z. anorg. Chem.*, 1937, **235**, 49–61.

37 Winkler, L. W., *Ber.*, 1901, **34**, 1409.

38 Yosim, S. J., Owens, B. B., *J. Chem. Phys.*, 1963, **39** (9), 2222–6.

Table 16.1 Properties of Nitric Oxide

		Ref.
Chemical formula	NO	(25)
Molecular weight	30.0057	(16)
Critical temperature	−92.9 °C	(25)
Critical pressure	64.85×10^5 N m^{-2}	(25)
Critical density	520.0 kg m^{-3}	(25)
Triple-point temperature	−163.6 °C	(25)
Triple-point pressure	0.2192×10^5 N m^{-2}	(25)
Boiling point at 1.0133×10^5 N m^{-2}	−151.75 °C	(25)
Melting point	−163.60 °C	(25)
Transition temperature	−220.51 °C	(17)
Density of solid at −195.0 °C	1556.0 kg m^{-3}	(4)
Coefficient of thermal expansion of gas at 30 °C and 1.0133×10^5 N m^{-2}	3.679×10^{-3} °C^{-1}	(8)
Heat of formation of gas at 25 °C	9.0351×10^7 J kg-mol^{-1}	(34)
Free energy of formation of gas at 25 °C	8.6751×10^7 J kg-mol^{-1}	(34)
Heat of fusion	2.3111×10^6 J kg-mol^{-1}	(10)
Heat of transition	5442.84 J kg-mol^{-1}	(17)
Entropy of gas at 25 °C	2.1079×10^5 J kg-mol^{-1} C^{-1}	(34)
Heat capacity ratio of gas at 15 °C and 1.0133×10^5 N m^{-2}	1.400	(26)
Velocity of sound at 27 °C and 1.0133×10^5 N m^{-2}	341.1 m s^{-1}	(14)
Diffusion in air at 80 °C and 1.0133×10^5 N m^{-2}	0.20×10^{-4} m^2 s^{-1}	(33)
Dielectric constant of gas at 0 °C and 1.0133×10^5 N m^{-2}	1.00044	(29)
Refractive index of liquid at −155 °C	1.3305	(27)
Threshold limit value	(25 ppm) 30.0×10^{-6} kg m^{-3}	(2)

16.1 Vapour Pressure of Nitric Oxide (9,15,20,23)

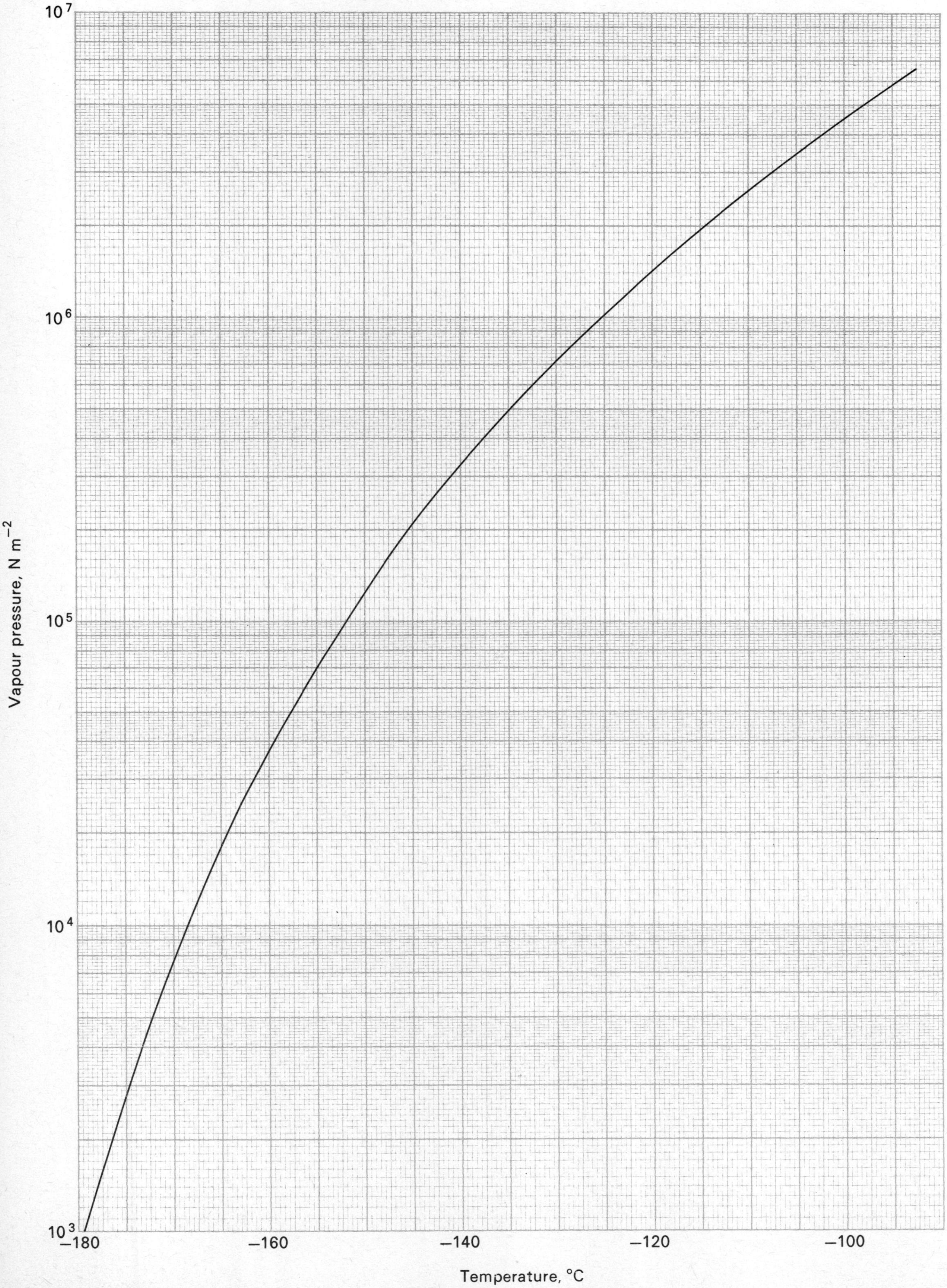

16.2 Orthobaric Density of Nitric Oxide (1,6, 23)

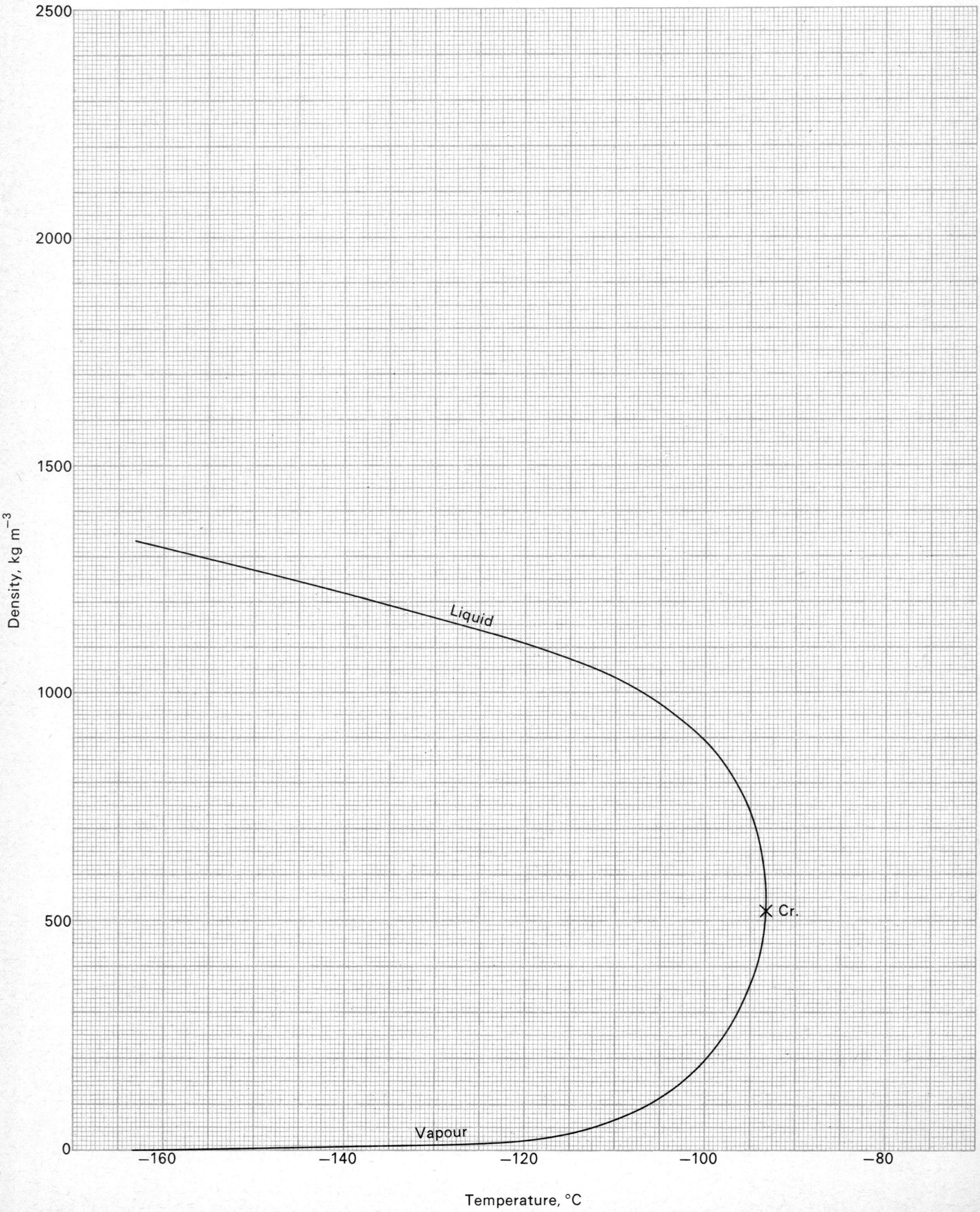

16.3 P-V-T Data for Nitric Oxide (5,23,35)

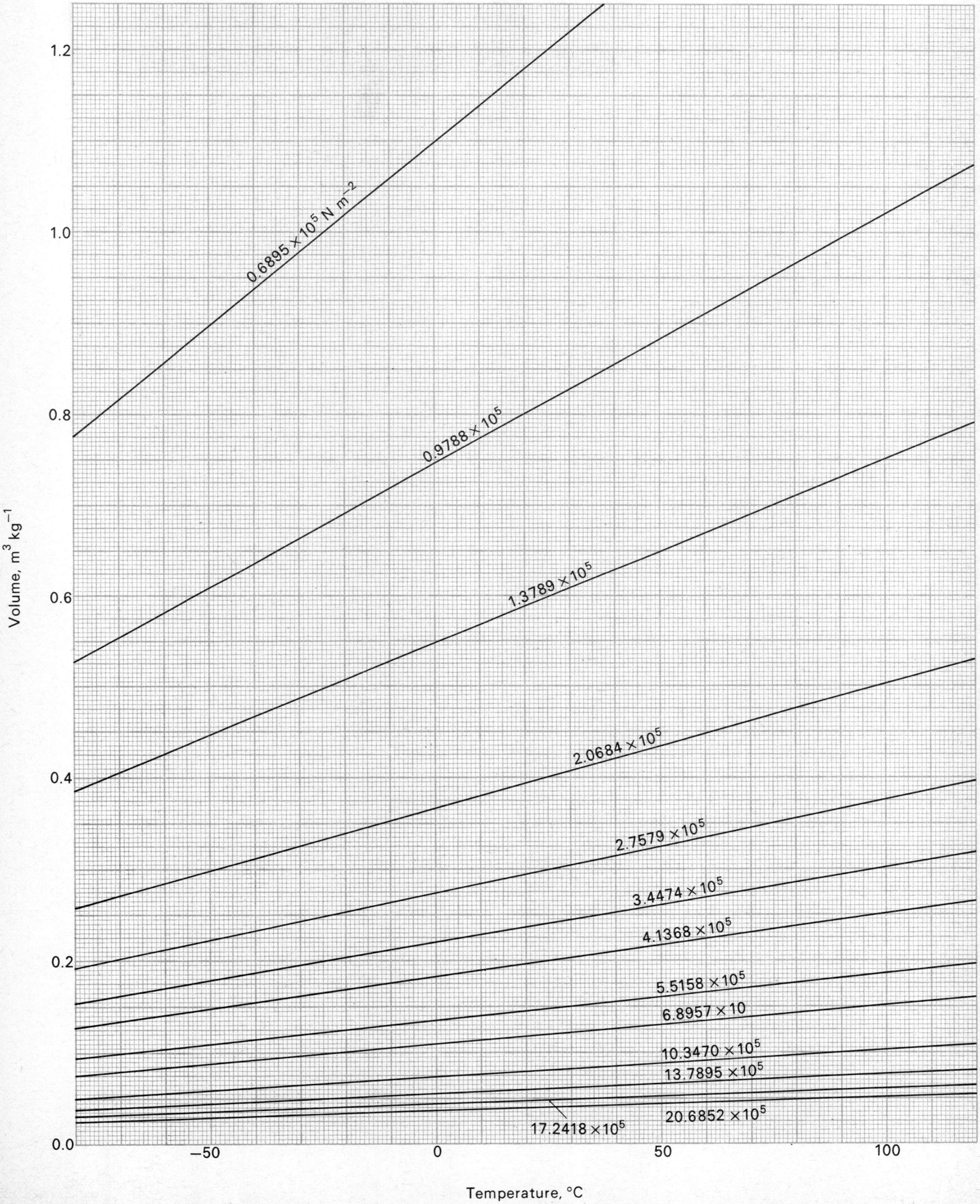

16.4 Latent Heat of Vaporisation of Nitric Oxide (17,21,28,38)

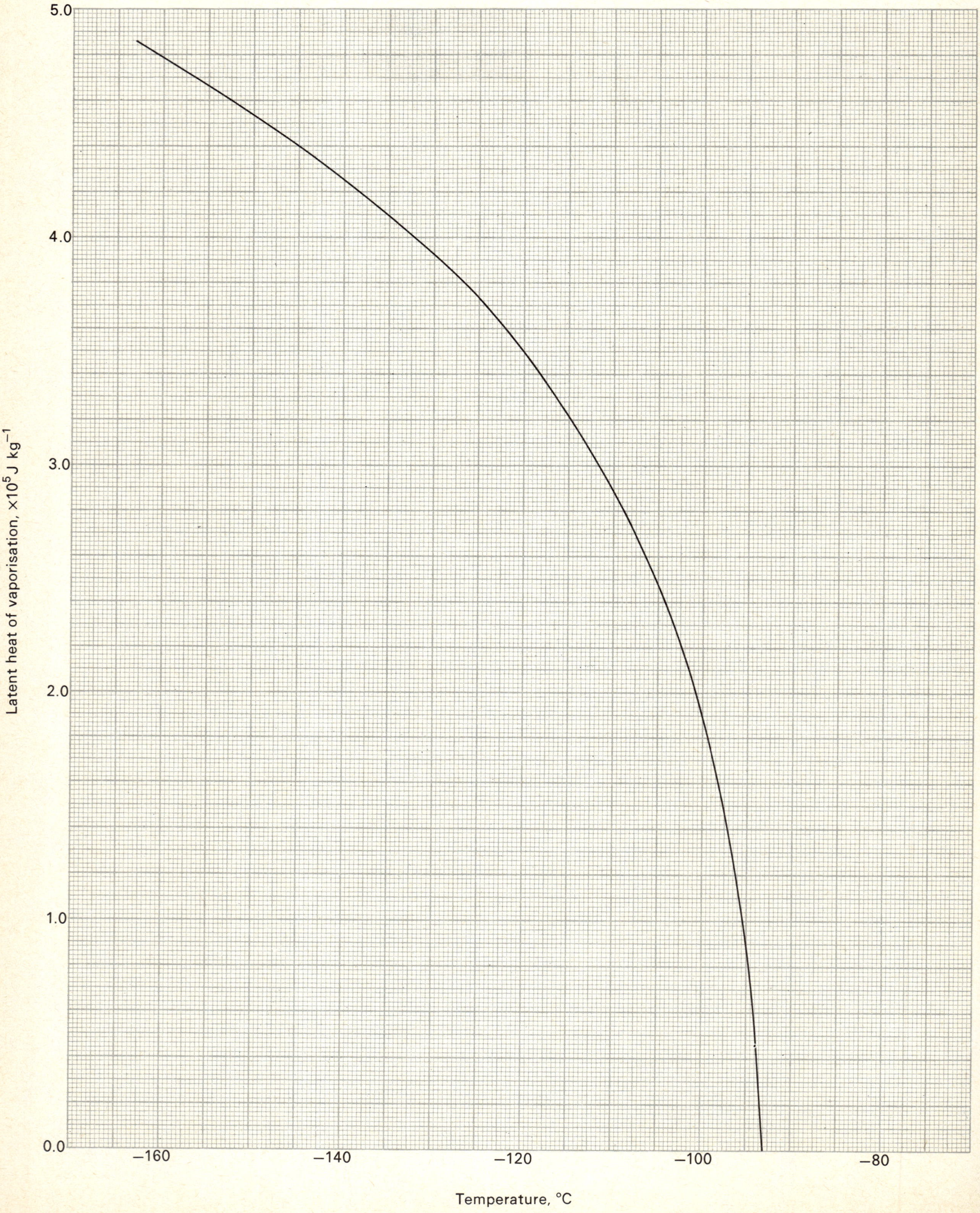

16.5 Heat Capacity of Liquid Nitric Oxide (10,17,22)

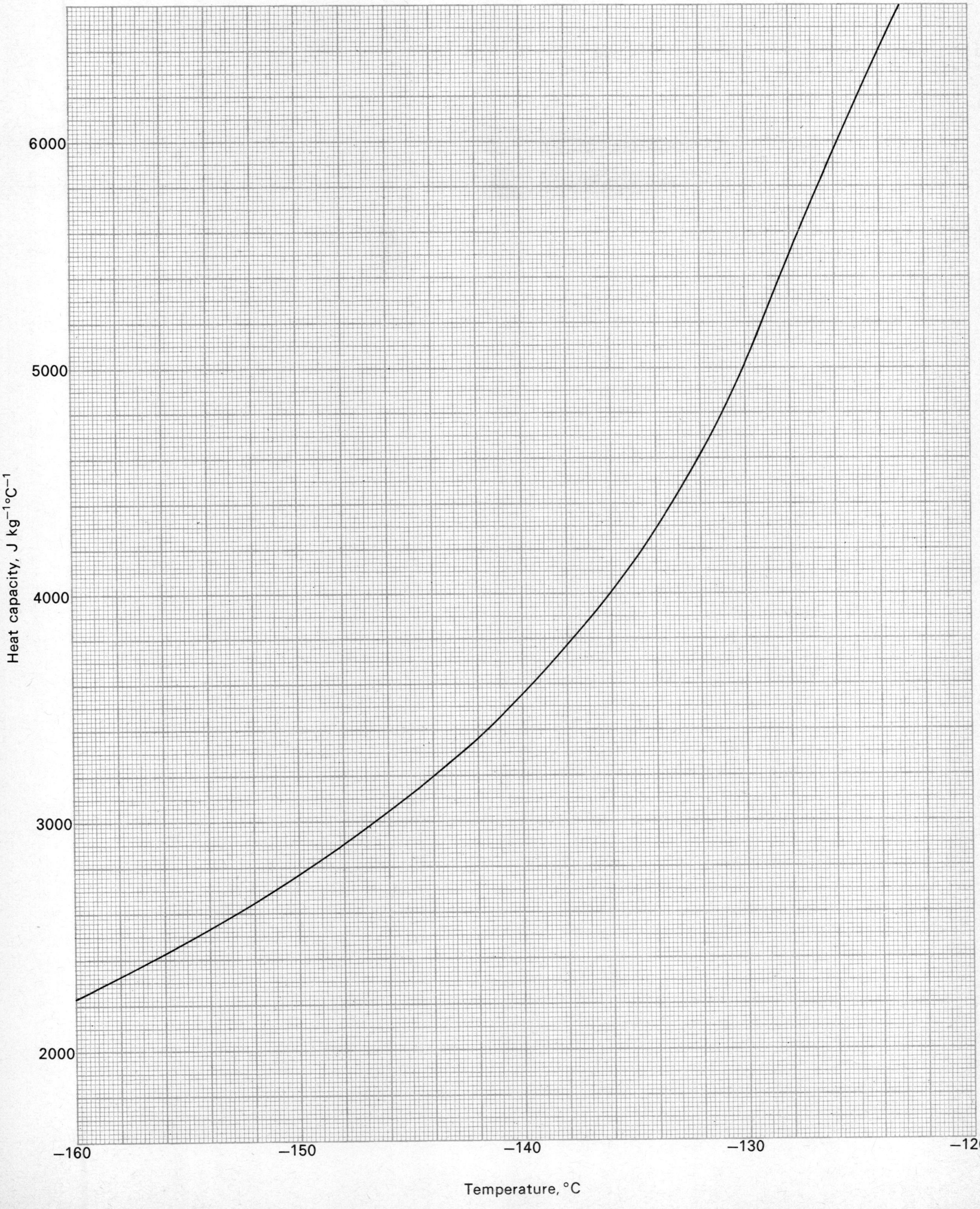

16.6 Heat Capacity of Gaseous Nitric Oxide at 1.0133×10^5 N m^{-2}(30,31,34)

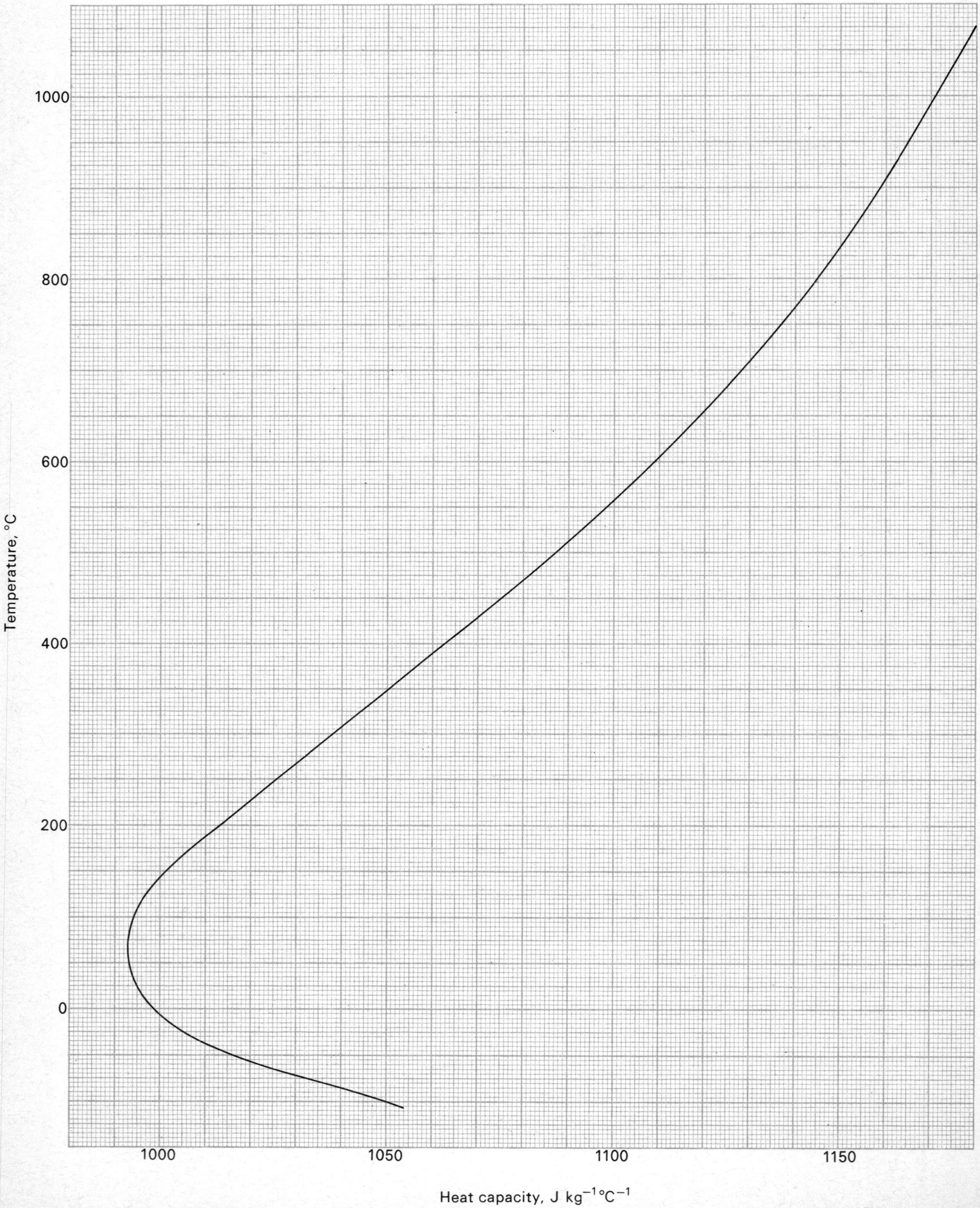

16.7 Viscosity of Gaseous Nitric Oxide at $1.0133 \times 10^5 N\ m^{-2}$ (13,19,25,31)

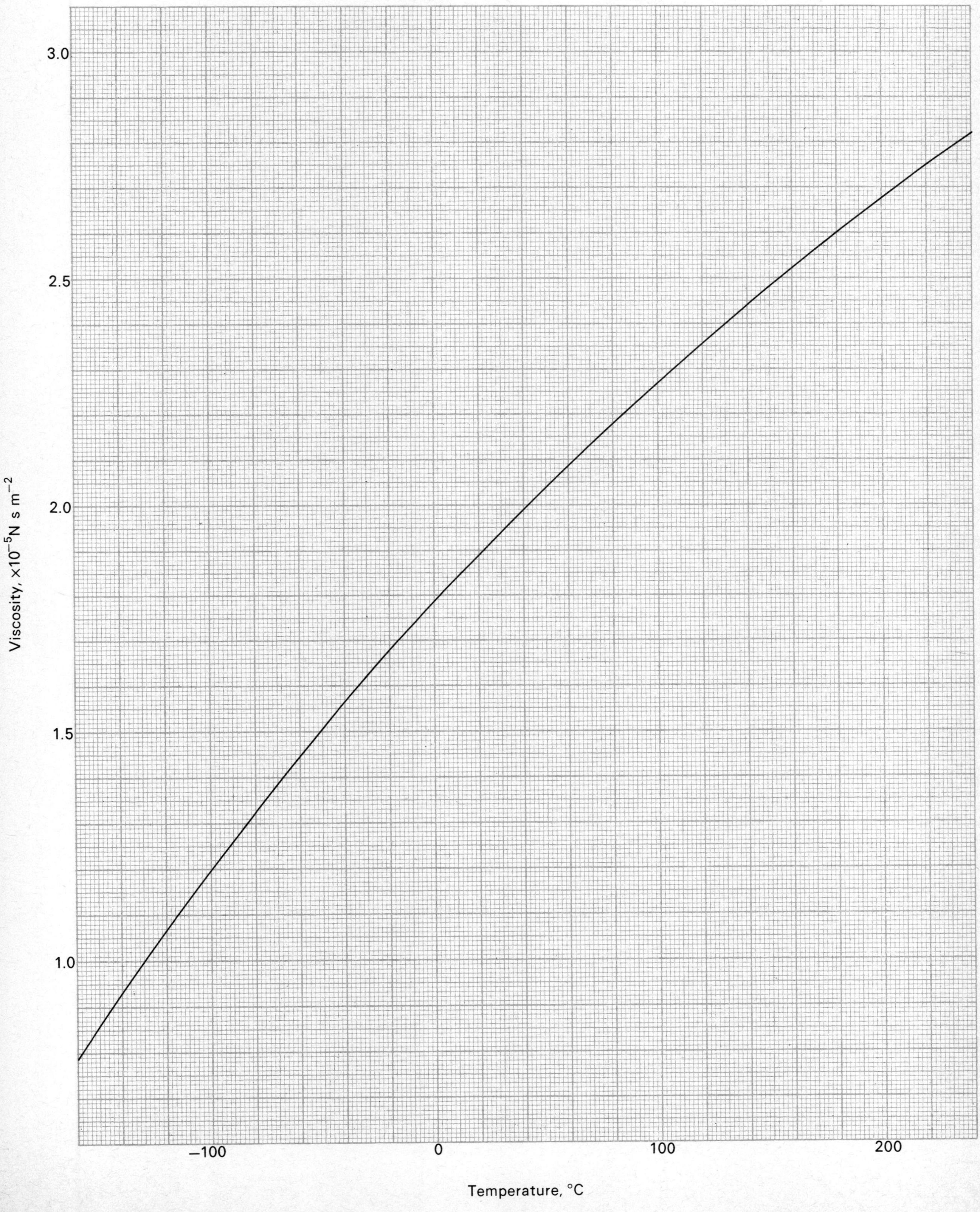

16.8 Thermal Conductivity of Liquid Nitric Oxide (18,32)

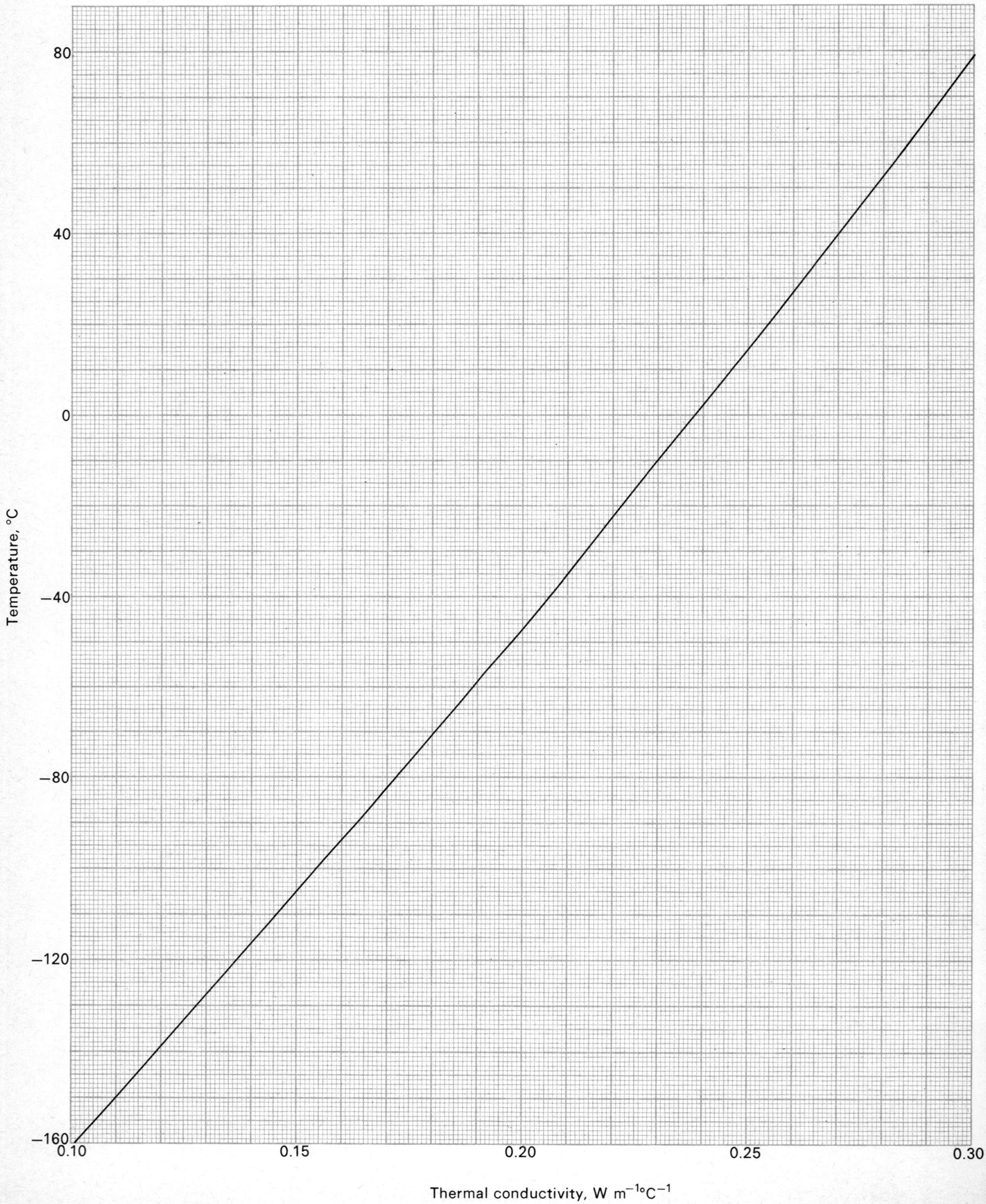

16.9 Thermal Conductivity of Gaseous Nitric Oxide at 1.0133 $\times 10^5$ N m^{-2} (7,11,18)

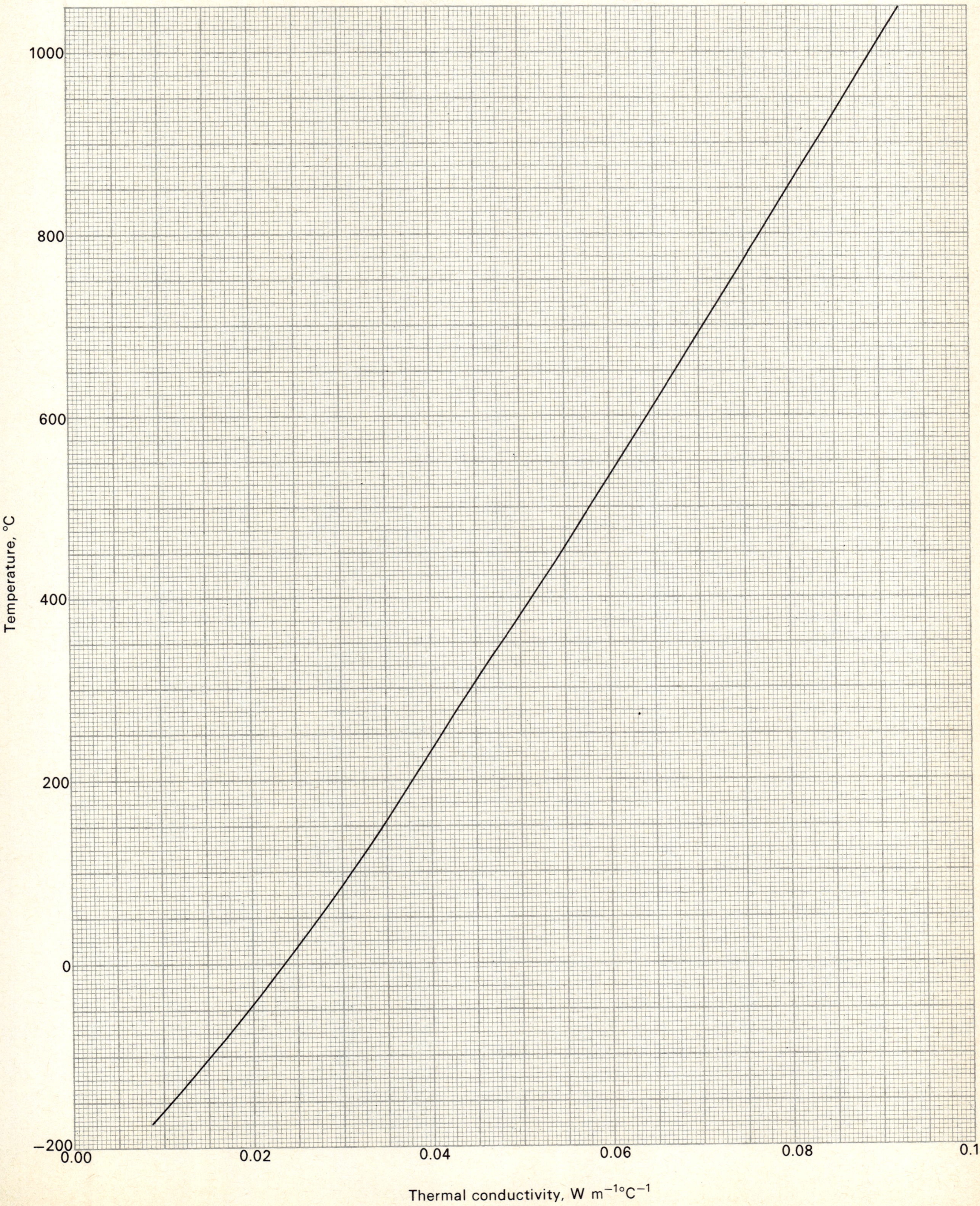

16.10 Surface Tension of Nitric Oxide (6)

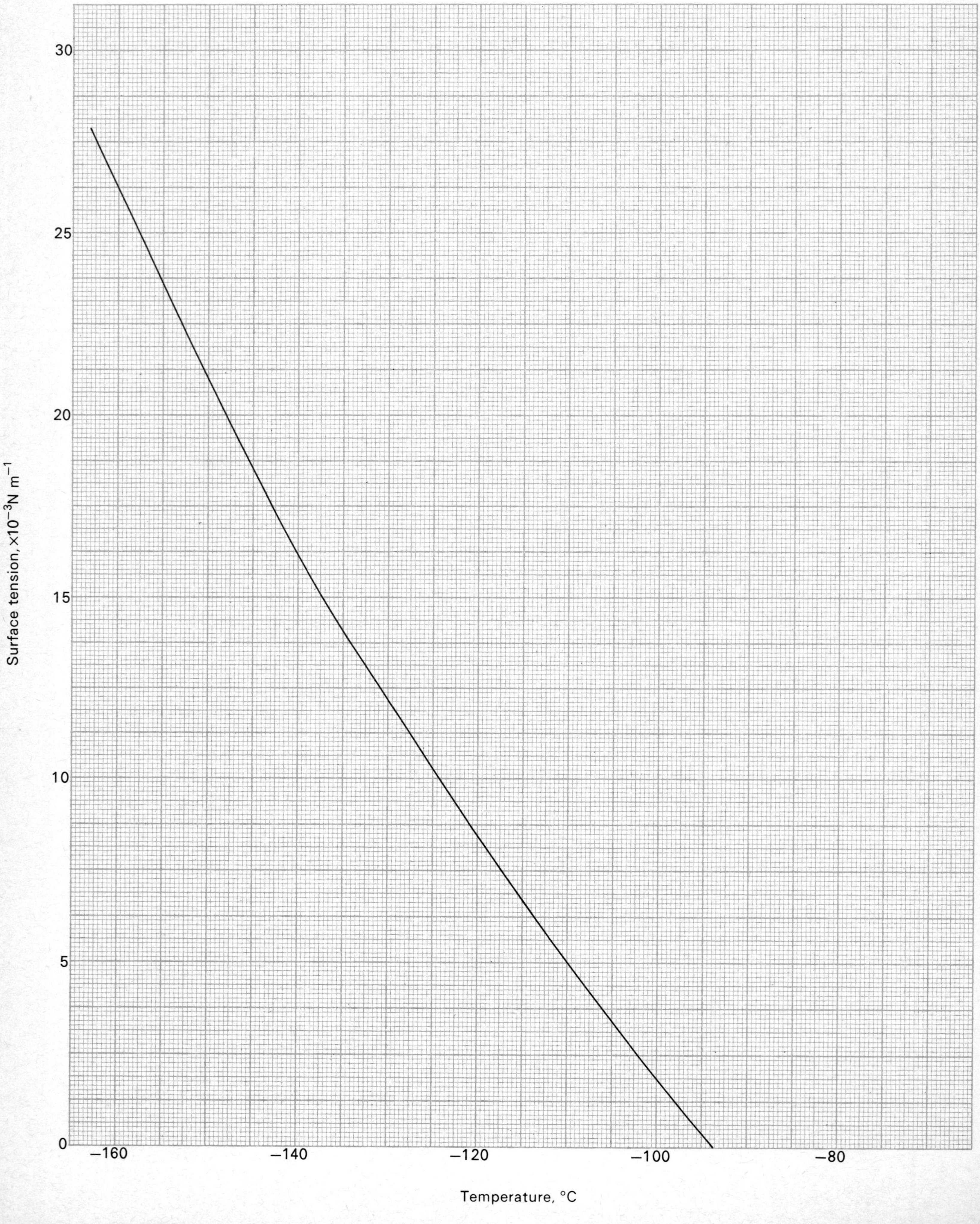

16.11 Solubility of Nitric Oxide in Water at $1.0133 \times 10^5 N\ m^{-2}$ Total Pressure (3,24,36,37)

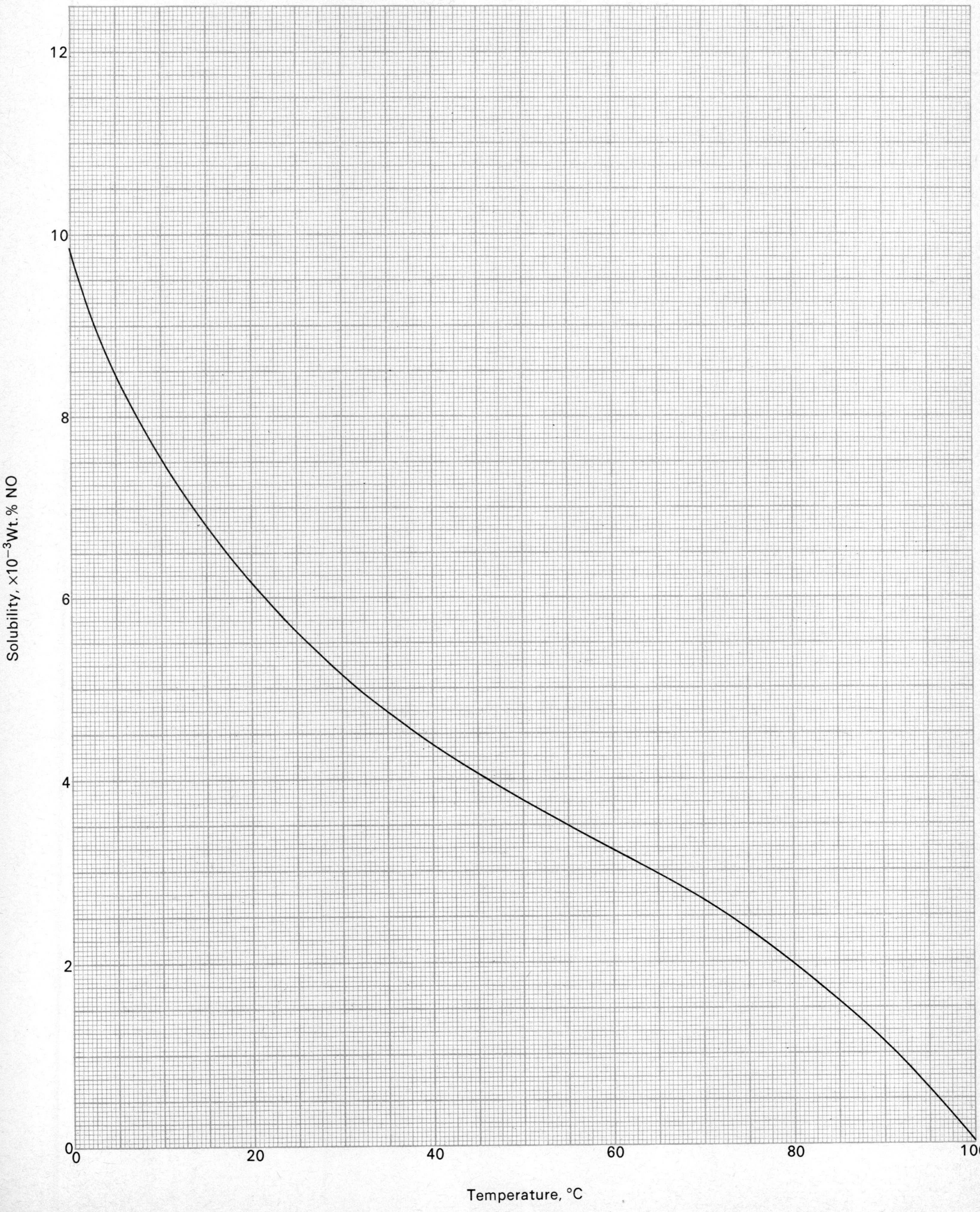

17 NITROUS OXIDE

Nitrous oxide is a colourless gas at room temperature and atmospheric pressure with a slightly sweet taste. The X-ray examination and the absorption spectrum show that the three atoms are in a line in the molecule, with the oxygen on one end, i.e. an unsymmetrical molecule. This is also indicated by the very small dipole moment (0.17 debye). N_2O is soluble in water and at low temperatures forms a hydrate, $N_2O.6H_2O$. Nitrous oxide resembles carbon dioxide in physical properties. There is a close agreement between the densities, vapour pressures, critical properties, molecular weights, etc. However CO_2 and N_2O are chemically dissimilar. The resemblance is due to the similar arrangement of the atoms in the molecules. The crystal lattice of solid N_2O is the same as that of solid CO_2.

In the cold, nitrous oxide is resistant to attack by O_2, O_3, H_2, the halogens, alkali metals, PH_3, H_2S, and aqua regia. At high temperatures it is a strong oxidizing agent, being reduced to elementary nitrogen. Gaseous N_2O can be liquefied at a pressure of 50 bars and at room temperature. It is used extensively as a low temperature refrigerant. Nitrous oxide is distributed in steel cylinders. The gas is employed in medical practice as an anaesthetic. It forms an explosive mixture with hydrogen.

The electrical conductivity of liquid nitrous oxide is identical with that of pure water at the same temperatures.

The physical, thermodynamic, and transport properties of nitrous oxide in the solid, liquid, and gaseous states have been compiled in the following books: Dreisbach (6), Gmelin (9), Landolt-Börnstein (20), Matheson (27), Mellor (29), and Pascal (32).

Most of the data presented in Table 17.1 and Figs. 17.1–17.12 have been adopted from Couch *et al.* (5).

References

1 Andrussow, L., *Diffusion, Viscosity and Conductivity of Gases*, Progr. Intern. Res. Thermodyn. Transport Properties, 2nd 2nd Symp., Princeton, Am. Soc. Mech. Eng., New York, 1962, pp. 279–87.

2 Biltz, W., Fischer, W., Wünnenberg, E., *Z. anorg. Chem.*, 1930, **193**, 321–57.

3 Blue, R. W., Giauque, W. F., *J. Am. Chem. Soc.*, 1935, **57**, 991–7.

4 Choy, P., Raw, C. J. G., *J. Chem. Phys.*, 1966, **45**, 1413.

5 Couch, E. J., Kobe, K. A., Hirth, L. J., *J. Chem. Eng. Data*, 1961, **6** (2), 229–37.

6 Dreisbach, R. R., *Physical Properties of Chemical Compounds – III*, Adv. in Chem. Ser. No. 29, Am. Chem. Soc., Washington, 1961, p. 470.

7 Edwards, D. G., *The Vapour Pressure of 30 Inorganic Liquids between One Atmosphere and the Critical Point*, Univ. of California, Lawrence Rad. Lab. UCRL 7167, Livermore, California, June 13, 1963, 148 pp.

8 Eucken, A., Donath, E., *Z. phys. Chem.*, 1926, **124**, 194–9.

9 *Gmelins Handbuch der anorganischen Chemie*, Syst. No. 4, Stickstoff, Verlag Chemie GmbH, Berlin, 1936, pp. 518–58.

10 Golubev, I. F., *Viscosity of Gases and Gas Mixtures*, Gosudarstvennoe Izdatel'stvo Fiziko-Matematicheskoi Literatury, Moskva, 1959, Table 4.

11 Gordon, J. S., WADC-TR-57-33, (AD 110735), 1957, 172 pp.

12 Hammick, D. L., *Phil. Mag.*, 1922, **44** (6), 593.

13 Hirth, L. J., *Gas Compressibility of Nitrous Oxide and of Sulphur Dioxide by the Burnett Method*, Univ. of Texas, Ph.D. dissertation in Chem. Eng., Austin, Texas, 1958.

14 Honig, R. E., Hook, H. O., Vapour Pressure Data for some Common Gases, *RCA Rev.*, 1960, **21**, 360–8.

15 IUPAC, Atomic Weights of the Elements 1971, *Pure & Appl. Chem.*, 1972, **30** (3–4), 638–49.

16 Jay, B. E., Wilson, R. H., Doty, V., Pingree, H., Hargis, B., *Ann. Chem.*, 1962, **34** (3), 414–8.

17 Johnston, H. L., Grilly, E. R., *J. Chem. Phys.*, 1946, **14**, 233–8.

18 Jossi, J. A., Stiel, L. I., Thodos, G., *AIChE Journal*, 1962, **8** (1), 59–63.

19 Kuneth, W., *Phys. Rev.*, 1922, **19** (2), 512.

20 *Landolt-Börnstein, Zahlenwerte und Funktionen aus Physik, Chemie, Astronomie, Geophysik, und Technik*, 7th ed., IV Band, Technik, 4 Teil, Bandteil a, Springer-Verlag, Berlin, 1967, pp. 725–7.

21 Lax, E., Synowietz, C., Eds., *Taschenbuch für Chemiker und Physiker*, 3rd ed., Springer-Verlag, Berlin, 1967.

22 Leduc, A., *Chem. Rev.*, 1929, **6**, 1–16.

23 Lenoir, J. M., Univ. Arkansas, Eng. Expt. Station Bull. 18, 1953, 48 pp.

24 Liley, P. E., *The Thermal Conductivity of 46 Gases at Atmospheric Pressure*, Proc. Symp. Thermophys. Prop. 4th, (T. R. Moszynski, Ed.) Am. Soc. Mech. Eng., New York, 1968, pp. 323–49.

25 Mannheimer, M., *Chem. Analyst*, 1956, **45** (1), 8–10.

26 Markham, A. E., Kobe, K. A., *J. Am. Chem. Soc.*, 1941, **63**, 449–54.

27 Matheson Gas Data Book, 4th ed., *Nitrous Oxide*, The Matheson Company, Inc., East Rutherford, N. J., 1966, pp. 387–91.

28 Mathews, J. F., *Chem. Rev.*, 1972, **72** (1), 71–100.

29 *Mellor's Comprehensive Treatise on Inorganic and Theoretical Chemistry*, Vol. VIII, Suppl. II, Nitrogen, Part II, Longmans, London, 1967, pp. 192–204.

30 McBride, B. J., Gordon, S., *J. Chem. Phys.*, 1961, **35**, 2198–2206.

31 McBride, B. J., Heimel, S., Ehlers, J. G., Gordon, S., *Thermodynamic Properties of 210 Substances Involving the First 18 Elements*, Natl. Aeron. Space Admin. NASA SP-3001, Washington, 1963.

32 Pascal, P., Ed., *Nouveau Traité de Chimie Minérale*, Vol. X, Azote, Phosphore, Masson et Cie, Éditeurs, Paris, 1956, pp. 597–605.

33 Pennington, R. E., Kobe, K. A., *J. Chem. Phys.*, 1954, **22** (8), 1442–7.

34 Quinn, E. L., Wernimont, G., *J. Am. Chem. Soc.*, 1930, **52**, 2723–30.

35 Raw, C. J. G., Ellis, C. P., *J. Chem. Phys.*, 1958, **28**, 1198–1200.

36 Richter, G. N., Sage, B. H., *J. Chem. Eng. Data*, 1963, **8** (2), 221–5.

37 Rohmann, H., *Ann. Phys.*, 1911, **34** (4), 1001.

38 Rossini, F. D., Wagman, D. D., Evans, W. H., Levine, S., Jaffe, I., *Selected Values of Chemical Thermodynamic Properties*, Natl. Bur. Std. Circular 500, Washington, Feb. 1, 1952.

39 Spangenberg, C. D., *Chem. Eng.*, 1951, **58** (9), 170–1.

40 Stull, D. R., Dir., *JANAF Thermochemical Tables*, Dow Chemical Co., Midland, Mich., 1965–8.

41 Svehla, R. A., *Estimated Viscosities and Thermal Conductivities of Gases at High Temperatures*, Lewis Res. Center, Techn. Rept. R-132, Cleveland, Ohio, 1962, 140 pp.

42 Washburn, E. W., Ed., *International Critical Tables of Numerical Data, Physics, Chemistry, and Technology*, Vol. VI, McGraw-Hill, New York, 1929, pp. 461–7.

Table 17.1 Properties of Nitrous Oxide

		Ref.
Chemical formula	N_2O	(21)
Molecular weight	44.0124	(15)
Critical temperature	36.41 °C	(28)
Critical pressure	72.45×10^5 N m^{-2}	(28)
Critical density	452.0 kg m^{-3}	(28)
Triple-point temperature	−90.81 °C	(21)
Triple-point pressure	0.8785×10^5 N m^{-2}	(21)
Boiling point at 1.0133×10^5 N m^{-2}	−88.47 °C	(21)
Melting point	−90.89 °C	(3)
Transition temperature	None	(9)
Density of solid at −195 °C	1606.0 kg m^{-3}	(2)
Coefficient of thermal expansion of liquid at boiling point	1.86×10^{-3} $°C^{-1}$	(12)
Heat of formation of gas at 25 °C	8.2103×10^7 J kg-mol^{-1}	(40)
Free energy of formation of gas at 25 °C	1.0425×10^7 J kg-mol^{-1}	(40)
Heat of sublimation at −137.65 °C	2.4376×10^7 J kg-mol^{-1}	(8)
Heat of fusion	6.5440×10^6 J kg-mol^{-1}	(38)
Heat of transition	None	(9)
Entropy of gas at 25 °C	2.2000×10^5 J kg-mol^{-1} $°C^{-1}$	(40)
Heat capacity ratio of gas at 15 °C and 1.0133×10^5 N m^{-2}	1.303	(22)
Velocity of sound at 0 °C and 1.0133×10^5 N m^{-2}	260.5 m s^{-1}	(42)
Diffusion in air at 80 °C and 1.0133×10^5 N m^{-2}	0.17×10^{-4} m^2 s^{-1}	(39)
Dielectric constant of gas at 0 °C and 1.0133×10^5 N m^{-2}	1.001129	(37)
Refractive index of gas at 20 °C and 1.0133×10^5 N m^{-2}	1.0005	(9, 29)

17.1 Vapour Pressure of Nitrous Oxide (5,7,14)

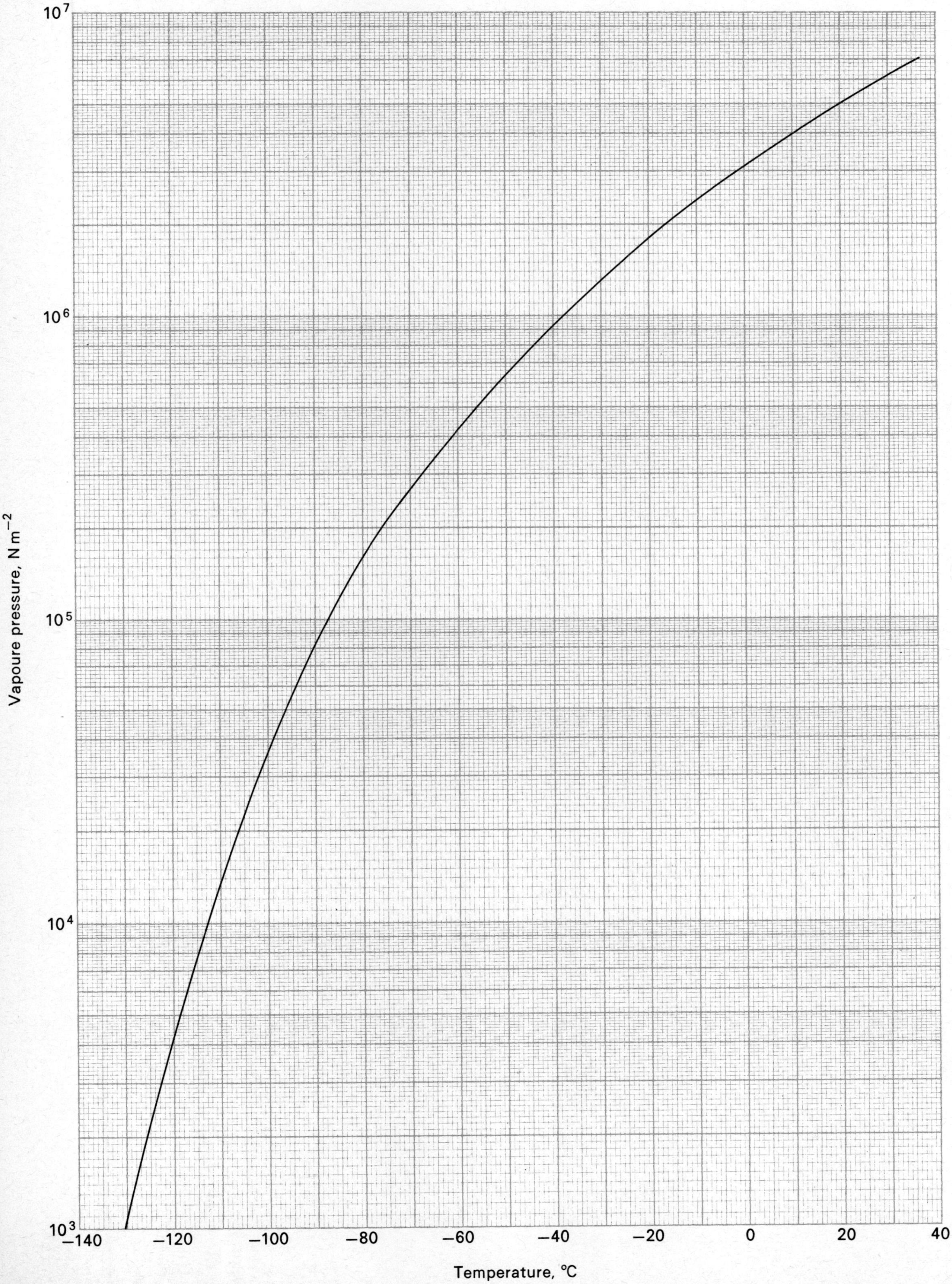

17.2 Orthobaric Density of Nitrous Oxide (5)

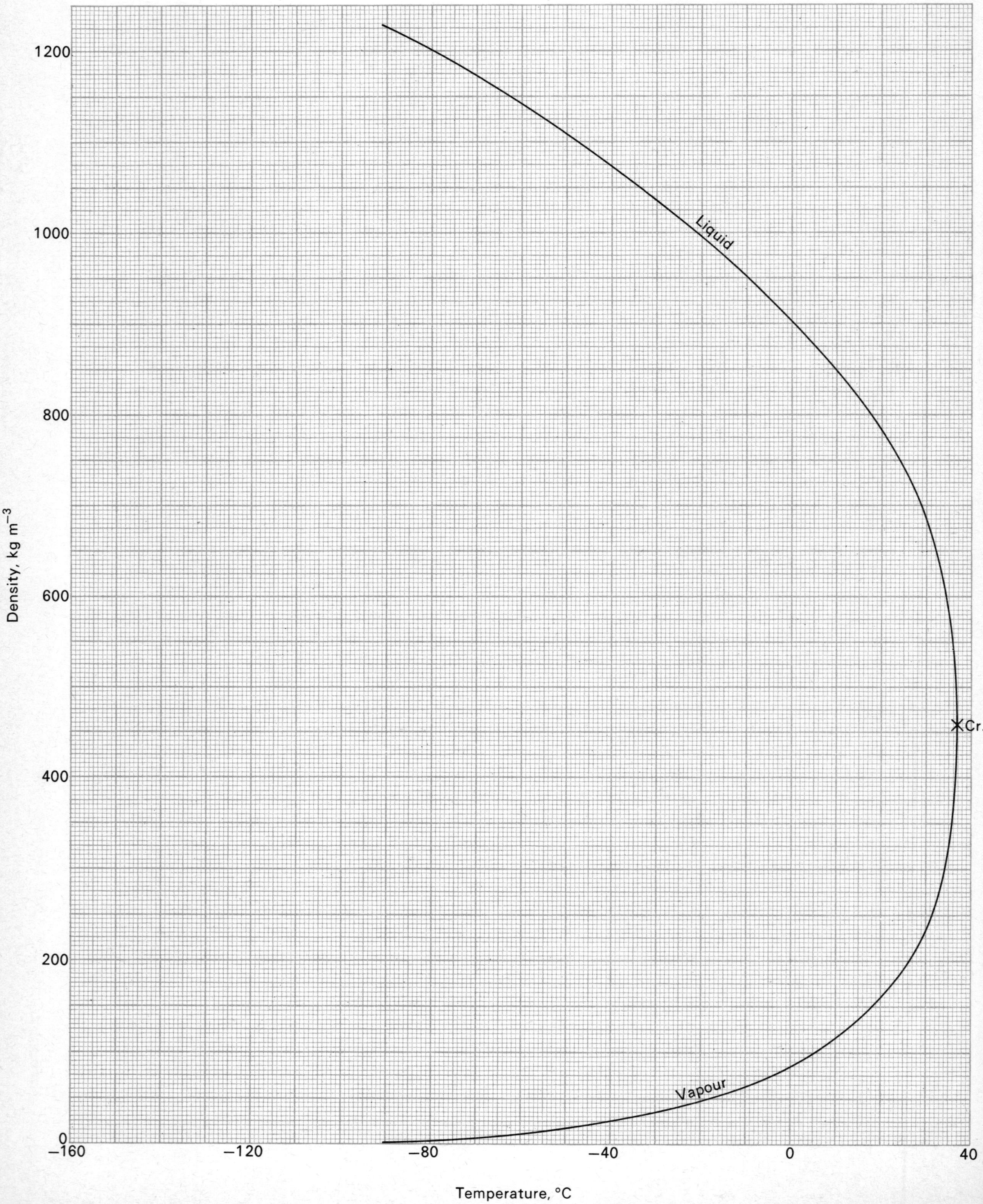

17.3 P-V-T Data for Nitrous Oxide (5,13)

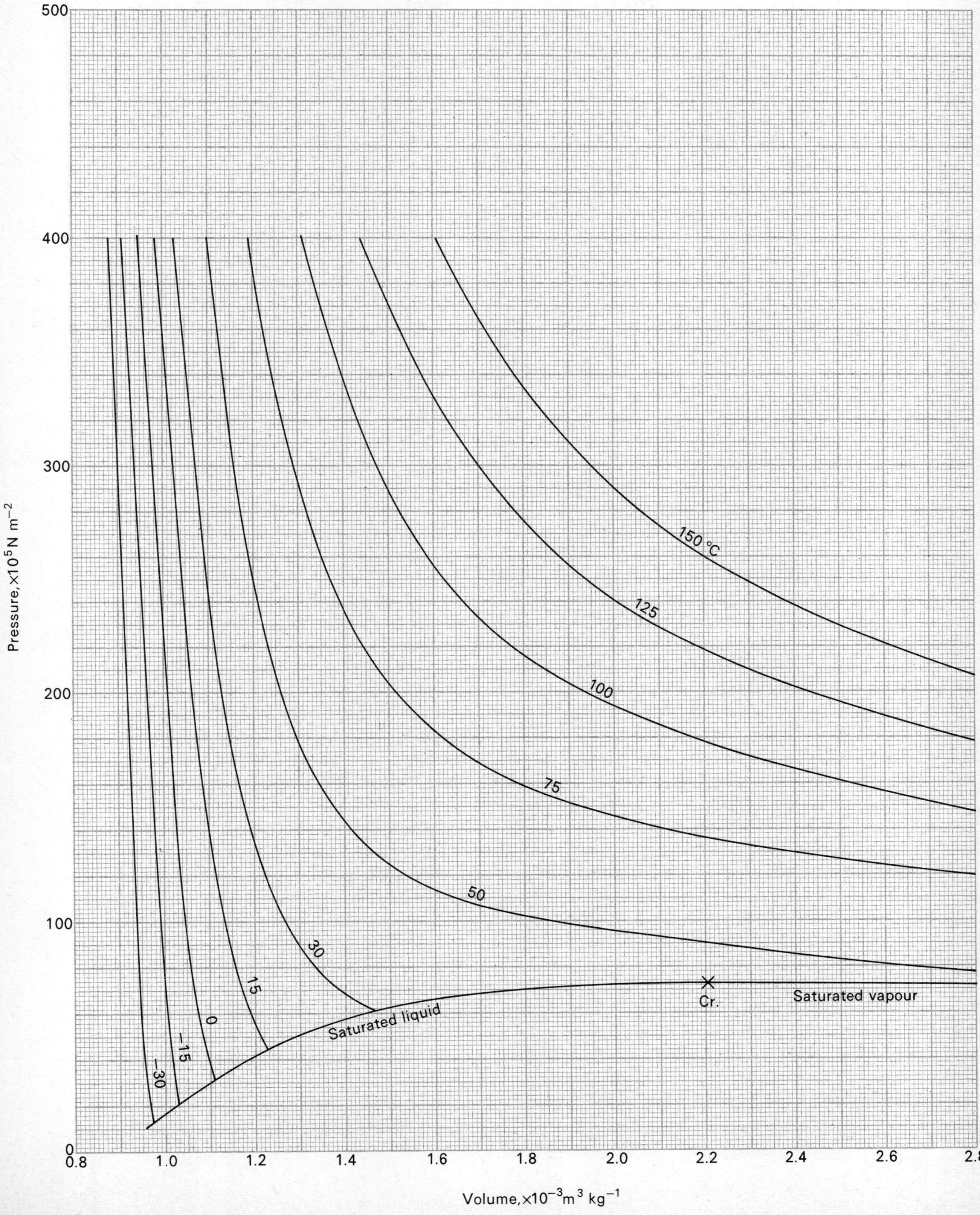

17.4 Latent Heat of Vaporisation of Nitrous Oxide (5,20,21,29,32,38)

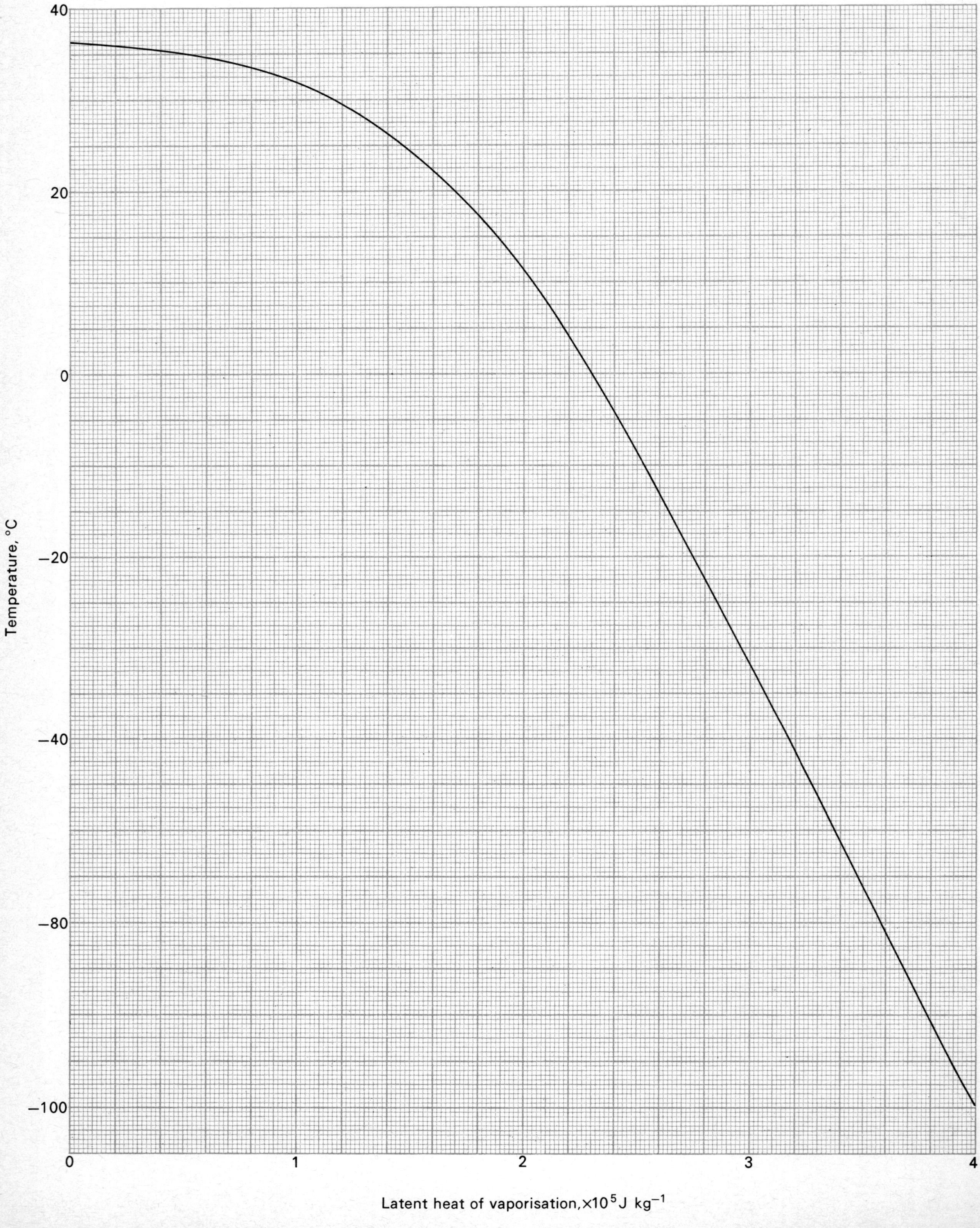

17.5 Heat Capacity of Liquid Nitrous Oxide (3)

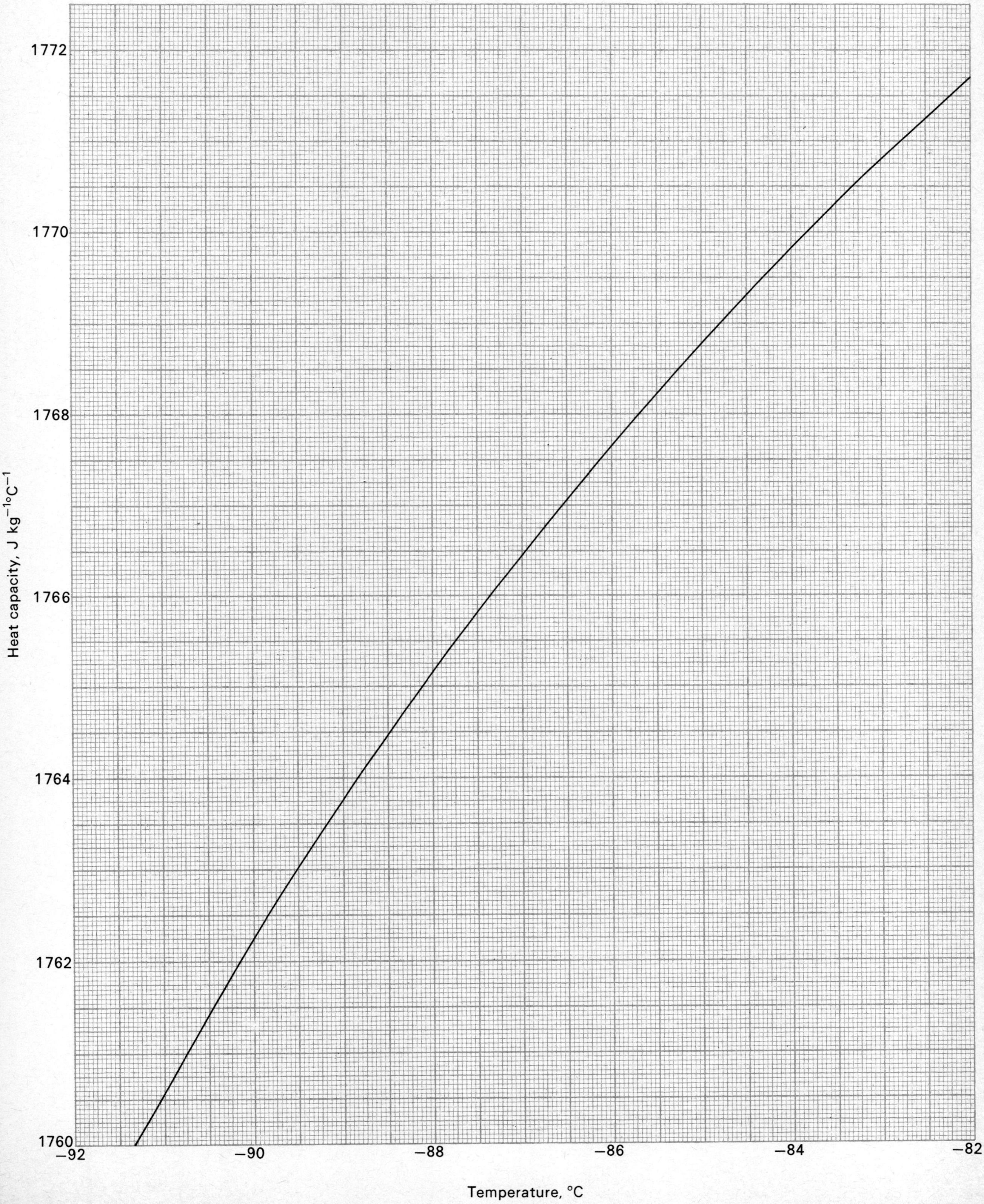

17.6 Heat Capacity of Gaseous Nitrous Oxide at 1.0133×10^5 N m^{-2}(11,30,31,33,40)

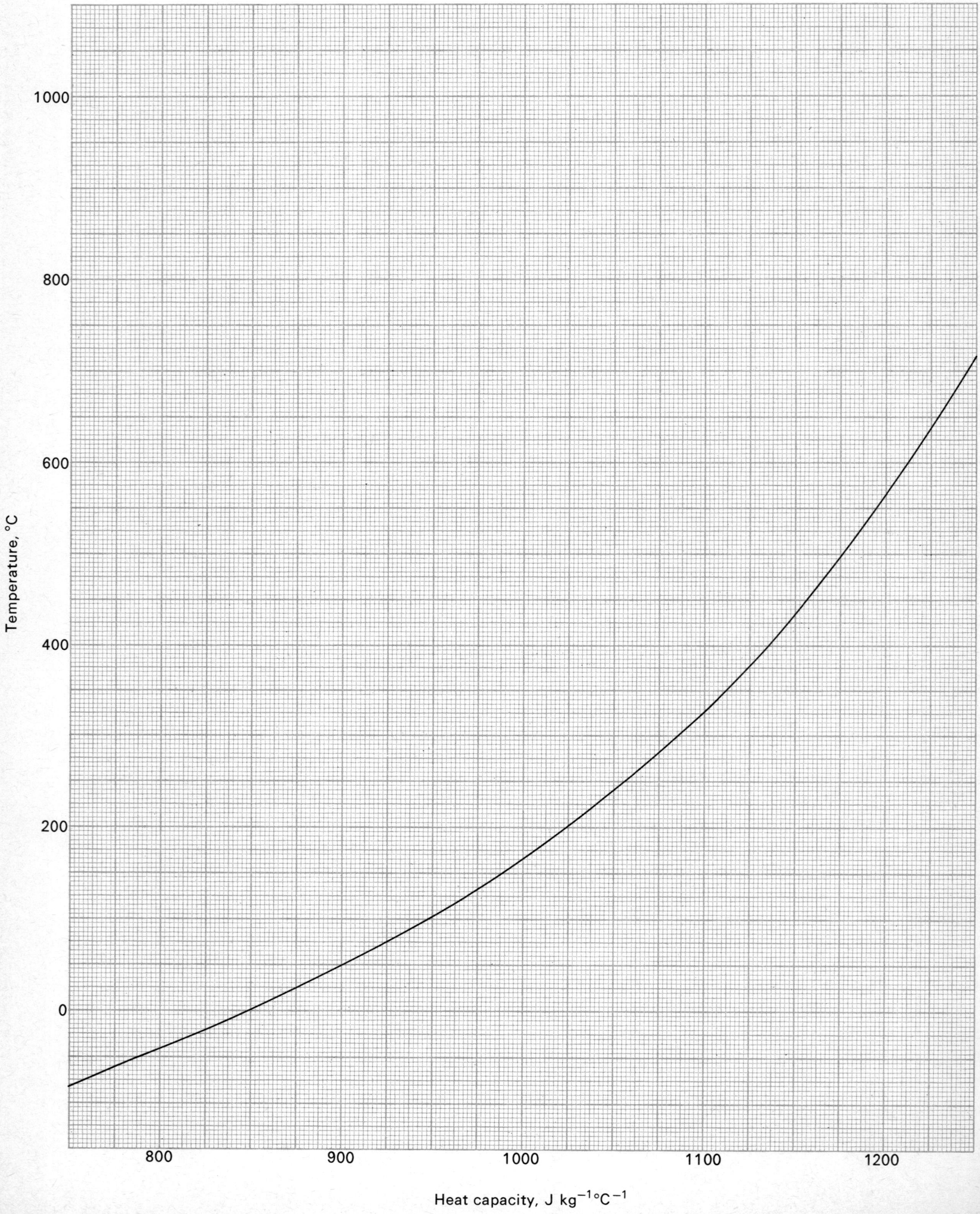

17.7 Viscosity of Liquid Nitrous Oxide (18)

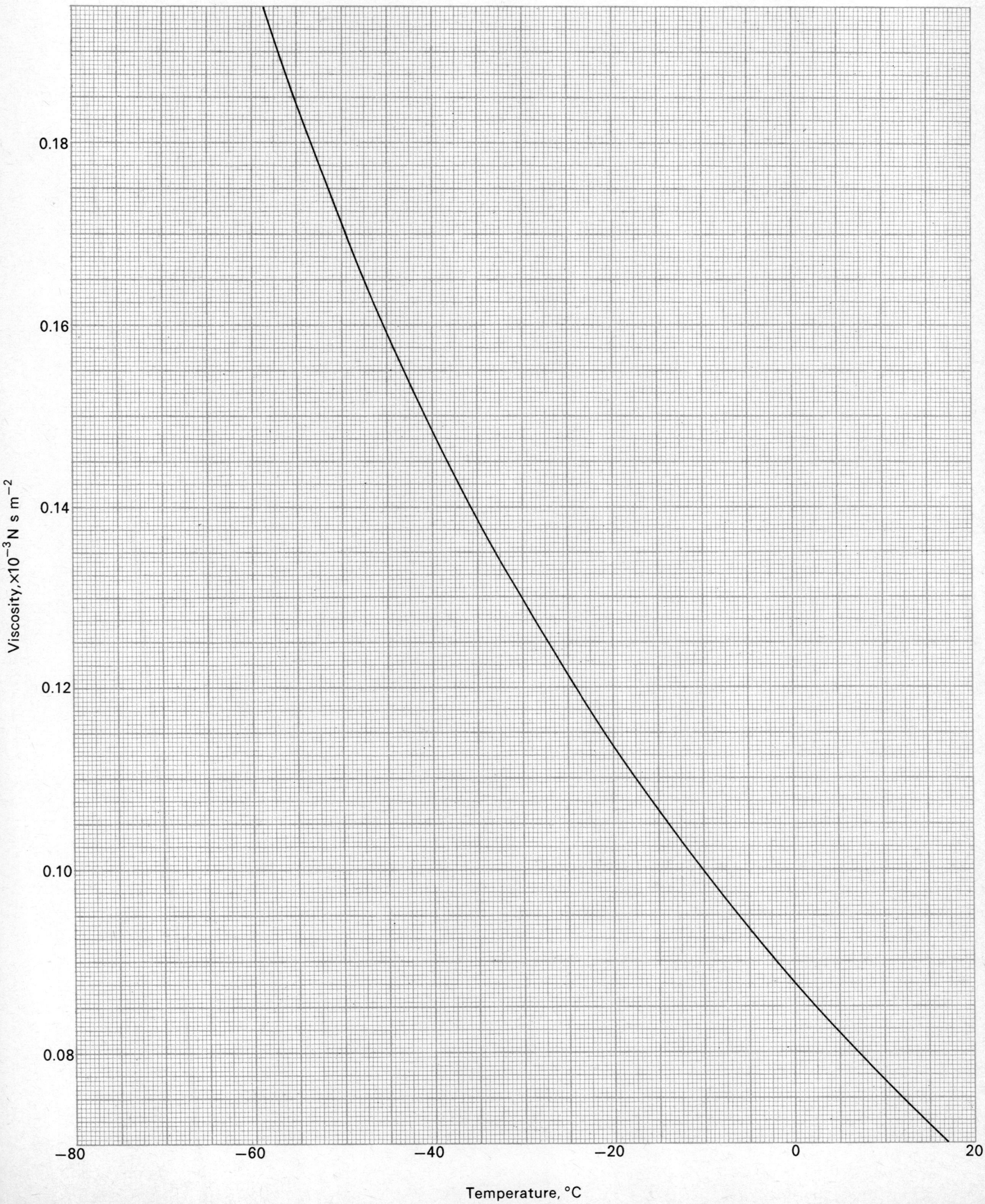

17.8 Viscosity of Gaseous Nitrous Oxide at 1.0133×10^5 N m^{-2} (1,10,18,35,41)

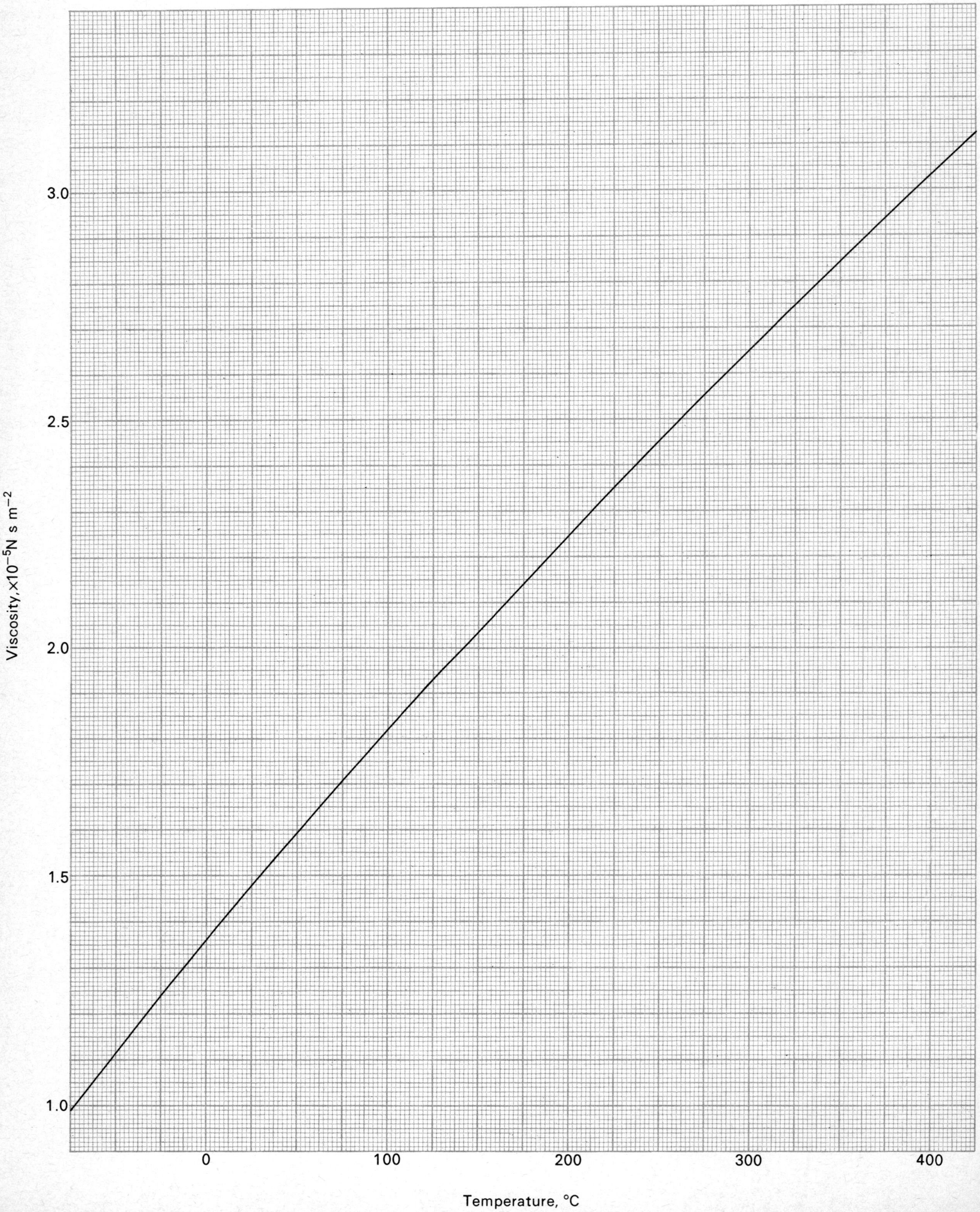

17.9 Thermal Conductivity of Liquid Nitrous Oxide (36)

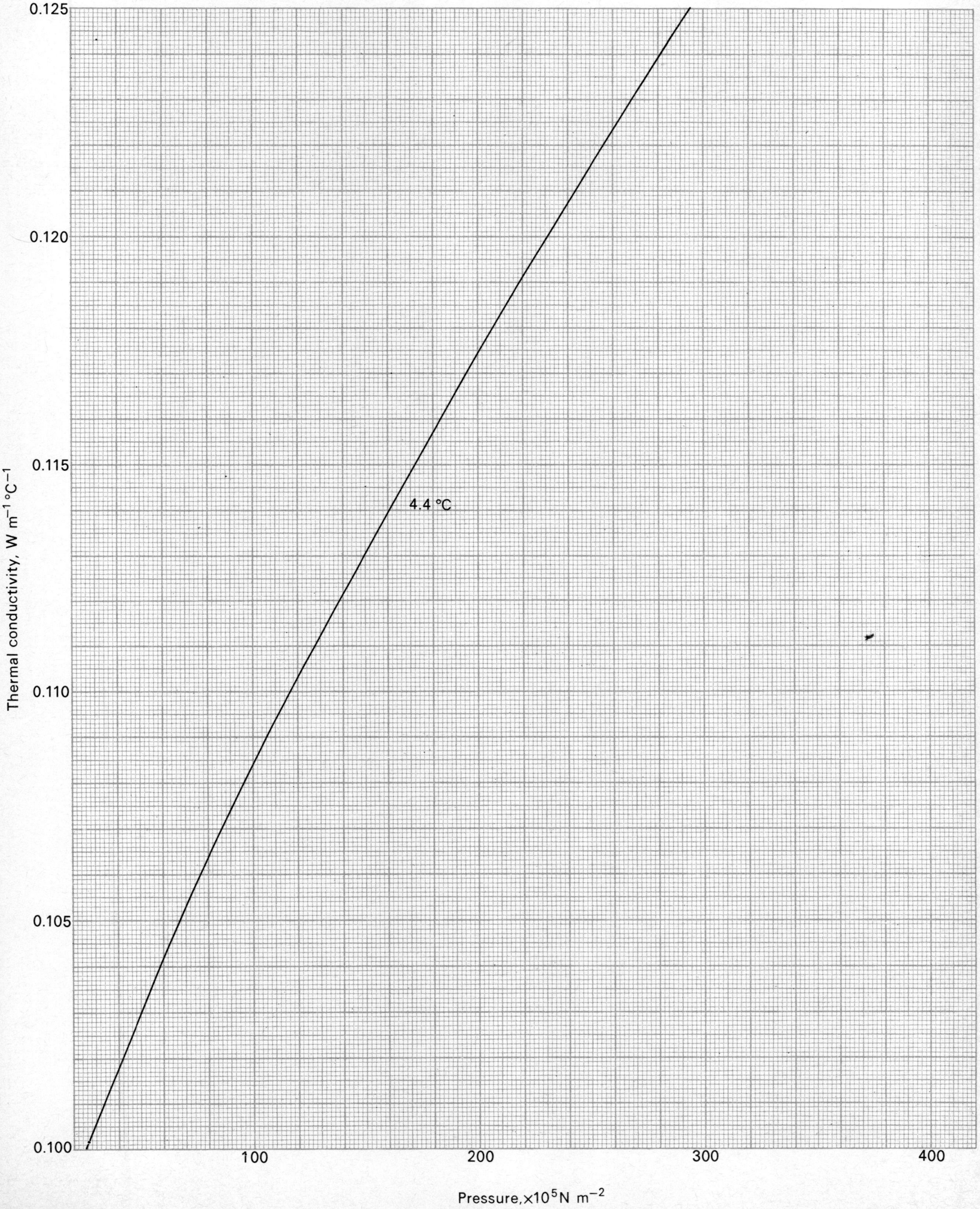

17.10 Thermal Conductivity of Gaseous Nitrous Oxide at $1.0133 \times 10^5 N\ m^{-2}$ (4,17,23,24,41)

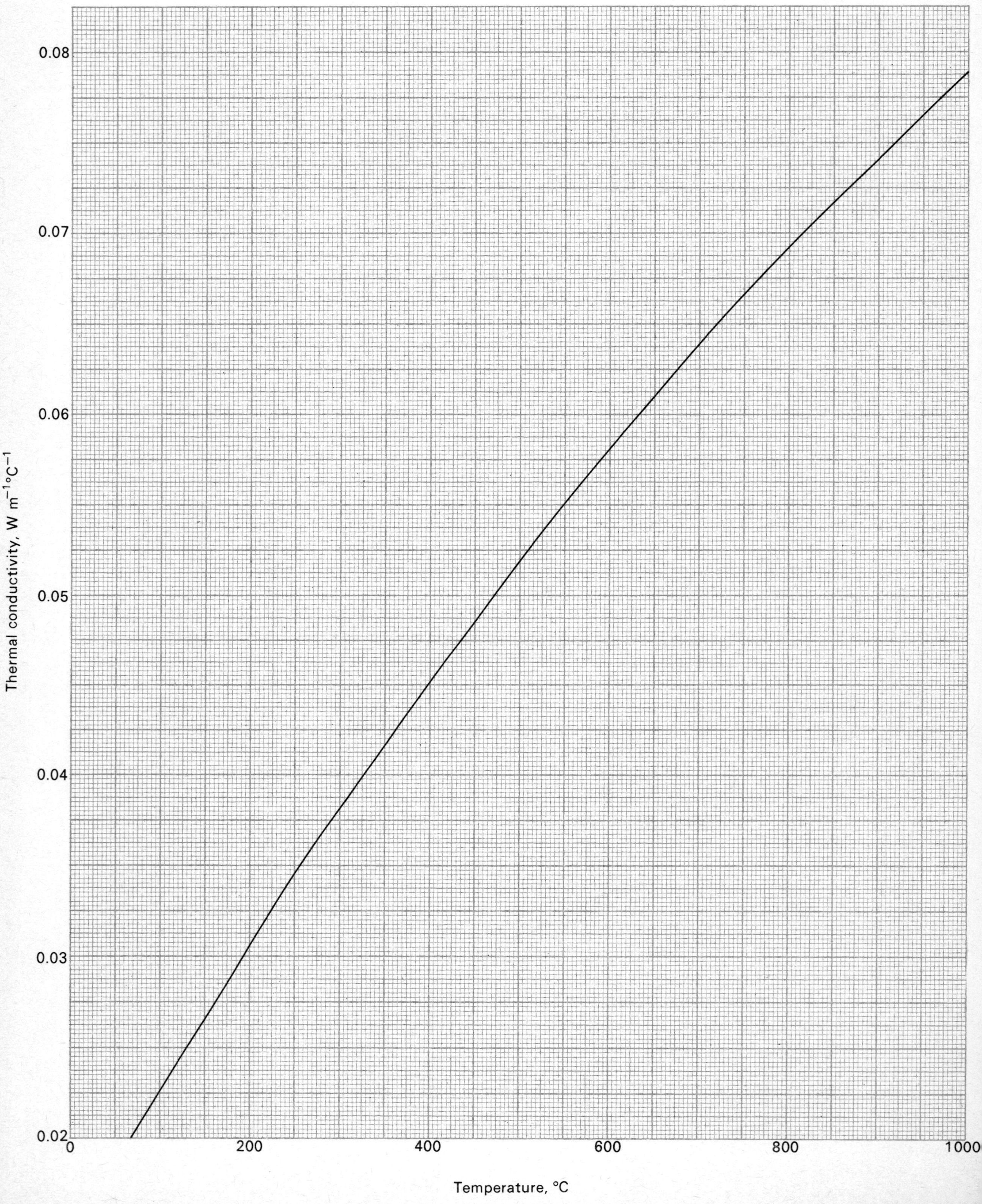

17.11 Surface Tension of Nitrous Oxide (34)

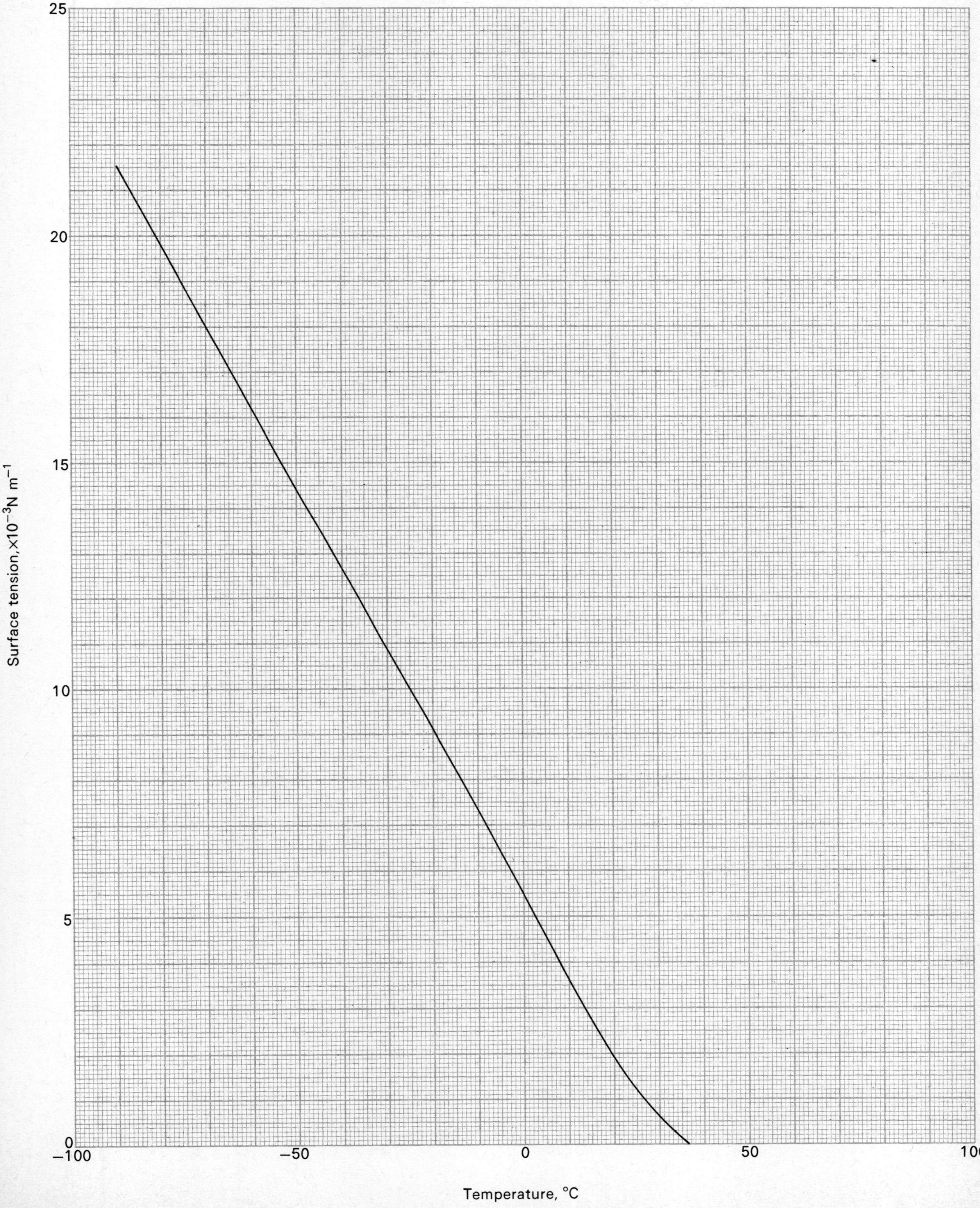

17.12 Solubility of Nitrous Oxide in Water at $1.0133 \times 10^5 N\ m^{-2}$ Partial Pressure (16,19,21,25,26)

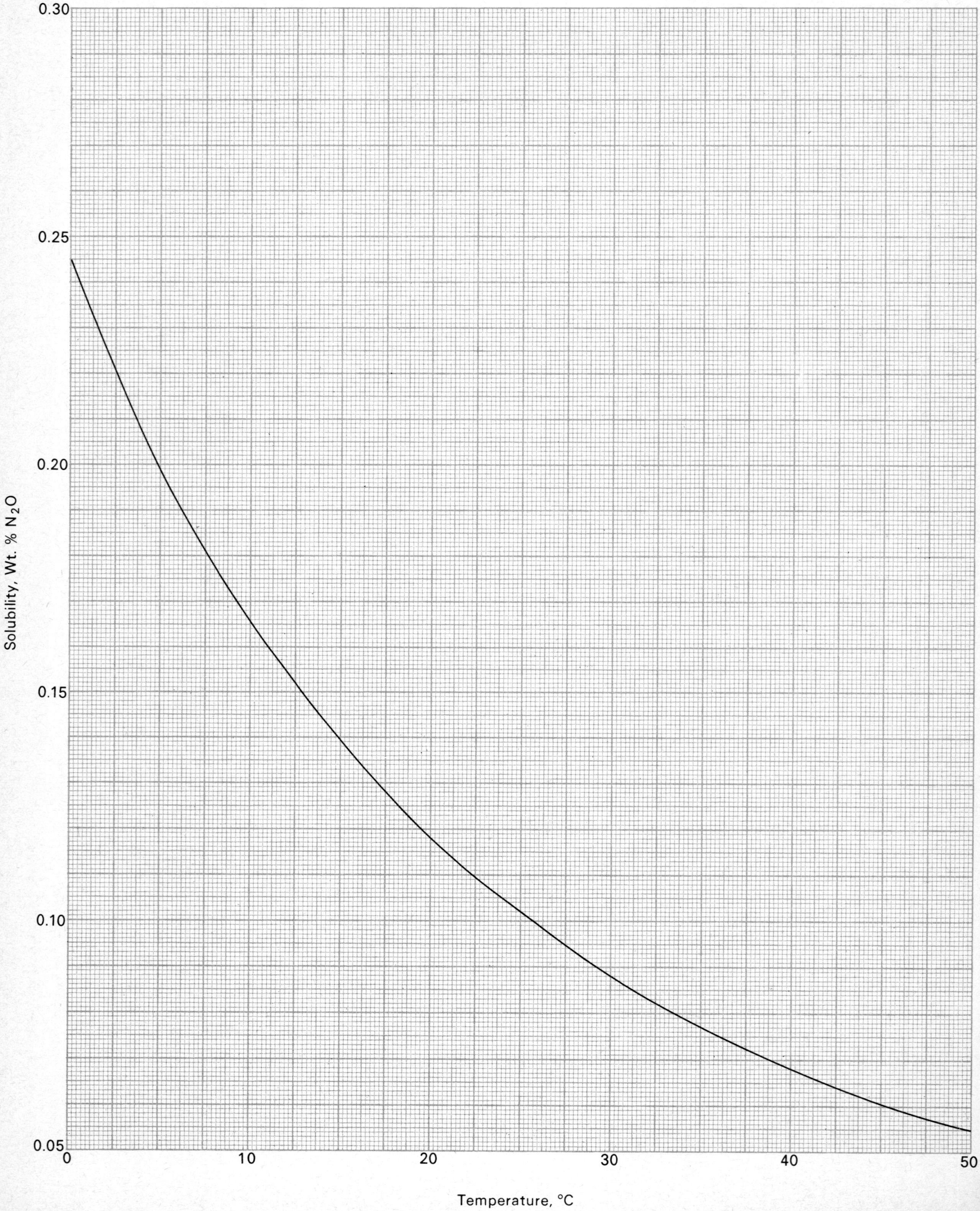

18 OXYGEN

Oxygen was discovered by Priestley in 1774. It is a colourless, odourless, tasteless gas at room temperature and atmospheric pressure, and is paramagnetic despite the fact that O_2 has an even number of electrons. This is because it has two unpaired electrons. Oxygen is very reactive, being able to combine with readily oxidizable compounds.

Liquid and solid oxygen are pale blue in colour. There are three modifications of oxygen in the solid phase: alpha, beta, and gamma forms. The crystal structures of the alpha and beta modifications are orthorhombic, while the gamma modification has a cubic structure. Solid oxygen has two transition points. Liquid oxygen has a very large coefficient of expansion, which might be due to the equilibrium

$$2\,O_2 \rightleftharpoons O_4$$

Oxygen is highly electronegative and combines directly with the majority of metals and nonmetals except the noble gases, halogens, and some of the noble metals. In the reactions oxygen acquires electrons by forming either covalent or ionic bonds.

Large scale production of oxygen is by fractional distillation of liquid air. The commercial applications of oxygen are in welding and cutting metals (oxy-acetylene torch). Distribution is in steel cylinders under high pressures. Further applications of interest are in medical practice, submarines, and fuel for space vehicles.

In connection with the U.S. space programme, the National Bureau of Standards investigated most of the physical, thermodynamic, and transport properties of solid, liquid, and gaseous oxygen at various temperatures and pressures. These publications (13, 33, 55, 56, 57, 58) formed the major part of this compilation. Further details, reviews, and selections, some with a comprehensive discussion are compiled in: Data Book and Buyers Guide (4), Gmelin (11), Landolt-Börnstein (22), Mullins *et al.* (32), Pascal (37), and Vasserman *et al.* (48, 49, 50, 51).

References

1 Boon, J. P., Thomaes, G., *Physica*, 1963, **29**, 208–14.
2 Burton, J. T. A., Ziebland, H., Ministry of Technology, Explosives Research and Development Establishment, Report No. 2/R/56, Waltham Abbey, 1956, 15 pp.
3 Clusius, K., *Z. phys. Chem. (Leipzig)*, 1929, **3B**, 41–79.
4 Data Book and Buyers Guide, Engineering Data Section, *Cryogenic and Ind. Gases*, 1971, **6** (3), 41–80.
5 De Bock, A., *Physica*, 1967, **34** (1), 49–52.
6 Dewar, J., *Proc. Roy. Soc.*, 1904, **73**, 251–61.
7 Engineering Science Data Units, *Specific Heats and Their Ratio as Functions of Temperature for Several Common Gases*, Royal Aeronautical Society, Data Sheet 00.01.08, London, 1965, 4 pp.
8 Engineering Science Data Units, *Approximate Data on the Viscosity of some Common Liquids*, Item No. 66024, London, July 1966, 26 pp.
9 Franck, E. U., *Z. Elektrochem.*, 1951, **55** (7), 636–43.
10 Geier, H., Schafer, K., *Allgem. Wärmetech.*, 1961, **10**, 70–5.
11 *Gmelins Handbuch der anorganischen Chemie*, Syst. No. 3, Sauerstoff, Lief. 3, Verlag Chemie GmbH, Weinheim, 1958, pp. 318–593.
12 Golubev, I. F., *Viscosity of Gases and Gas Mixtures*, Gosudarstvennoe Izdatel'stvo Fiziko-Mathematicheskoi Literatury, Moskva, 1959, Table 4.
13 Goodwin, R. D., Weber, L. A., *J. Res. Natl. Bur. Std.*, 1969, **73A**, 1–36.
14 Hamman, G., *Ann. Phys.*, Ser. 5, 1938, **32**, 593.
15 Honig, R. E., Hook, H. O., Vapour Pressure Data for some Common Gases, *RCA Rev.*, 1960, **21**, 360–8.
16 IUPAC, Atomic Weights of the Elements 1971, *Pure & Appl. Chem.*, 1972, **30** (3–4), 638–49.
17 Ibrahim, S. H., Kuloor, N. R., *Brit. Chem. Eng.*, 1961, **6** (12), 862–3.
18 Ivanova, Z. A., *Teploenergetika*, 1967, **14** (10), 74–7.
19 Kanda, E., *Bull. Chem. Soc. Japan*, 1937, **12**, 469–72.
20 Kestin, J., Leidenfrost, W., *The Effect of Moderate Pressures on the Viscosity of Five Gases*, Progr. Internat. Res. Thermodyn. Transport Properties, 2nd Symp. Thermophys. Properties (Masi, J. F., Tsai, D. H., Eds.) Am. Soc. Mech. Eng., Academic Press, New York, 1962, pp. 321–38.
21 Landine, R. C., *Water and Sewage Works*, 1971, **118** (8), 242–4.
22 *Landolt-Börnstein, Zahlenwerte und Funktionen aus Physik, Chemie, Astronomie, Geophysik, und Technik*; Technik, 4 Teil, Bandteil a, Springer-Verlag, Berlin, 1967, pp. 228–35.
23 Lange, N. A., Ed., *Handbook of Chemistry*, 10th ed., McGraw-Hill, New York, 1967.
24 Langen, A., *Z. Electrochem.*, 1919, **25**, 25–45.
25 Lax, E., Synowietz, C., Eds., *Taschenbuch für Chemiker und Physiker*, 3rd ed., Springer-Verlag, Berlin, 1967.
26 Leduc, A., *Chem. Rev.*, 1929, **6**, 1–16.
27 Lewis, B., Von Elbe, G., *J. Am. Chem. Soc.*, 1933, **55**, 511–19.
28 Maitland, G. C., Smith, E. B., *J. Chem. Eng. Data*, 1972, **17** (2), 150–6.
29 Maryott, A. A., Smith, E. R., *Table of Dielectric Constants of Pure Liquids*, Natl. Bur. Std. Circular 514, Washington, Aug. 10, 1951, 44 pp.
30 Mathews, J. F., *Chem. Rev.*, 1972, **72** (1), 71–100.
31 Montgomery, J. B., DeVries, T., *J. Am. Chem. Soc.*, 1942, **64**, 2372–5.
32 Mullins, J. C., Ziegler, W. T., Kirk, B. S., *The Thermodynamic Properties of Oxygen from 20 to 100 K*, Georgia Inst. of Technology, Eng. Expt. Station Techn. Rept. No. 2, March 1, 1962.
33 McCarty, R. D., Weber, L. A., *Thermophysical Properties of Oxygen from the Freezing Liquid Line to 600°R for Pressure to 5000 PSIA*, Natl. Bur. Std. Note 384, Washington, July 1971, 189 pp.
34 Okamura, T., *Tohoku Imp. Univ. Science Rept.*, 1933, **22**, 519–36.
35 Orlova, M. P., *Russ. J. Phys. Chem.*, 1966, **40** (12), 1603–7.
36 Pao, C.-S., *Phys. Rev.*, 1943, **64** (2), 60–74.
37 Pascal, P., Ed., *Nouveau Traité de Chimie Minérale*, Vol. XIII, Part 1, Oxygene, Ozone, Oxydes, Eau Oxygénée, La combustion, Soufre, Sélénium, Tellure, Polonium; Masson et Cie, Éditeurs, Paris, 1960, pp. 75–147.
38 Prosad, S., *Brit. J. Appl. Phys.*, 1952, **3**, 58.
39 Rossini, F. D., Wagman, D. D., Evans, W. H., Levine, S., Jaffe, I., *Selected Values of Chemical Thermodynamic Properties*, Natl. Bur. Std. Circular 500, Washington, Feb. 1, 1952.
40 Rudenko, N. S., *Zh. Expt. Theor. Fiz.*, 1939, **9**, 1078–80.
41 Saji, Y., Kobayashi, S., *Cryogenics*, 1964, **4**, 136–40.
42 Schaefer, C. A., Thodos, G., *AIChE Journal*, 1959, **5** (3), 367–72.
43 Stull, D. R., Dir., *JANAF Thermochemical Tables*, Dow Chemical Co., Midland, Mich., 1965–8.
44 Tsederberg, N. V., Timrot, D. L., *Soviet Phys., Tech. Phys.*, 1957, **1**, 1791–7.
45 Van Itterbeek, A., de Bock, A., Verhaegen, L., *Physica*, 1949, **15** (2), 624–6.
46 Van Itterbeek, A., Van Paemel, O., *Physica*, 1941, **8** (2), 133–43.
47 Van Paemel, O., *Verh. Kon. Vlaamsche Acad. Kl. Wetensch.*, 1941, **3** (3), 1–59.

48 Vasserman, A. A., Kazavchinskii, Ya. Z., Rabinovich, V. A., *Thermophysical Properties of Air and Air Components*, Izdatel'stvo Nauk, Moskva, 1966.
49 Vasserman, A. A., Rabinovich, V. A., *Zh. Fiz. Khim.*, 1968, **152** (7), 8–35.
50 Vasserman, A. A., Rabinovich, V. A., *Thermophysical Properties of Liquid Air and Its Components*, Izdatel'stvo Komiteta Standartov, Mer i Izmeritel'nykh Priborov pri Sovete Ministrov SSSR, Moskva, 1968.
51 Vasserman, A. A., Rabinovich, V. A., *Teplo Massoperenos*, 1968, 7, 47–57.
52 Verschaffelt, J. E., *Meded. natuurkundig Labor. Univ. Gent*, 1925, **2**, 1–9.
53 Wacker, P. F., Cheney, R. K., Scott, R. B., *J. Res. Natl. Bur. Std.*, 1947, **38**, 651–9.
54 Washburn, E. W., Ed., *International Critical Tables of Numerical Data, Physics, Chemistry, and Technology*, Vol. VI, McGraw-Hill, New York, 1929, pp. 461–7.
55 Weber, L. A., *Phys. Letters*, 1969, **30A** (7), 390–1.
56 Weber, L. A., The P–V–T Surface of Oxygen in the Critical Region; Densities of Saturated Liquid and Vapour, in *Advances in Cryogenic Engineering*, (K. D. Timmerhous, Ed.) Vol. 15, Plenum Press, New York, 1970, B-4, pp. 50–7.
57 Weber, L. A., *J. Res. Natl. Bur. Std.*, 1970, **74A** (1), 93–129.
58 Weber, L. A., *Phys. Rev.*, 1970, **2A** (6), 2379–88.
59 Winkler, L. W., *Z. phys. Chem.*, 1891, **14**, 3606.
60 Yosim, S. J., *J. Chem. Phys.*, 1964, **40** (10), 3069–75.
61 Ziebland, H., Die Wärmeleitfähigkeit von Stickstoff, Sauerstoff und Argon in flüssigem und gasformigem Zustand, *Dechema Monographien*, 1959, **32** (451–76), 74.
62 Ziebland, H., Burton, J. T. A., *Brit. J. Appl. Phys.*, 1955, **6**, 416–20.

Table 18.1 Properties of Oxygen

		Ref.
Chemical formula	O_2	(25)
Molecular weight	31.9980	(16)
Critical temperature	−118.57 °C	(30)
Critical pressure	50.43×10^5 N m^{-2}	(30)
Critical density	436.0 kg m^{-3}	(30)
Triple-point temperature	−218.80 °C	(25)
Triple-point pressure	152.0 N m^{-2}	(25)
Boiling point at 1.0133×10^5 N m^{-2}	−182.97 °C	(25)
Melting point	−218.40 °C	(25)
Transition temperatures	−249.26 °C −229.35 °C	(15) (15)
Density of solid at −252.5 °C	1425.6 kg m^{-3}	(6)
Compressibility (adiabatic) of liquid at −198.15 °C	0.780×10^{-9} m^2 N^{-1}	(45)
Coefficient of thermal expansion of liquid at boiling point	3.95×10^{-3} °C^{-1}	(60)
Heat of formation of gas at 25 °C	0.0 J kg-mol^{-1}	(43)
Free energy of formation of gas at 25 °C	0.0 J kg-mol^{-1}	(43)
Heat of sublimation at −219.05 °C	8.2082×10^6 J kg-mol^{-1}	(24)
Heat of fusion	4.4506×10^5 J kg-mol^{-1}	(39)
Heat of transitions	9.3784×10^4 J kg-mol^{-1} 7.4358×10^5 J kg-mol^{-1}	(15) (15)
Entropy of gas at 25 °C	2.0515×10^5 J kg-mol^{-1} °C^{-1}	(43)
Heat capacity ratio of gas at 15 °C and 1.0133×10^5 N m^{-2}	1.401	(26)
Velocity of sound at 0 °C and 1.0133×10^5 N m^{-2}	316.2 m s^{-1}	(54)
Diffusion in air at 0 °C and 1.0133×10^5 N m^{-2}	1.80×10^{-5} m^2 s^{-1}	(17)
Dielectric constant of liquid at −193.0 °C	1.507	(29)
Electrical conductivity of liquid at −183.15 °C and 4.01×10^5 V m^{-1}	3×10^{-19} ohm^{-1} m^{-1}	(36)
Refractive index of liquid at −181.0 °C	1.221	(25)

18.1 Vapour Pressure of Oxygen (33,56,57)

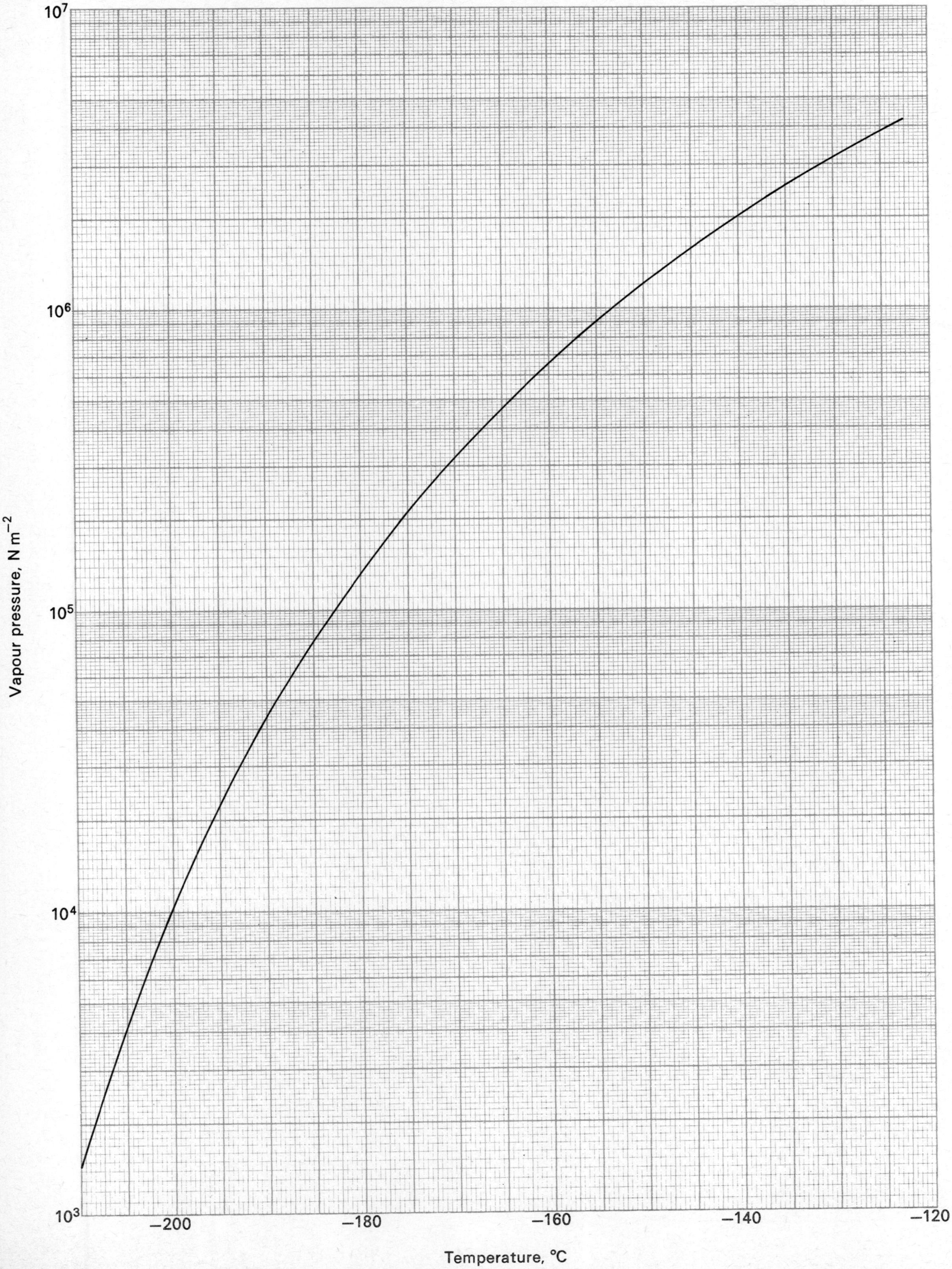

18.2 Orthobaric Density of Oxygen (23,33,41,55, 56,57,58)

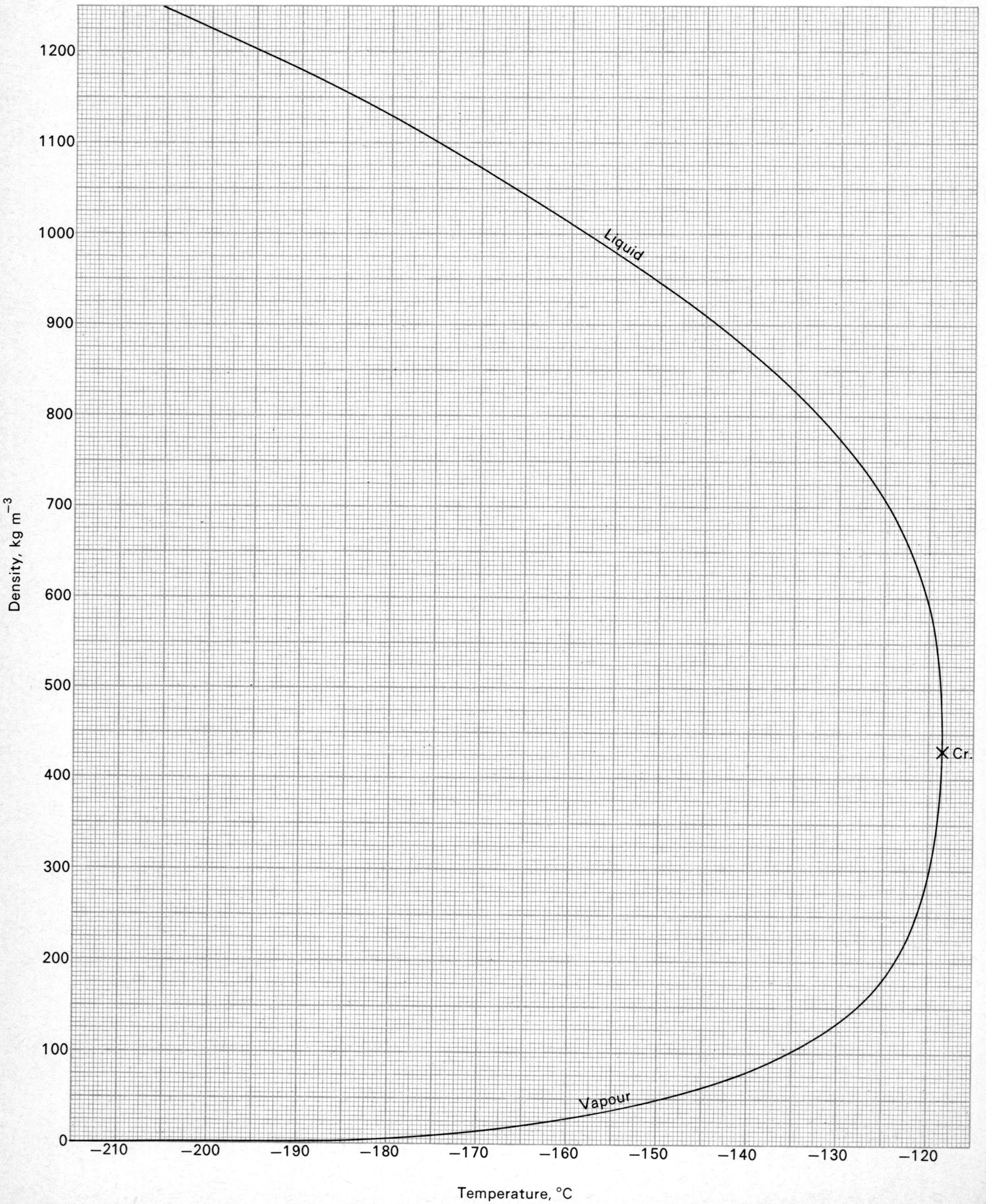

18.3 P-V-T Data for Oxygen (33,48,49,50,56,57)

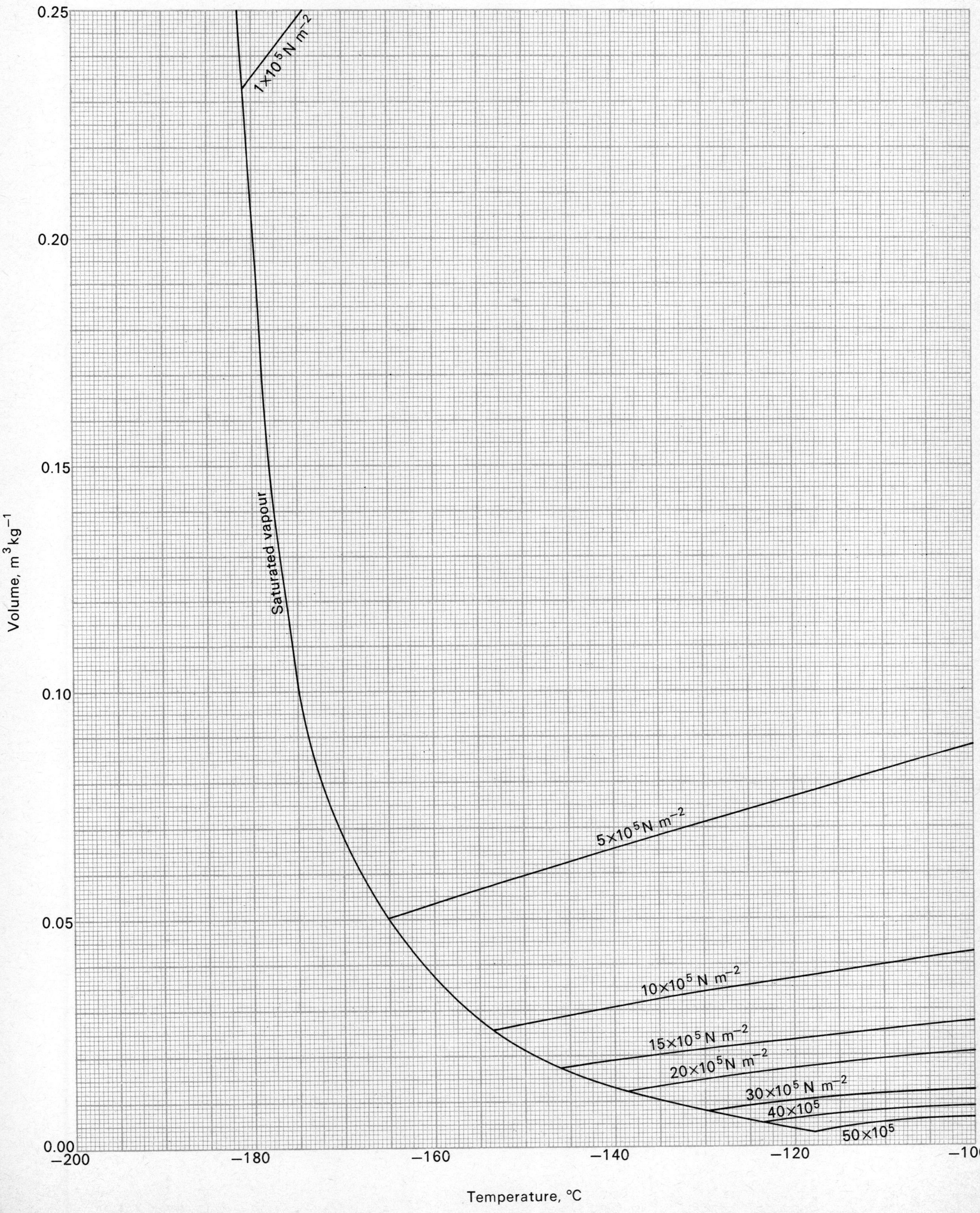

18.4 Latent Heat of Vaporisation of Oxygen (33,48,50)

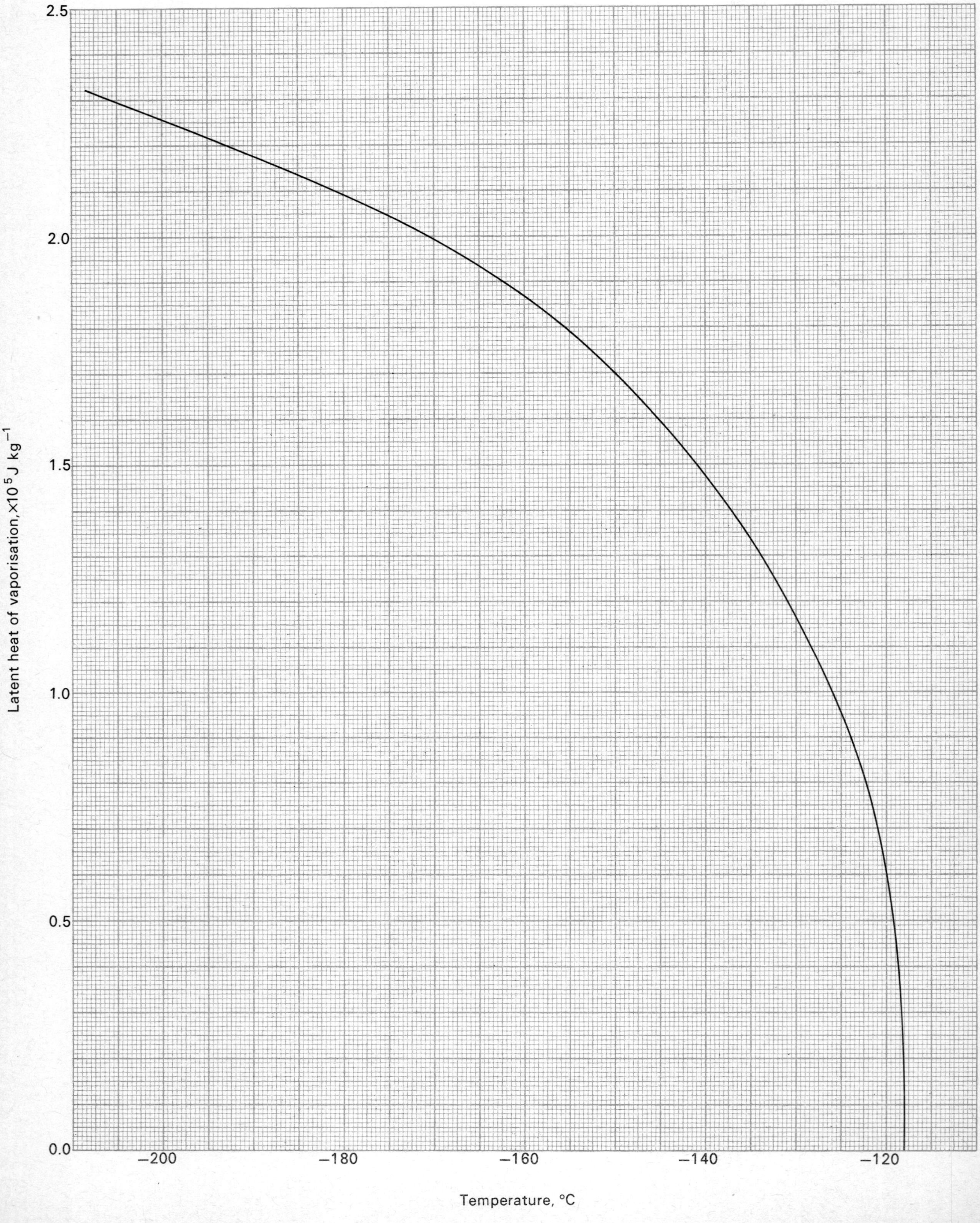

18.5 Heat Capacity of Liquid Oxygen (3,13,33,35)

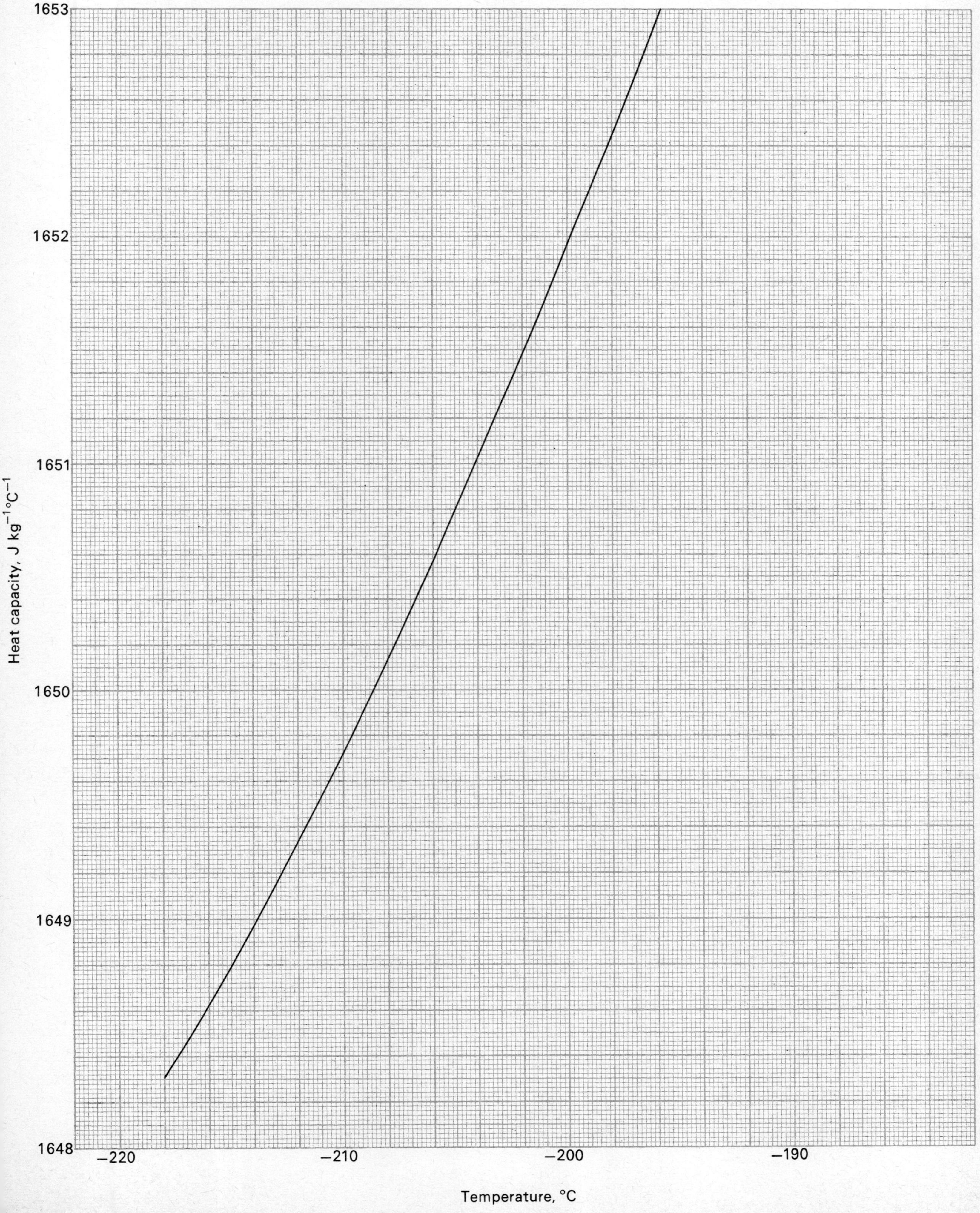

18.6 Heat Capacity of Gaseous Oxygen at $1.0133 \times 10^5 N\ m^{-2}$ (13,27,31,34,43,53)

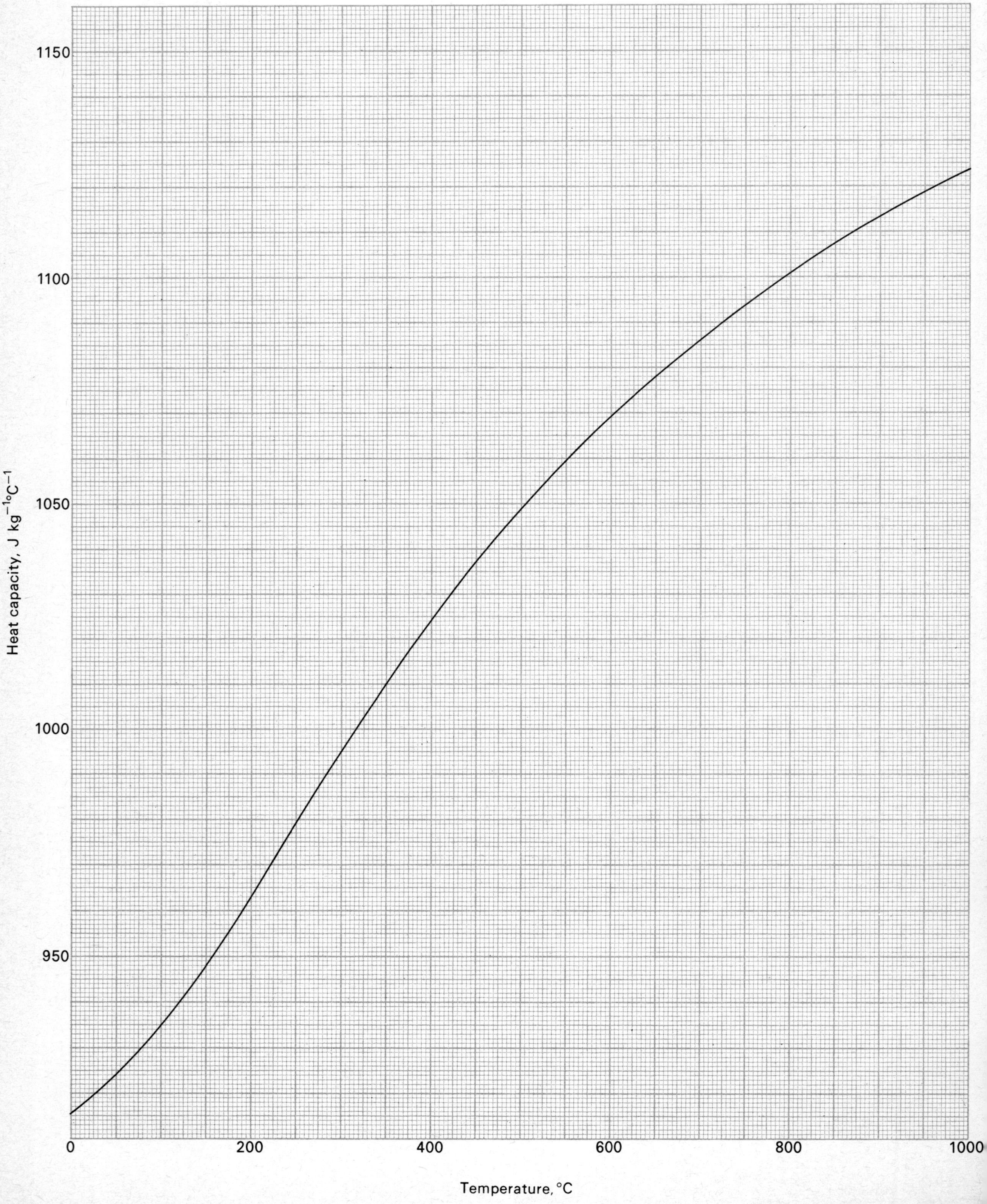

18.7 Heat Capacity Ratio (C_p/C_v) of Gaseous Oxygen at 1.0133×10^5 N m^{-2}(7,33,57)

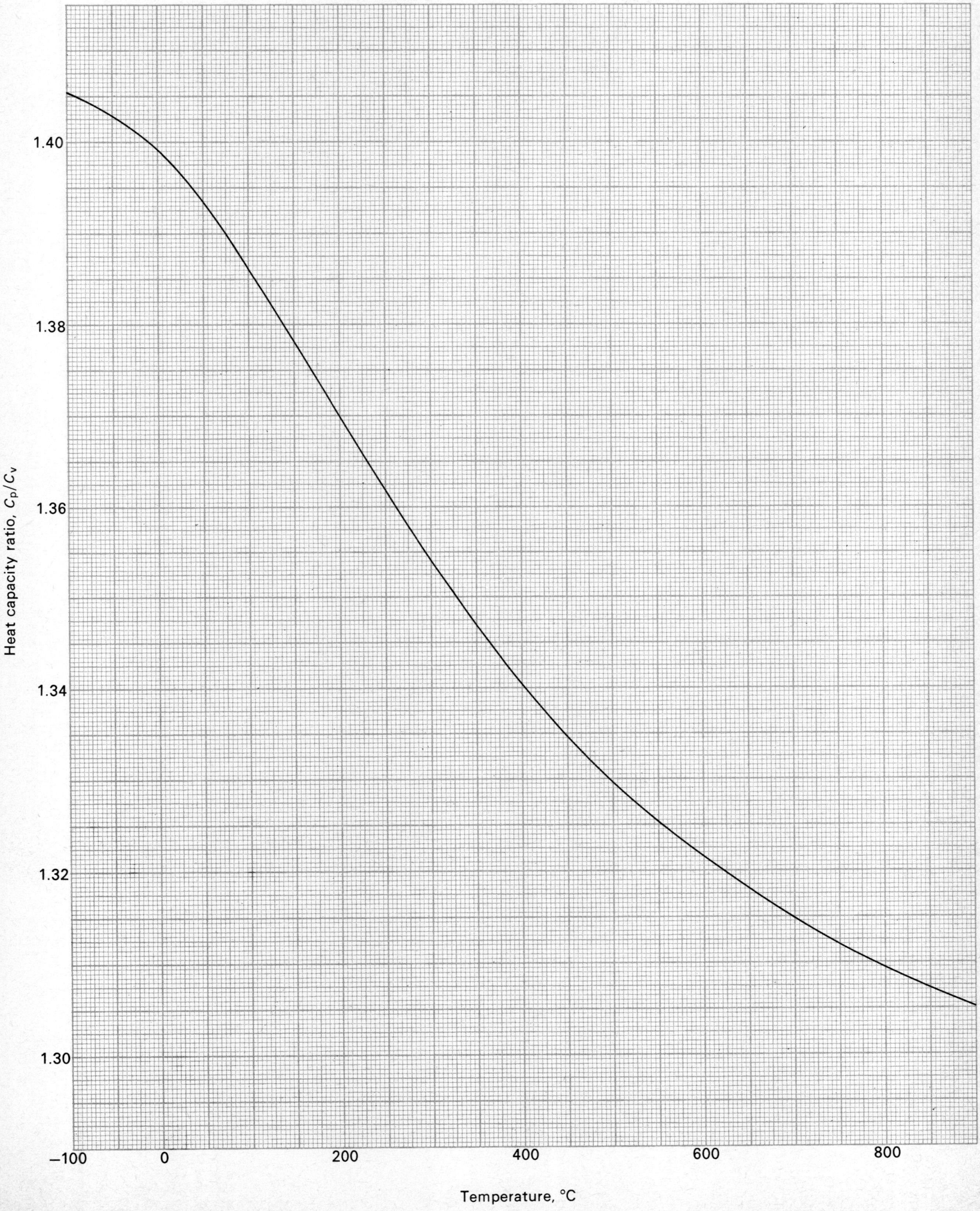

18.8 Viscosity of Liquid Oxygen (1,5,8,33,40,41,46,47)

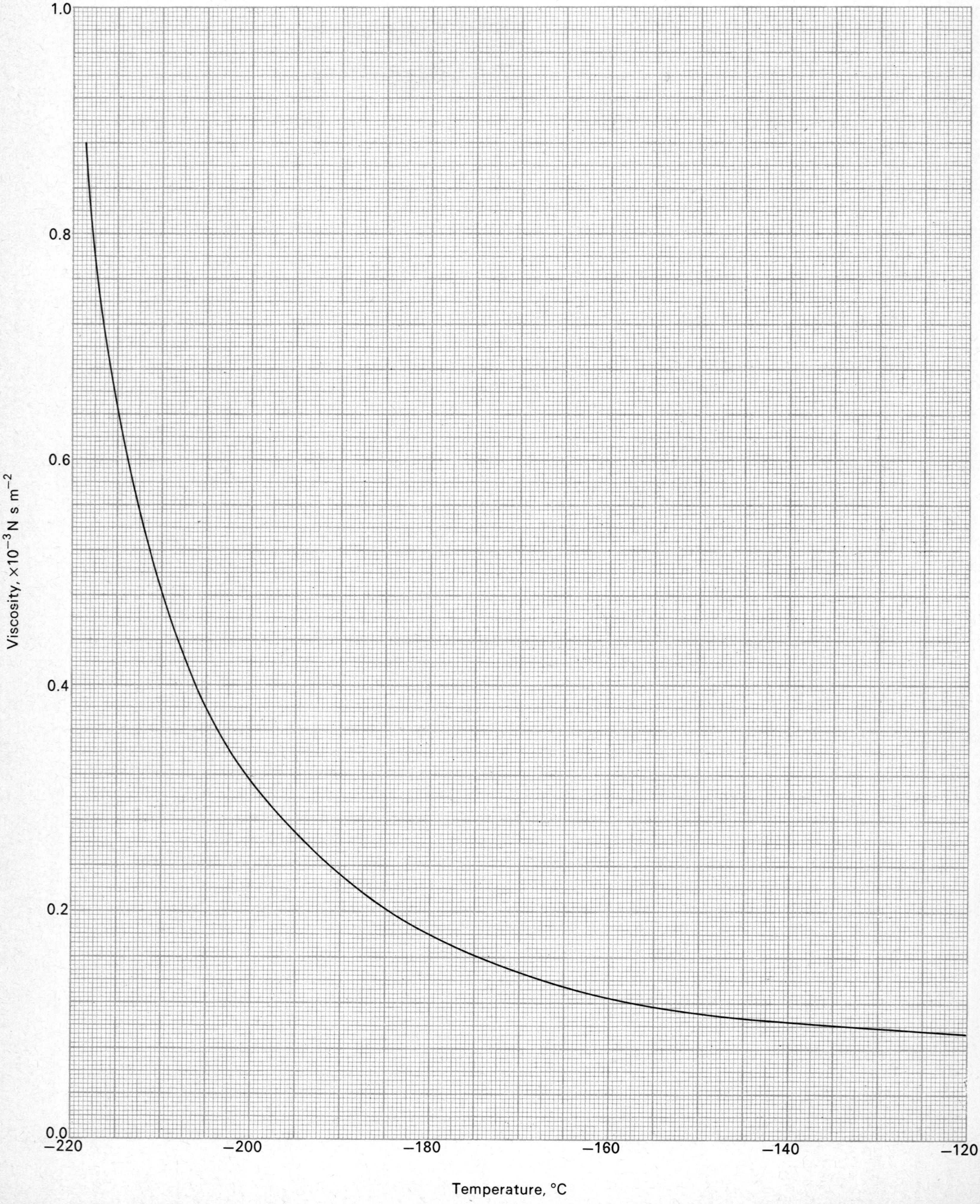

18.9 Viscosity of Gaseous Oxygen at $1.0133 \times 10^5 N\ m^{-2}$ (12,20,28,48,50)

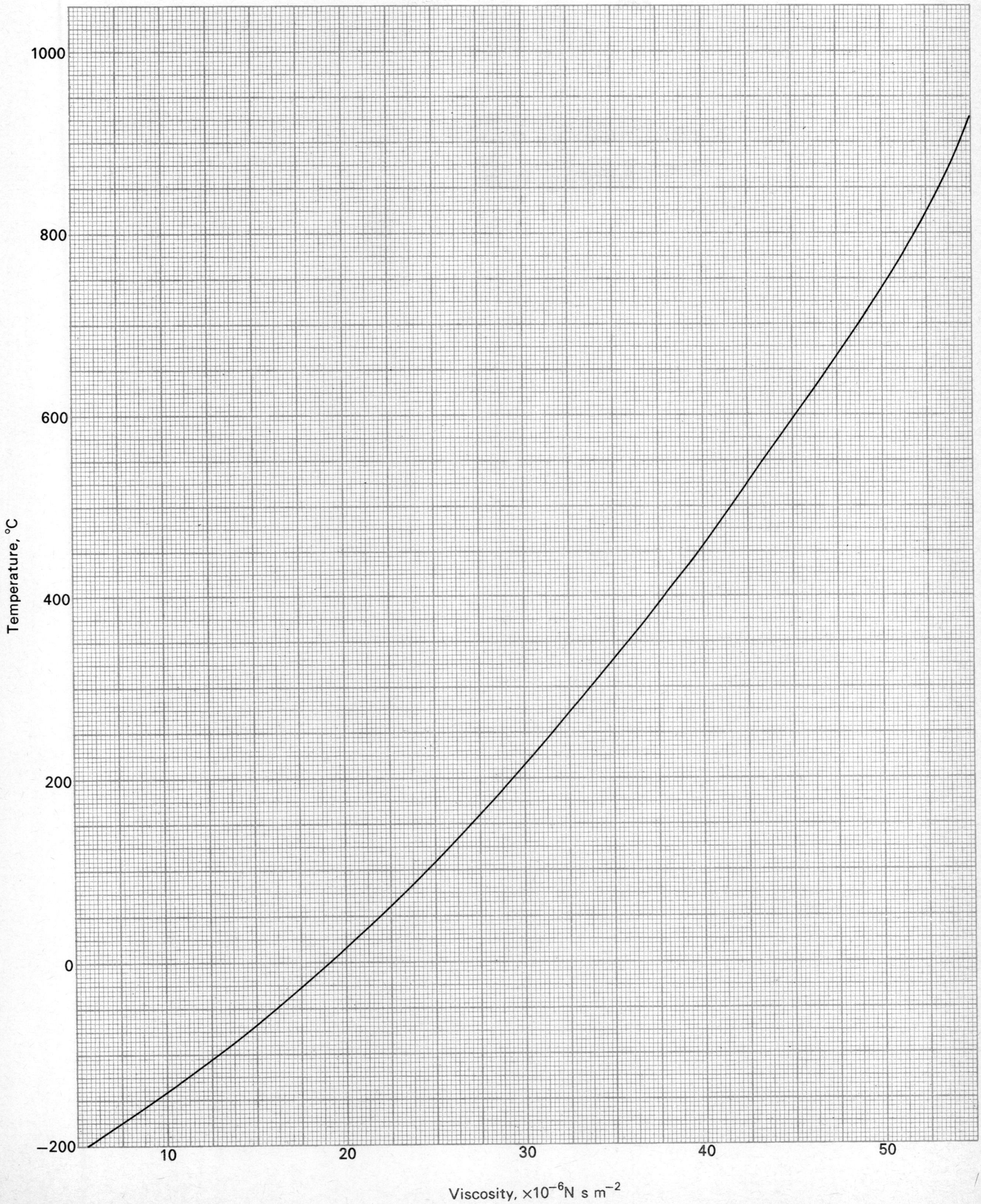

18.10 Thermal Conductivity of Liquid Oxygen (2,14,18,38,44,61,62)

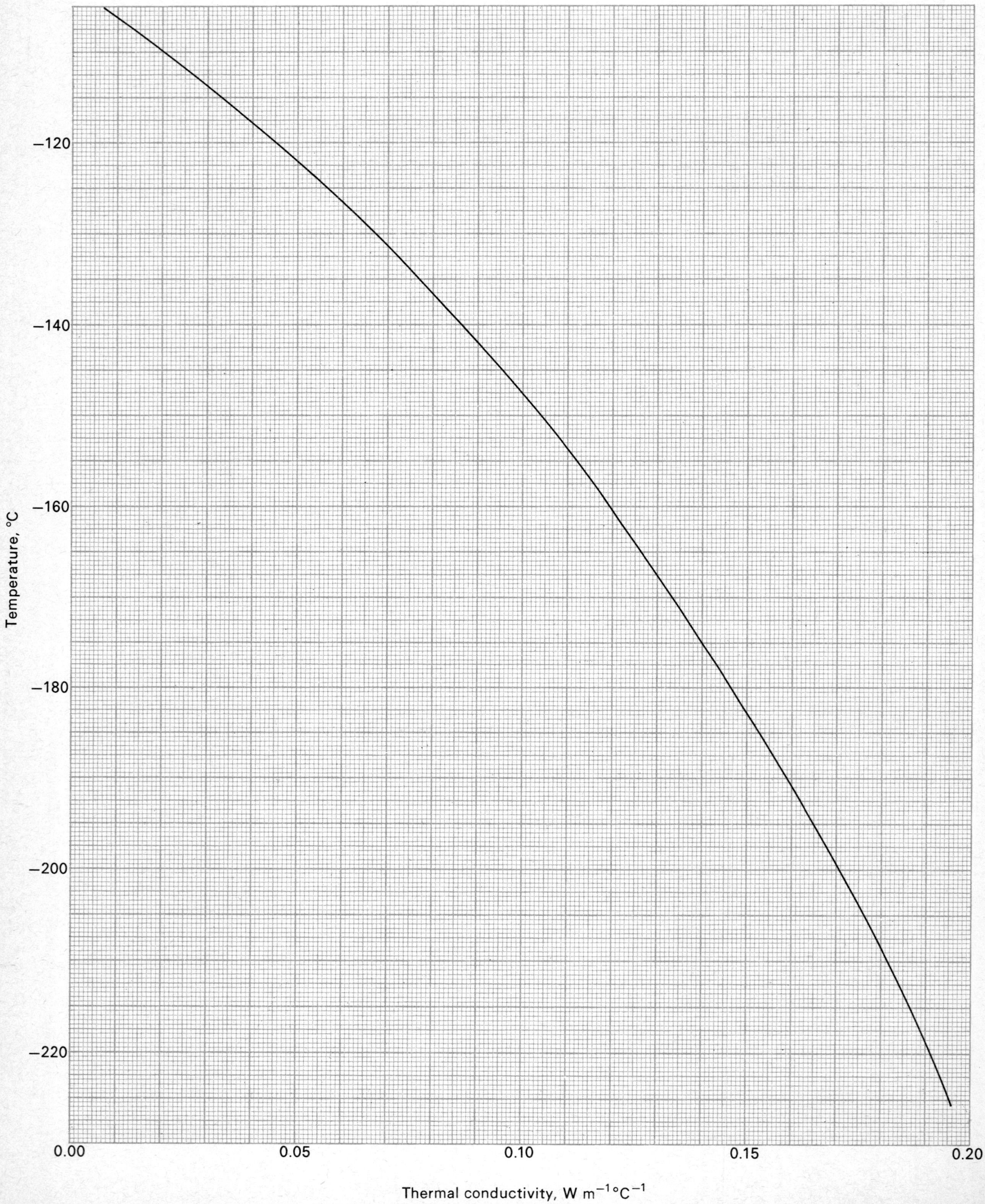

18.11 Thermal Conductivity of Oxygen (9,10,18,33,42,61)

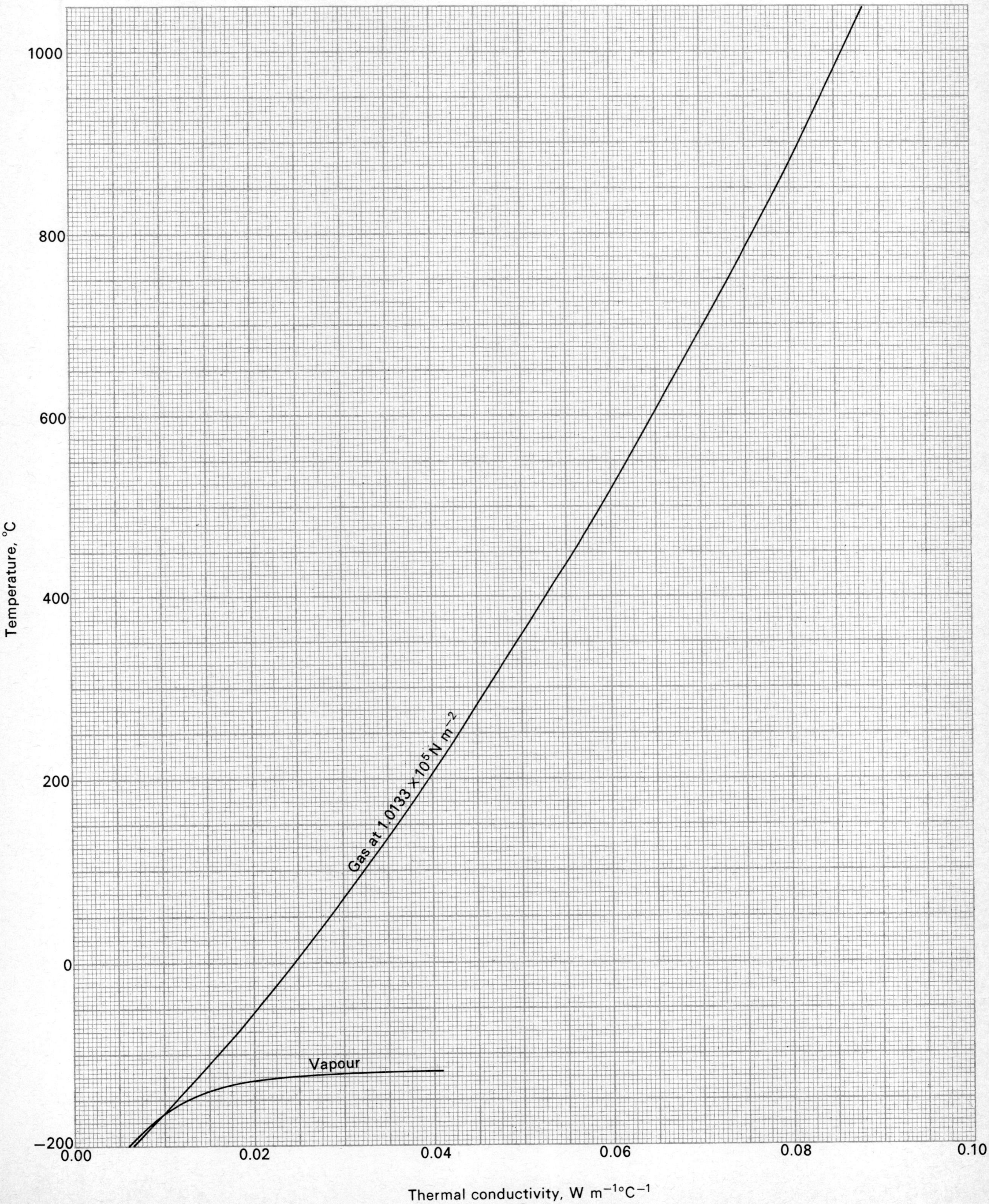

18.12 Surface Tension of Oxygen (19,33,41,52)

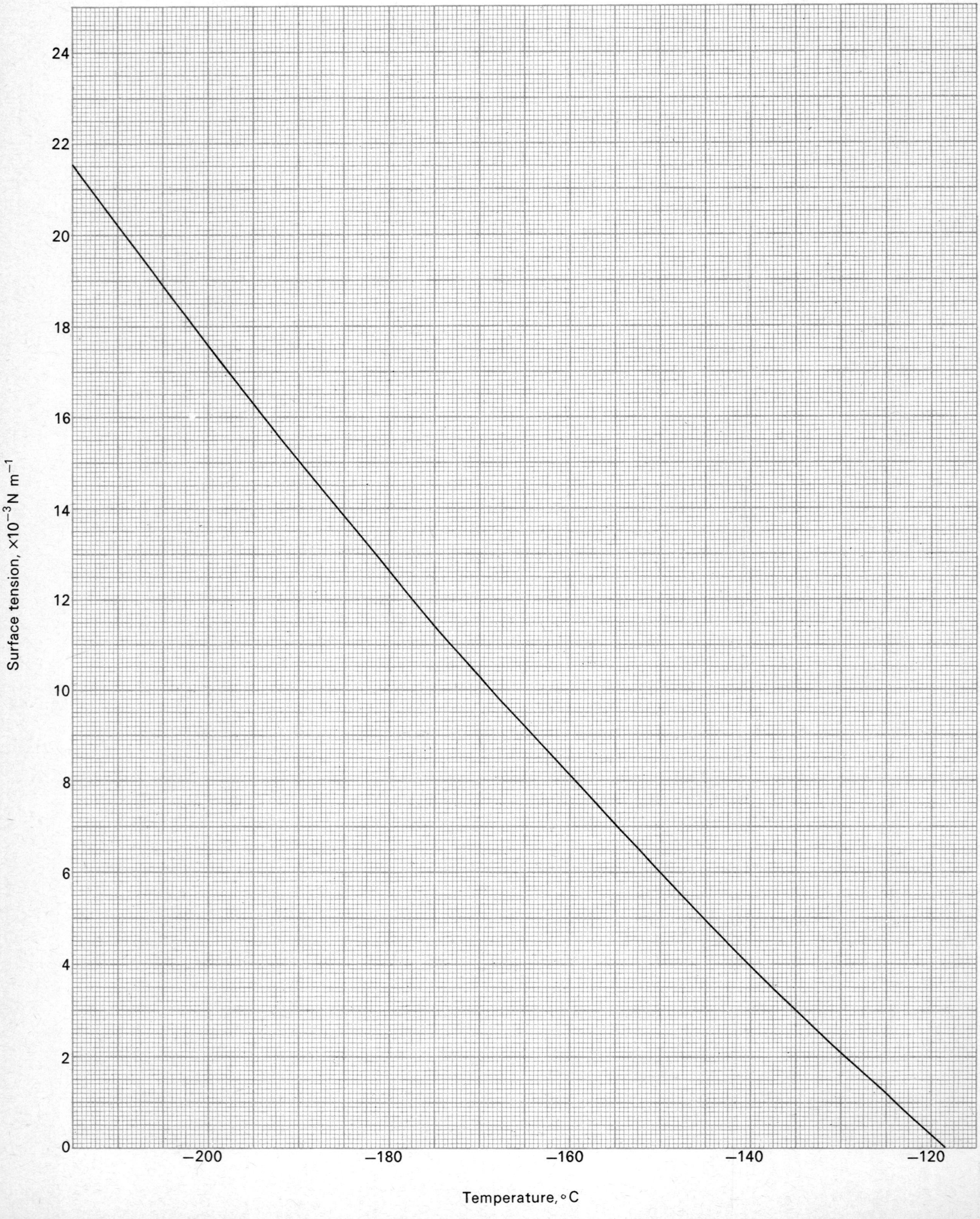

18.13 Solubility of Oxygen in Water at 1.0133×10^5 N m^{-2} Total Pressure (21,23,59)

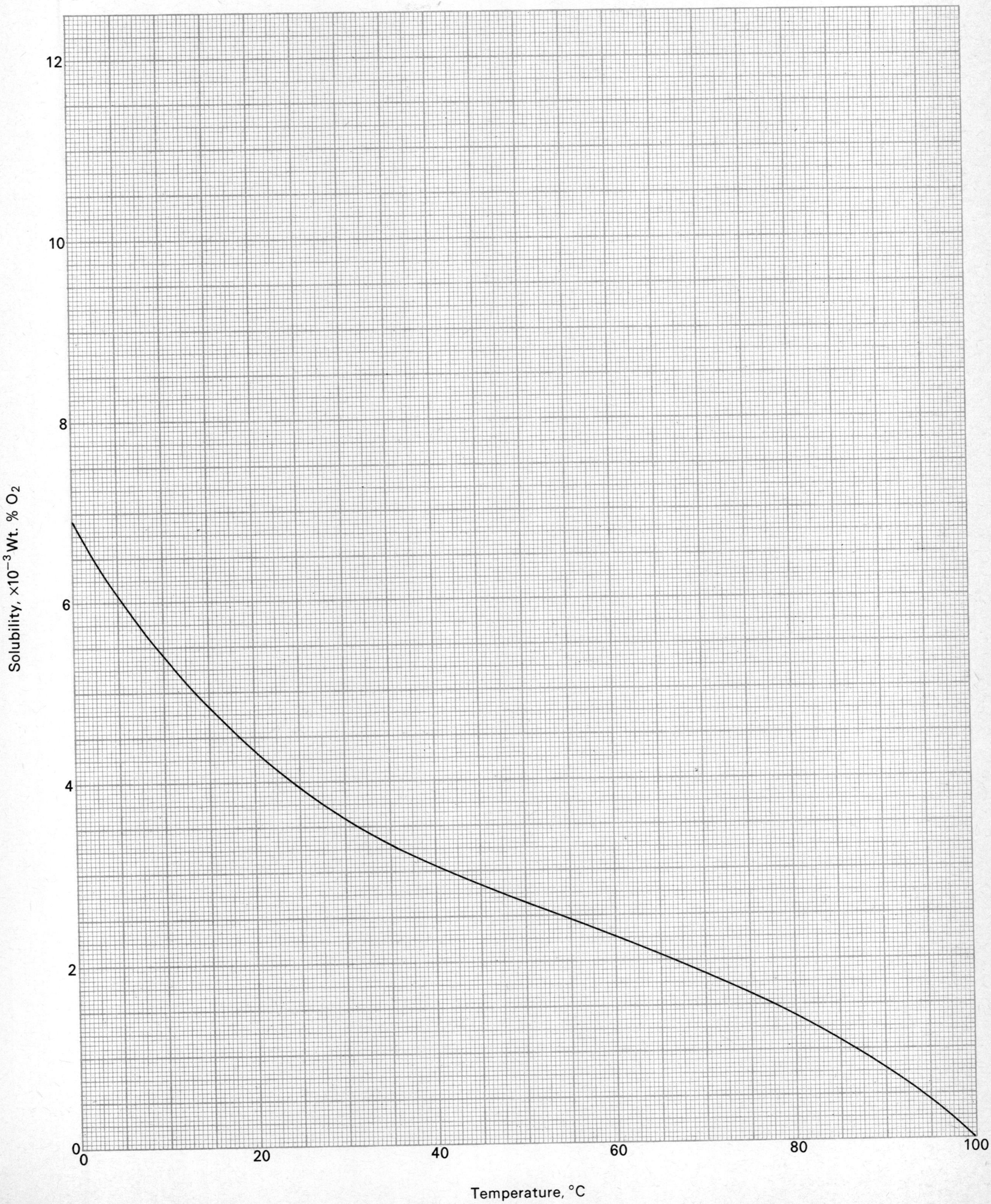

19 OZONE

Ozone is a blue coloured gas at ambient temperature and atmospheric pressure, with high colour intensity. It has a very characteristic pungent odour. The molecule appears to be V-shaped with an angle of 127°. The triatomic ozone molecule is paramagnetic and endothermic. In the gaseous phase ozone is a very powerful oxidizing agent, second only to fluorine. In most of its reactions, it is reduced to molecular oxygen. The gas is fairly stable and condenses to an indigo-blue liquid, which on further cooling solidifies to deep blue-violet crystals.

Being an endothermic compound, under rapid heating both the solid and liquid are explosive, and precautions have to be taken during the warming processes. The gas is more soluble in water than is oxygen.

Besides the use of ozone in bleaching of fabrics, papers, silks, etc., the main application in recent years is as a powerful propellant in high-energy chemistry. Further practical aspects are covered by Diaper (6) and Thorp (24).

In 1961, Streng (22) carried out a survey of the physical properties of ozone. The selected values are presented in Tables 19.1. Further reviews and compilations of interest are available in: Gmelin (9), Hersh *et al.* (11), Mellor (15), and Pascal (18).

References

1 American Conference of Government Industrial Hygienists, *Documentation of Threshold Limit Values*, revised edition, Committee on Threshold Limit Values, Cincinnati, Ohio 45202, 1966, 203 pp.

2 Birdsall, C. M., Jenkins, A. C., DiPaolo, F. S., Beattie, J. A., Apt, C. M., *J. Chem. Phys.*, 1955, **23** (3), 441–52.

3 Brabets, R. I., McDonough, J. M., *J. Chem. Phys.*, 1957, **27** (4), 880–2.

4 Briner, E., Perrottet, E., *Helv. Chim. Acta*, 1939, **22**, 397–404.

5 Cleveland, F. F., *J. Chem. Phys.*, 1952, **20** (2), 337–8.

6 Diaper, E. W. T., *Chem. Tech.*, 1972 (June), 368–75.

7 Engineering Science Data Units, *Approximate Data on the Viscosity of some Common Liquids*, Item No. 66024, London, July 1966, 26 pp.

8 Epprecht, G. W., *Z. ang. Math. Phys. (Basel)*, 1950, **1**, 138–44.

9 *Gmelins Handbuch der anorganischen Chemie*, Syst. No. 3, Sauerstoff, Lief. 4, Verlag Chemie GmbH, Weinheim, 1960, pp. 1110–46.

10 Heather, R. C., *J. Inst. Water Eng.*, 1948, **4**, 358.

11 Hersh, C. K., Berger, A. W., Brown, J. R. C., Physical Properties of Liquid Ozone–Oxygen Mixtures: Density, Viscosity, and Surface Tension, Adv. in Chem. Ser. No. 21, *Ozone Chemistry and Technology*, Am. Chem. Soc., Washington, 1959, pp. 22–7.

12 IUPAC, Atomic Weights of the Elements 1971, *Pure & Appl. Chem.*, 1972, **30** (3–4), 638–49.

13 Jacobs, A., Univ. of Marburg, *Dissertation*, Marburg, 1904.

14 Lax, E., Synowietz, C., Eds., *Taschenbuch für Chemiker und Physiker*, 3rd ed., Springer-Verlag, Berlin, 1967.

15 *Mellor's Comprehensive Treatise on Inorganic and Theoretical Chemistry*, Vol. I, H, O, Longmans, London, 1922, pp. 877–961.

16 Olszewski, C., *Phil. Mag.*, 1895, **38** (5), 188–212.

17 Owens, E. J., Thodos, G., *AIChE Journal*, 1960, **6** (4), 676–81.

18 Pascal, P., Ed., *Nouveau Traite de Chimie Minérale*, Vol. XIII, Masson et Cie, Éditeurs, Paris, 1960, pp. 243–312.

19 Riesenfeld, E. H., Beja, M., *Medd. Vetenskapsakad. Nobelinst., Stockholm*, 1923, **6** (7), 1–27.

20 Riesenfeld, E. H., Beja, M., *Z. anorg. Chem.*, 1924, **132**, 179–200.

21 Schwab, G.-M., *Z. phys. Chem.*, 1924, **110**, 599–625.

22 Streng, A. G., *J. Chem. Eng. Data*, 1961, **6** (3), 431–6.

23 Stull, D. R., Dir., *JANAF Thermochemical Tables*, Dow Chemical Co., Midland, Mich., 1965–8.

24 Thorp, C. E., *Bibliography of Ozone Technology*, Vol. 2, Armour Research Foundation, Illinois Inst. of Technology, Chicago, Ill., 1955.

25 Von Mailfert, A., *Compt. Rend.*, 1894, **119**, 951.

26 Waterman, T. E., *Thermal Conductivity of Liquid Ozone*, Symp. on Thermal Properties, Thermodynamic and Transport Properties of Gases, Liquids, and Solids, (Touloukian, Y. S., Chairman), Am. Soc. Mech. Eng., New York, 1959, pp. 301–3.

27 Yastrebov, V. V., *Russ. J. Phys. Chem.*, 1960, **34** (1), 21–3.

Table 19.1 Properties of Ozone

		Ref.
Chemical formula	O_3	(14)
Molecular weight	47.9970	(12)
Critical temperature	−12.1 °C	(14)
Critical pressure	55.73×10^5 N m^{-2}	(14)
Critical density	540.0 kg m^{-3}	(14)
Triple-point temperature	−193.0 °C	(22)
Triple-point pressure	1.14 N m^{-2}	(22)
Boiling point at 1.0133×10^5 N m^{-2}	−111.9 °C	(14)
Melting point	−192.7 °C	(14)
Transition temperature	None	(9)
Density of solid at −195.8 °C	1728.0 kg m^{-3}	(22)
Coefficient of thermal expansion of liquid at boiling point	2.5×10^{-3} °C^{-1}	(21)
Heat of formation of gas at 25 °C	1.4277×10^8 J kg-mol^{-1}	(23)
Free energy of formation of gas at 25 °C	1.6291×10^8 J kg-mol^{-1}	(23)
Heat of fusion	2.0934×10^6 J kg-mol^{-1}	(22)
Heat of transition	None	(9)
Entropy of gas at 25 °C	2.3898×10^5 J kg-mol^{-1} °C^{-1}	(23)
Heat capacity ratio of gas at 25 °C and 1.0133×10^5 N m^{-2}	1.29	(13)
Dielectric constant of gas at 0 °C and 1.0133×10^5 N m^{-2}	1.00190	(8)
Refractive index of liquid at −140 °C and 5893 Å	1.2226	(16)
Threshold limit value	(0.1 ppm) 0.2×10^{-6} kg m^{-3}	(1)

19.1 Vapour Pressure of Ozone (20,22)

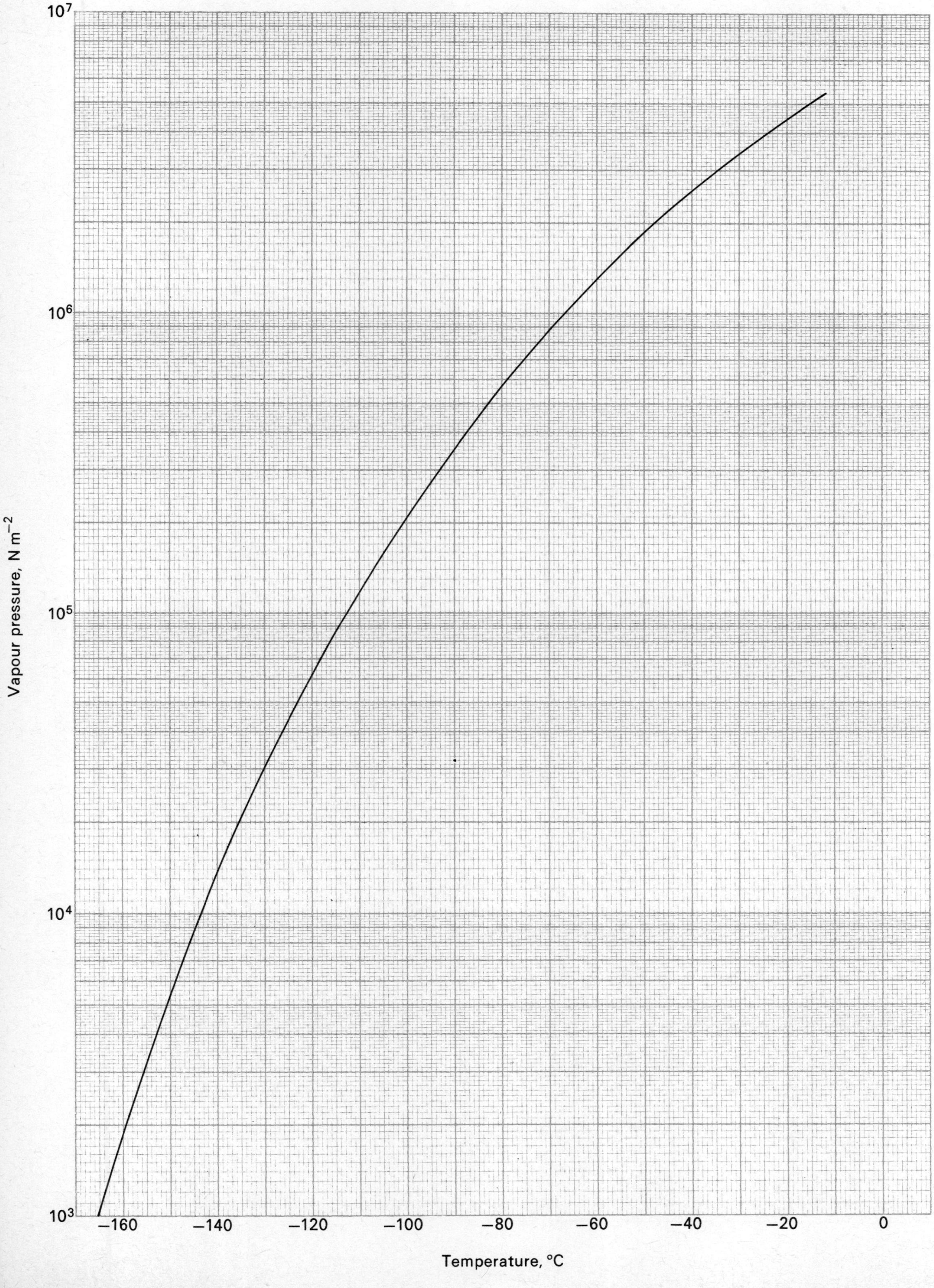

19.2 Orthobaric Density of Ozone (3,22)

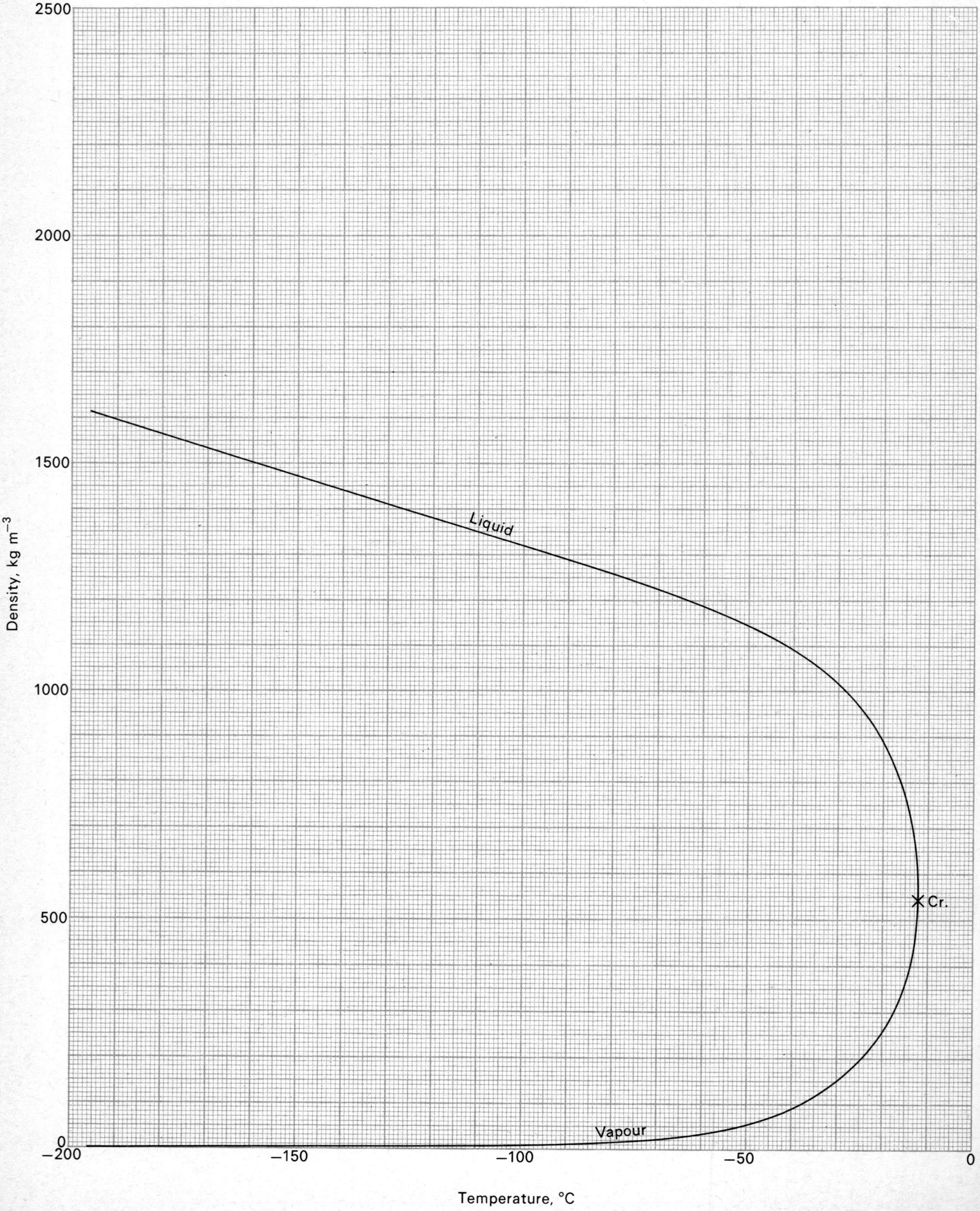

19.3 P-V-T Data for Ozone (2)

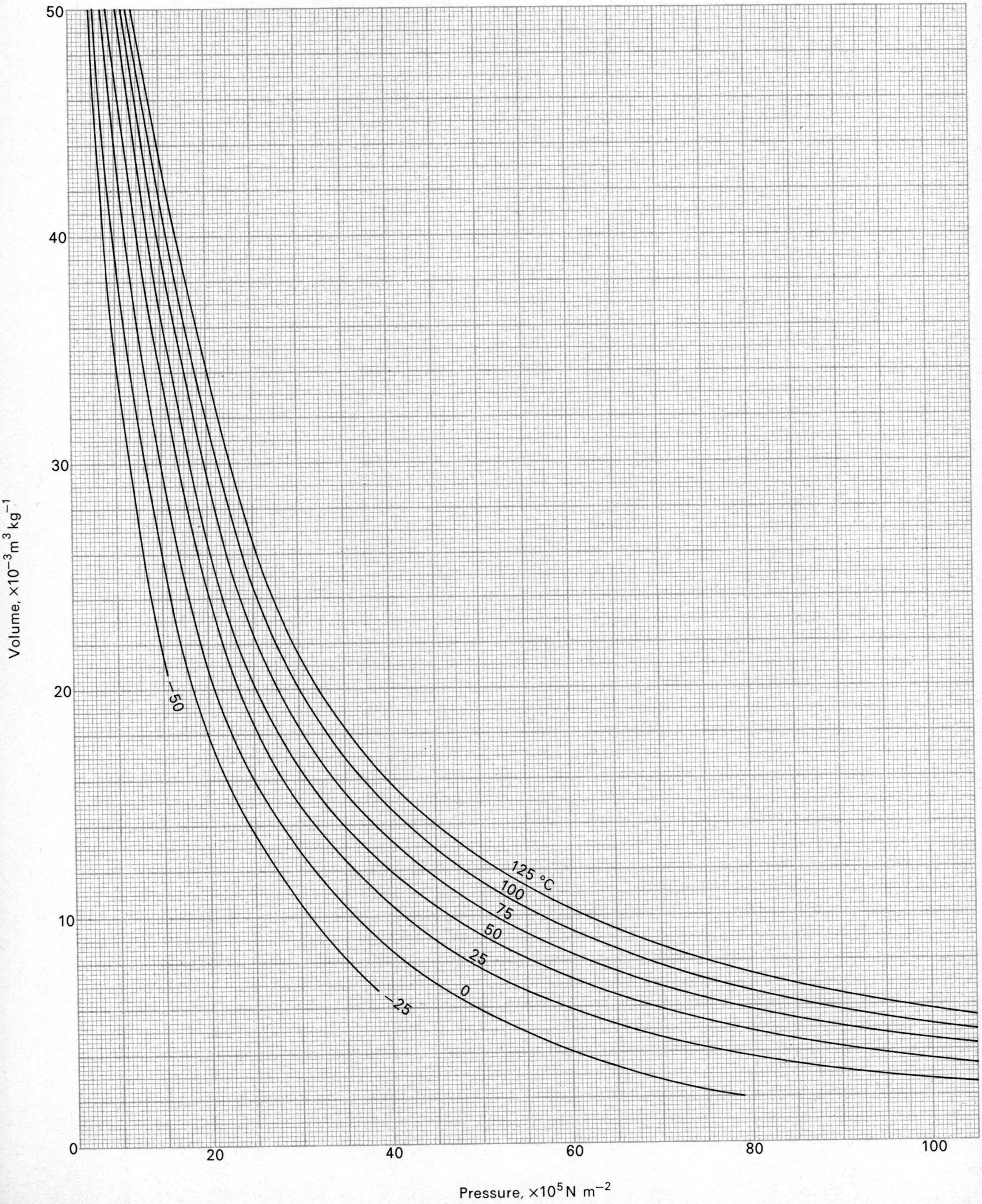

19.4 Latent Heat of Vaporisation of Ozone (19,20)

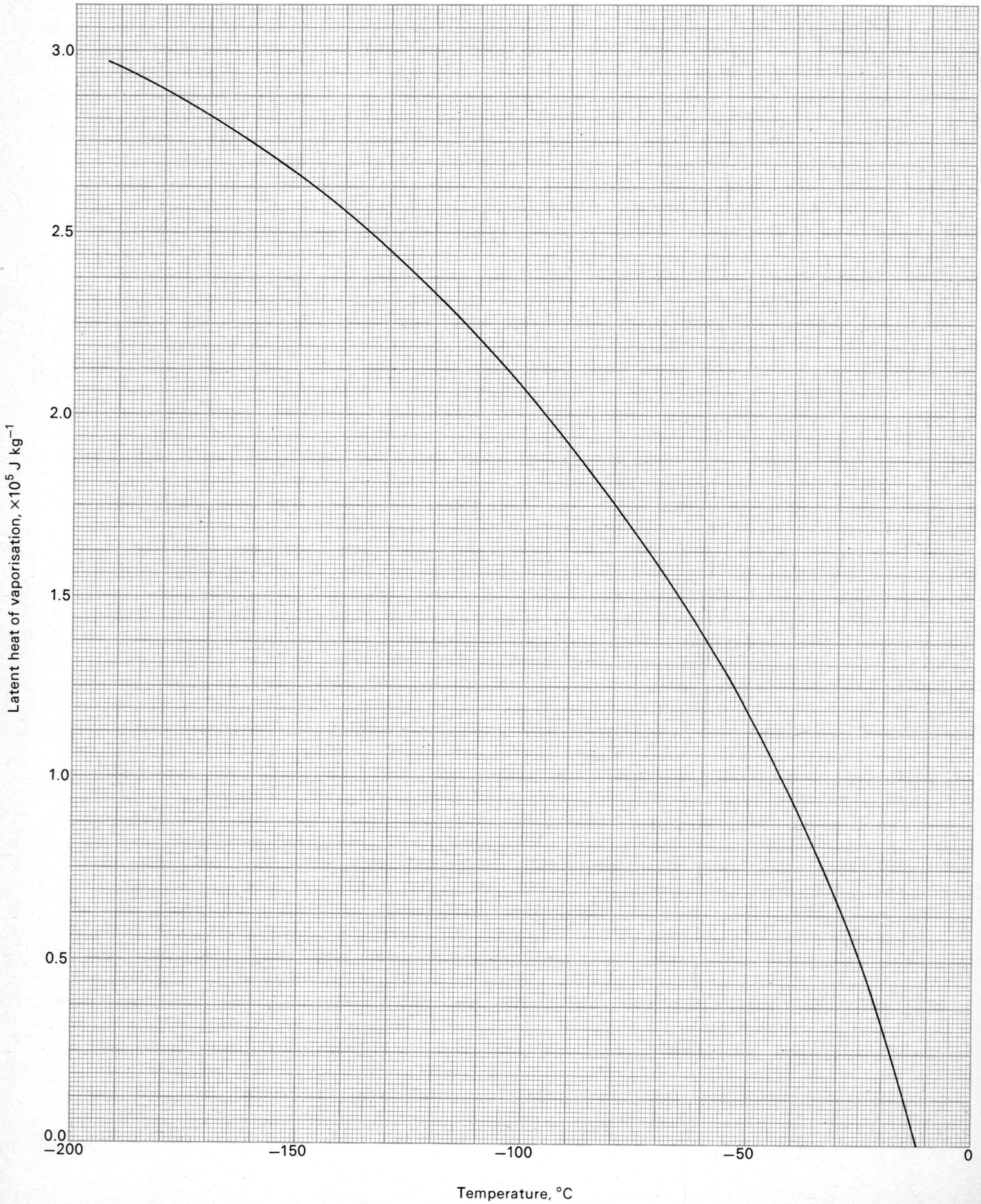

19.5 Heat Capacity of Liquid Ozone (22)

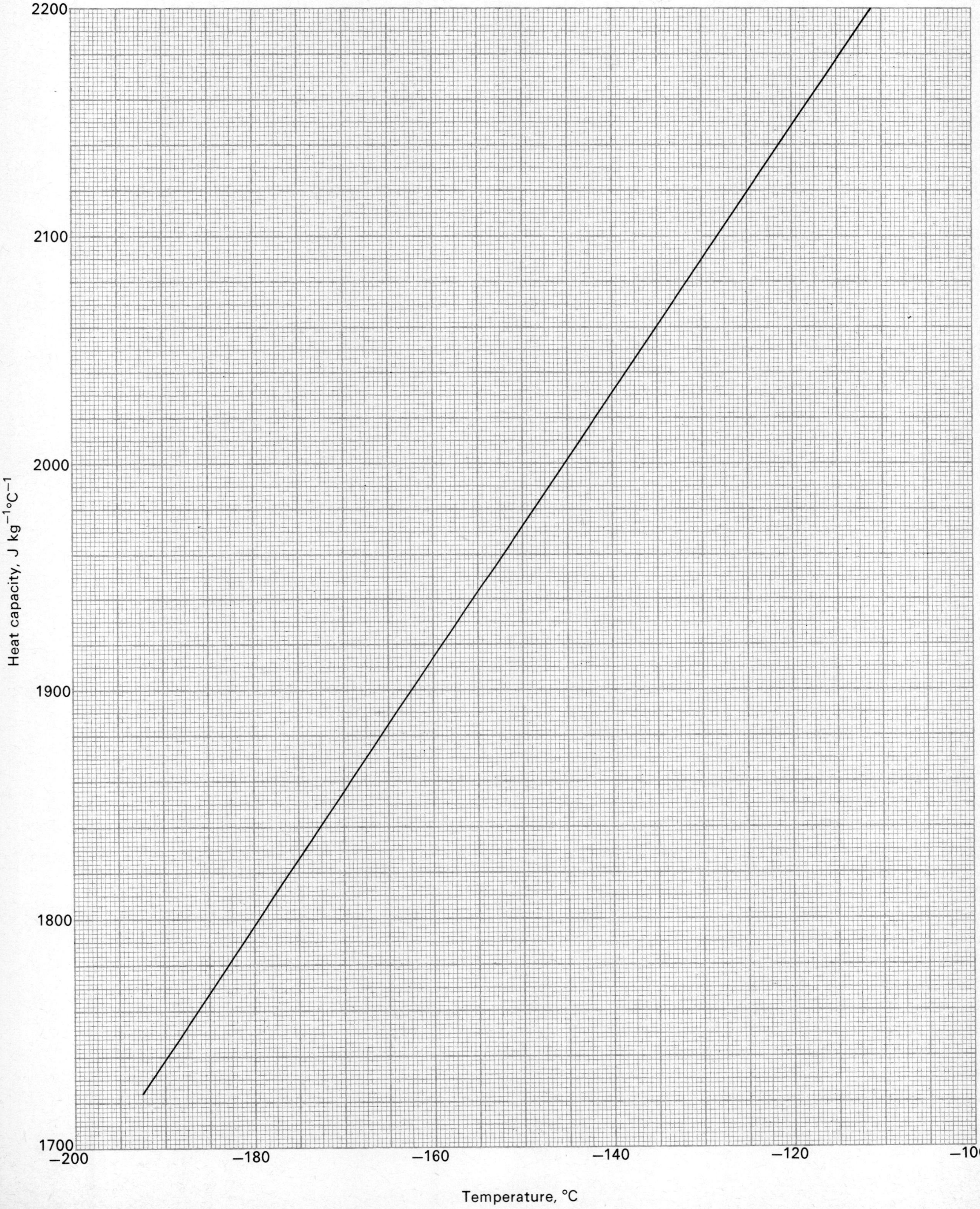

19.6 Heat Capacity of Gaseous Ozone at 1.0133×10^5 N m^{-2} (5,23)

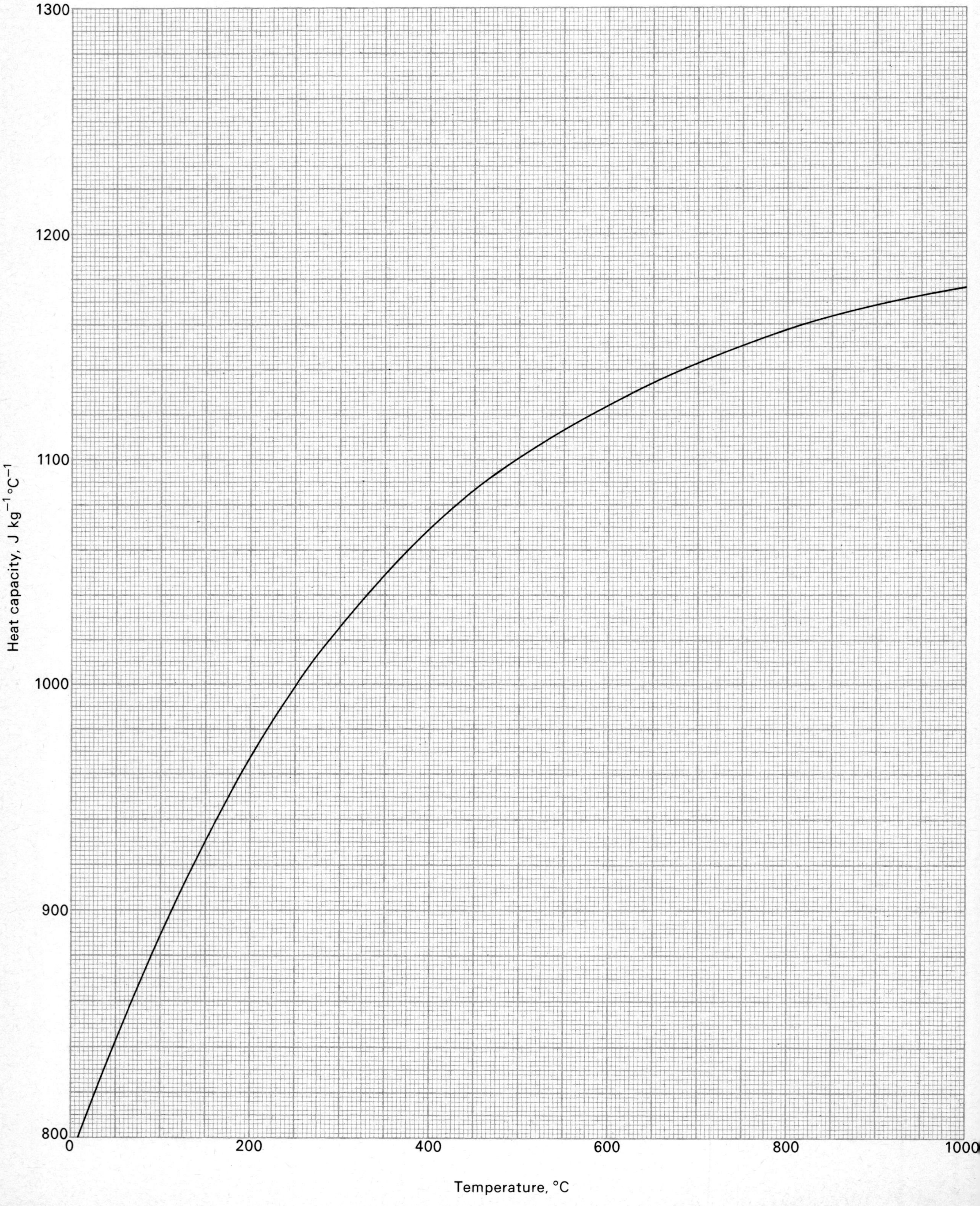

19.7 Viscosity of Liquid Ozone (7,22)

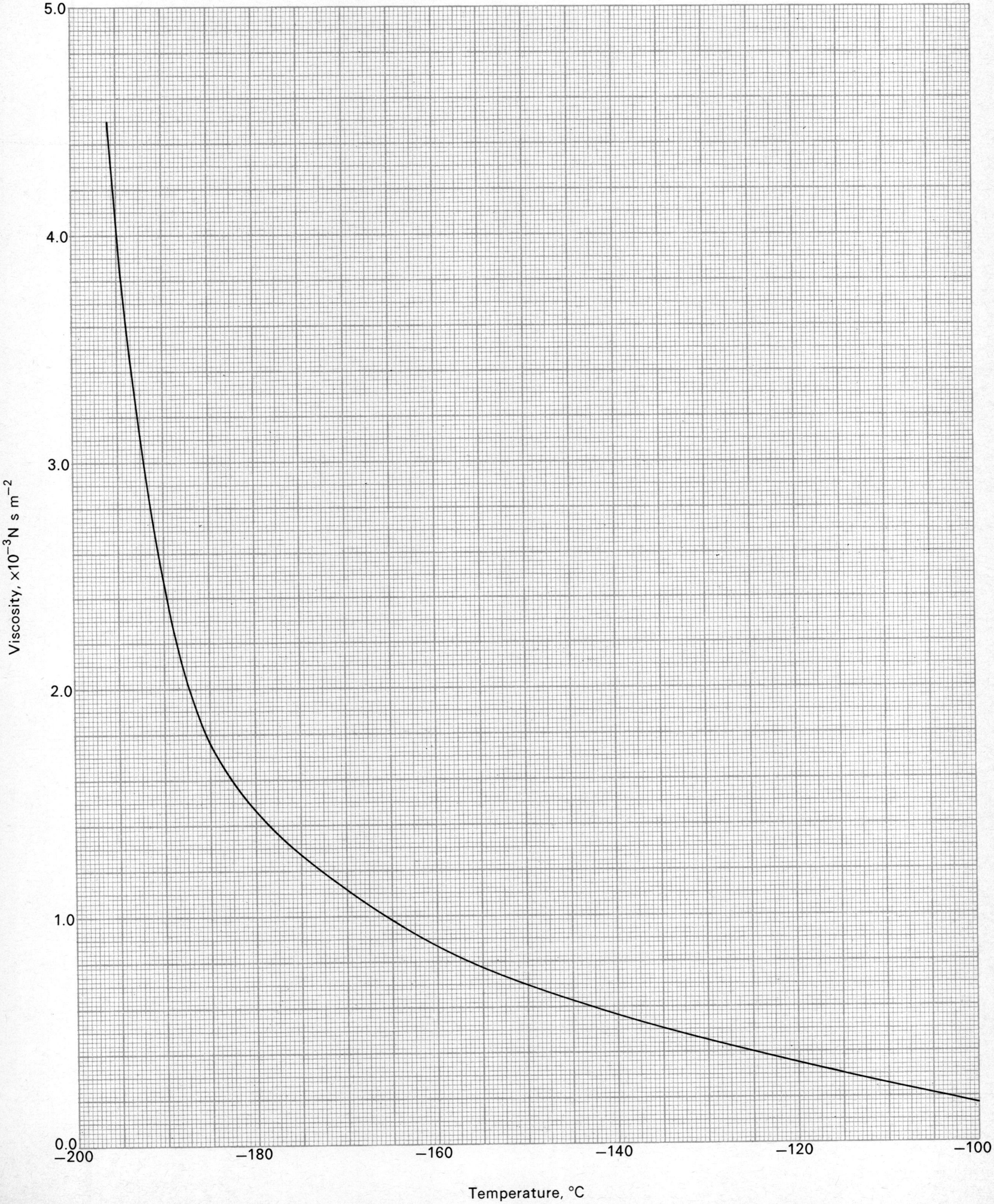

19.8 Viscosity of Gaseous Ozone at 1.0133×10^5 N m^{-2}(22)

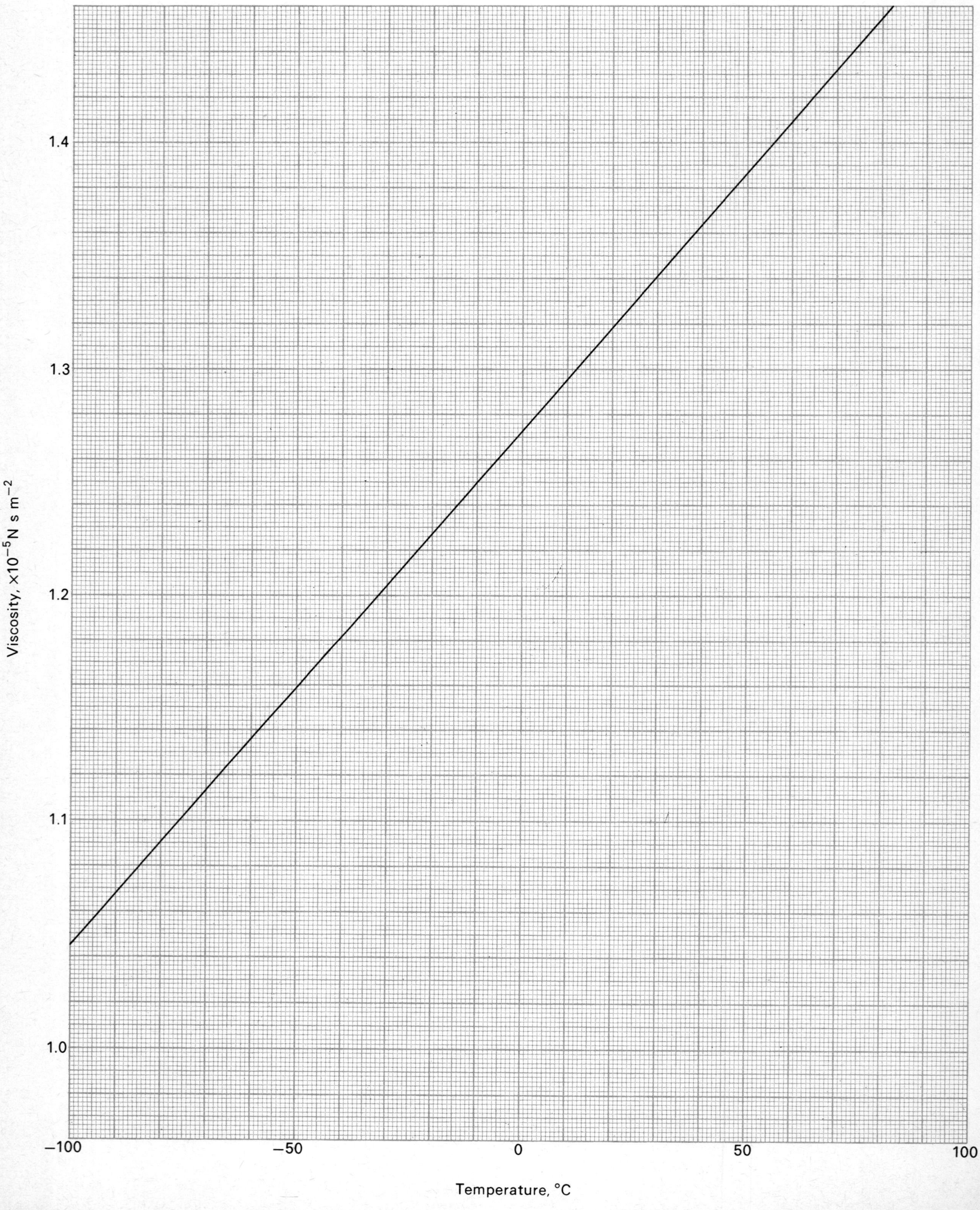

19.9 Thermal Conductivity of Liquid Ozone (14,26)

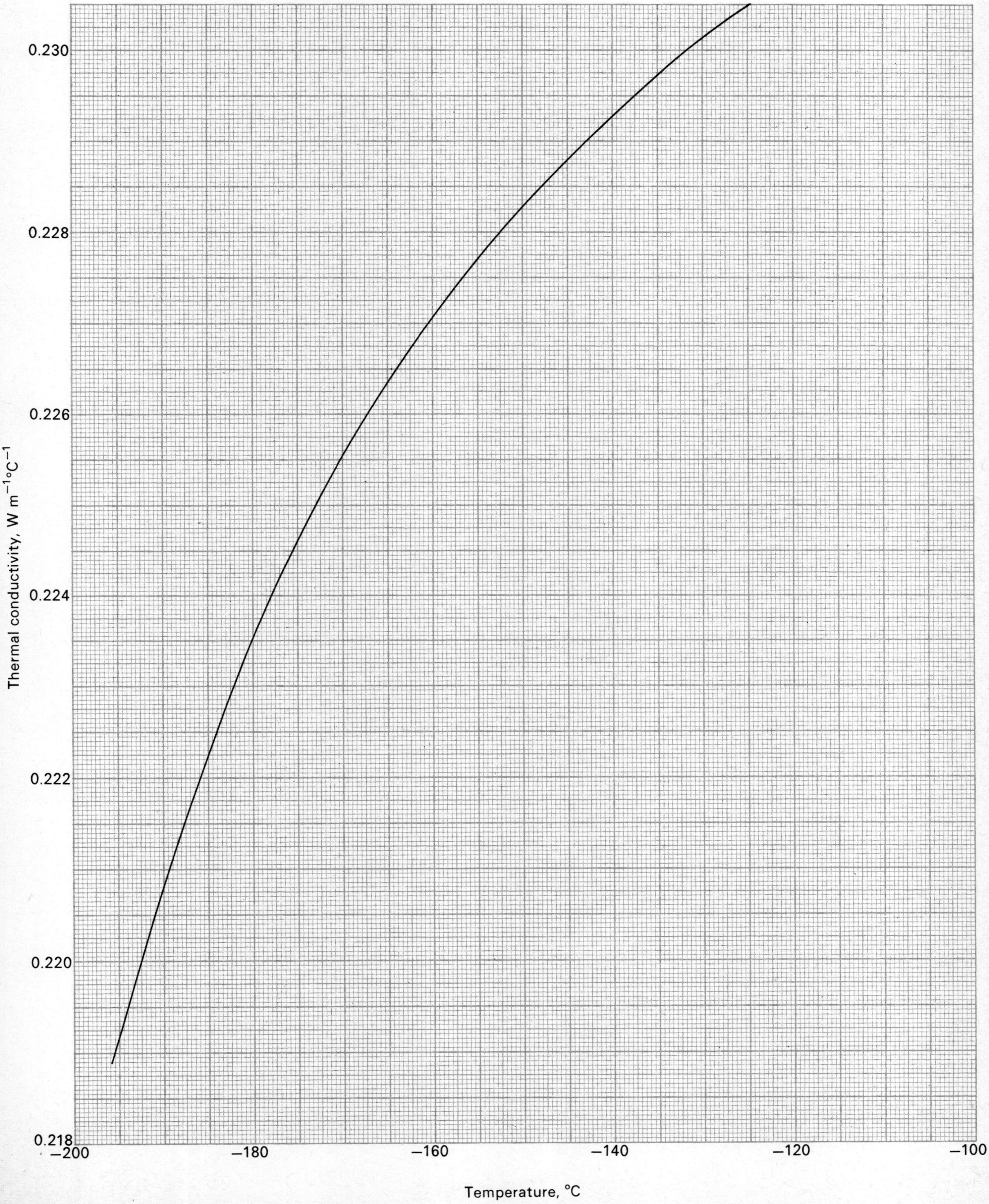

19.10 Thermal Conductivity of Gaseous Ozone at $1.0133 \times 10^5 N\ m^{-2}$(17, 27)

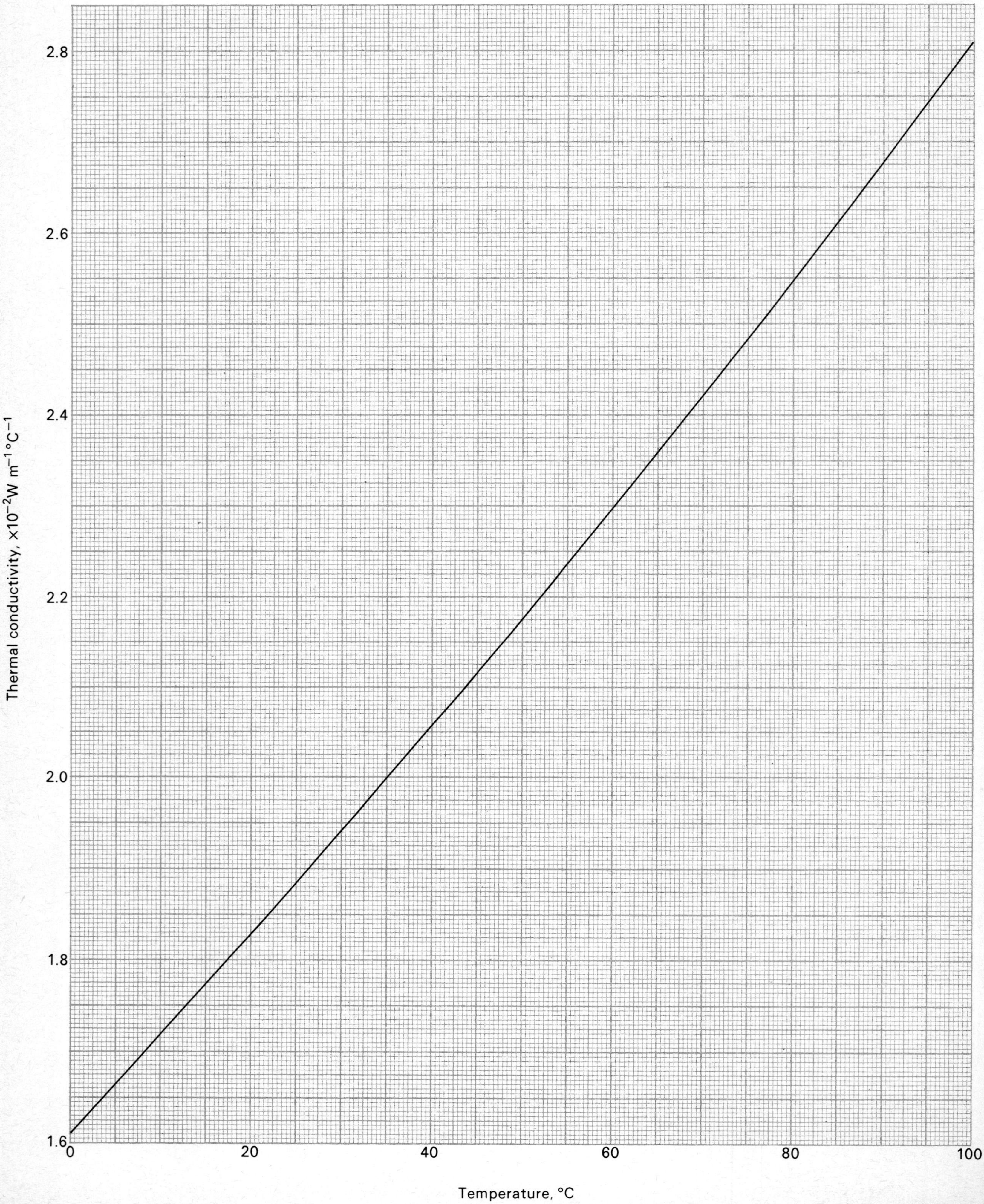

19.11 Surface Tension of Ozone (22)

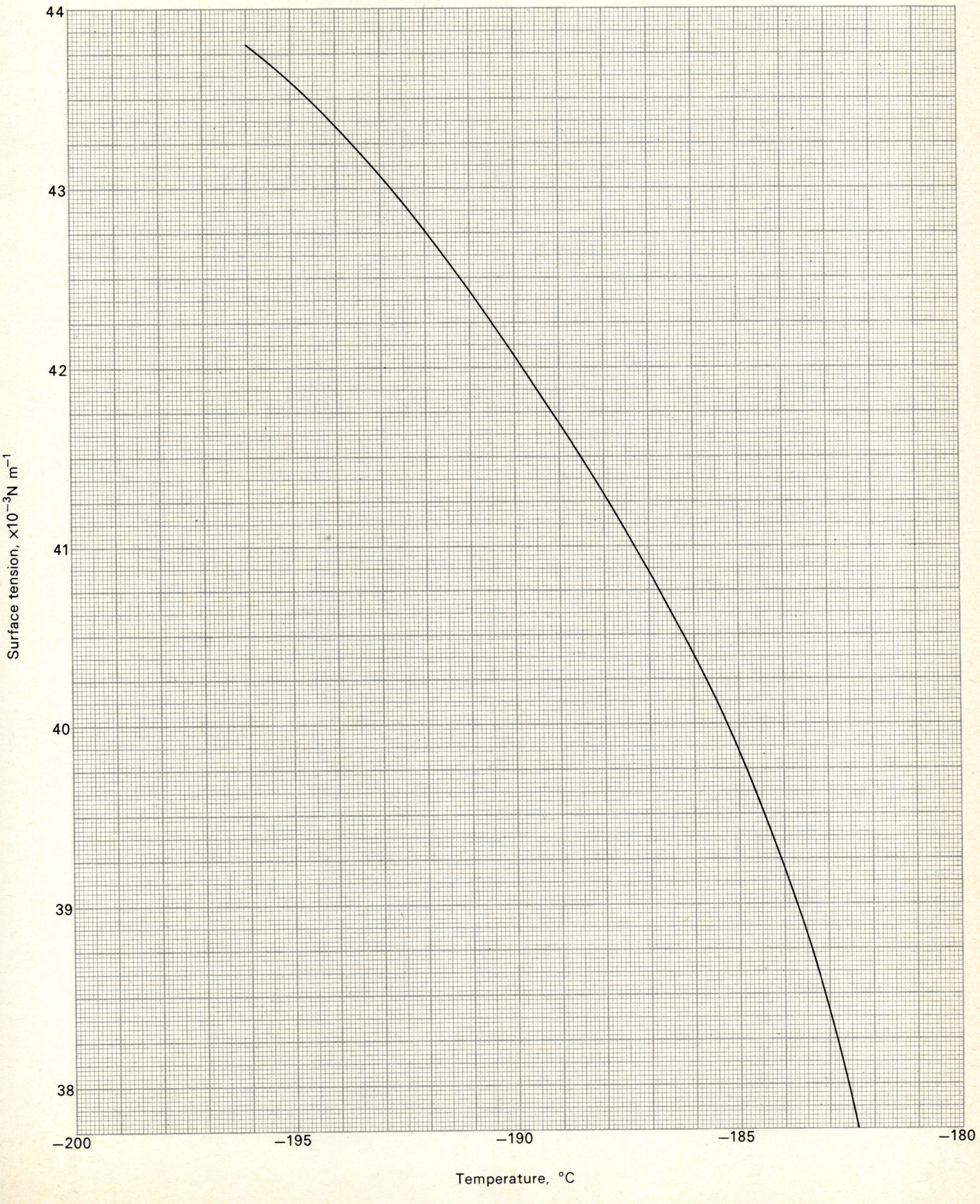

19.12 Solubility of Ozone in Water at 1.0133×10^5 N m^{-2} Total Pressure (4,10,22, 25)

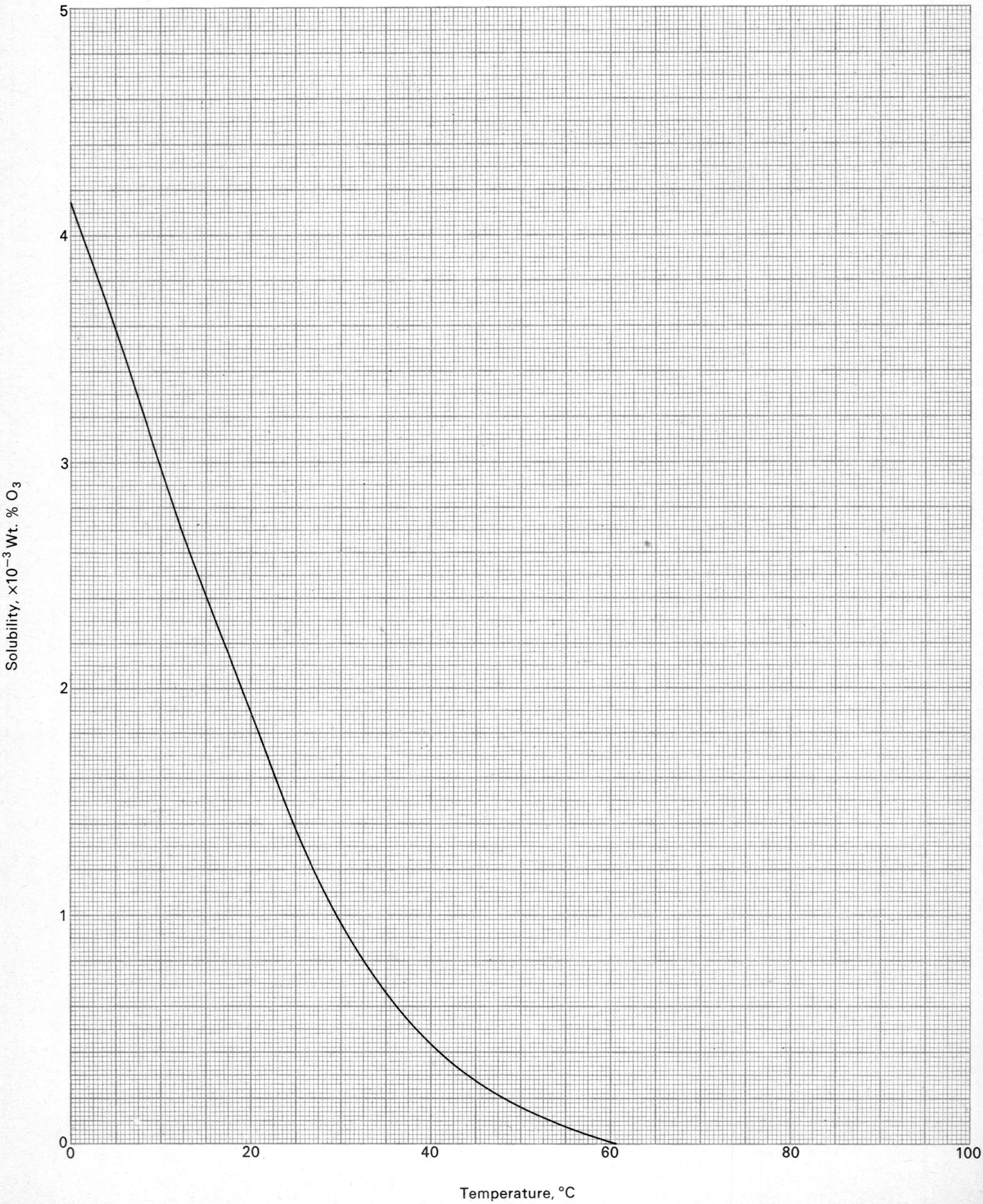

20 PHOSPHORIC ACID

Pure phosphoric acid is a white solid at room temperature and atmospheric pressure. The crystalline solid has a monoclinic crystal structure. In the solid phase the molecule contains four oxygen atoms, arranged tetrahedrally around the central phosphorus atom. The solid has three modifications at various temperatures. The behaviour of solid and liquid acids indicates the strong effect of hydrogen bonding.

Phosphoric acid (or orthophosphoric acid) is a tribasic acid, having three dissociation constants:

$K_1 = 1.1 \times 10^{-2}$
$K_2 = 5.6 \times 10^{-8}$
$K_3 = 1.2 \times 10^{-12}$

and is a very weak oxidizing agent.

The syrupy liquid acid has a strong tendency to be supercooled; therefore, the solutions remain liquid below the temperatures given by the equilibrium diagram. The solid crystals are stable. However, on heating above the melting point, it loses water, becoming first the pyro- and then the meta-acid. Orthophosphoric acid is very soluble in water.

Besides the multi-volume handbooks: Gmelin (10), Mellor (17), and Pascal (19), there are two publications of interest on the physical, thermodynamic, and transport properties of orthophosphoric acid in various phases, by Farr (9) and Monsanto (18).

The most comprehensive investigations have been carried out on the liquid density of the acid. In addition to the citations on Fig. 20.2, Biltz and Hülsmann (2) and Christensen and Reed (6) might be of further interest.

Gaseous phosphoric acid does not exist. Most of the liquid properties at higher temperatures have been obtained by extrapolations or from thermodynamic calculations.

The data in Table 20.1 and Figs. 20.1–20.9 have been selected from the most reliable sources and are accurate enough for most chemical and engineering calculations. Some properties have been correlated from several sources.

References

1 American Conference of Government Industrial Hygienists, *Documentation of Threshold Limit Values*, revised edition, Committee on Threshold Limit Values, Cincinnati, Ohio 45202, 1966, 203 pp.
2 Biltz, W., Hülsmenn, O., *Z. anorg. Chem.*, 1932, **207**, 377–84.
3 Bricke, E. V., Pestov, N. E., *Trudy naucnogo Inst. Udobrenijam Insektofungisidam*, 1929, **59**, 5–169.
4 Brown, E. H., Whitt, C. D., *Ind. Eng. Chem.*, 1952, **44**, 615–18.
5 Cherbuliez, E., Leber, J-P., *Helv. Chim. Acta*, 1950, **33**, 2264–7.
6 Christensen, J. H., Reed, R. B., *Ind. Eng. Chem.*, 1955, **47**, 1277–80.
7 Egan, E. P., Luff, B. B., *Ind. Eng. Chem.*, 1955, **47**, 1280–1.
8 Egan, E. P., Wakefield, Z. T., *J. Phys. Chem.*, 1957, **61**, 1500–4.
9 Farr, T. D., *Phosphorus—Properties of the Element and some of Its Compounds*, Tenn. Valley Authority, Chem. Eng. Rept. No. 8, 1950.
10 *Gmelins Handbuch der anorganischen Chemie*, Syst. No. 16, Phosphor, Teil C, Verlag Chemie GmbH, Weinheim, 1965, pp. 155–68.
11 Greenwood, N. N., Thompson, A., *J. Chem. Soc.*, 1959, 3485–92.
12 Holmes, W. S., *Trans. Faraday Soc.*, 1962, **58**, 1916–25.
13 IUPAC, Atomic Weights of the Elements 1971, *Pure & Appl. Chem.*, 1972, **30** (3–4), 638–49.
14 Kondratenko, Z. V., Fedorchenko, I. G., *Russ. J. Inorg. Chem.*, 1959, **4** (5), 447–9.
15 Kondratenko, Z. V., Fedorchenko, I. G., Kovalev, I. A., *J. Appl. Chem. USSR*, 1967, **40**, 1882–5.
16 Lax, E., Synowietz, C., Eds., *Taschenbuch für Chemiker und Physiker*, 3rd ed., Springer-Verlag, Berlin, 1967.
17 *Mellor's Comprehensive Treatise on Inorganic and Theoretical Chemistry*, Vol. VIII, Suppl. III, Phosphorus, Longmans, London, 1971, pp. 669–717.
18 Monsanto Chemical Company, *Phosphoric Acid. Physical and Chemical Properties*, Techn. Bull. No. P-26, St. Louis, Mo., 1946, 12 pp.
19 Pascal, P., Ed., *Nouveau Traité de Chimie Minérale*, Vol. X, Azote, Phosphore, Masson et Cie, Éditeurs, Paris, 1956, pp. 817–19.
20 Rass, W. H., Jones, R. M., *J. Am. Chem. Soc.*, 1925, **47**, 2165–70.
21 Smith, J. P., Brown, W. E., Lehr, J. R., *J. Am. Chem. Soc.*, 1955, **77**, 2728–30.
22 Stull, D. R., Dir., *JANAF Thermochemical Tables*, Dow Chemical Co., Midland, Mich., 1965–8.
23 Takeda, T., *J. Chem. Soc. Japan*, 1939, **60**, 885–94.
24 Takeda, T., *J. Sci. Hiroshima Univ.*, 1953, **15**, 151–60.
25 Turnbull, A. G., *J. Chem. Eng. Data*, 1965, **10** (2), 118–19.
26 Wakefield, Z. T., Luff, B. B., Reed, R. B., *J. Chem. Eng. Data*, 1972, **17** (4), 420–3.

Table 20.1 Properties of Phosphoric Acid

		Ref.
Chemical formula	H_3PO_4	(16)
Molecular weight	97.9935	(13)
Boiling point at 1.0133×10^5 N m^{-2}	255.3 °C	(3)
Melting point	42.35 °C	(20)
Transition temperature	None	(10)
Density of solid at 10 °C	1990.0 kg m^{-3}	(21)
Coefficient of thermal expansion of solid at −137 °C	0.190×10^{-3} C^{-1}	(16)
Heat of formation of liquid at 25 °C	-1.2511×10^9 J kg-mol^{-1}	(22)
Heat of formation of crystal at 25 °C	-1.2799×10^9 J kg-mol^{-1}	(12)
Free energy of formation of liquid at 25 °C	-1.1071×10^9 J kg-mol^{-1}	(22)
Free energy of formation of crystal at 25 °C	-1.1072×10^9 J kg-mol^{-1}	(22)
Heat of fusion	1.2979×10^7 J kg-mol^{-1}	(22)
Heat of transition	None	(10)
Entropy of liquid at 25 °C	1.4654×10^5 J kg-mol^{-1} °C^{-1}	(22)
Entropy of crystal at 25 °C	1.1062×10^5 J kg-mol^{-1} °C^{-1}	(22)
Electrical conductivity of liquid at 25 °C	4.675 ohm^{-1} m^{-1}	(11)
Refractive index of liquid at 20 °C	1.4503	(11)
Threshold limit value	1.0×10^{-6} kg m^{-3}	(1)

20.1 Vapour Pressure of Phosphoric Acid (4,5,10,17)

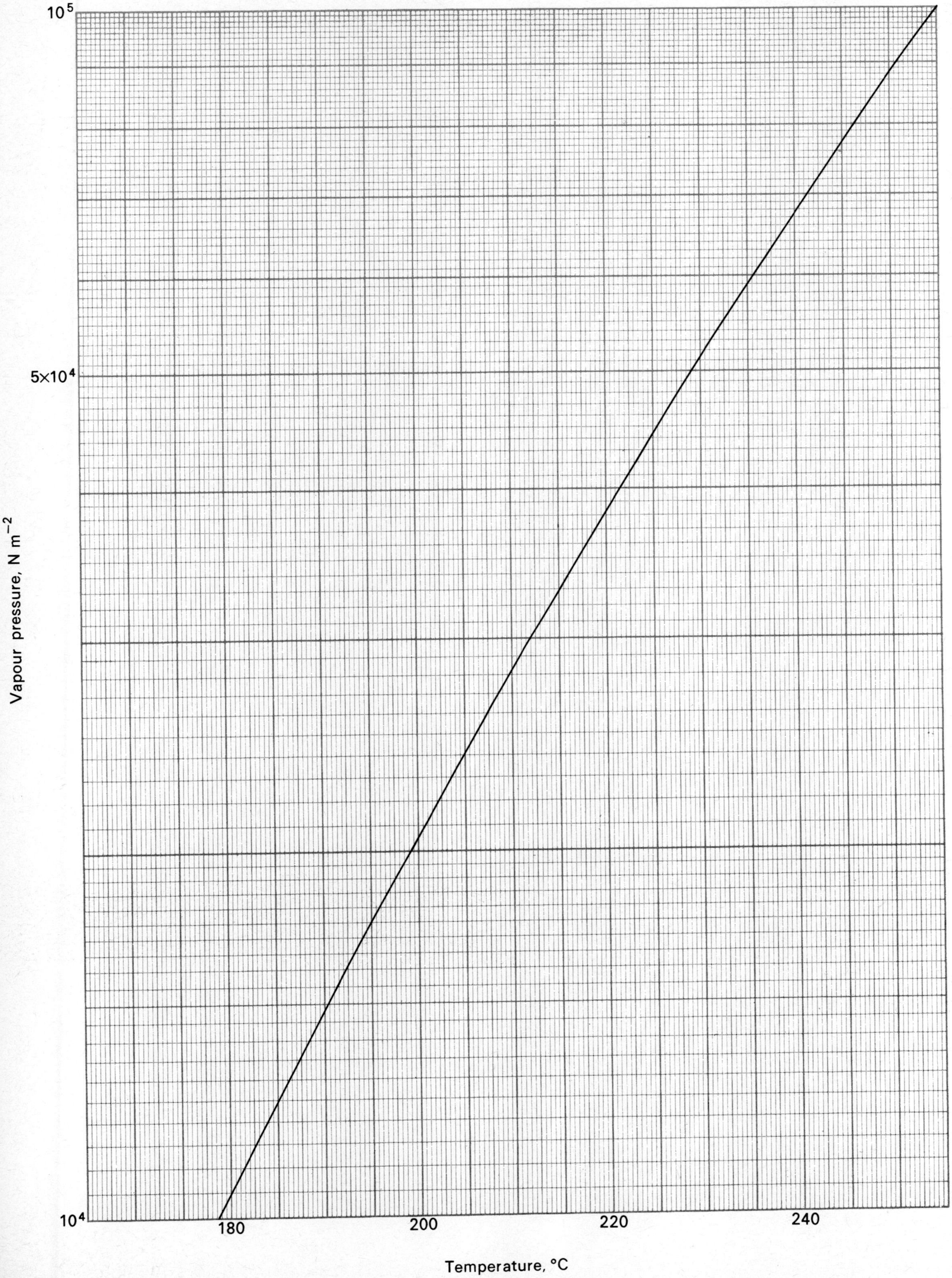

20.2 Liquid Density of Phosphoric Acid (7,11,14,15)

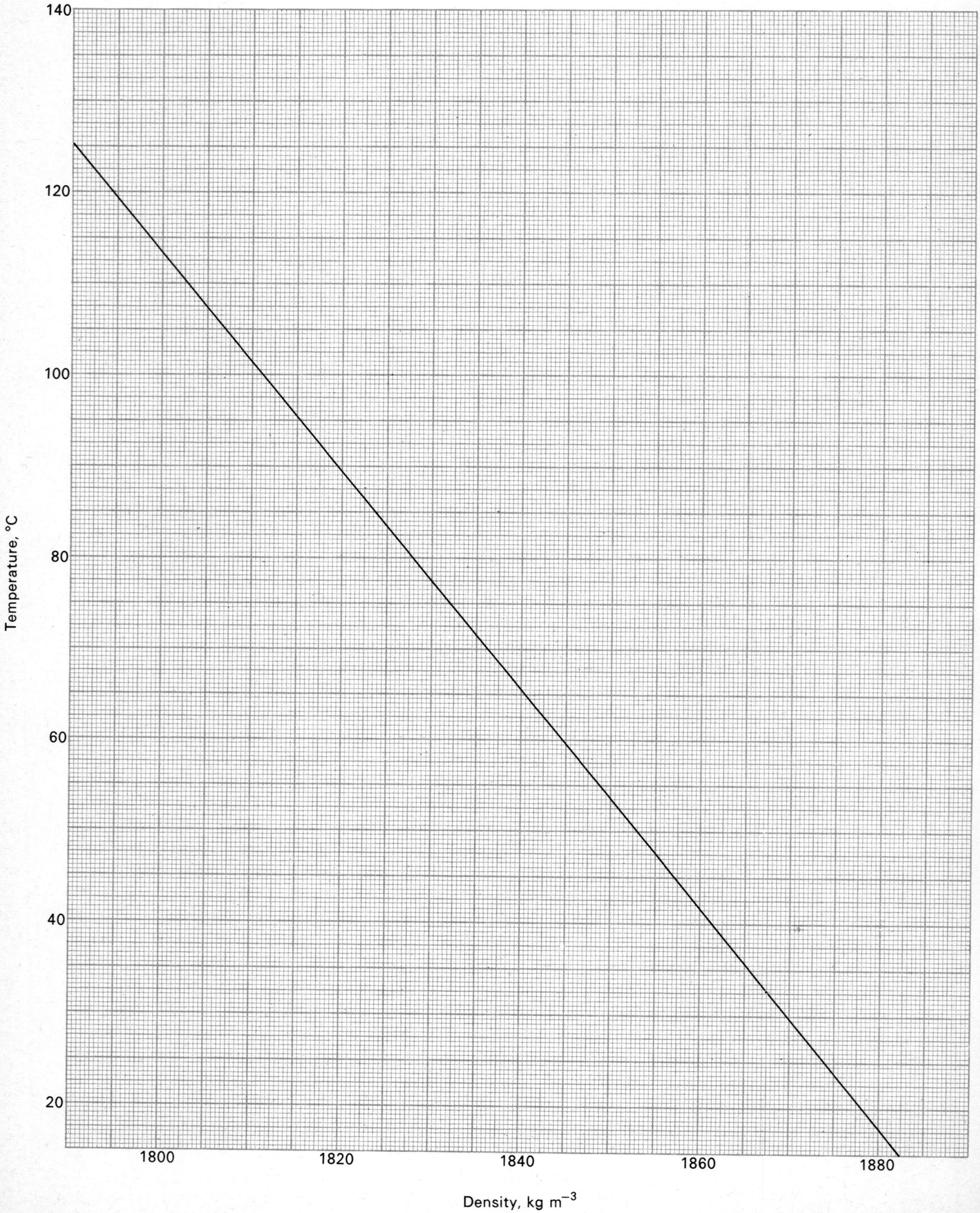

20.3 Latent Heat of Vaporisation of Phosphoric Acid (4,10)

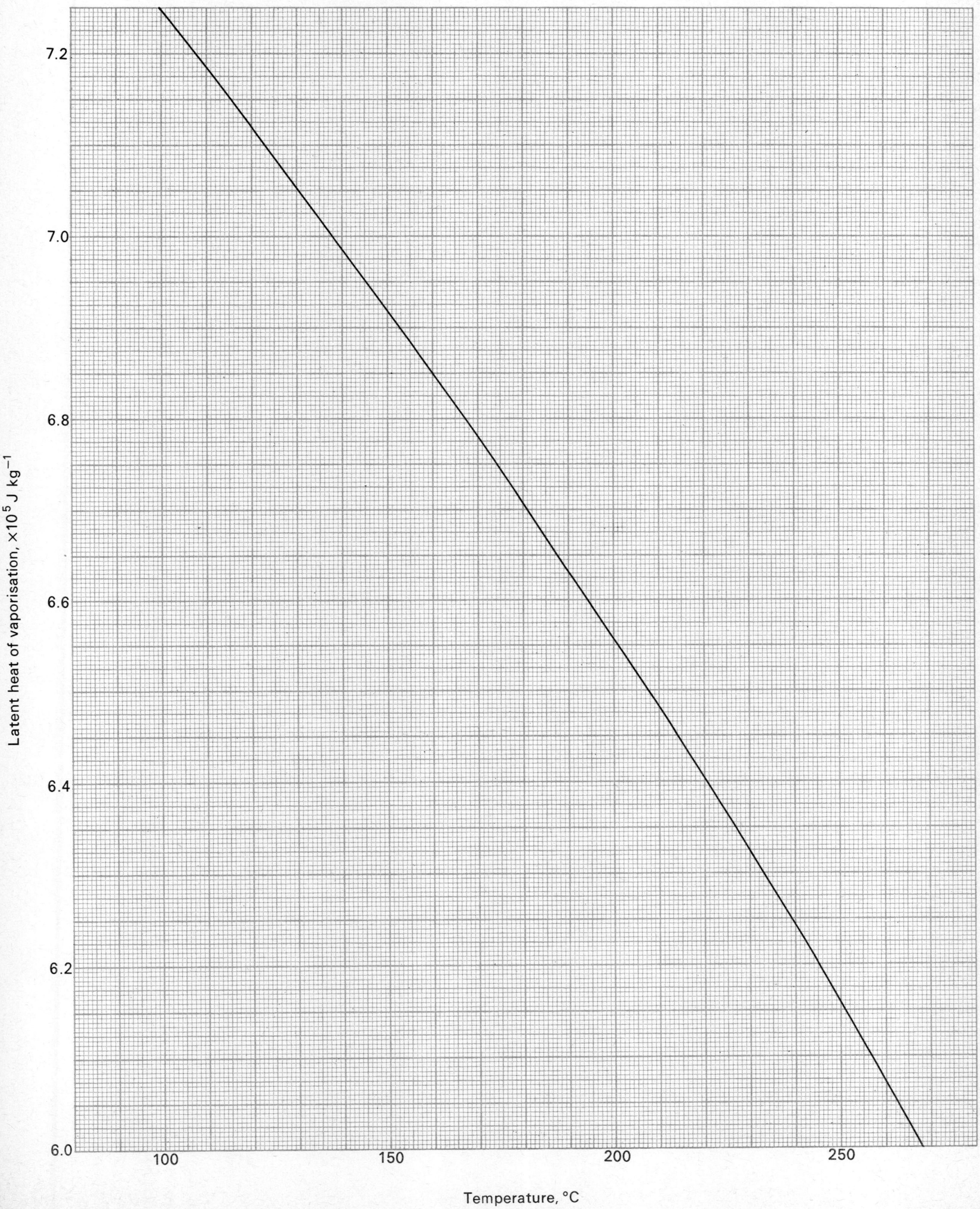

20.4 Heat Capacity of Solid and Liquid Phosphoric Acid (8,22,26)

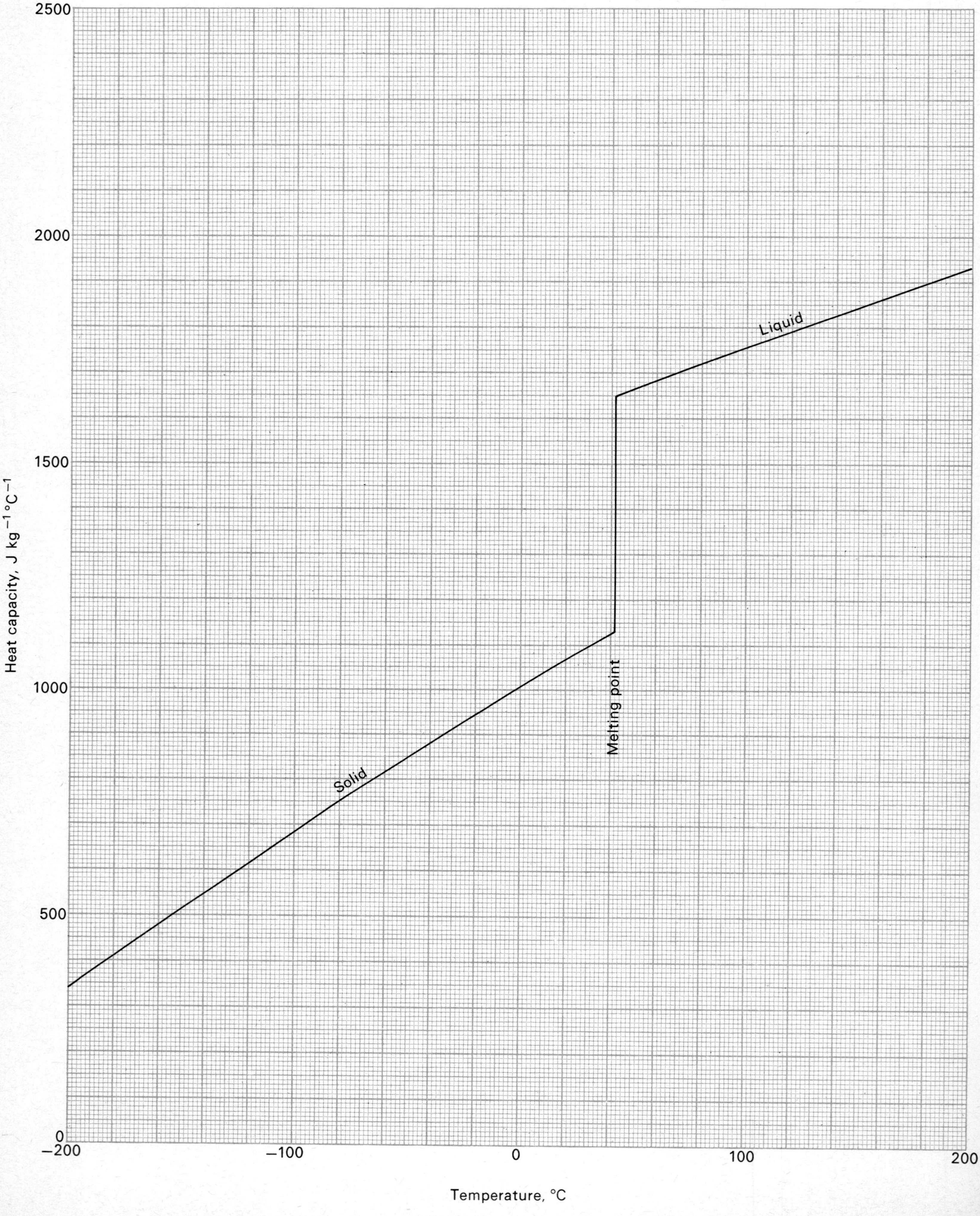

20.5 Viscosity of Liquid Phosphoric Acid (14,15)

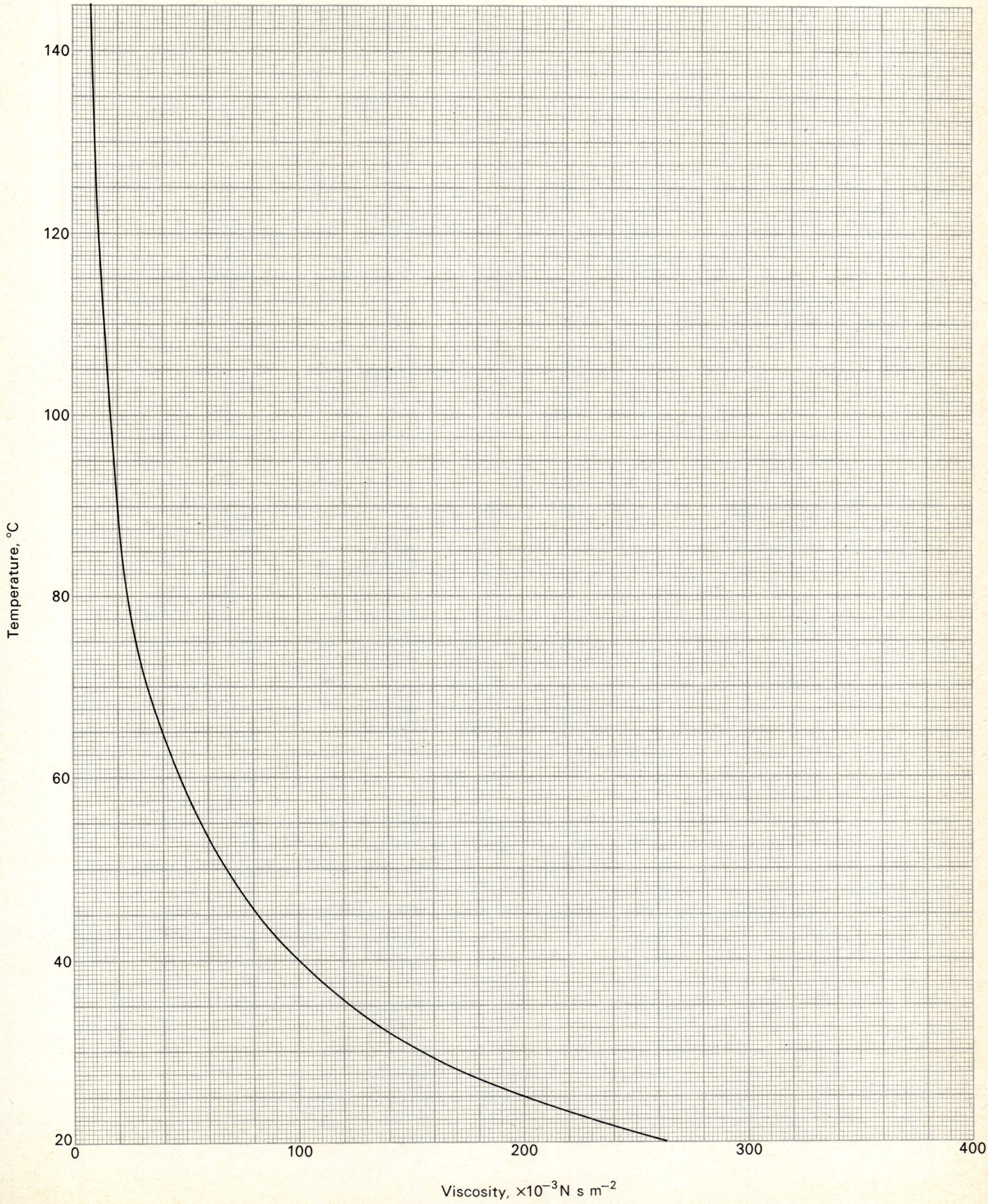

20.6 Thermal Conductivity of Liquid Phosphoric Acid (25)

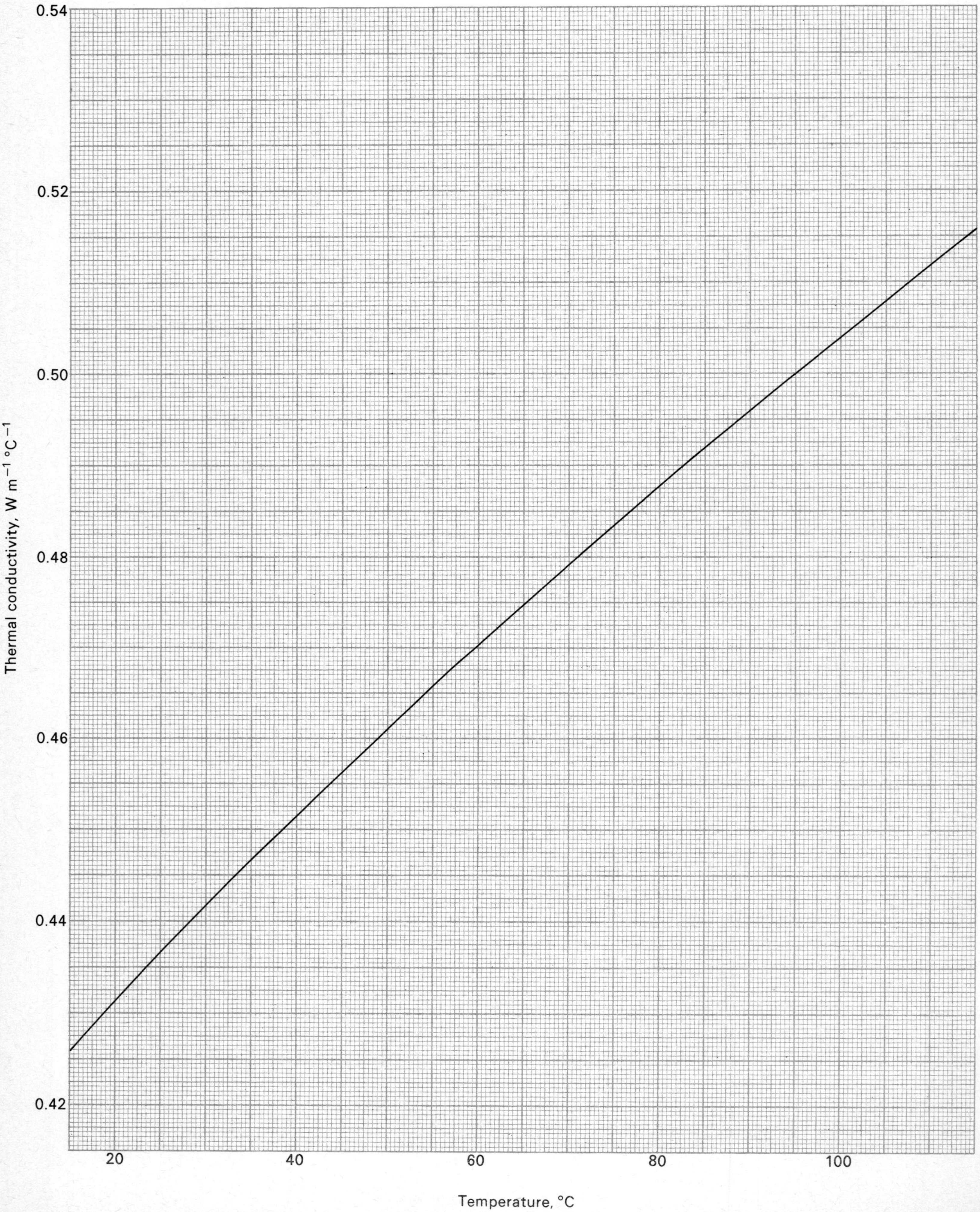

20.7 Surface Tension of Phosphoric Acid (23,24)

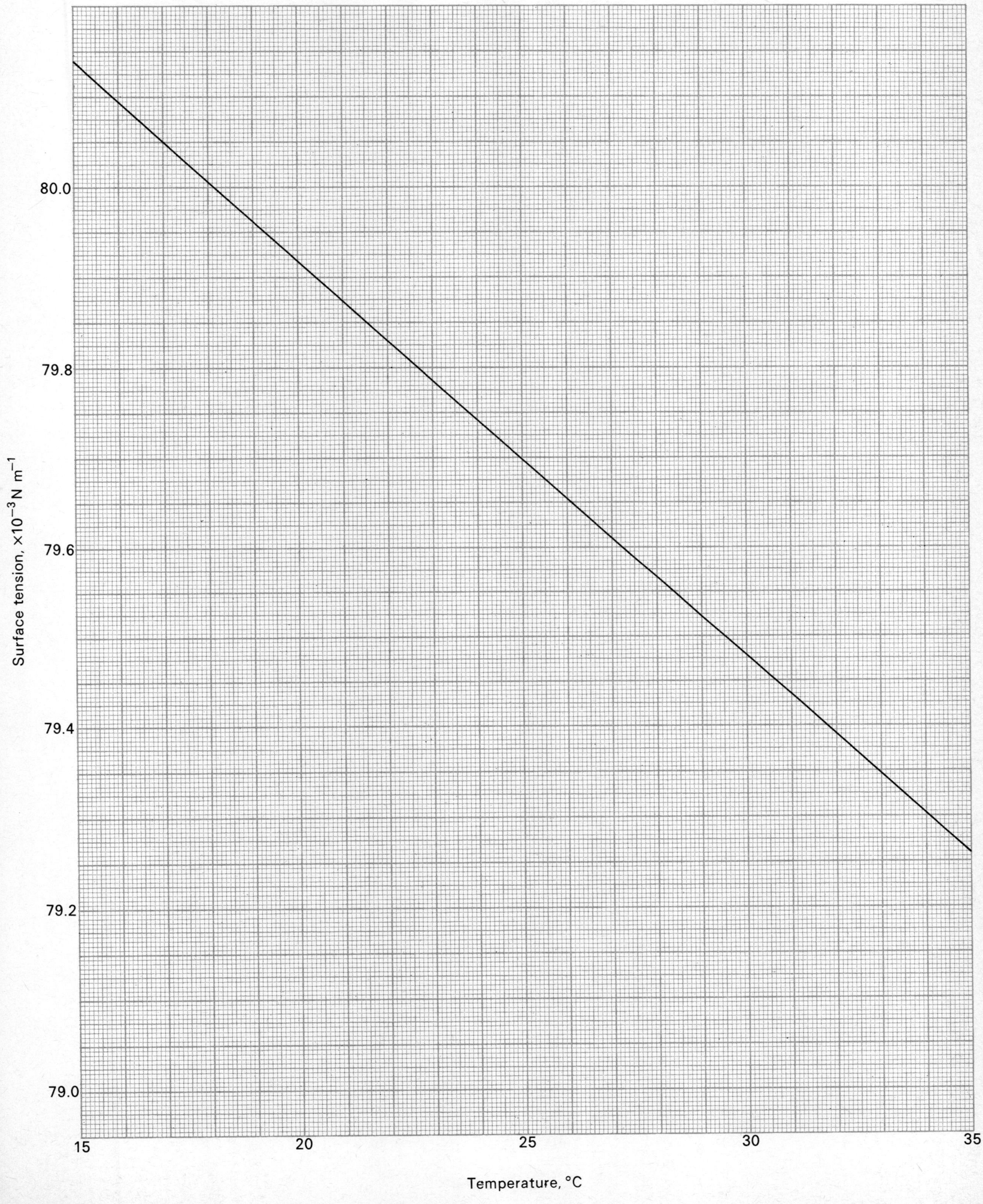

20.8 Phase Diagram for H_3PO_4-H_2O System (4,18,20)

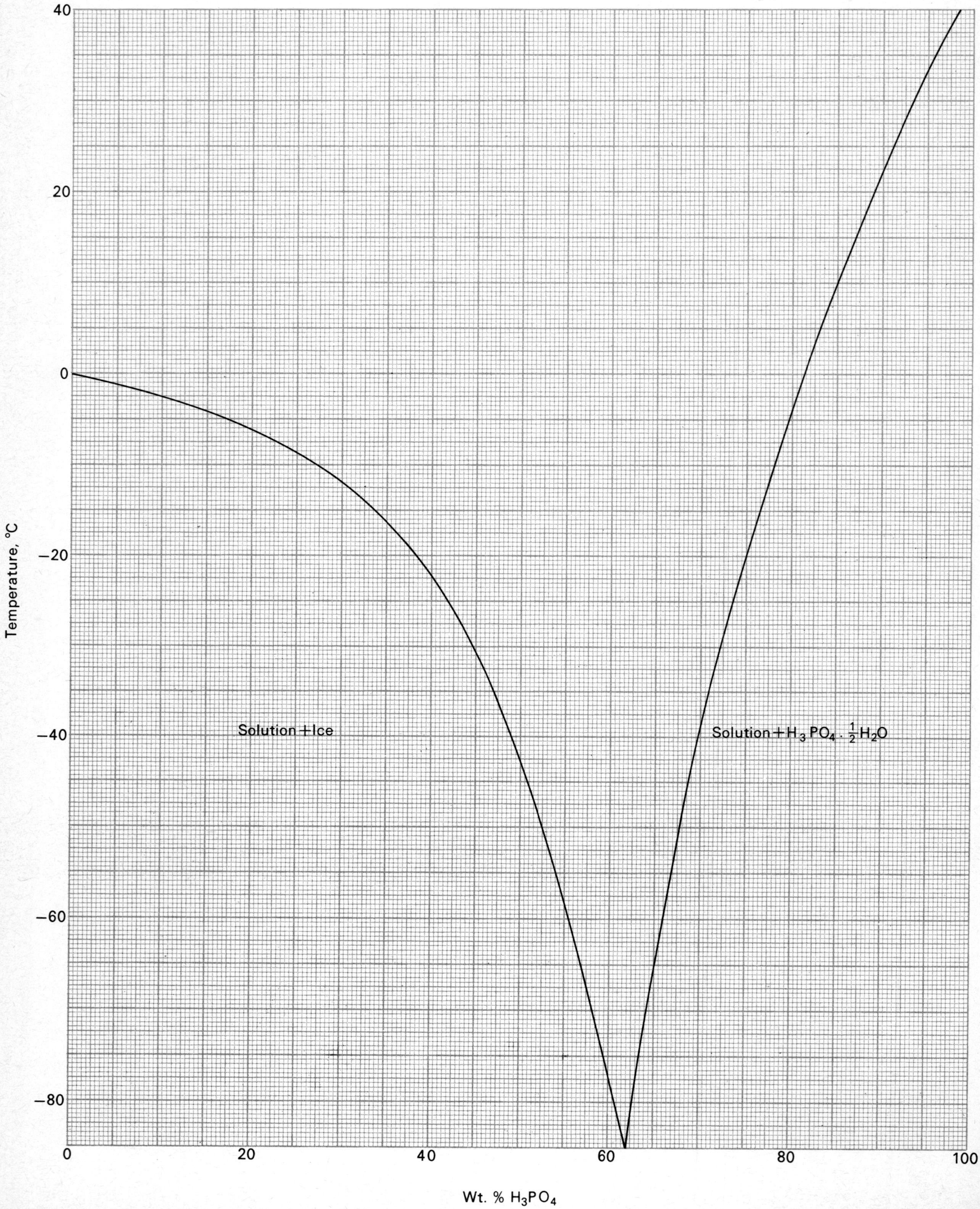

20.9 Electrical Conductivity of Liquid Phosphoric Acid (11)

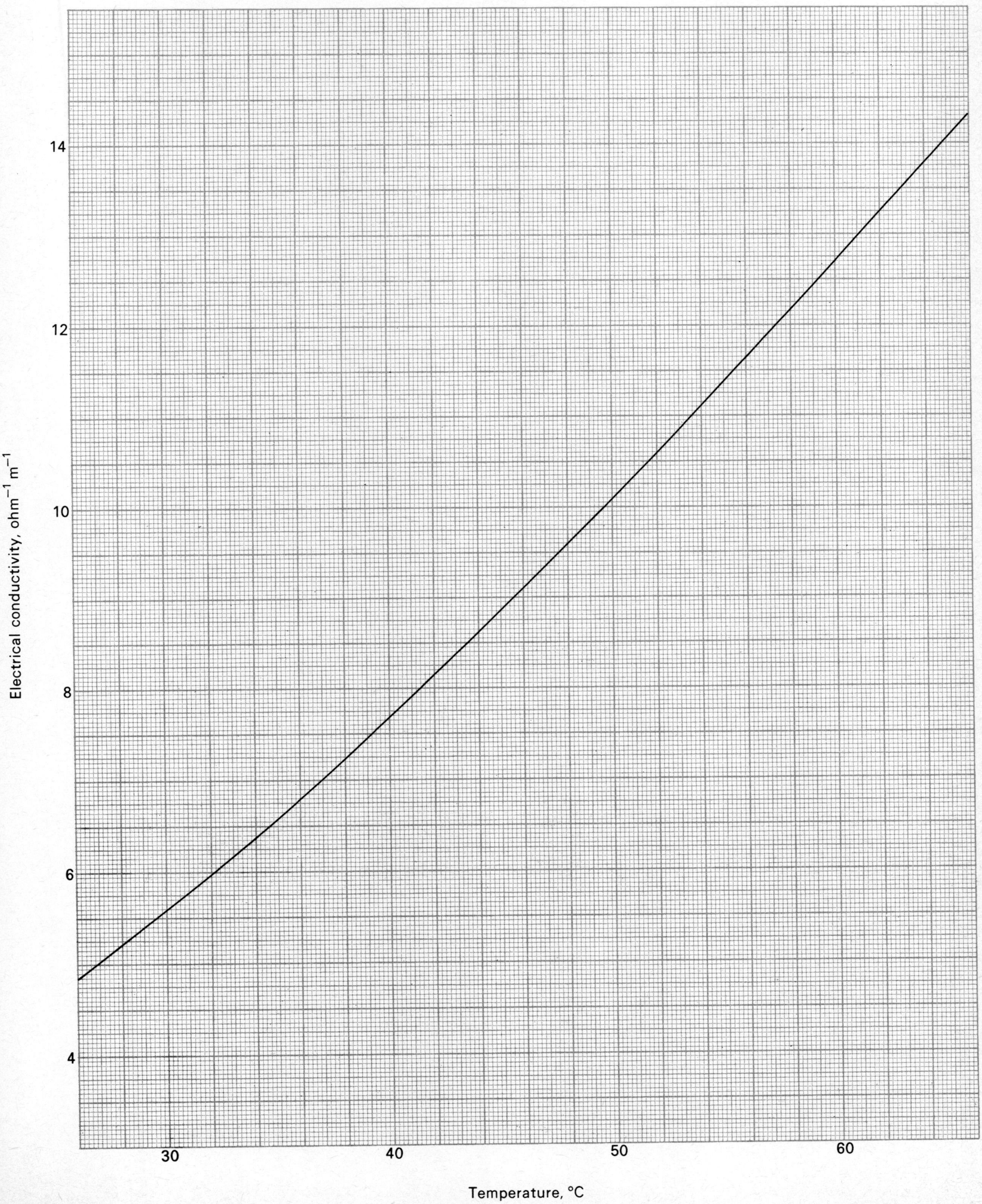

21 SULPHUR DIOXIDE

Sulphur dioxide is a colourless gas of pungent odour, causing suffocation and choking. It is a highly polar gas and moderately soluble in water. The dissolved gas can be easily expelled by boiling. The dipole moment of the molecule shows that it is not linear.

The gas may be liquefied at 20°C and by a pressure of 3 bars. Because of the ease of liquefaction, it is used in refrigeration processes. During evaporation, it causes great temperature depression, particularly at low pressures. The distribution of commercial sulphur dioxide is as a liquid in steel cylinders. Liquid sulphur dioxide is completely miscible with benzene, but only partially miscible with water. The pure liquid is a non-conductor. It is not associated in the liquid phase, but there are strong intermolecular attractions, which are indicated by the high boiling point. The liquid is an excellent solvent for most substances, particularly for salts, whose solutions have a characteristic colour.

Solid SO_2 is polymorphic, and according to measurements of the heat capacity, it does not show any phase transition. The well established crystalline structure is rhombic.

Commercial applications of sulphur dioxide are in the chemical industry as a bleaching agent for sensitive materials, and as an important intermediate in the manufacture of sulphuric acid.

The physical, thermodynamic, and transport properties of solid, liquid, and gaseous SO_2 at various temperatures and pressures have been compiled in the following publications: Burow (4), Gmelin (12), Helwig (15), Hirth (16), Kang (20), and Pascal (35).

The selected properties of SO_2 are presented in Table 21.1 and Figs. 21.1–21.14. These data were mostly correlated from the cited sources.

References

1 American Conference of Government Industrial Hygienists, *Documentation of Threshold Limit Values*, revised edition, Committee on Threshold Limit Values, Cincinnati, Ohio 45202, 1966, 203 pp.
2 Baker, C. B., de Haas, N., *Phys. Fluids*, 1964, **7**, 1400–2.
3 Biltz, W., Hülsmann, O., *Z. anorg. Chem.*, 1932, **207**, 377–84.
4 Burow, D. F., Liquid Sulphur Dioxide, in *The Chemistry of Nonaqueous Solvents*, Vol. III, (Lagowski, J. J., Ed.), Academic Press, New York, 1970, pp. 137–83.
5 Edwards, D. G., *The Vapour Pressure of 30 Inorganic Liquids between One Atmosphere and the Critical Point*, Univ. of California, Lawrence Rad. Lab., UCRL 7167, Livermore, California, June 13, 1963, 148 pp.
6 Engineering Science Data Units, *Approximate Data on the Viscosity of some Common Liquids*, Item No. 66024, London, July 1966, 26 pp.
7 Evans, W. H., Wagman, D. D., *J. Res. Natl. Bur. Std.*, 1952, **49**, 141–8.
8 Faltin, H., *Technik*, 1960, **15**, 592–6.
9 Fürstenau, R., *Ann. Phys.*, 1908, **27** (4), 735–72.
10 Fürstenau, R., *Verh. phys. Ges.*, 1908, **10**, 968–79.
11 Giauque, W. F., Stephenson, C. C., *J. Am. Chem. Soc.*, 1938, **40**, 1389–94.
12 *Gmelins Handbuch der anorganischen Chemie*, Syst. No. 9, Schwefel, Teil B, Lief. 1, Verlag Chemie, GmbH, Weinheim, 1953, pp. 204–62.
13 Golubev, I. F., *Viscosity of Gases and Gas Mixtures*, Gosudarstvennoe Izdatel'stvo Fiziko-Matematicheskoi Literatury, Moskva, 1959, Table 4.
14 Greene, E. F., Toennies, J. P., *Chemical Reactions in Shock Waves*, Edward Arnold, London, 1964, pp. 333–4.
15 Hellwig, L. R., *Pressure–Volume–Temperature Properties of Sulphur Dioxide*, Univ. of Texas, Ph.D. dissertation in Chem. Eng., Austin, Texas, 1955.
16 Hirth, L. J., *Gas Compressibility of Nitrous Oxide and of Sulphur Dioxide by the Burnett Method*, Univ. of Texas, Ph.D. dissertation in Chem. Eng., Austin, Texas, 1958.
17 Hudson, J. C., *J. Chem. Soc.*, 1925, **127**, 1332.
18 IUPAC, Atomic Weights of the Elements 1971, *Pure & Appl. Chem.*, 1972, **30** (3–4), 638–49.
19 Ibrahim, S. H., Kuloor, N. R., *Brit. Chem. Eng.*, 1961, **6** (12), 862–3.
20 Kang, T. L., *Thermodynamic Properties of Sulphur Dioxide*, Univ. of Texas, Ph.D. dissertation in Chem. Eng., Austin, Texas, 1960.
21 Kang, T. L., Hirth, L. J., Kobe, K. A., McKetta, J. J., *J. Chem. Eng. Data*, 1961, **6** (2), 220–6.
22 Kang, T. L., McKetta, J. J., *J. Chem. Eng. Data*, 1961, **6** (2), 227–8.
23 Kardos, A., *Z. ges. Kälte-Ind.*, 1934, **41**, 1–35.
24 Kelley, K. K., *The Free Energies of Vaporization and Vapour Pressures of Inorganic Substances*, Bureau of Mines Bull. 383, U.S. Government Printing Office, Washington, 1935, 132 pp.
25 Kelley, K. K., *Heats of Fusion of Inorganic Substances*, Bureau of Mines Bull. 393, U.S. Government Printing Office, Washington, 1936, 166 pp.
26 Kobe, K. A., Long, E. G., *Petrol. Ref.*, 1950, **29** (1), 126–30.
27 Koch, B., *Grundlagen des Wärmeaustausches*, Verlagsanstalt H. Beucke & Sohne, Berlin, 1950.
28 Lange, A., *Z. ang. Chem.*, 1899, pp. 300–5.
29 Lange, N. A., Ed., *Handbook of Chemistry*, 10th ed., McGraw-Hill, New York, 1967.
30 Lax, E., Synowietz, C., Eds., *Taschenbuch für Chemiker und Physiker*, 3rd ed., Springer-Verlag, Berlin, 1967.
31 Leduc, A., *Chem. Rev.*, 1929, **6**, 1–16.
32 Mahenc, J., Fournes, L., *Bull. Soc. Chim. France*, 1969, **7**, 2277–80.
33 Maryott, A. A., Smith, E. R., *Table of Dielectric Constants of Pure Liquids*, Natl. Bur. Std. Circular 514, Washington, Aug. **10**, 1951, 44 pp.
34 Meyer, G. R., Thodos, G., *J. Chem. Eng. Data*, 1962, **7** (4), 532–6.
35 Pascal, P., Ed., *Nouveau Traité de Chimie Minérale*, Vol. XIII, Part 2, Masson et Cie, Éditeurs, Paris, 1960, pp. 1204–28.
36 Rao, R. V. G., Nammalvar, T., *Z. phys. Chem. Neue Folge*, 1971, **75**, 77–87.
37 Ribaud, G., *Publ. Sci. Tech. Ministre Air (France)*, 1952, **266**, 169 pp.
38 Riedel, L., *Bull. Internat. Inst. Refrig.*, 1939, **20** (4), (Annex No. 5), B1–10.
39 Rossini, F. D., Wagman, D. D., Evans, W. H., Levine, S., Jaffe, I., *Selected Values of Chemical Thermodynamic Properties*, Natl. Bur. Std. Circular 500, Washington, Feb. 1, 1952.
40 Scheuer, O., *Anz. Wien. Akad.*, 1911, **48**, 304–8.
41 Seyer, W. F., Peck, W. S., *J. Am. Chem. Soc.*, 1930, **52**, 14–23.
42 Stowe, V. M., *J. Am. Chem. Soc.*, 1929, **51**, 410–15.
43 Stull, D. R., Dir., *JANAF Thermochemical Tables*, Dow Chemical Co., Midland, Mich., 1965–8.
44 Yosim, S. J., *J. Chem. Phys.*, 1964, **40** (10), 3069–75.

Table 21.1 Properties of Sulphur Dioxide

		Ref.
Chemical formula	SO_2	(30)
Molecular weight	64.0580	(18)
Critical temperature	157.5°C	(21)
Critical pressure	78.84×10^5 N m^{-2}	(21)
Critical density	525.0 kg m^{-3}	(21)
Triple-point temperature	−75.5°C	(30)
Triple-point pressure	1674.39 N m^{-2}	(30)
Boiling point at 1.0133×10^5 N m^{-2}	−10.02°C	(30)
Melting point	−73.15°C	(25)
Transition temperature	None	(12)
Density of solid at −191.0°C	1928.0 kg m^{-3}	(3)
Compressibility of liquid at 13°C	1.32×10^{-9} m^2 N^{-1}	(28)
Coefficient of thermal expansion of liquid at boiling point	1.70×10^{-3} C^{-1}	(44)
Heat of formation of gas at 25°C	-2.9704×10^8 J kg-mol^{-1}	(43)
Free energy of formation of gas at 25°C	-3.0036×10^8 J kg-mol^{-1}	(43)
Heat of sublimation at 25°C	-7.7999×10^6 J kg-mol^{-1}	(24)
Heat of fusion	7.4064×10^6 J kg-mol^{-1}	(39)
Heat of transition	None	(12)
Entropy of gas at 25°C	2.4828×10^5 J kg-mol^{-1} °C^{-1}	(43)
Heat capacity ratio of gas at 15°C and 1.0133×10^5 N m^{-2}	1.29	(31)
Velocity of sound at 15°C and 1.0133×10^5 N m^{-2}	224.1 m s^{-1}	(14)
Diffusion in air at 0°C and 1.0133×10^5 N m^{-2}	1.22×10^{-5} m^2 s^{-1}	(19)
Dielectric constant of liquid at 20°C	14.1	(33)
Electrical conductivity of liquid at 35°C	1.5×10^{-6} ohm^{-1} m^{-1}	(29)
Refractive index of liquid at 13°C and 5889 Å	1.357	(4)
Threshold limit value	(5 ppm) 1.3×10^{-5} kg m^{-3}	(1)

21.1 Vapour Pressure of Sulphur Dioxide (5,11,21,36)

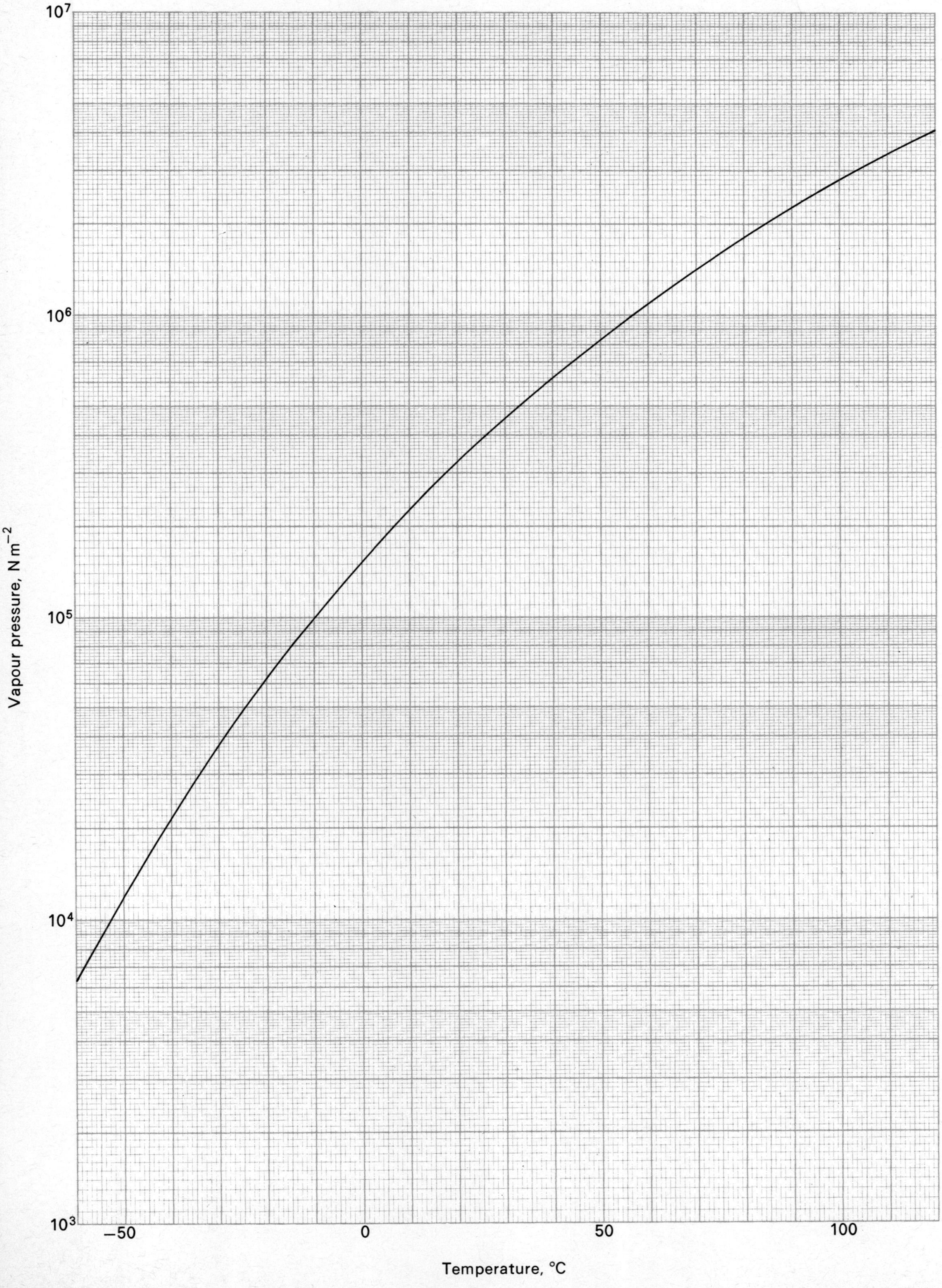

21.2 Orthobaric Density of Sulphur Dioxide (21,36)

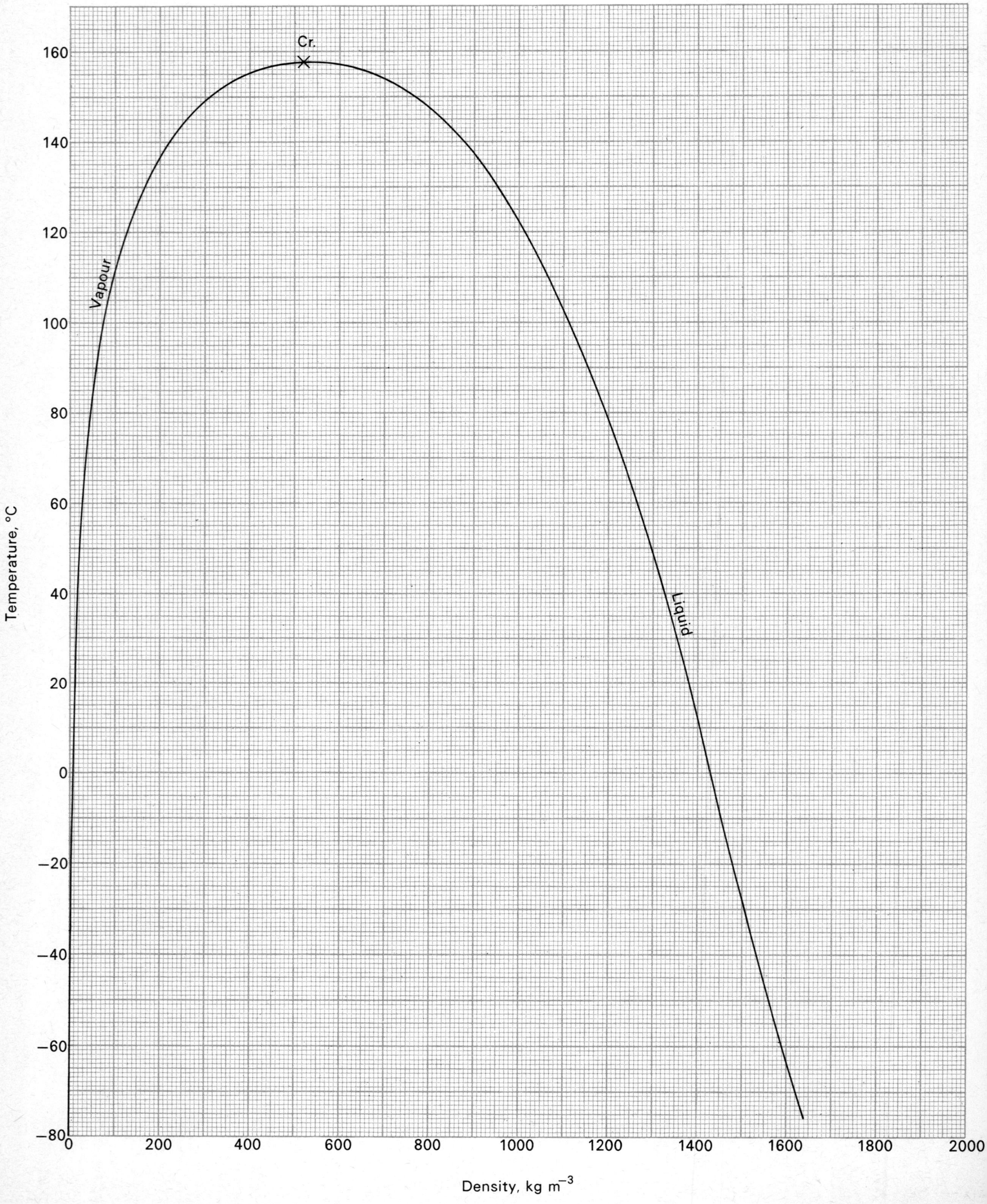

21.3 P-V-T Data for Sulphur Dioxide (15,16,21,22)

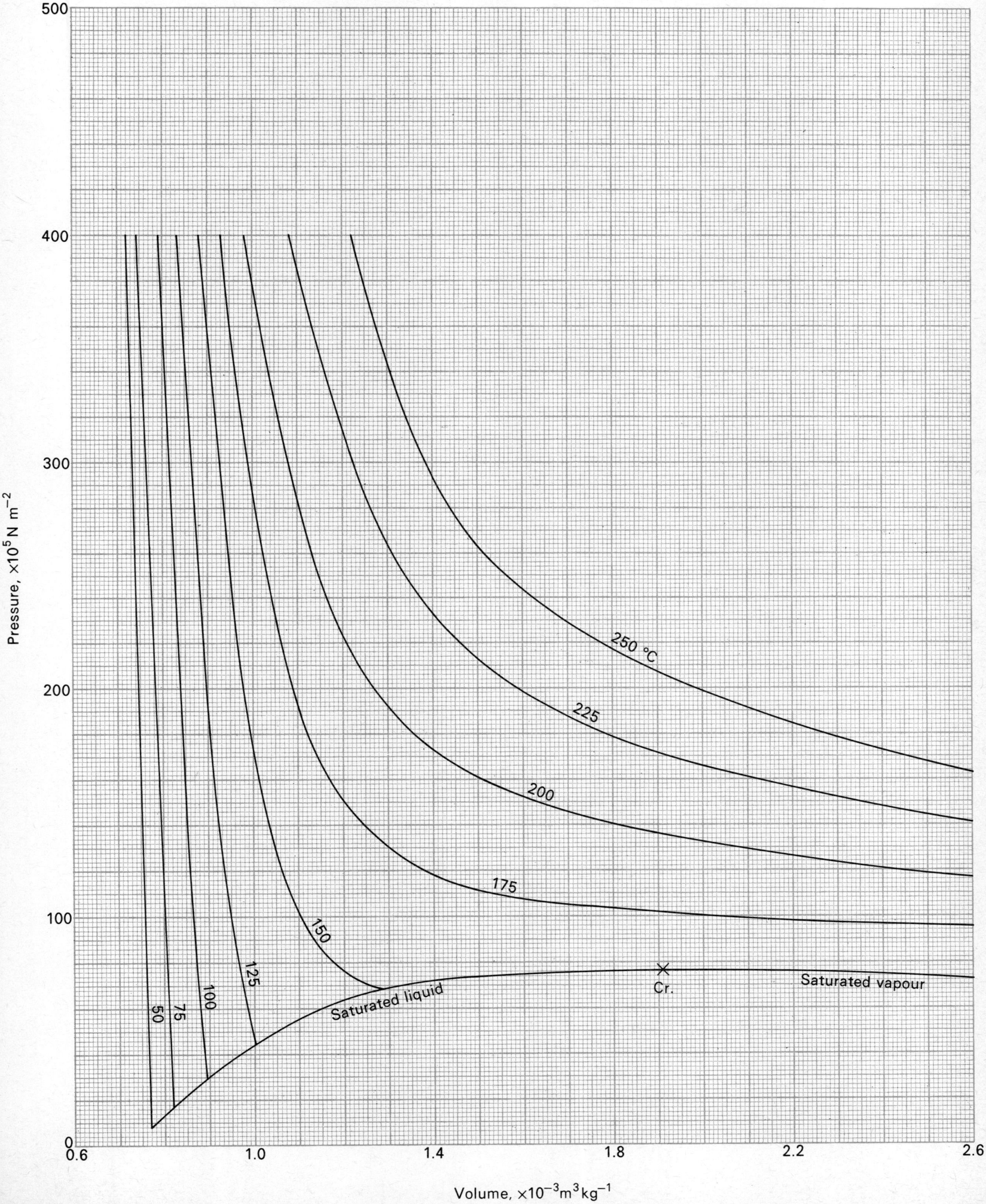

21.4 Latent Heat of Vaporisation of Sulphur Dioxide (20,21,22)

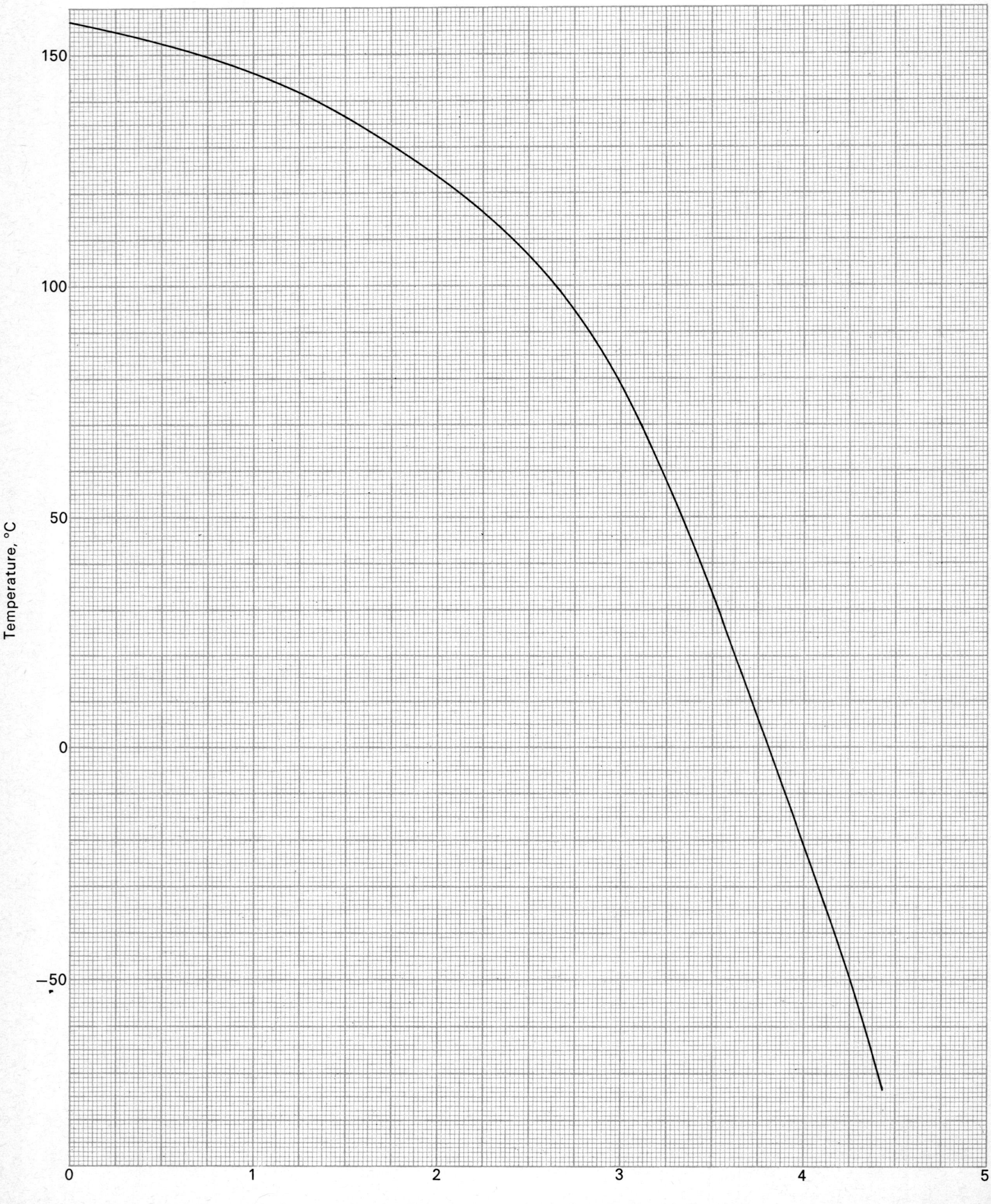

21.5 Heat Capacity of Liquid Sulphur Dioxide (11,20,27,38)

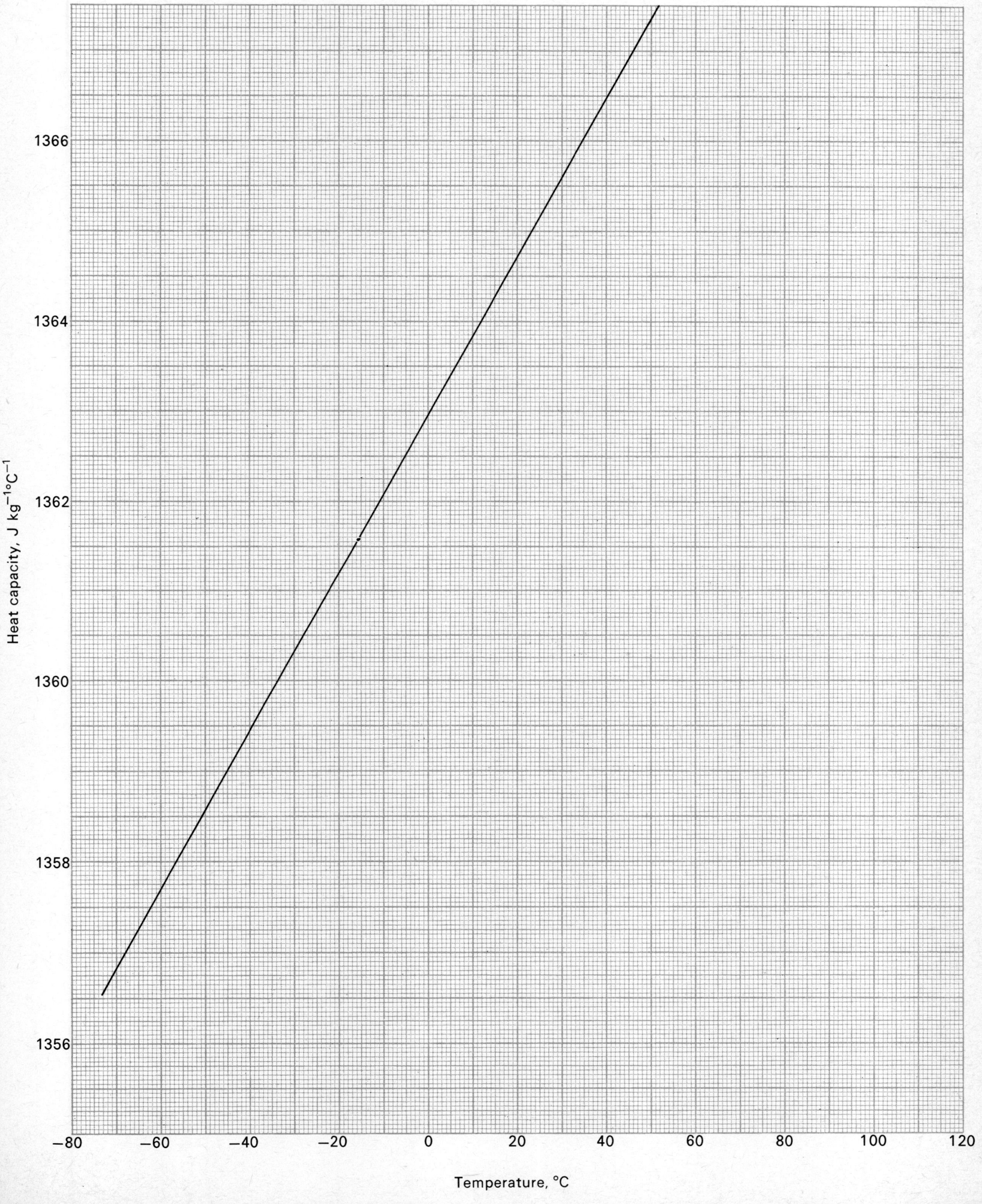

21.6 Heat Capacity of Gaseous Sulphur Dioxide at $1.0133 \times 10^5 N\ m^{-2}$(7,8,26,37)

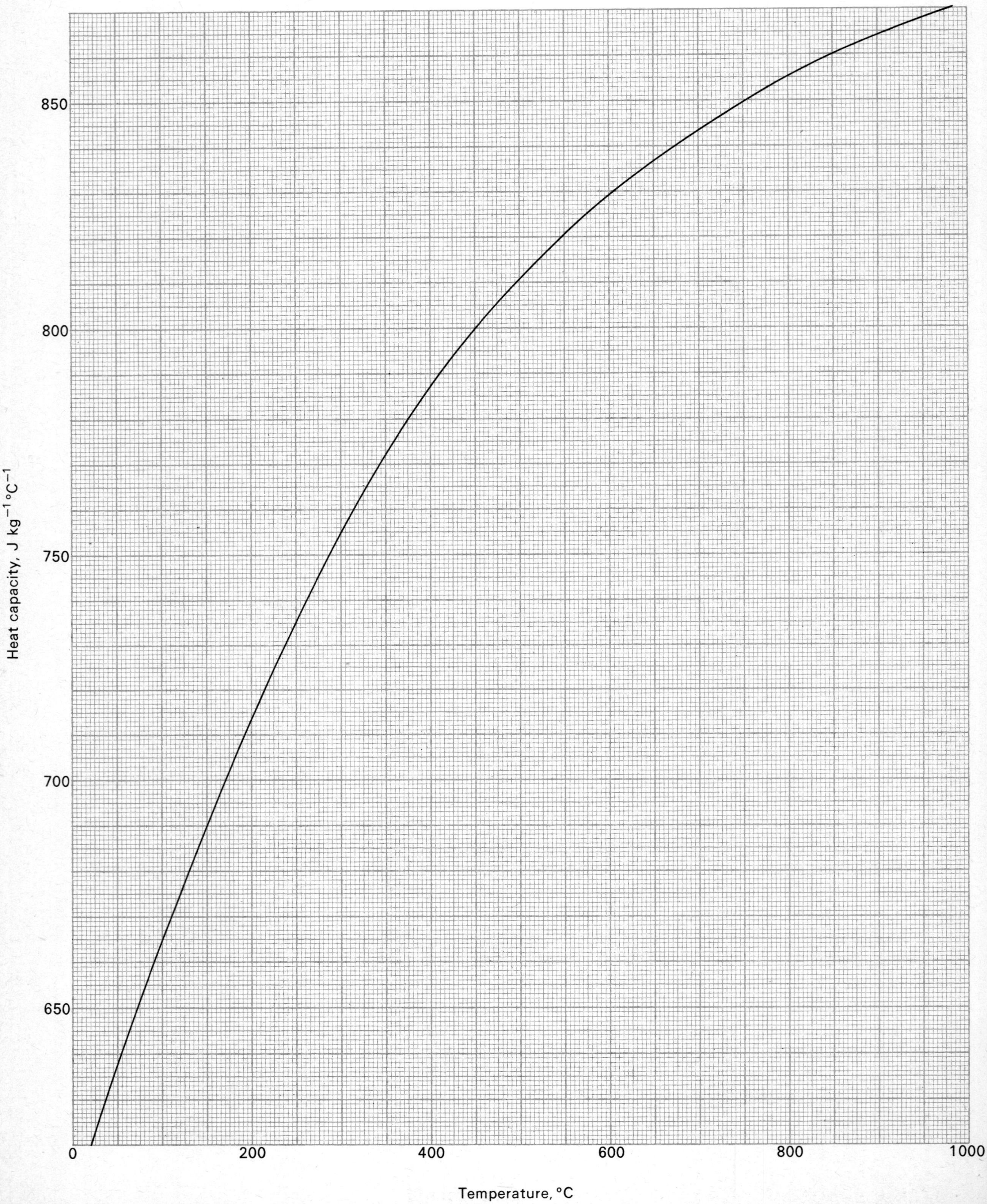

21.7 Heat Capacity Ratio (C_p/C_v) of Gaseous Sulphur Dioxide at 1.0133×10^5 N m^{-2} (9,10)

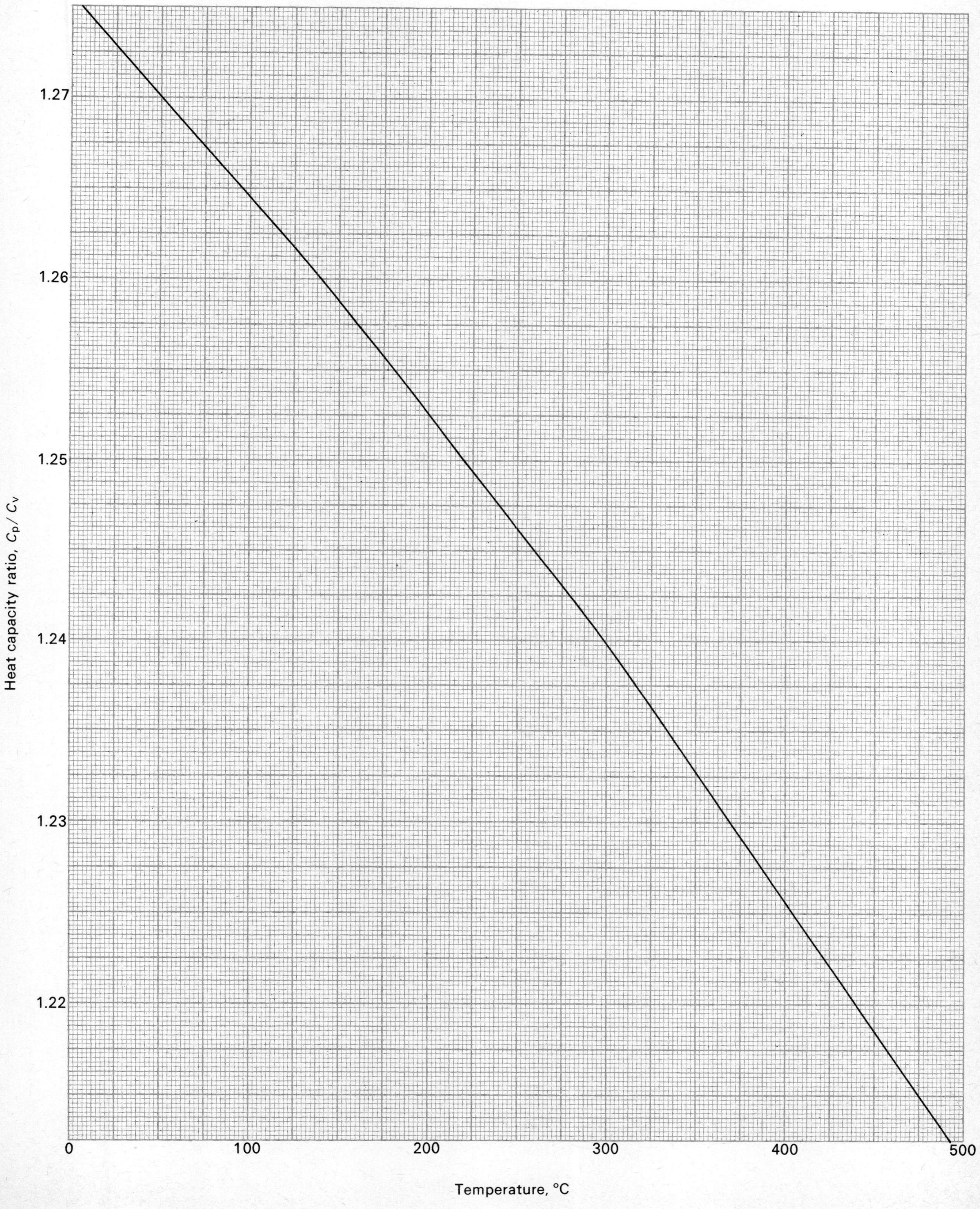

21.8 Viscosity of Liquid Sulphur Dioxide (6,34)

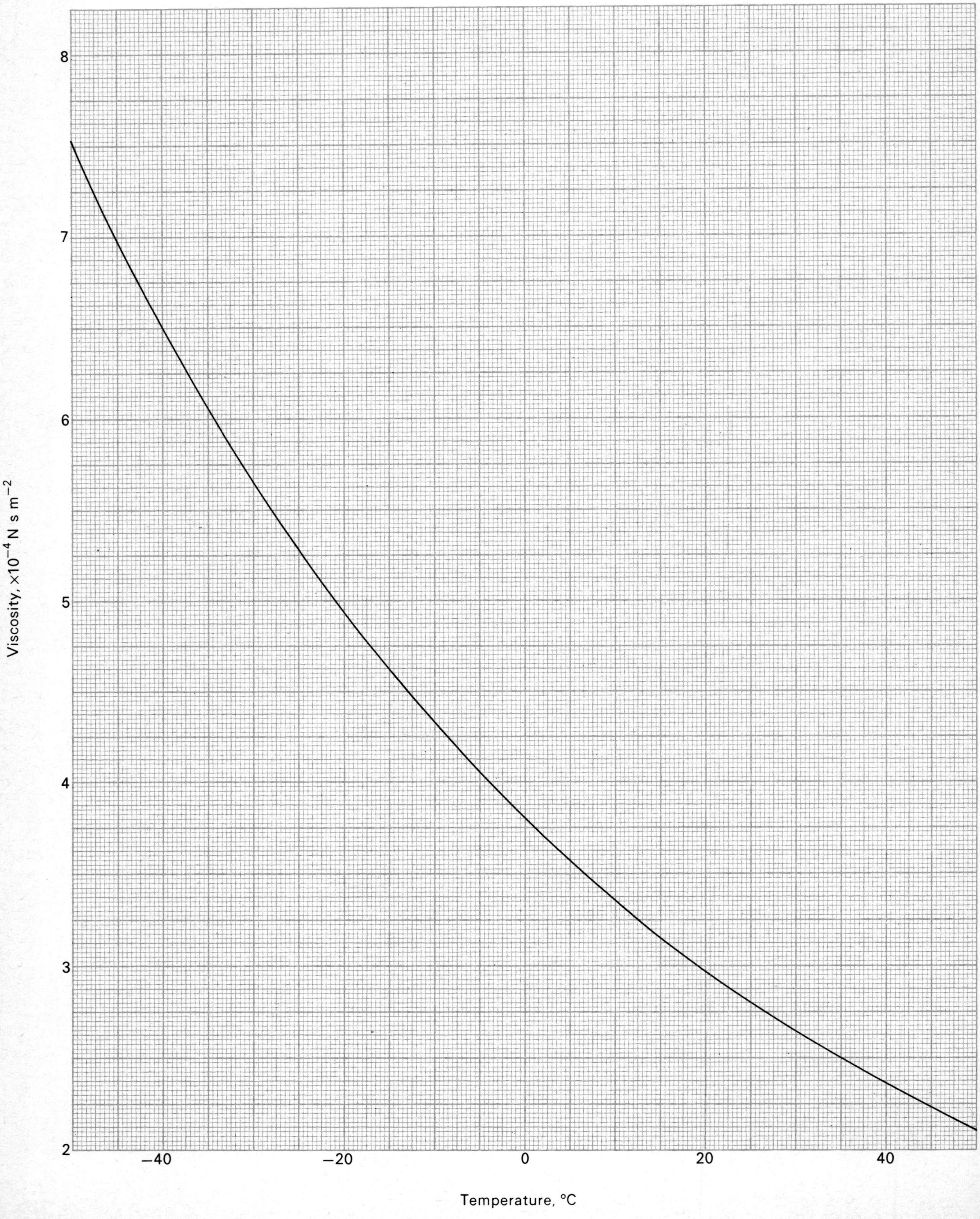

21.9 Viscosity of Gaseous Sulphur Dioxide at $1.0133 \times 10^5 N\ m^{-2}$ (13,34)

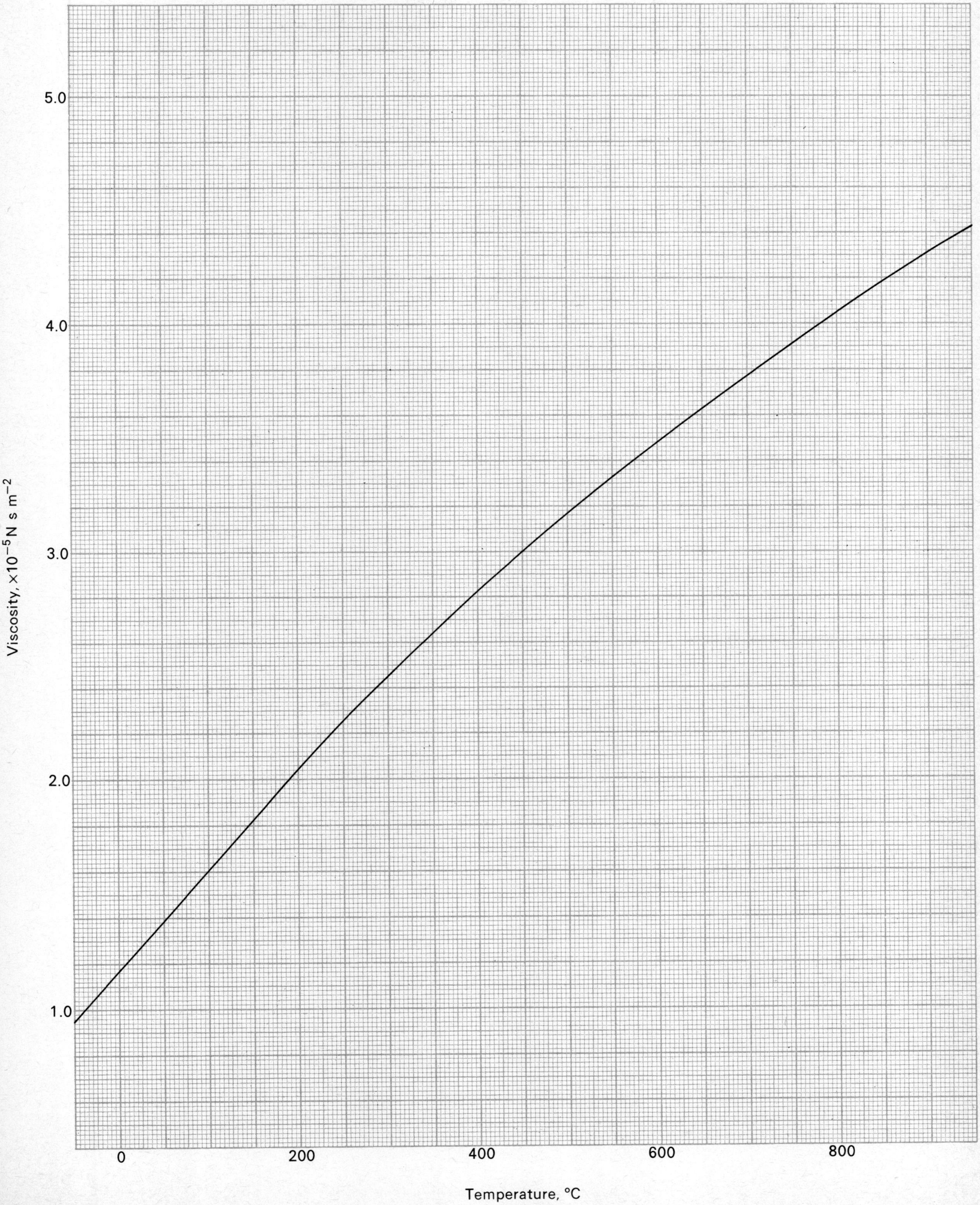

21.10 Thermal Conductivity of Liquid Sulphur Dioxide (23,27,34)

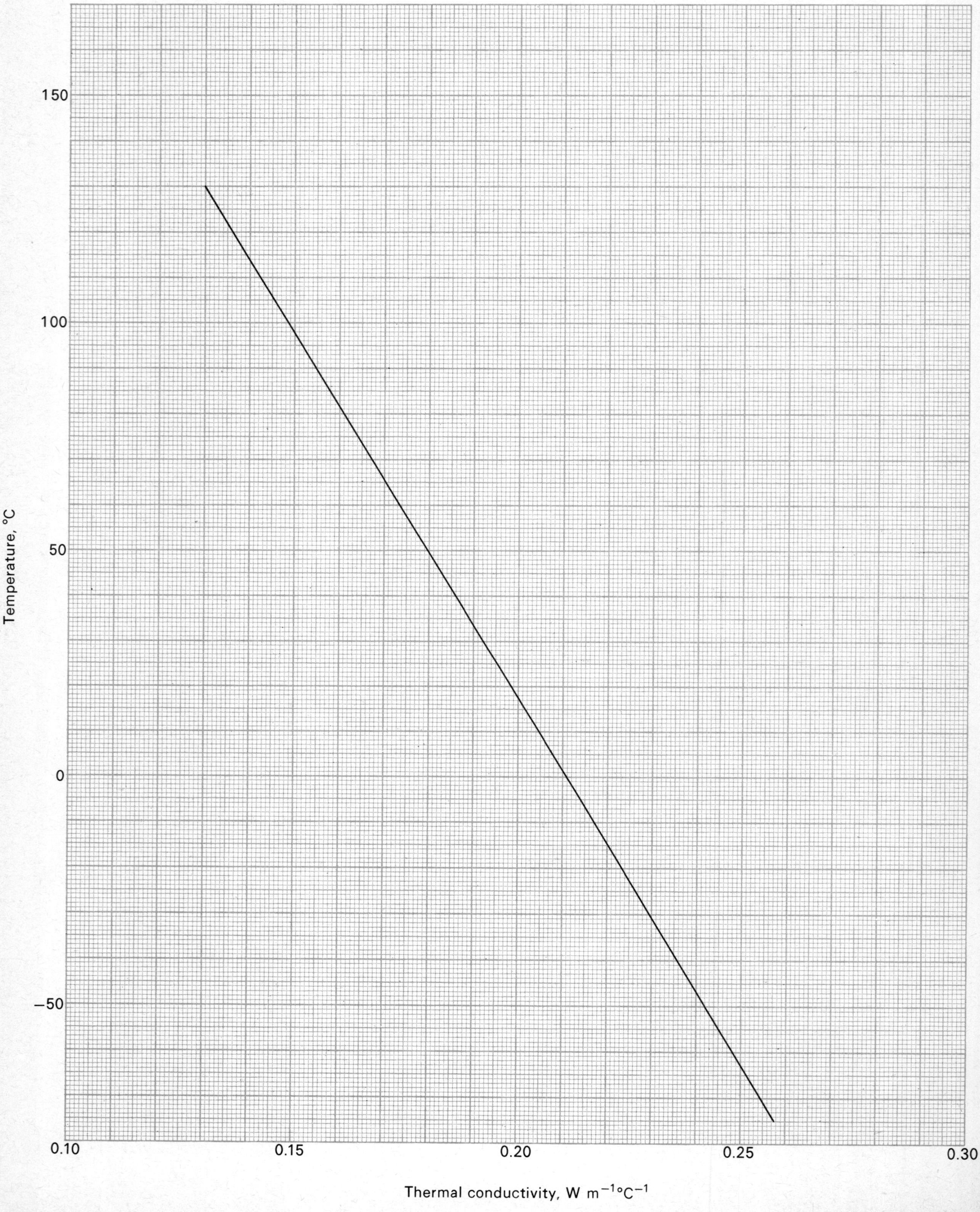

21.11 Thermal Conductivity of Gaseous Sulphur Dioxide at $1.0133 \times 10^5 N\ m^{-2}$(2,34)

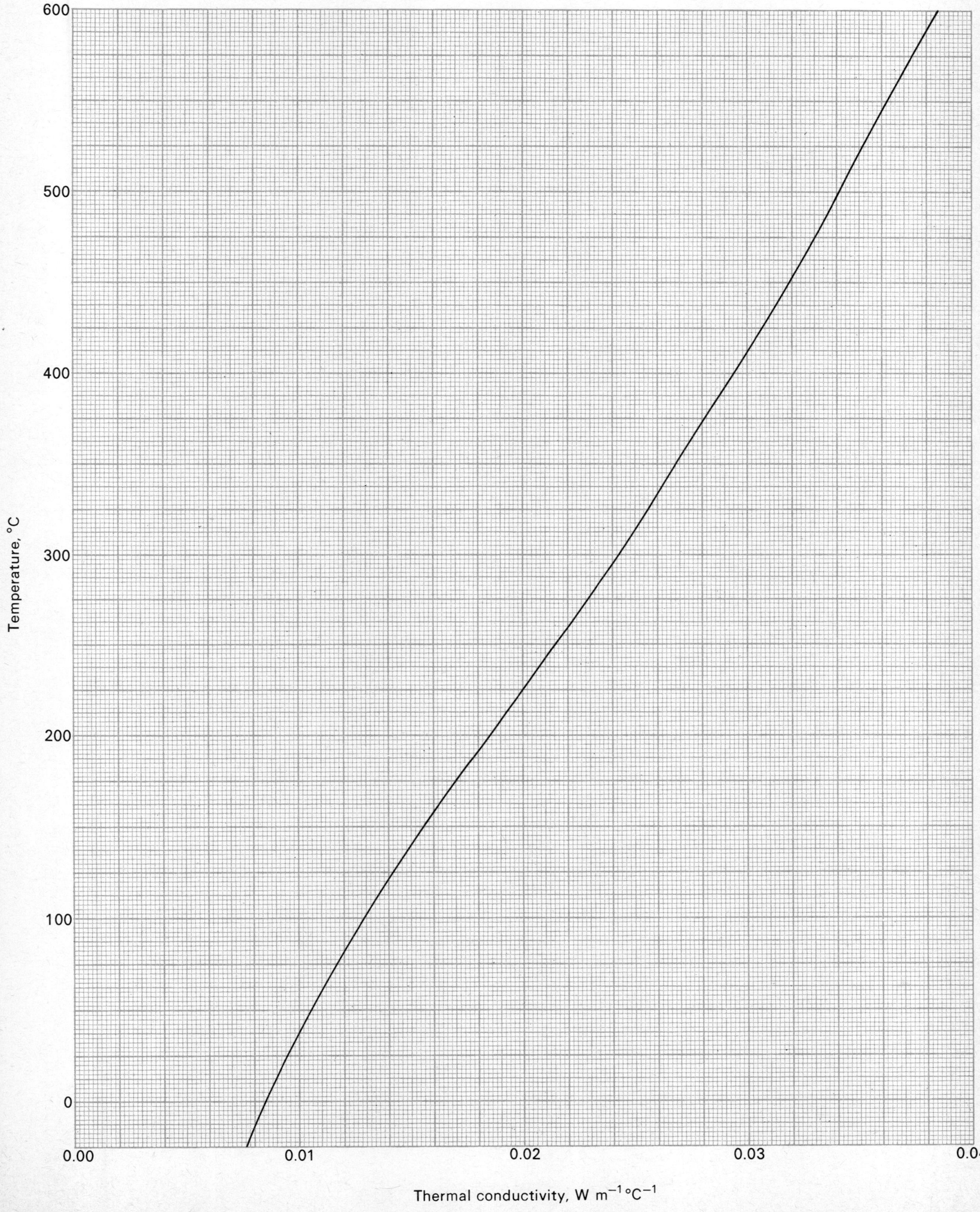

21.12 Surface Tension of Sulphur Dioxide (40,41,42)

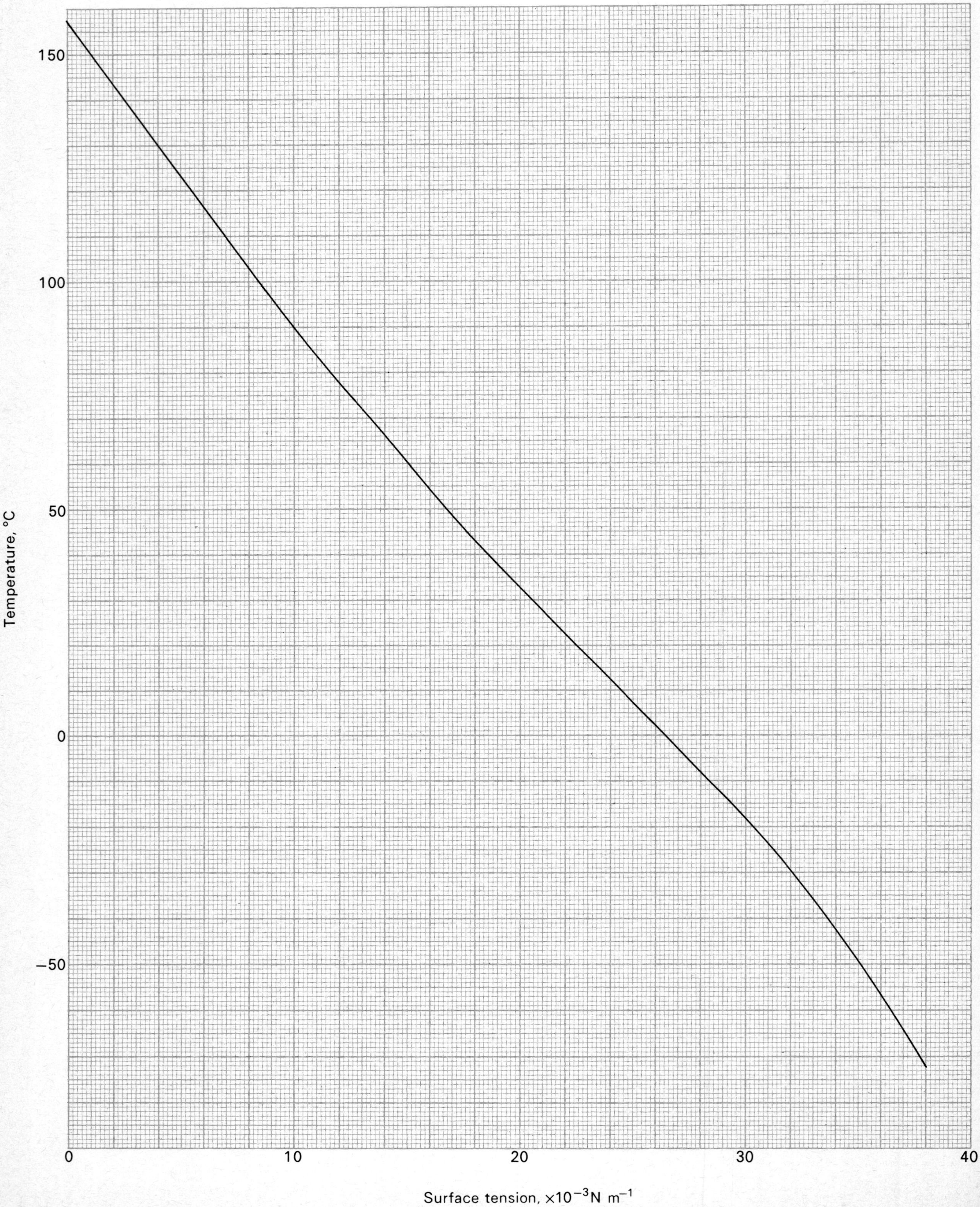

21.13 Solubility of Sulphur Dioxide in Water at 1.0133×10^5 N m^{-2} Total Pressure (17,29)

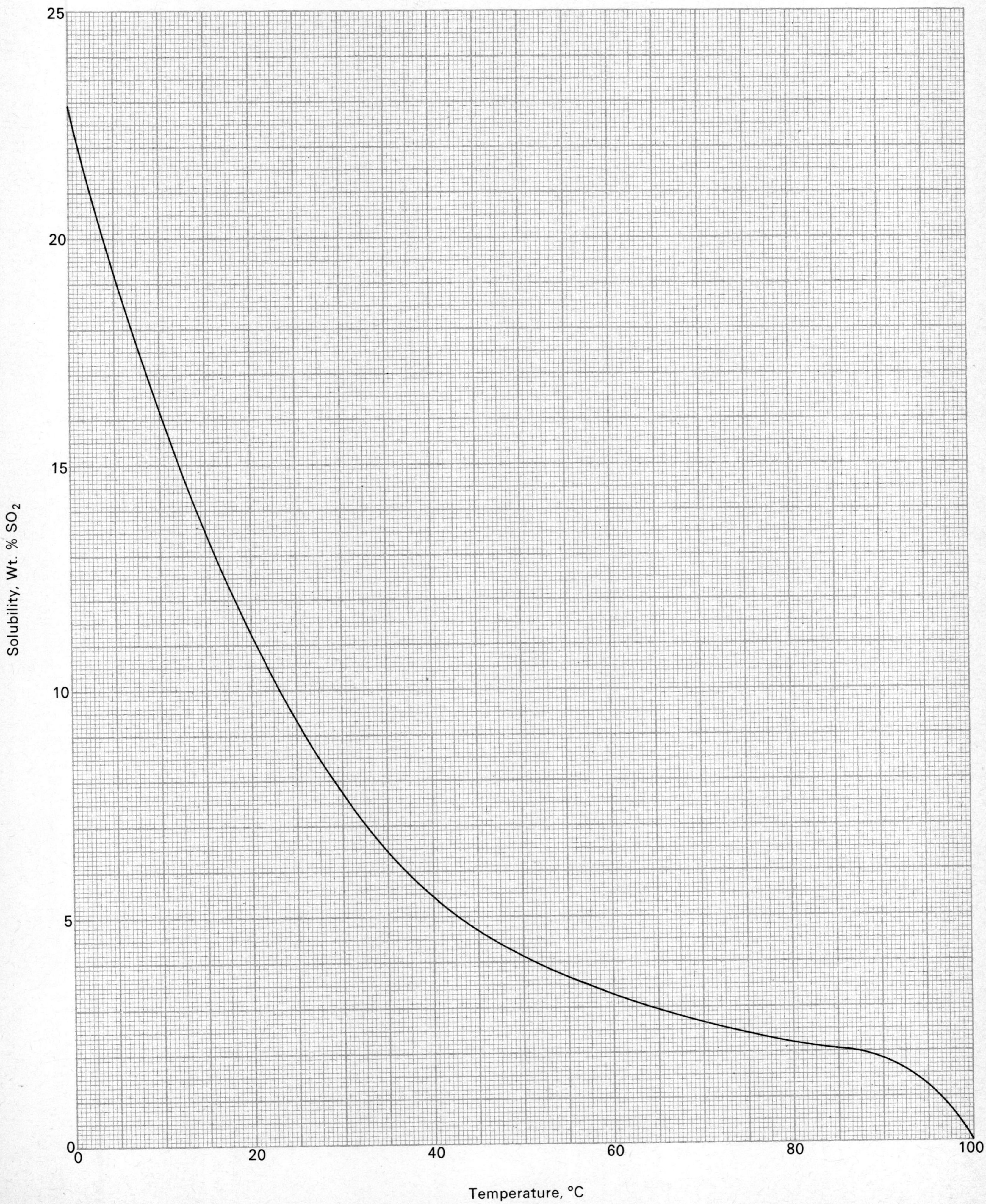

21.14 Electrical Conductivity of Liquid Sulphur Dioxide (32)

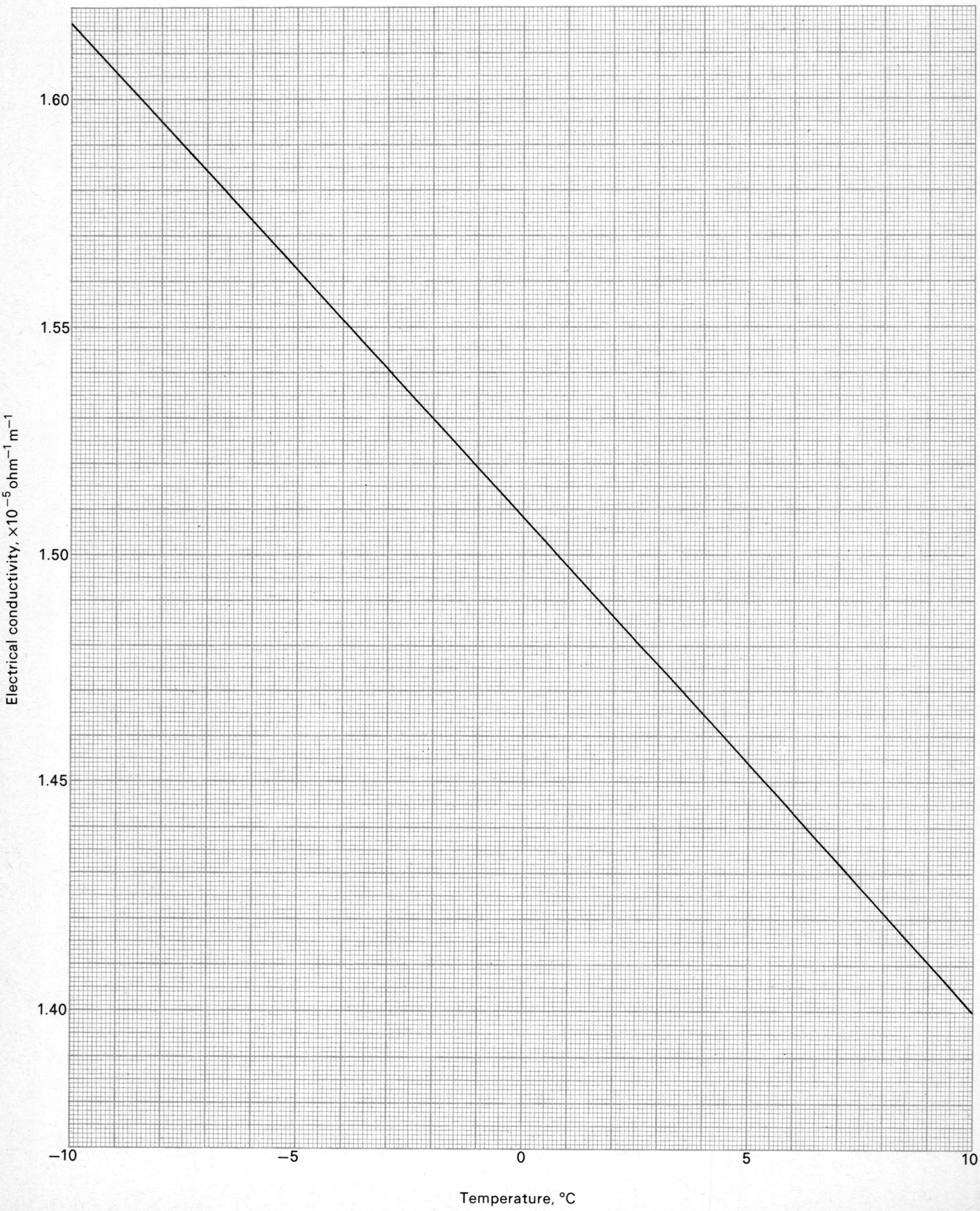

22 SULPHUR TRIOXIDE

Sulphur trioxide is a colourless liquid at room temperature and atmospheric pressure. The liquid has an abnormally high coefficient of thermal expansion of 1/448 per degree. The liquid phase is an equilibrium mixture of monomer and trimer modifications. Its readiness for polymerization in the liquid state makes precise physicochemical measurements difficult. The presence of two forms in the condensed phase accounts for the unusual phenomena observed in the vapour pressure, which suddenly increases after freezing the liquid, i.e. the vapour pressure is greater just below the melting point than that of liquid SO_3 just above the melting point. The liquid does not conduct electricity.

Solid sulphur trioxide is a colourless trimorphic crystalline substance, possessing three solid phases: alpha-SO_3 has a polymeric cross-linked structure, resembling needles of ice; the beta-SO_3 modification consists of polymeric molecules, forming white asbestos-like crystals with a silky lustre; and gamma-SO_3 or the colloidal form consists of cyclic trimer and monomer molecules.

The vapour of SO_3 is monomeric at high temperatures and normal pressures, and shows a dipole moment of zero at about 100°C, so that the molecule must be planar and symmetrical. This is supported by the infra-red and Raman spectra. The vapour readily combines with water forming sulphuric acid and its hydrates. SO_3 is reasonably stable to heat; however, in a red hot tube it decomposes into SO_2 and O_2. With moist air it forms dense white fumes of a sweet taste and suffocating odour. Chemically, it is a powerful oxidizing agent. The solution of SO_3 in concentrated H_2SO_4 is known as oleum.

Most of the thermophysical properties of solid, liquid, and gaseous sulphur trioxide have been compiled in Gmelin (12), Landolt-Börnstein (18), and Pascal (29). Further investigations of interest are presented by Colwell (8) and Stuckey and Secoy (37), the latter, dealing with the critical temperature of the SO_3–H_2O system. Due to the narrow temperature interval between the melting and boiling points of SO_3, only one value has been presented for the liquid heat capacity (see Table 22.1). This value can be used with confidence in the narrow temperature interval of the liquid state.

References

1 Abercromby, D. C., Hyne, R. A., Tiley, P. F., *J. Chem. Soc.*, 1963, 5832–8.
2 Abercromby, D. C., Tiley, P. F., *J. Chem. Soc.*, 1963, 4902–4.
3 Askel'rod, Yu. V., Ramm, V. M., Surkov, E. I., *J. Appl. Chem. USSR*, 1964, **37** (6), 1203–7.
4 Berthoud, A., *Helv. Chim. Acta*, 1922, **5**, 513–32.
5 Berthoud, A., *J. Chim. Phys.*, 1923, **20**, 77–86.
6 Bodenstein, M., Pohl, W., *Z. Elektrochem.*, 1905, **11**, 373–84.
7 Campbell, A. N., Smith, N. O., *Trans. Faraday Soc.*, 1937, **33**, 545–51.
8 Colwell, J. H., *The Physical Properties of* SO_3, Univ of Washington, Ph.D. thesis, Washington, 1961, 63 pp.
9 Edwards, D. G., *The Vapour Pressure of 30 Inorganic Liquids between One Atmosphere and the Critical Point*, Univ. of California, Lawrence Rad. Lab. UCRL 7167, Livermore, California, June 13, 1963, 148 pp.
10 Gable, C. M., Betz, H. F., Maron, S. H., *J. Am. Chem. Soc.*, 1950, **72**, 1445–8.
11 Gardy, H., Bachellier, J., Roget, J., *Prediction of Physical Magnitudes Necessary for Chemical Engineering Calculations. II. Estimation of the Heats of Vaporization of Sulphur Dioxide and Sulphur Trioxide*, Compt. Rend. 78th Congr. socs. savantes Paris et depts., Sect. sci. 1953, pp. 517–26.
12 *Gmelins Handbuch der anorganischen Chemie*, Syst. No. 9, Schwefel, Teil B, Lief. 1, Verlag Chemie GmbH, Weinheim, 1953, pp. 337–53.
13 Grau, R., Roth, W. A., *Z. anorg. Chem.*, 1930, **188**, 173–85.
14 Hülsmann, O., Biltz, W., *Z. anorg. Chem.*, 1934, **218**, 369–78.
15 Hyne, R. A., Tiley, P. F., *J. Chem. Soc.*, 1961, 2348–52.
16 IUPAC, Atomic Weights of the Elements 1971, *Pure & Appl. Chem.*, 1972, **30** (3–4), 638–49.
17 Kobe, K. A., Long, E. G., *Petrol. Ref.*, 1950, **29** (1), 126–30.
18 *Landolt-Börnstein, Zahlenwerte und Funktionen aus Physik, Chemie, Astronomie, Geophysik, und Technik;* II Band, 4 Teil, Kalorische Zustandgrössen, Springer-Verlag, Berlin, 1961, p. 445.
19 Lax, E., Synowietz, C., Eds., *Taschenbuch für Chemiker und Physiker*, 3rd ed., Springer-Verlag, Berlin, 1967.
20 Lichty, D. M., *J. Am. Chem. Soc.*, 1912, **34**, 1440–8.
21 Luchinskii, G. P., *Zh. Fiz. Khim.*, 1939, **12**, 280–5.
22 Maryott, A. A., Smith, E. R., *Table of Dielectric Constants of Pure Liquids*, Natl. Bur. Std. Circular 514, Washington, Aug. 10, 1951, 44 pp.
23 Mathews, J. F., *Chem. Rev.*, 1972, **72** (1), 71–100.
24 Miles, F. D., Niblock, H., Wilson, G. L., *Trans. Faraday Soc.*, 1940, **36**, 345–56.
25 Moghadame, P. E., *Rev. Inst. Franc. Petrole et Ann. Combustibles Liquides*, 1957, **12** (1), 58–66.
26 Nasini, R., *Ber.*, 1882, **15**, 2878–94.
27 Orlicek, A. F., Pöll, H., *Hilfsbuch für Mineralöltechniker 1*, Springer-Verlag, Wien, 1951.
28 Owens, E. J., Thodos, G., *AIChE Journal*, 1960, **6** (4), 676–81.
29 Pascal, P., Ed., *Nouveau Traité de Chimie Minérale*, Vol. XIII, Part 2, Oxygene, Ozone, Oxydes, Eau Oxygénée, La combustion, Soufre, Selenium, Tellure, Polonium; Masson et Cie, Éditeurs, Paris, 1960, pp. 1266–78.
30 Popiel, R., *J. Chem. Eng. Data*, 1964, **9** (2), 269–71.
31 Rossini, F. D., Wagman, D. D., Evans, W. H., Levine, S., Jaffe, I., *Selected Values of Chemical Thermodynamic Properties*, Natl. Bur. Std. Circular 500, Washington, Feb. 1, 1952.
32 Schenck, R., *Lieb. Ann.*, 1901, **316**, 1–17.
33 Smith, A., Schoenmaker, P., *J. Chem. Soc.*, 1926, 1108–27.
34 Spangenberg, C. D., *Chem. Eng.*, 1951, **58** (9), 170–1.
35 Stiel, L. I., Thodos, G., *AIChE Journal*, 1964, **10** (1), 26–30.
36 Stockmayer, W. H., Kavanagh, G. M., Mickley, H. S., *J. Chem. Phys.*, 1944, **12**, 408–12.
37 Stuckey, J. E., Secoy, C. H., *J. Chem. Eng. Data*, 1963, **8** (3), 386–9.
38 Stull, D. R., Dir., *JANAF Thermochemical Tables*, Dow Chemical Co., Midland, Mich., 1965–8.
39 Thomas, L. H., *J. Chem. Soc.*, 1959, 2132–52.
40 Walden, P., *Z. anorg. Chem.*, 1900, **25**, 209–26.
41 Westrik, R., MacGillavry, C. H., *Rec. trav. chim.*, 1941, **60**, 794–810.

Table 22.1 Properties of Sulphur Trioxide

		Ref.
Chemical formula	SO_3	(19)
Molecular weight	80.0570	(16)
Critical temperature	217.8°C	(23)
Critical pressure	82.08×10^5 N m^{-2}	(23)
Critical density	630.0 kg m^{-3}	(23)
Triple-point temperature (gamma)	16.8°C	(8)
Triple-point pressure (gamma)	0.2113×10^5 N m^{-2}	(33)
Boiling point at 1.0133×10^5 N m^{-2}	44.8°C	(33)
Melting point (gamma)	16.8°C	(33)
Transition temperature	−183.0°C	(8)
Density of solid (gamma) at −10.0°C	2290.0 kg m^{-3}	(41)
Coefficient of thermal expansion of liquid at 18°C	2.005×10^{-3} °C^{-1}	(20)
Heat capacity of liquid at 30.0°C	3223.8 J kg^{-1} °C^{-1}	(12)
Heat of formation of gas at 25°C	-3.9603×10^8 J kg-mol^{-1}	(38)
Free energy of formation of gas at 25°C	-3.7062×10^8 J kg-mol^{-1}	(38)
Heat of sublimation (gamma) at triple-point temperature	4.9825×10^7 J kg-mol^{-1}	(8)
Heat of fusion	1.9678×10^6 J kg-mol^{-1}	(31)
Heat of transition	1.0467×10^7 J kg-mol^{-1}	(8)
Entropy of gas at 25°C	2.5684×10^5 J kg-mol^{-1} C^{-1}	(38)
Heat capacity ratio of gas at 27°C and 1.0133×10^5 N m^{-2}	1.333	(6, 36)
Diffusion in air at 80°C and 1.0133×10^5 N m^{-2}	0.13×10^{-4} m s^{-1}	(34)
Dielectric constant of liquid at 18°C	3.11	(22)
Electrical conductivity of liquid	None	(30, 40)
Refractive index of liquid at 20°C and 6567 Å	1.4077	(26)

22.1 Vapour Pressure of Sulphur Trioxide (2,3, 9,15,24)

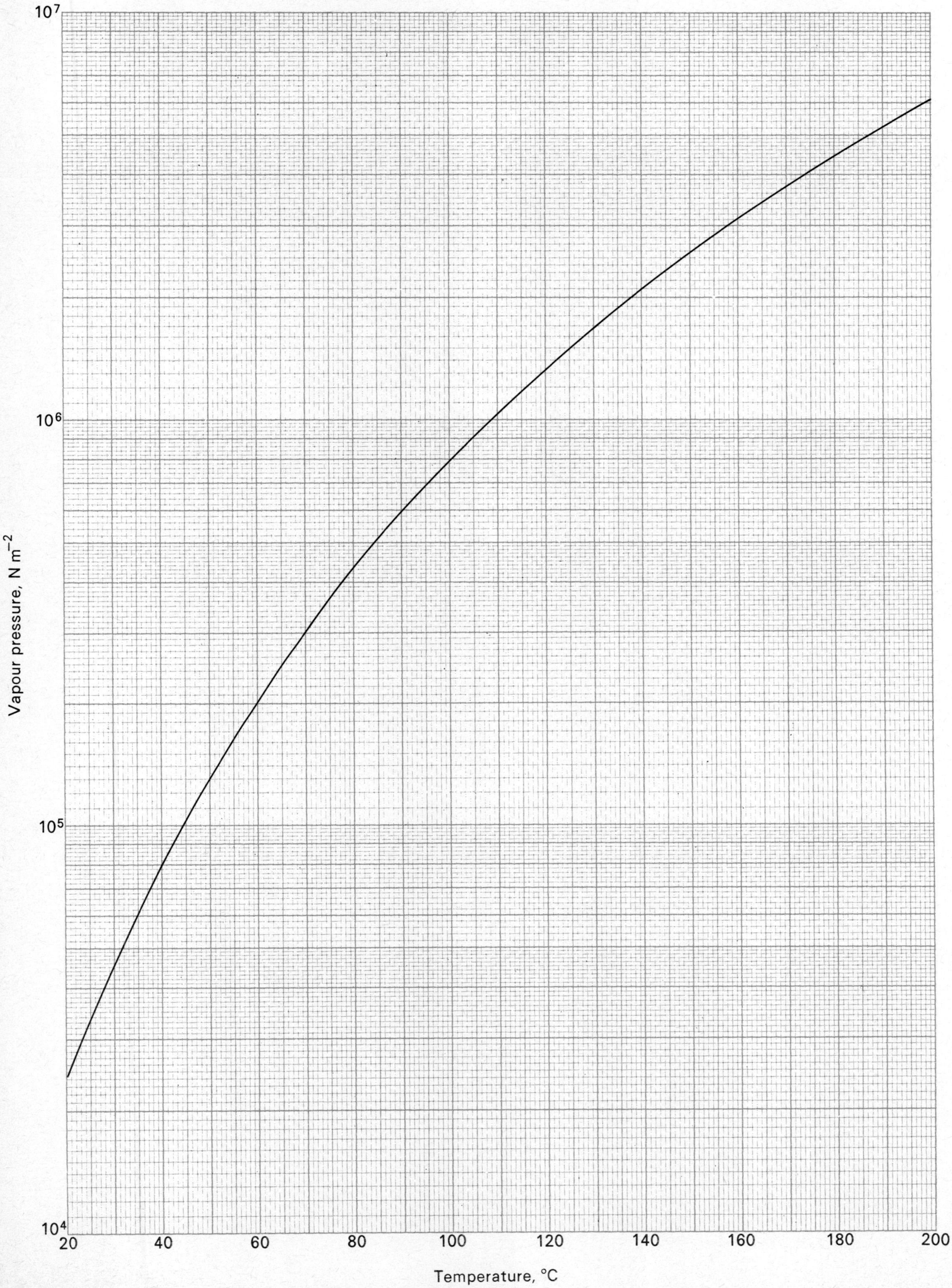

22.2 Orthobaric Density of Sulphur Trioxide (4,5,19,32)

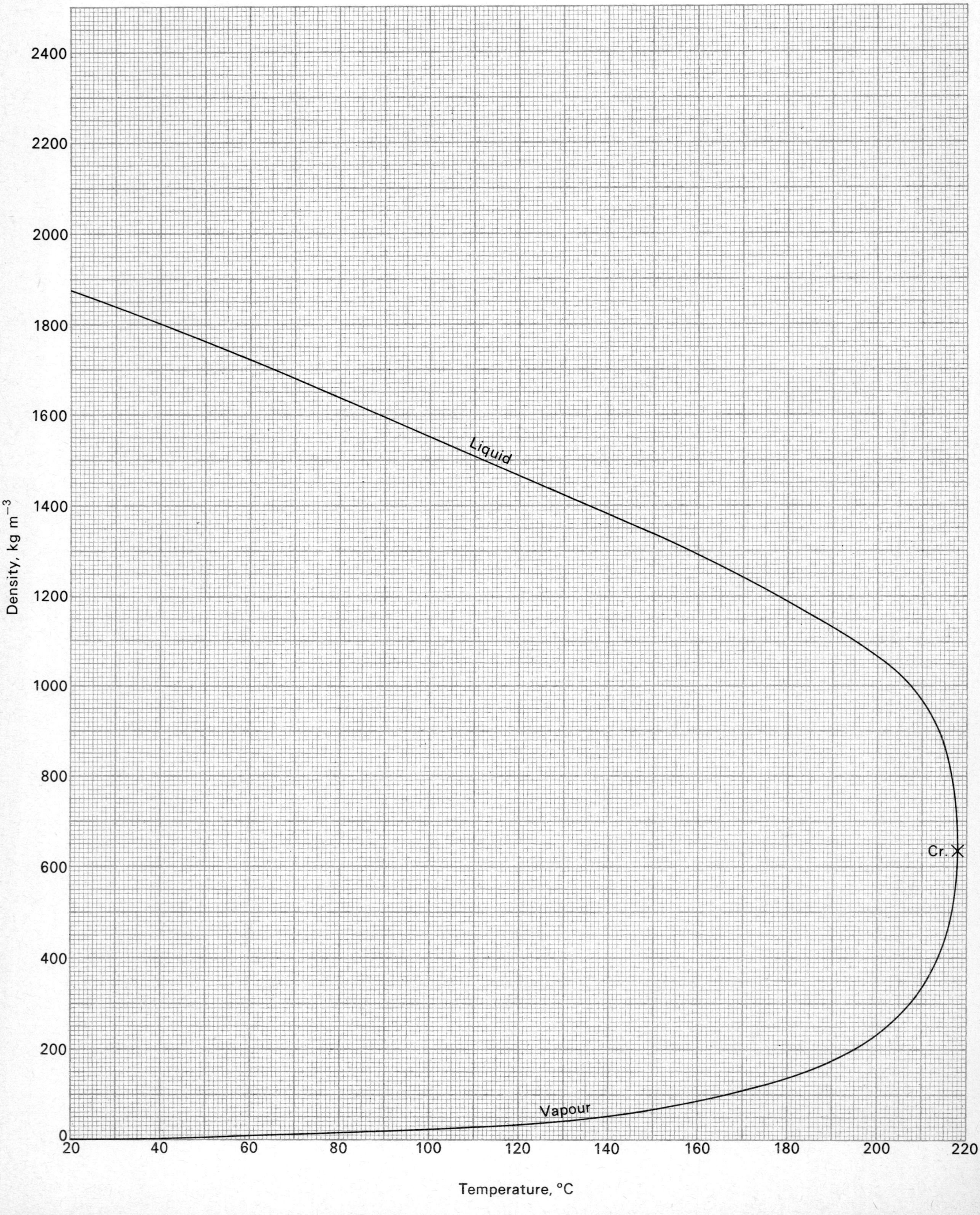

22.3 Latent Heat of Vaporisation of Sulphur Trioxide (11,13,15,33,39)

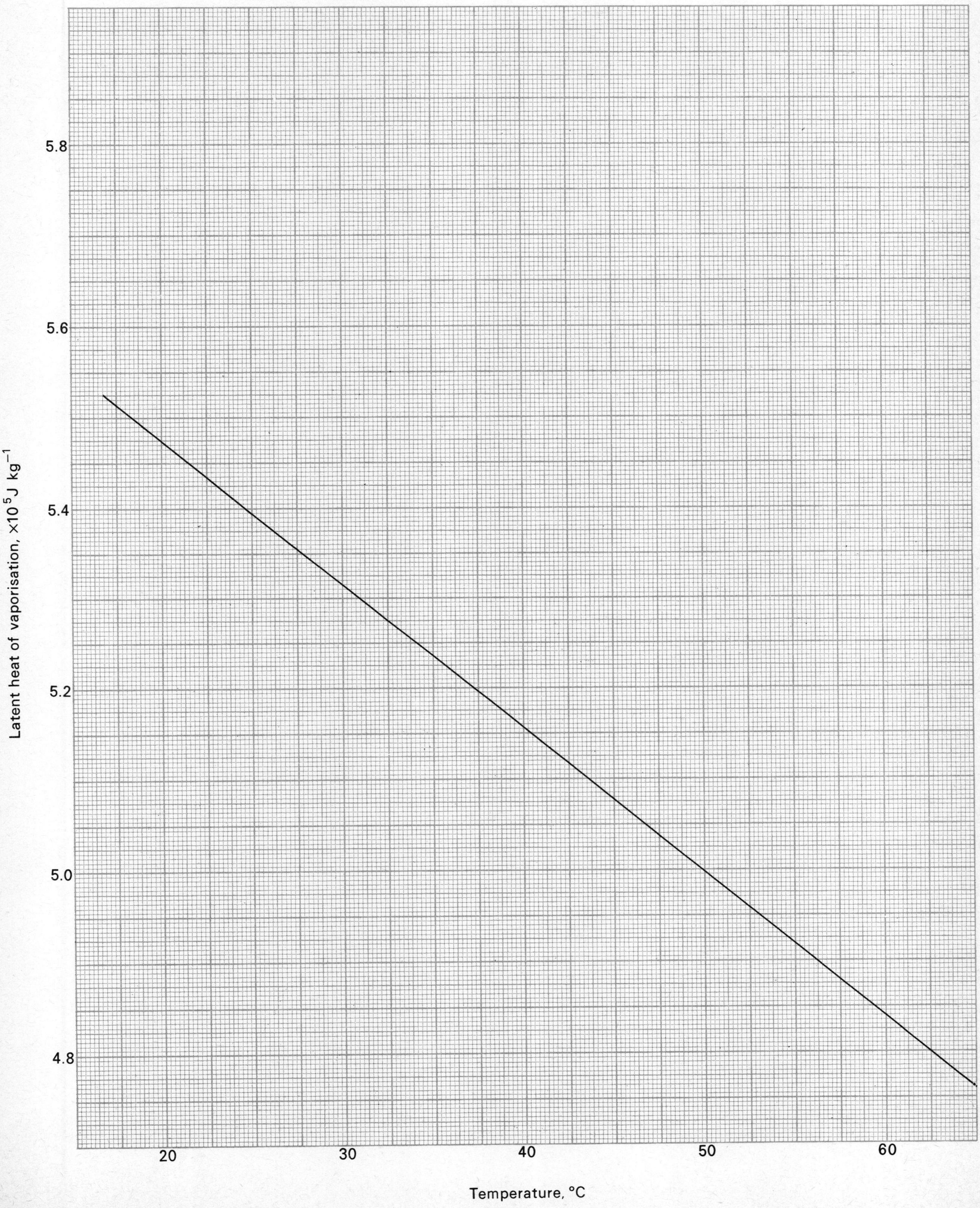

22.4 Heat Capacity of Gaseous Sulphur Trioxide at 1.0133×10^5 N m^{-2} (17,38)

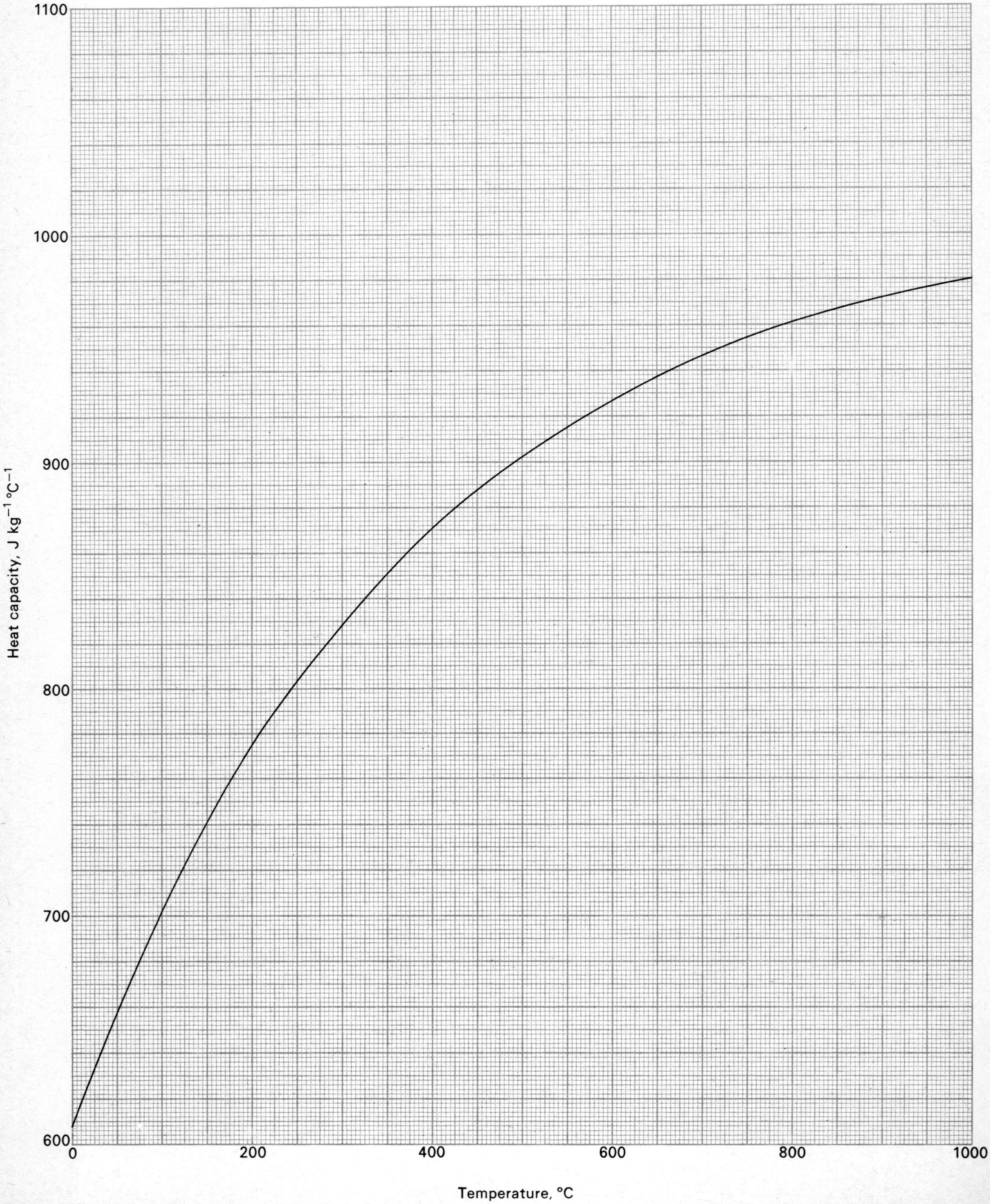

22.5 Viscosity of Liquid Sulphur Trioxide (1,15,21)

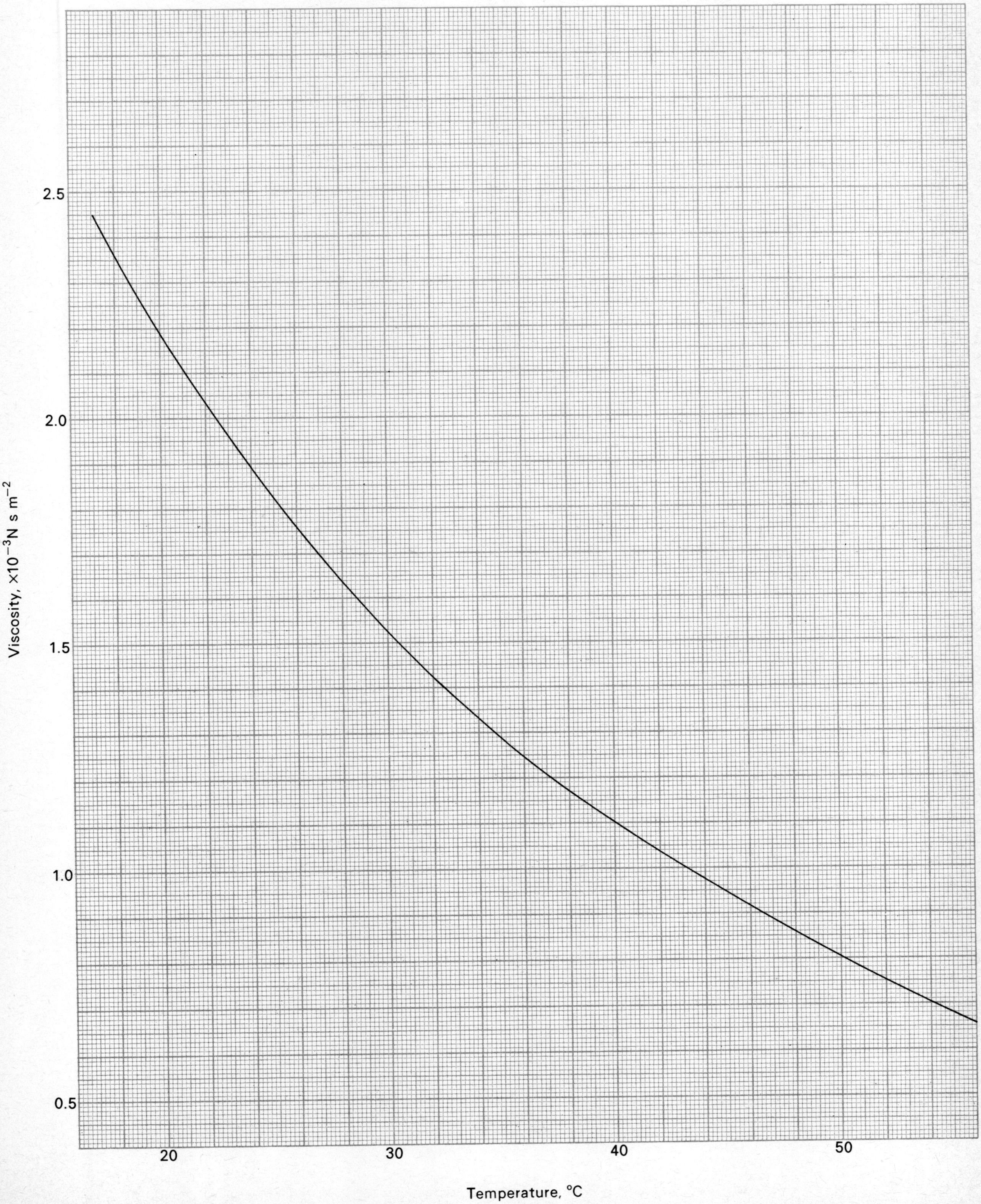

22.6 Viscosity of Gaseous Sulphur Trioxide at 1.0133×10^5 N m^{-2}(12,27,29)

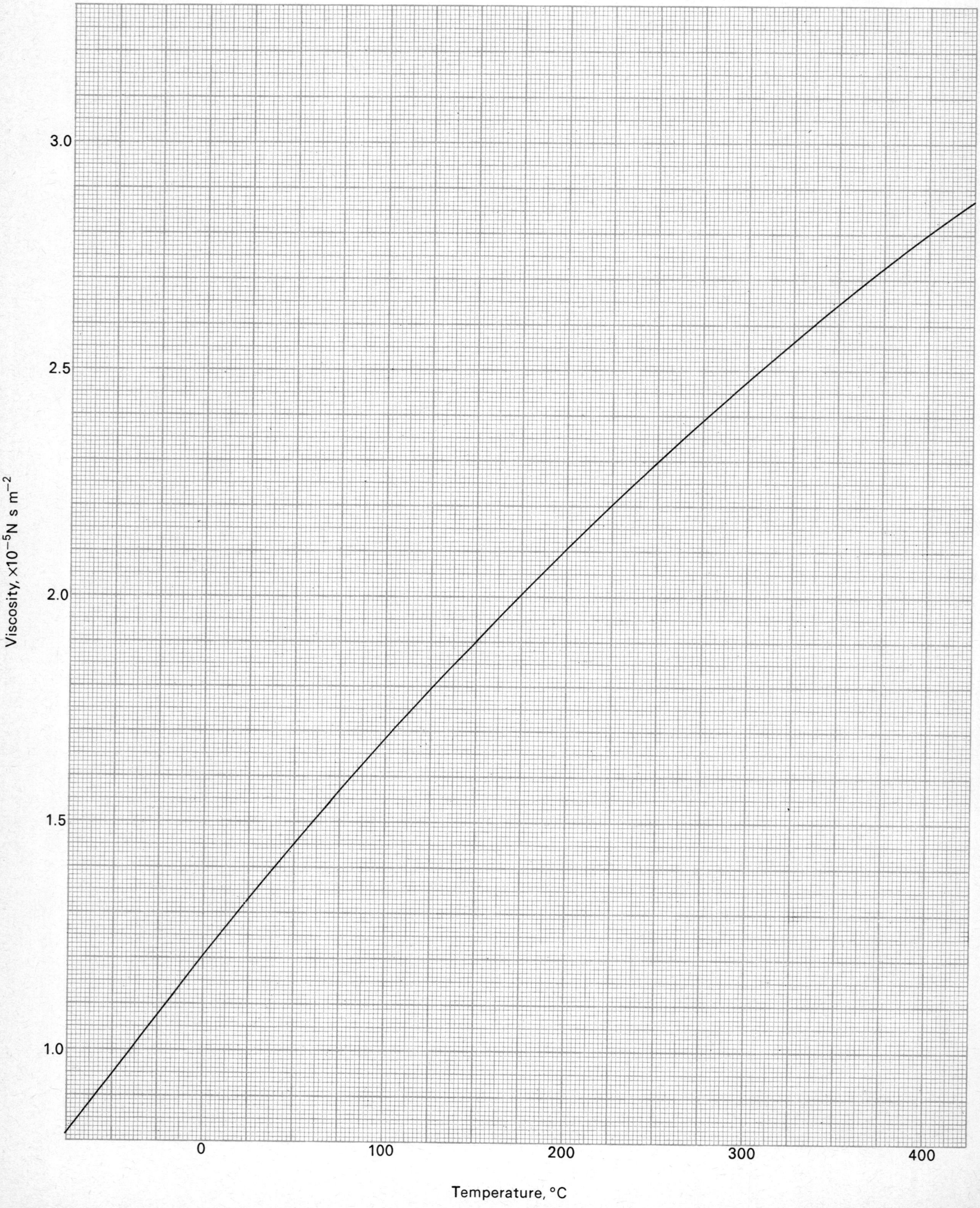

22.7 Thermal Conductivity of Liquid Sulphur Trioxide (35)

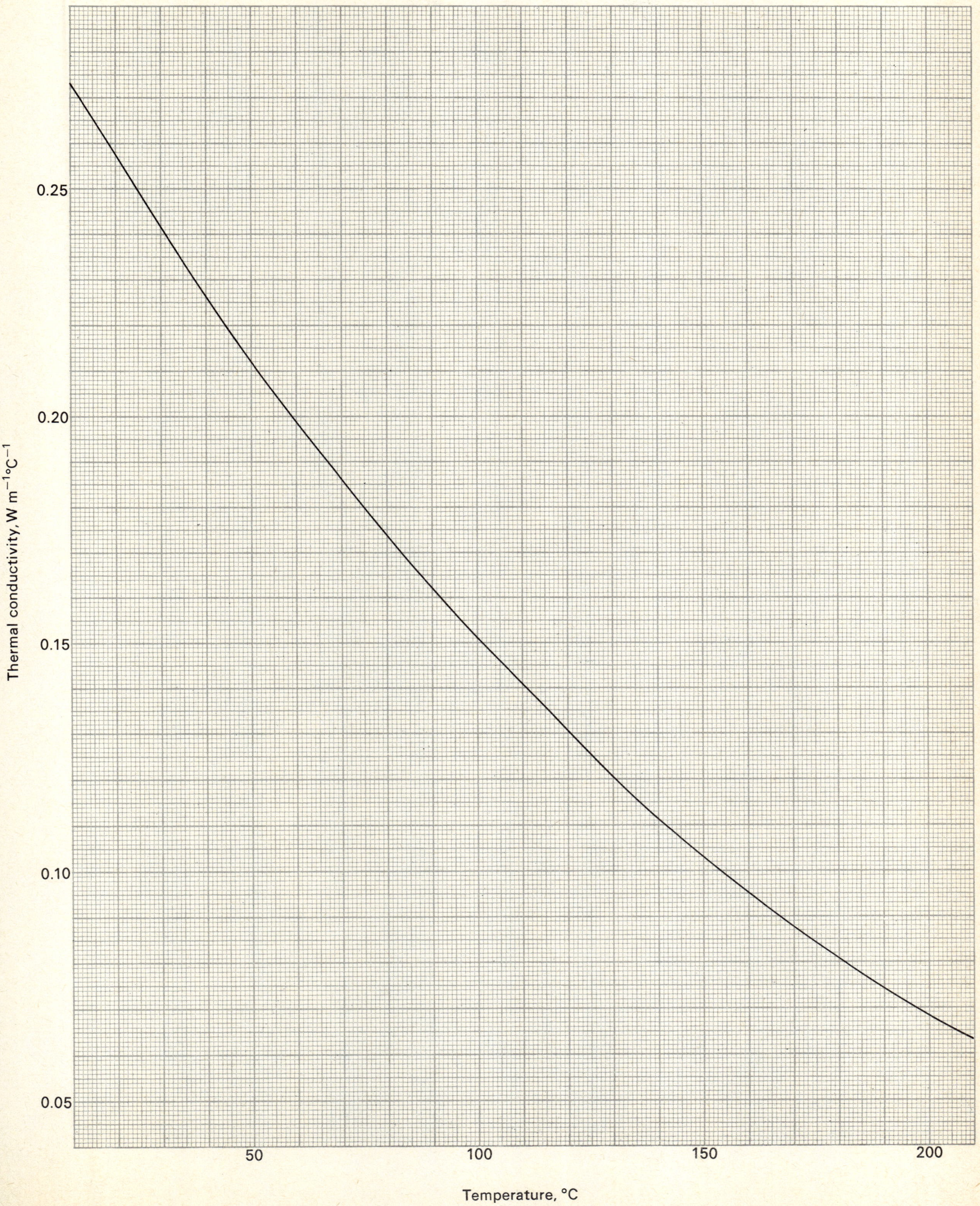

22.8 Thermal Conductivity of Gaseous Sulphur Trioxide at 1.0133×10^5 N m^{-2} (25,28)

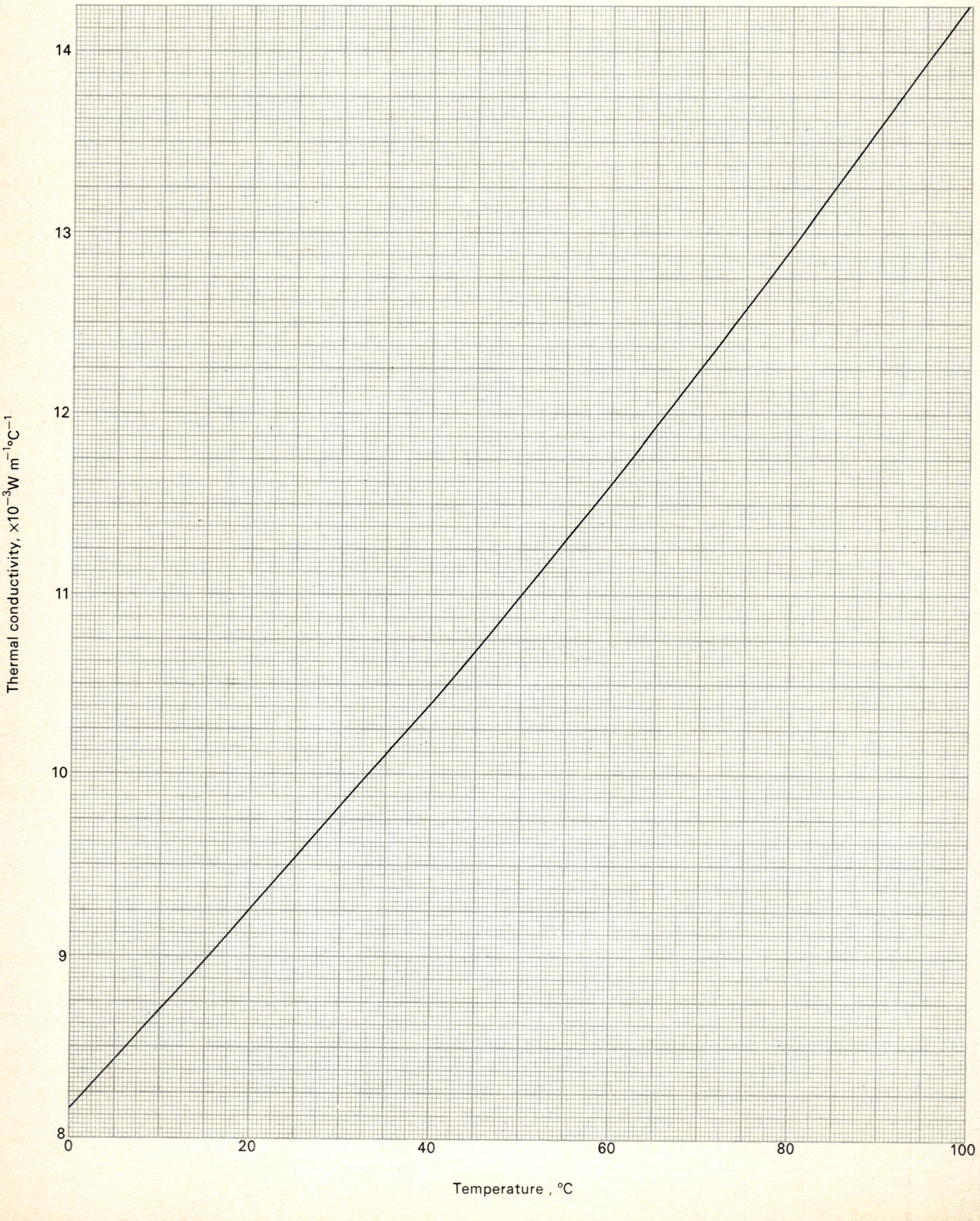

22.9 Surface Tension of Sulphur Trioxide (4,5,7,32)

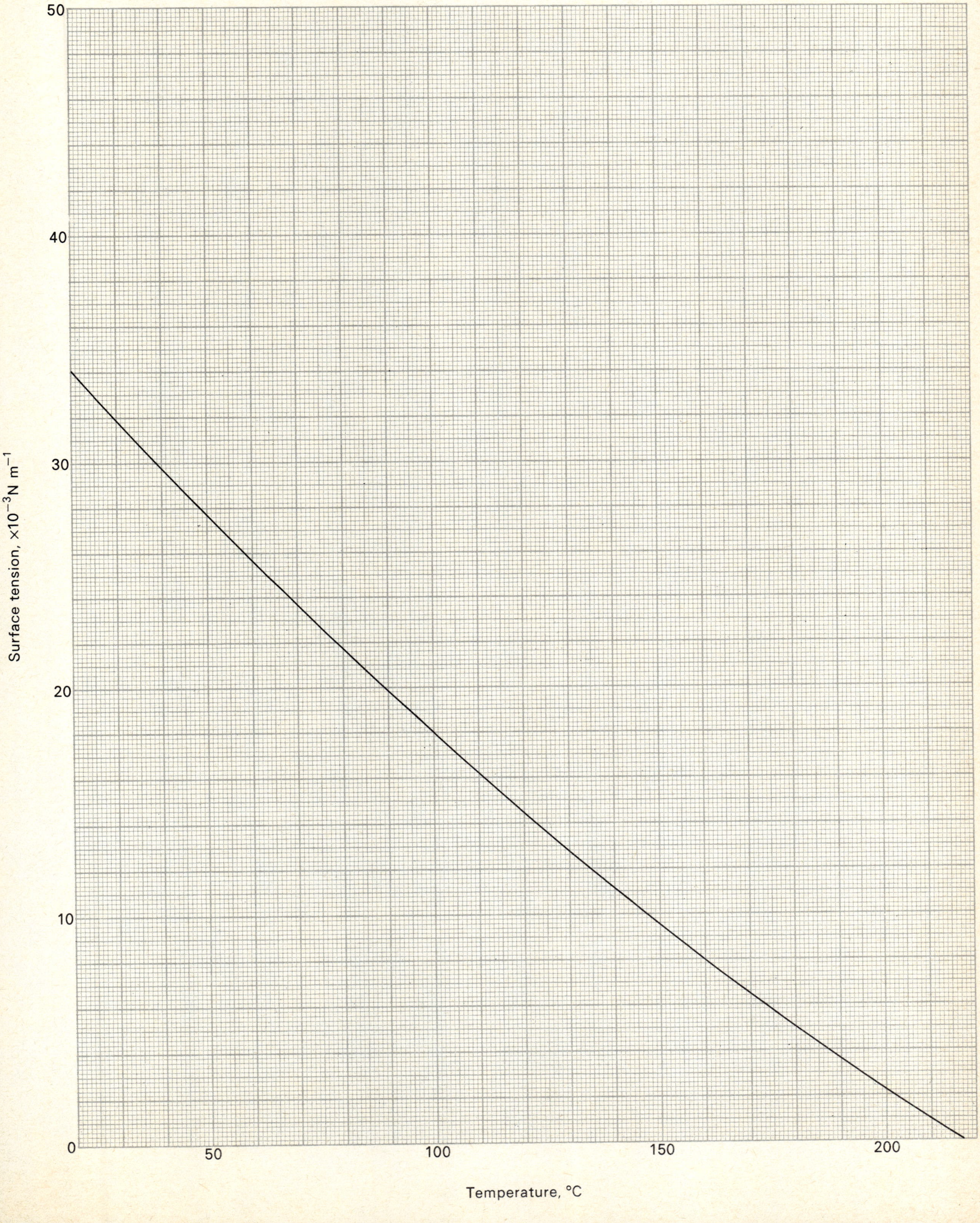

22.10 Phase Diagram for SO_3-H_2O System (10,14)

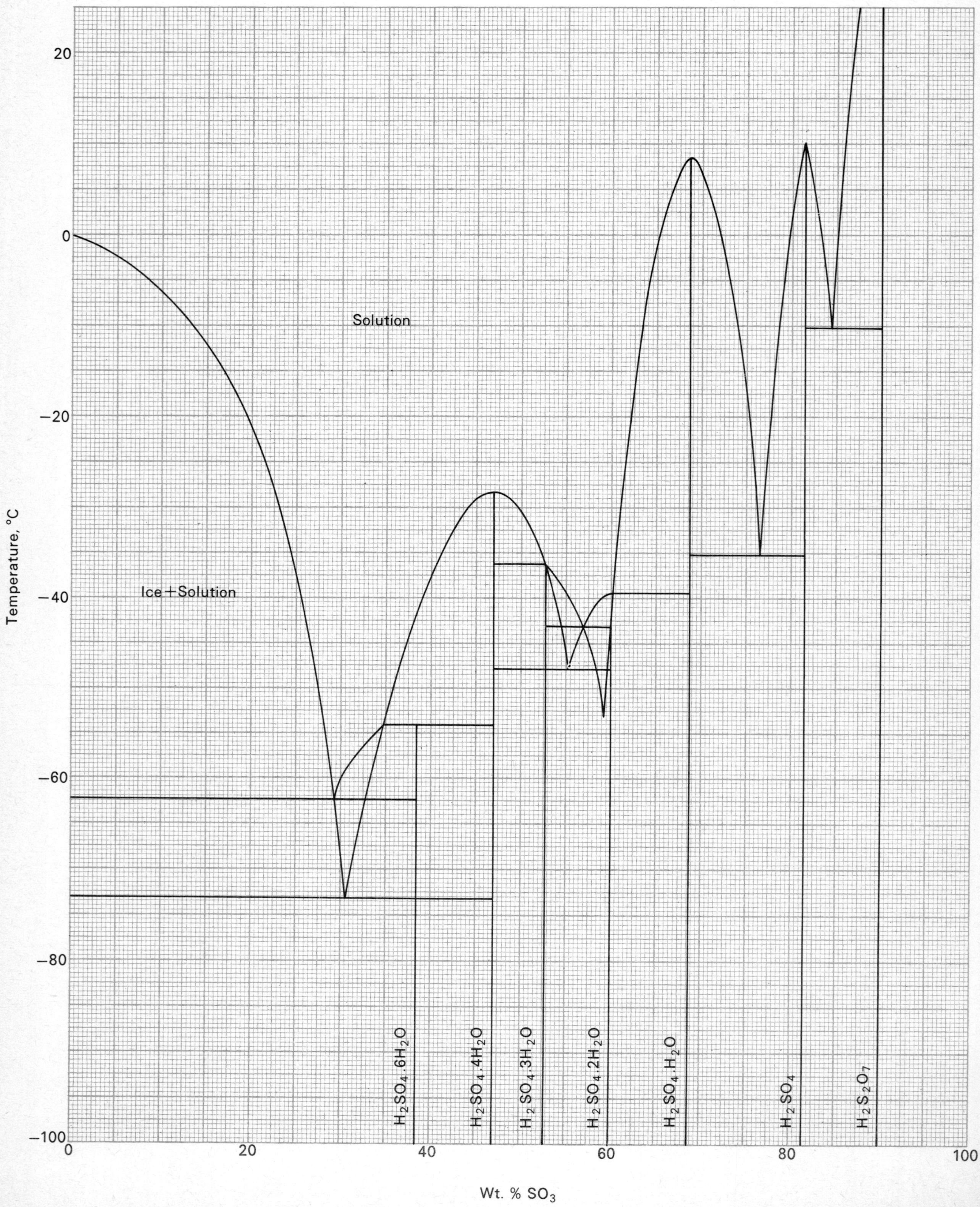

23 SULPHURIC ACID

Sulphuric acid is a colourless liquid of syrupy consistency at room temperature and atmospheric pressure. It is a covalent compound; however, the pure liquid behaves like a fused salt, showing definite electrical conductivity. H_2SO_4 is a strong dibasic acid and a vigorously active chemical compound. On heating the liquid begins to fume, giving off SO_3. It is a good solvent for many compounds.

Liquid H_2SO_4 solidifies at a low temperature. The system SO_3—H_2O forms several solid phases in various proportions:

monohydrate, H_2SO_4
double hydrate, $H_2SO_4.H_2O$
triple hydrate, $H_2SO_4.2H_2O$
quintuple hydrate, $H_2SO_4.4H_2O$

Sulphuric acid forms an azeotrope (constant boiling point solution) with water, consisting of 98.3% by weight of H_2SO_4. This is the strongest acid and can be obtained by distillation. Concentrated H_2SO_4 can be mixed with water in all proportions, with evolution of heat during the mixing and a contraction in volume. The pure acid can be prepared by the addition of SO_3 to concentrated H_2SO_4. It has a strong tendency to take up water.

The vapour of sulphuric acid decomposes into SO_3 and H_2O during heating. The dissociation is complete at 450°C.

Sulphuric acid is a very important industrial substance. It is a strong acid, having oxidizing powers, and is used as a dehydrating agent.

Various properties of H_2SO_4 are compiled in multi-volume handbooks, such as: Gmelin (11), and Pascal (23). There are several publications of interest on H_2SO_4 technology, presenting some selections of its thermophysical data: Bump and Sibbitt (6), Fairle (8), Fasullo (9), MCA (22), and Szubinski (29).

References

1 American Conference of Government Industrial Hygienists, *Documentation of Threshold Limit Values*, revised edition, Committee on Threshold Limit Values, Cincinnati, Ohio 45202, 1966, 203 pp.
2 Baranova, A. I., *J. Appl. Chem. USSR*, 1958, **31**, 161–71.
3 Beckman, E., *Z. phys. Chem.*, 1905, **53**, 129.
4 Biltz, W., Hülsmann, O., *Z. anorg. Chem.*, 1932, **207**, 377–84.
5 Bright, N. F., Hutchison, H., Smith, D., *J. Soc. Chem. Ind.*, 1946, **65**, 385–8.
6 Bump, T. R., Sibbitt, W. L., *Ind. Eng. Chem.*, 1955, **47** (8), 1665–70.
7 Engineering Science Data Units, *Approximate Data on the Viscosity of some Common Liquids*, Item No. 66024, London, July 1966, 26 pp.
8 Fairlie, A. M., *Sulphuric Acid Manufacture*, 4th printing, Reinhold Publ. Corp., New York, 1949, 669 pp.
9 Fasullo, O. T., *Sulphuric Acid: Use and Handling*, McGraw-Hill, New York, 1965.
10 Filippov, L. P., *Vestnik Moskov. Univ.*, **9** (6), *Ser. Fiz.-Mat. i Estestven. Nauk.*, 1954, **4**, 59–61.
11 *Gmelins Handbuch der anorganischen Chemie*, Syst. No. 9, Schwefel, Teil B, Lief. 2, Verlag Chemie GmbH, Weinheim, 1960, pp. 613–50.
12 Gmitro, J. I., Vermeulen, T., *Vapour-Liquid Equilibria for Aqueous Sulphuric Acid*, Univ. of California, Lawrence Rad. Lab. UCRL-10886, Berkeley, California, June 24, 1963, 81 pp.
13 Greenwood, N. N., Thompson, A., *J. Chem. Soc.*, 1959, 3485–92.
14 Hantzsch, A., Dürigen, F., *Z. phys. Chem.*, 1928, **136**, 1–17.
15 Hetherington, G., Hub, D. R., Nichols, M. J., Robinson, P. L., *J. Chem. Soc.*, 1955, 3300–1.
16 Hülsmann, O., *Z. anorg. Chem.*, 1934, **218**, 369–78
17 IUPAC, Atomic Weights of the Elements 1971, *Pure & Appl. Chem.*, 1972, **30** (3–4), 638–49.
18 Kobeko, P. P., Kuvsinskij, E. V., Siskin, N.I., *Acta Phys. Chem. USSR*, 1937, **6**, 954.
19 Kohlraush, W., *Wied. Ann.*, 1882, **17**, 69–85.
20 Lange, N. A., Ed., *Handbook of Chemistry*, 10th ed., McGraw-Hill, New York, 1967.
21 Lax, E., Synowietz, C., Eds., *Taschenbuch für Chemiker und Physiker*, 3rd ed., Springer-Verlag, Berlin, 1967.
22 Manufacturing Chemists Association, *Sulphuric Acid*, Chemical Safety Data Sheet SD-20, Washington, 1963.
23 Pascal, P., Ed., *Nouveau Traité de Chimie Minérale*, Vol. XIII, Part 2, Oxygene, Ozone, Oxydes, Eau Oxygénée, La combustion, Soufre, Selenium, Tellure, Polonium, Masson et Cie, Éditeurs, Paris, 1960, pp. 1322–67.
24 Pickering, S. U., *J. Chem. Soc.*, 1890, **57**, 64–184.
25 Rossini, F. D., Wagman, D. D., Evans, W. H., Levine, S., Jaffe, I., *Selected Values of Chemical Thermodynamic Properties*, Natl. Bur. Std. Circular 500, Washington, Feb. 1, 1952.
26 Rubin, T. R., Giauque, W. F., *J. Am. Chem. Soc.*, 1952, **74**, 800–4.
27 Sabinina, L., Terpugow, L., *Z. phys. Chem.*, 1935, **173A**, 237–41.
28 Stull, D. R., Dir., *JANAF Thermochemical Tables*, Dow Chemical Co., Midland, Mich., 1965–8.
29 Szubinski, H.-W., *Die Schwefelsäurefabrikation*, Friedr. Vieweg. & Sohn, Braunschweig, 1961, 488 pp.
30 Tartakowskaja, W., Bondarenko, J., Jameljanowa, L., *Acta Phys. Khim. USSR*, 1937, **6**, 609–24.
31 Thomas, J. S., Barker, W. F., *J. Chem. Soc.*, 1925, **127**, 2820–31.
32 Thomas, J. S., Ramsay, A. G., *J. Chem. Soc.*, 1923, **123** (II), 3256–70.
33 Usanovic, M., Sumarokova, T., Udovenko, V., *Zh. Obsh. Khim.*, 1939, **9**, 1967–75.
34 Usanovic, M., Sumarokova, T., Udovenko, V., *Acta Phys. Khim. USSR*, 1939, **11**, 505–20.
35 Walden, P., *Z. phys. Chem.*, 1909, **65**, 129–225.
36 Zeisberg, F. C., *Chem. Metallurg. Eng.*, 1922, **27** (1), 22–6.

Table 23.1 Properties of Sulphuric Acid

		Ref.
Chemical formula	H_2SO_4	(21)
Molecular weight	98.0718	(17)
Triple-point temperature	None	(11)
Triple-point pressure	None	(11)
Boiling point at 1.0133×10^5 N m^{-2}	*c.* 309 °C	(31)
Melting point	10.37 °C	(26)
Transition temperature	None	(11)
Density of solid at −191 °C	2144.0 kg m^{-3}	(4)
Coefficient of thermal expansion of liquid at 18 °C	5.7×10^{-4} °C^{-1}	(19)
Heat of formation of liquid at 25 °C	-8.1186×10^8 J kg-mol^{-1}	(28)
Free energy of formation of liquid at 25 °C	-6.9011×10^8 J kg-mol^{-1}	(28)
Heat of fusion	9.8808×10^6 J kg-mol^{-1}	(25)
Heat of transition	None	(11)
Entropy of liquid at 25 °C	1.5696×10^5 J kg-mol^{-1} °C^{-1}	(28)
Dielectric constant of solid at −113.15 °C	54.0	(18)
Electrical conductivity of liquid at 25 °C	1.045 ohm^{-1} m^{-1}	(15)
Refractive index of liquid at 20 °C	1.41828	(14)
Threshold limit value	1.0×10^{-6} kg m^{-3}	(1)

23.1 Vapour Pressure of Sulphuric Acid (2,8,9,11,12,23,29,32)

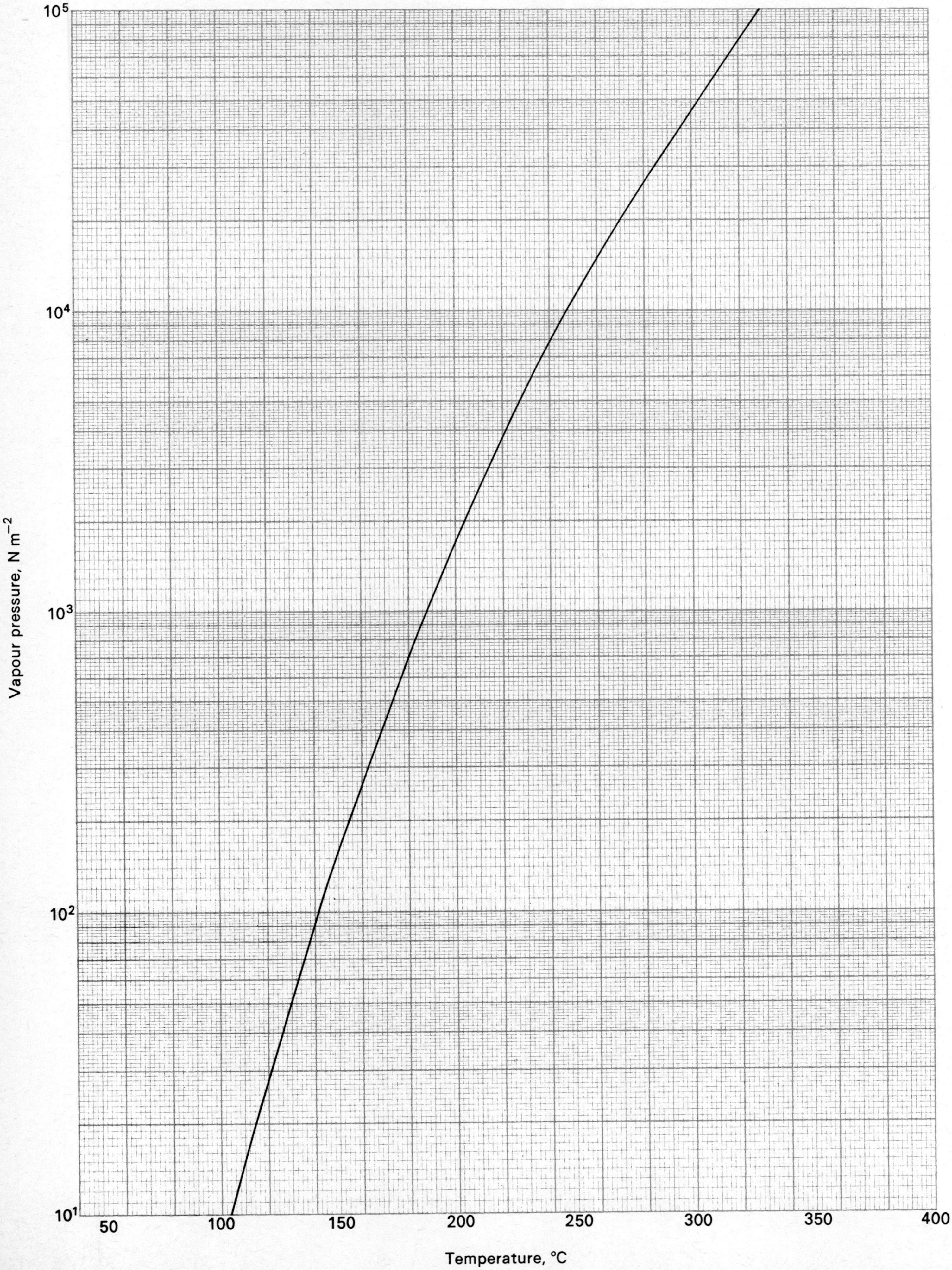

23.2 Liquid Density of Sulphuric Acid (5,13,21,24)

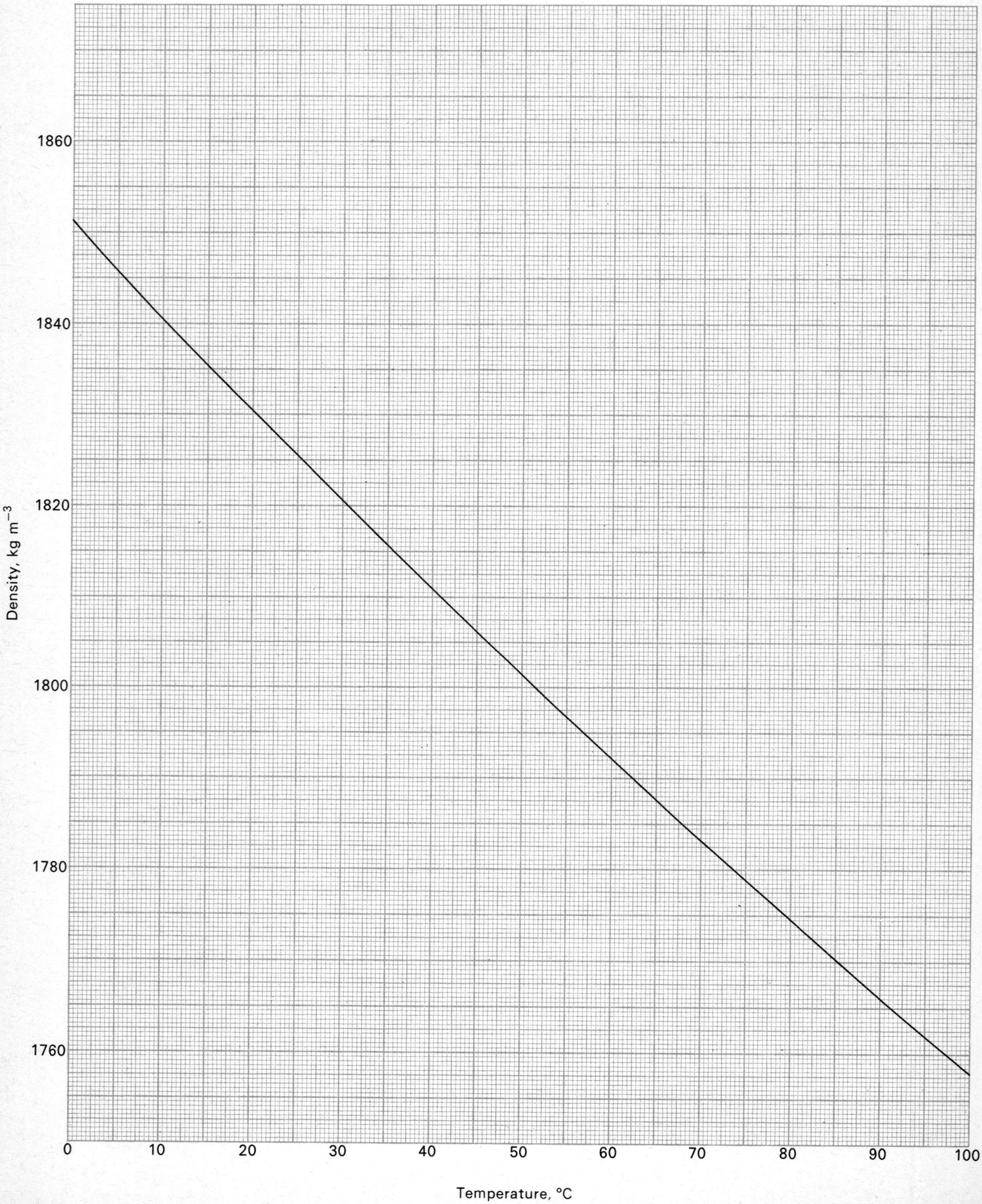

23.3 Latent Heat of Vaporisation of Sulphuric Acid (2,3,20,32,36)

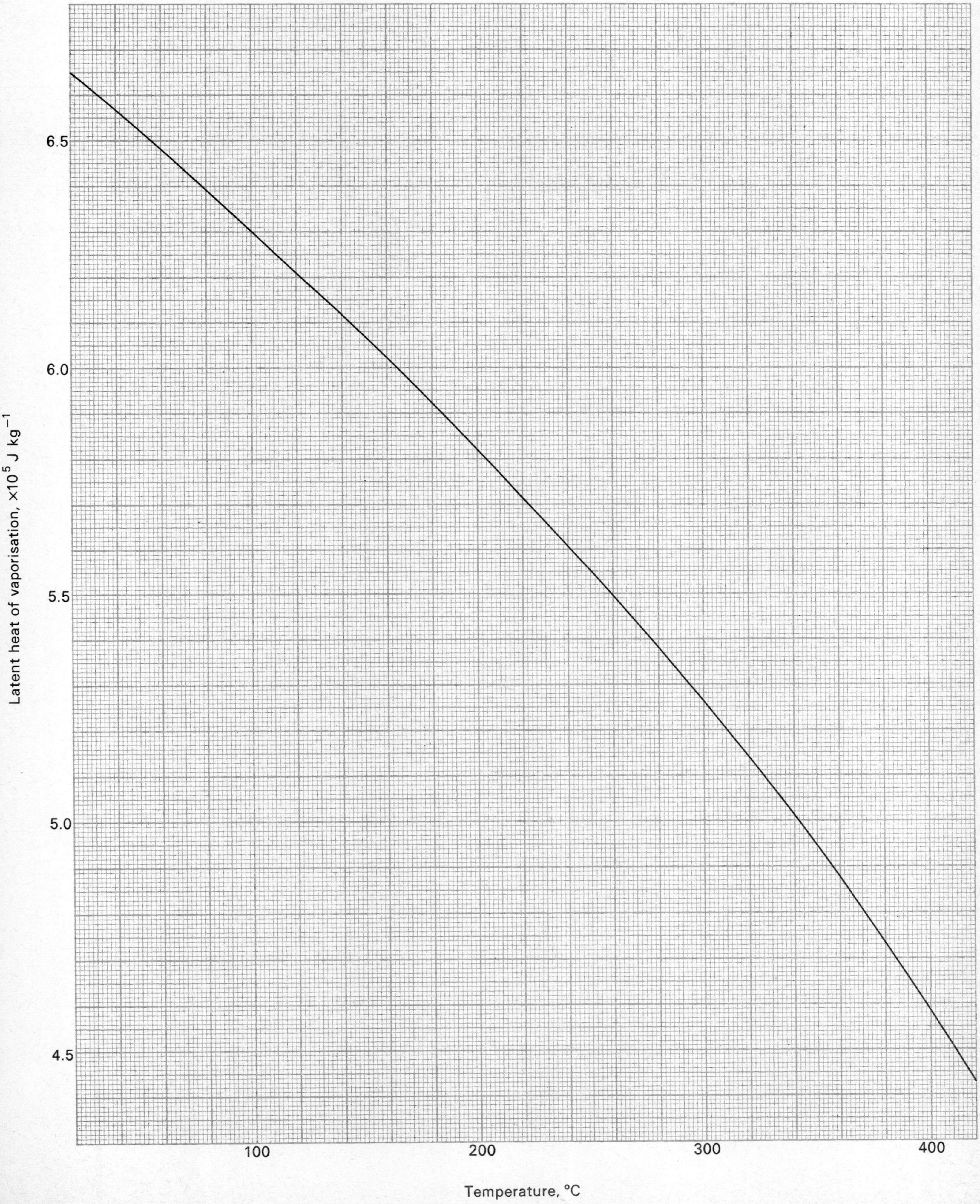

23.4 Heat Capacity of Liquid Sulphuric Acid (11,23,28)

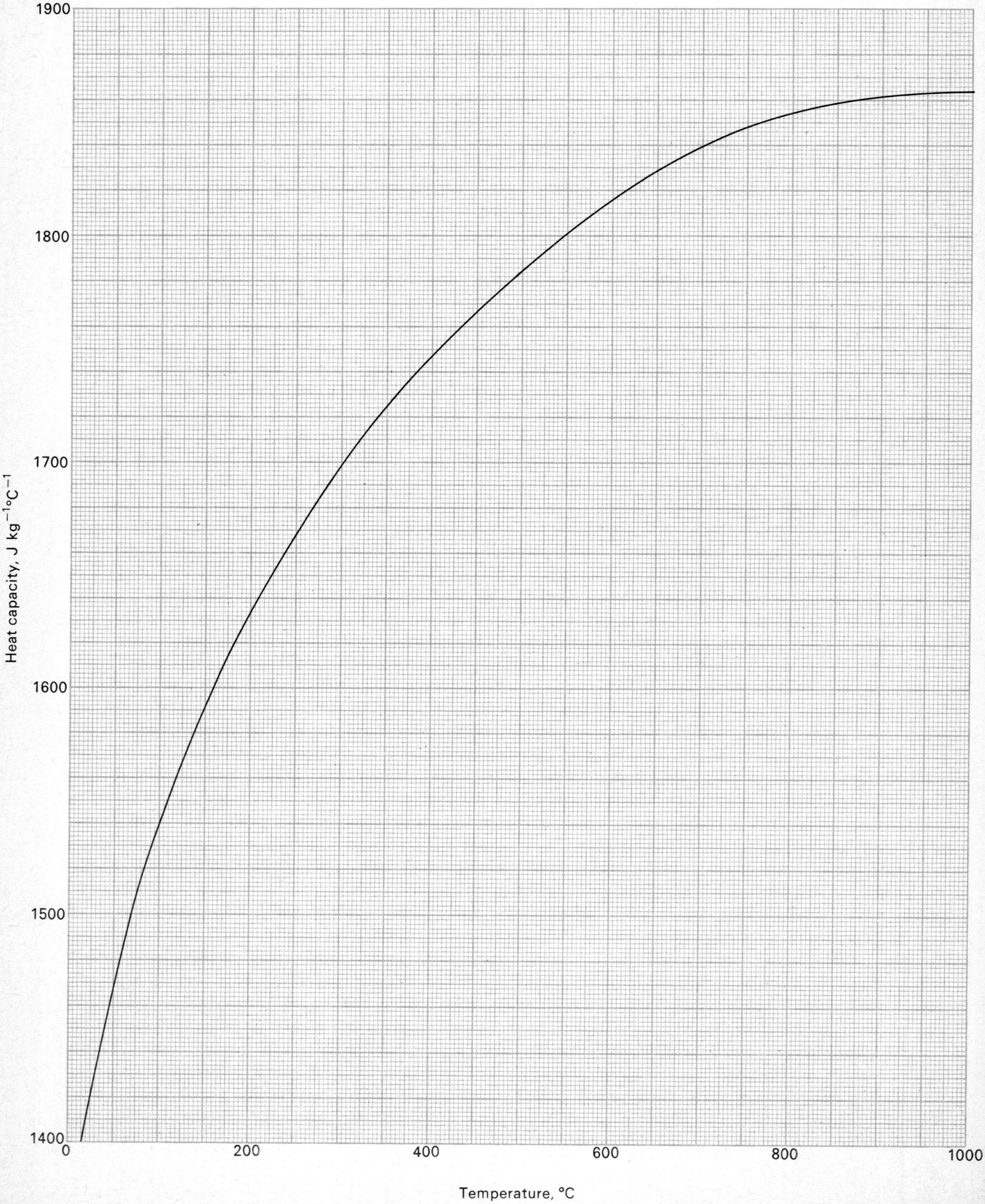

23.5 Viscosity of Liquid Sulphuric Acid (5,7,13,30,33,34)

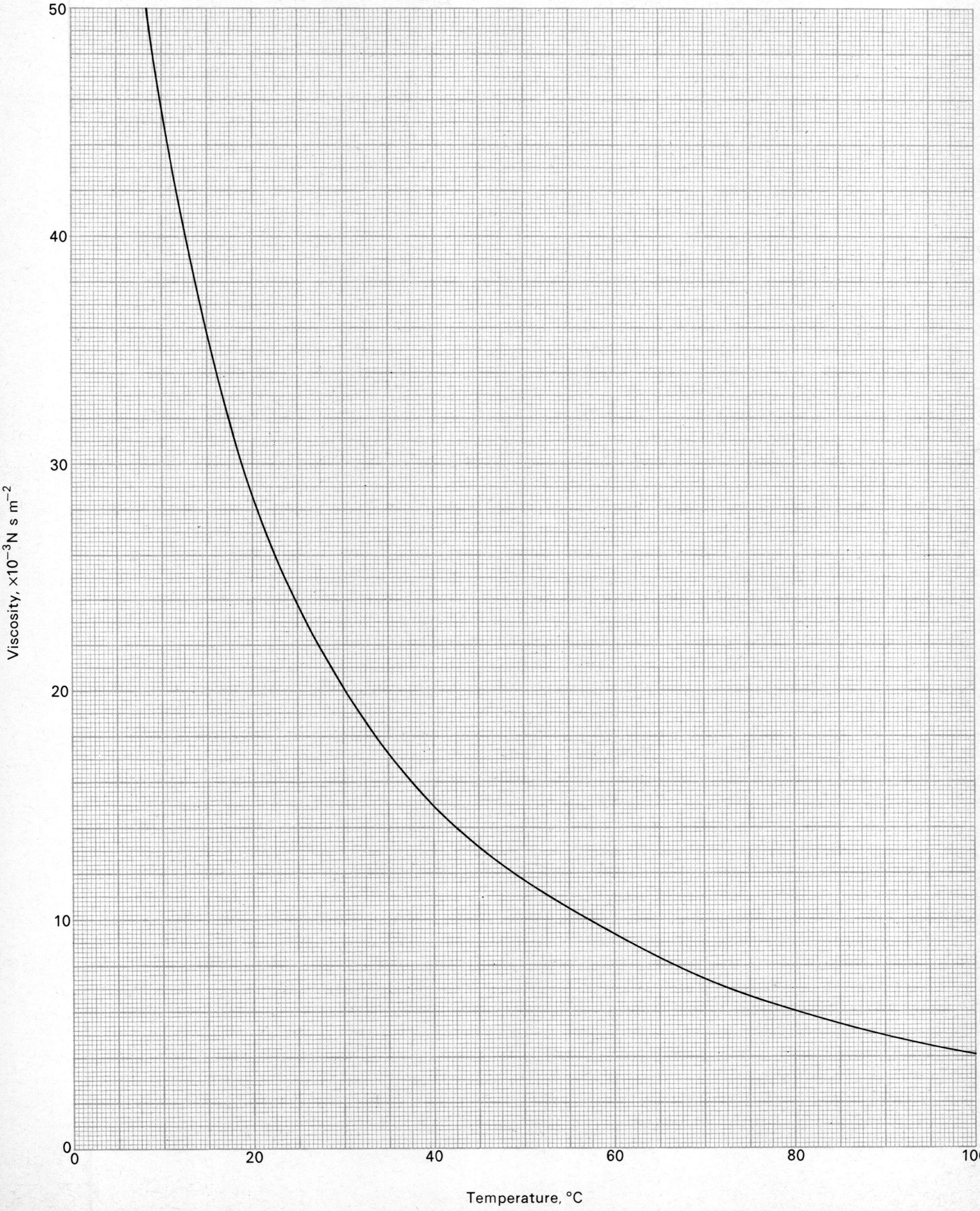

23.6 Thermal Conductivity of Liquid Sulphuric Acid (10)

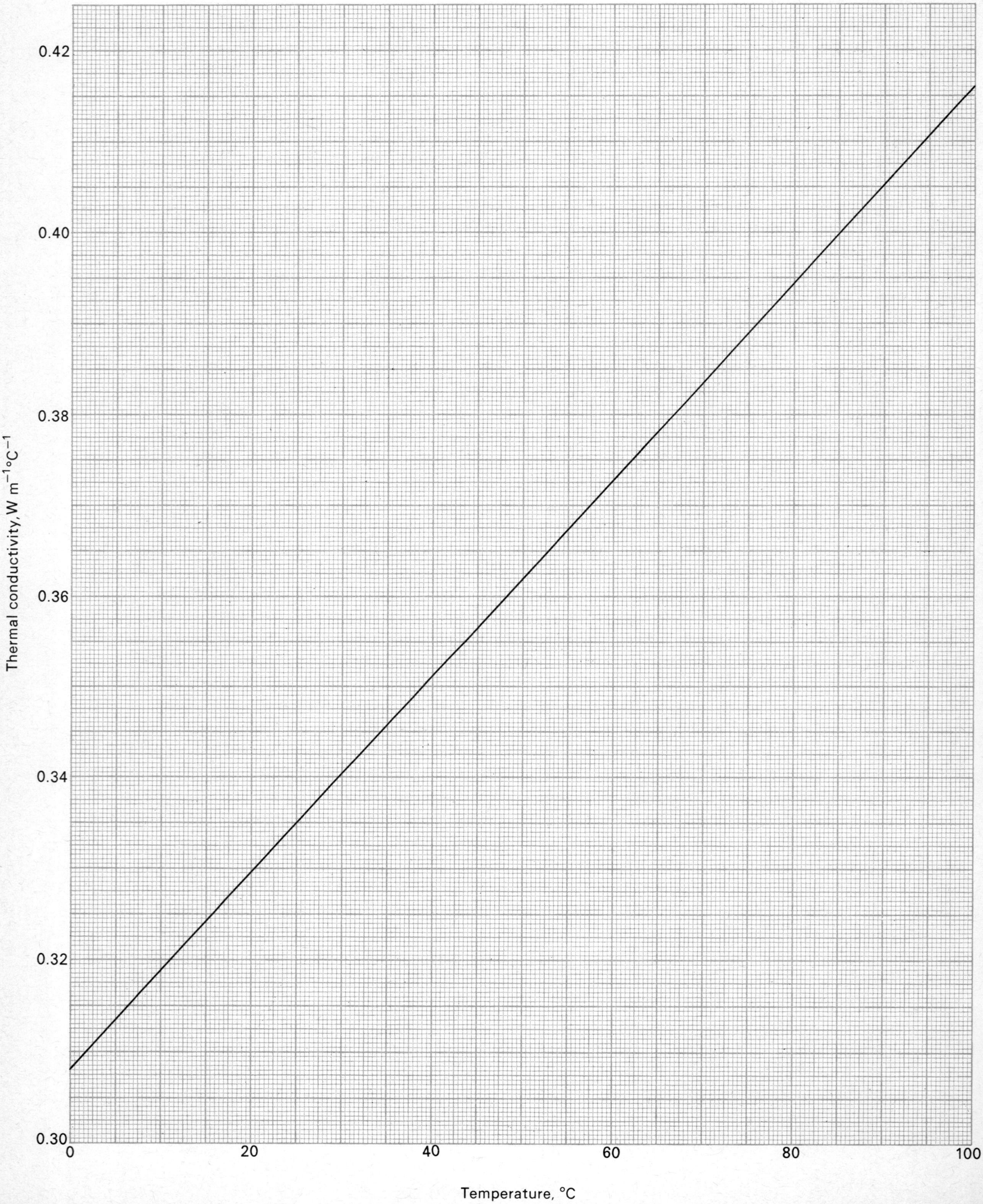

23.7 Surface Tension of Sulphuric Acid (27,33,34,35)

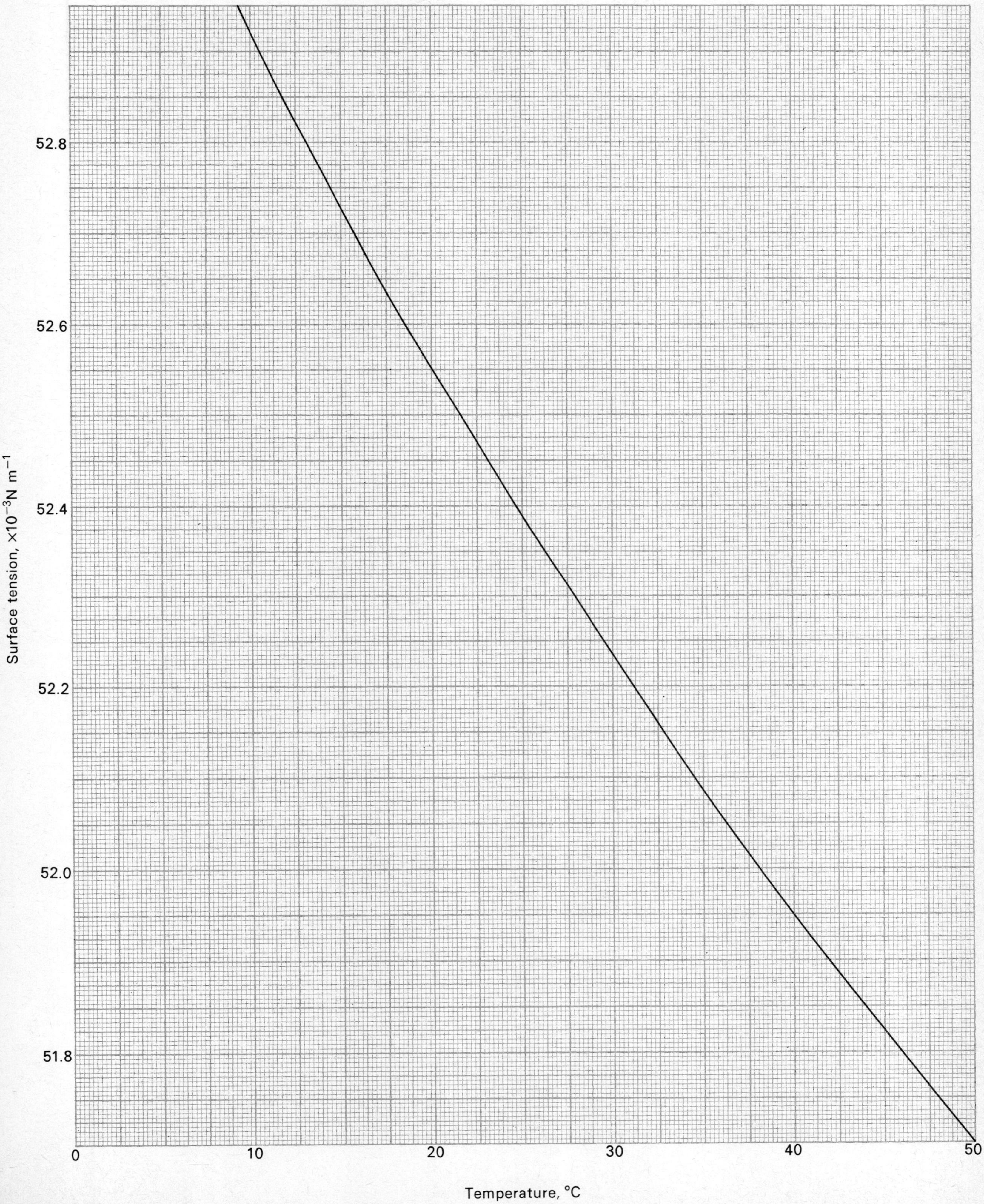

23.8 Electrical Conductivity of Liquid Sulphuric Acid (13)

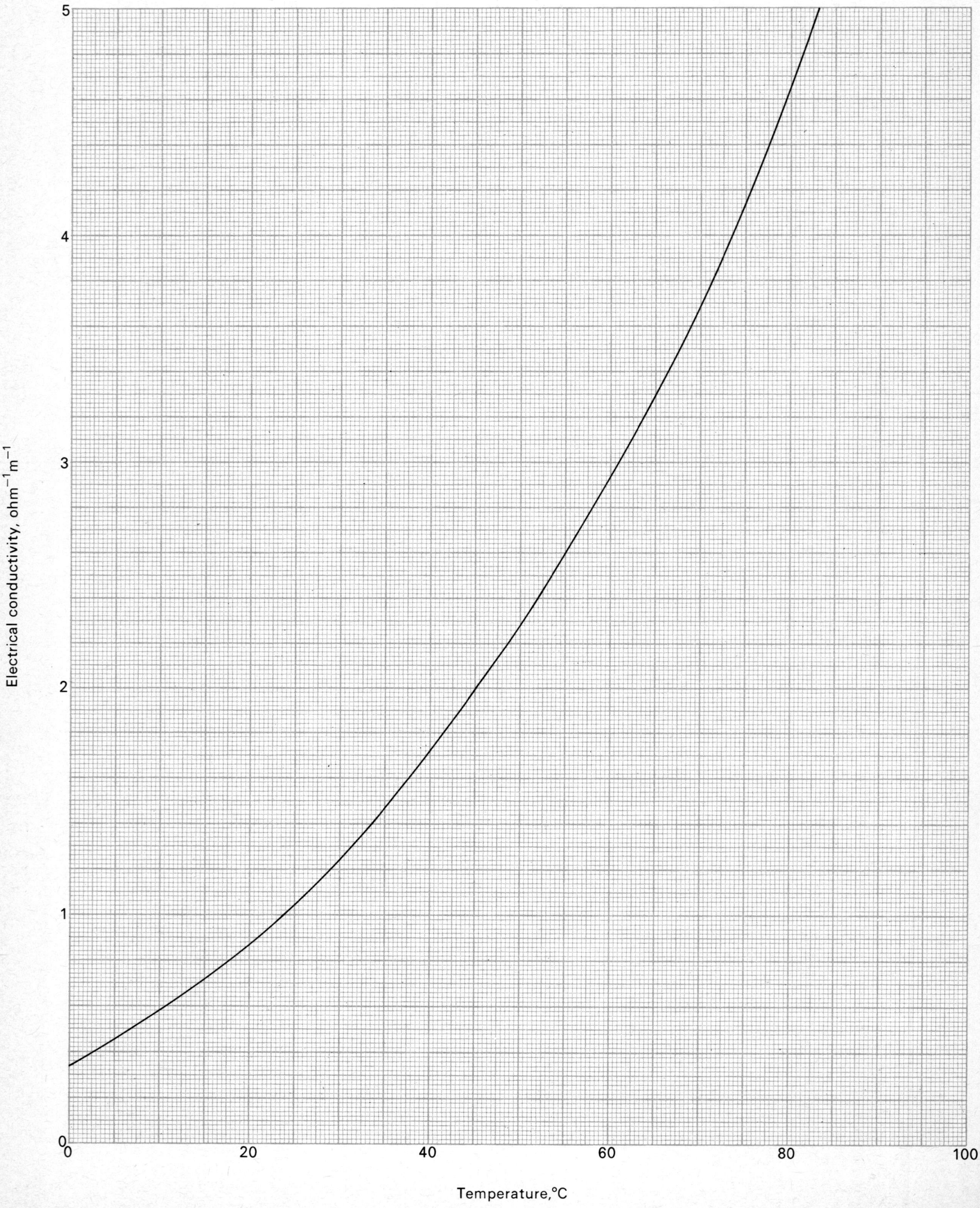

23.9 Phase Diagram for H_2SO_4-H_2O System (16)

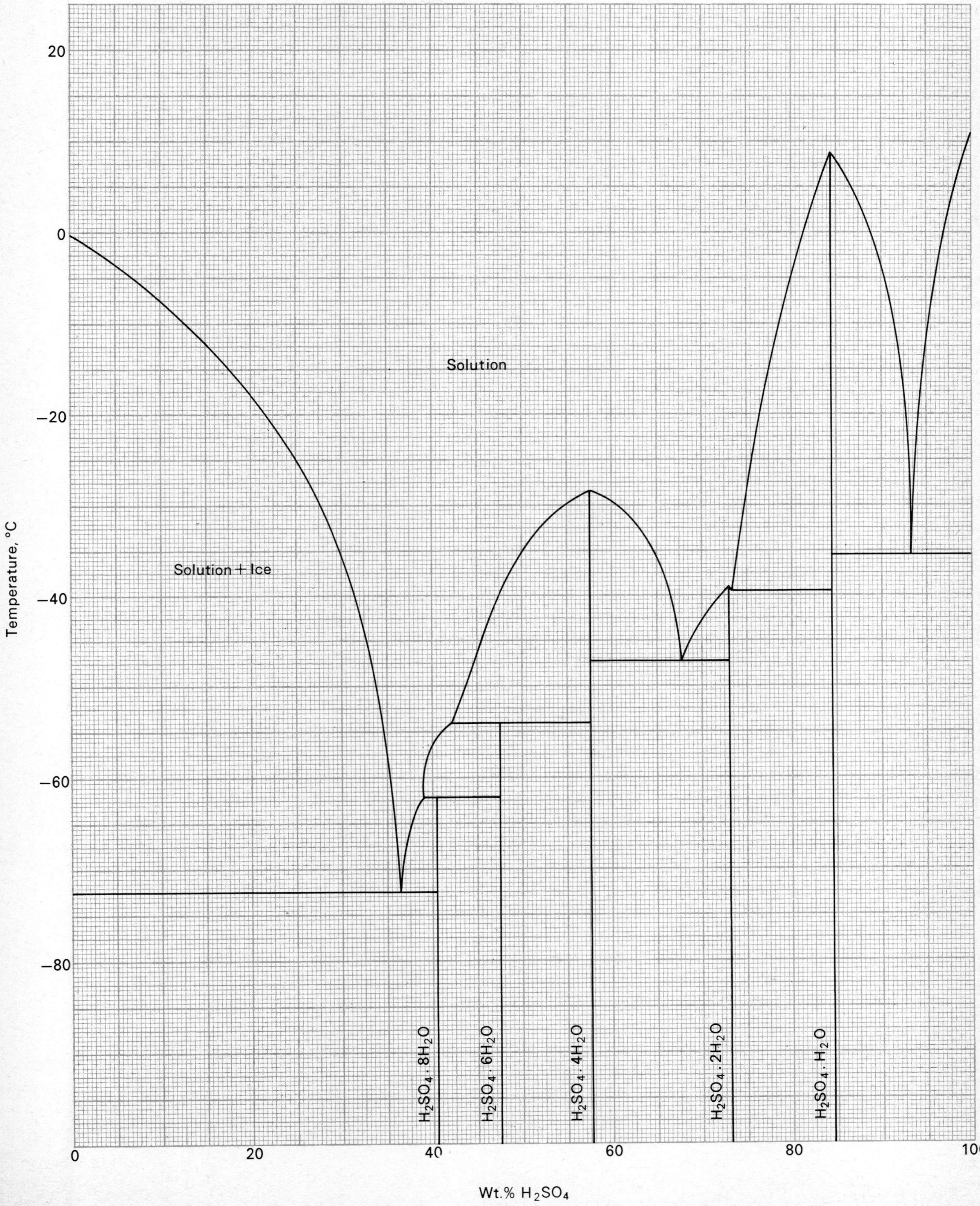

24 WATER

Water is the most intensively studied of all known chemical substances. Many of its physical properties have been adopted as the basis of the international standards, e.g. triple point, boiling point, density, viscosity, etc. Water is one of the unusual liquids, whose density is greater in the liquid (at 4°C) than in the solid phase (ice). The reason is that the molecules are looser packed in the ice than in the liquid.

Ice forms several allotropic modifications at high pressures. In ordinary ice, four oxygen atoms surround each oxygen atom, and the hydrogen atom is in line between two oxygen atoms (hydrogen bond).

Liquid water is highly polar which is indicated by its high dielectric constant. Its physical properties give the impression that water has physical and chemical anomalies (polywater).

Although the study of the properties of water has been very extensive, there are still numerous mysteries (14, 33). There are several good works on the physical, chemical, thermodynamic, and transport properties of water and steam, e.g. (6, 8, 9, 20, 24, 25, 26, 38, 39, 45, 47, 49, 50). Schmidt (39) includes a very comprehensive list of references, covering all relevant sources of physical, thermodynamic, and transport properties of water and steam. A recently published book (14) deals exclusively with its physics and physical chemistry, and is highly recommended for scientists dealing with the properties and behaviour of water and steam.

Physical, thermodynamic, and transport properties of water are in good agreement in various sources and therefore do not require a rigorous selection. The data in the following table and figures have been collected from the most reliable sources in the world literature.

References

1 Blank, G., *Wärme-Stoffübertragung*, 1969, **2**, 53.
2 Brain, T. J. S., *The Thermal Conductivity of Steam at Atmospheric Pressure*, Univ. of Glasgow, Mech. Eng. Dept., Techn. Report No. 26, Glasgow, 1968.
3 Brewer, A. D., Hutchins, J. E. C., *Nature*, 1966, **210** (5042), 1257–8.
4 Bruges, E. A., *Transport Properties of Water and Steam—Thermal Conductivity*, Univ. of Glasgow, Mech. Eng. Dept., Techn. Report No. 9(B), Glasgow, 1963.
5 Bruges, E. A., Gibson, M. R., *The Dynamic Viscosity of Compressed Water to 10 kbar and Steam to 1500°C*, Univ. of Glasgow, Mech. Eng. Dept., Techn. Report No. 25, Glasgow, 1968.
6 Burnham, C. W., Holloway, J. R., Davis, N. F., *Thermodynamic Properties of Water to 1000°C and 10 000 bars*, Pennsylvania State Univ., Dept. of Geochem. and Mineralogy, College of Earth and Mineral Sciences, University Park, Pennsylvania, 1968.
7 Cappi, J., *The Viscosity of Water at High Pressure*, Univ. of London, Doctoral Dissertation, London, 1964.
8 Dorsey, N. E., *Properties of Ordinary Water Substance*, Hafner Publ. Co., New York, 1968 (Reprint).
9 Eisenberg, D., Kauzmann, W., *The Structure and Properties of Water*, Oxford University Press, Oxford, 1969.
10 Electrical Research Association, *1967 Steam Tables*, Edward Arnold, London, 1967.
11 Engineering Science Data Units, *Approximate Data on the Viscosity of Some Common Liquids*, Item No. 66024, London, July 1966, 26 pp.
12 Engineering Science Data Units, *Specific Heats and Their Ratio as Functions of Temperature for Several Common Gases*, Royal Aeronautical Society, Data Sheet 00.01.08, London, 1965, 4 pp.
13 Fano, L., Hubbell, J. H., Beckett, C. W., NACA, TN 3273 (AD 103415), 1956, 1–61.
14 Franks, F., Ed., *Water. A Comprehensive Treatise*, Vol. 1, The Physics and Physical Chemistry of Water, Plenum Press Corp., New York, 1972, 596 pp.
15 Ginnings, D. C., Furukawa, G. T., *J. Am. Chem. Soc.*, 1953, **75**, 522–7.
16 Golubev, I. F., *Viscosity of Gases and Gas Mixtures*, Gosudarstvennoe Izdatel'stvo Fiziko-Matematicheskoi Literatury, Moskva, 1959, Table 4.
17 Grigull, U., Bach, J., *Brennstoff-Wärme-Kraft*, 1966, **18** (2), 73–5.
18 Hacker, P. T., *Experimental Values of the Surface Tension of Supercooled Water*, National Advisory Committee for Aeronautics, NACA TN 2510, Washington, 1951.
19 Heck, R. C. H., *Mech. Eng.*, 1941, **63**, 126–35.
20 Hilsenrath, J., Beckett, C. W., Benedict, W. S., Fano, L., Hoge, H. J., Masi, J. F., Nuttal, R. L., Touloukian, Y. S., Woolley, H. W., Natl. Bur. Std., Circular 564, Washington, 1955, 1–488.
21 Holzapfel, W. B., *J. Chem. Phys.*, 1969, **50** (10), 4424–8.
22 IUPAC, Atomic Weights of the Elements 1971, *Pure & Appl. Chem.*, 1972, **30** (3–4), 638–49.
23 Jones, J., Taylor, E. F., Vogel, R. C., *J. Am. Chem. Soc.*, 1948, **70**, 966–77.
24 Juza, J., Kmonicek, V., Sifner, O., *An Equation of State for Water and Steam*, Academia, Nakledatelstvi Ceskoslovenske Akademie, VED, Praha, 1966.
25 Keenan, J. H., Keyes, F. G., Hill, P. G., Moore, J. G., *Steam Tables*, Wiley, New York, 1969.
26 Keyes, F. G., 5th Internat. Conf. Properties of Steam, London, 1956.
27 Keyes, F. G., Smith, L. B., Gerry, H. T., *Mech. Eng.*, 1934, **56**, 87–92.
28 Koch, B., Fritz, W., *Wärme und Kältetech.*, 1940, **42**, 113–17.
29 Lange, N. A., Ed., *Handbook of Chemistry*, 10th ed., McGraw-Hill, New York, 1967.
30 Latto, B., *The Viscosity of Steam at Atmospheric Pressure*, Univ. of Glasgow, Mech. Eng. Dept., Techn. Report No. 16, Glasgow, 1965.
31 Lax, E., Synowietz, C., Eds., *Taschenbuch für Chemiker und Physiker*, 3rd ed., Springer-Verlag, Berlin, 1967.
32 Le Neindre, B., Bury, P., Tufeu, R., Johannin, P., Vodar, B., *Resultats expérimentaux sur la conductivité thermique de l'eau lourde en phase liquide, jusqu'á une température de 370°C, et leurs discussions*, Centre National de la Recherche, Laboratoire des Hautes Pressions, Paris, 1968.
33 Levelt Sengers, J. M. H., *Thermodynamic Anomalies near the Critical Point of Steam*, 7th Internat. Conf. Properties of Steam, Tokyo, 1968.
34 Masi, J. F., *Trans. Am. Soc. Mech. Eng.*, 1954, **76**, 1067–74.
35 O'Connell, J. P., Prausnitz, J. M., *Ind. Eng. Chem.*, 1968, **60** (1), 36–65.
36 Osborne, N. S., Stimson, H. F., Ginnings, D. C., *J. Res. Natl. Bur. Std.*, 1939, **23**, 197–260.
37 Rossini, F. D., Wagman, D. D., Evans, W. H., Levine, S., Jaffe, I., *Selected Values of Chemical Thermodynamic Properties*, Natl. Bur. Std., Circular 500, Washington, Feb. 1, 1952.
38 Salomon, M., *Thermodynamic Properties of Liquid H_2O and D_2O and their Mixtures*, National Aeronautics and Space Administration, NASA TN D-5223, Washington, 1969.
39 Schmidt, E., *Properties of Water and Steam in SI Units*, Springer-Verlag, Berlin, 1969.
40 Schmidt, E., Sellschopp, W., *Forsch. Gebiete Ingenierw.*, 1932, **3**, 277–86.
41 Schrock, V. E., Gott, R. E., Starkman, E. S., WADC TR-56-104, (AD 103097), 1956, 1–144.

42 Stull, D. R., Dir., *JANAF Thermochemical Tables*, Dow Chemical Co., Midland, Mich., 1965–8.

43 Sugisaki, H., Suga, H., Seki, S., *Bull. Chem. Soc. Japan*, 1967, **40**, 2984.

44 Tanaka, K., Sasaki, M., Hattori, H., Kobayashi, Y., Haishima, K., Sato, K., Tashiro, M., *Viscosity of Steam at High Pressures and High Temperatures*, Resources Res. Inst., JCPS Report No. 10, Kawaguchi, Saitama, Japan, 1964.

45 Theiss, R. V., Thodos, G., *J. Chem. Eng. Data*, 1963, **8** (3), 390–5.

46 Timrot, D. L., Vargaftik, N. B., *J. Tech. Phys. USSR*, 1940, **10**, 1063–73.

47 Vargaftik, N. B., Smirnova, E. V., *Soviet Phys., Tech. Phys.*, 1957, **1**, 1221–31.

48 Vaughan, W. E., *Digest of Lit. on Dielectrics*, 1969, **33**, 21–83.

49 Vukalovich, M. P., *Thermodynamic Properties of Water and Steam*, VEB Verlag Technik, Berlin, 1958.

50 Walker, W. A., *Thermodynamic Properties of Liquid Water to 1 kbar*, U.S. Naval Ordnance Laboratory, NOL-TR 66–217, Washington, 1967.

51 Wilson, W. D., *J. Acoust. Soc. Am.*, 1961, **33**, 374.

Table 24.1 Properties of Water

		Ref.
Chemical formula	H_2O	(22)
Molecular weight	18.0148	(22)
Critical temperature	373.91 °C	(1)
Critical pressure	2.205×10^7 N m^{-2}	(1)
Critical density	315.0 kg m^{-3}	(31)
Triple-point temperature	0.01 °C	(31)
Triple-point pressure	615.066 N m^{-2}	(31)
Boiling point at 1.0133×10^5 N m^{-2}	100.0 °C	(31)
Melting point at 1.0133×10^5 N m^{-2}	0.0 °C	(31)
Transition temperature	−138.15 °C	(43)
Density of ice at 0 °C and 1.0133×10^5 N m^{-2}	918.0 kg m^{-3}	(31)
Compressibility of liquid at 10 °C	48.0×10^{-11} m^2 N^{-1}	(35)
Coefficient of thermal expansion of liquid at 25 °C	2.5632×10^{-4} °C^{-1}	(23)
Heat of formation of gas at 25 °C	-2.4199×10^8 J kg-mol^{-1}	(42)
Free energy of formation of gas at 25 °C	-2.2877×10^8 J kg-mol^{-1}	(42)
Heat of sublimation at −298.15 °C	4.7260×10^7 J kg-mol^{-1}	(31)
Heat of fusion at 0.0 °C	6.0135×10^6 J kg-mol^{-1}	(37)
Heat of transition	-1.5910×10^5 J kg-mol^{-1}	(43)
Entropy of gas at 25 °C	1.8885×10^5 J kg-mol^{-1} °C^{-1}	(42)
Heat capacity ratio of gas at 100 °C and 1.0133×10^5 N m^{-2}	1.324	(29)
Velocity of sound at 0 °C and 1.0133×10^5 N m^{-2}	1403.0 m s^{-1}	(51)
Diffusion in air at 8 °C and 1.0133×10^5 N m^{-2}	2.40×10^{-5} m^2 s^{-1}	(31)
Dielectric constant of liquid at 25 °C and 1.0133×10^5 N m^{-2}	78.40	(48)
Electrical conductivity of liquid at 18 °C	4×10^{-6} ohm^{-1} m^{-1}	(29)
Refractive index of liquid at 25 °C and 1.0133×10^5 N m^{-2}	1.333	(31)
Threshold limit value	None	(31)

24.1 Vapour Pressure of Water (10,31,39)

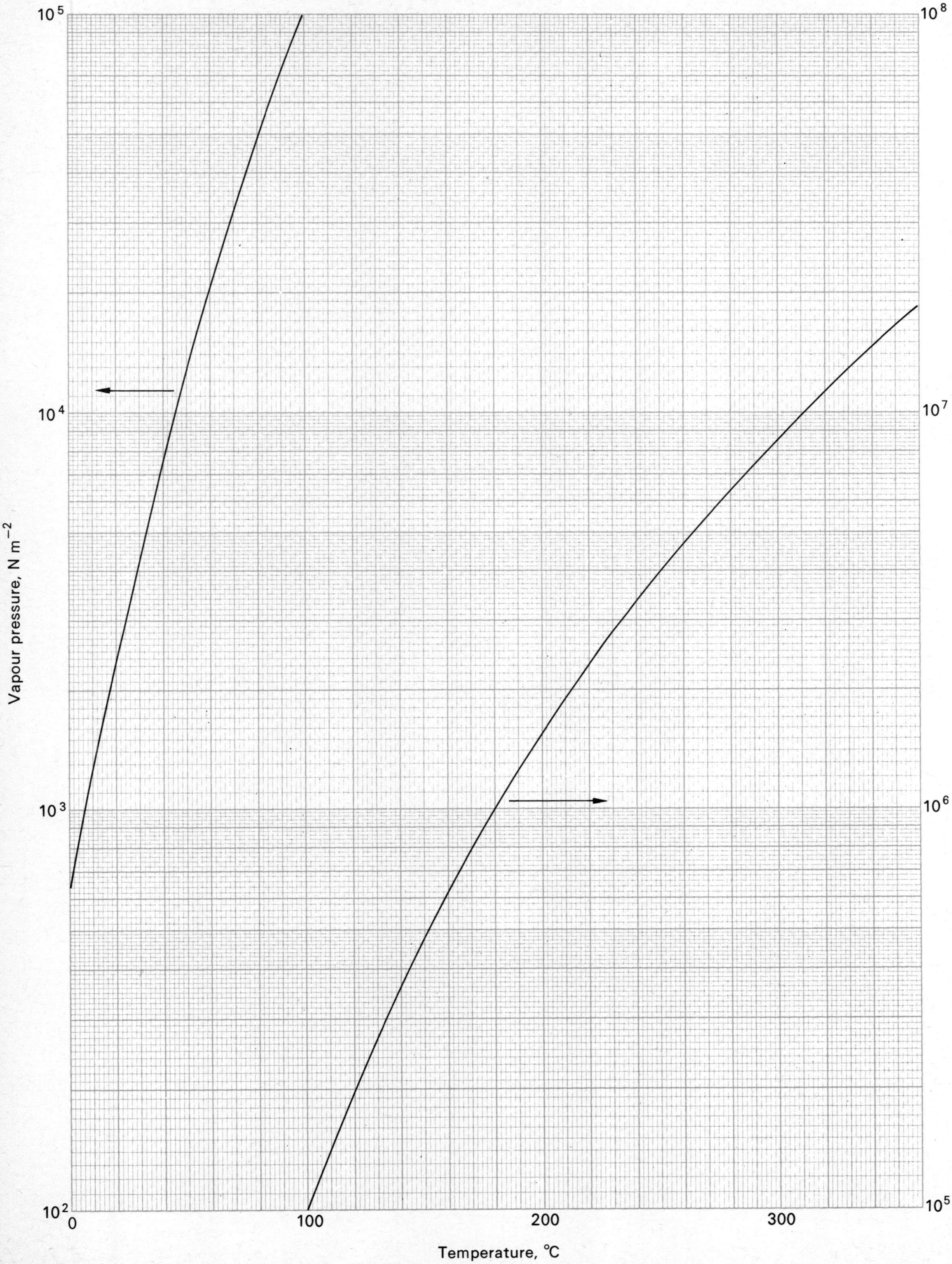

24.2 Orthobaric Density of Water (10,31,39)

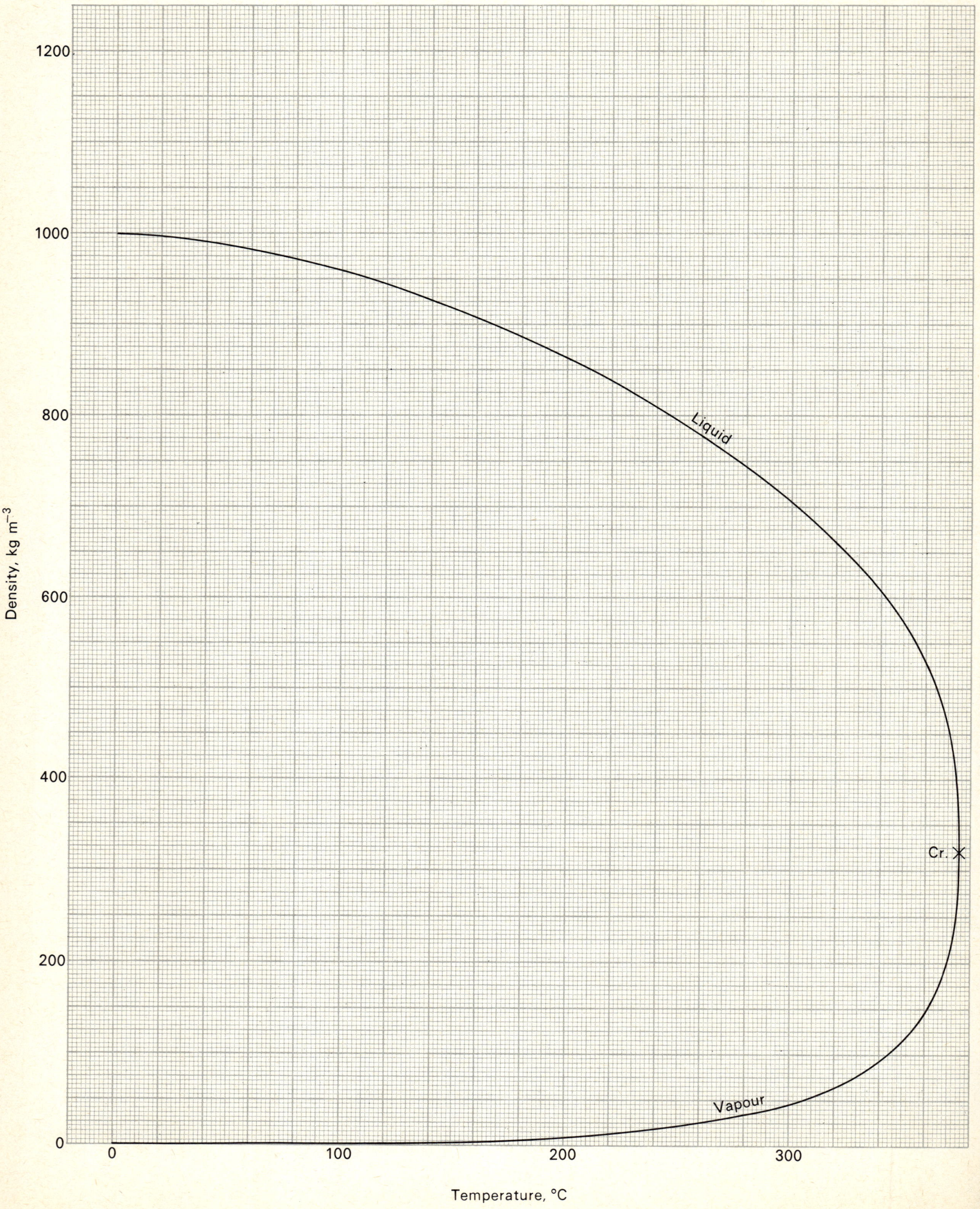

24.3 P-V-T Data for Water (10,31,39)

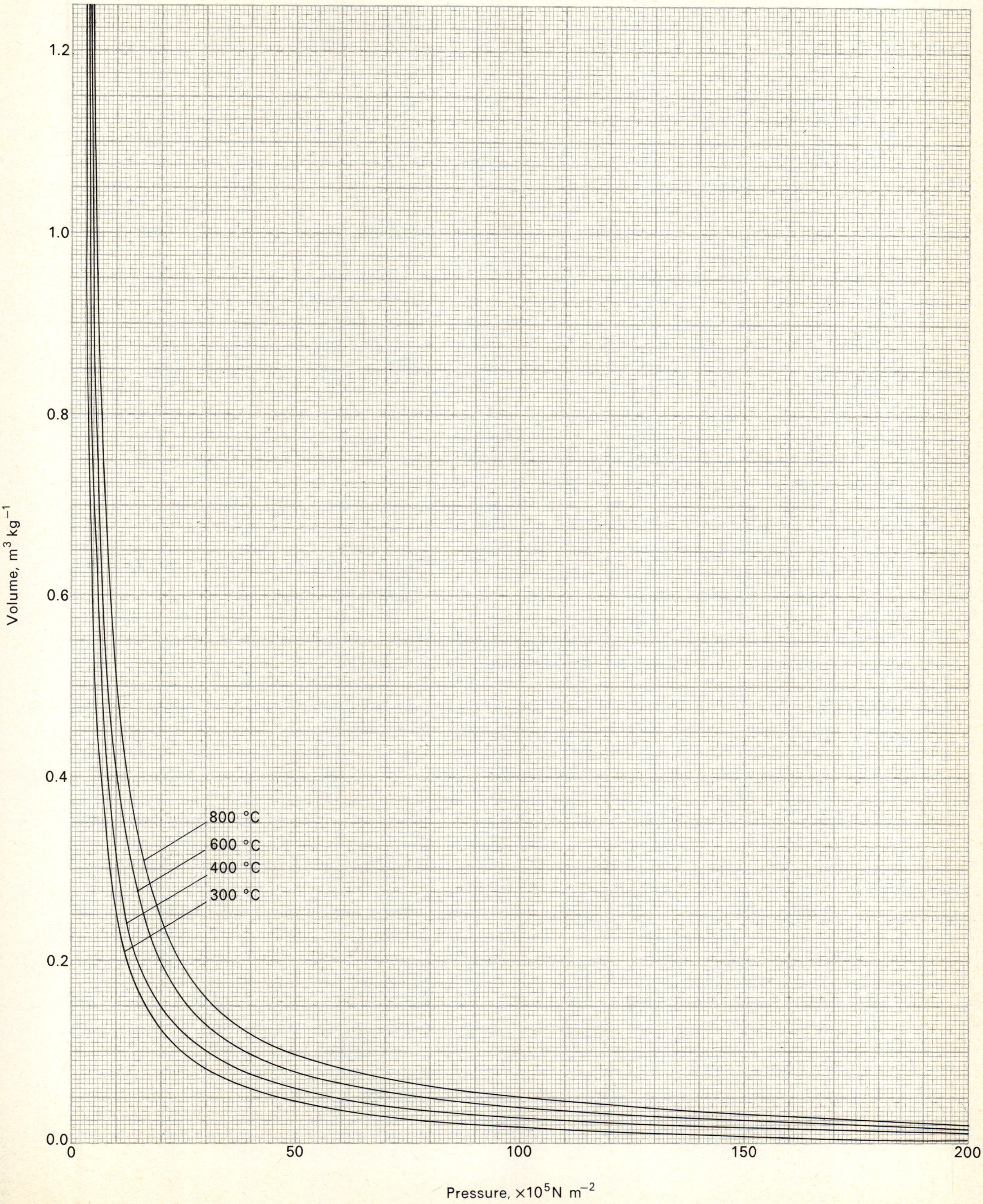

24.4 Latent Heat of Vaporisation of Water (10,31,39)

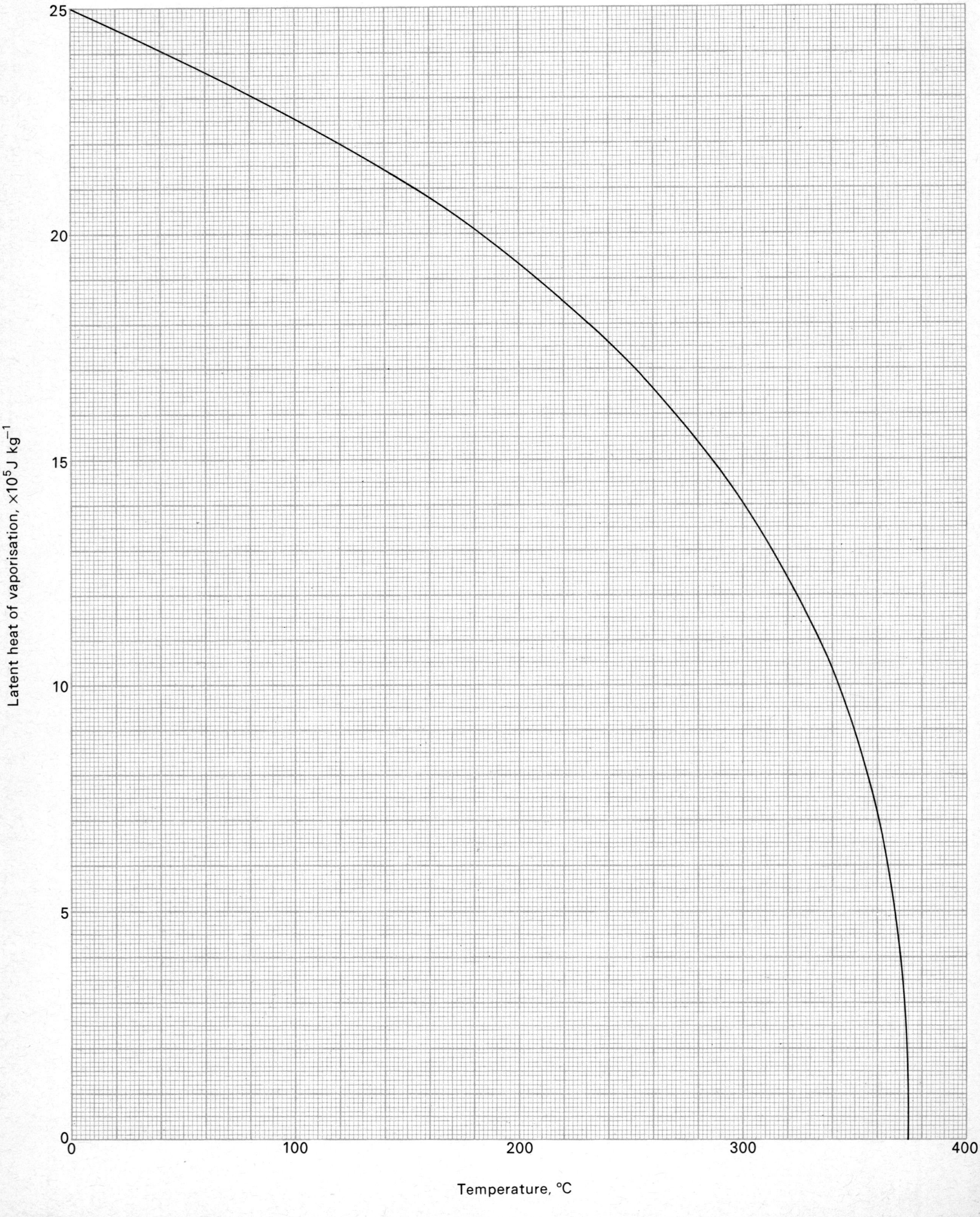

24.5 Heat Capacity of Liquid Water (15,36,41)

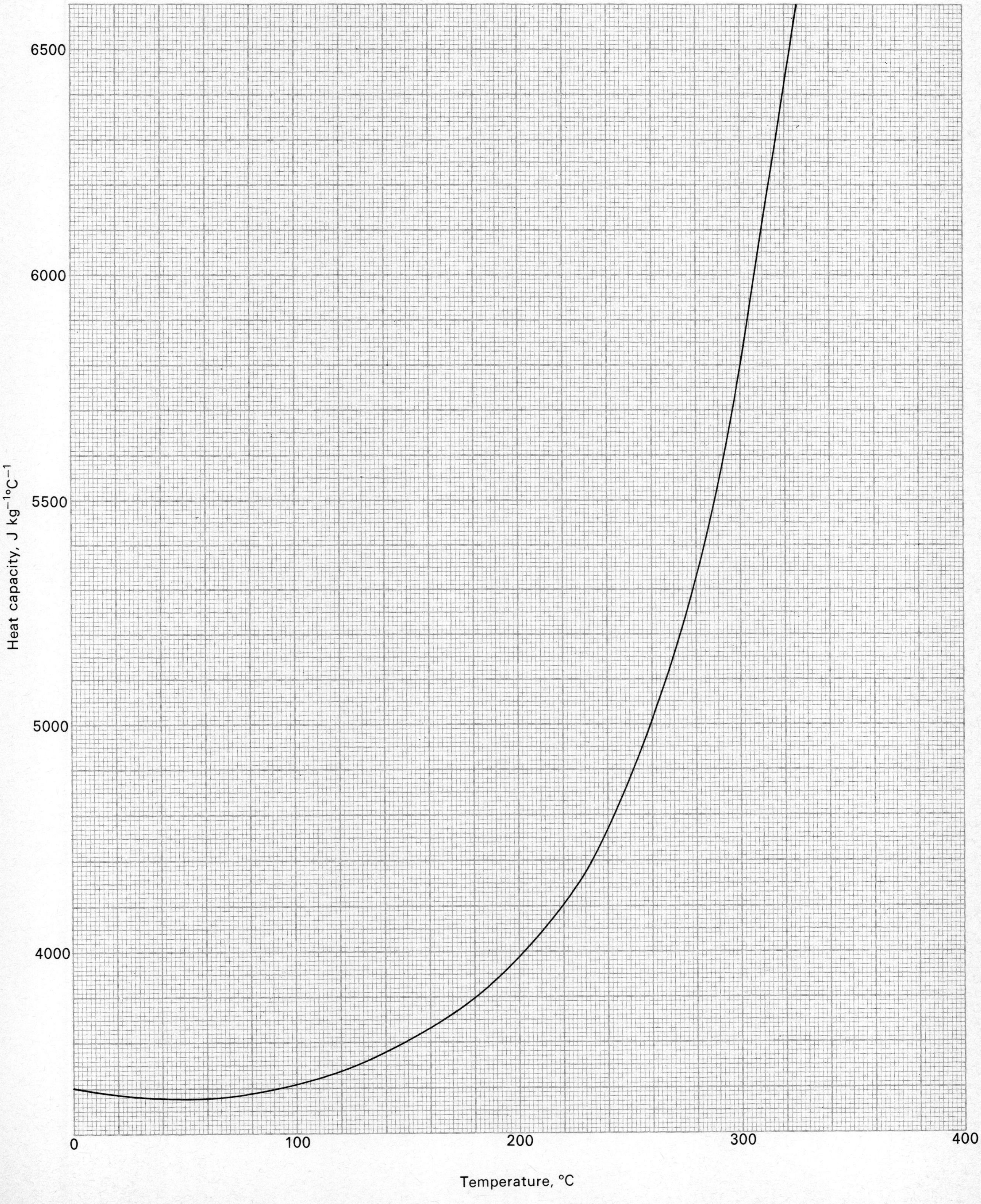

24.6 Heat Capacity of Gaseous Water at 1.0133×10^5 N m^{-2}(13,19,34)

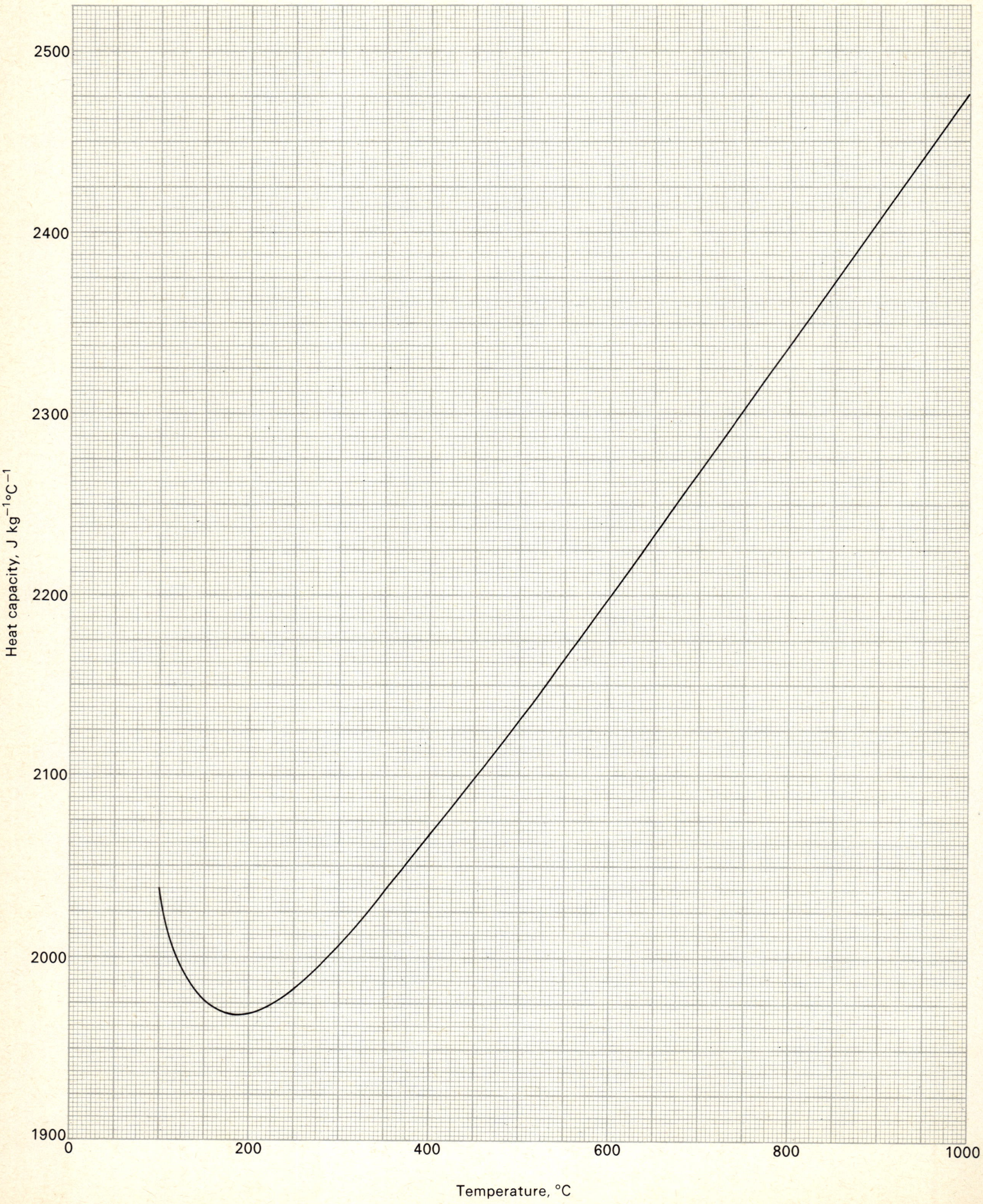

24.7 Heat Capacity Ratio (C_p/C_v) of Gaseous Water at 1.0133×10^5 N m^{-2} (12)

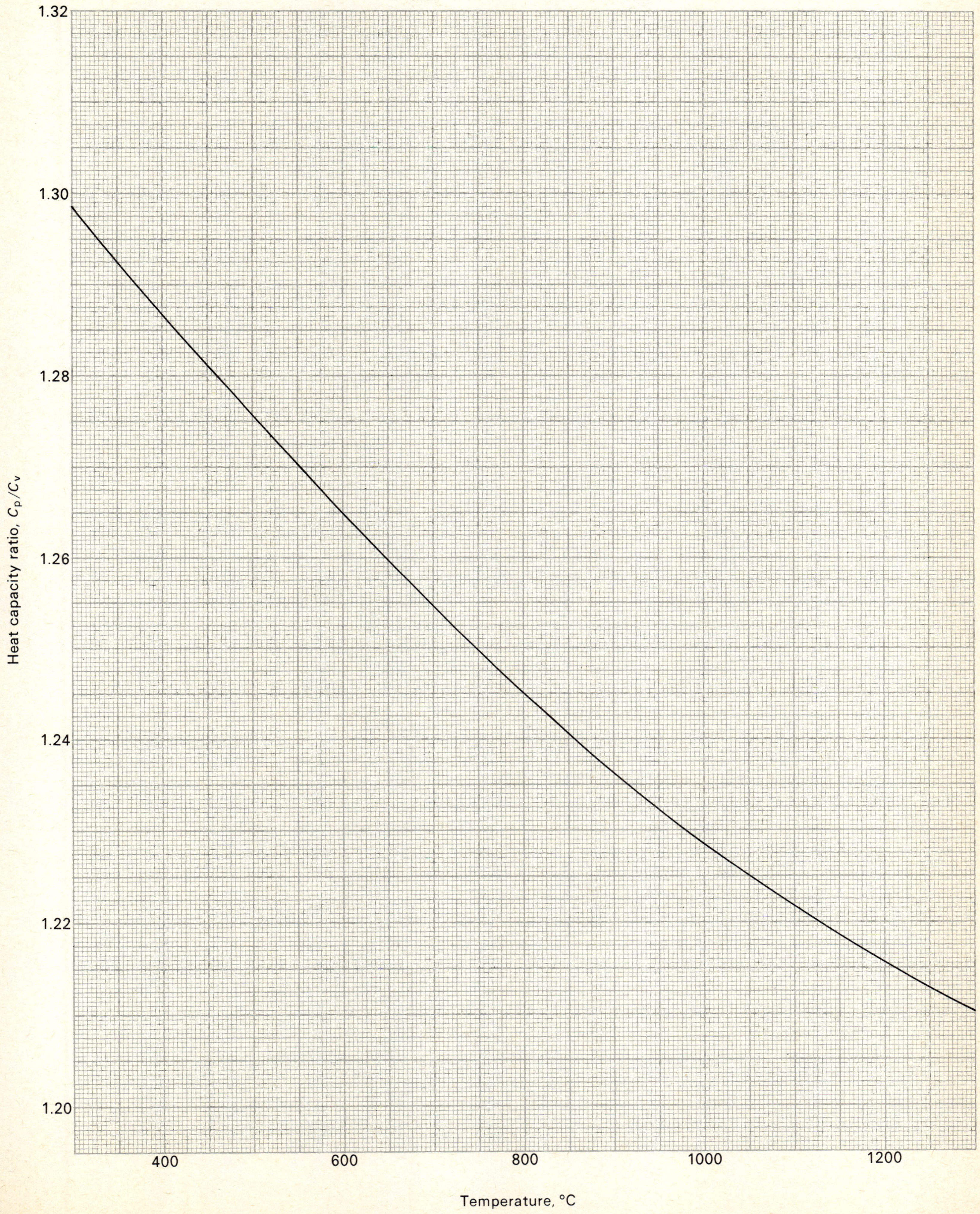

24.8 Viscosity of Liquid Water (5,7,11,39)

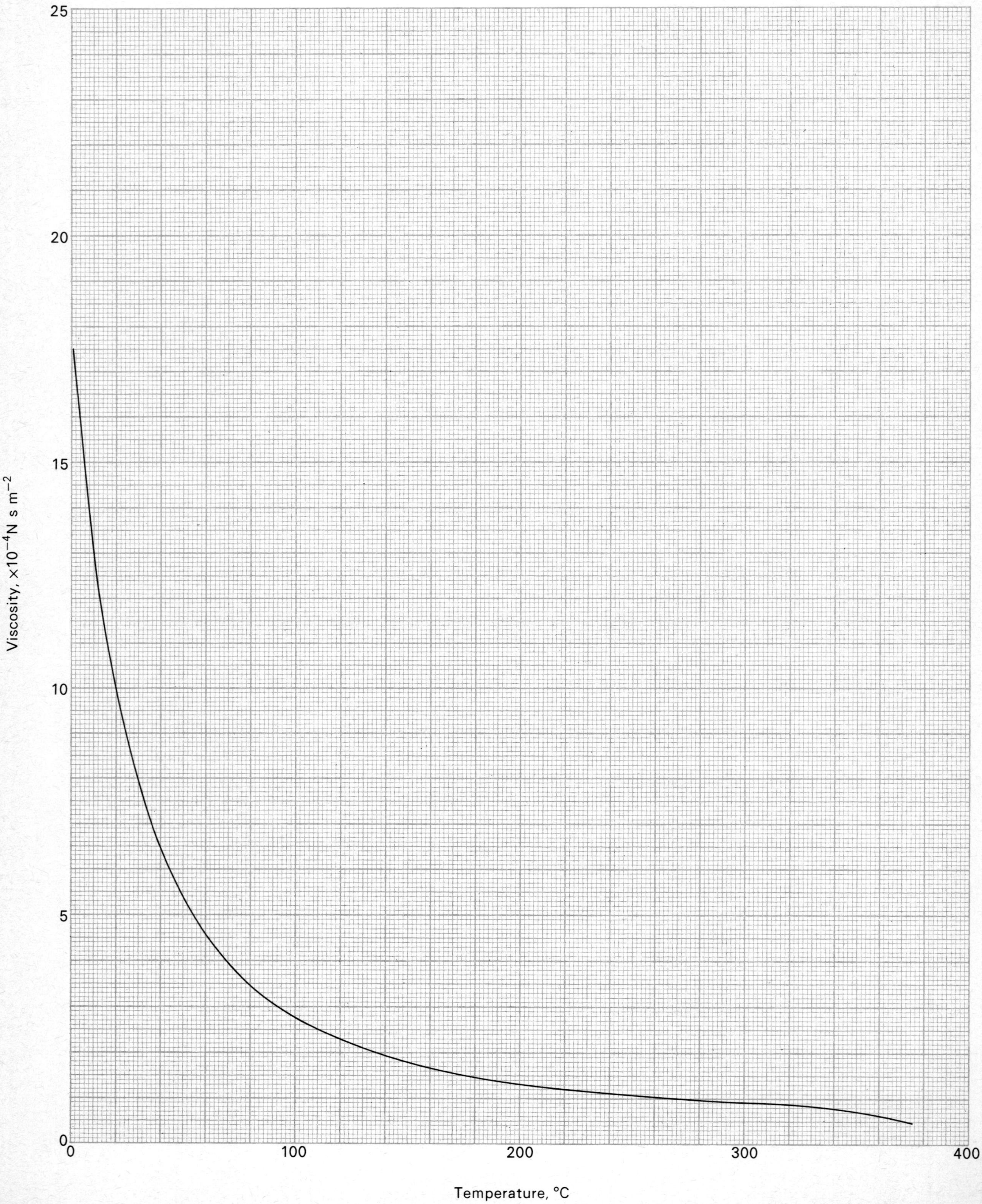

24.9 Viscosity of Gaseous Water at 1.0133×10^5 N m^{-2} (16,30,39,44)

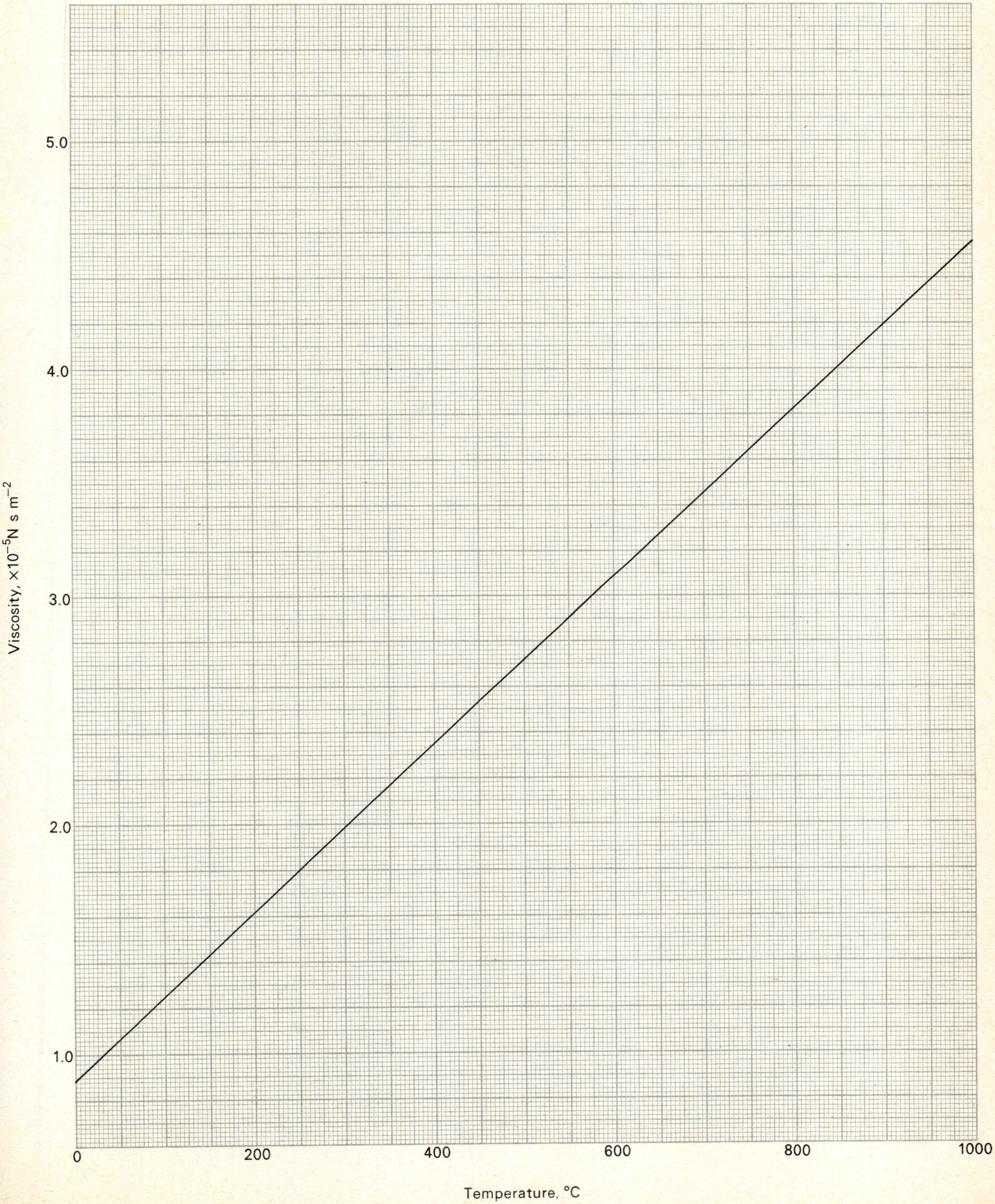

24.10 Thermal Conductivity of Liquid Water (Saturated) (4,32,40,46)

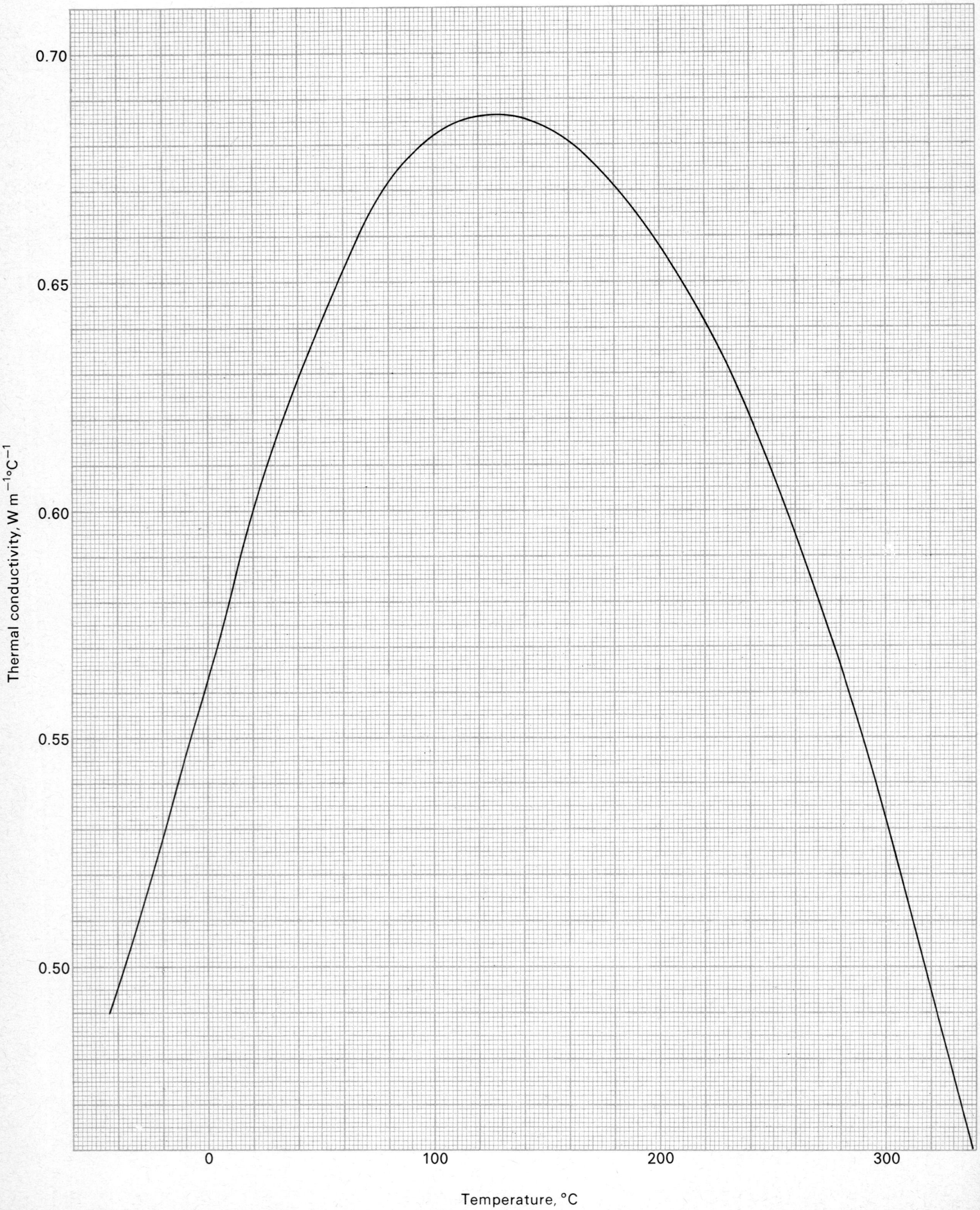

24.11 Thermal Conductivity of Gaseous Water at 1.0133×10^5 N m^{-2} (2,4,10,27,28)

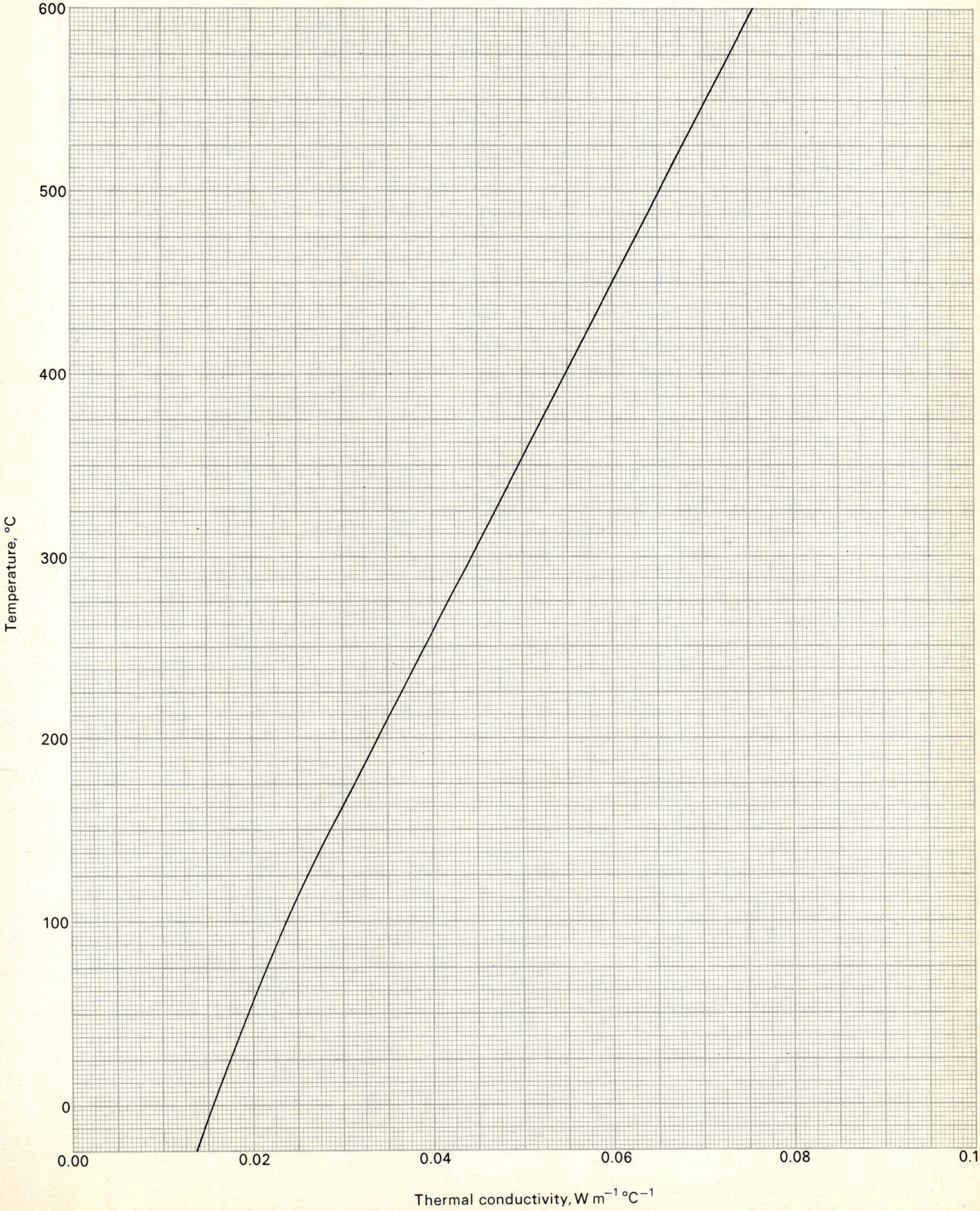

24.12 Surface Tension of Water (10,17,18,39)

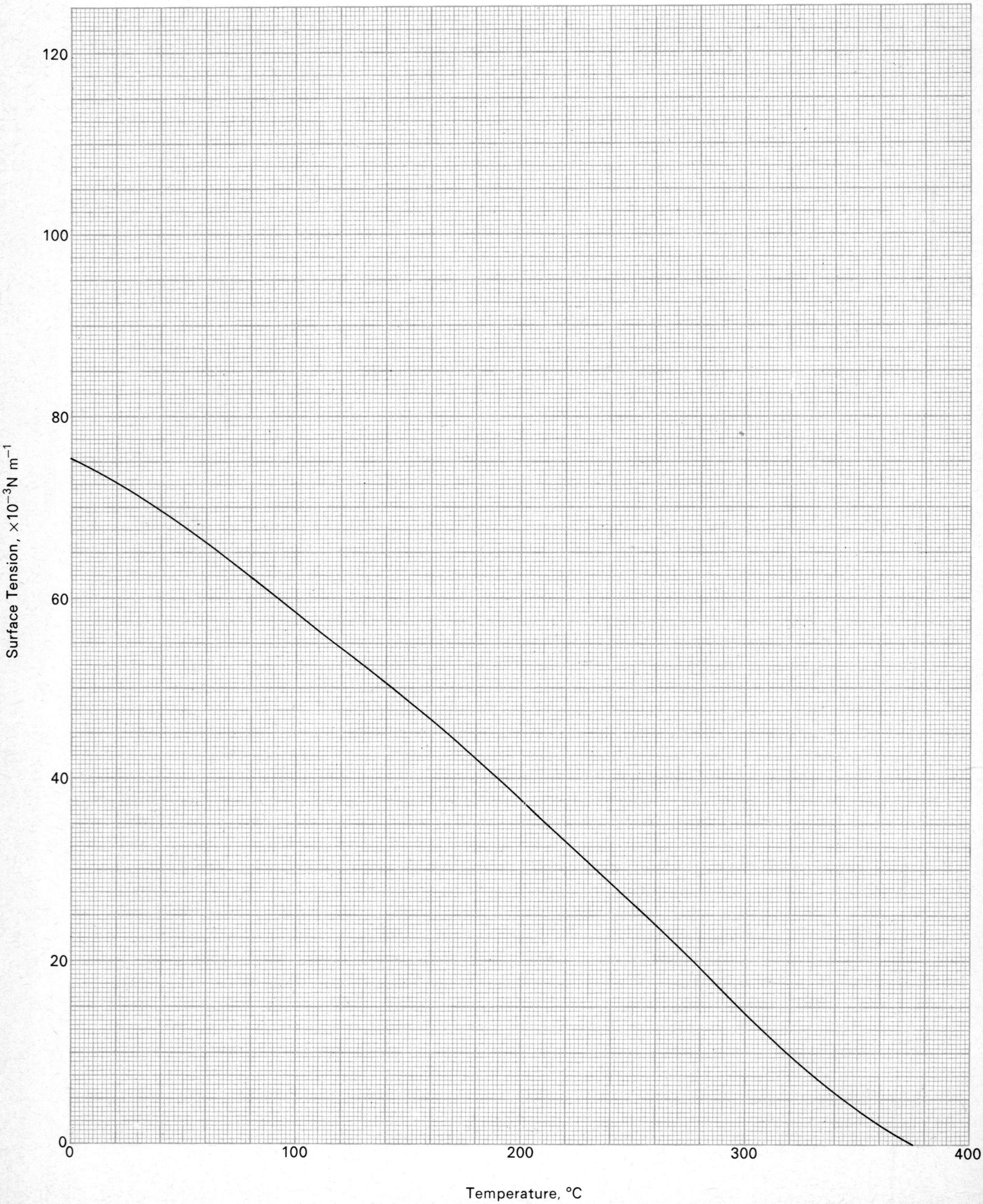

24.13 Electrical Conductivity of Water (21)

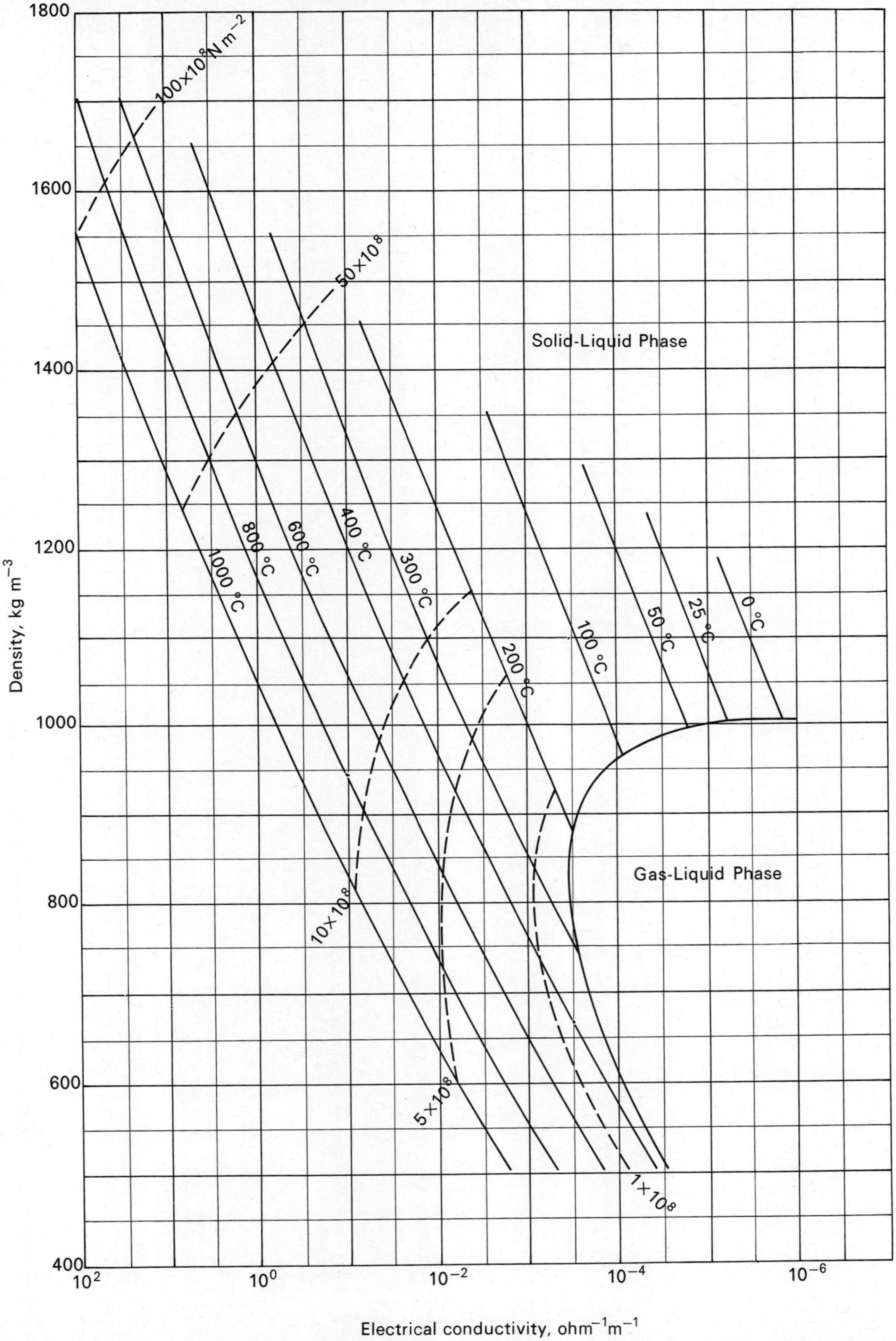

25 AMMONIA

Ammonia is a colourless gas with a characteristic smell at room temperature and atmospheric pressure. Experimental evidence shows that the ammonia molecule is pyramidal with the nitrogen atom at the apex. The angle H—N—H is 116° and the N—H distances are 1.016 Å (according to absorption spectra). The nitrogen atom vibrates (having two positions) below and above the plane of the hydrogen atoms. The dipole moment of NH_3 gas is 1.46×10^{-18} e.s.u.

Ammonia does not burn in air, but in oxygen it burns to H_2O and N_2. On heating the molecule dissociates. It is extremely soluble in water, forming solutions of ammonium hydroxide, which is a weak base.

Liquid ammonia is also colourless and resembles water more than any other substance. It is associated in the condensed phase, which is due to the unshared electrons. The high Trouton constant (23.4) also indicates the high degree of association. Liquid ammonia, being an excellent solvent, dissolves practically all ordinary acids.

Solid ammonia crystallises in a slightly deformed cubic close-packed structure. Each nitrogen atom is surrounded by six hydrogen atoms. Each unit cell contains four NH_3 molecules. The small temperature coefficient and low dielectric constant indicate that there is no rotation in the solid state.

Ammonia is distributed as a liquid in iron cylinders under pressure. The major applications of ammonia are in the manufacture of fertilizers, salts, and nitric acid. It is also used in refrigeration cycles.

The physical, thermodynamic, and transport properties of ammonia at various temperatures and pressures are well established. There are several compilations on the evaluated, correlated, and selected properties as follows: Canjar and Manning (7), Din (14), Gmelin (19), Harrison (25), Jacolin (32), Jandler (33), Keyes (36), Landolt-Börnstein (39), Matheson (46), Mayhew (48), Mellor (49), Natl. Bur. Std. (52), Pascal (55), Tans (66), and Tsoiman (69). In 1968, Haar (24) at the National Bureau of Standards carried out a comprehensive work on the thermodynamic properties of ammonia as an ideal gas, being a basis for real gas correlation of the properties. Furthermore, its viscosity has been investigated at high pressures (9, 32). The equation of state for NH_3 has been presented and discussed by Ramelho and Fizello (58) and Tsiklis (68).

References

1 American Conference of Government Industrial Hygienists, *Documentation of Threshold Limit Values*, revised edition, Committee on Threshold Limit Values, Cincinnati, Ohio 45202, 1966, 203 pp.
2 Baker, C. E., Brokaw, R. S., *J. Chem. Phys.*, 1965, **43**, 3519–28.
3 Baldi, G., Specchia, V., *Chim. Ind. (Milan)*, 1971, **53** (10), 929–33.
4 Blagoi, Yu. P., *Zh. Fiz. Khim.*, 1968, **42** (5), 1075–8.
5 Bowen, D. E., Thompson, J. C., *J. Chem. Eng. Data*, 1968, **13** (2), 206–8.
6 Callear, A. B., Robb, J. C., *Trans. Faraday Soc.*, 1955, **51**, 630–8.
7 Canjar, L. N., Manning, F. S., Thermodynamic Properties of Ammonia, in *Thermodynamic Properties and Reduced Correlations for Gases*, Gulf Publ. Co., Houston, Texas, 1967, pp. 119–25.
8 Carmichael, L. T., Sage, B. H., *Ind. Eng. Chem.*, 1952, **44** (11), 2728.
9 Carmichael, L. T., Sage, B. H., *J. Chem. Eng. Data*, 1963, **8** (3), 400–4.
10 Clifford, I. L., Hunter, E., *J. Phys. Chem.*, 1933, **37**, 101–18.
11 Cragoe, C. S., *Refrig. Eng.*, 1925/26, **12**, 136–8.
12 Cragoe, C. S., Meyers, C. H., *J. Am. Chem. Soc.*, 1920, **42**, 222–7.
13 Davis, D. S., *Chem. Proc. Eng.*, 1958, **39** (9), 325.
14 Din, F., Ed., *Thermodynamic Functions of Gases*, Vol. 1, Ammonia, Carbon Dioxide, Carbon Monoxide, Butterworths, London, 1962, 175 pp.
15 Edwards, D. G., *The Vapour Pressure of 30 Inorganic Liquids between One Atmosphere and the Critical Point*, Univ. of California, Lawrence Rad. Lab., UCRL 7167, Livermore, California, June 13, 1963, 148 pp.
16 Engineering Science Data Units, *Approximate Data on the Viscosity of some Common Liquids*, Item No. 66024, London, July 1966, 26 pp.
17 Eucken, A., Donath, E., *Z. phys. Chem.*, 1926, **124**, 194–9.
18 Eucken, A., Karwat, E., *Z. phys. Chem.*, 1924, **112**, 467–85.
19 *Gmelins Handbuch der anorganischen Chemie*, Syst. No. 4, Stickstoff, Verlag Chemie, GmbH, Berlin, 1936, pp. 408–79.
20 Golubev, I. F., *Viscosity of Gases and Gas Mixtures*, Gosudarstvennoe Izdatel'stvo Fiziko-Matematicheskoi Literatury, Moskva, 1959, Table 4.
21 Golubev, I. F., Sokolova, V. P., *Teploenergetika*, 1964, **11** (9), 64–7.
22 Groenier, W. S., Thodos, G., *J. Chem. Eng. Data*, 1960, **5** (3), 285–8.
23 Groenier, W. S., Thodos, G., *J. Chem. Eng. Data*, 1961, **6** (2), 240–4.
24 Haar, L., *J. Res. Natl. Bur. Std.*, 1968, **72A** (2), 207–16.
25 Harrison, R. H., *The Thermodynamic Properties of Ammonia*, Univ. of Texas, M.S. thesis in Chem. Eng., Austin, Texas, 1952.
26 Heide, R., *Luft- und Kältetechnik*, 1971, **7** (3), 128–30.
27 Honig, R. E., Hook, H. O., Vapour Pressure Data for some Common Gases, *RCA Rev.*, 1960, **21**, 360–8.
28 Horvata, L. J., *Bull. Soc. Chim. Beograd.*, 1969, **34** (5–7), 359–65.
29 IUPAC, Atomic Weights of the Elements 1971, *Pure & Appl. Chem.*, 1972, **30** (3–4), 638–49.
30 Ibrahim, S. H., Kuloor, N. R., *Brit. Chem. Eng.*, 1961, **6** (12), 862–3.
31 Ivassaki, H., Takahachi, M., *Rev. Phys. Chem. Japan*, 1968, **38** (1), 18.
32 Jacolin, O., *Proprietes Physiques et Chimiques de L'Ammoniac*, Commissariat à l'Energie Atomique, Bibliographie No. 57, Paris, 1965, 68 pp.
33 Jandler, J., *Anorganische und allgemeine Chemie in flüssigem Ammoniak*, Vol. I, Part 1, Friedr. Vieweg & Sohn, Braunschweig, 1966.
34 Kardos, A., *Z. Ver. deutsch. Ing.*, 1933, **77**, 1158.
35 Kardos, A., *Z. ges. Kälte-Ind.*, 1934, **41**, 1–35.
36 Keyes, F. G., Brownlee, R. B., *Thermodynamic Properties of Ammonia*, Wiley, New York, 1916.
37 Kobe, K. A., Harrison, R. H., *Petrol. Ref.*, 1954, **33** (11), 161–4.
38 Koch, B., *Grundlagen des Wärmeaustausches*, Verlagsanstalt H. Beucke & Sohne, Berlin, 1950.
39 *Landolt-Börnstein, Zahlenwerte und Funktionen aus Physik, Chemie, Astronomie, und Technik*; 6th ed., Technik, 4 Teil, Bandteil a, Springer-Verlag, Berlin, 1967, pp. 586–602.
40 Lange, N. A., Ed., *Handbook of Chemistry*, 10th ed., McGraw-Hill, New York, 1967.
41 Lax, E., Synowietz, C., Eds., *Taschenbuch für Chemiker und Physiker*, 3rd ed., Springer-Verlag, Berlin, 1967.
42 Leduc, A., *Chem. Rev.*, 1929, **6**, 1–16.
43 Mapstone, G. E., *Aust. Chem. Proc. Eng.*, 1971, **24** (7), 15.
44 Maryott, A. A., Smith, E. R., *Table of Dielectric Constants of Pure Liquids*, Natl. Bur. Std. Circular 514, Washington, Aug. 10, 1951, 44 pp.
45 Masi, J. F., *Trans. Am. Soc. Mech. Eng.*, 1954, **76**, 1067–74.
46 Matheson Gas Data Book, 4th ed., *Ammonia*, The Matheson Company, Inc., East Rutherford, N. J., 1966, pp. 13–21.

47 Mathews, J. F., *Chem. Rev.*, 1972, **72** (1), 71–100.
48 Mayhew, Y. R., Rogers, G. F. C., *Thermodynamic and Transport Properties of Fluids*, 2nd ed., Blackwell, Oxford, 1972, 20 pp.
49 *Mellor's Comprehensive Treatise on Inorganic and Theoretical Chemistry*, Vol. VIII, Suppl. 1, Nitrogen, Part I, Longmans, London, 1964, pp. 276–327.
50 Missenard, A., *Thermal Conductivity of Solids, Liquids, and Gases and Their Mixtures*, Éditions Eyrolles, Paris, 1965, 576 pp.
51 McKelvy, E. C., Taylor, C. S., *Natl. Bur. Std., Sci. Paper*, 1922/23, **18**, 688.
52 Natl. Bur. Std., *Tables of Thermodynamic Properties of Ammonia*, 1st ed., Circular 142, Washington, 1923, 7–30.
53 Needham, D. P., Ziebland, H., *Inter. J. Heat and Mass Transfer*, 1965, **8**, 1387–414.
54 Overstreet, R., Giauque, W. F., *J. Am. Chem. Soc.*, 1937, **59**, 254–9.
55 Pascal, P., Ed., *Nouveau Traité de Chimie Minérale*, Vol. X, Azote, Phosphore, Masson et Cie, Éditeurs, Paris, 1956, 89–125.
56 Pickering, S. F., *Natl. Bur. Std., Sci. Paper*, 1926/27, **21**, 604.
57 Popov, V. A., Manzhelii, V. G., Bagatskii, M. I., *J. Low Temp. Phys.*, 1971, **5** (4), 427–33.
58 Ramalho, R. S., Frizelle, W. G., *J. Chem. Eng. Data*, 1965, **10** (4), 366–70.
59 Richter, G. N., Sage, B. H., *J. Chem. Eng. Data*, 1964, **9** (1), 75–8.
60 Rossini, F. D., Wagman, D. D., Evans, W. H., Levine, S., Jaffe, I., *Selected Values of Chemical Thermodynamic Properties*, Natl. Bur. Std. Circular 500, Washington, Feb. 1, 1952.
61 Sellschopp, W., *Z. Ver. Deutsch. Ing.*, 1935, **79**, 69.
62 Smith, C., *Proc. Royal Soc.*, 1912, **87A**, 369.
63 Smits, A., Swart, E., Bruin, P., *J. Chem. Soc.*, 1929, 2723.
64 Stern, H., *Ind. Eng. Chem.*, 1952, **44** (10), 2442–4.
65 Stull, D. R., Dir., *JANAF Thermochemical Tables*, Dow Chemical Co., Midland, Mich., 1965–8.
66 Tans, A. M. P., *Chem. Eng.*, 1962 (July 9), **71**, 162.
67 Thatte, V. N., *Phil. Mag.*, 1929, **7** (7), 890.
68 Tsiklis, D. S., *J. Phys. Chem.*, 1969, **43** (3), 406–7.
69 Tsoiman, G. I., *Teplofiz. Kharakter. Veshchestv.*, 1968, **1**, 190–3.
70 Van Cleave, A. B., Maass, O., *Can. J. Res.*, 1935, **13B**, 140–8.
71 Varlashkin, P. G., Thompson, J. C., *J. Chem. Eng. Data*, 1963, **8** (4), 526.
72 Washburn, E. W., Ed., *International Critical Tables of Numerical Data, Physics, Chemistry, and Technology*, Vol. VI, McGraw-Hill, New York, 1929, pp. 461–7.

Table 25.1 Properties of Ammonia

		Ref.
Chemical formula	NH_3	(41)
Molecular weight	17.0304	(29)
Critical temperature	132.4°C	(47)
Critical pressure	112.78×10^5 N m^{-2}	(47)
Critical density	235.0 kg m^{-3}	(47)
Triple-point temperature	−77.74°C	(41)
Triple-point pressure	6077.2 N m^{-2}	(41)
Boiling point at 1.0133×10^5 N m^{-2}	−33.35°C	(12)
Melting point	−77.75°C	(63)
Transition temperature	None	(19)
Density of solid at −185°C	836.0 kg m^{-3}	(51)
Compressibility of liquid at −51.0°C	63.8×10^{-11} m^2 N^{-1}	(5, 52)
Coefficient of thermal expansion of liquid at −25°C	1.95×10^{-3} °C^{-1}	(67)
Heat of formation of gas at 25°C	-4.5929×10^7 J kg-mol^{-1}	(65)
Free energy of formation of gas at 25°C	-1.6161×10^7 J kg-mol^{-1}	(65)
Heat of sublimation at −104.15°C	3.1327×10^7 J kg-mol^{-1}	(17)
Heat of fusion	5.6564×10^6 J kg-mol^{-1}	(60)
Heat of transition	None	(19)
Entropy of gas at 25°C	1.9272×10^5 J kg-mol^{-1} C^{-1}	(65)
Heat capacity ratio of gas at 15°C and 1.0133×10^5 N m^{-2}	1.31	(42)
Velocity of sound at 0°C and 1.0133×10^5 N m^{-2}	414.8 m s^{-1}	(4, 72)
Diffusion in air at 0°C and 1.0133×10^5 N m^{-2}	1.98×10^{-5} m^2 s^{-1}	(30)
Dielectric constant of liquid at 25°C	16.9	(44)
Electrical conductivity of liquid at 20°C	1.22×10^{-8} ohm^{-1} m^{-1}	(28)
Refractive index of liquid at 16.5°C	1.325	(62)
Threshold limit value	(50 ppm) 3.5×10^{-5} kg m^{-3}	(1)

25.1 Vapour Pressure of Ammonia (10,15,27,64)

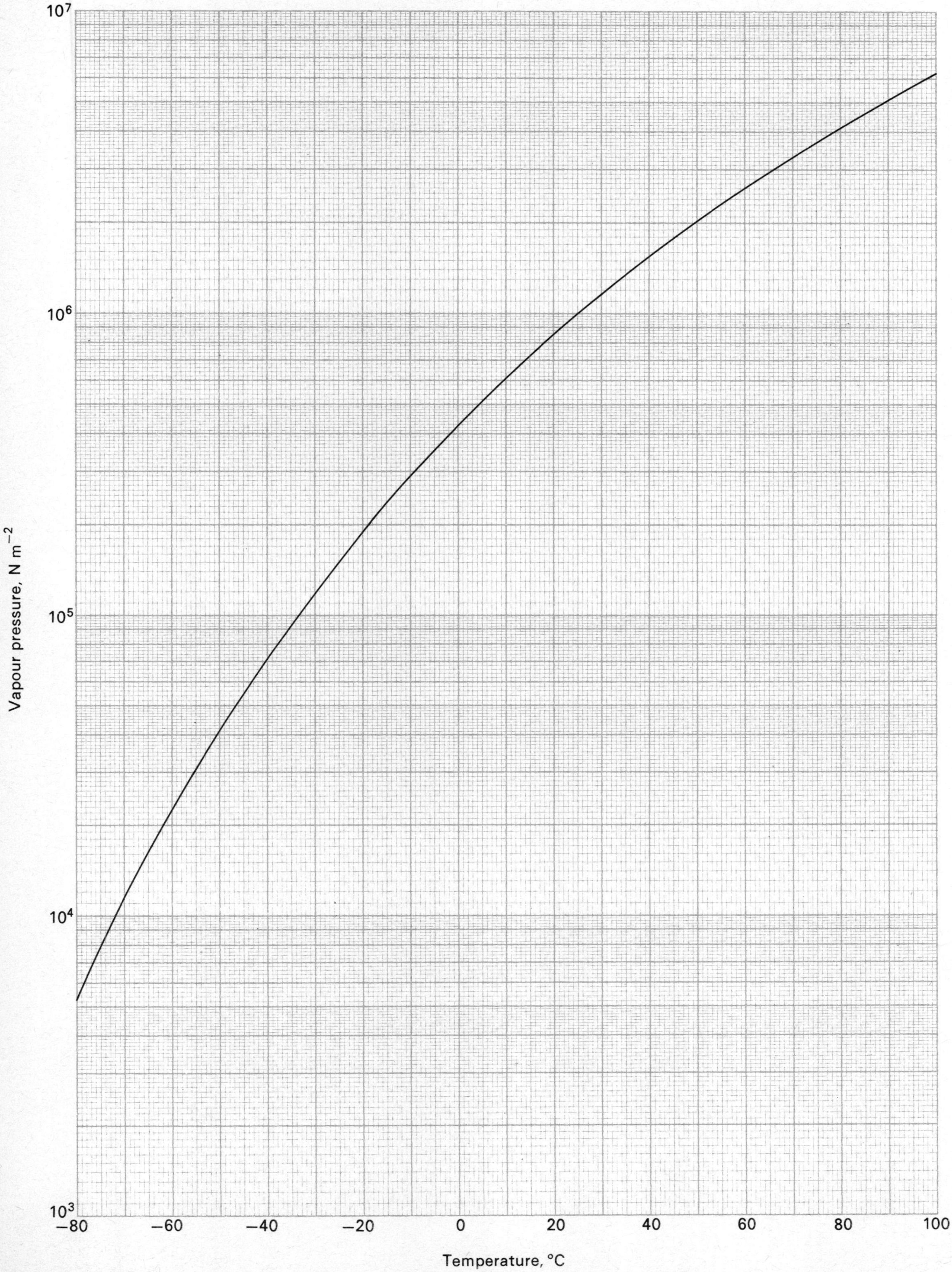

25.2 Orthobaric Density of Ammonia (22,40,41)

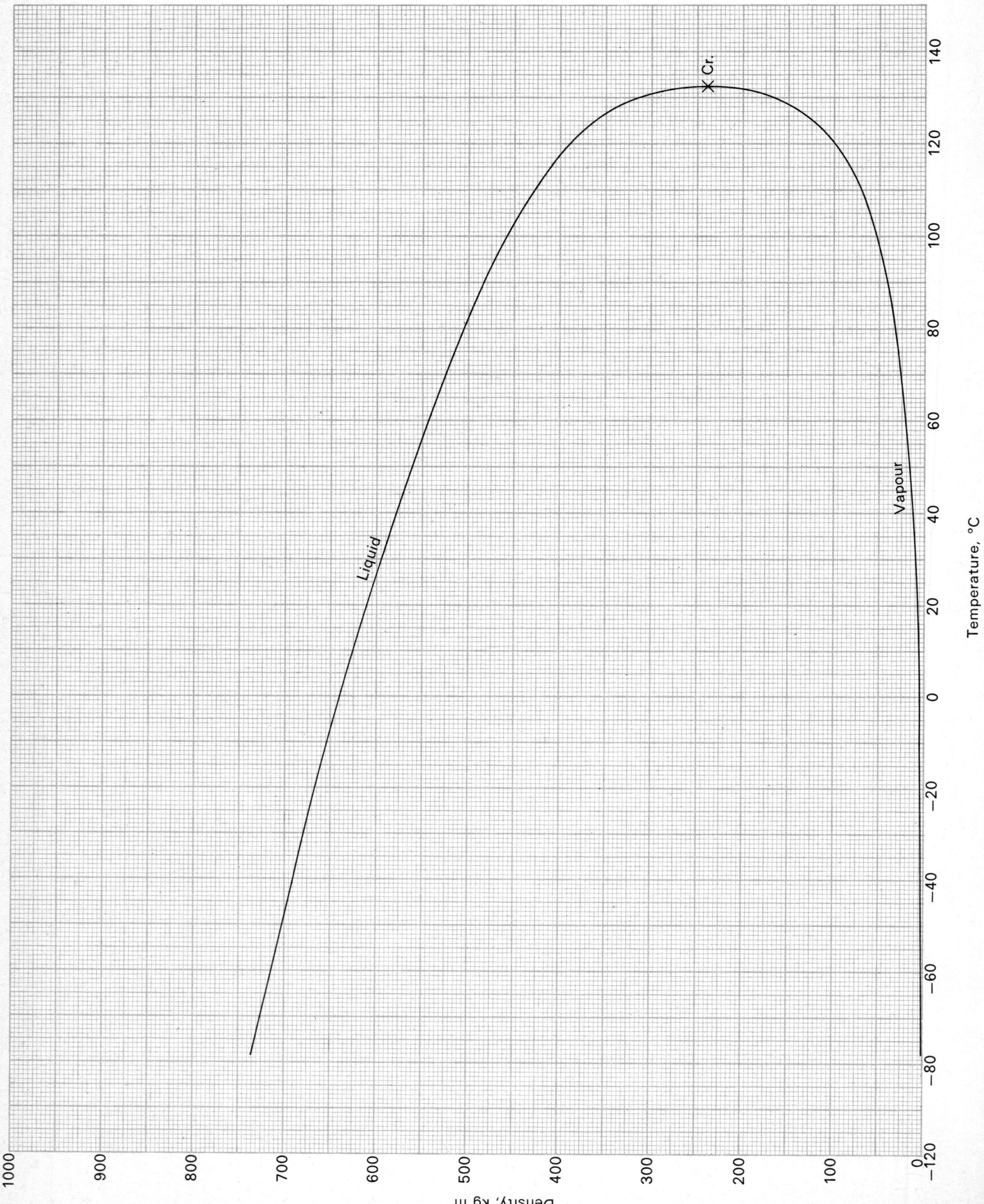

25.3 P-V-T Data of Ammonia (22,25,39,41)

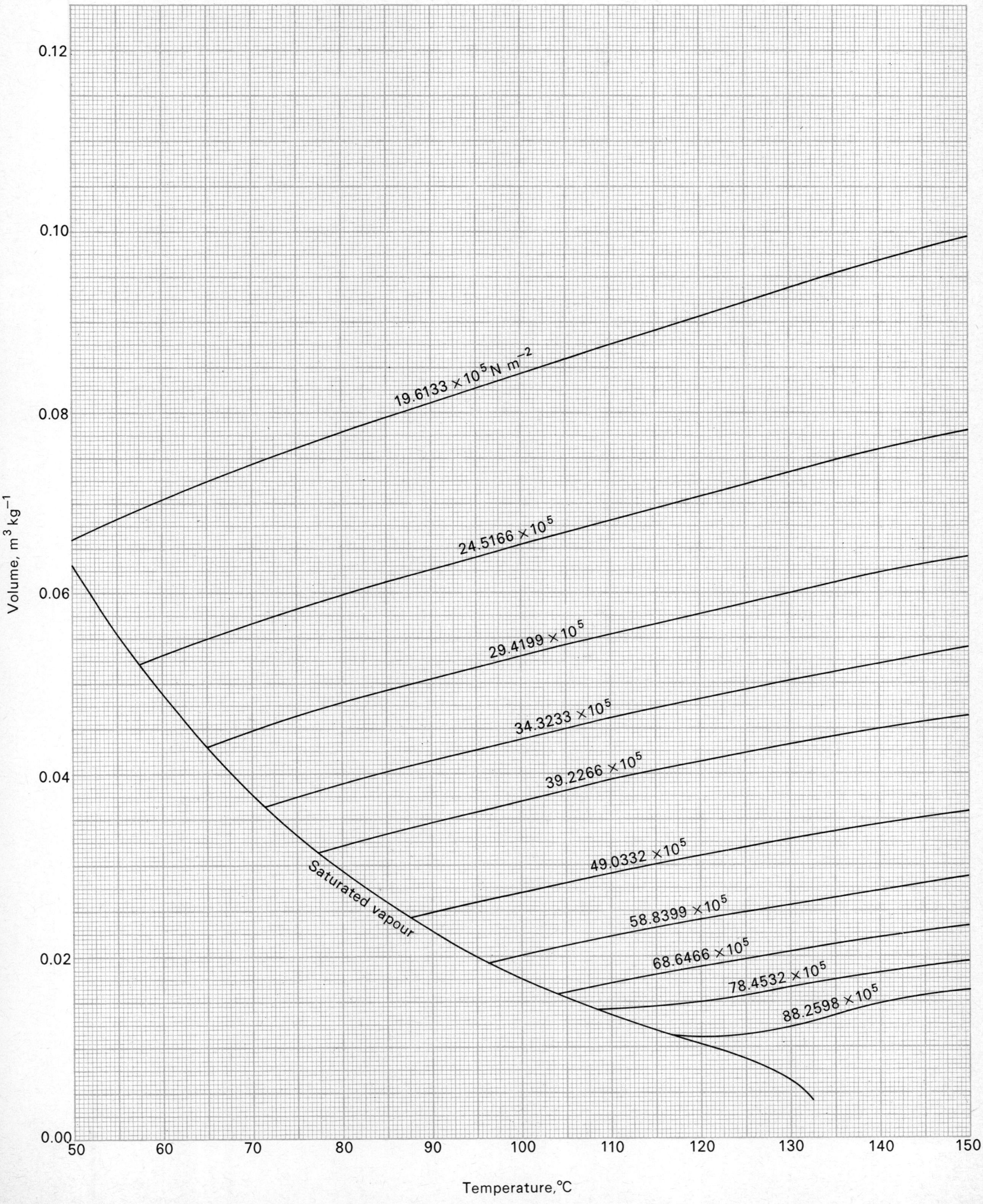

25.4 Latent Heat of Vaporisation of Ammonia (14,19,25,32,39,52)

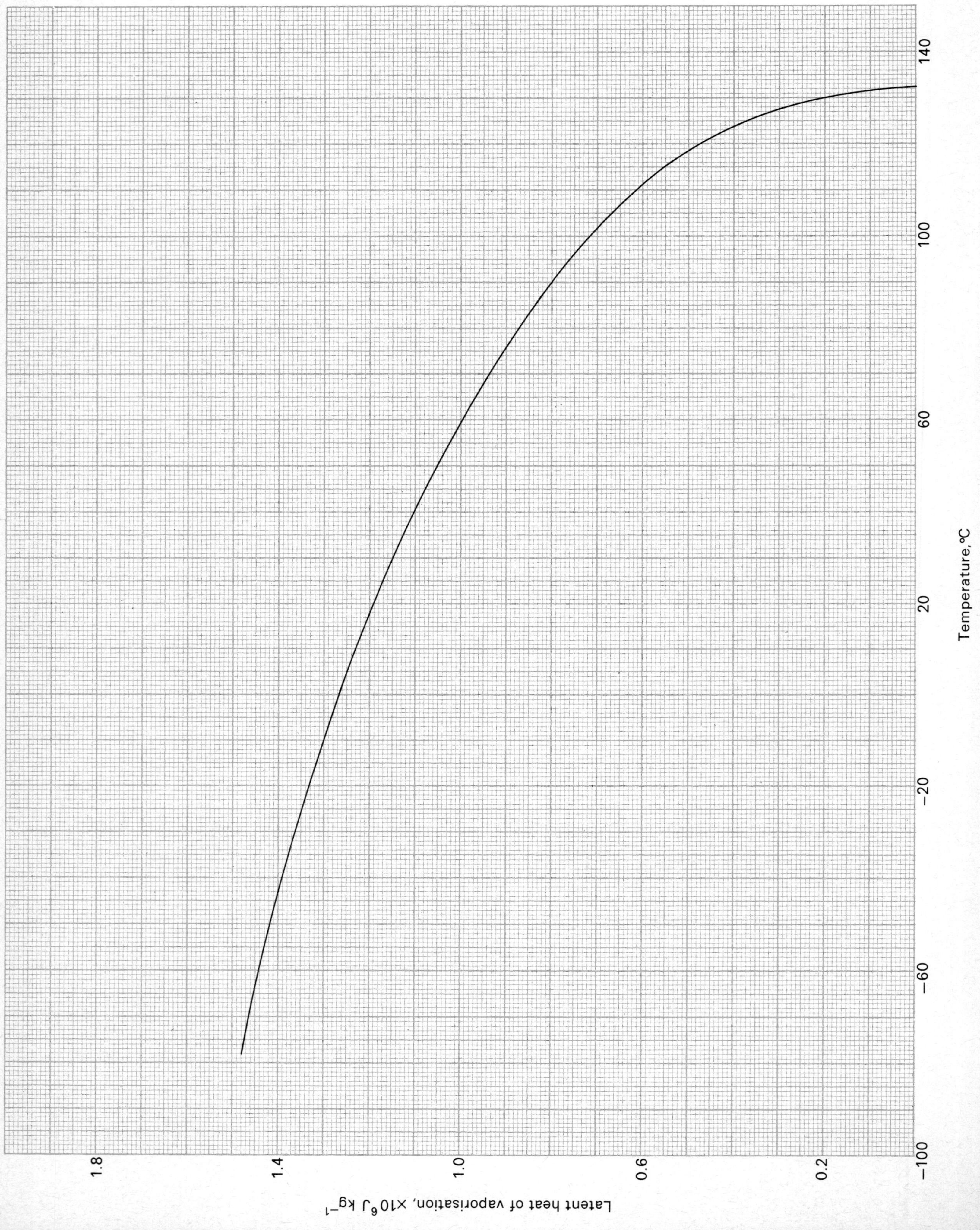

25.5 Heat Capacity of Liquid Ammonia (38,52,54,57)

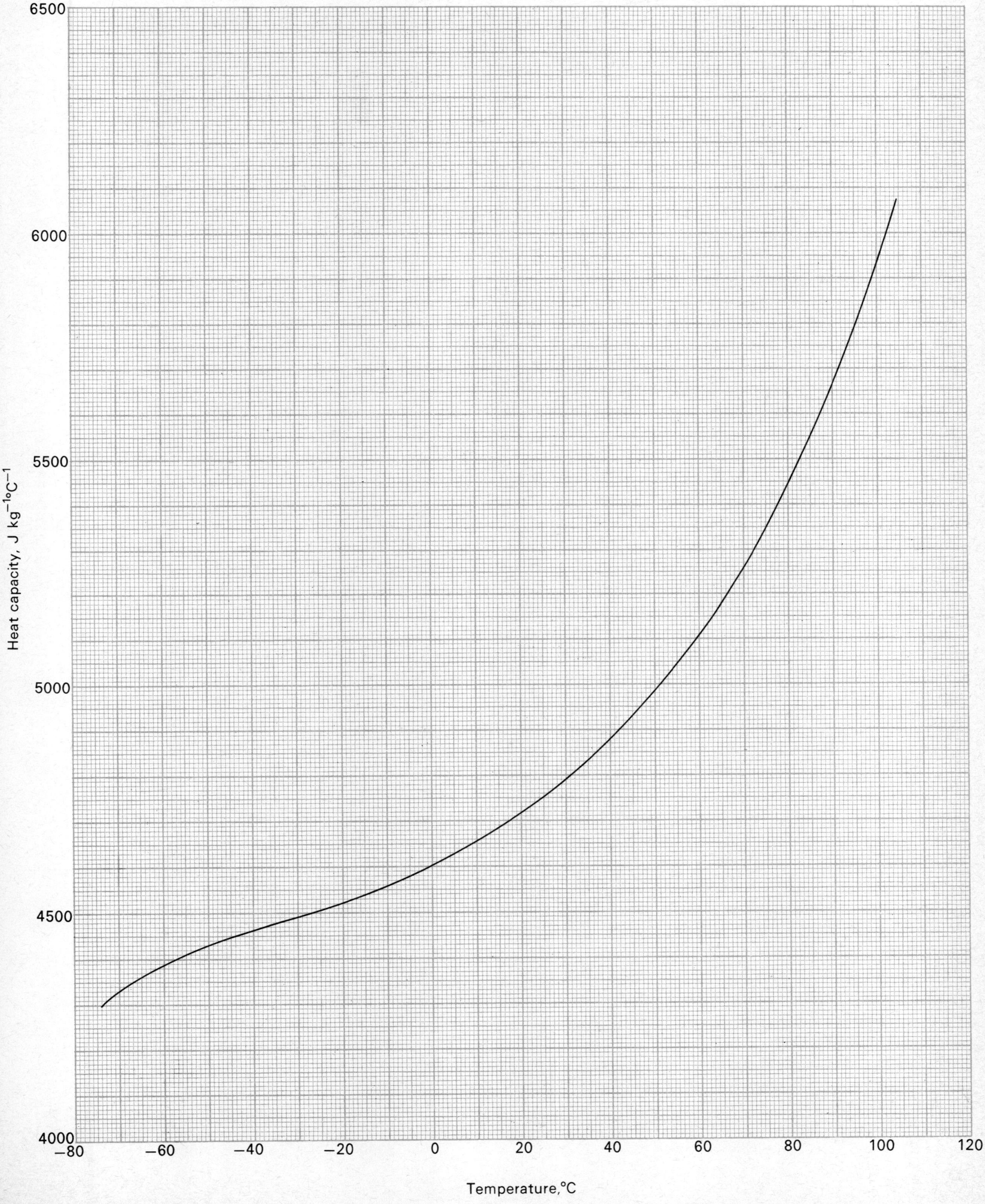

25.6 Heat Capacity of Gaseous Ammonia at $1.0133 \times 10^5 N\ m^{-2}$ (14,45,65)

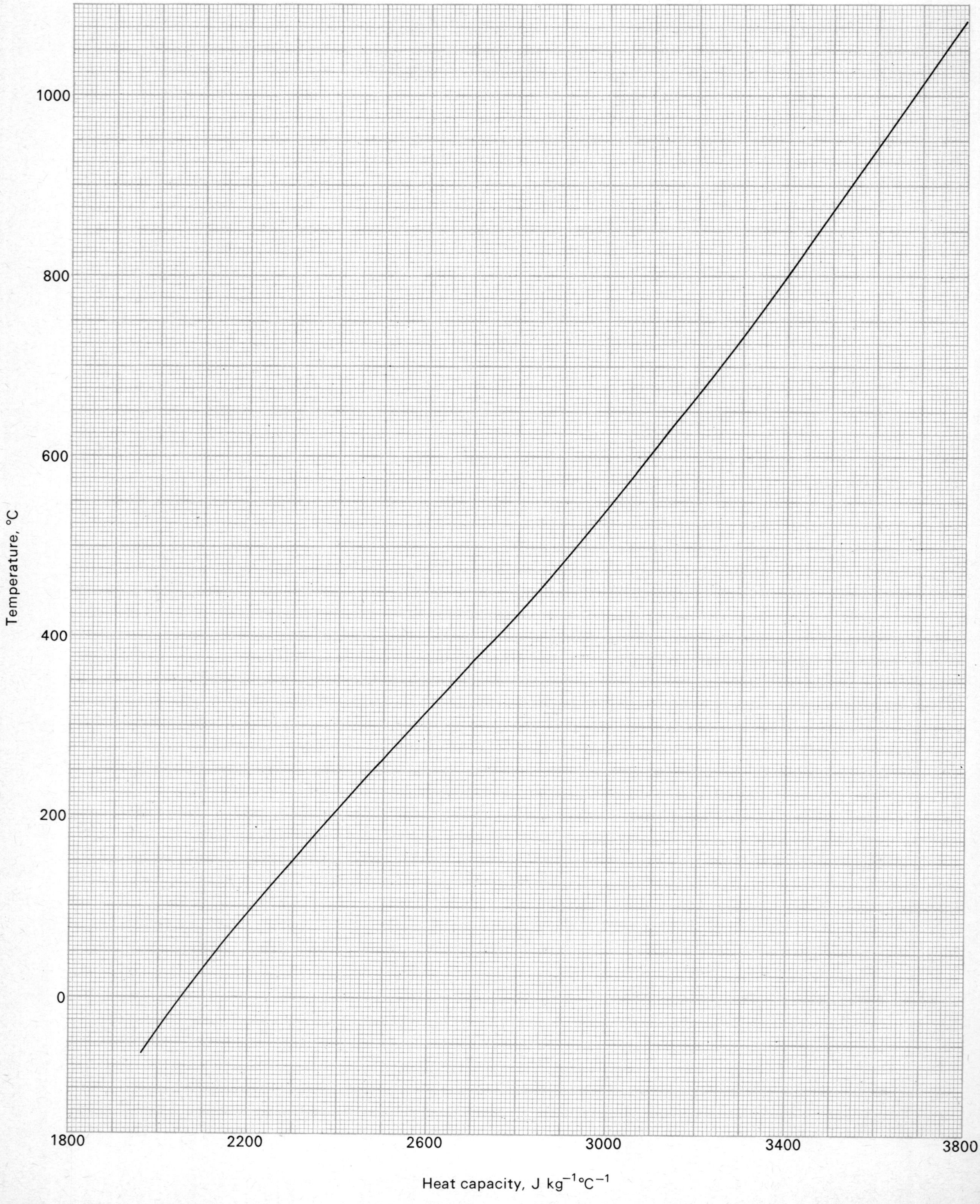

25.7 Heat Capacity Ratio C_p/C_v of Gaseous Ammonia (11)

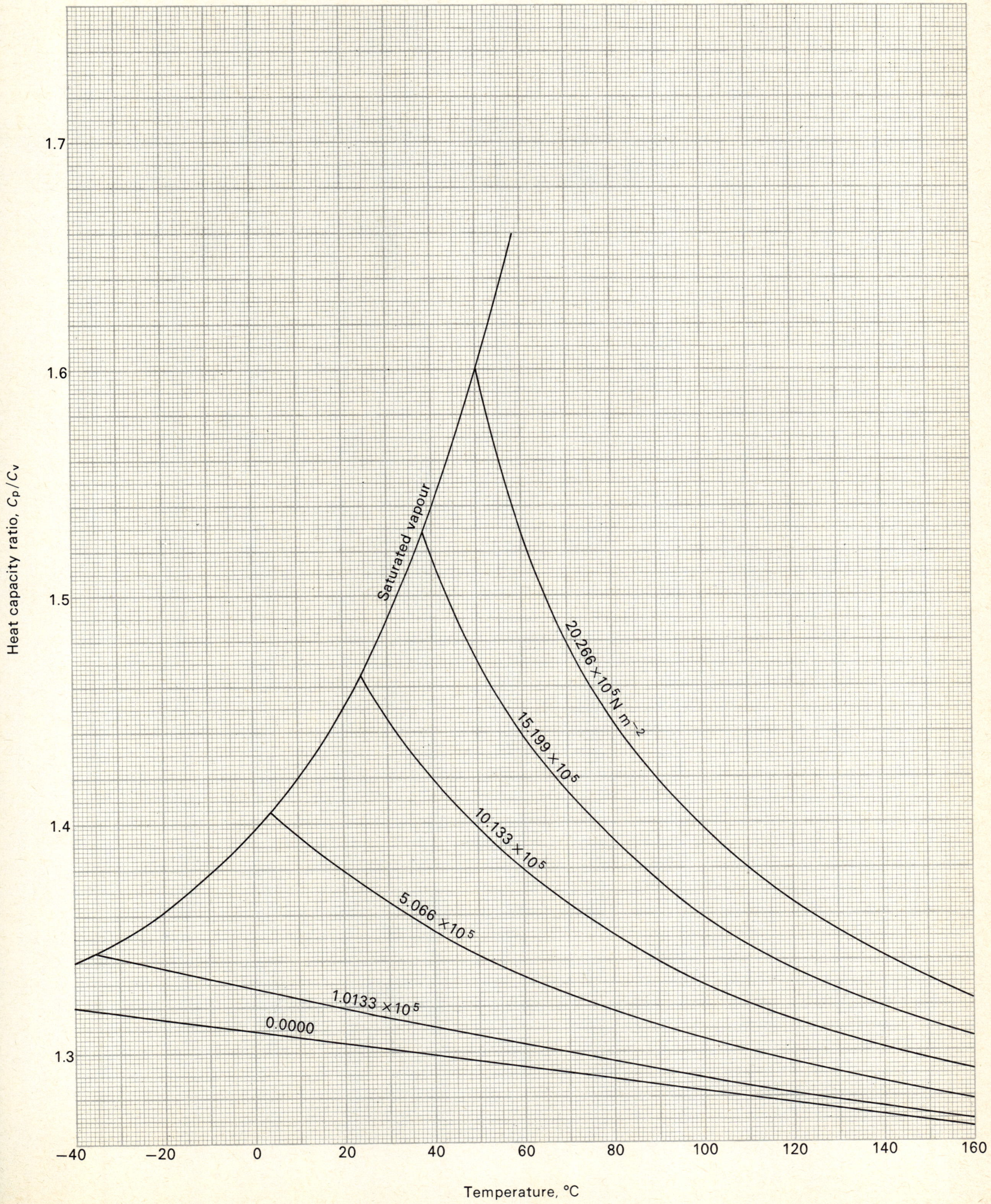

25.8 Viscosity of Liquid Ammonia (16,23,26,33,43)

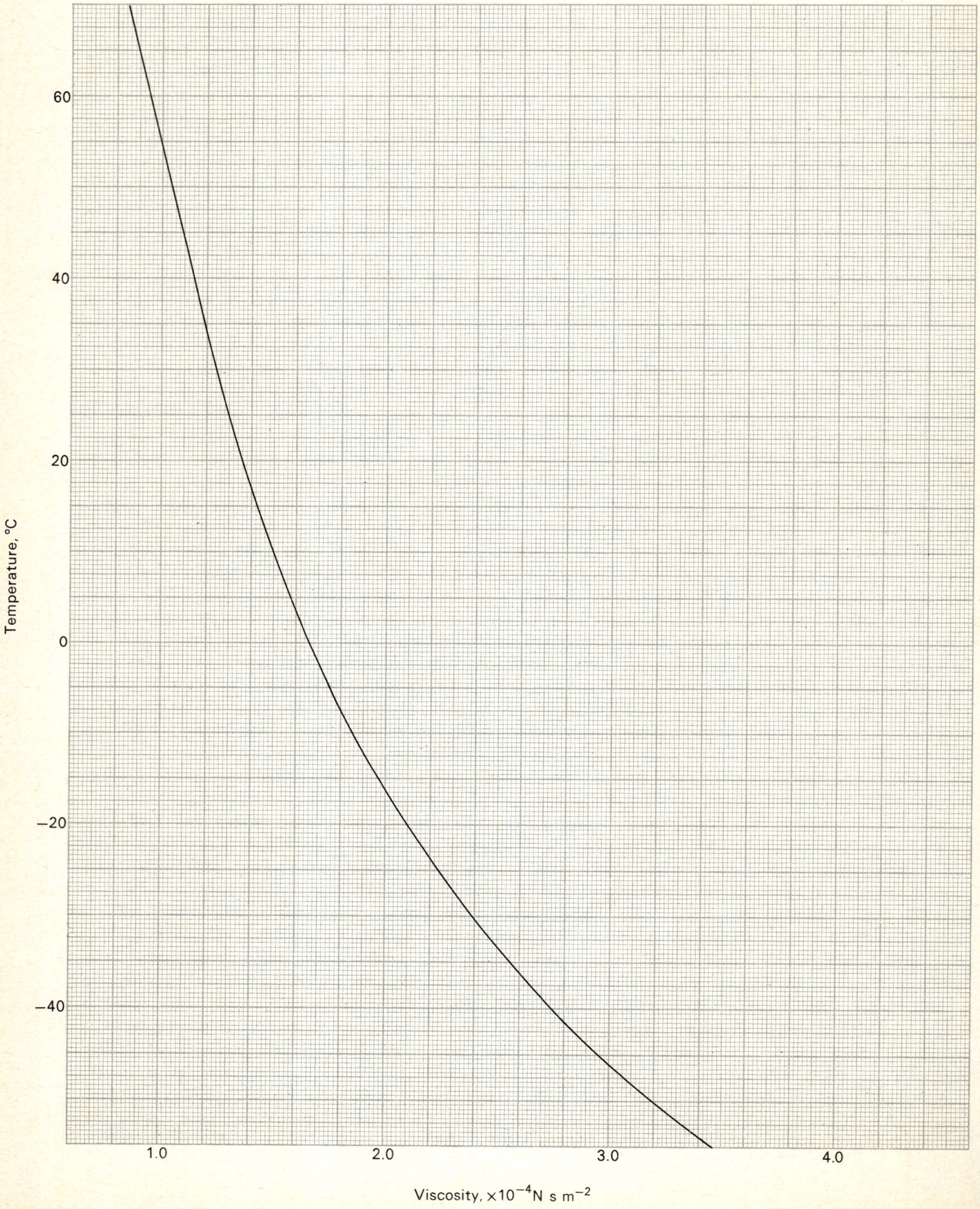

25.9 Viscosity of Gaseous Ammonia at $1.0133 \times 10^5 N\ m^{-2}$ (20,23,26,49,70)

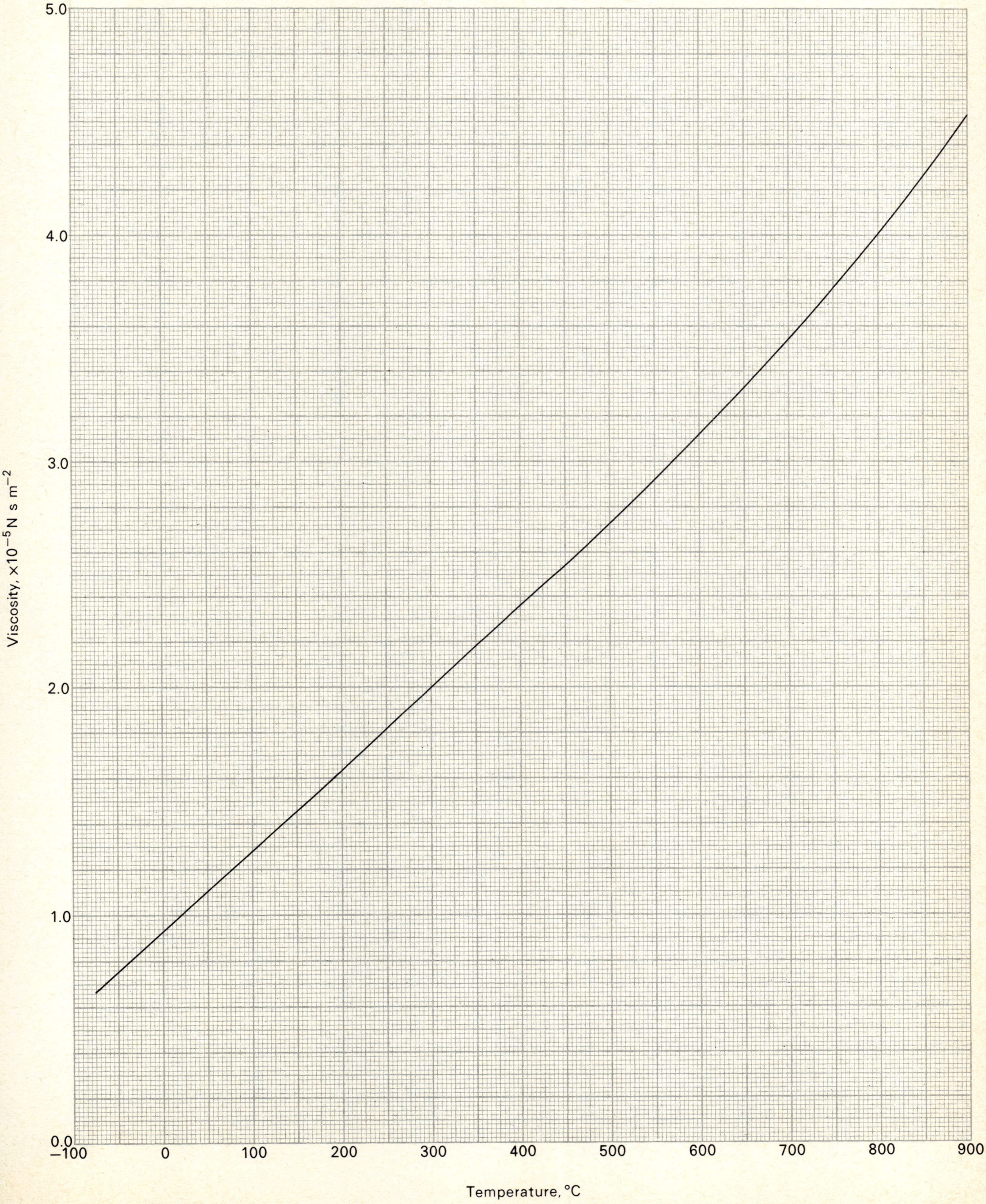

25.10 Thermal Conductivity of Liquid Ammonia (23,26,34,35,59,61)

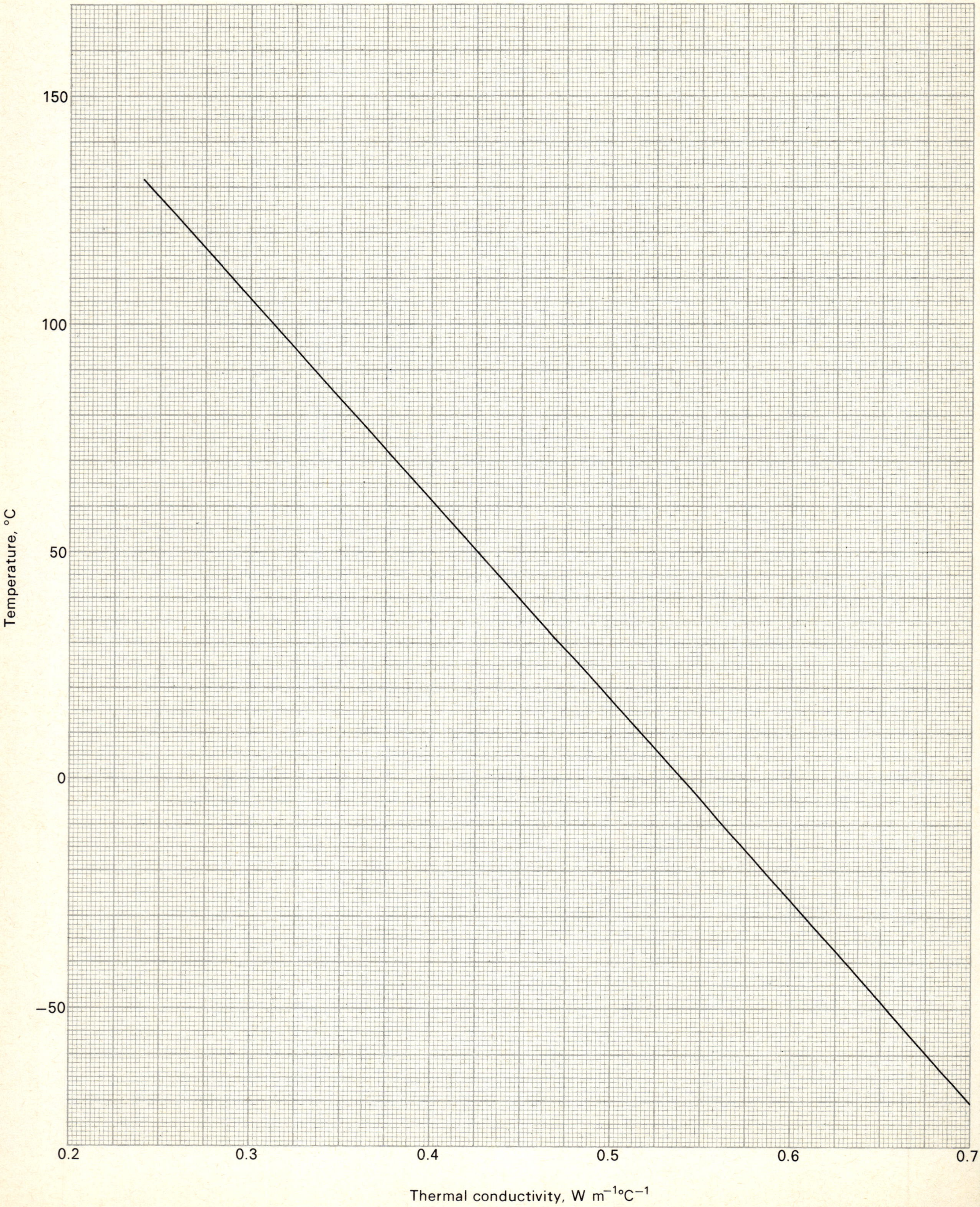

25.11 Thermal Conductivity of Gaseous Ammonia at $1.0133 \times 10^5 N\ m^{-2}$ (2,6,21,26,50,53)

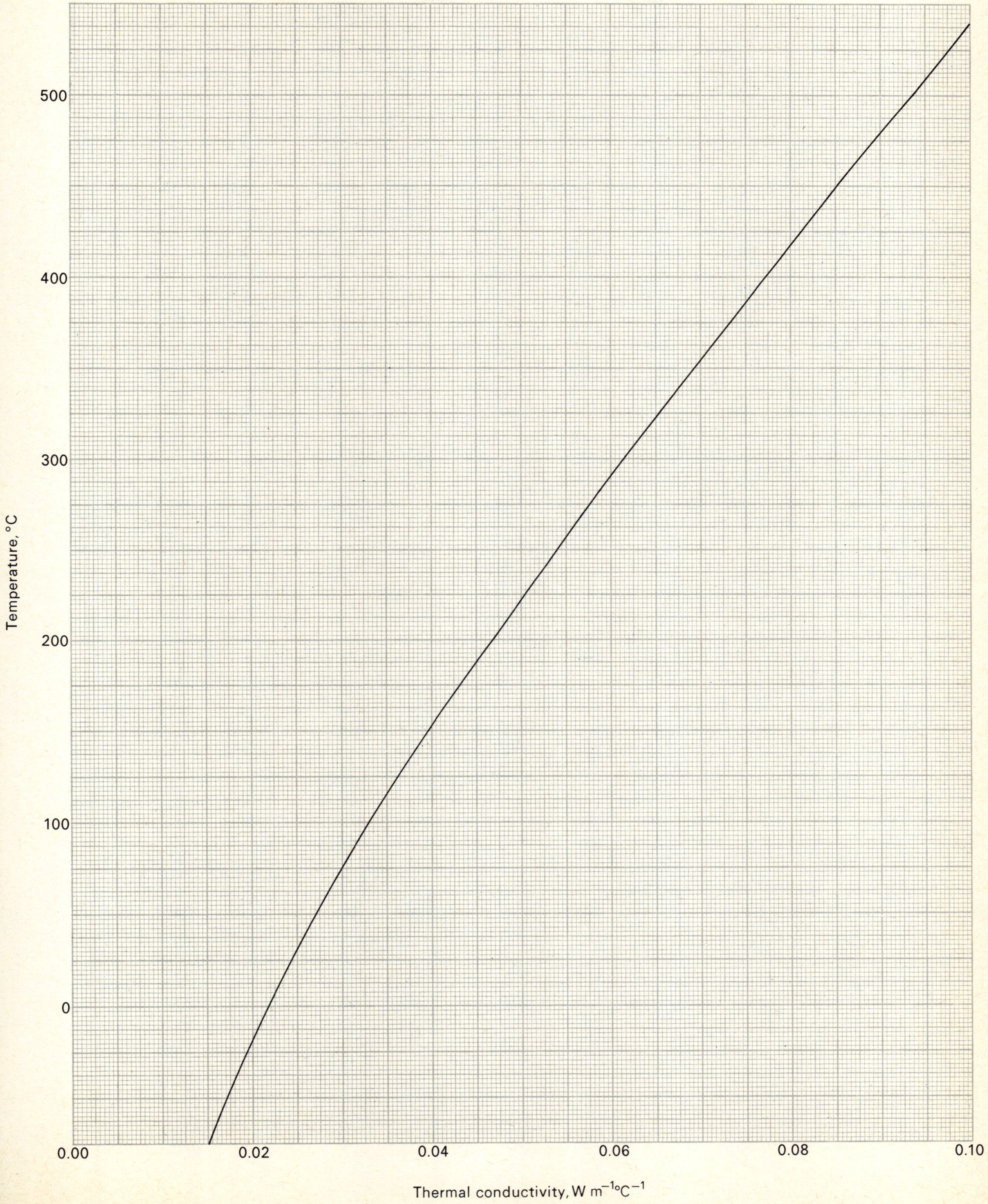

25.12 Surface Tension of Ammonia (19,26,33,55)

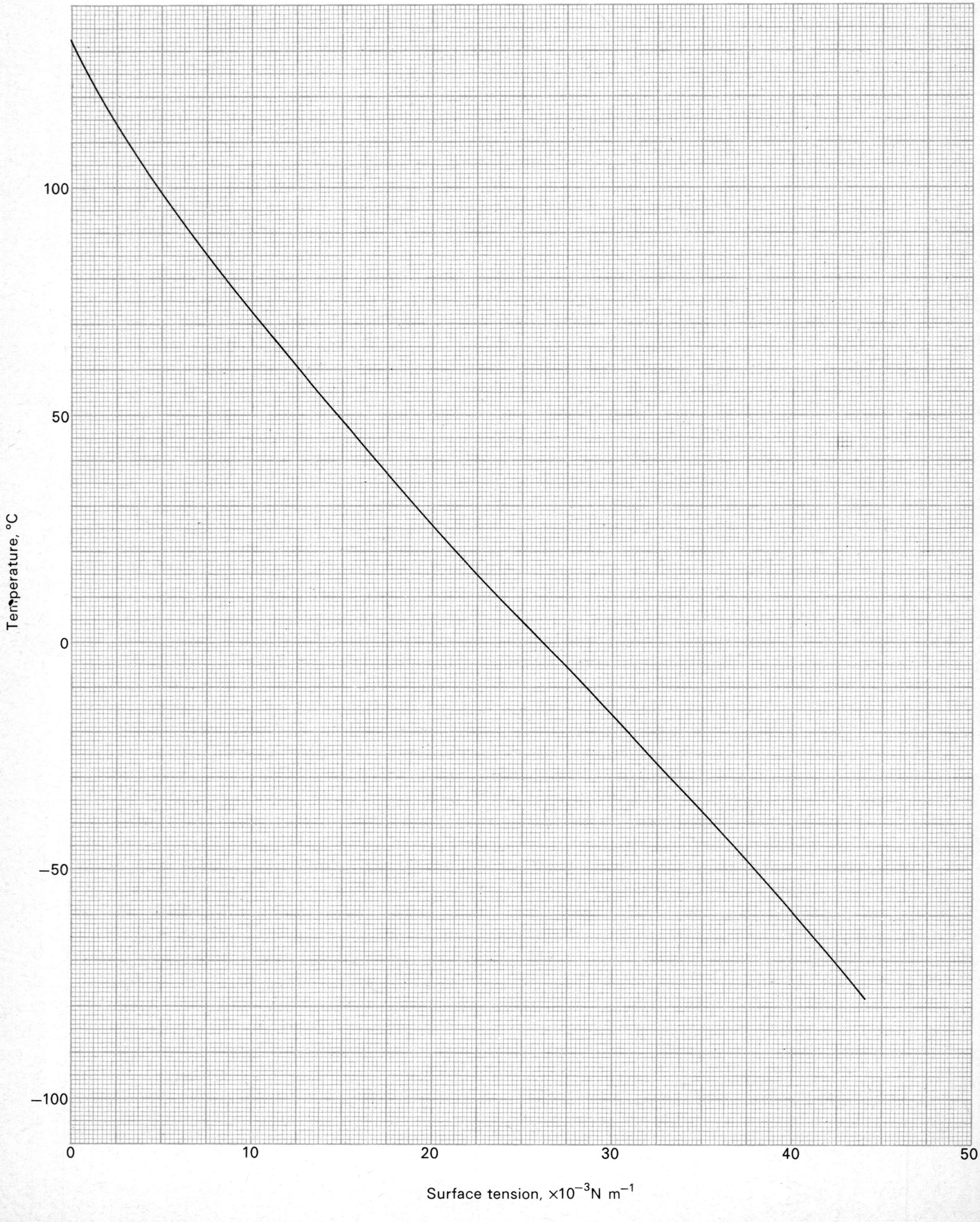

25.13 Electrical Conductivity of Liquid Ammonia (28)

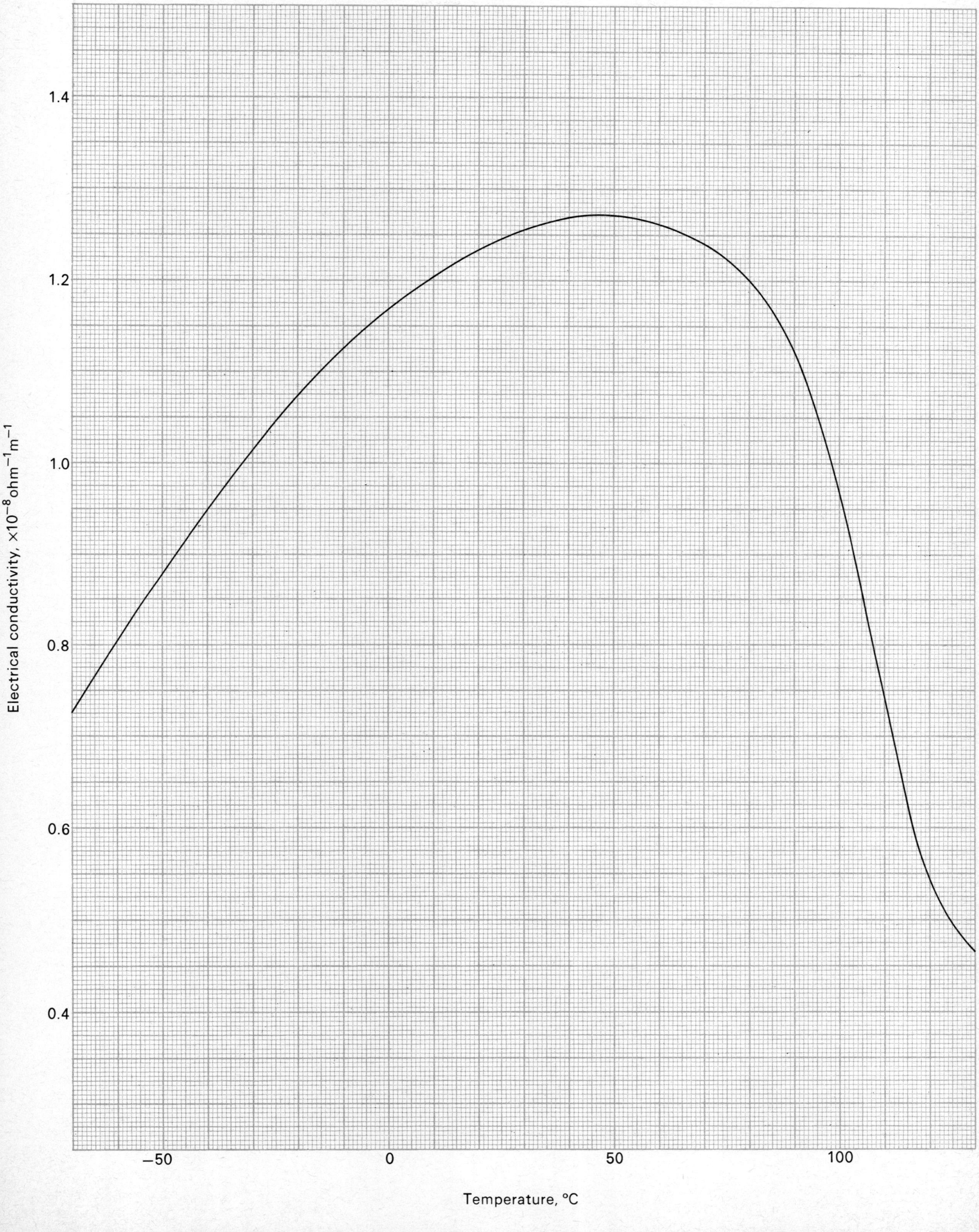

25.14 Solubility of Ammonia in Water at $1.0133 \times 10^5 N\ m^{-2}$ Total Pressure (3,13,40,41)

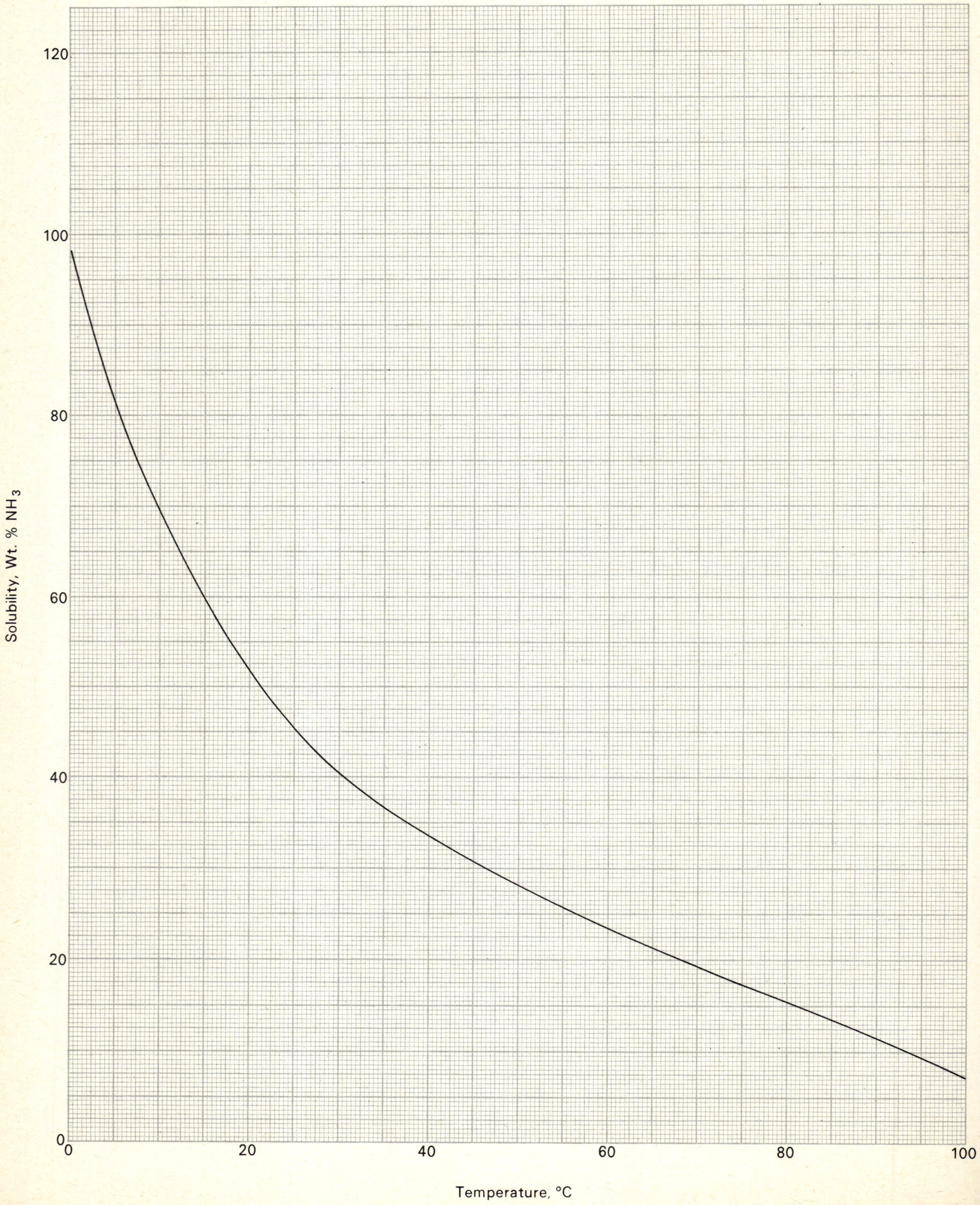

26 DEUTERIUM

Deuterium was discovered by Urey and co-workers at Columbia University in 1931. It is one of the isotopes of hydrogen with mass number of 2 and atomic mass of 2.016. It occurs in natural hydrogen, water, and other compounds with hydrogen contents. The average abundance is 0.015 mole %. It can be separated from hydrogen by electrolysis, chemical exchange, gas chromatography, and distillation.

The physical properties of D_2 are somewhat different from those of normal hydrogen. However, like hydrogen, it exists in ortho and para forms. Deuterium shows nuclear spin isomerism and the spin is 1 instead of $\frac{1}{2}$ in the case of hydrogen, and therefore the ortho modification is more stable than the para form at low temperatures, in contrast to hydrogen. The ratio of ortho to para is 2:1 for D_2 and 3:1 for H_2.

The dissociation energy is greater for deuterium than for the hydrogen molecule, because the electronic binding energy is lower in D_2. Hence deuterium is more stable than hydrogen. In reactions, D_2 requires greater activation energy than hydrogen. The most important factor contributing to the different bond energies is the large difference in zero-point energy between a bond to deuterium and a bond to hydrogen.

The differences in the physical and chemical properties of D_2 and H_2 are greater than those for isotopes of other elements, because of the large relative difference in mass.

Deuterium finds applications in biological processes, chemical reactions, and as a moderator for neutrons. In chemical reactions, it reacts more slowly than hydrogen with corresponding compounds, due to the large mass and zero-point energy differences.

The thermophysical properties of deuterium have been determined since 1932, and the agreement between the various investigations is quite good. The compilations on various properties are available in the following works: Guizouan (22), Landolt-Börnstein (29), Pascal (42), and Prydz *et al.* (44). In 1965, Dimiduk (13) compiled a bibliography on the thermodynamic properties of several relevant gases including deuterium.

There is no reliable data on the solubility of D_2 in water, but it is expected that the solubility of D_2 is lower in water than that of hydrogen at the same temperature and pressure.

References

1 Andrusov, L., *J. Chim. Phys.*, 1952, **49**, 599–604.
2 Andrussow, L., *Diffusion, Viscosity, and Conductivity of Gases*, Progr. Intern. Res. Thermodyn. Transport Properties, 2nd Symp., Thermophys. Properties, (J. F. Masi, D. H. Tsai, Eds.), Princeton Univ., Am. Soc. Mech. Eng., Academic Press, New York, 1962, pp. 279–87.
3 Arnold, R. D., Hoge, H. T., *J. Chem. Phys.*, 1950, **18** (9), 1295.
4 Bartholomé, H. E., Eucken, A., *Z. Elektrochem.*, 1936, **42**, 547–51.
5 Barua, A. K., Afzal, M., Flynn, G. P., Ross, T., *J. Chem. Phys.*, 1964, **41** (2), 374–8.
6 Bigeleisen, J., *J. Chim. Phys.*, 1963, **60** (1–2), 35–43.
7 Chelton, D. B., Mann, D. B., OTS, Univ. of California, Lawrence Rad. Lab. Rept. UCRL-3421, 1956, 116 pp.
8 Chelton, D. B., Mann, D. B., WADC-TR-59-8, (AD 208155), 1959, 144 pp.
9 Clusius, K., Bartholome, E., *Nachr. Ges. Wiss. Göttingen Math-Phys. Klasse*, 1934, **1**, 29–39.
10 Clusius, K., Bartholome, E., *Z. phys. Chem.*, 1935, **30B**, 237–57.
11 Clusius, K., Bartholome, E., *Z. Elektrochem.*, 1935, **41**, 487.
12 Daney, D. E., *Cryogenics*, 1971, **11** (4), 290–7.
13 Dimiduk, P. W., *Bibliography of References for the Thermophysical Properties of He, H, D, Ne, Ar, N, O, CO, CO_2, CH_4, etc. . . .*, U.S. Air Force Systems Command, Res. Tech. Div., Techn. Rept. AFML 65–338, Washington, 1965, 140 pp.
14 Engineering Science Data Units, *Approximate Data on the Viscosity of Some Common Liquids*, Item No. 66024, London, July 1966, 26 pp.
15 Eyring, H., Henderson, D., Ree, T., *Thermodynamic and Transport Properties of Liquids*, Progr. Internat. Res. Thermodyn. Transport Properties, 2nd Symp., Thermophys. Properties (J. F. Masi, D. H. Tsai, Eds.), Princeton Univ., Am. Soc. Mech. Eng., Academic Press, New York, 1962, pp. 340–51.
16 Friedman, A. S., Hilsenrath, J., Natl. Bur. Std. Rept. 3163, Washington, 1954, 12 pp.
17 Friedman, A. S., Trzeiciak, M., Johnston, H. L., *J. Am. Chem. Soc.*, 1954, **76**, 1552–3.
18 Golubev, I. F., *Viscosity of Gases and Gas Mixtures*, Gosudarstvennoe Izdatel'stvo Fiziko-Matematicheskoi Literatury, Moskva, 1959, Table 4.
19 Grigor'ev, V. N., Rudenko, N. S., *Zh. Eksperim. Teor. Fiz.*, 1964, **47** (1), 92–6.
20 Grilly, E. R., *J. Am. Chem. Soc.*, 1951, **73**, 843–6.
21 Grosh, J., Jhon, M. S., Ree, T., Eyring, H., *Proc. Natl. Acad. Sci. U.S.*, 1967, **58** (6), 2196–2203.
22 Guizouarn, L., *Molecular Physical Properties of the Hydrogen Isotopes and of Their Isotopic Combinations*, Commis. Energ. At.(Fr.), Serv. Doc. Ser. Bibliogr. No. 87, Paris, 1967, 49 pp.
23 Hall, L. A., *Thermal Conductivity of 10 Cryogenic Liquids—A Bibliography*, Natl. Aeron. Space Admin., Accession No. 65-17283, Rept. No. NASA-CR-50292, Washington, 1963, 21 pp.
24 Henderson, D., Kim, S., Eyring, H., *Proc. Natl. Acad. Sci. U.S.*, 1962, **48**, 1753–9.
25 IUPAC, Atomic Weights of the Elements 1971, *Pure & Appl. Chem.*, 1972, **30** (3–4), 638–49.
26 Johnston, H. L., Long, E. A., *J. Chem. Phys.*, 1934, **2**, 389–95.
27 Kerrisk, J. F., Rogers, J. D., Hammel, E. F., LADC 5791, 1963, 21 pp.
28 Klipping, G., Schmidt, F., *Kältetechnik*, 1965, **17** (12), 382–4.
29 *Landolt-Börnstein, Zahlenwerte und Funktionen aus Physik, Chemie, Astronomie, und Technik*, 7th ed., IV Band, Technik, 4 Teil, Bandteil a, Springer-Verlag, Berlin, 1967, 206–28.
30 Lax, E., Synowietz, C., Eds., *Taschenbuch für Chemiker und Physiker*, 3rd ed., Springer-Verlag, Berlin, 1967.
31 Lenoir, J. M., Univ. Arkansas, Eng. Expt. Station Bull. 18, 1953, 48 pp.
32 Liley, P. E., *The Thermal Conductivity of 46 Gases at Atmospheric Pressure*, Proc. Symp. Thermophys. Prop. 4th, (T. R. Moszynski, Ed.), Am. Soc. Mech. Eng., New York, 1968, pp. 323–49.
33 Maryott, A. A., Smith, E. R., *Table of Dielectric Constants of Pure Liquids*, Natl. Bur. Std. Circular 514, Washington, Aug. 10, 1951, 44 pp.
34 Mathews, J. F., *Chem. Rev.*, 1972, **72** (1), 71–100.
35 Megaw, H. D., *Phil. Mag.*, 1939, **28**, 129.
36 Megaw, H. D., Simon, F., *Nature*, 1936, **138**, 244.
37 Michels, A., DeGraaff, W., Wolkers, G. J., *Physica*, 1959, **25**, 1097–124.
38 Michels, A., DeGraaff, W., Wolker, G. J., *Appl. Sci. Res.*, 1963, **12A**, 9–32.
39 Michels, A., Goudeket, M., *Physica*, 1941, **8** (3), 353–60.
40 Michels, A., Schipper, A. C. J., Rintoul, W. H., *Physica*, 1953, **19**, 1011.
41 Mittelhauser, H. M., Thodos, G., *Cryogenics*, 1964, **4**, 368–73.
42 Pascal, P., Ed., *Nouveau Traite de Chimie Minérale*, Vol. I, Generalités, Air, Hydrogene, Deuterium, Tritium, Helium, et Gas Inerts, Masson et Cie, Éditeurs, Paris, 1956, pp. 720–47.
43 Powers, R. W., Mattox, R. W., Johnston, H. L., *J. Am. Chem. Soc.*, 1954, **76**, 5974.

44 Prydz, R., Timmerhaus, K. D., Stewart, R. B., The Thermodynamic Properties of Deuterium, *Adv. in Cryogenic Eng.*, 1968, **13**, 384–96.
45 Rossini, F. D., Wagman, D. D., Evans, W. H., Levine, S., Jaffe, I., *Selected Values of Chemical Thermodynamic Properties*, Natl. Bur. Std. Circular 500, Washington, Feb. 1, 1952.
46 Rudenko, N. S., Konareva, V. G., *Zh. Fiz. Khim.*, 1963, **37**, 1964 and 2761.
47 Rudenko, N. S., Konareva, V. G., *Zh. Fiz. Khim.*, 1964, **38**, 270.
48 Rudenko, N. S., Slyusar, V. P., *Russ. J. Phys. Chem.*, 1968, **42** (1), 126–7.
49 Stiel, L. I., Thodos, G., *AIChE Journal*, 1964, **10** (1), 26–30.
50 Ubbink, J. B., *Physica*, 1948, **14**, 165–74.
51 Van Itterbeek, A., *Physica*, 1940, **7**, 325–8.
52 Van Itterbeek, A., Van Paemel, O., *Physica*, 1940, **7**, 208 and 262.
53 Van Itterbeek, A., Van Paemel, O., *Physica*, 1941 **8** (2), 133–43.
54 Willis, W. L., *Cryogenics*, 1966, **6** (5), 279–84.
55 Woolley, H. W., Scott, R. B., Brickwedde, F. G., *J. Res. Natl. Bur. Std.*, 1948, **41** (5), 379–475.

Table 26.1 Properties of Deuterium (normal)

		Ref.
Chemical formula	D_2	(30)
Molecular weight	4.0320	(25)
Critical temperature	−234.9 °C	(34)
Critical pressure	16.52×10^5 N m^{-2}	(34)
Critical density	66.9 kg m^{-3}	(34)
Triple-point temperature	−254.5 °C	(30)
Triple-point pressure	0.1707×10^5 N m^{-2}	(30)
Boiling point at 1.0133×10^5 N m^{-2}	−249.48 °C	(15)
Melting point	−254.42 °C	(15)
Density of solid at −268.95 °C	51.12 kg m^{-3}	(12, 35, 36)
Compressibility of liquid at −252.82 °C	1.0136×10^{-8} m^2 N^{-1}	(17)
Heat of formation of gas at 25 °C	0.0 J kg-mol^{-1}	(45)
Free energy of formation of gas at 25 °C	0.0 J kg-mol^{-1}	(45)
Heat of sublimation at triple-point temperature	1.5240×10^6 J kg-mol^{-1}	(20)
Heat of fusion	1.9680×10^5 J kg-mol^{-1}	(45)
Entropy of gas at 25 °C	1.4487×10^5 J kg-mol^{-1} °C^{-1}	(45)
Heat capacity ratio of gas at −75 °C and 1.0133×10^5 N m^{-2}	1.401	(29)
Velocity of sound at 25 °C and 1.0133×10^5 N m^{-2}	930.0 m s^{-1}	(29)
Diffusion in air at 27.7 °C and 1.0133×10^5 N m^{-2}	0.565×10^{-4} m^2 s^{-1}	(30)
Dielectric constant of liquid at −253 °C	1.277	(33)
Electrical conductivity of liquid at −253.6 °C	1.35×10^{-14} ohm^{-1} m^{-1}	(54)
Refractive index of gas at 0 °C and 1.0133×10^5 N m^{-2}	1.000138	(11)

26.1 Vapour Pressure of Deuterium (3,6,20,21,28,41,44,55)

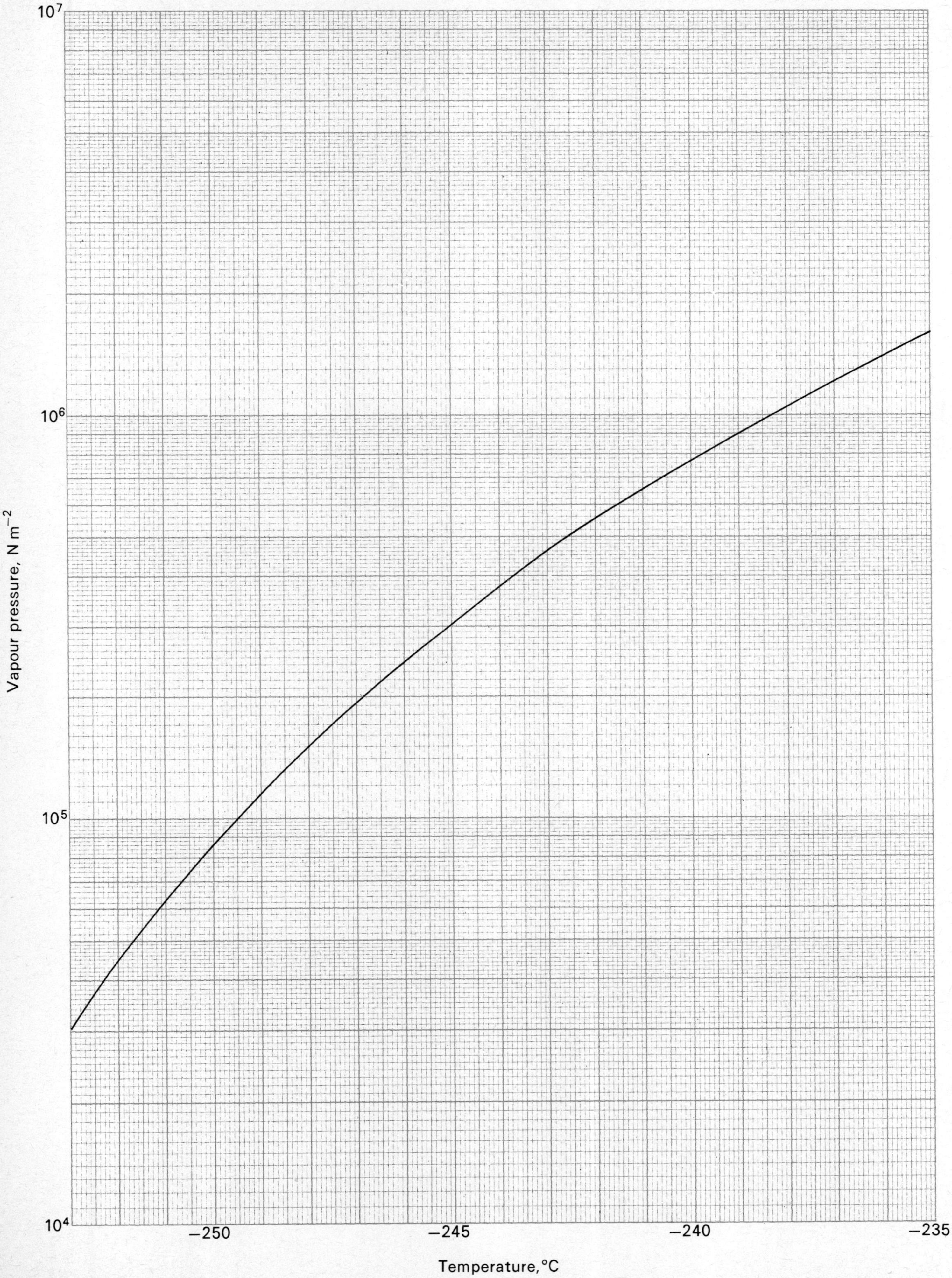

26.2 Orthobaric Density of Deuterium (44,48,55)

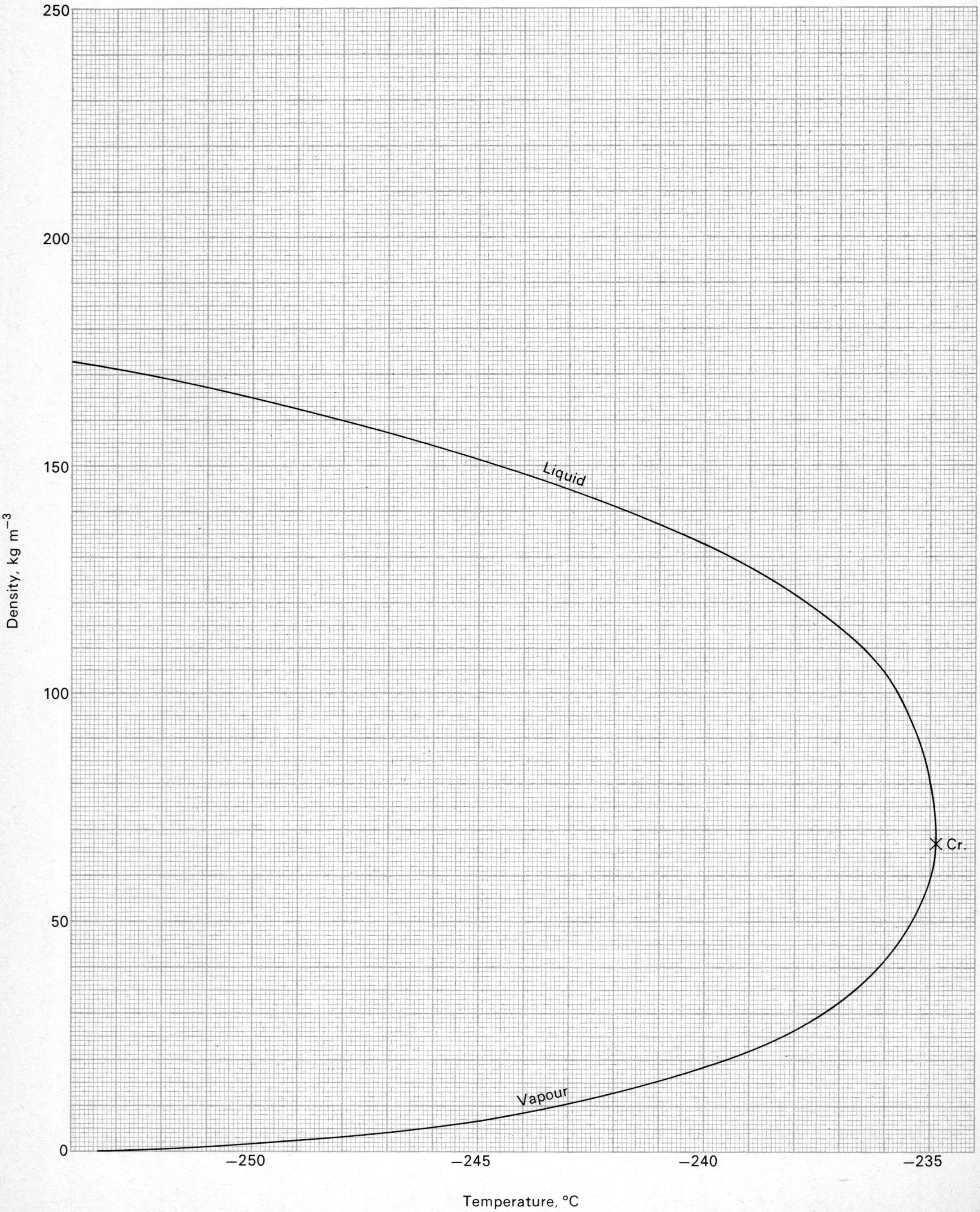

26.3 P-V-T Data for Deuterium (5,17,39,44)

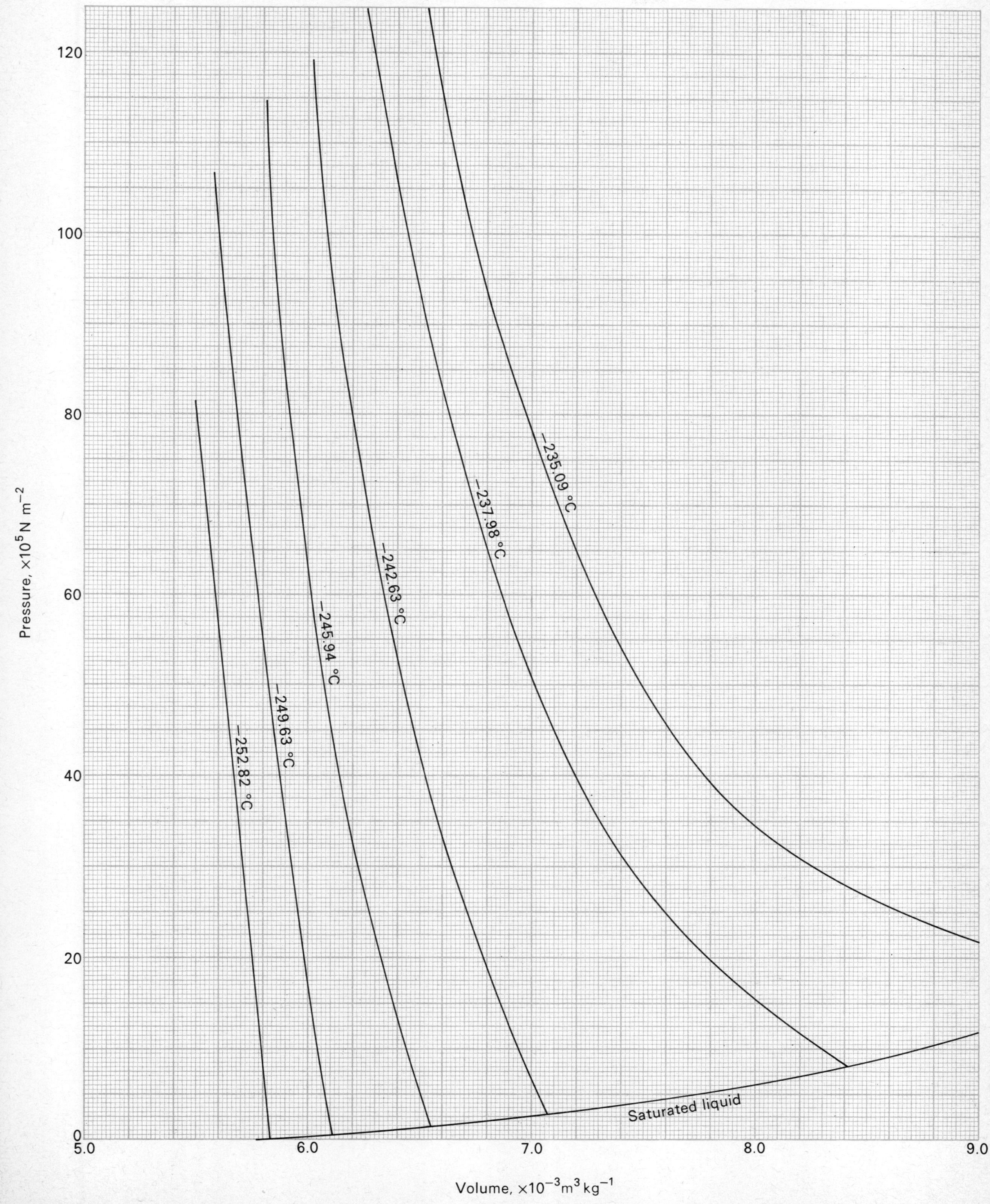

26.4 Latent Heat of Vaporisation of Deuterium (10,20,29,42)

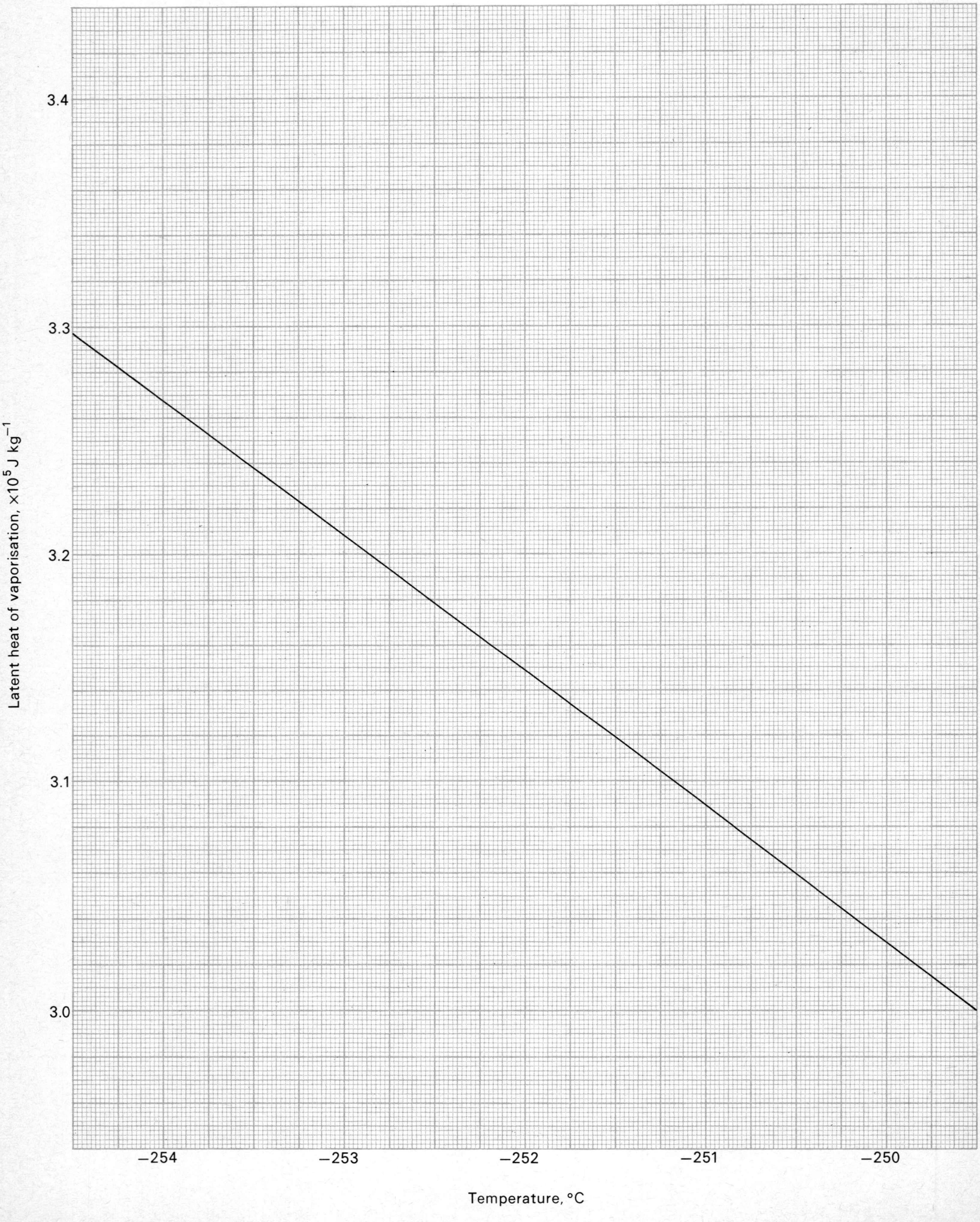

26.5 Heat Capacity of Liquid Deuterium (7,8,10,11,24,55)

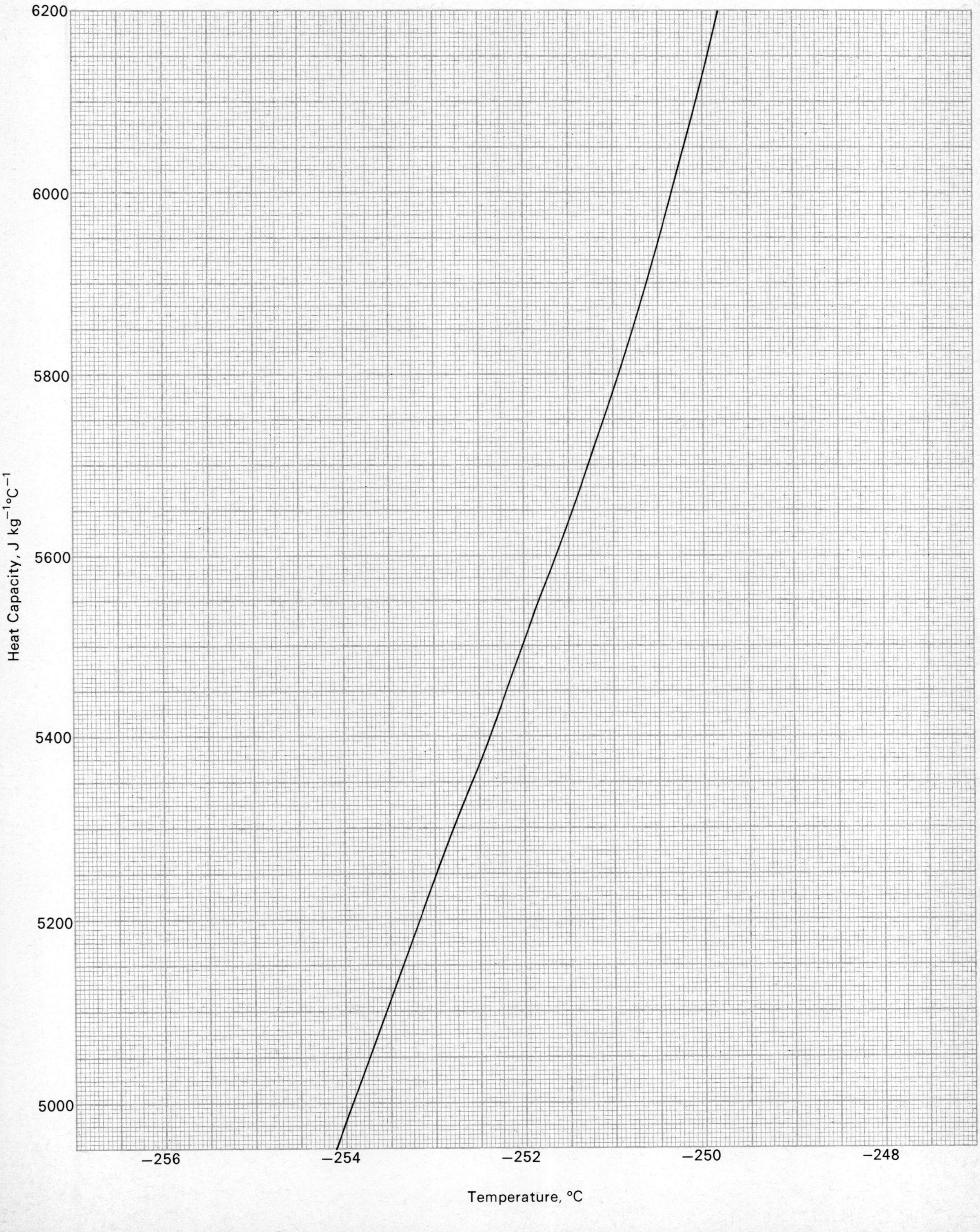

26.6 Heat Capacity of Gaseous Deuterium at $1.0133 \times 10^5 N\ m^{-2}$ (26,37,38,55)

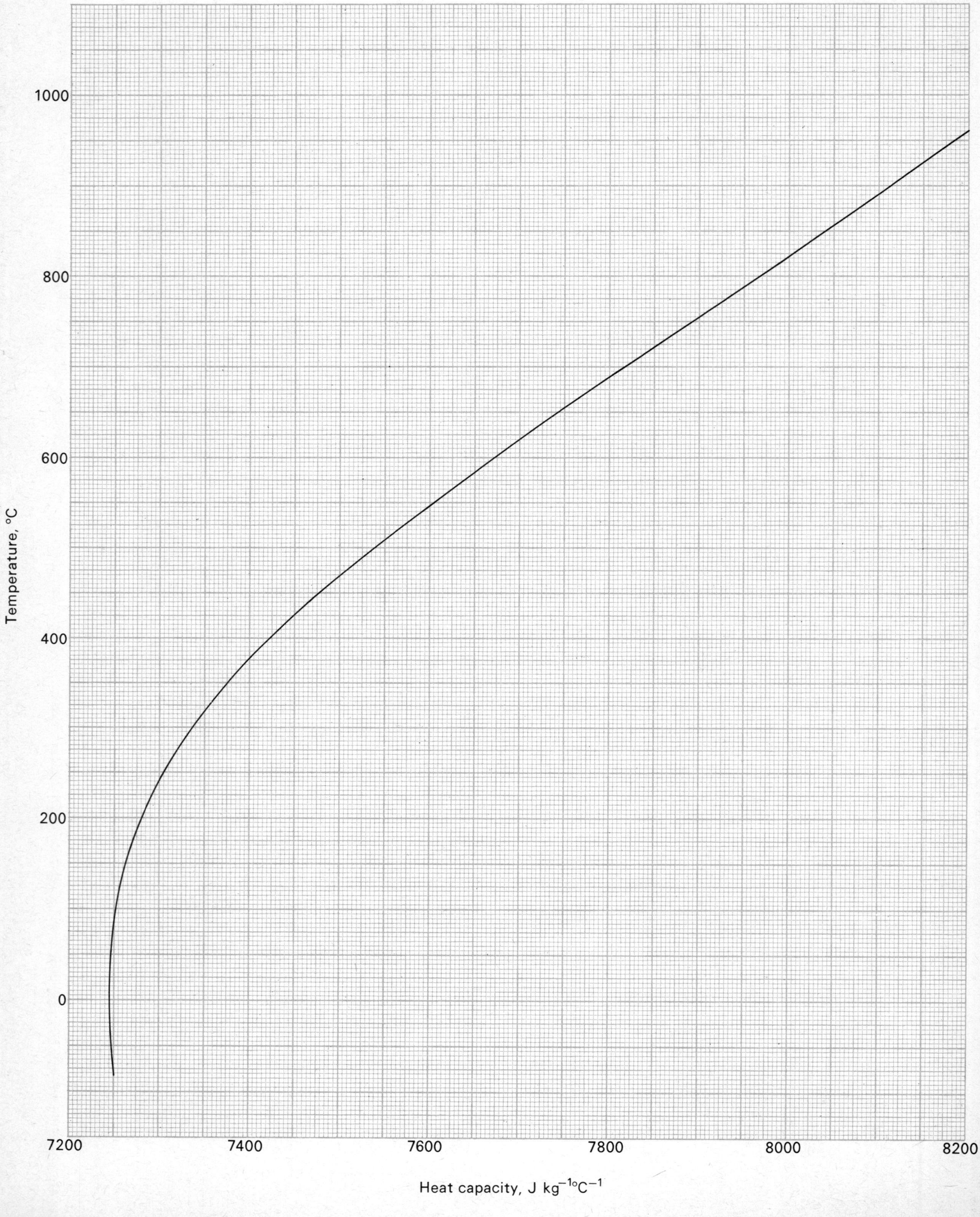

26.7 Viscosity of Liquid Deuterium (14,46,47,52,53)

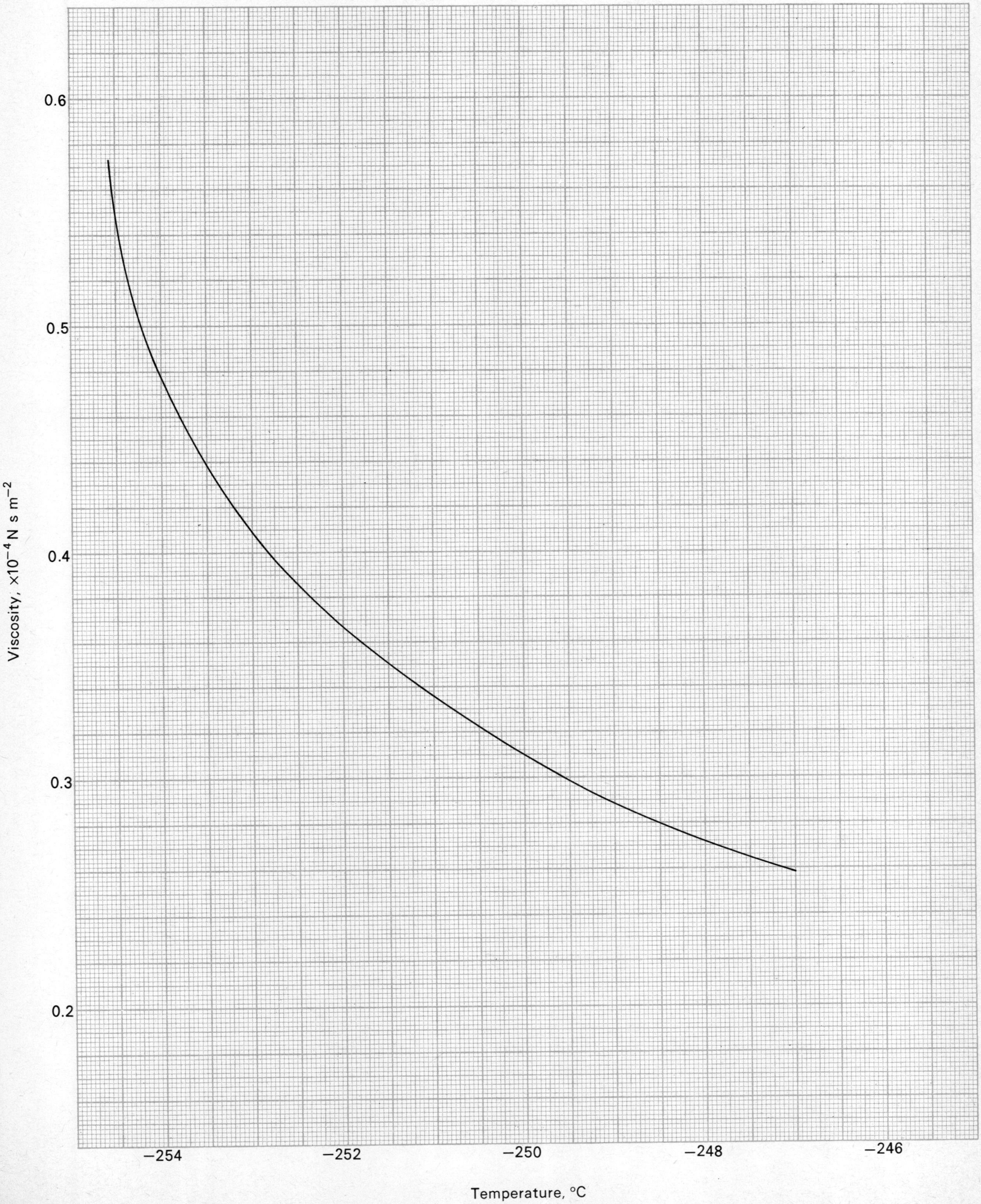

26.8 Viscosity of Gaseous Deuterium (2,5,18,40)

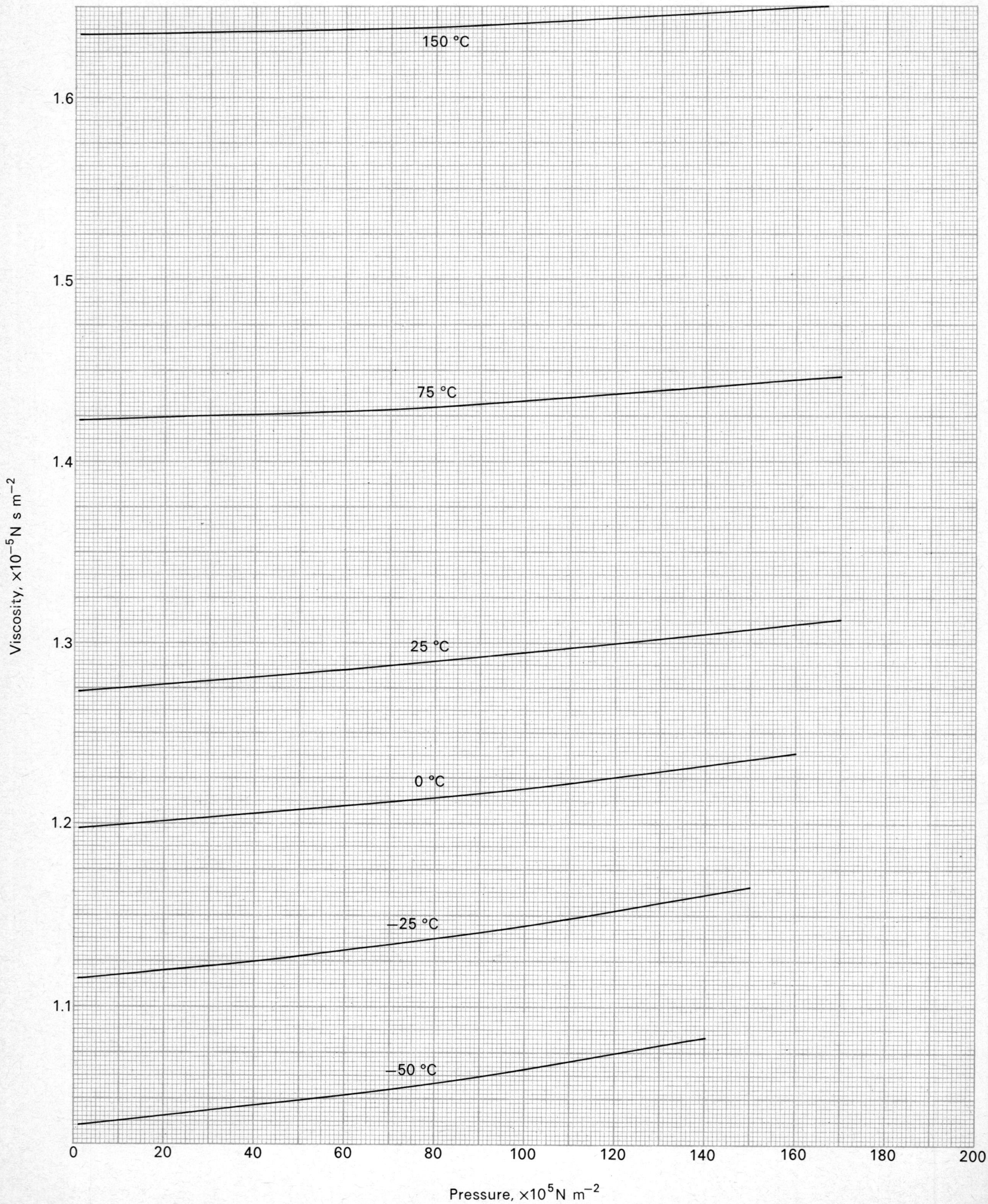

26.9 Thermal Conductivity of Liquid Deuterium (16,23,43)

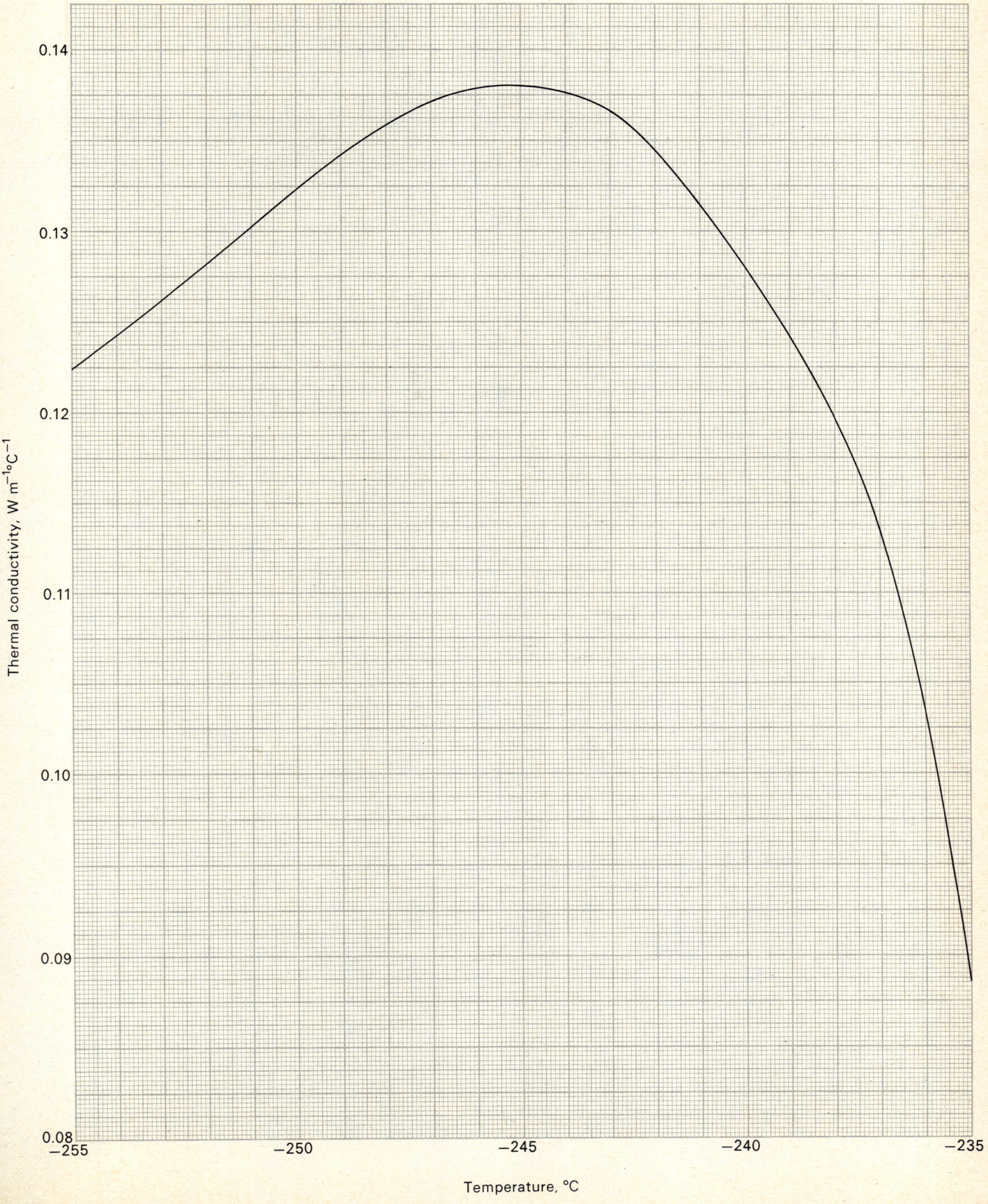

26.10 Thermal Conductivity of Deuterium (2,23,31,32,37,49,50)

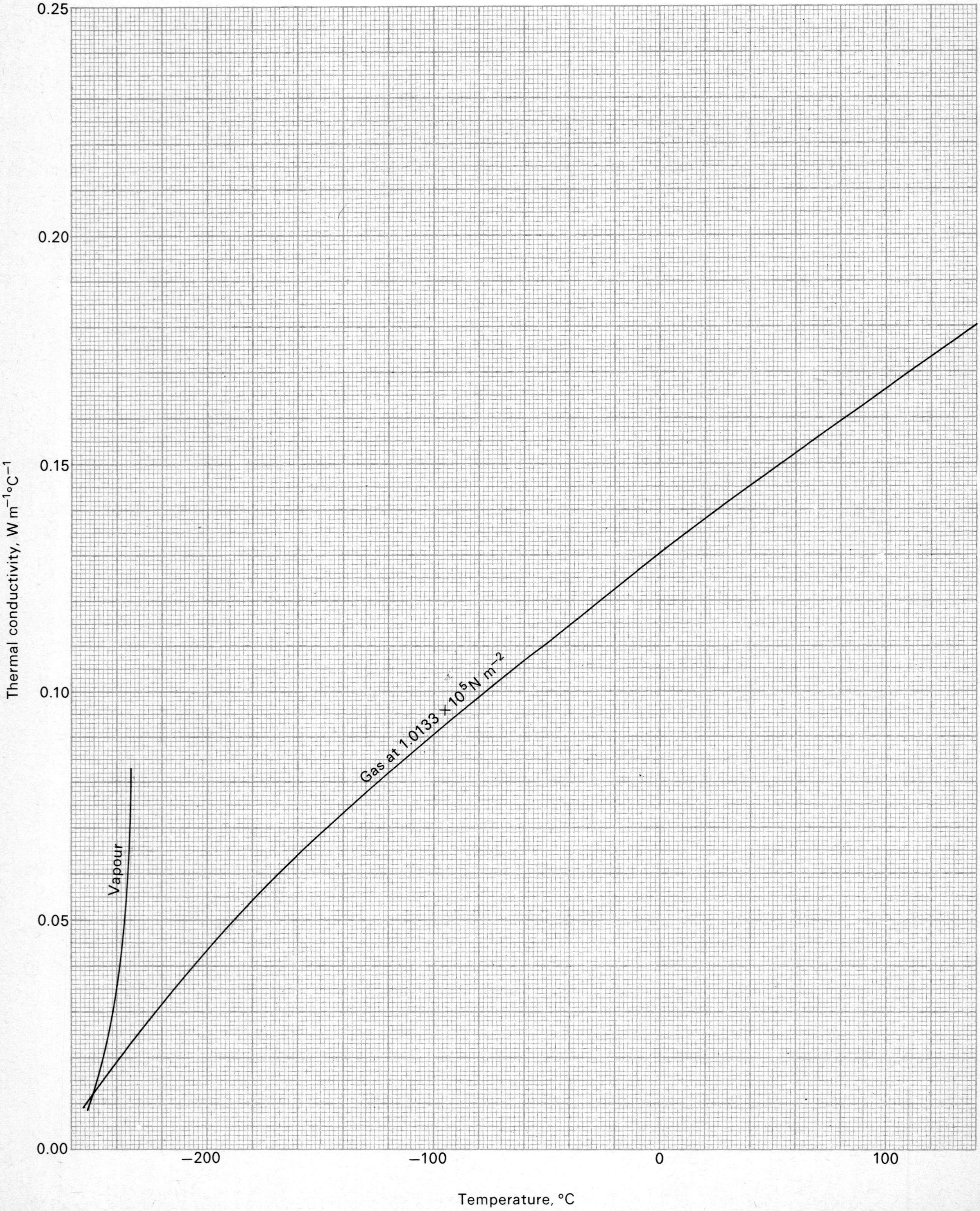

26.11 Surface Tension of Deuterium (19,51)

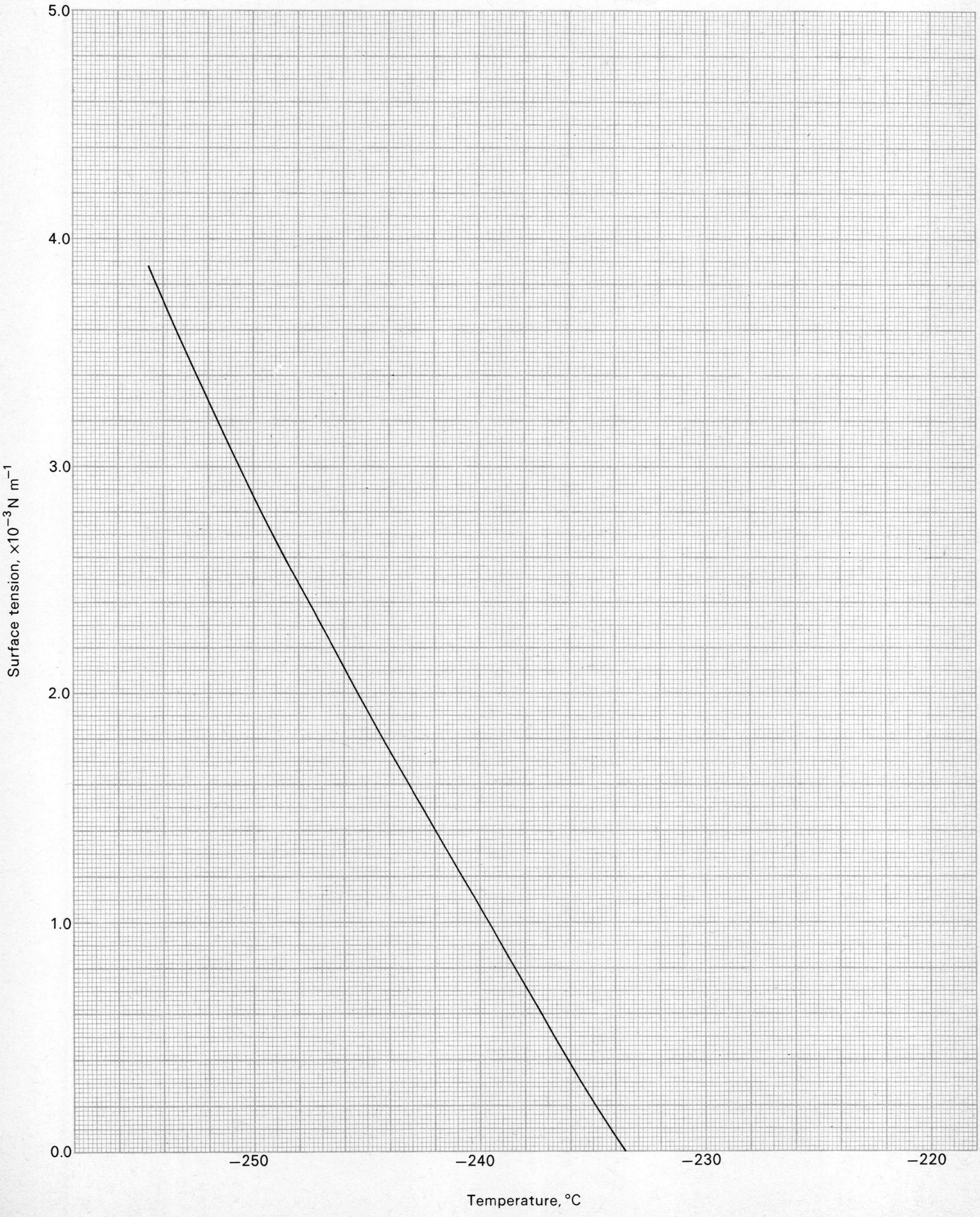

27 HYDRAZINE

Hydrazine was first obtained by T. Curtius in 1887 from organic substances. Anhydrous hydrazine is a colourless, hygroscopic, toxic liquid, having an odour resembling that of ammonia. As a solvent, it behaves like liquid ammonia. The high dipole moment (1.35 debye) and the Raman spectrum are consistent with an unsymmetrical structure for the molecule, as is the high dielectric constant (53.0 at 23°C).

Since hydrazine is a highly polar substance, it is miscible only with polar solvents such as water, the alcohols, ammonia, and the amines; it is insoluble in nonpolar solvents such as the hydrocarbons and the halogenated hydrocarbons. Hydrazine is monomeric in the vapour state and a highly associated substance in the liquid state; aggregation of hydrazine molecules into larger units accounts for some of its unusual properties. Several of its physical properties are similar to those of water. Hydrazine is an endothermic compound and thermodynamically unstable with respect to decomposition into ammonia and nitrogen above 250°C. For studies of thermal decompositions, high-purity hydrazine is necessary; this may be prepared in a special apparatus (30), taking full account of safety precautions.

Hydrazine is extremely soluble in water from which it is obtained as a monohydrate $N_2H_4.H_2O$ (m.p. = −40°C; b.p. = 120°C).

The physical, thermodynamic, and transport properties of anhydrous hydrazine have been summarized in several works: Audrieth and Ogg (4), Clark (11), Das and Kuloor (12), Dreisbach (14), Gallant (18), Gmelin (20), Kirk-Othmer (26), Kobe and Harrison (27), Mellor (34), and Pascal (37).

There has been no experimental investigation of P–V–T relations for anhydrous hydrazine at various temperatures and pressures. A single value for liquid thermal conductivity has been obtained from Fisons Limited, Harston, Cambridge (see Acknowledgement), which agrees with the data predicted by a method of Stiel and Thodos (29). The evaluated result is presented in Fig. 27.9.

References

1 Ahlert, R. C., Bauerle, G. L., Lecce, J. V., *J. Chem. Eng. Data*, 1962, **7** (1), 158–60.
2 Ahlert, R. C., Younts, C. A., *J. Chem. Eng. Data*, 1968, **13** (3), 402–5.
3 American Conference of Government Industrial Hygienists, *Documentation of Threshold Limit Values*, revised edition, Committee on Threshold Limit Values, Cincinnati, Ohio 45202, 1966, 203 pp.
4 Audrieth, L. F., Ogg, B. A., *The Chemistry of Hydrazine*, Wiley, New York, 1951, 244 pp.
5 Baglin, F. G., Bush, S. F., During, J. R., *Far Infrared Spectra and Space Group of Crystalline Hydrazine and Hydrazine-d_4*, Natl. Bur. Std. Spec. Publ. 301, Washington, 1969, 437–40.
6 Baker, N. B., Gilbert, E. C., *J. Am. Chem. Soc.*, 1940, **62**, 2479–80.
7 Barrick, L. D., Drake, G. W., Lochte, H. L., *J. Am. Chem. Soc.*, 1936, **58**, 160–2.
8 Beck, F., *Wien. Chem. Ztg.*, 1943, **46**, 18–22.
9 Bernard, M. L., Goudeau, J. C., *J. Chim. Phys.*, 1965, **62** (7–8), 732–5.
10 Callear, A. B., Robb, J. C., *Trans. Faraday Soc.*, 1955, **51**, 630–8.
11 Clark, C. C., *Hydrazine*, 1st ed., Mathieson Chemical Corp., Baltimore, Maryland, 1953, 133 pp.
12 Das, T. R., Kuloor, N. R., *J. Indian Inst. Sci.*, 1968, **50**, 13–25.
13 De Bruyn, L., *Rec. Trav. Chim.*, 1896, **15**, 174–84.
14 Dreisbach, R. R., *Physical Properties of Chemical Compounds*, Vol. III, Adv. in Chem. Ser. No. 29, Am. Chem. Soc., Washington, 1961, p. 463.
15 Engineering Science Data Units, *Approximate Data on the Viscosity of some Common Liquids*, Item. No. 66024, London, July 1966, 26 pp.
16 Eucken A., Krome, H., *Z. phys. Chem.*, 1940, **45B** (3), 175–92.
17 Fresenius, W., Karweil, J., *Z. phys. Chem.*, 1939, **44B** (1), 1–13.
18 Gallant, R. W., *Hydrocarbon Processing*, 1969, **48** (8), 117–26.
19 Giguere, P. A., Rundle, R. E., *J. Am. Chem. Soc.*, 1941, **63**, 1135–7.
20 *Gmelins Handbuch der anorganischen Chemie*, Syst. No. 4, Stickstoff, Verlag Chemie GmbH, Berlin, 1936, pp. 307–20.
21 Gray, P., Wright, P. G., *Combustion and Flame*, 1960, **4**, 95–6.
22 Haws, J. L., Harden, D. G., *J. Spacecraft Rockets*, 1965, **2** (6), 972–4.
23 Hieber, W., Woerner, A., *Z. Elektrochem.*, 1934, **40**, 252–6.
24 Hill, T. C. H., Sumner, J. F., *J. Chem. Soc.*, 1951, 838–40.
25 IUPAC, Atomic Weights of the Elements 1971, *Pure & Appl. Chem.*, 1972, **30** (3–4), 638–49.
26 *Kirk-Othmer Encyclopedia of Chemical Technology*, 2nd ed., Vol. 11, Hydrazine, Interscience Publ., New York, 1966, pp. 164–7.
27 Kobe, K. A., Harrison, R. H., *Petrol. Ref.*, 1954, **33** (11), 161–4.
28 *Landolt-Börnstein, Zahlenwerte und Funktionen aus Physik, Chemie, Astronomie, Geophysik, und Technik*, II Band, 4th Teil, Kalorische Zustandgrössen, Springer-Verlag, Berlin, 1961, p. 446.
29 Lax, E., Synowietz, C., Eds., *Taschenbuch für Chemiker und Physiker*, 3rd ed., Springer-Verlag, Berlin, 1967.
30 Lucien, H. W., *J. Chem. Eng. Data*, 1962, **7** (4), 541–2.
31 Maryott, A. A., Smith, E. R., *Table of Dielectric Constants of Pure Liquids*, Natl. Bur. Std. Circular 514, Washington, Aug. 10, 1951, 44 pp.
32 Mason, D. M., Wilcox, O. W., Sage, B. H., *J. Phys. Chem.*, 1952, **56**, 1008–10.
33 Mathews, J. F., *Chem. Rev.*, 1972, **72** (1), 71–100.
34 *Mellor's Comprehensive Treatise on Inorganic and Theoretical Chemistry*, Vol. VIII, Suppl. II, Nitrogen, Part 2, Longmans, London, 1967, pp. 69–114.
35 Mohr, P. H., Andrieth, L. F., *J. Phys. Colloid. Chem.*, 1949, **53**, 901–6.
36 Papousek, D., *Coll. Czech. Chem. Comm.*, 1958, **23**, 1767–73.
37 Pascal, P., Ed., *Nouveau Traité de Chimie Minérale*, Vol. X, Azote, Phosphore, Masson et Cie, Éditeurs, Paris, 1956, pp. 542–66.
38 Scott, D. W., Oliver, G. D., Gross, M. E., Hubbard, W. N., Huffman, H. M., *J. Am. Chem. Soc.*, 1949, **71**, 2293–5.
39 Stiel, L. I., Thodos, G., *AIChE Journal*, 1964, **10** (1), 26–30.
40 Stull, D. R., Dir., *JANAF Thermochemical Tables*, Dow Chemical Co., Midland, Mich., 1965–8.
41 Sutton, J. R., *A Method of Calculating the Viscosities of Polar Gases*, Progr. Intern. Res. Thermodyn. Transport Properties, Paper, Symp. Thermophysical Prop. 2nd., Am. Soc. Mech. Eng., New York, 1962, pp. 266–70.
42 Walden, P., Hilgert, H., *Z. phys. Chem.*, 1944, **165A**, 241–71.
43 Yungman, V. S., Gurvich, L. V., Rtishcheva, N. P., *Tr. Gos. Inst. Prikl. Khim.*, 1962, **49**, 20–37.

Table 27.1 Properties of Hydrazine

		Ref.
Chemical formula	N_2H_4	(29)
Molecular weight	32.0450	(25)
Critical temperature	380.0°C	(33)
Critical pressure	146.9×10^5 N m^{-2}	(33)
Critical density	231.0 kg m^{-3}	(1, 26)
Boiling point at 1.0133×10^5 N m^{-2}	113.5°C	(29)
Melting point	1.54°C	(29)
Transition temperature	None	(5, 20)
Density of solid at −5.0°C	1146.0 kg m^{-3}	(8)
Compressibility (adiabatic) of liquid at 25°C	23.34×10^{-11} m^2 N^{-1}	(36)
Heat of formation of gas at 25°C	9.5417×10^7 J kg-mol^{-1}	(40)
Free energy of formation of gas at 25°C	1.5864×10^8 J kg-mol^{-1}	(40)
Heat of fusion	1.2665×10^7 J kg-mol^{-1}	(38)
Heat of transition	None	(5, 20)
Heat of combustion at 25°C and 1.0133×10^5 N m^{-2}	6.1378×10^8 J kg-mol^{-1}	(26, 34)
Entropy of gas at 25°C	2.3877×10^5 J kg-mol^{-1} °C^{-1}	(40)
Velocity of sound at 25°C and 1.0133×10^5 N m^{-2}	2068.9 m s^{-1}	(36)
Dielectric constant of liquid at 20°C	52.9	(31)
Electrical conductivity of liquid at 25°C	2.4×10^{-3} ohm^{-1} m^{-1}	(42)
Refractive index of liquid at 25°C	1.46444	(7)
Threshold limit value	(1 ppm) 1.3×10^{-6} kg m^{-3}	(3)

27.1 Vapour Pressure of Hydrazine (13,17,19,22,23,38)

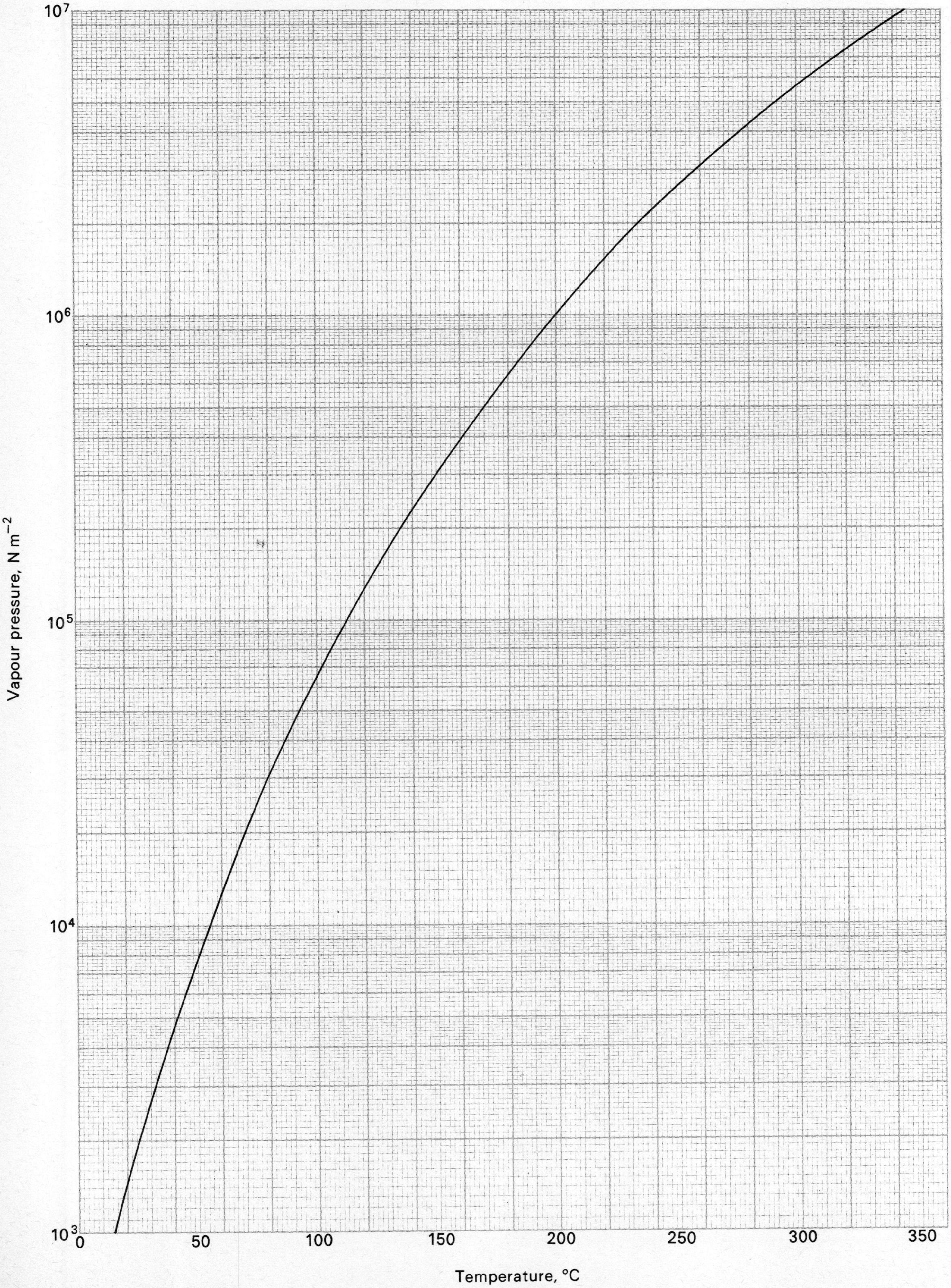

27.2 Orthobaric Density of Hydrazine (1,19)

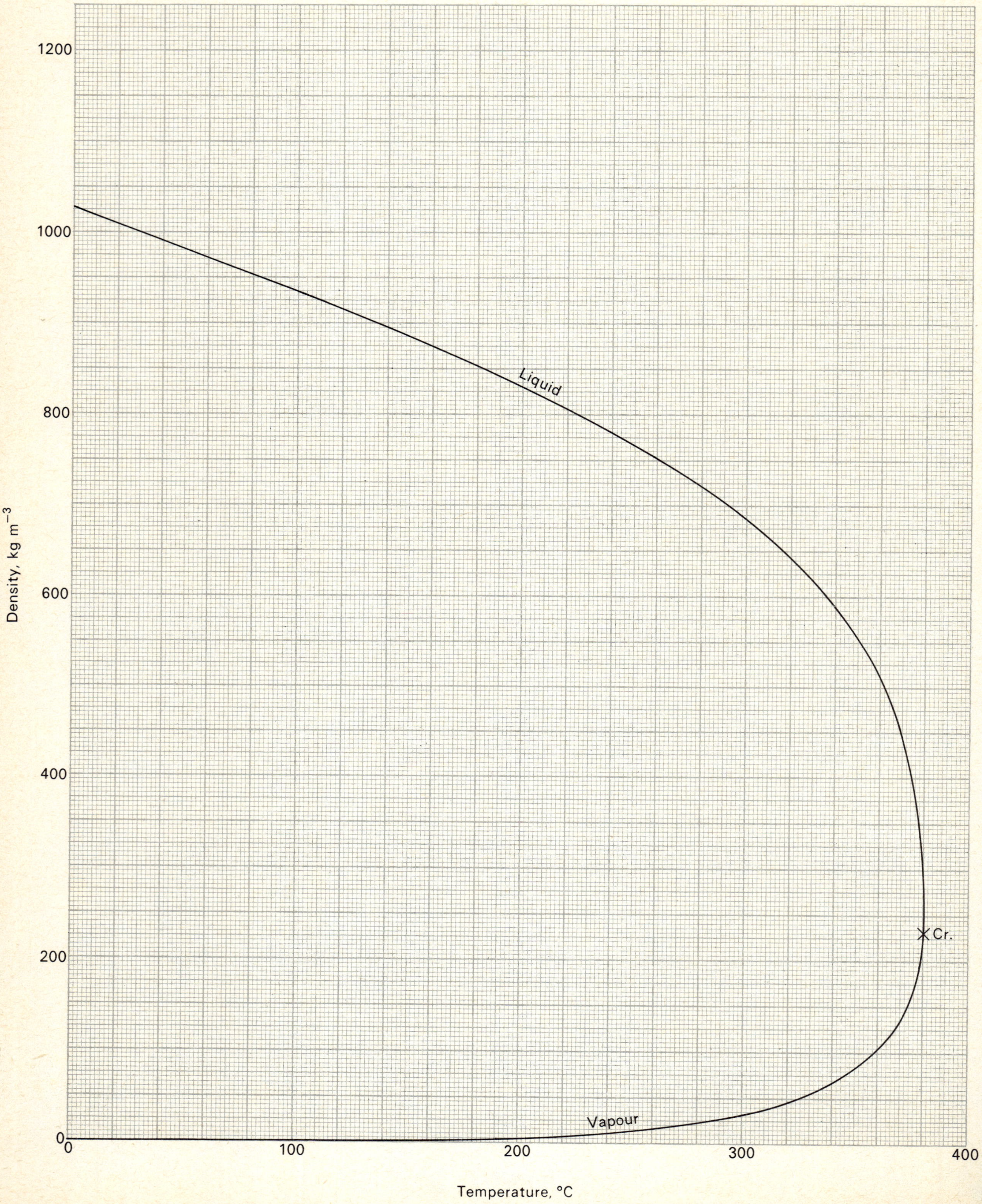

27.3 Latent Heat of Vaporisation of Hydrazine (17,19,38)

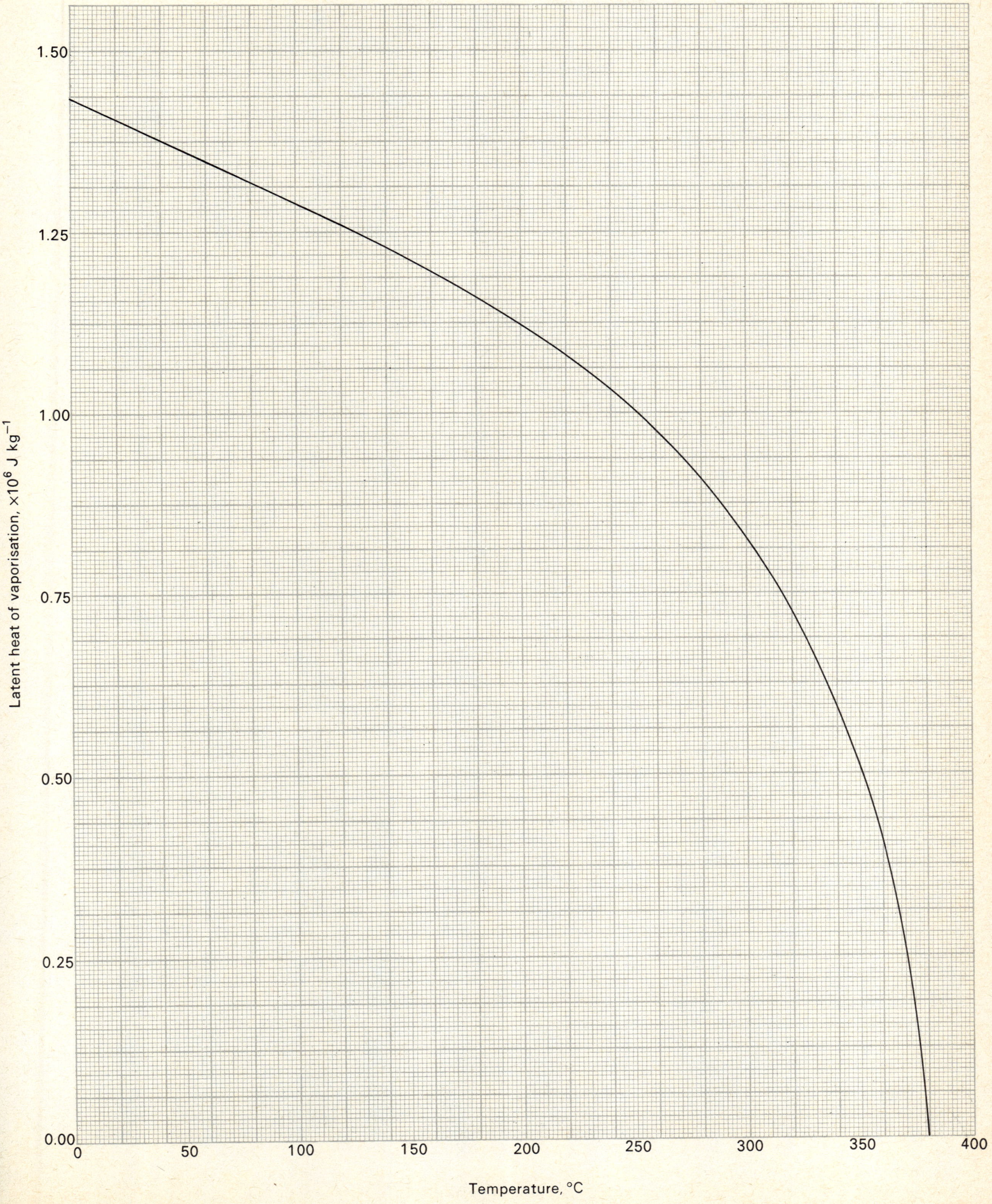

27.4 Heat Capacity of Solid and Liquid Hydrazine (2,16,38,40)

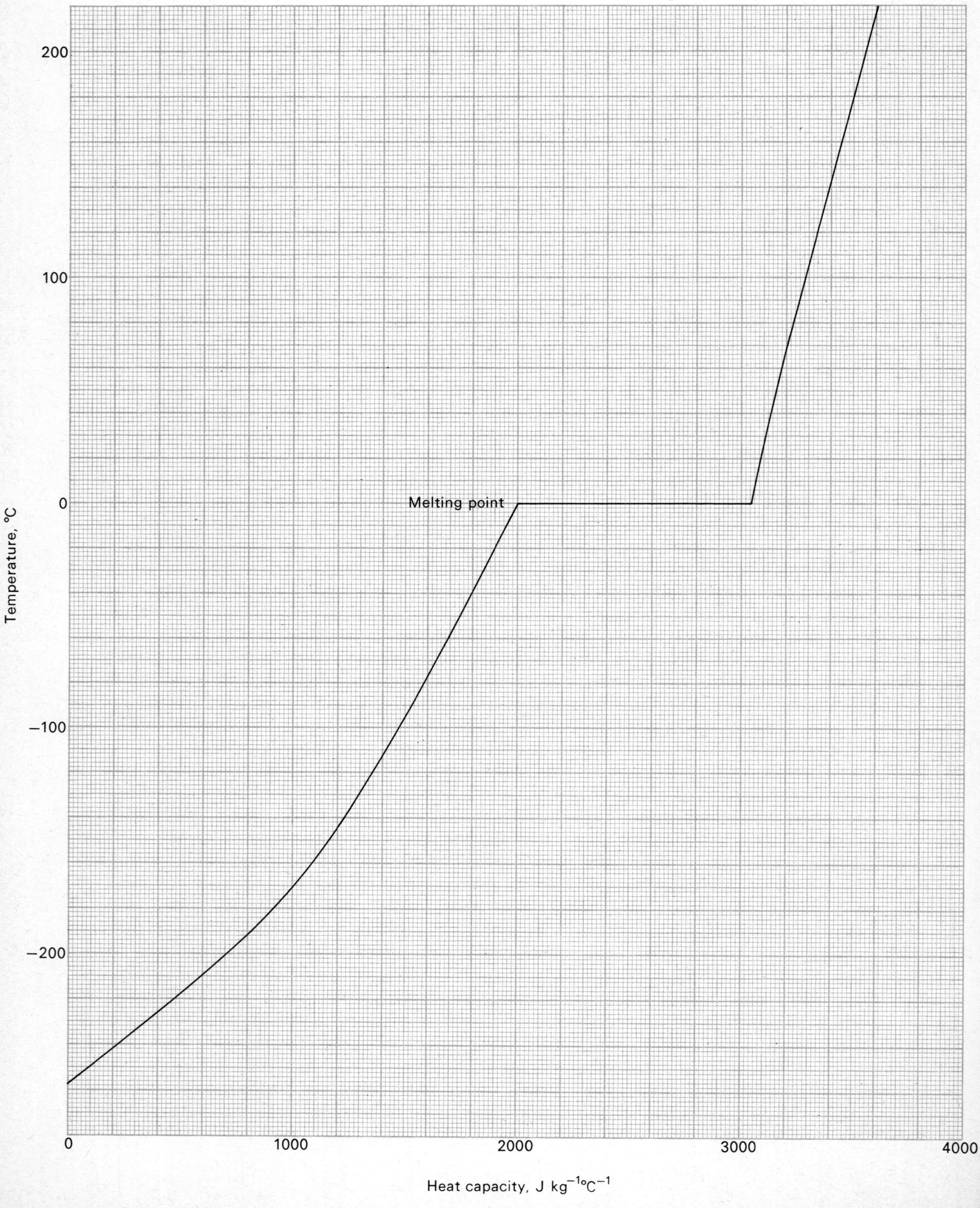

27.5 Heat Capacity of Gaseous Hydrazine at 1.0133×10^5 N m^{-2} (22,27,28,40,43)

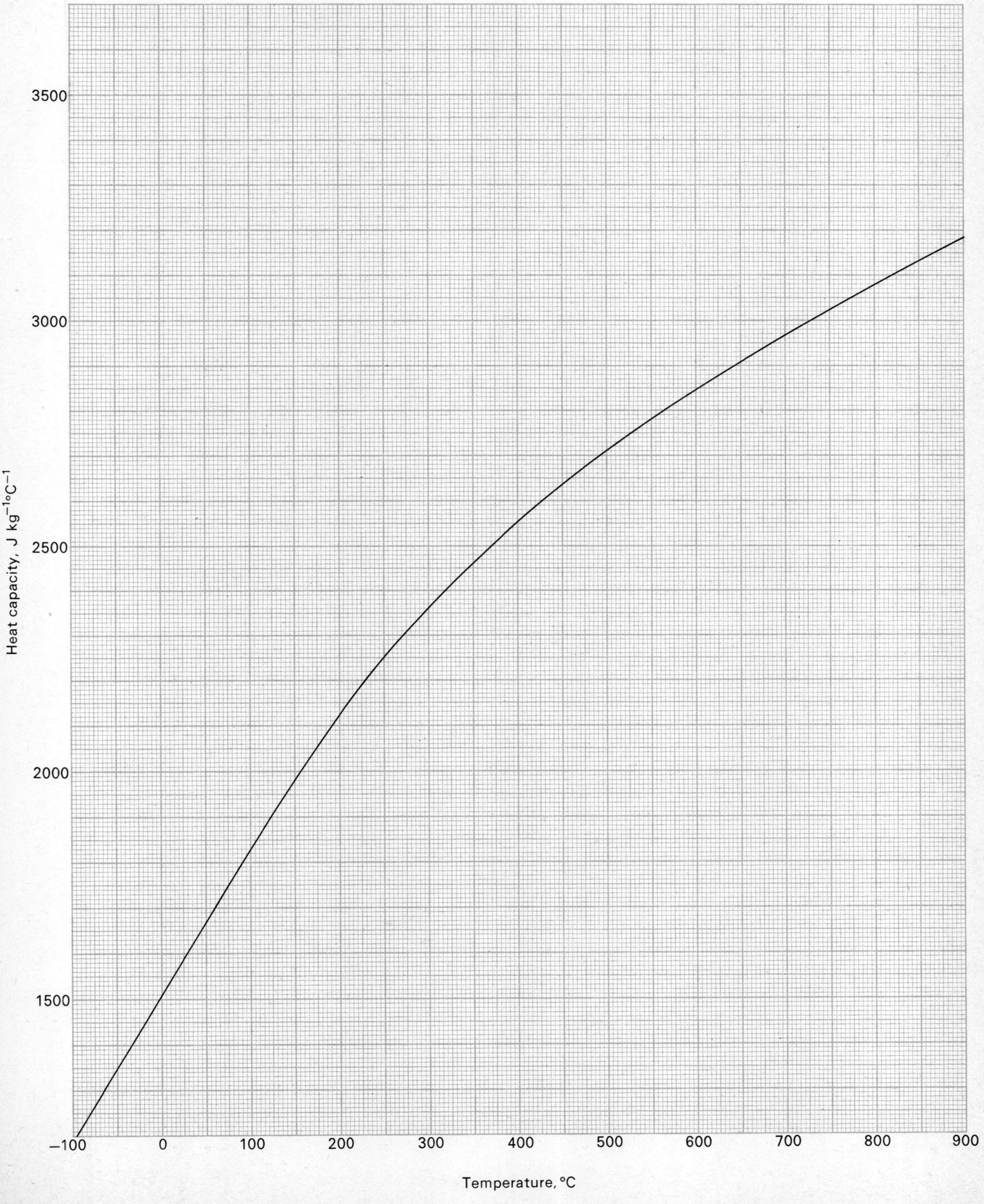

27.6 Viscosity of Liquid Hydrazine (1,15)

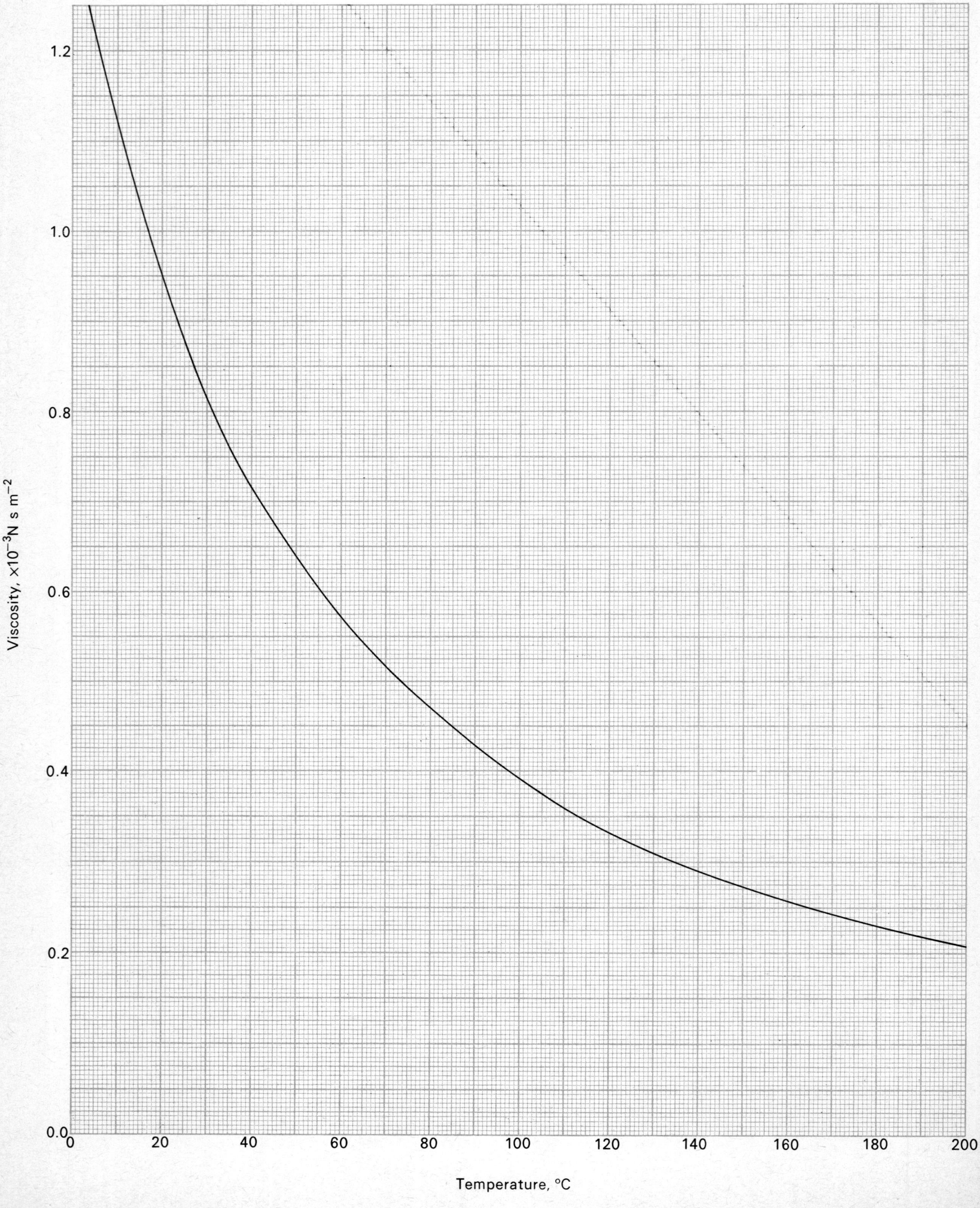

27.7 Viscosity of Liquid Hydrazine at High Pressure (32)

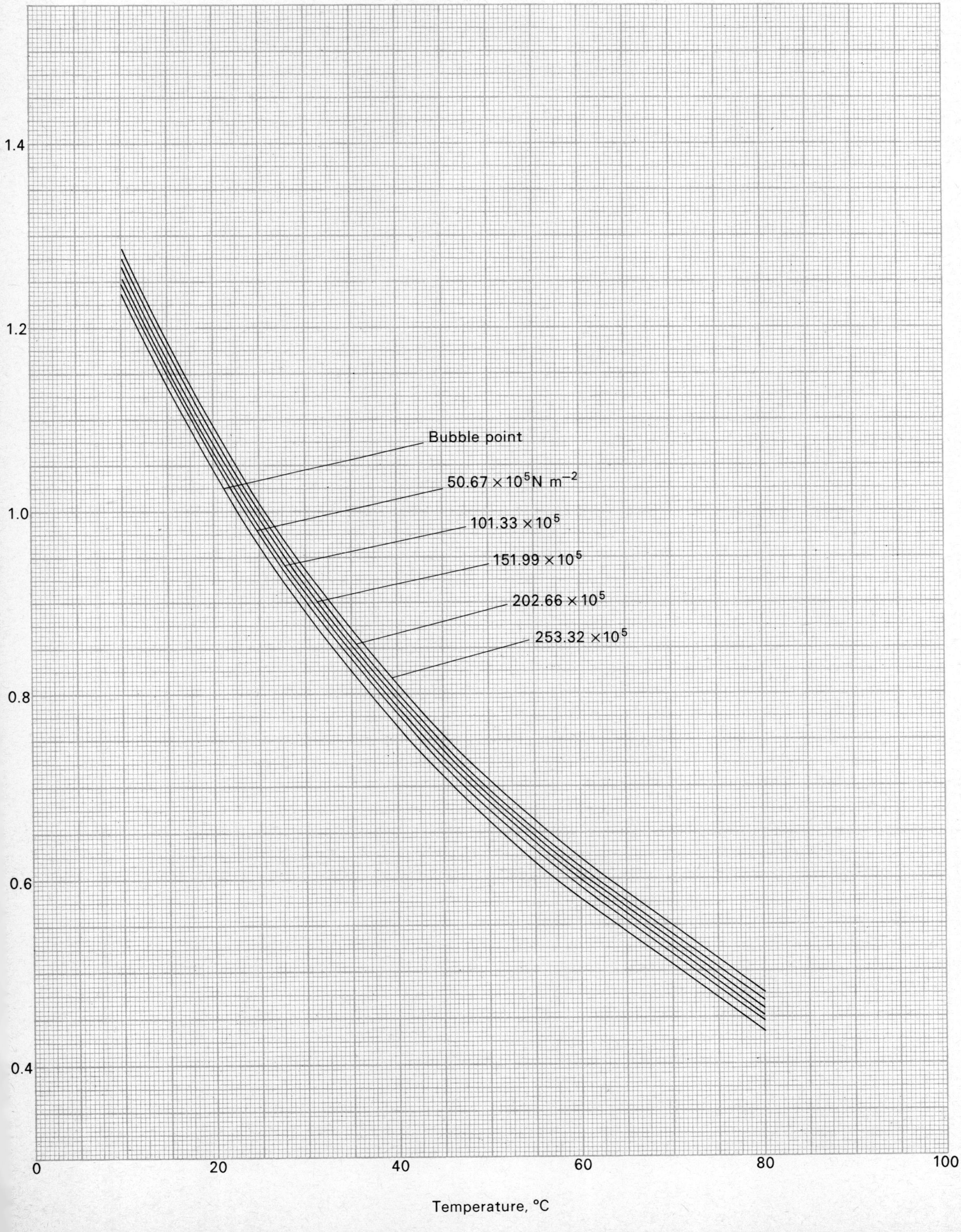

27.8 Viscosity of Gaseous Hydrazine at $1.0133 \times 10^5 N\ m^{-2}$ (18,41)

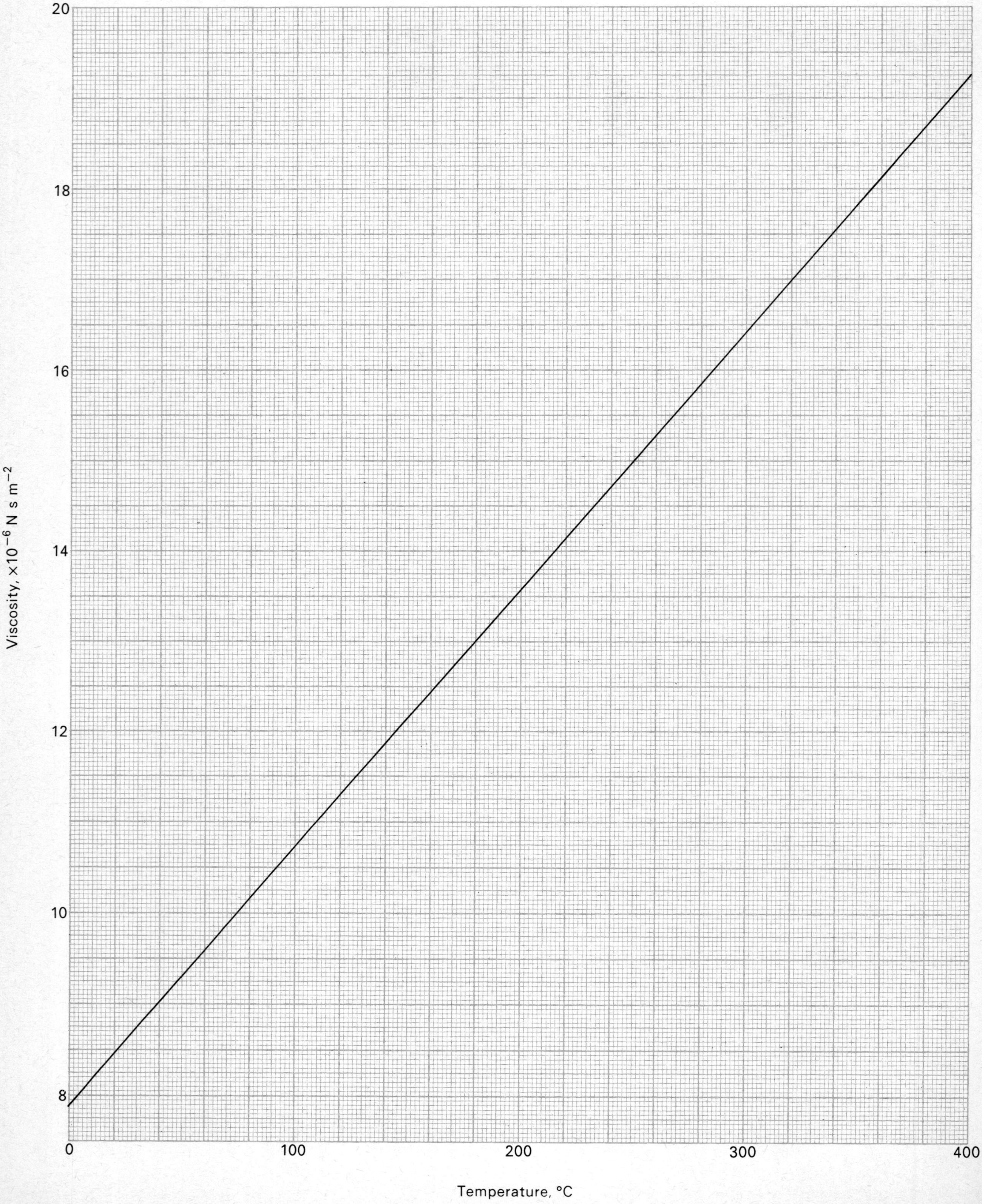

27.9 Thermal Conductivity of Liquid Hydrazine (39)

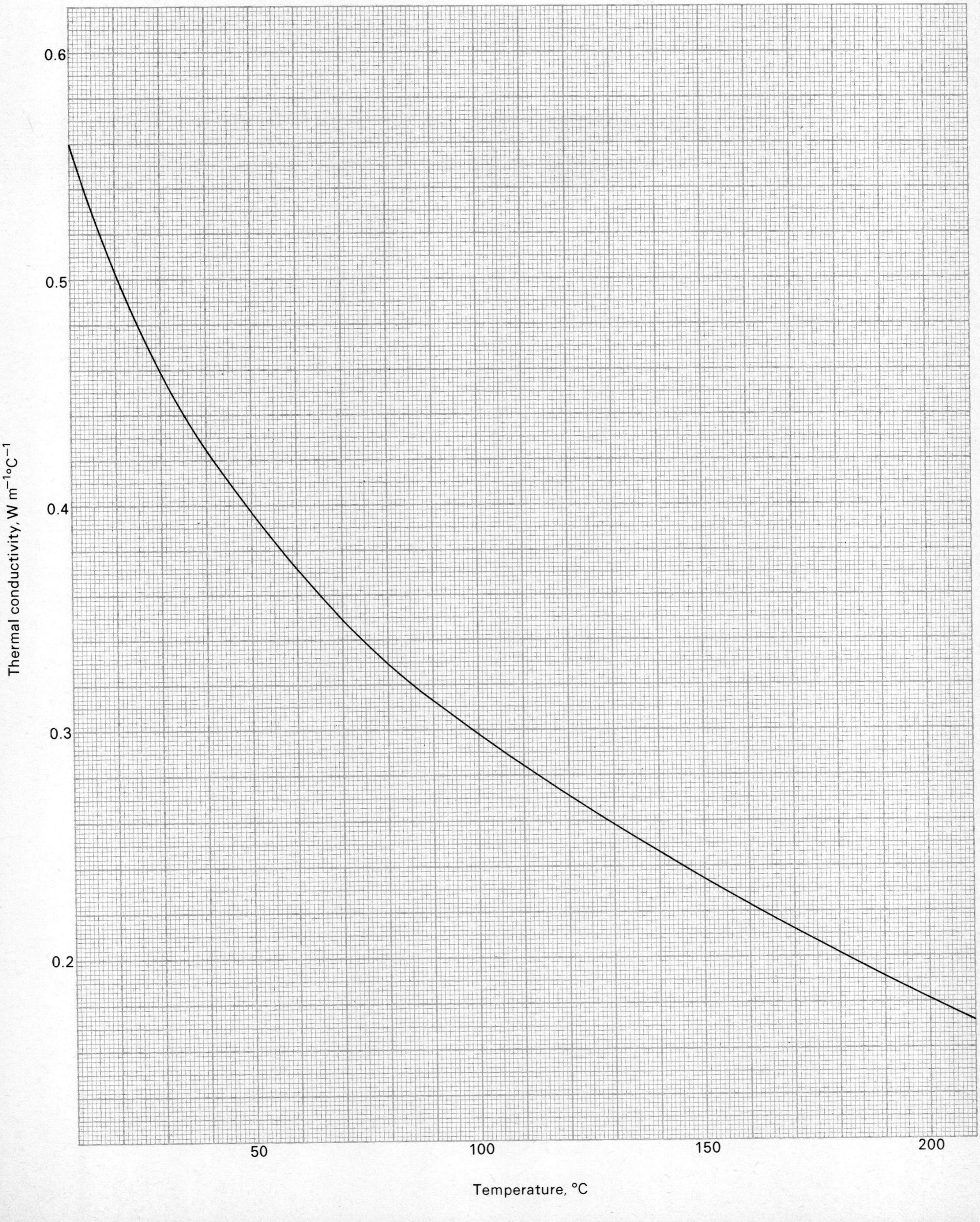

27.10 Thermal Conductivity of Gaseous Hydrazine at 1.0133×10^5 N m^{-2} (10,21)

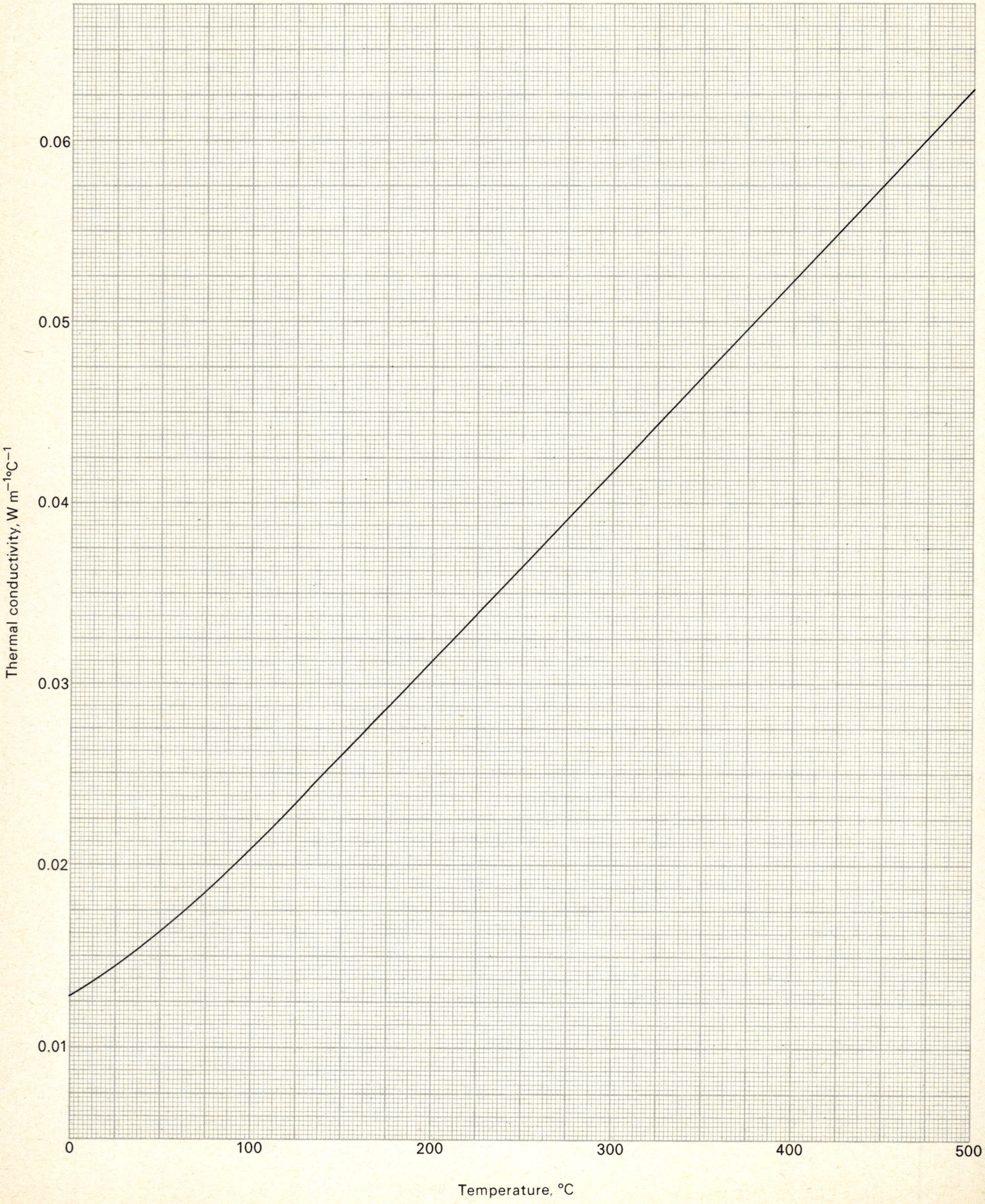

27.11 Surface Tension of Hydrazine (6,9,29,34)

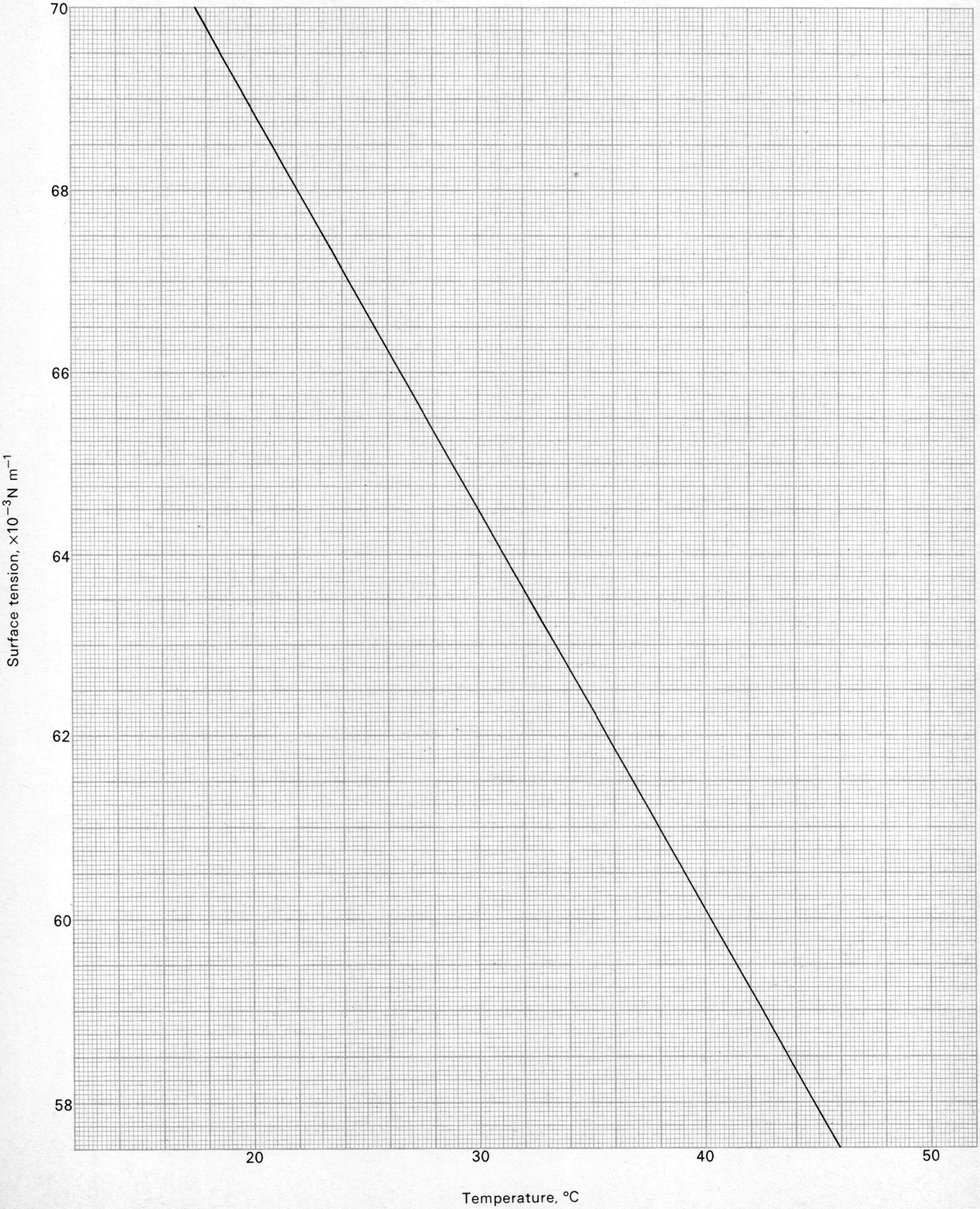

27.12 Phase Diagram for N_2H_4-H_2O System (24,35)

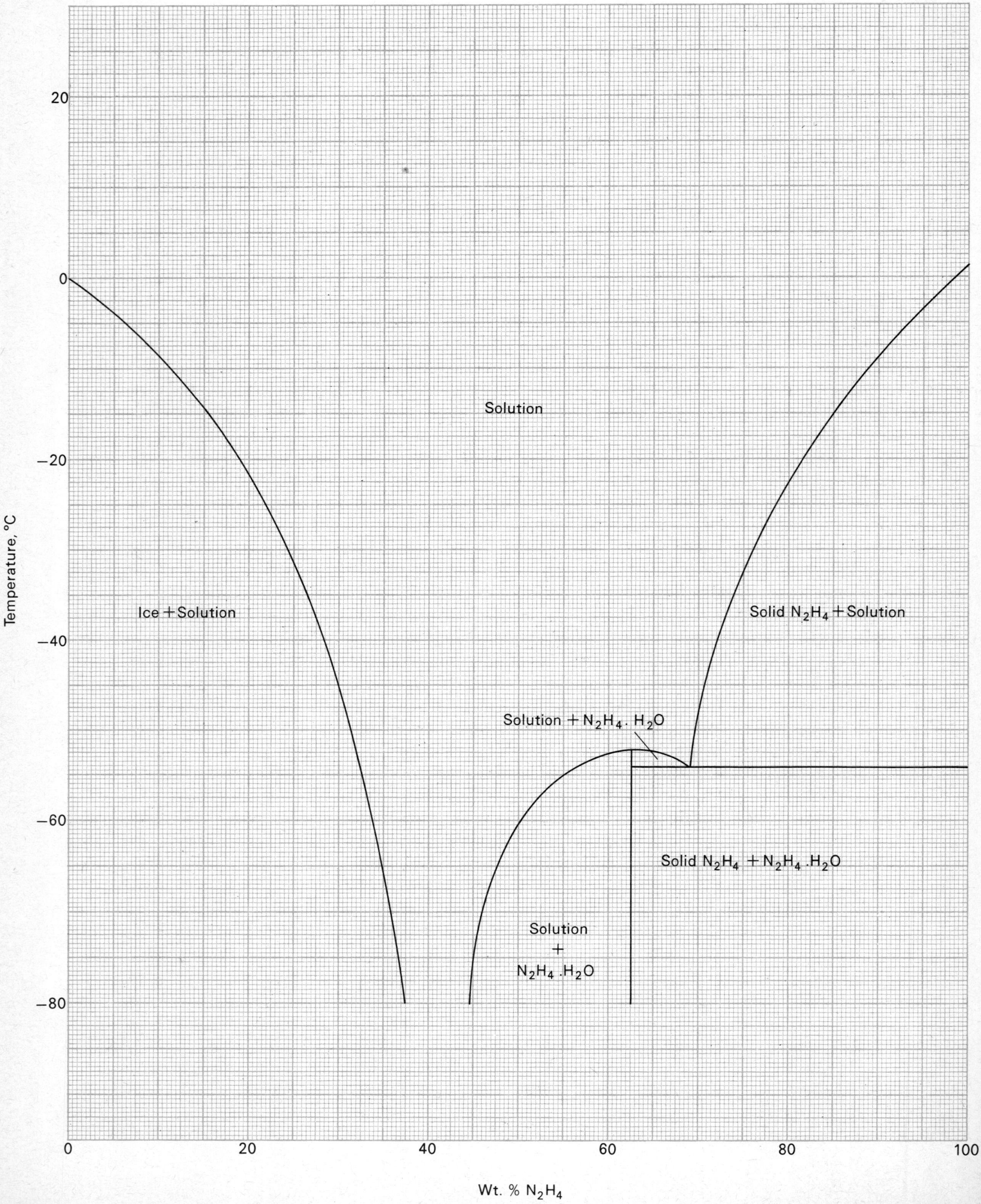

28 HYDROGEN

Hydrogen is a colourless, odourless, and tasteless gas at ambient temperature and atmospheric pressure. It has the lowest molecular weight and the lowest liquid and solid densities of all substances. Its melting and boiling points are the second lowest after helium. Normal hydrogen is a mixture of the ortho and para modifications. At room temperature and atmospheric pressure 25% para and 75% ortho hydrogen are in equilibrium. However, at low temperatures the concentration of para increases, as more of the molecules go into the zero quantum state.

The differences between the two forms are due to rotational energy. Ortho H_2 has the odd quantum states, while para H_2 has the even ones. Therefore, the specific heat of ortho is lower than that of para form. In the ortho molecule the spins are parallel, while in the para molecule the spins are antiparallel. The thermal conductivities are also different. Otherwise, the physical properties of ortho and para forms differ only slightly. Because of the low density of H_2, it diffuses faster than most gases, e.g. four times as fast as oxygen.

On cooling, liquid hydrogen solidifies to a transparent mass like ice with a foamy surface. The crystal belongs to the cubic system, either trisoctahedral or hexoctahedral.

The major industrial application of hydrogen is in the manufacture of synthetic ammonia (Haber process), methanol, and other liquid fuels. Hydrogen is also used in high temperature welding (oxy-hydrogen flame).

There are several excellent publications on various properties of H_2, particularly by the National Bureau of Standards, whose investigations have been considered as the most reliable sources, heavy weight being given to them in the correlations. The physical, thermodynamic, and transport properties of hydrogen have been surveyed and presented in the following works: Bizjak (3), Data Book and Buyers Guide (7), Esel'son, *et al.* (14), Farkas (16), Gmelin (19), Hilsenrath *et al.* (24), Landolt-Börnstein (31, 32), Malkov (39), Mellor (43), McBride *et al.* (45), Pascal (46), Roder *et al.* (50), Serdyuk (53), Sindt and Mann (54), and Tans (59).

In this presentation only the properties of normal hydrogen have been compiled. Various properties of para H_2, in addition to the above citations, can be found in the following sources: (8, 12, 16, 20, 21, 31, 32, 46, 50, 54). Some further references of interest are: P–V–T data (4), viscosity (10, 23, 36, 52), and electrical conductivity (47).

References

1 American Conference of Government Industrial Hygienists, *Documentation of Threshold Limit Values*, revised edition, Committee on Threshold Limit Values, Cincinnati, Ohio 45202, 1966, 203 pp.

2 Barua, A. K., Afzal, M., Flynn, G. P., Ross, J., *J. Chem. Phys.*, 1964, **41** (2), 374–8.

3 Bizjak, F., *Physical Properties of Hydrogen*, U.S. Gov. Res. Dev. Rept. AE61-0726, (AD 677550), Clearinghouse, Springfield, Va., July, 26, 1961, 18 pp.

4 Brief, A., Jaffe, J., *J. Chem. Eng. Data*, 1970, **15** (1), 114–19.

5 Clusius, K., Hiller, K., *Z. phys. Chem.*, 1929, **4B**, 158–68.

6 Crommelin, K. O., *Comm. Leiden* **157***a*, *Akad. Amsterdam Vessl*,. 1913, **22**, 214.

7 Data Book and Buyers Guide, Engineering Data Section, *Cryogenic and Ind. Gases*, 1971, **6** (3), 41–80.

8 De Palma, J. V., Thodos, G., *J. Chem. Eng. Data*, 1966, **11** (1), 31–7.

9 Diller, D. E., *J. Chem. Phys.*, 1965, **42**, 2089.

10 Diller, D. E., *J. Chem. Phys.*, 1968, **49** (7), 3096–3105.

11 Diller, D. E., Roder, H. M., Thermal Conductivity Measurements of Fluid Hydrogen at 17 to 200 K and Pressures to 10 MN/m^2, in *Advances in Cryogenic Engineering*, Vol. 15 (Timmerhouse, K. D., Ed), Plenum Press, New York, 1970, pp. 58–64.

12 Dwyer, R. F., Cook, G. A., Berwaldt, O. E., *J. Chem. Eng. Data*, 1966, **11** (3), 351–3.

13 Engineering Science Data Units, *Specific Heats and Their Ratio as Functions of Temperature for Several Common Gases*, Royal Aeronautical Society, Data Sheet 00.01.08, London, 1965, 4 pp.

14 Essel'son, B. N., Blagoi, Yu. P., Grigor'ev, V. N., Manzhelii, V. G., Mikhailenko, S. A., Neklyudov, N. P., *Properties of Liquid and Solid Hydrogen*, Komitet Standartov, Mer i Izmeritel'nykh Priborov pri Sovete Ministrov SSSR, Moskva, 1969, 123 pp.

15 Eucken, A., *Verh. phys. Ges.*, 1916, **18**, 17.

16 Farkas, A., *Orthohydrogen, Parahydrogen, and Heavy Hydrogen*, Macmillan, New York, 1935.

17 Friedman, A. S., Hilsenrath, J., Natl. Bur. Std. Rept. 3163, Washington, 1954, 12 pp.

18 Geier, H., Schafer, K., *Allgem. Wärmetech.*, 1961, **10**, 70–5.

19 *Gmelins Handbuch der anorganischen Chemie*, Syst. No. 2, Wasserstoff, Verlag Chemie GmbH, Berlin, 1927, pp. 55–212.

20 Goodwin, R. D., *J. Res. Natl. Bur. Std.*, 1966, **70A** (6), 541–4.

21 Goodwin, R. D., Diller, D. E., Roder, H. M., Weber, L. A., *J. Res. Natl. Bur. Std.*, 1963, **67A** (2), 173–92.

22 Gordon, J. S., WADC-TR-57-33 (AD 110735), 1957, 172 pp.

23 Gracki, J. A., *J. Chem. Phys.*, 1969, **51** (9), 3856–63.

24 Hilsenrath, J., Beckett, C. W., Benedict, W. S., Fano, L., Hoge, H. J., Masi, J. F., Nuttal, R. L., Touloukian, Y. S., Woolley, H. W., *Tables of Thermal Properties of Gases*, Natl. Bur. Std. Circular 564, Washington, 1955, 488 pp.

25 Hilsenrath, J., Touloukian, Y. S., *Trans. Am. Soc. Mech. Eng.*, 1954, **76**, 967–85.

26 IUPAC, Atomic Weights of the Elements 1971, *Pure & Appl. Chem.*, 1972, **30** (3–4), 638–49.

27 Ibrahim, S. H., Kuloor, N. R., *Brit. Chem. Eng.*, 1961, **6** (12), 862–3.

28 Johnston, H. L., Clarke, J. T., Rifkin, E. B., Keer, E. C., *J. Am. Chem. Soc.*, 1950, **72**, 3933–8.

29 Johnston, H. L., Keller, W. E., Friedman, A. S., *J. Am. Chem. Soc.*, 1954, **76**, 1482–8.

30 Keyes, F. G., Techn. Rept. 37, Project Squid (ATI 16173), 1952, 33 pp.

31 *Landolt-Börnstein, Zahlenwerte und Funktionen aus Physik, Chemie, Astronomie, Geophysik, und Technik*, 7th ed., 4 Teil, Kalorische Zustandgrössen, Springer-Verlag, Berlin, 1961, p. 400.

32 *Landolt-Börnstein, Zahlenwerte und Funktionenaus Physik, Chemie, Astronomie, Geophysik, und Technik*, 7th ed., 4 Teil, Bandteil a, Technik, Springer-Verlag, Berlin, 1967, pp. 206–28.

33 Lange, N. A., Ed., *Handbook of Chemistry*, 10th ed., McGraw-Hill, New York, 1967.

34 Lax, E., Synowietz, C., Eds., *Taschenbuch für Chemiker und Physiker*, 3rd ed., Springer-Verlag, Berlin, 1967.

35 Leduc, A., *Chem. Rev.*, 1929, **6**, 1–16.

36 Lyusternik, V. E., *High Temperature*, 1969, **7** (2), 335–7.

37 Maitland, G. C., Smith, E. B., *J. Chem. Eng. Data*, 1972, **17** (2), 150–6.

38 Malkov, M. P., Ed., *Zhidkii Vodorod*, (*Liquid Hydrogen*), A Collection of Articles, Mir, Moskva, 1964.

39 Maryott, A. A., Smith, E. R., *Table of Dielectric Constants of Pure Liquids*, Natl. Bur. Std. Circular 514, Washington, Aug. 10, 1951, 44 pp.

40 Mathews, J. F., *Chem. Rev.*, 1972, **72** (1), 71–100.

41 Medvedev, V. A., Dedikov, Yu. D., Astrov, D. N., *Zh. Fiz. Khim.*, 1967, **41** (8), 1480.

42 Meissner, M., *Verh. phys. Ges.*, 1913, **15**, 540.
43 *Mellor's Comprehensive Treatise on Inorganic and Theoretical Chemistry*, Vol. I, H, O, Longmans, London, 1922, pp. 270–343.
44 Morrison, T., Billett, F., *J. Chem. Soc.*, 1948, 2033–5.
45 McBride, B. J., Heimel, S., Ehlers, J. G., Gordon, S., Natl. Aeron. and Space Admin. Report SP-3001, 1963, 328 pp.
46 Pascal, P., Ed., *Nouveau Traité de Chimie Minérale*, Vol. I, Géneralites, Air, Eau, Hydrogéne, Deutérium, Tritium, Helium et Gas Inertes, Masson et Cie, Éditeurs, Paris, 1956, pp. 572–604.
47 Plantikow, U., Steinberger, S., *Z. Phys.*, 1970, **231**, 109–18.
48 Powers, R. W., Mattox, R. W., Johnston, H. L., *J. Am. Chem. Soc.*, 1954, **76**, 5972–3.
49 Roder, H. M., Diller, D. E., *J. Chem. Phys.*, 1970, **52** (11), 5928–49.
50 Roder, H. M., Weber, L. A., Goodwin, R. D., *Thermodynamic and Related Properties of Parahydrogen from the Triple-Point to 100 K and Pressures to 340 Atmospheres*, Natl. Bur. Std. Monograph 94, Washington, Aug. 10, 1965, 115 pp.
51 Rossini, F. D., Wagman, D. D., Evans, W. H., Levine, S., Jaffe, I., *Selected Values of Chemical Thermodynamic Properties*, Natl. Bur. Std. Circular 500, Washington, Feb. 1, 1952.
52 Rudenko, N. S., Slyusar, V. P., *Ukr. Fiz. Zh.*, 1968, **13** (6), 917–21.
53 Serdyuk, L. S., *Russ. J. Phys. Chem.*, 1969, **43** (2), 263–5.
54 Sindt, C. F., Mann, D. B., *Temperature–Entropy Diagram for Parahydrogen, Triple-Point Region*, Natl. Bur. Std. Techn. Rept. 343, Washington, June 22, 1966, 6 pp.
55 Sinke, G. C., AR-15-59, (Ad 214587), 1959.
56 Smith, A. L., Hallett, N. C., Johnston, H. L., *J. Am. Chem. Soc.*, 1954, **76**, 1486–8.
57 Stull, D. R., Dir., *JANAF Thermochemical Tables*, Dow Chemical Co., Midland, Mich., 1965–8.
58 Svehla, R. A., *Estimated Viscosities and Thermal Conductivities of Gases at High Temperatures*, Lewis Res. Center Techn. Rept. R-132, Cleveland, Ohio, 1962, 140 pp.
59 Tans, A. M. P., *Brit. Chem. Eng.*, 1960, **5** (1), 52.
60 Van Itterbeek, A., Verbeke, O., Theewes, F., Staes, K., De Boelpaep, J., *Physica*, 1964, **30**, 1238–44.
61 Vargaftik, N. B., Filippov, L. P., Tarzimanov, A. A., Yurchak, R. P., *Thermal Conductivity of Gases and Liquids*, Teploprovodnost' Gazov i Zhidkostey, Moskva, 1970, 185 pp.
62 Wiebe, R., Gaddy, V. L., *J. Am. Chem. Soc.*, 1934, **56**, 76.
63 Willis, W. L., *Cryogenics*, 1966, **6** (5), 279–84.
64 Woolley, H. W., Scott, R. B., Brickwedde, F. G., *J. Res. Natl. Bur. Std.*, 1948, **41** (5), 379–475.
65 Yosim, S. J., *J. Chem. Phys.*, 1964, **40** (10), 3069–75.

Table 28.1 Properties of Hydrogen (normal)

		Ref.
Chemical formula	H_2	(34)
Molecular weight	2.0158	(26)
Critical temperature	−240.17 °C	(40)
Critical pressure	12.93×10^5 N m^{-2}	(40)
Critical density	31.4 kg m^{-3}	(40)
Triple-point temperature	−259.2 °C	(34)
Triple-point pressure	7221.0 N m^{-2}	(34)
Boiling point at 1.0133×10^5 N m^{-2}	−252.77 °C	(14)
Melting point	−259.14 °C	(14)
Transition temperature	None	(14, 19)
Density of solid at −262.0 °C	80.77 kg m^{-3}	(6)
Compressibility of liquid at −256 °C	1.978×10^{-8} m^2 N^{-1}	(15)
Coefficient of thermal expansion of liquid at boiling point	1.58×10^{-2} °C^{-1}	(65)
Heat of formation of gas at 25 °C	0.0 J kg-mol^{-1}	(57)
Free energy of formation of gas at 25 °C	0.0 J kg-mol^{-1}	(57)
Heat of sublimation at −273.15 °C	7.96×10^5 J kg-mol^{-1}	(31)
Heat of fusion	1.1720×10^5 J kg-mol^{-1}	(51)
Heat of transition	0.0 J kg-mol^{-1}	(14, 19)
Entropy of gas at 25 °C	1.3067×10^5 J kg-mol^{-1} °C^{-1}	(57)
Heat capacity ratio of gas at 15 °C and 1.0133×10^5 N m^{-2}	1.410	(35)
Velocity of sound at 0 °C and 1.0133×10^5 N m^{-2}	1261.7 m s^{-1}	(34)
Diffusion in air at 0 °C and 1.0133×10^5 N m^{-2}	6.12×10^{-5} m^2 s^{-1}	(27)
Dielectric constant of liquid at −252.75 °C	1.228	(39)
Electrical conductivity of liquid at −253.7 °C	2.18×10^{-18} ohm^{-1} m^{-1}	(63)
Refractive index of liquid at −255 °C	1.1087	(42)

28.1 Vapour Pressure of Hydrogen (Normal) (14,38,60,64)

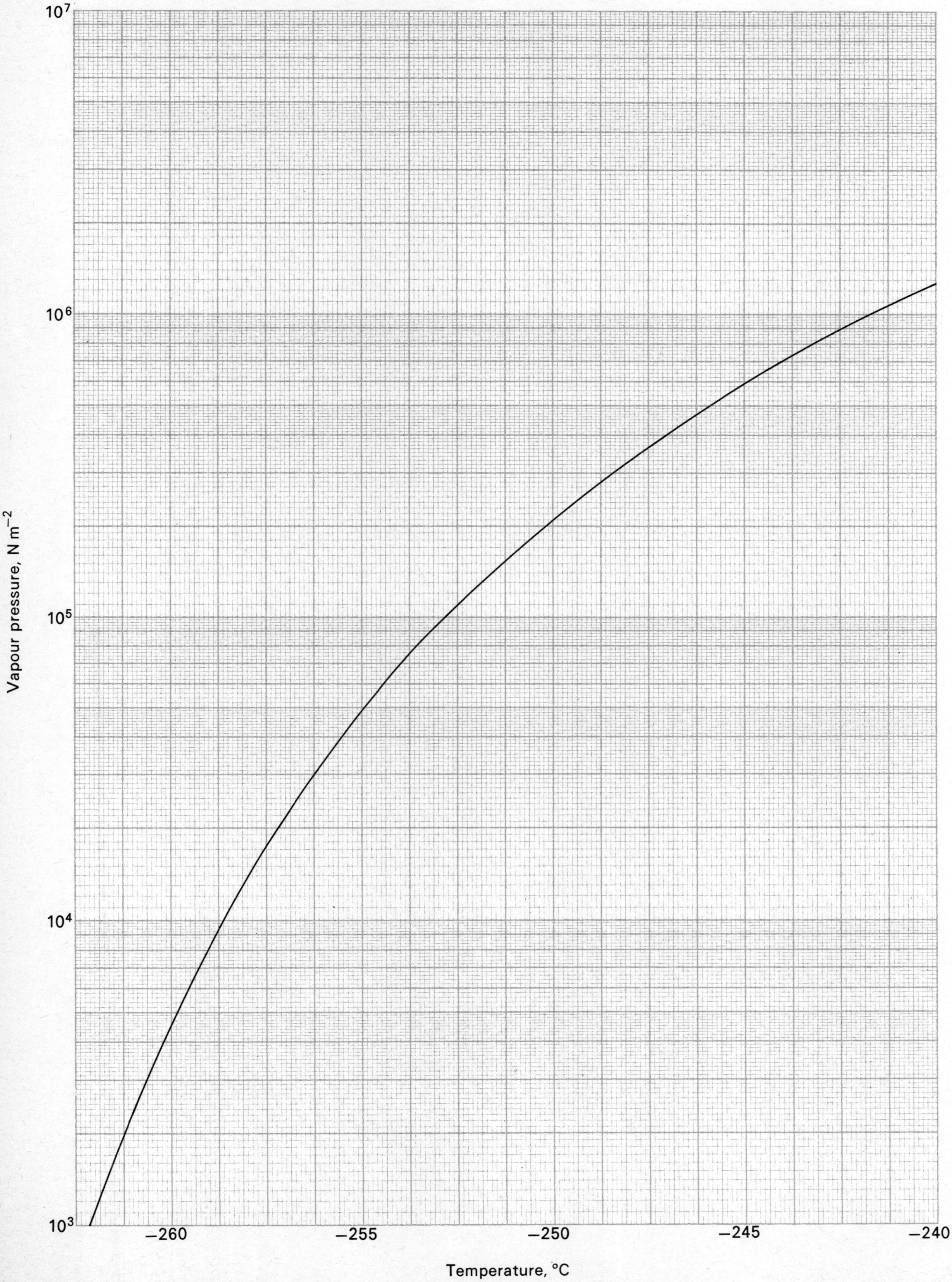

28.2 Orthobaric Density of Hydrogen (Normal) (3,14,33)

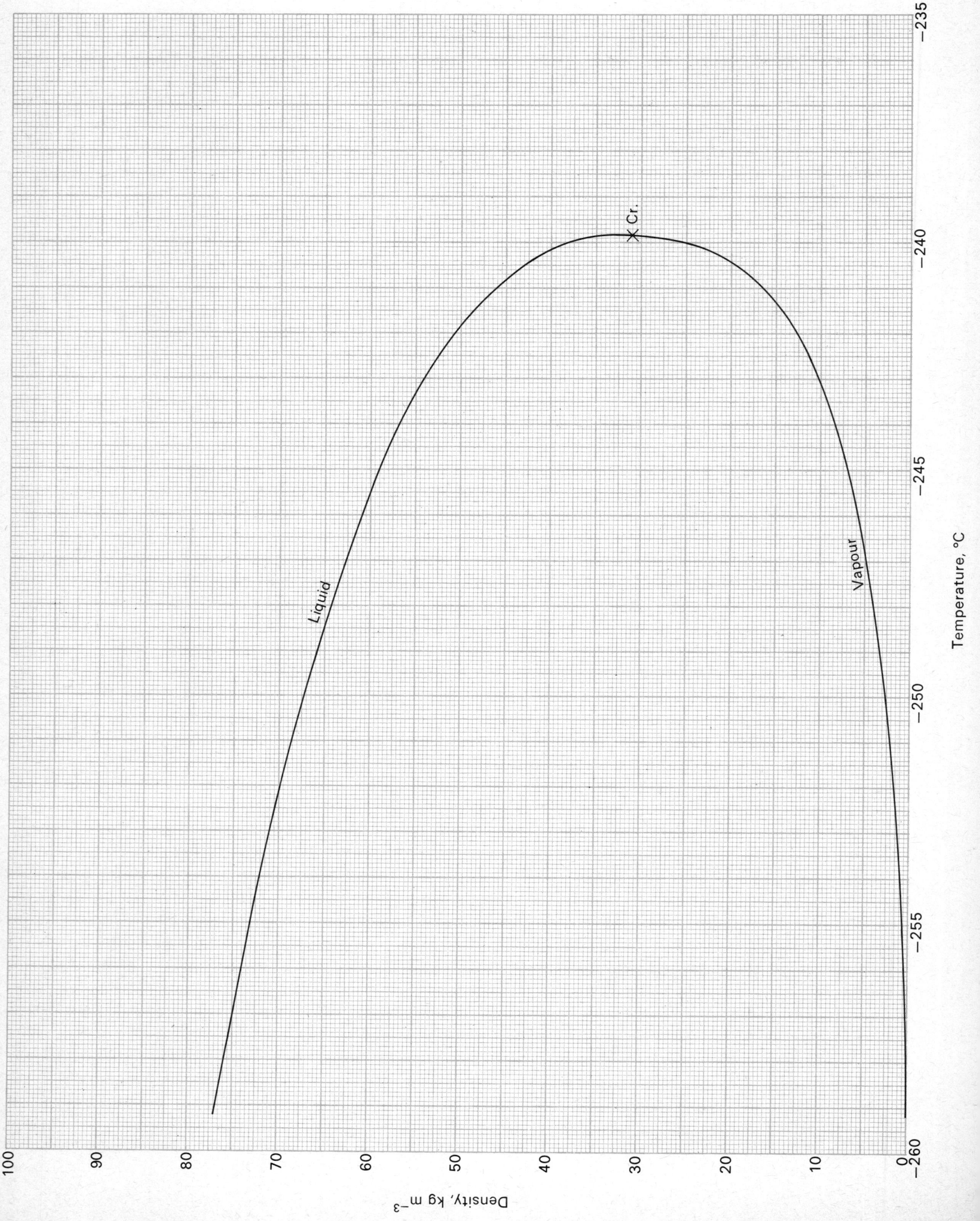

28.3 P-V-T Data for Hydrogen (Normal) (14,29)

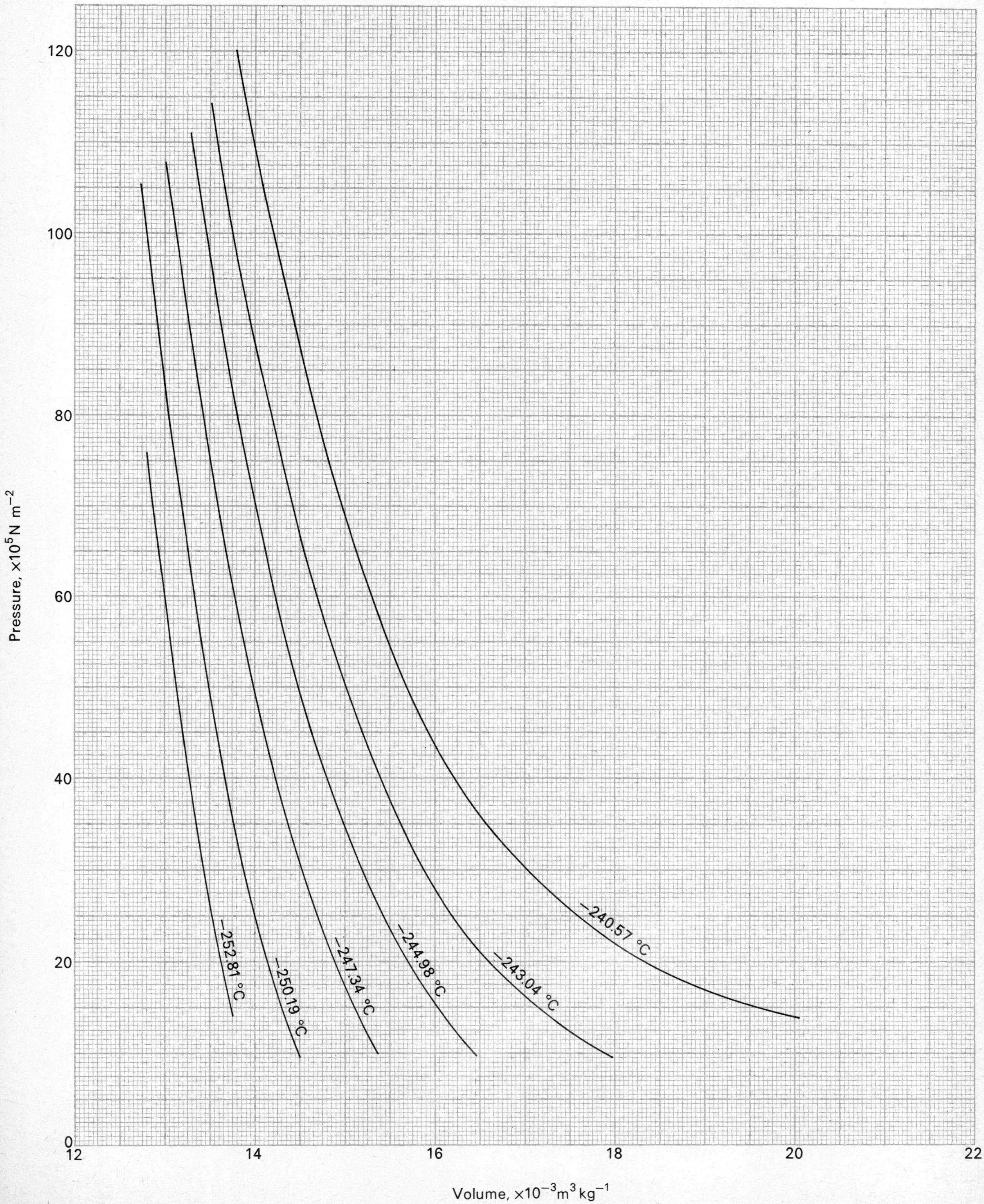

28.4 Latent Heat of Vaporisation of Hydrogen (Normal) (14,50)

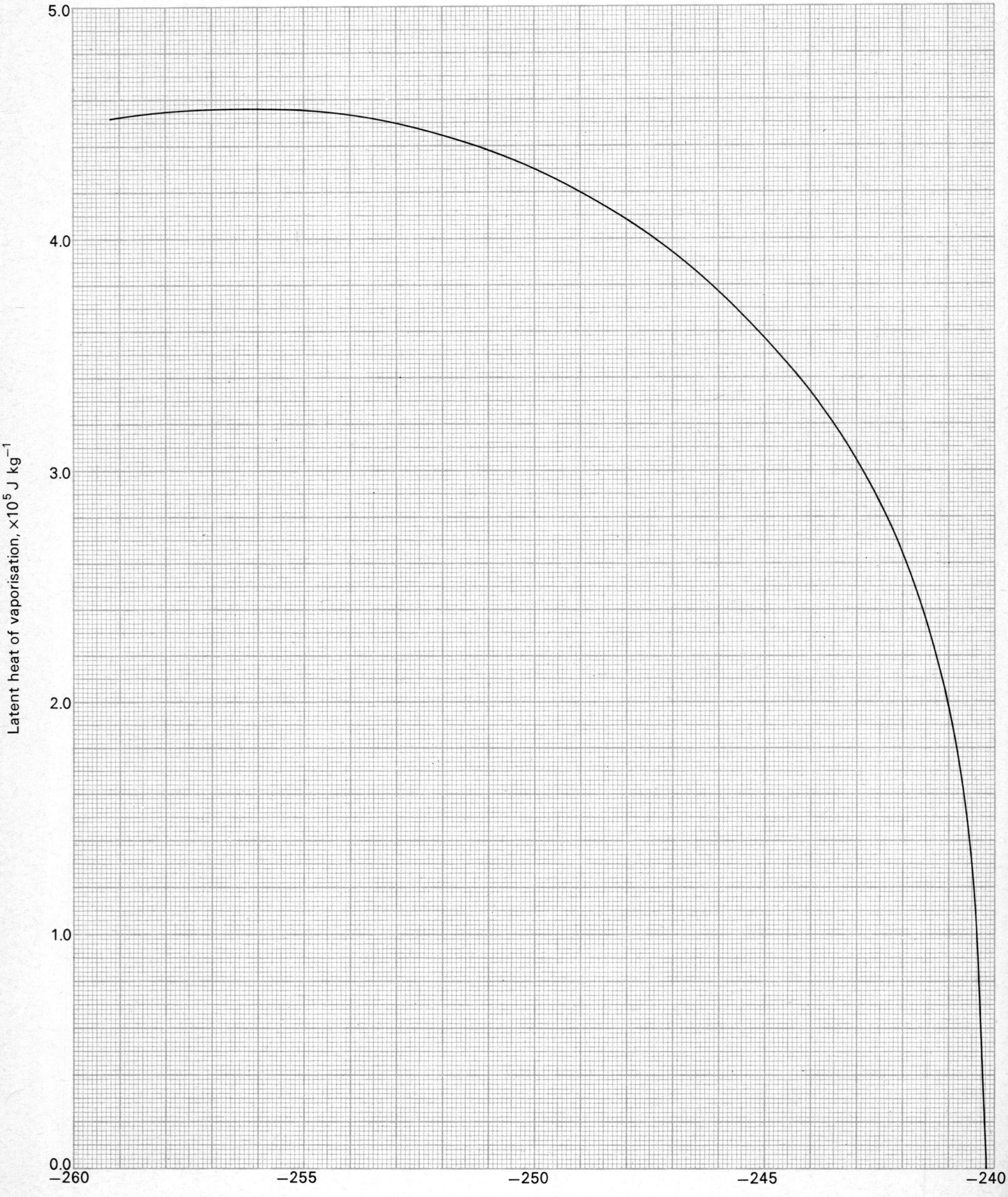

28.5 Heat Capacity of Liquid Hydrogen (Normal) (5,28,56,64)

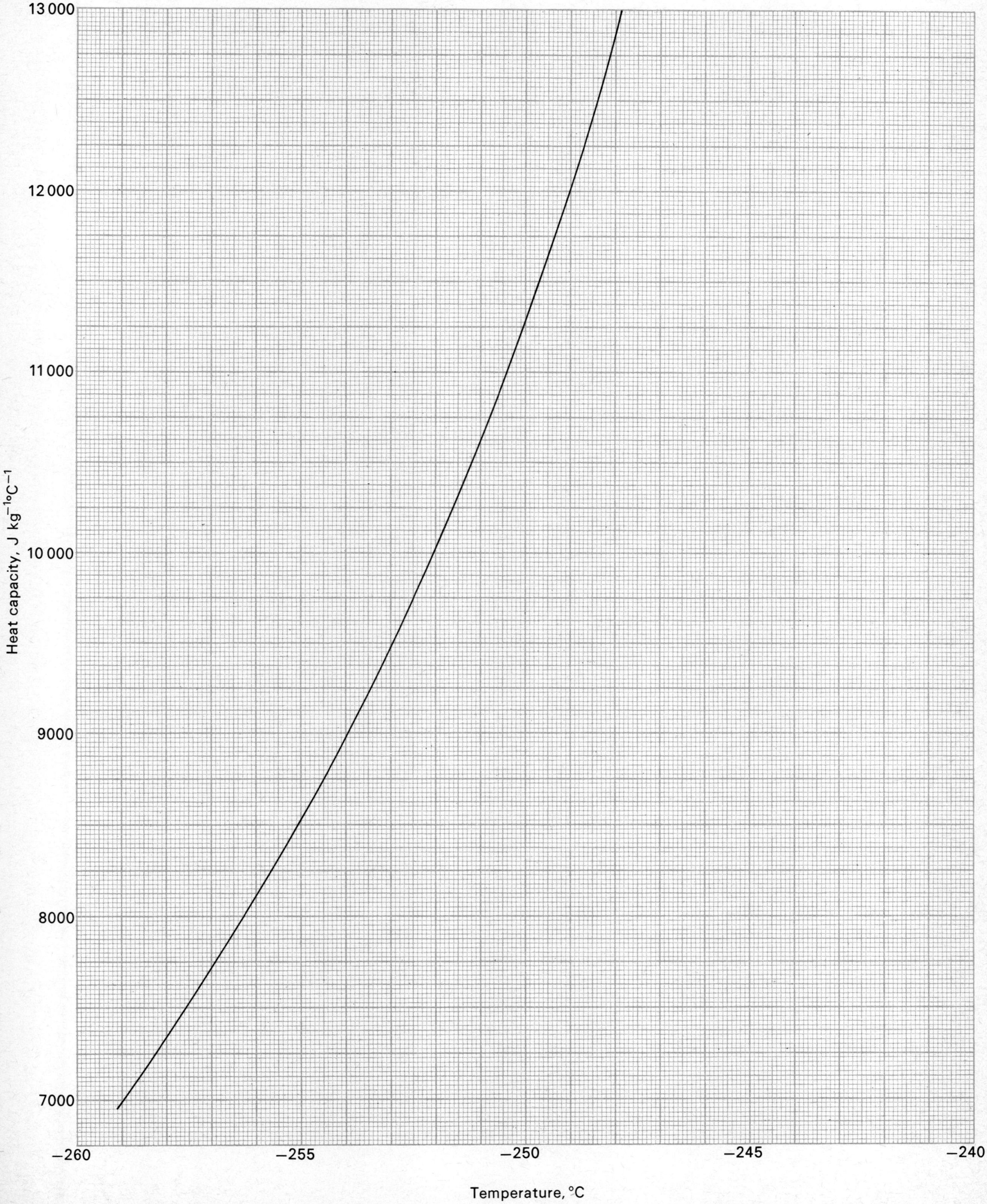

28.6 Heat Capacity of Gaseous Hydrogen (Normal) at $1.0133 \times 10^5 N\ m^{-2}$ (3,22,24,45,55,57,64)

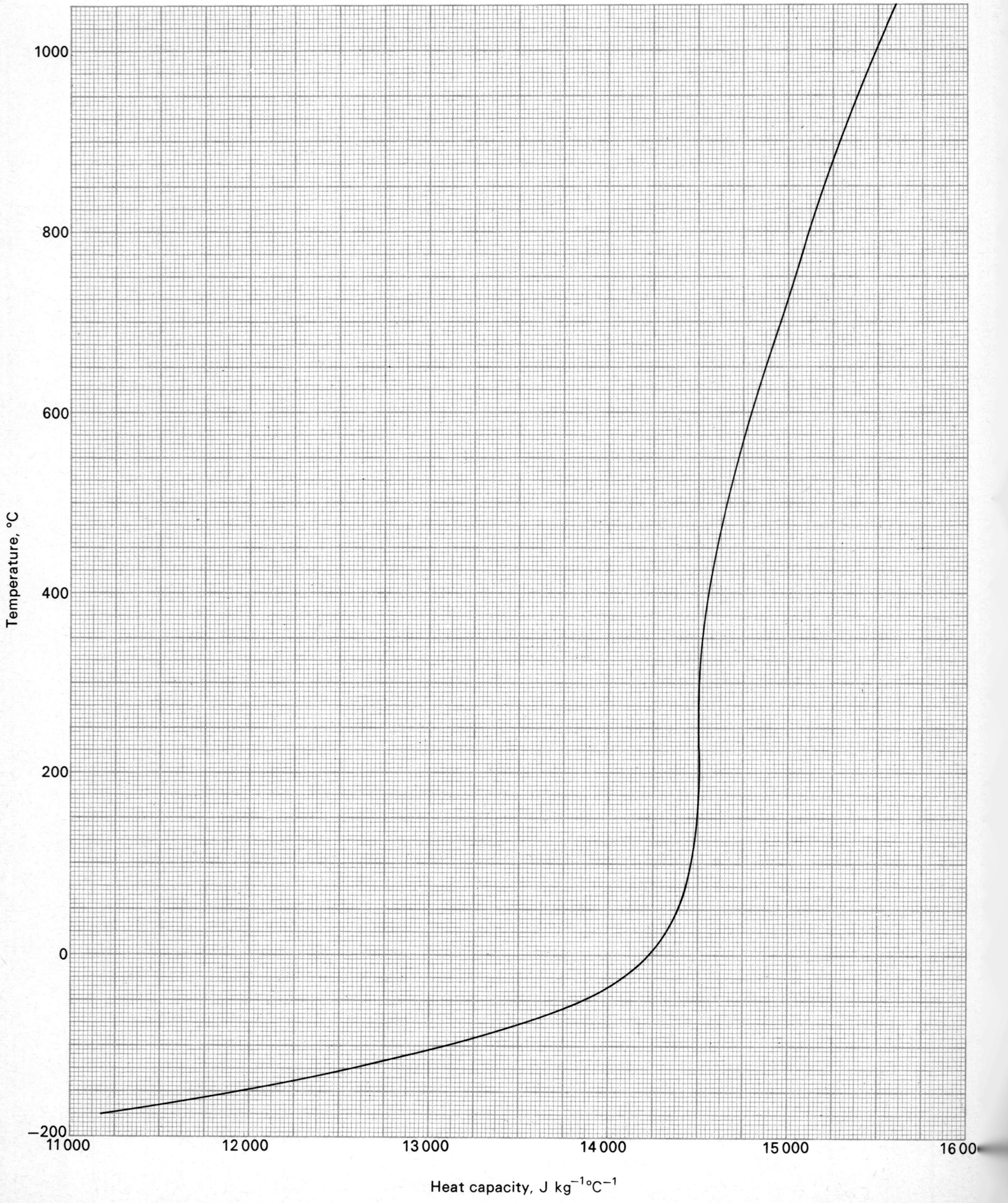

28.7 Heat Capacity Ratio (C_p/C_v) of Gaseous Hydrogen at 1.0133×10^5 N m^{-2} (13)

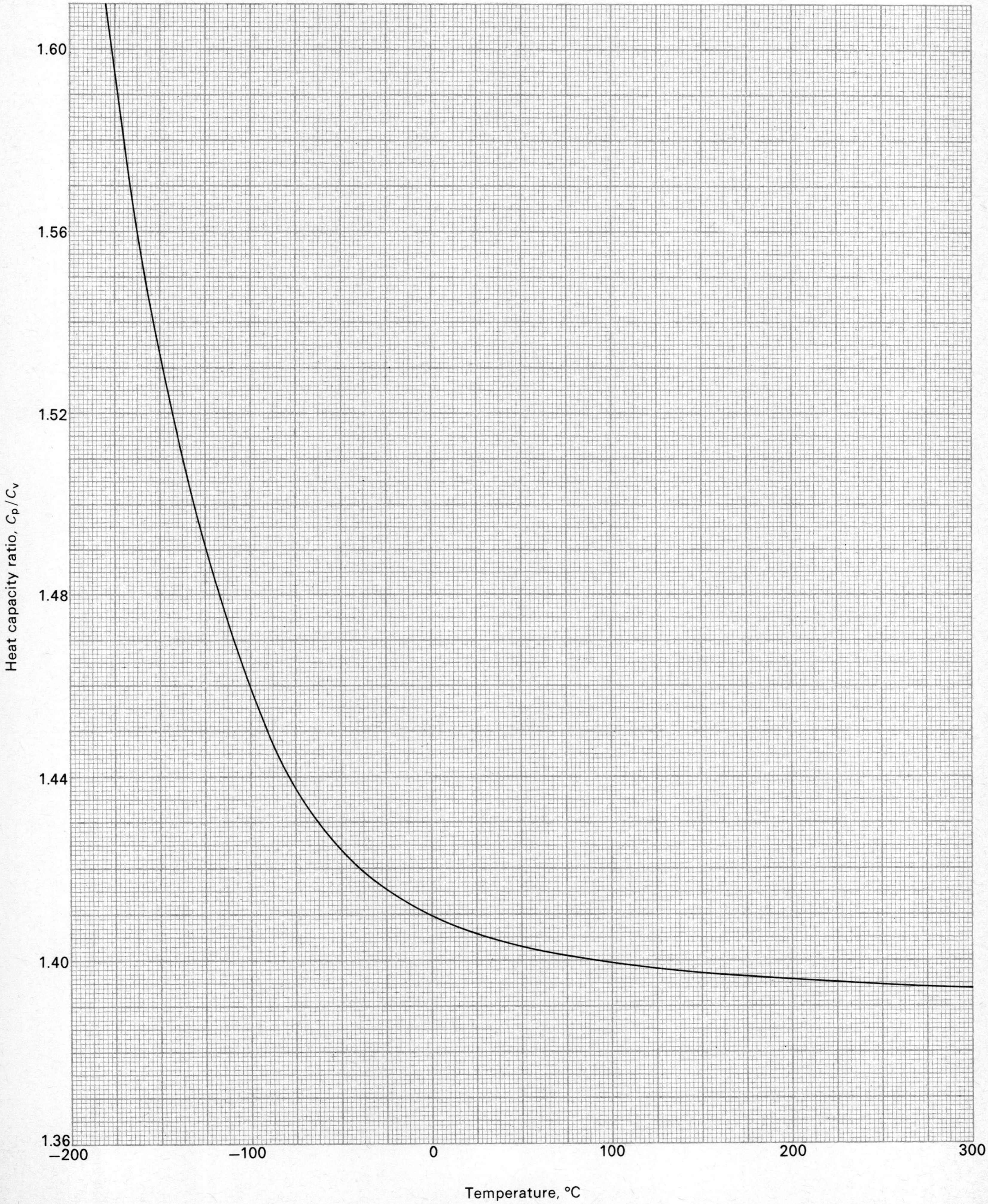

28.8 Viscosity of Liquid Hydrogen (Normal) (3,9,14)

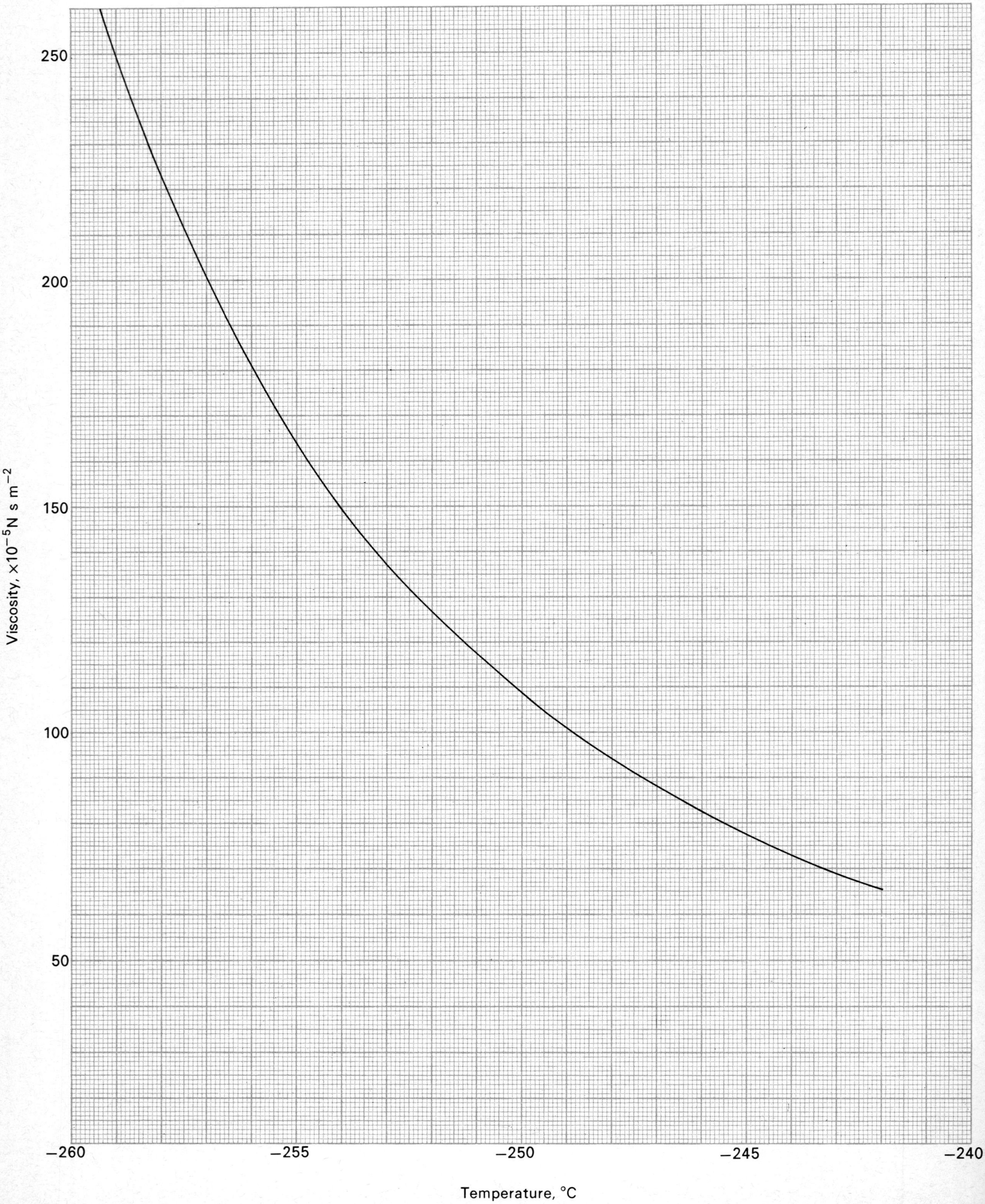

28.9 Viscosity of Gaseous Hydrogen (Normal) at 1.0133×10^5 N m^{-2} (2,3,14,37)

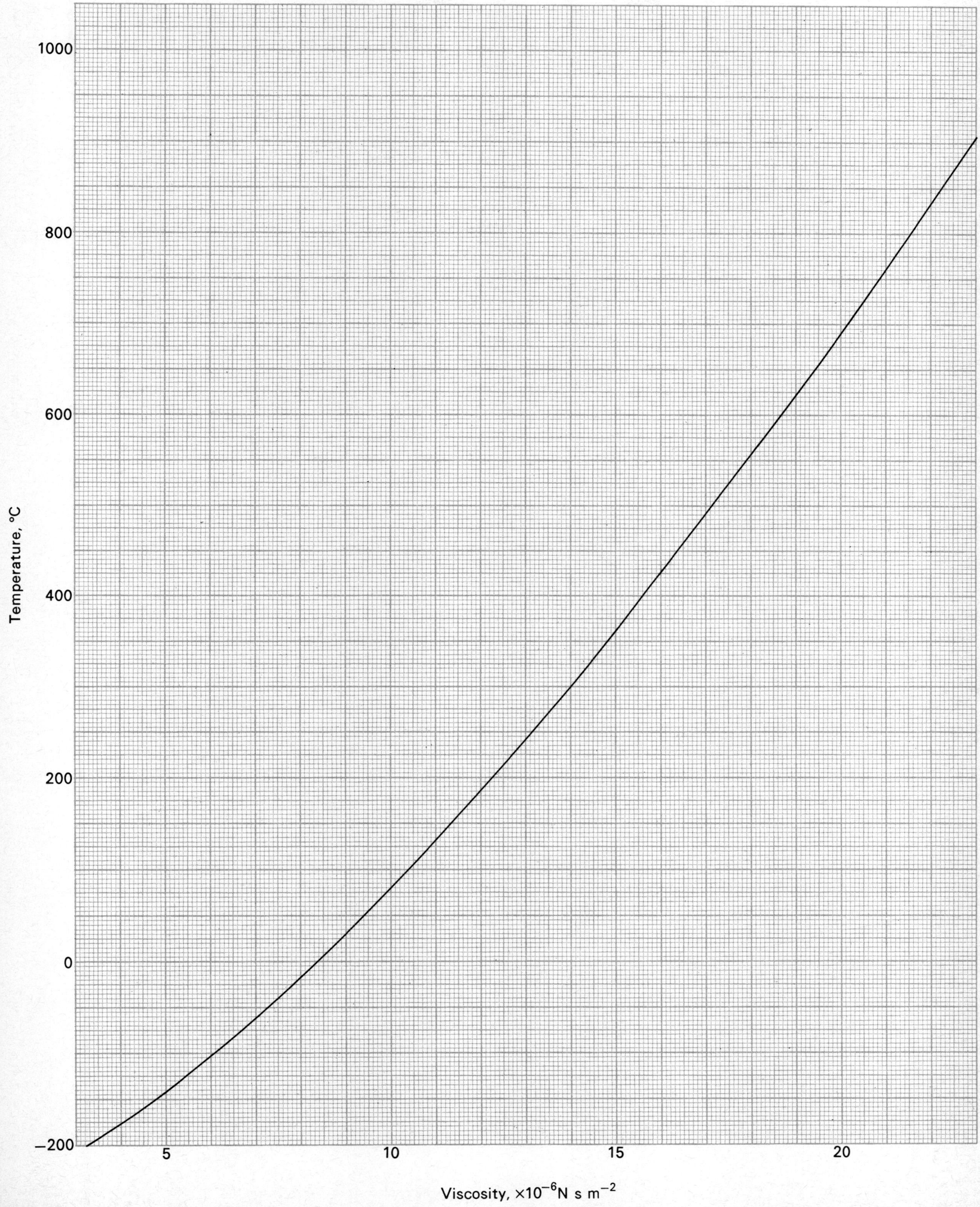

28.10 Thermal Conductivity of Liquid Hydrogen (Normal) (3,14,17,48,61)

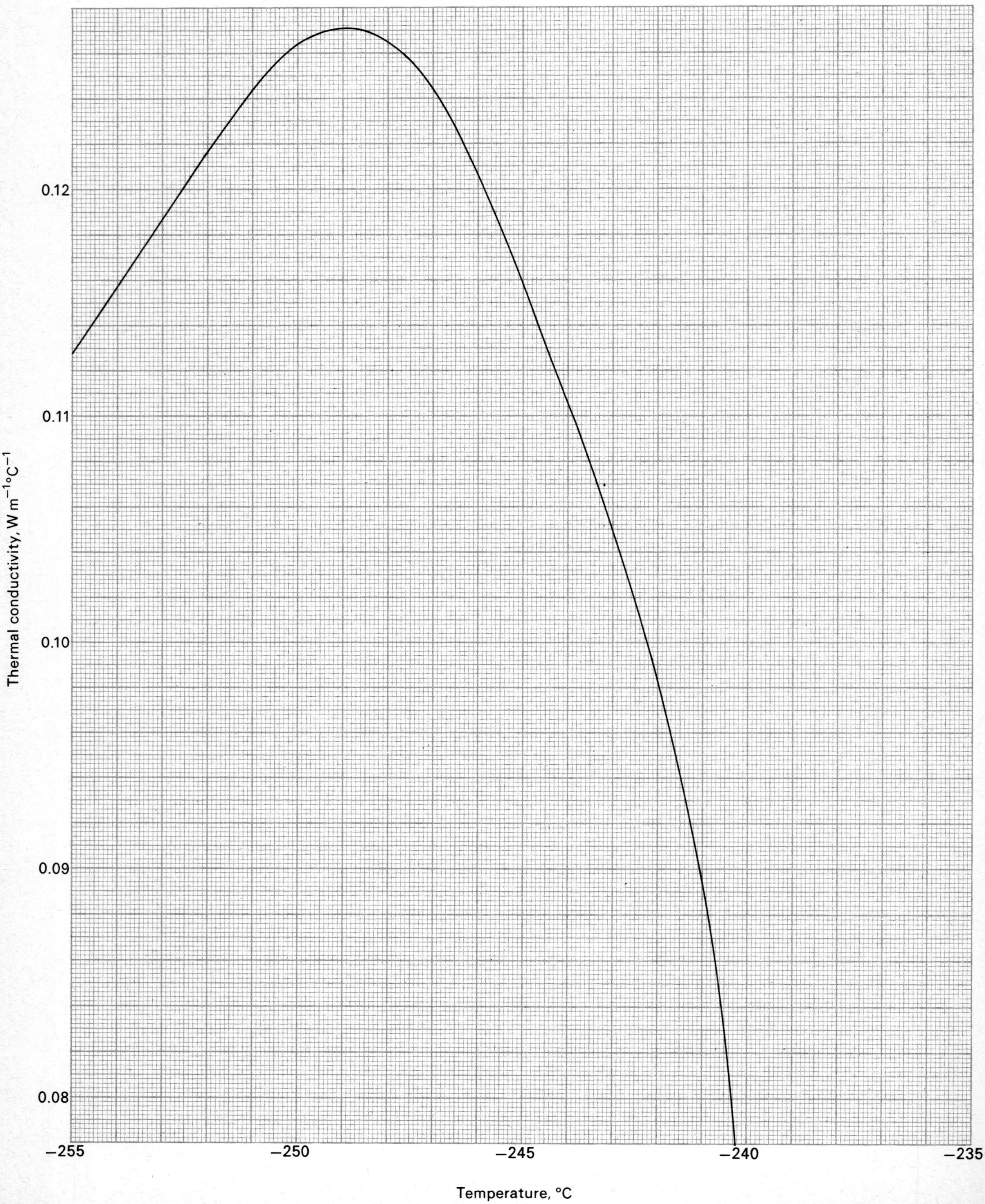

28.11 Thermal Conductivity of Gaseous Hydrogen (Normal) at $1.0133 \times 10^5 N\ m^{-2}$ (3,11,18,24,25,30,58,61)

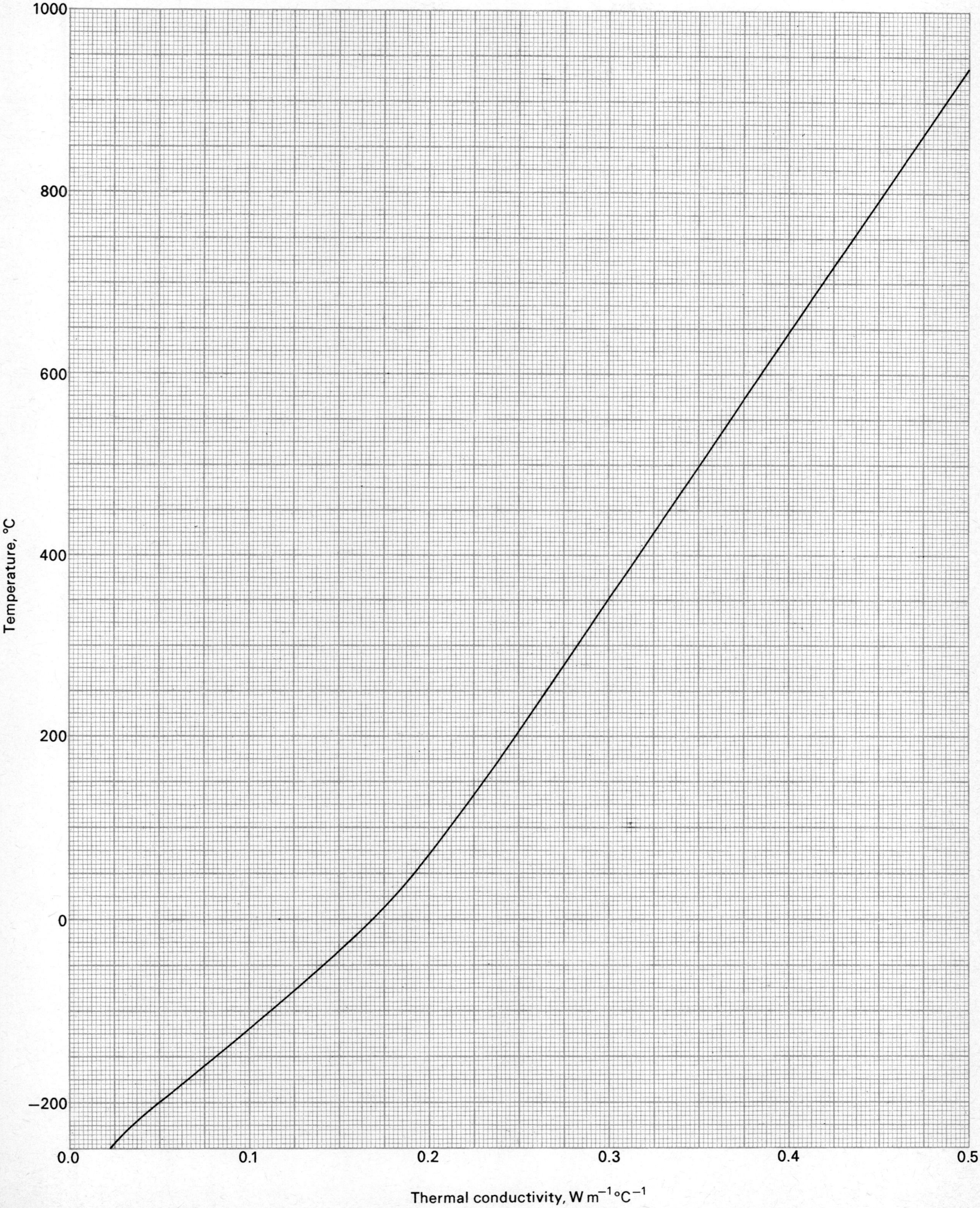

28.12 Surface Tension of Hydrogen (Normal) (14)

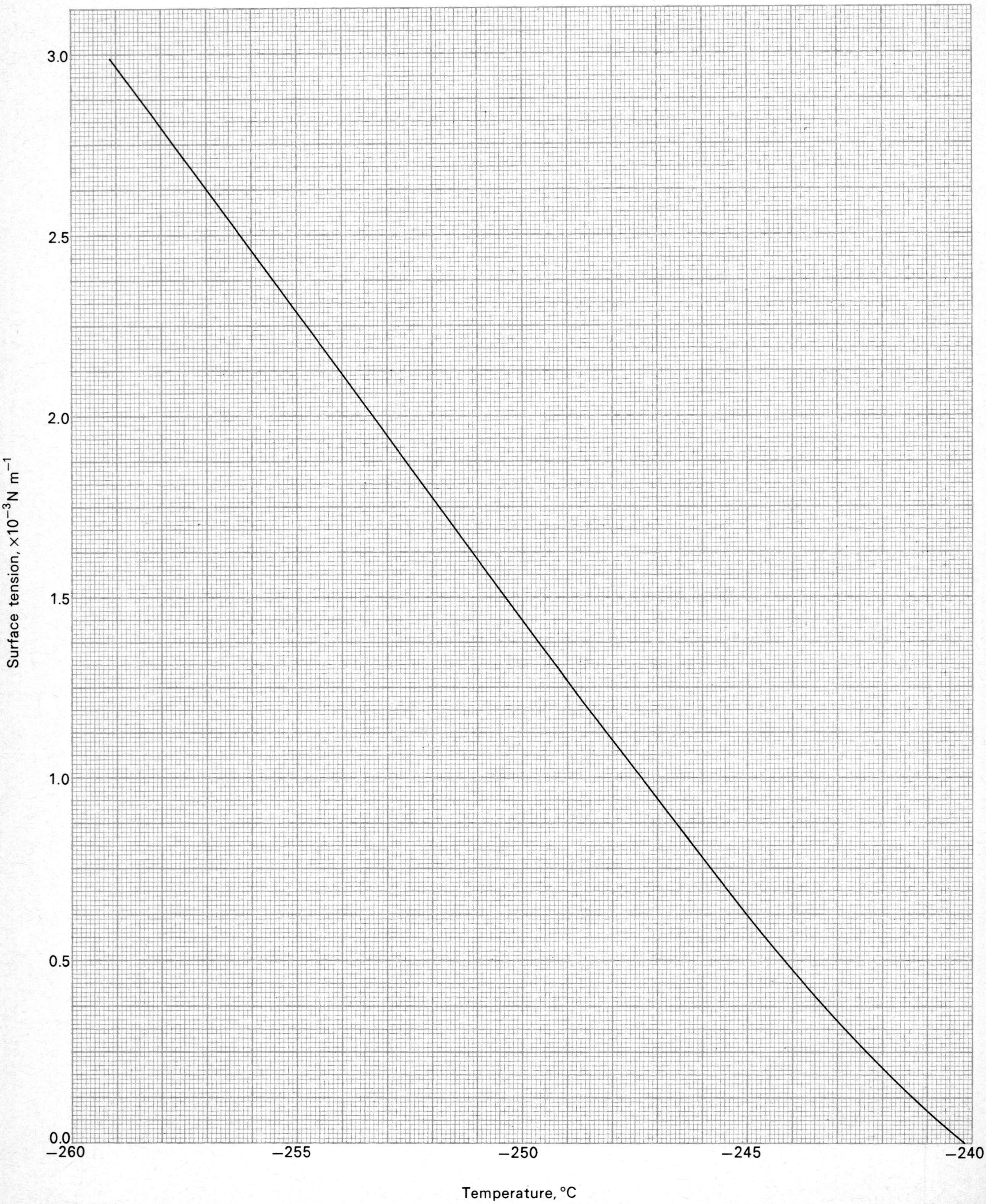

28.13 Solubility of Hydrogen in Water at $1.0133 \times 10^5 \text{N m}^{-2}$ Total Pressure (33,34,44,62)

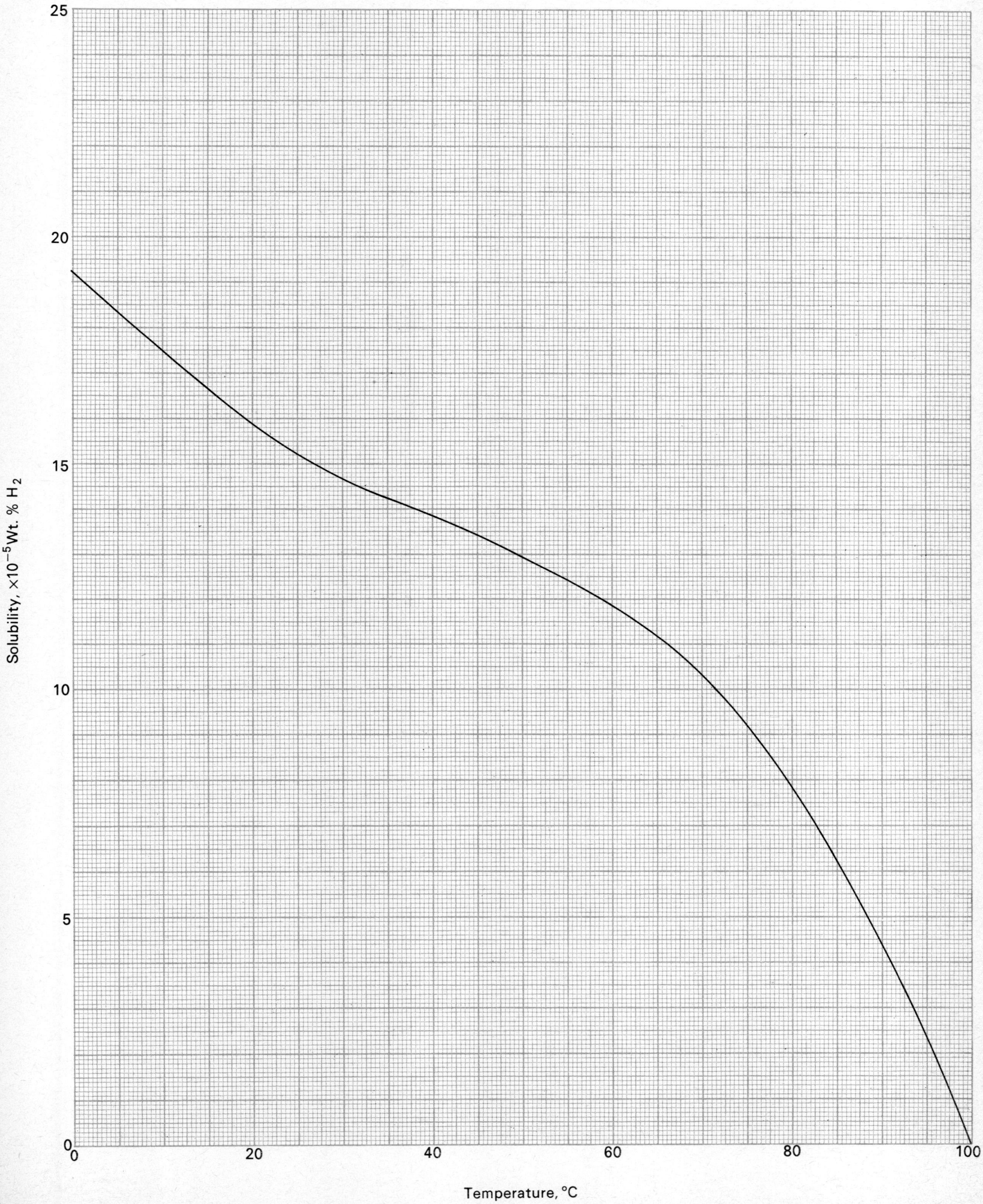

29 HYDROGEN CYANIDE

Hydrogen cyanide is a colourless, very volatile, and strongly poisonous liquid at room temperature and atmospheric pressure. The gas has an odour of bitter almonds. It burns in air and a mixture with oxygen may explode. Liquid hydrogen cyanide is unstable and decomposes into elements. It has a very high dielectric constant. In the gaseous state the molecules are associated, about 18% at the normal boiling point. The high boiling point is explained by the dipole attraction. Liquid HCN is a good solvent for salts.

Gaseous hydrogen cyanide is readily soluble in water, forming a weak acid. The solution is slowly hydrolysed to ammonia formate.

The boiling point, dielectric constant, heat of vaporisation, etc., of HCN are very much higher than would be expected of a compound of similar molecular dimensions. These differences are the result of association of HCN molecules into linear polymers as a consequence of hydrogen bonding. Furthermore, there is a connection between the hydrogen bonding and the viscosity of the liquid. If the bond is very strong, greater energy will be required to break these bonds in addition to overcoming the dipole and van der Waals forces.

On cooling, the liquid solidifies to crystalline hydrogen cyanide. There are two modifications of the crystalline structure: tetragonal and rhombic. In the crystalline state the HCN molecules do not rotate.

The thermophysical and transport properties of HCN have been compiled in the following publications: Gallant (11), Gmelin (13), Grüttner (16), Kirk-Othmer (19), Pascal (25), Sinozaki and Hara (30), and Ullmann (33).

The liquid thermal conductivity of HCN has been predicted, using a method recommended by Stiel and Thodos (31) and the estimated values are presented in Fig. 29.9. For the reliability of the estimation method, see the discussion under 'Fluorine'.

References

1 American Conference of Government Industrial Hygienists, *Documentation of Threshold Limit Values*, revised edition, Committee on Threshold Limit Values, Cincinnati, Ohio 45202, 1966, 203 pp.

2 Braune, H., Asche, T., *Z. phys. Chem.*, 1931, **14B**, 18–26.

3 Braune, H., Linke, R., *Z. phys. Chem.*, 1930, **148A**, 195–214.

4 Bredig, G., Teichmann, L., *Z. Elektrochem.*, 1925, **31**, 449–53.

5 Coates, J. E., Davies, R. H., *J. Chem. Soc.*, 1950, 1194–9.

6 Coates, J. E., Hartshorne, N. H., *J. Chem. Soc.*, 1931, 657–65.

7 Coates, J. E., Taylor, E. G., *J. Chem. Soc.*, 1936, 1245–56.

8 Edwards, D. G., *The Vapour Pressure of 30 Inorganic Liquids between One Atmosphere and the Critical Point*, Univ. of California, Lawrence Radiation Lab., UCRL 7167, Livermore, California, June 13, 1963, 148 pp.

9 Franck, E. U., *Z. Electrochem.*, 1951, **55** (7), 636–43.

10 Fredenhagen, K., Dahmlos, J., *Z. anorg. Chem.*, 1929, **179**, 77–81.

11 Gallant, R. W., *Hydrogen Processing*, 1969, **48** (7), 135–41.

12 Giauque, W. F., Ruehrwein, R. A., *J. Am. Chem. Soc.*, 1939, **61**, 2626–33.

13 *Gmelins Handbuch der anorganischen Chemie*, System No. 14, Kohlenstoff, Teil D1, Verlag Chemie GmbH., Weinheim, 1971 120–242.

14 Golubev, I. F., *Viscosity of Gases and Gas Mixtures*, Gosudarstvennoe Izdatel'stvo Fiziko-Matematicheskoi Literatury, Moskva, 1959, Table 4.

15 Greene, E. F., Toennies, J. P., *Chemical Reactions in Shock Waves*, Edward Arnold, London, 1964, 333–4.

16 Grüttner, B., Dove, M. F. A., Clifford, A. F., *Chemistry in Anhydrous, Prototropic Inorganic Solvents*, Part 1. Inorganic Chemistry in Liquid Hydrogen Cyanide and Liquid Hydrogen Fluoride, Pergamon Press, Oxford, 1971, 315 pp.

17 Horiuchi, J., Tanabe, K., *J. Res. Inst. Catalysis, Hokkaido Univ.*, 1951, **1**, 117–31.

18 IUPAC, *Pure & Appl. Chem.*, 1972, **30** (3–4), 638–49.

19 *Kirk-Othmer Encyclopedia of Chemical Technology*, 2nd ed., Vol. 6, Hydrogen Cyanide, Interscience Publ., New York, 1965, 574–85.

20 Klotz, I. M., Miller, D. K., *J. Am. Chem. Soc.*, 1947, **69**, 2557–8.

21 Lange, J., Berga, J., *Monatsh. Chem.*, 1950, **81**, 921–38.

22 Lange, N. A., Ed., *Handbook of Chemistry*, 10th ed., McGraw-Hill, New York, 1967.

23 Lax, E., Synowietz, C., Eds., *Taschenbuch für Chemiker und Physiker*, 3rd ed., Springer-Verlag, Berlin, 1967.

24 Partington, J. R., Carroll, M. F., *Phil. Mag.*, 1925, **49** (6), 665–80.

25 Pascal, P., Ed., *Nouveau Traité de Chimie Minérale*, Vol. VIII, Carbone, Masson et Cie, Éditeurs, Paris, 1968, 901–29.

26 Ribaud, M. G., *Publ. Sci. Tech. Min. Air (France)*, 1952, (266), 119–35.

27 Rossini, F. D., Wagman, D. D., Evans, W. H., Levine, S., Jaffe, I., *Selected Values of Chemical Thermodynamic Properties*, Natl. Bur. Std. Circular 500, Washington, Feb. 1, 1952.

28 Senftleben, H., *Z. Angew. Phys.*, 1964, **17**, 86–7.

29 Sinozaki, H., Hara, R., *Proc. Imp. Acad. (Japan)*, 1927, **3** (9), 590–4.

30 Sinozaki, H., Hara, R., Hydrocyanic Acid II. Determination of the Densities of Gaseous and Liquid Hydrocyanic Acid, and some Thermodynamic Calculations, *Tech. Repts. Tohoku Imp. Univ.*, 1929, **8** (3), 19–64.

31 Stiel, L. I., Thodos, G., *AIChE Journal*, 1964, **10** (1), 26–30.

32 Stull, D. R., Dir., *JANAF Thermochemical Tables*, Dow Chemical Co., Midland, Mich., 1965–8.

33 *Ullmanns Encyklopädie der technischen Chemie*, 3rd ed., Vol. 5, Cyanwasserstoff, Urban & Schwarzenberg, München, 1954, 628–41.

34 Werner, O., *Z. phys. Chem.*, 1929, **4B**, 371–92.

35 Woolf, W. L., *J. Acous. Soc. Amer.*, 1943, **15**, 83–8.

Table 29.1 Properties of Hydrogen Cyanide

		Ref.
Chemical formula	HCN	(22)
Molecular weight	27.0256	(18)
Critical temperature	183.6 °C	(4)
Critical pressure	53.9076×10^5 N m^{-2}	(4)
Critical density	195.0 kg m^{-3}	(4)
Triple-point temperature	−13.296 °C	(12)
Triple-point pressure	0.1872×10^5 N m^{-2}	(12)
Boiling point at 1.0133×10^5 N m^{-2}	25.7 °C	(27)
Melting point at 1.0133×10^5 N m^{-2}	−13.3 °C	(6)
Transition temperature	−102.8 °C	(12)
Density of solid at −40 °C	925.0 kg m^{-3}	(34)
Coefficient of thermal expansion of liquid at 18 °C	1.93×10^{-3} °C^{-1}	(23)
Heat of formation of gas at 25 °C	1.3063×10^8 J kg-mol^{-1}	(27)
Free energy of formation of gas at 25 °C	1.2020×10^8 J kg-mol^{-1}	(32)
Heat of sublimation at triple-point temperature	3.6551×10^7 J kg-mol^{-1}	(21)
Heat of fusion	8.4113×10^6 J kg-mol^{-1}	(12)
Heat of transition	1.6747×10^4 J kg-mol^{-1}	(27)
Entropy of gas at 25 °C	2.0184×10^5 J kg-mol^{-1} °C^{-1}	(32)
Heat capacity ratio of gas at 65 °C and 1.0133×10^5 N m^{-2}	1.257	(24)
Velocity of sound at 65 °C and 1.0133×10^5 N m^{-2}	438.0 m s^{-1}	(15)
Diffusion in air at 0 °C and 1.0133×10^5 N m^{-2}	0.173×10^{-4} m^2 s^{-1}	(20)
Dielectric constant of gas at 50.75 °C and 1.0133×10^5 N m^{-2}	1.02332	(2)
Electrical conductivity of liquid at 18 °C	0.90×10^{-5} ohm^{-1} m^{-1}	(7)
Refractive index of liquid at 10 °C and 1.0133×10^5 N m^{-2}	1.2675	(22)
Threshold limit value	(10 ppm) 11×10^{-6} kg m^{-3}	(1)

29.1 Vapour Pressure of Hydrogen Cyanide (8,26)

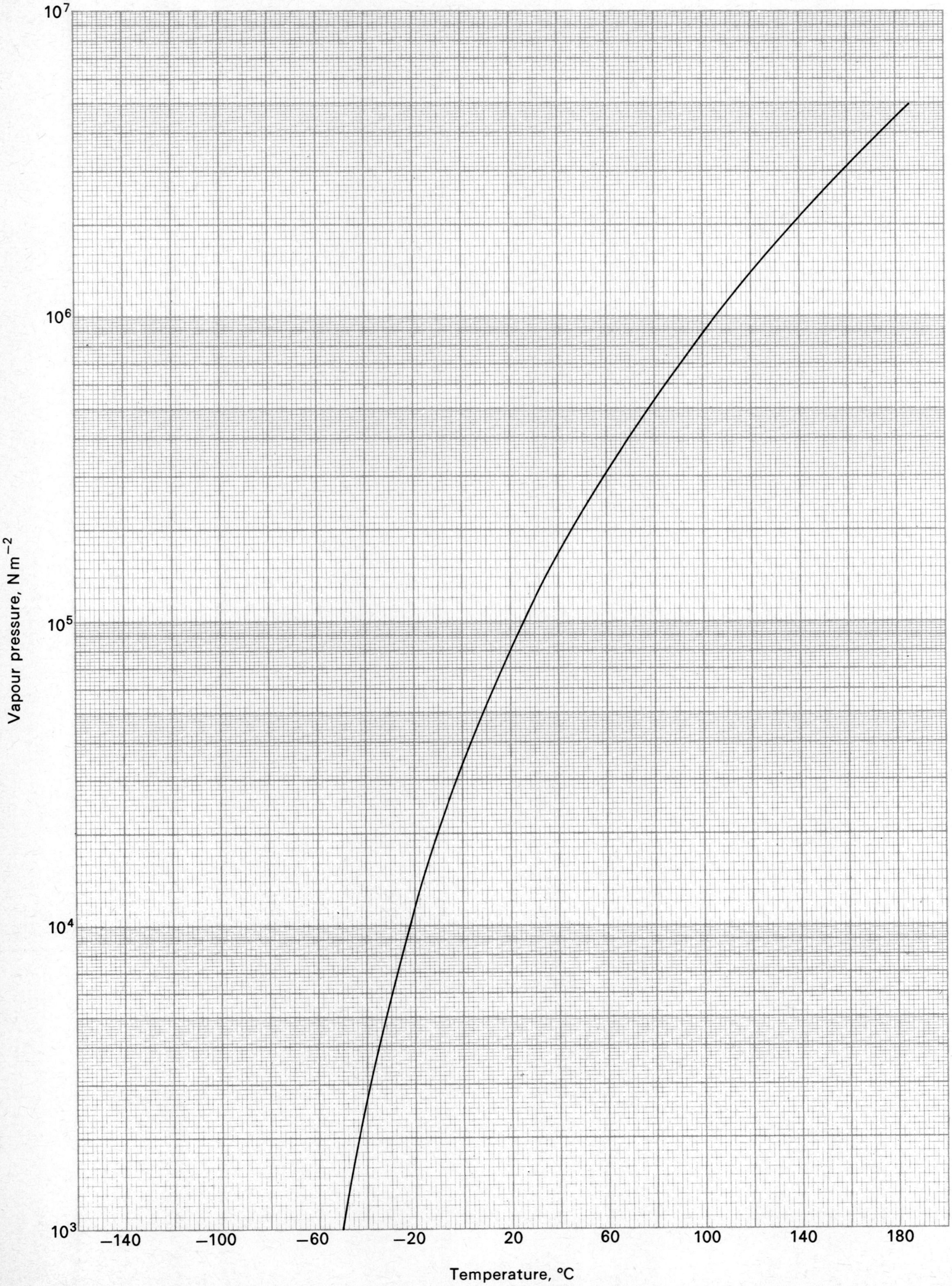

29.2 Orthobaric Density of Hydrogen Cyanide (4,11)

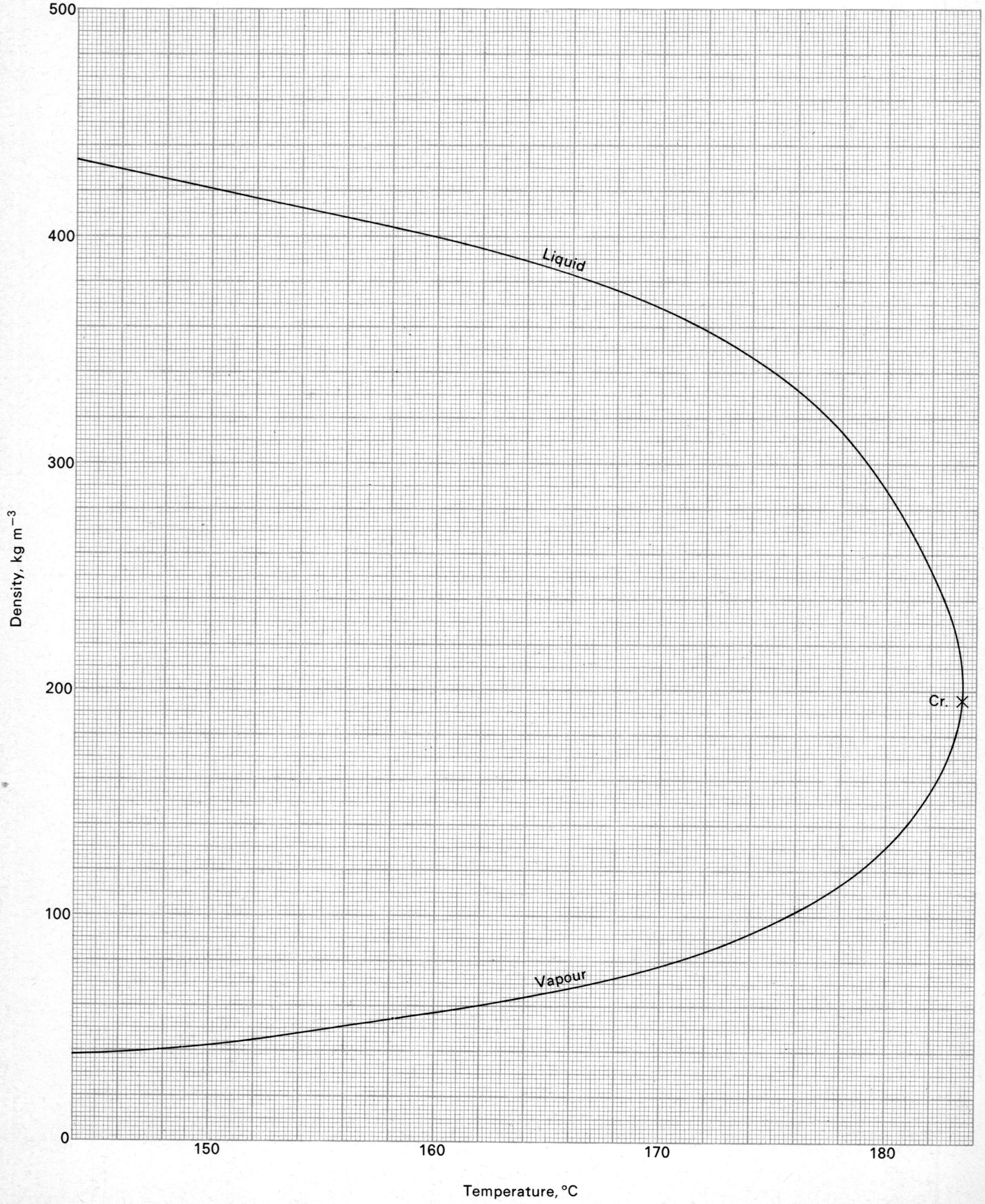

29.3 P-V-T Data for Hydrogen Cyanide (29,30)

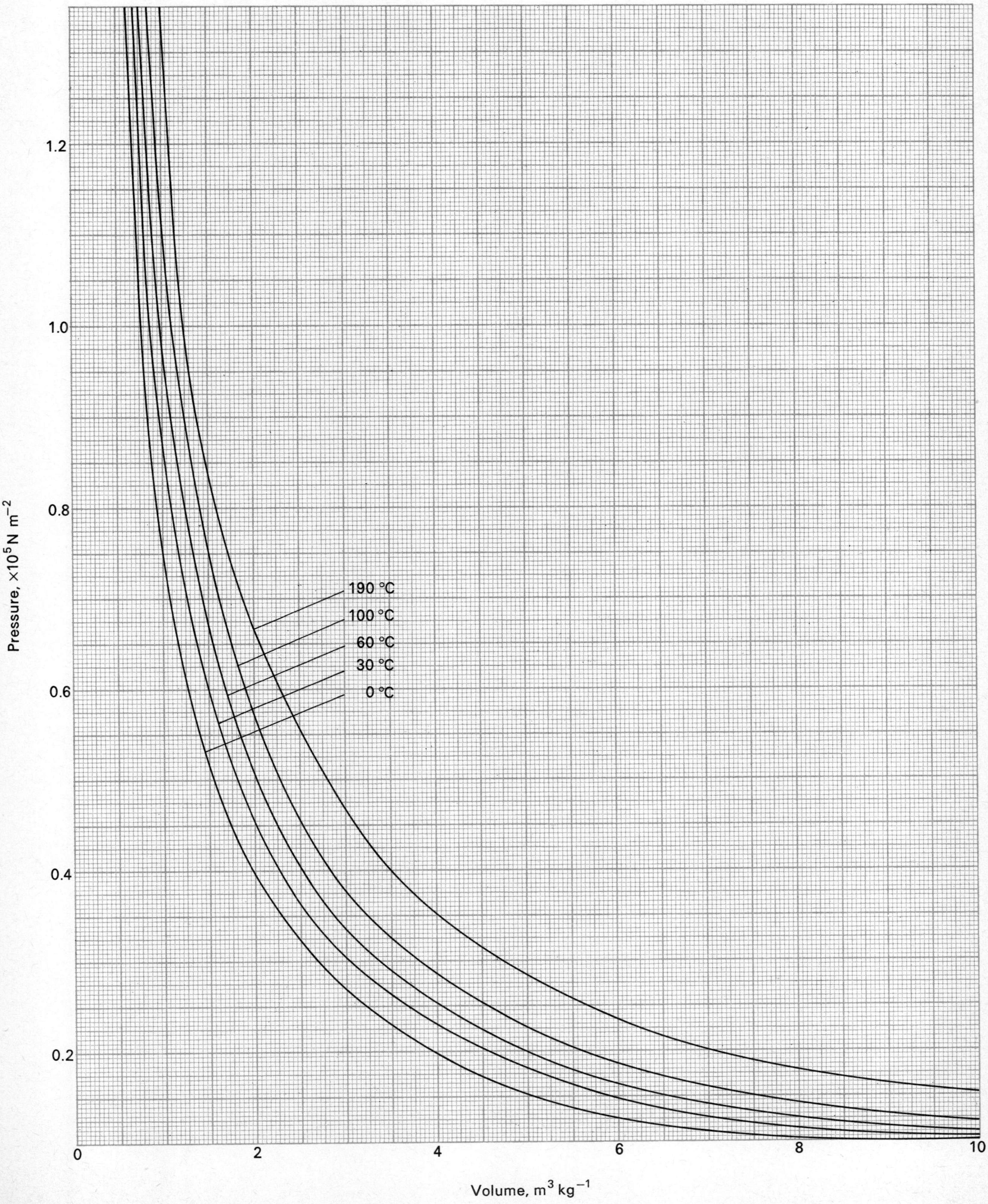

29.4 Latent Heat of Vaporisation of Hydrogen Cyanide (8,11,27,30)

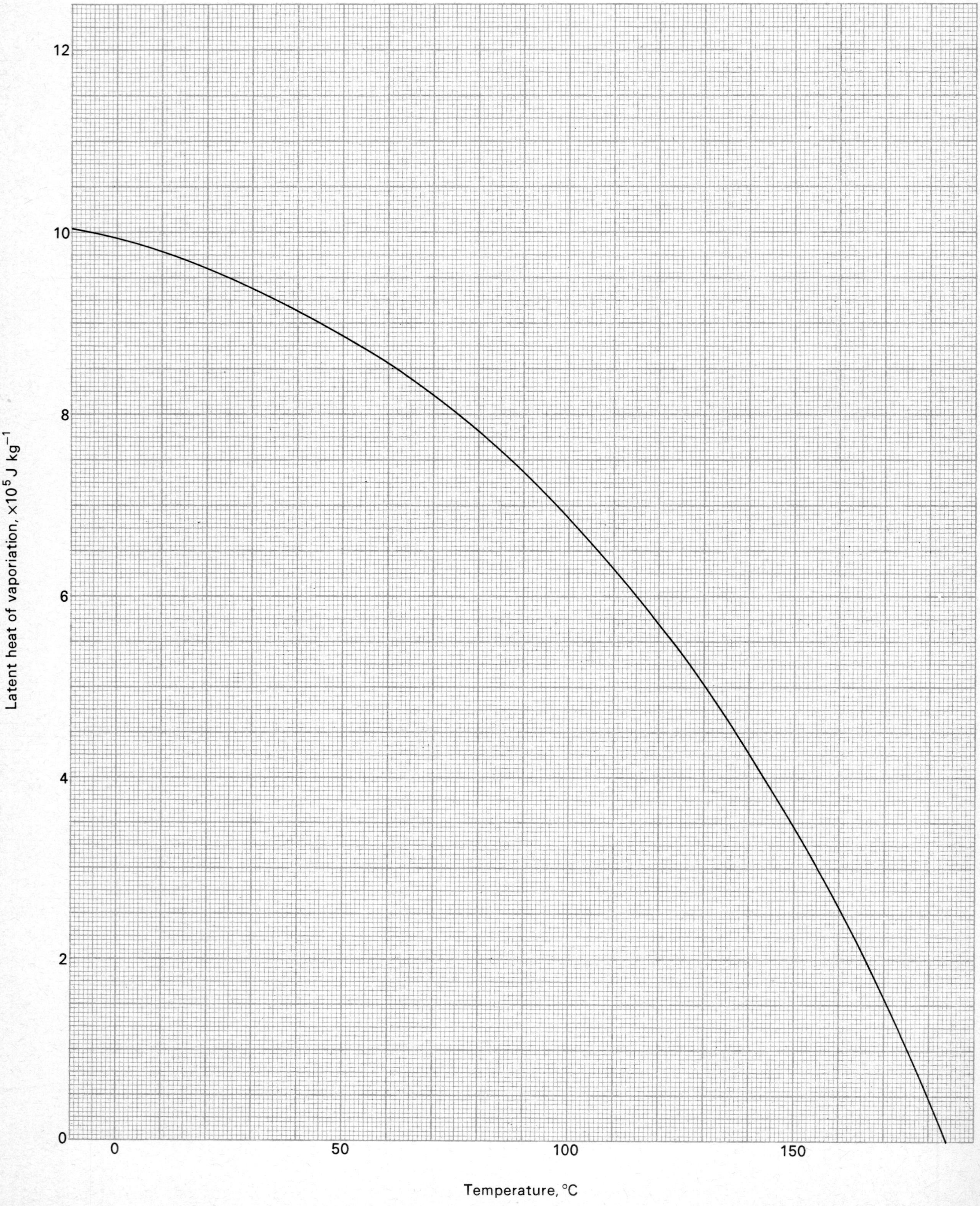

29.5 Heat Capacity of Solid and Liquid Hydrogen Cyanide (12)

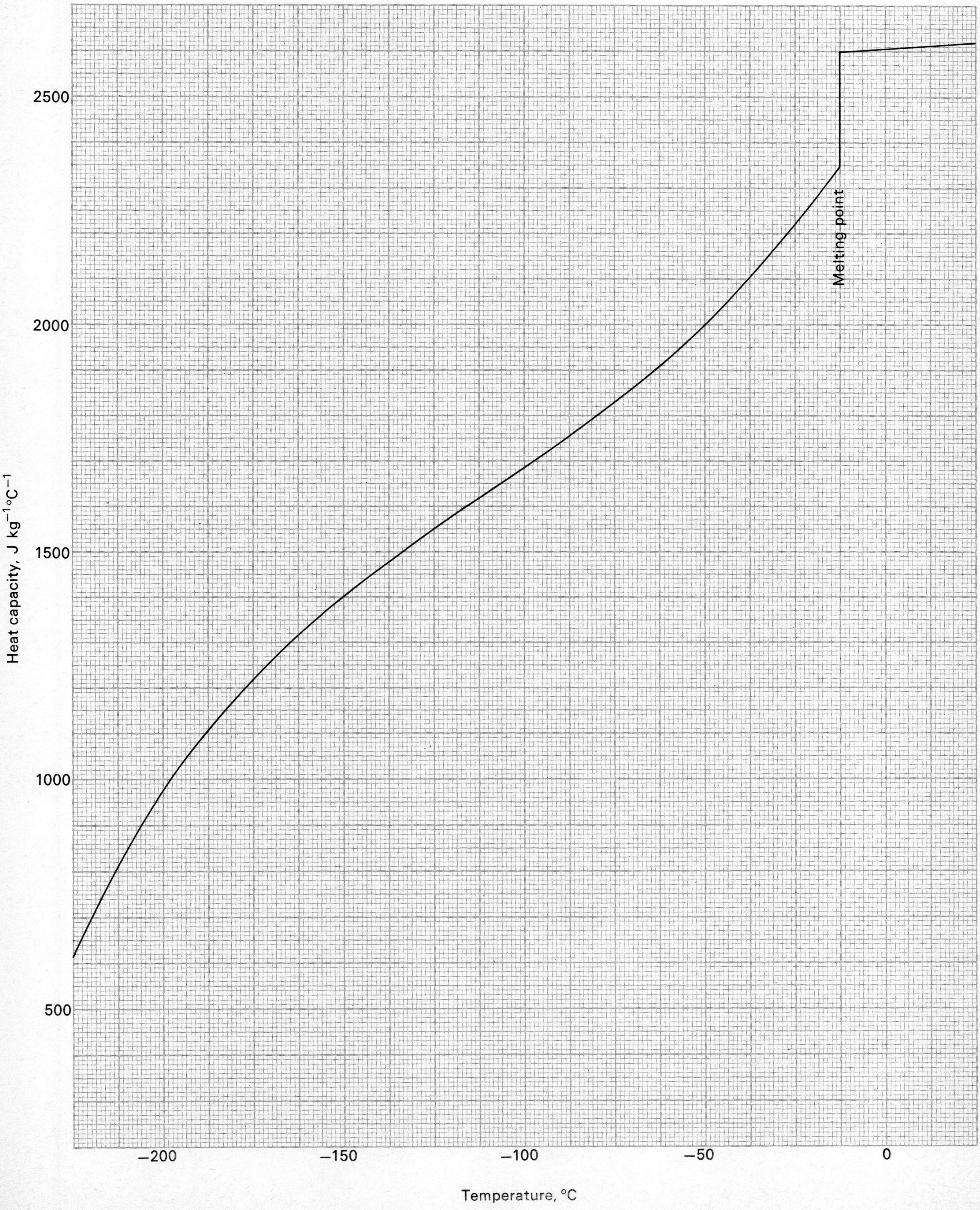

29.6 Heat Capacity of Gaseous Hydrogen Cyanide at $1.0133 \times 10^5 N\ m^{-2}$ (32)

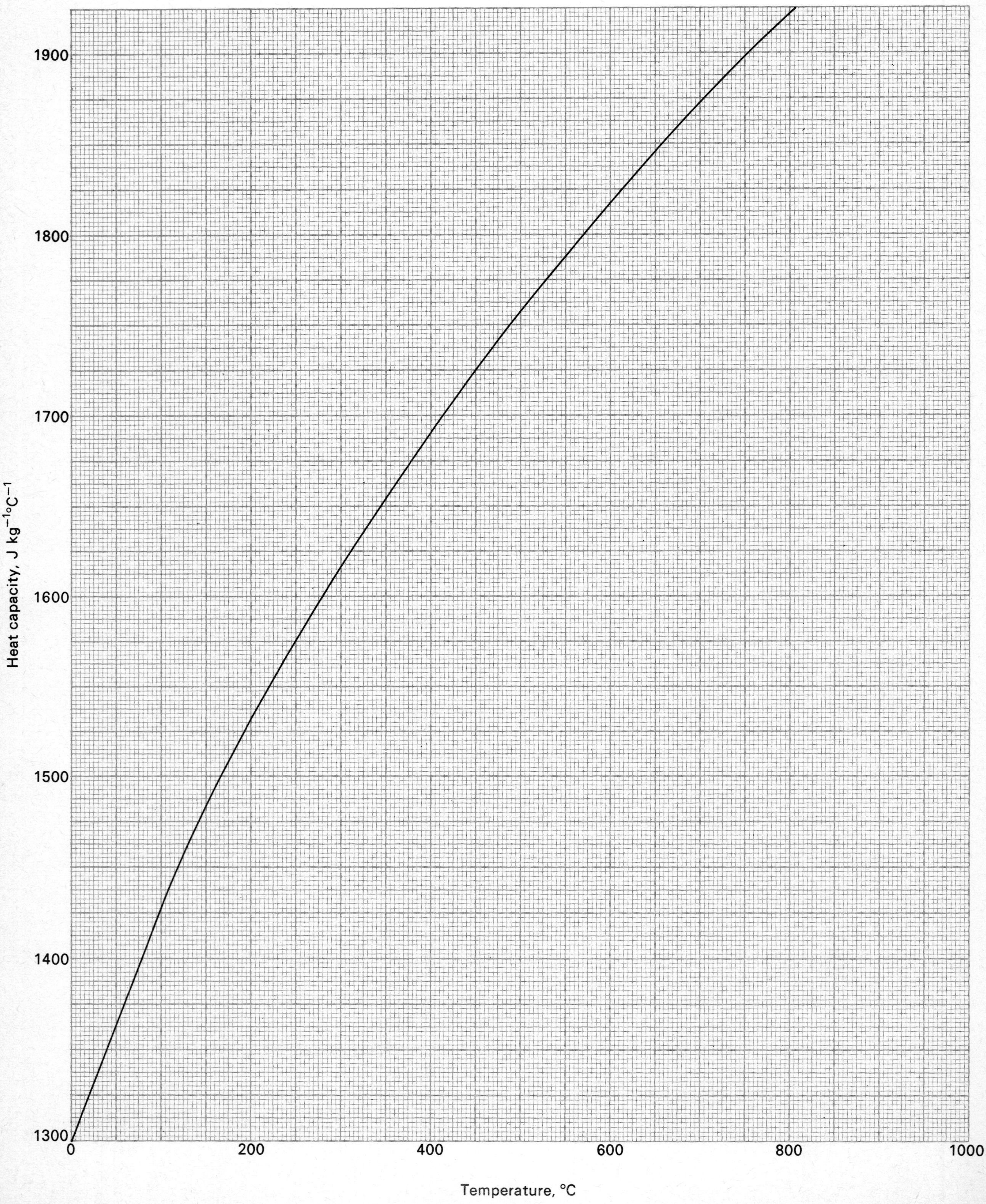

29.7 Viscosity of Liquid Hydrogen Cyanide (5)

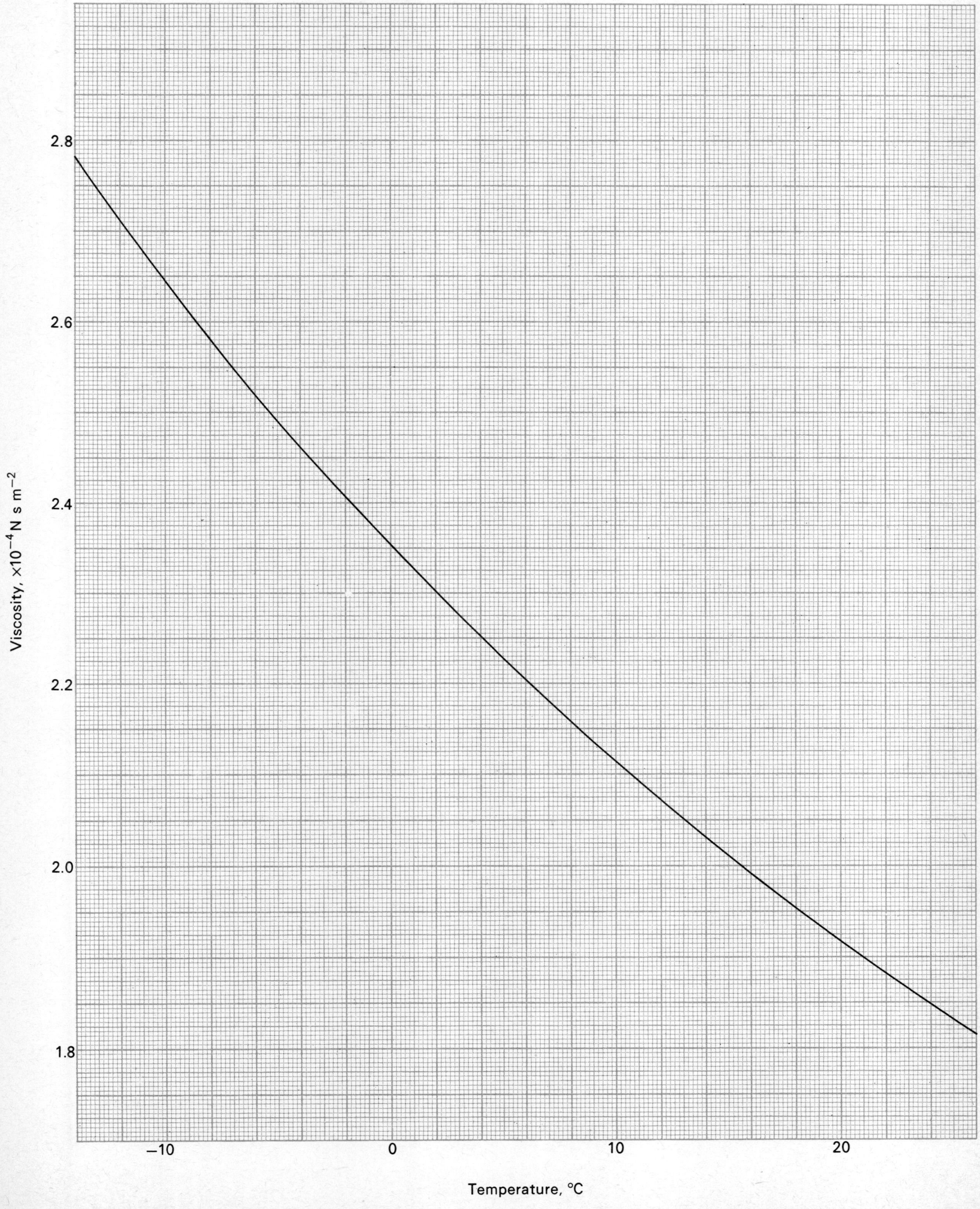

29.8 Viscosity of Gaseous Hydrogen Cyanide at 1.0133×10^5 N m^{-2} (3,14)

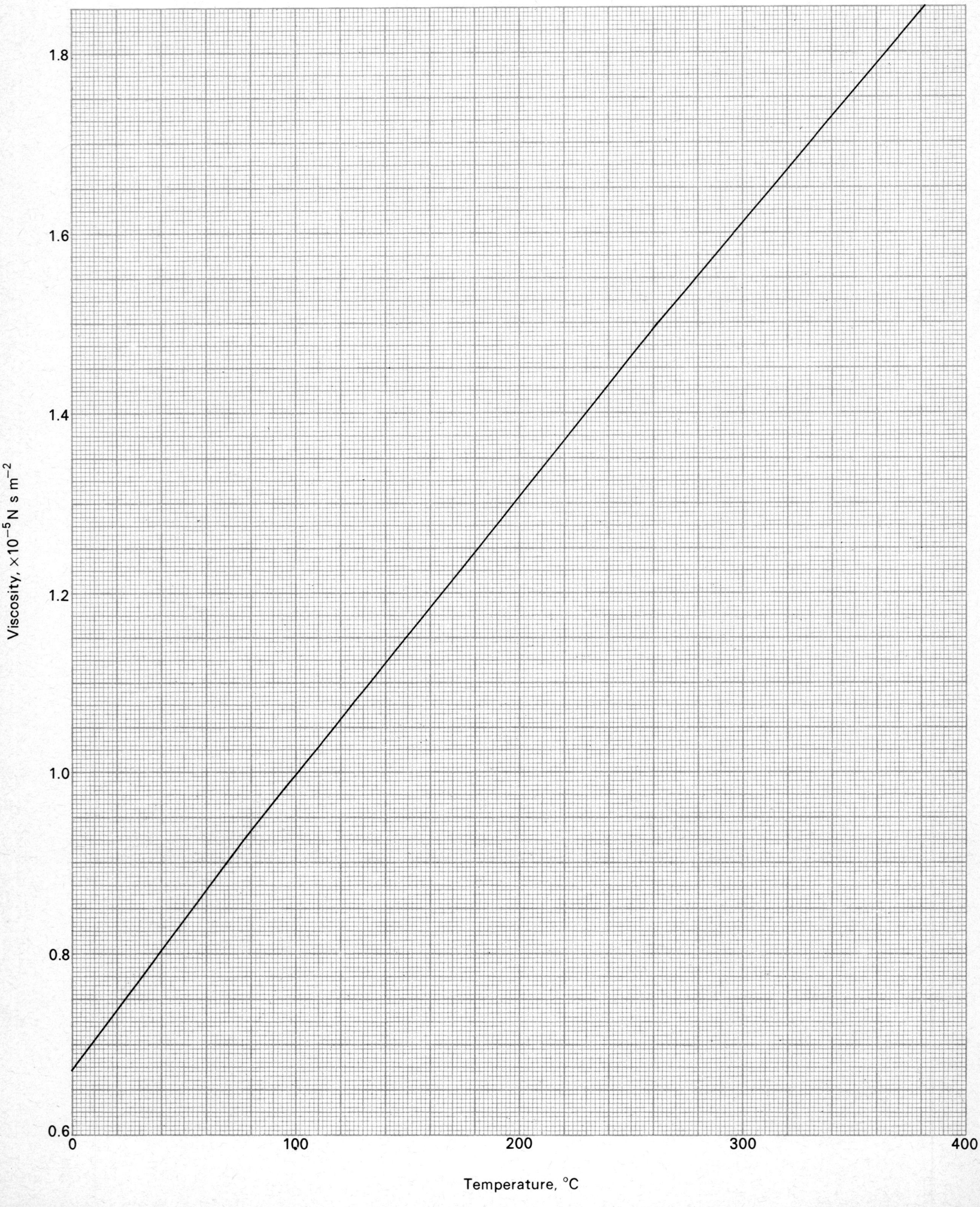

29.9 Thermal Conductivity of Liquid Hydrogen Cyanide (31)

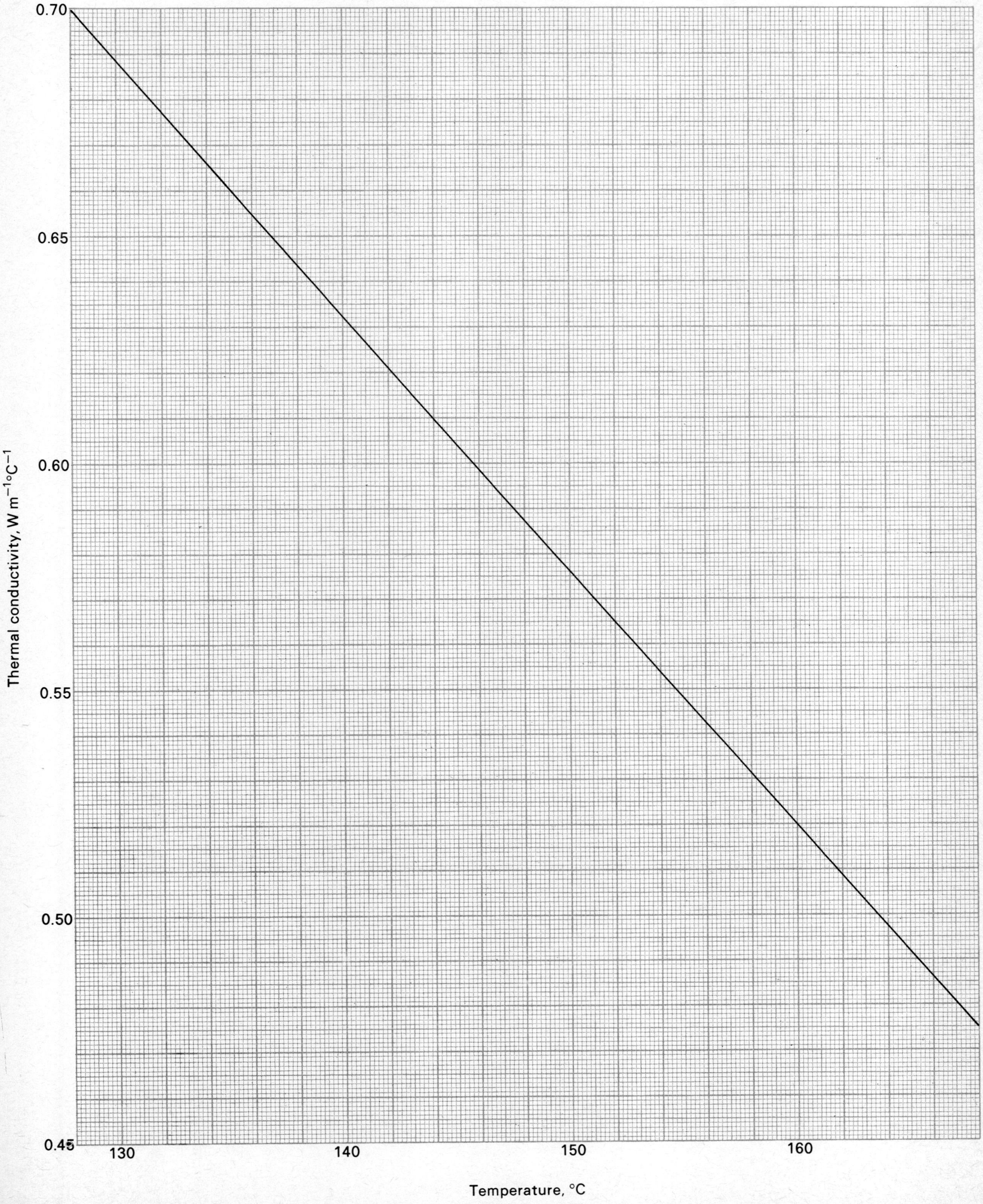

29.10 Thermal Conductivity of Gaseous Hydrogen Cyanide at $1.0133 \times 10^5 N\ m^{-2}$(9,28)

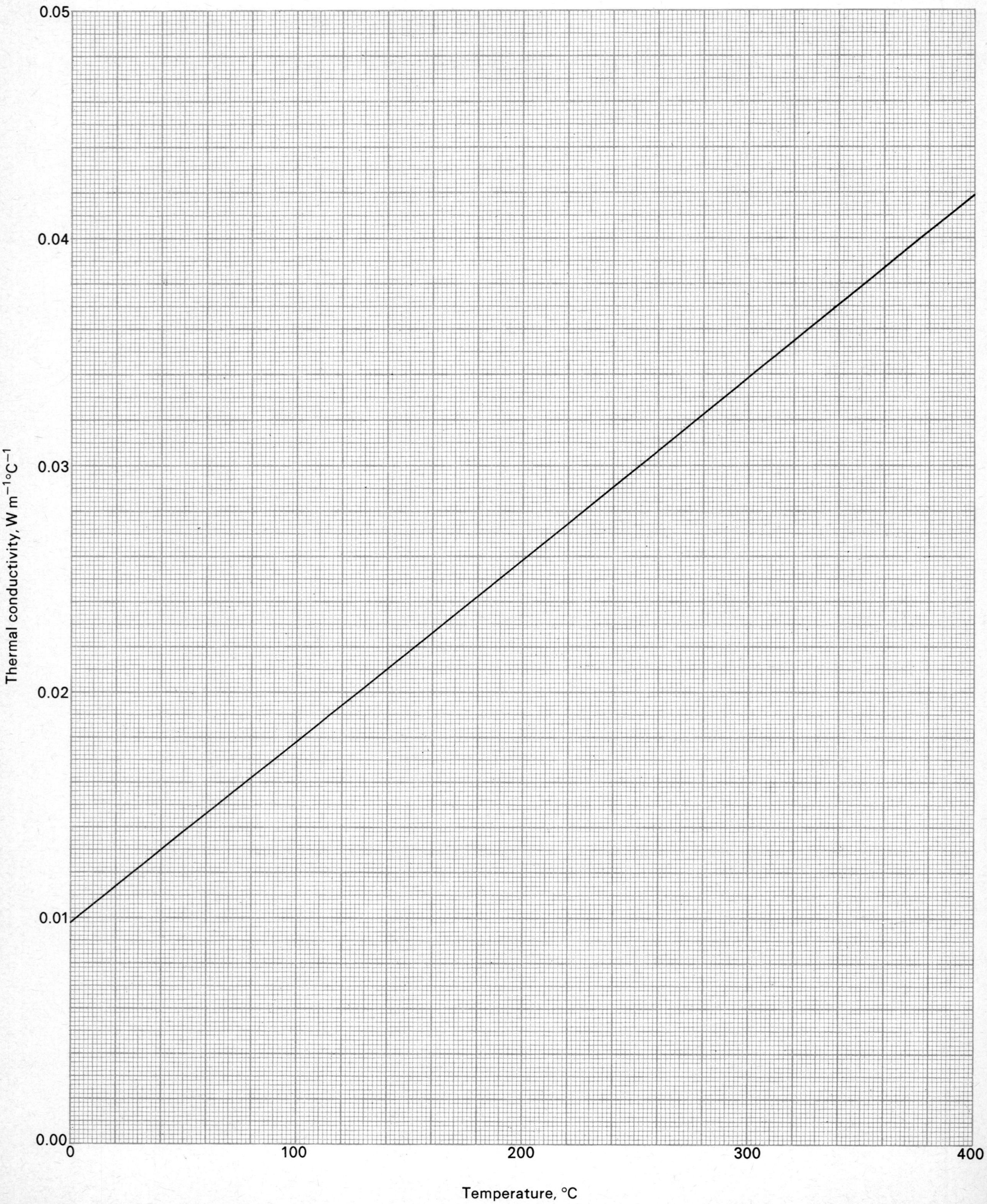

29.11 Surface Tension of Hydrogen Cyanide (4,5)

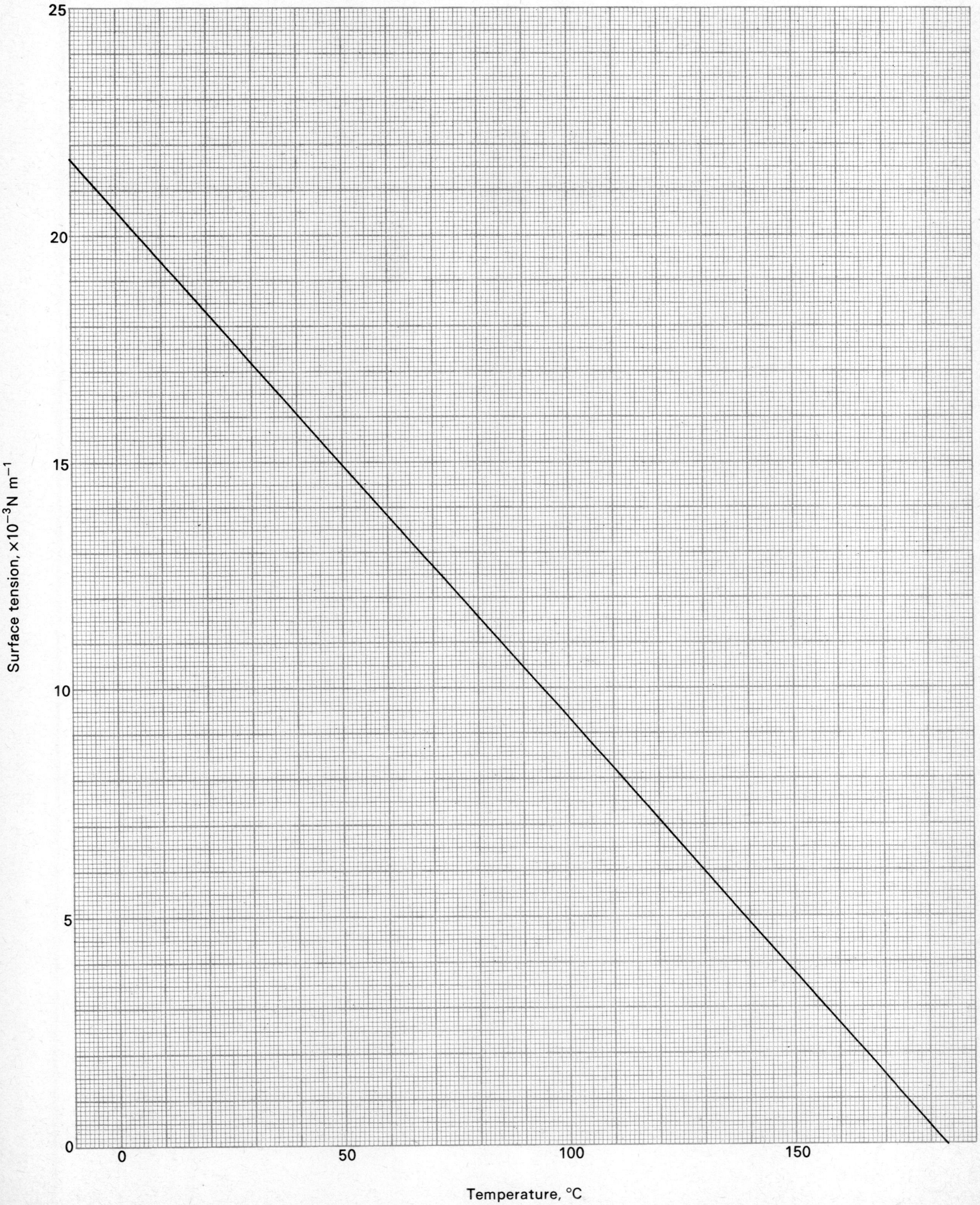

29.12 Solubility of Hydrogen Cyanide in Water at 1.0133×10^5 N m^{-2} Partial Pressure (17)

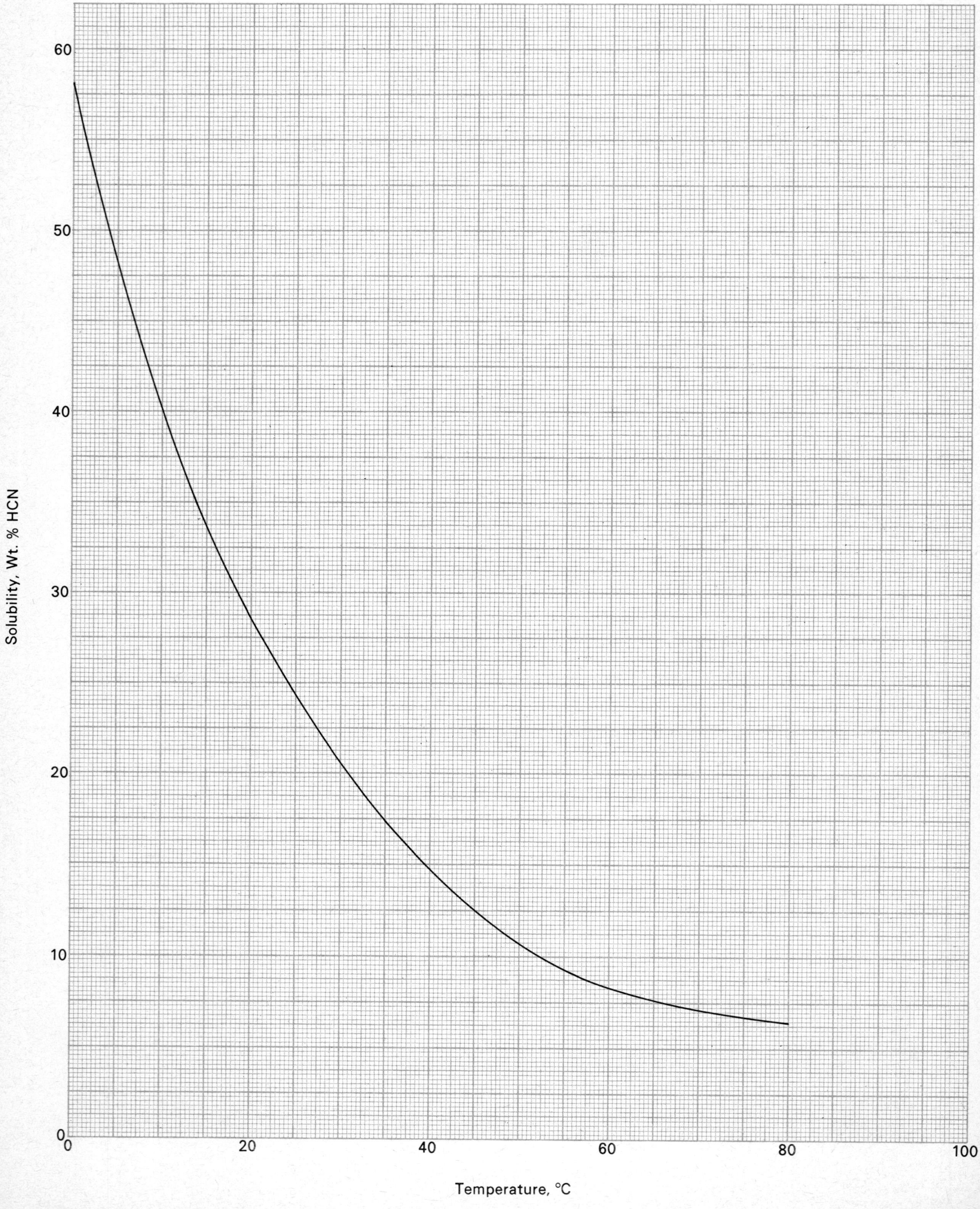

30 HYDROGEN SULPHIDE

Hydrogen sulphide is a colourless gas at ambient temperature and atmospheric pressure with a characteristic odour and toxic effect. On cooling, the vapour liquefies into a colourless non-associated liquid with a low dielectric constant of 8.3 at −78.6°C. Liquid hydrogen sulphide is relatively inactive and it shows the properties of a protonated solvent, which agrees well with its non-associated character. The liquid provides solvolyzing properties for compounds having covalent or ionic crystal structures. The infra-red and Raman spectra indicate that the H_2S molecule is non-linear, having a V-shape with an angle of 109°.

On heating the H_2S molecule decomposes above 400°C. Hydrogen sulphide burns in excess of air, forming SO_2 and H_2O. The gas is soluble in water, giving a weak dibasic acid. With water it forms a solid hydrate, $H_2S.6H_2O$.

Solid hydrogen sulphide is trimorphic, having two transition points. The structure of the transparent solid is a close-packed assemblage of H_2S molecules. One of the more pronounced crystalline structures is a face-centred cubic form, having four sulphur atoms in each unit cell.

Various properties of H_2S in solid, liquid, and gaseous states have been compiled in the following publications: Fehér (12), Gmelin (15), Landolt-Börnstein (26), Pascal (40), and Sage and Lacey (44).

There is no experimental data on the thermal conductivity of liquid hydrogen sulphide and an estimation method recommended by Stiel and Thodos (46) has been applied. The reliability of the data predicted by this technique is discussed under 'Fluorine'.

References

1 Alpert, N. L., *Phys. Rev.*, 1949, **75**, 398–410.

2 American Conference of Government Industrial Hygienists, *Documentation of Threshold Limit Values*, revised edition, Committee on Threshold Limit Values, Cincinnati, Ohio 45202, 1966, 203 pp.

3 Andrussow, L., Progr. Intern. Res. Thermodyn. Transport Properties, 2nd Symp., Princeton, Am. Soc. Mech. Eng., New York, 1962, pp. 279–87.

4 Barna, A. K., Manna, A., Mukhopadhyay, P., *J. Chem. Phys.*, 1968, **49** (5), 2422–5.

5 Biltz, W., Lemke, A., *Z. anorg. Chem.*, 1932, **203**, 321–9.

6 Bleekrode, L., *Proc. Roy. Soc.*, 1884, **37**, 339–51.

7 Clusius, K., Frank, A., *Z. phys. Chem.*, 1936, **34B**, 420–31.

8 Correia, P., Schraumm, B., Schaefer, K., *Ber. Bunsenges. Phys. Chem.*, 1968, **72** (3), 393–6.

9 Devyatykh, G. G., Zorin, A. D., Runovskaya, I. V., *Dokl. Phys. Chem.*, 1969, **188** (5), 663–4.

10 Edwards, D. G., *The Vapour Pressure of 30 Inorganic Liquids between One Atmosphere and the Critical Point*, Univ. of California, Lawrence Rad. Lab., UCRL 7167, Livermore, California, June 13, 1963, 148 pp.

11 Evans, W. H., Wagman, D. D., *J. Res. Natl. Bur. Std.*, 1952, **49**, 141–8.

12 Fehér, F., Liquid Hydrogen Sulphide, in *The Chemistry of Nonaqueous Solvents*, Vol. III (J. J. Lagowski, Ed.), Academic Press, New York, 1970, pp. 219–43.

13 Felsing, W. A., Drake, G. W., *J. Am. Chem. Soc.*, 1936, **58**, 1714–17.

14 Giauque, W. F., Blue, R. W., *J. Am. Chem. Soc.*, 1936, **58**, 831–7.

15 *Gmelins Handbuch der anorganischen Chemie*, Syst. No. 9, Schwefel, Teil B, Lief. 1, Verlag Chemie GmbH, Weinheim, 1953, pp. 14–45.

16 Golubev, I. F., *Viscosity of Gases and Gas Mixtures*, Gosudarstvennoe Izdatel'stvo Fiziko-Matematicheskoi Literatury, Moskva, 1959, Table 4.

17 Haar, L., Bradley, J. C., Friedman, A. S., Natl. Bur. Std. Rept. 4067, Washington, 1955, 22 pp.

18 Haar, L., Bradley, J. C., Friedman, A. S., *J. Res. Natl. Bur. Std.*, 1955, **55**, 285–90.

19 Holleman, E. M., *J. Phys. Chem.*, 1969, **73** (11), 3700–3.

20 IUPAC, Atomic Weights of the Elements 1971, *Pure & Appl. Chem.*, 1972, **30** (3–4), 638–49.

21 Kamiike, O., Kataoka, S., Inaba, S., *J. Chem. Soc. Japan*, 1942, **63**, 1007–14.

22 Karapet'yants, M. Kh., *Zh. Fiz. Khim.*, 1958, **32**, 556–68.

23 Kelley, K. K., *The Free Energies of the Vaporisation and Vapour Pressures of Inorganic Substances*, Bureau of Mines, Bull. 383, U.S. Government Printing Office, Washington, 1935, 132 pp.

24 Klemenc, A., *Z. Elektrochem.*, 1932, **38**, 592–5.

25 Klemenc, A., Baukowski, O., *Z. anorg. Chem.*, 1932, **208**, 348–66.

26 *Landolt-Börnstein, Zahlenwerte und Funktionen aus Physik, Chemie, Astronomie, und Technik*, 7th ed., Technik, 4 Teil, Bandteil a, Springer-Verlag, Berlin, 1967, pp. 266–9.

27 Lange, N. A., Ed., *Handbook of Chemistry*, 10th ed., McGraw-Hill, New York, 1967.

28 Lax, E., Synowietz, C., Eds., *Taschenbuch für Chemiker und Physiker*, 3rd ed., Springer-Verlag, Berlin, 1967.

29 Leduc, A., *Chem. Rev.*, 1929, **6**, 1–16.

30 Lenoir, J. M., Univ. Arkansas, Eng. Expt. Station Bull. 18, 1953, 48 pp.

31 Lewis, L. C., Fredricks, W. J., *J. Chem. Eng. Data*, 1968, **13** (4), 482–5.

32 Maryott, A. A., Smith, E. R., *Table of Dielectric Constants of Pure Liquids*, Natl. Bur. Std. Circular 514, Washington, Aug. 10, 1951, 44 pp.

33 Masuda, H., Yorizane, M., *Kagaku Kogaku*, 1968, **32**, 820–2.

34 Mathews, J. F., *Chem. Rev.*, 1972, **72** (1), 71–100.

35 Monteil, J. M., *J. Chim. Phys.*, 1969, **66** (10), 1673–5.

36 McBride, B. J., Heimel, S., Ehlers, J. G., Gordon, S., Natl. Aeron. Space Admin. Report SP-3001, 1963, 328 pp.

37 McIntosh, D., Steele, B. D., *Proc. Roy. Soc.*, 1904, **73**, 450–3.

38 McIntosh, D., Steele, B. D., *Phil. Trans.*, 1905, **205A**, 99–167.

39 McIntosh, D., Steele, B. D., *Z. phys. Chem.*, 1906, **55**, 129–50.

40 Pascal, P., Ed., *Nouveau Traité de Chimie Minérale*, Vol. XIII, Part 1, Oxygene, Ozone, Oxydes, Eau Oxygénée, La combustion, Soufre, Sellenium, Tellure, Polonium, Masson et Cie, Éditeurs, Paris, 1960, pp. 959–76.

41 Reamer, H. H., Sage, B. H., Lacey, W. N., *Ind. Eng. Chem.*, 1950, **42** (1), 140–3.

42 Robinson, P. L., Scott, W. E., *J. Chem. Soc.*, 1932, 972–9.

43 Rossini, F. D., Wagman, D. D., Evans, W. H., Levine, S., Jaffe, I., *Selected Values of Chemical Thermodynamic Properties*, Natl. Bur. Std. Circular 500, Washington, Feb. 1, 1952.

44 Sage, B. H., Lacey, W. N., *Some Properties of the Lighter Hydrocarbons, Hydrogen Sulphide and Carbon Dioxide*, Am. Petrol. Inst., New York, 1955, pp. 30–7.

45 Sinke, G. C., AR-15-59 (AD 214587), 1959.

46 Stiel, L. I., Thodos, G., *AIChE Journal*, 1964, **10** (1), 26–30.

47 Stull, D. R., Dir., *JANAF Thermochemical Tables*, Dow Chemical Co., Midland, Mich., 1965–8.

48 Svehla, R. A., *Estimated Viscosities and Thermal Conductivities of Gases at High Temperatures*, Lewis Res. Center, Techn. Rept. R-132, Cleveland, Ohio, 1962, 140 pp.

49 Washburn, E. W., Ed., *International Critical Tables of Numerical Data, Physics, Chemistry, and Technology*, Vol. VI, McGraw-Hill, New York, 1929, pp. 461–7.

50 West, J. R., *Chem. Eng. Progr.*, 1948, **44** (4), 287–92.

51 Winkler, L. W., *Z. phys. Chem.*, 1906, **55**, 350.

52 Wright, R. H., Maass, O., *Can. J. Res.*, 1932, **6**, 94.

Table 30.1 Properties of Hydrogen Sulphide

		Ref.
Chemical formula	H_2S	(28)
Molecular weight	34.0768	(20)
Critical temperature	100.0 °C	(34)
Critical pressure	89.37×10^5 N m^{-2}	(34)
Critical density	346.0 kg m^{-3}	(34)
Triple-point temperature	−85.7 °C	(28)
Triple-point pressure	0.2267×10^5 N m^{-2}	(28)
Boiling point at 1.0133×10^5 N m^{-2}	−60.35 °C	(31)
Transition temperatures	−169.55 °C −146.95 °C	(1) (1)
Density of solid at −195 °C	1217.0 kg m^{-3}	(5)
Coefficient of thermal expansion of liquid at −73 °C	1.74×10^{-3} C^{-1}	(42)
Heat of formation of gas at 25 °C	-2.0432×10^7 J kg-mol^{-1}	(47)
Free energy of formation of gas at 25 °C	-3.3076×10^7 J kg-mol^{-1}	(47)
Heat of sublimation at −190 °C	2.4289×10^7 J kg-mol^{-1}	(23)
Heat of fusion	2.3781×10^6 J kg-mol^{-1}	(43)
Heats of transition	1.5324×10^6 J kg-mol^{-1} 4.5220×10^5 J kg-mol^{-1}	(43) (43)
Entropy of gas at 25 °C	2.0578×10^5 J kg-mol^{-1} °C^{-1}	(47)
Heat capacity ratio of gas at 15 °C and 1.0133×10^5 N m^{-2}	1.32	(29)
Velocity of sound at 0 °C and 1.0133×10^5 N m^{-2}	289.3 m s^{-1}	(49)
Diffusion in water at 25 °C	1.36×10^{-9} m^2 s^{-1}	(21)
Dielectric constant of liquid at −78.5 °C	9.05	(32)
Electrical conductivity of liquid at −61 °C	1.0×10^{-9} ohm^{-1} m^{-1}	(27)
Refractive index of liquid at 18.5 °C	1.390	(6)
Threshold limit value	(10 ppm) 1.5×10^{-5} kg m^{-3}	(2)

30.1 Vapour Pressure of Hydrogen Sulphide (10,24,25,44)

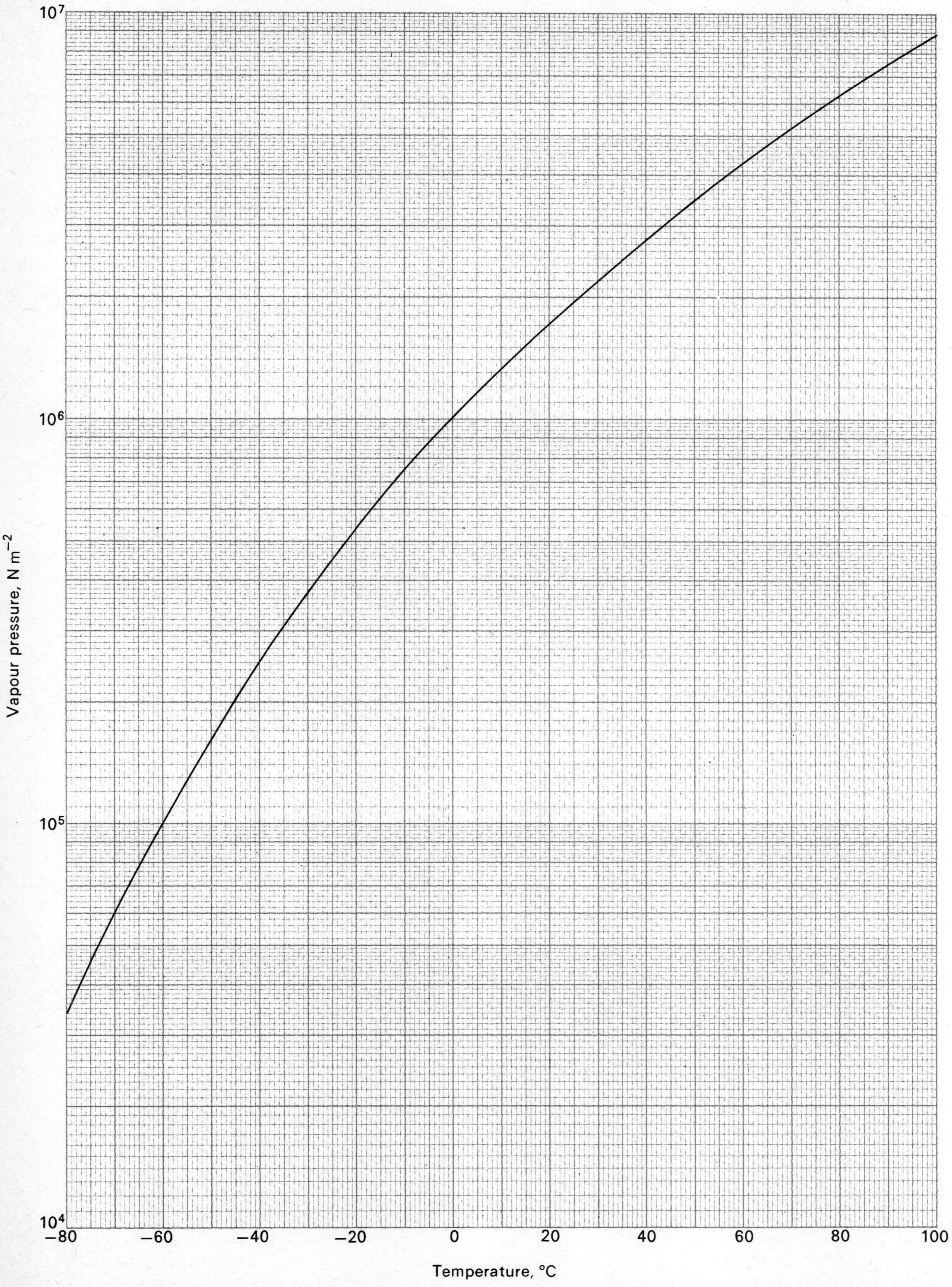

30.2 Orthobaric Density of Hydrogen Sulphide (19,26,28,44)

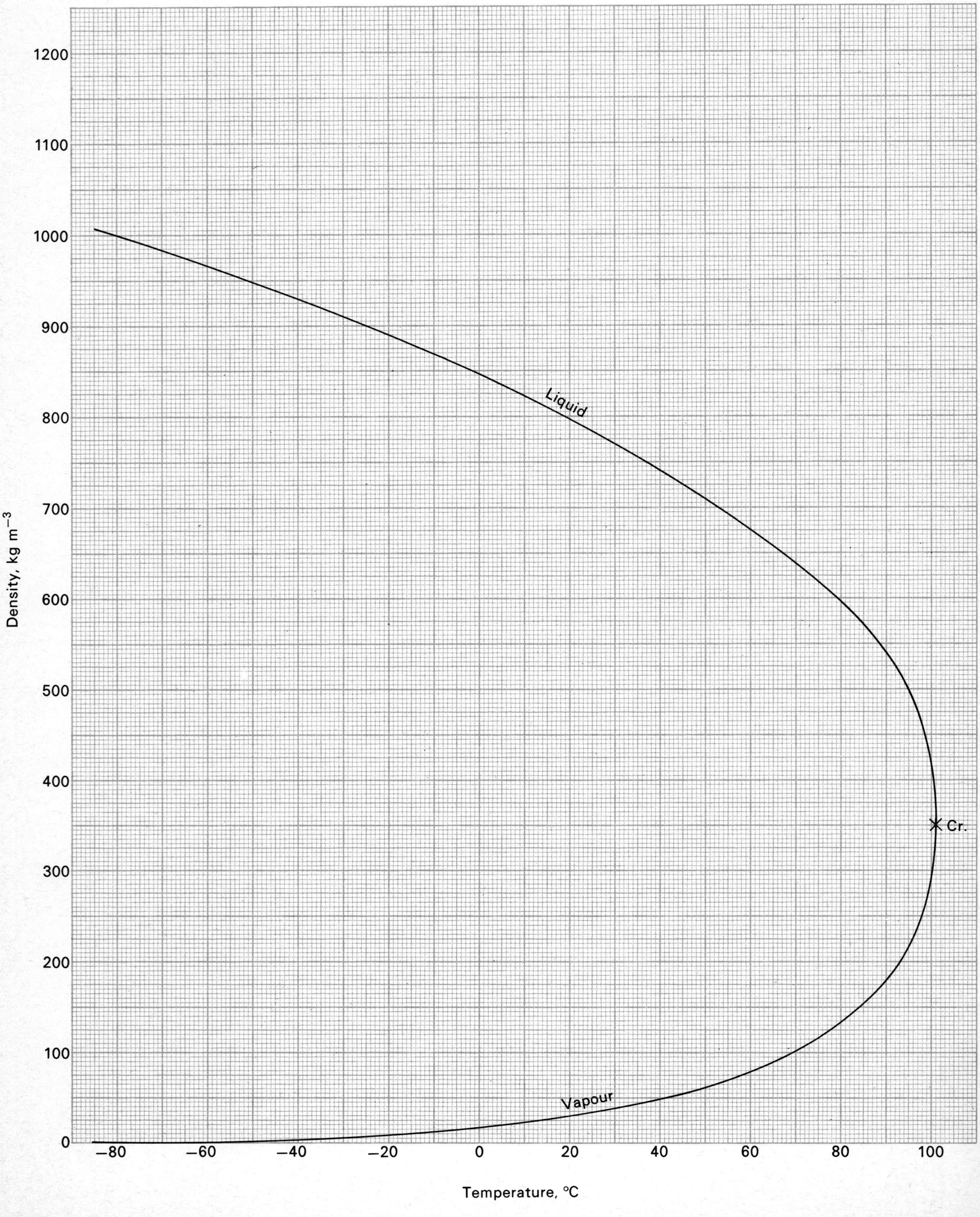

30.3 P-V-T Data for Hydrogen Sulphide (31,33,41,44)

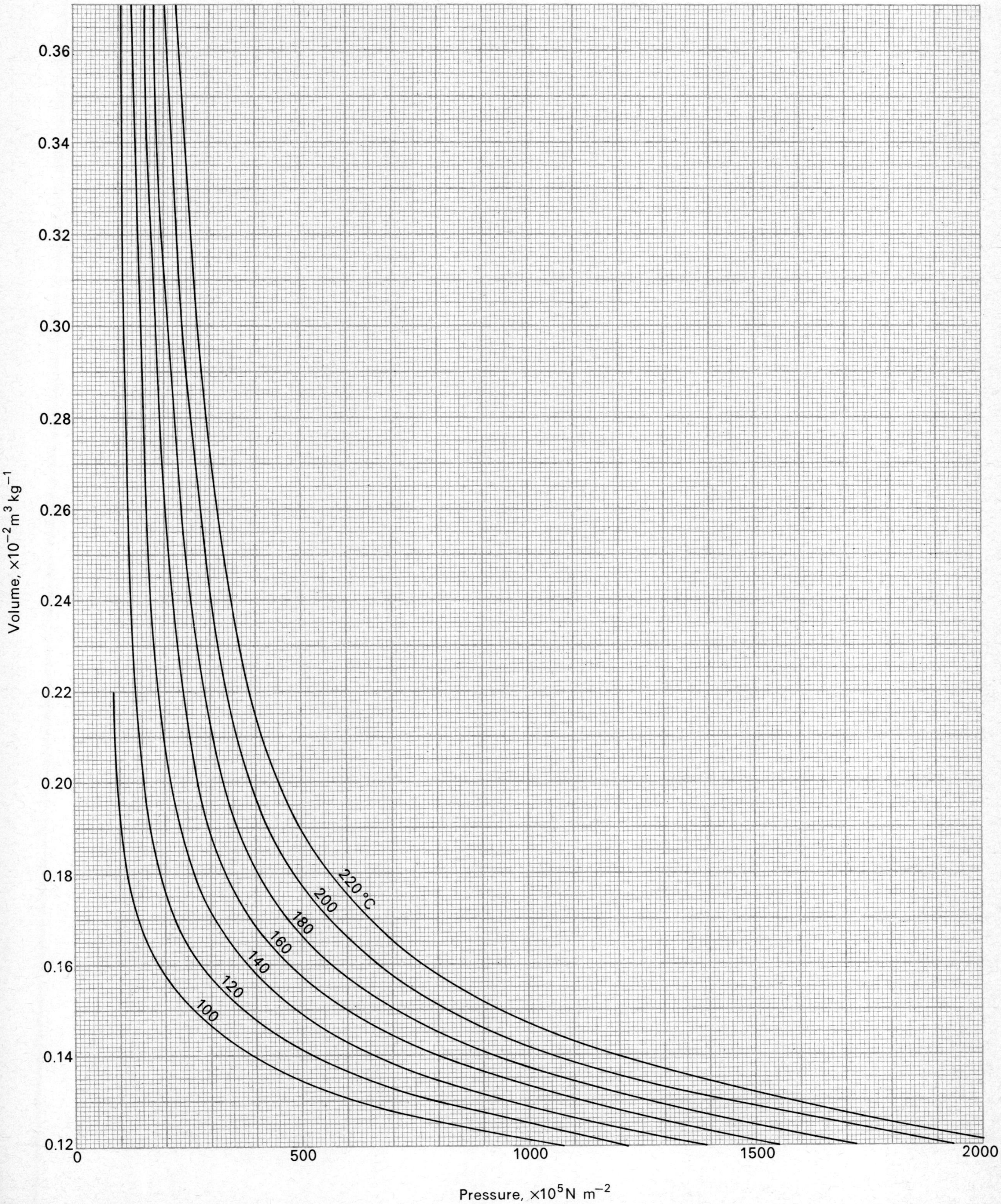

30.4 Latent Heat of Vaporisation of Hydrogen Sulphide (15,22,26)

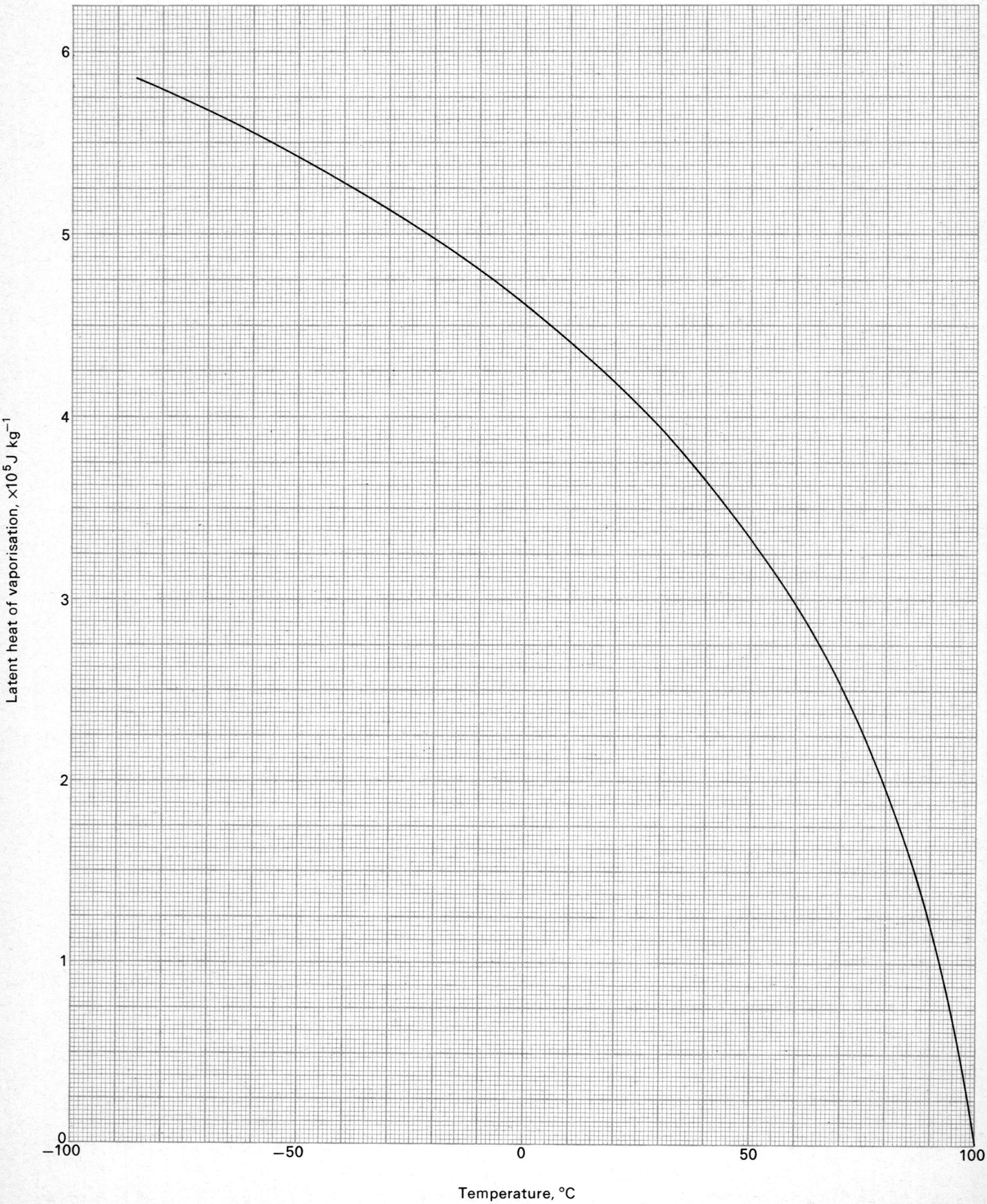

30.5 Heat Capacity of Liquid Hydrogen Sulphide (7,14)

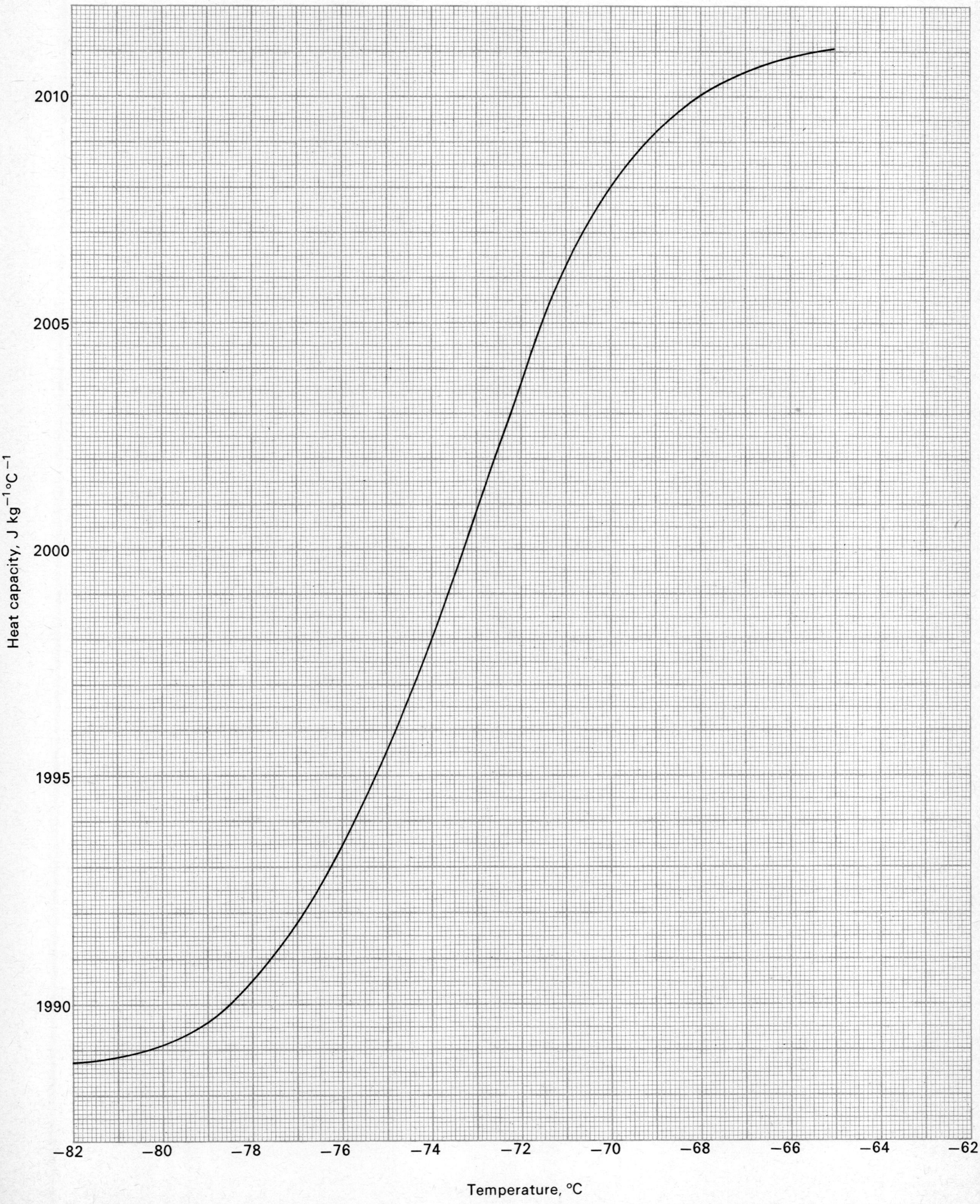

30.6 Heat Capacity of Gaseous Hydrogen Sulphide at 1.0133×10^5N m^{-2} (11,13,17,18,36,45,47,50)

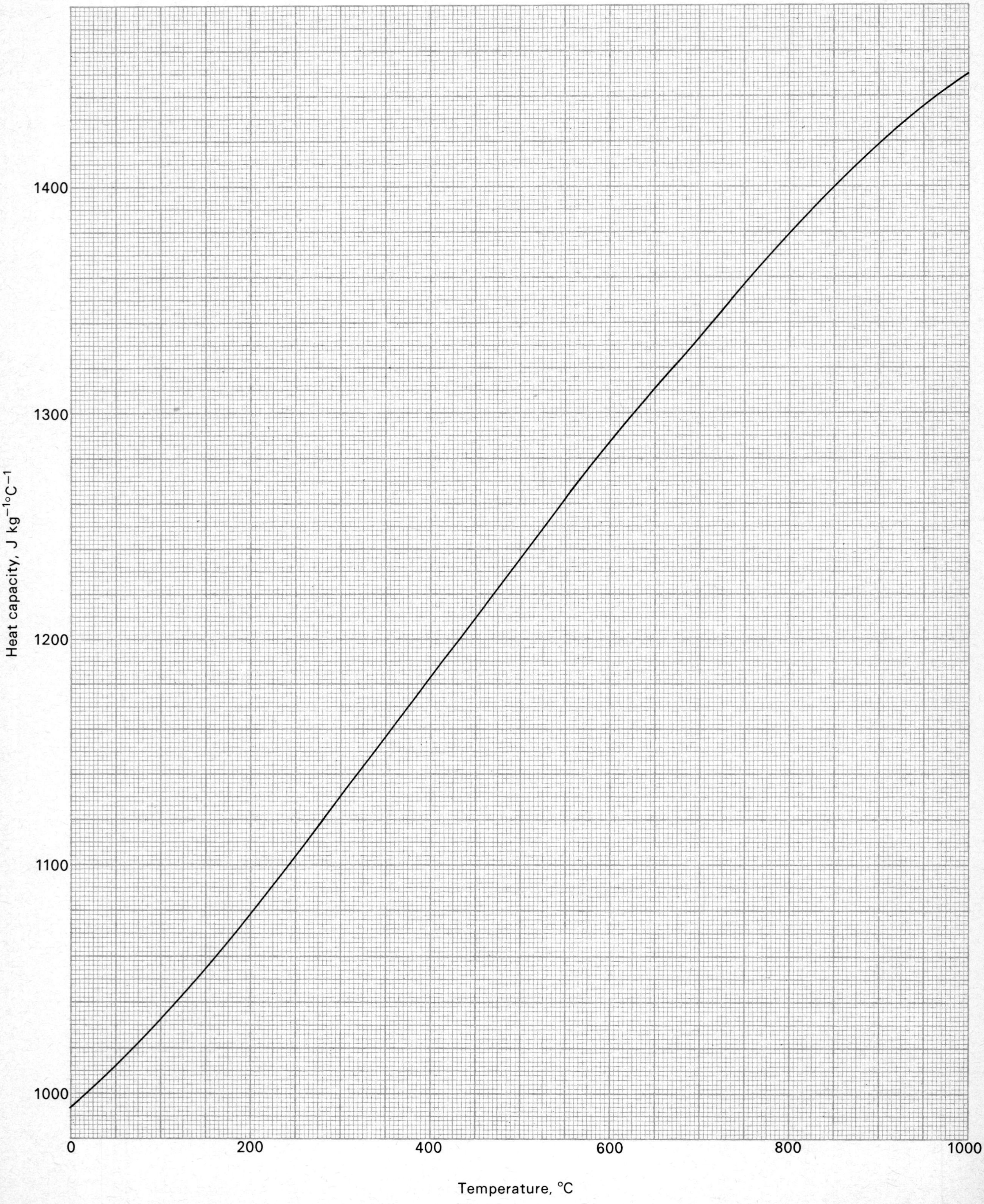

30.7 Viscosity of Liquid Hydrogen Sulphide (37,38,39)

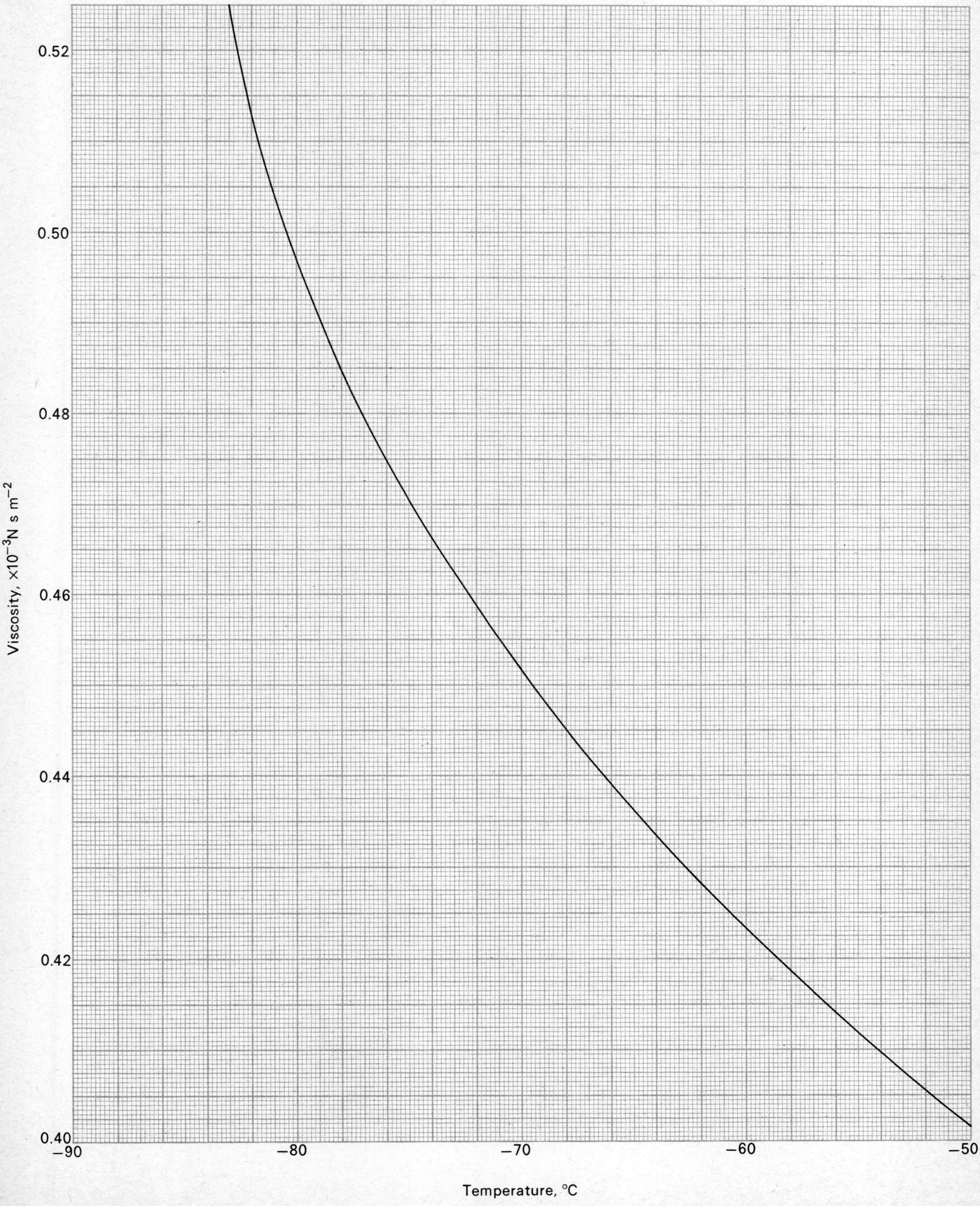

30.8 Viscosity of Gaseous Hydrogen Sulphide at $1.0133 \times 10^5 N\ m^{-2}$ (16,35)

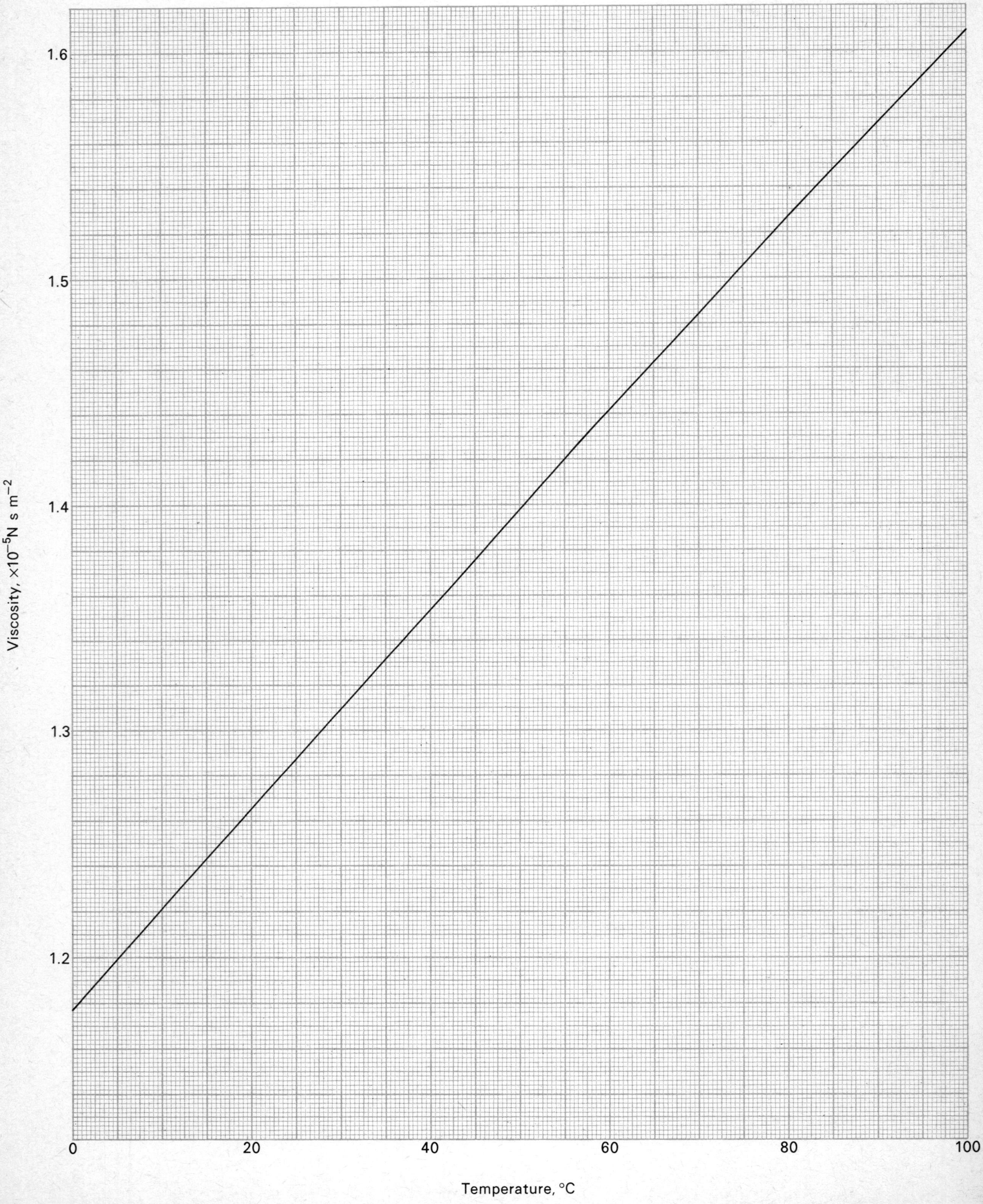

30.9 Thermal Conductivity of Liquid Hydrogen Sulphide (46)

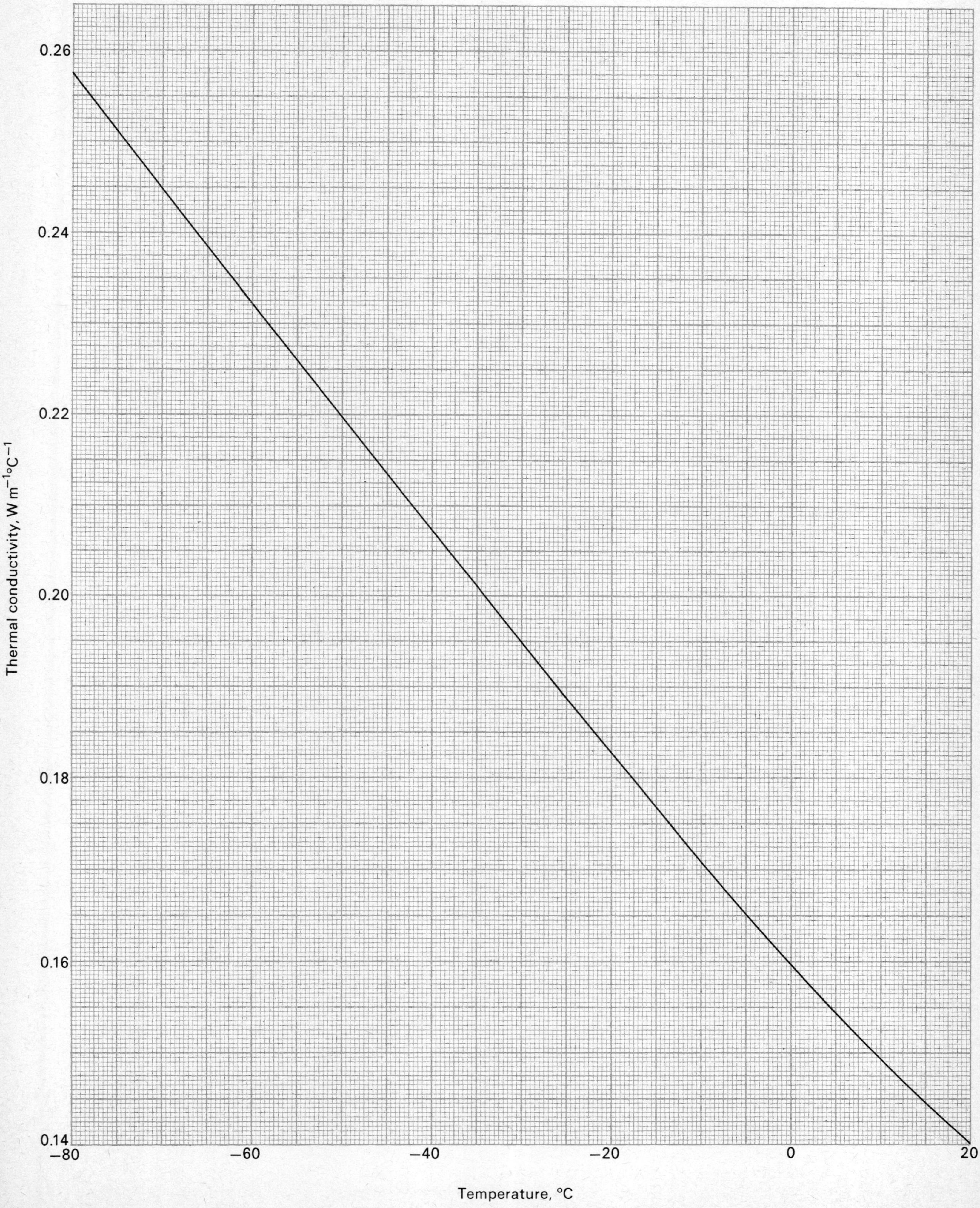

30.10 Thermal Conductivity of Gaseous Hydrogen Sulphide at $1.0133 \times 10^5 N\ m^{-2}$ (3,8,30,48)

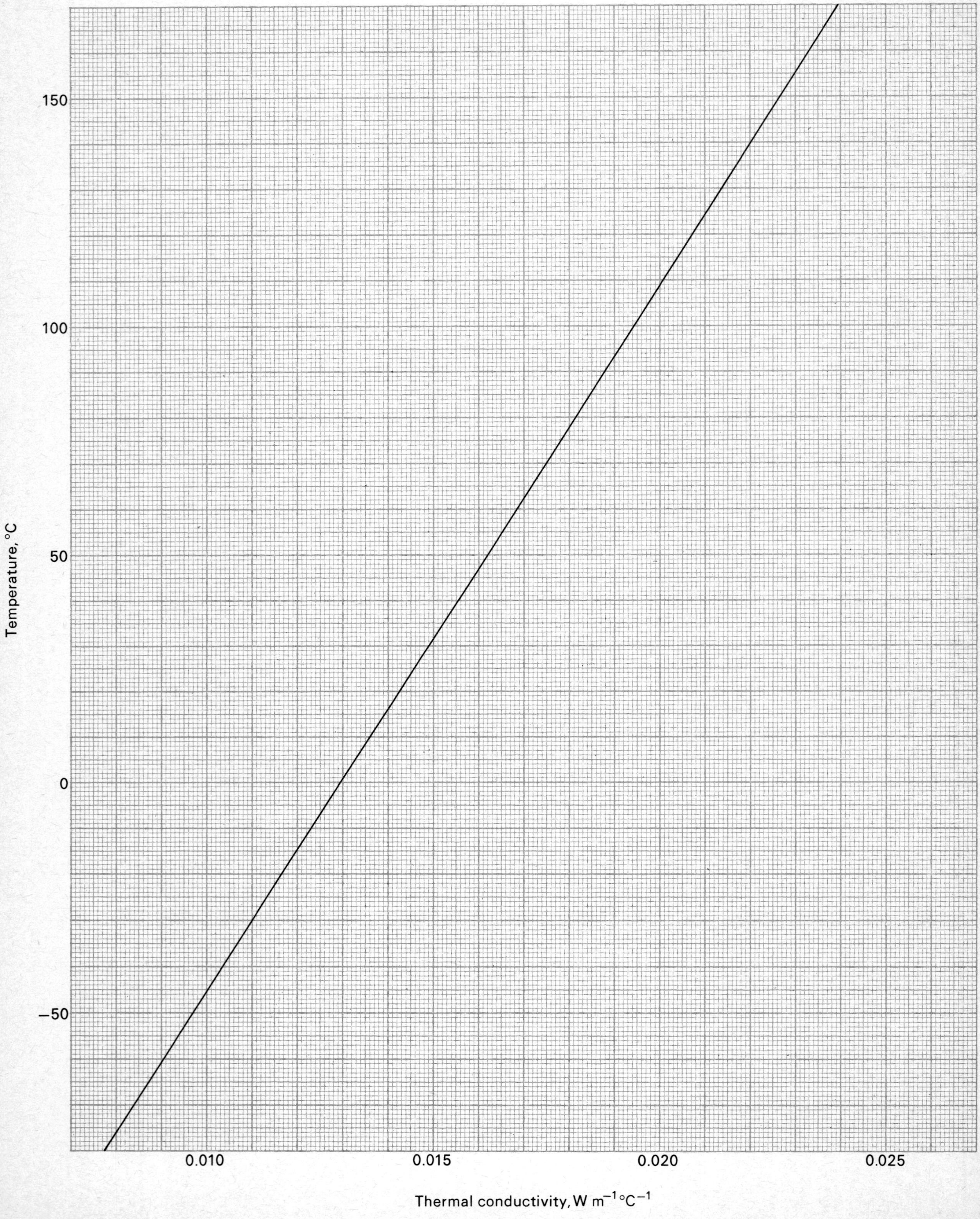

30.11 Surface Tension of Hydrogen Sulphide (9,15)

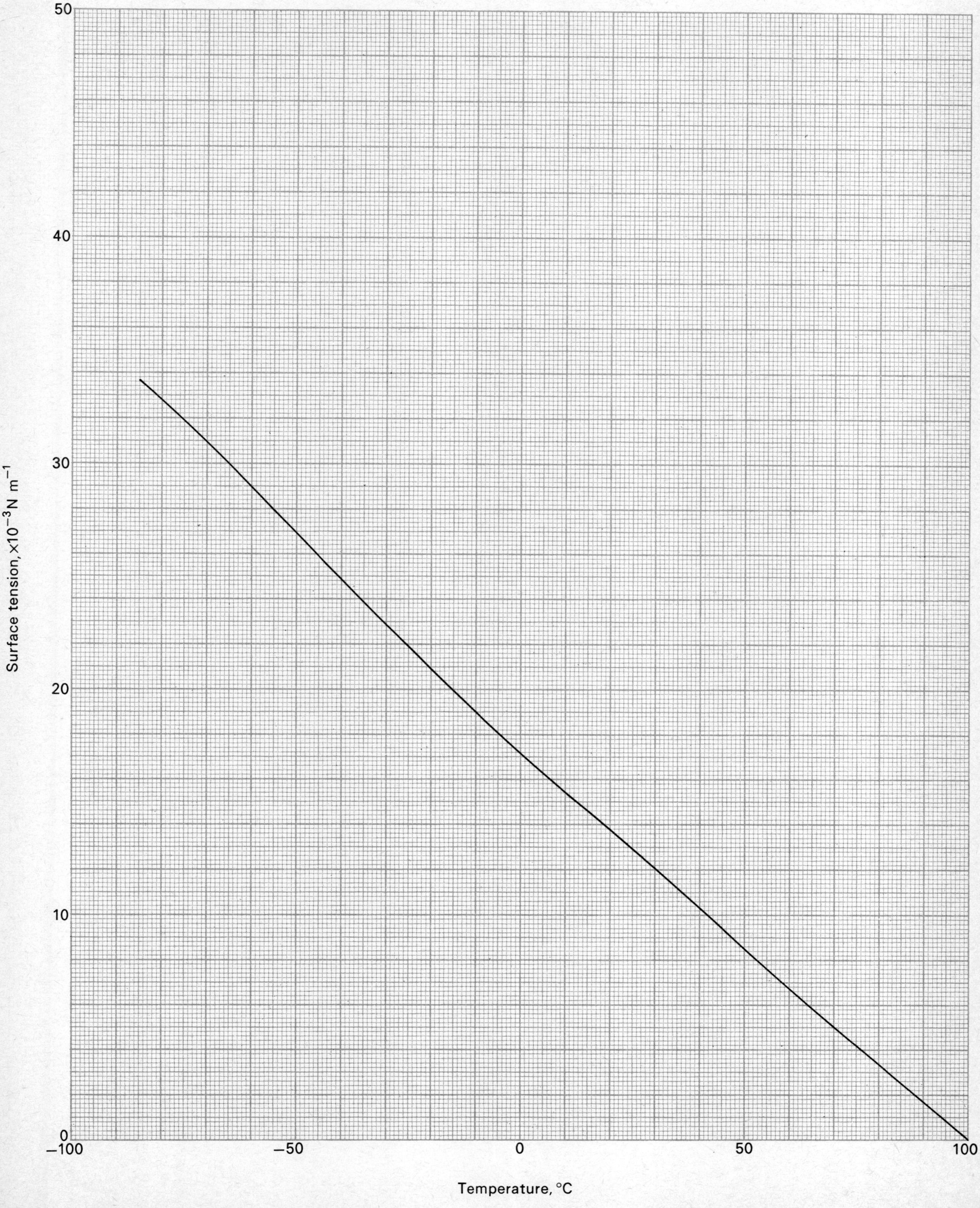

30.12 Solubility of Hydrogen Sulphide in Water at 1.0133×10^5 N m^{-2} Total Pressure (27,28,51,52)

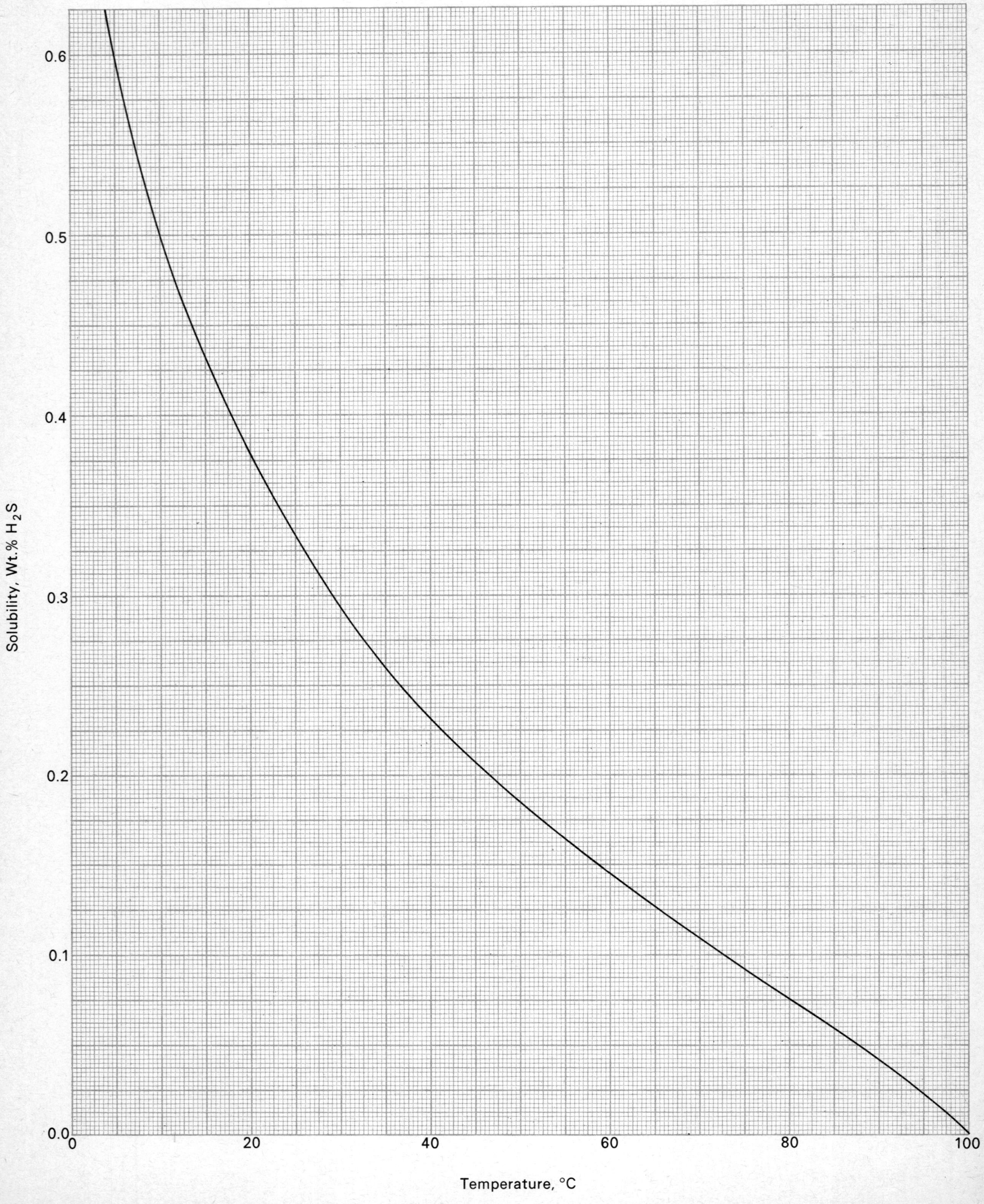

31 NITROGEN

Nitrogen is a colourless and tasteless gas at room temperature and atmospheric pressure. Free nitrogen contributes about 78% of the air, which is the most frequently used source for its large scale manufacture. Chemically, it is an inert gas at ambient temperature, due to the large heat of formation of the triple N≡N link. The nitrogen molecule is very stable and even at 8000°C only about 40% dissociates into atomic nitrogen. The molecule is diamagnetic.

Liquid nitrogen consists of diatomic molecules, shown by the scattering of neutrons and by X-ray diffraction measurements. Regarding the existence of ortho–para states of nitrogen, there is some disagreement among the investigators.

On cooling, liquid nitrogen crystallises in two enantiotropic modifications. In the alpha-form, it has a cubic structure, with four molecules in each unit cell, while in the beta-form the hexagonal structure has a close sphere packing, with two molecules in each unit cell. Solid nitrogen has only one transition point.

Nitrogen has the same number of extranuclear electrons as carbon monoxide, so that there are great similarities in the physicochemical properties of these two molecules.

The physical, thermodynamic, and transport properties of nitrogen in the solid, liquid, and gaseous states are rather consistent in the published literature. There are several good compilations on various properties as follows: ASHRAE (1), Brahinsky and Neel (7), Canjar and Manning (9), Data Book and Buyers Guide (13), Din (14), Friedman (19), Gmelin (22), Grabau and Brahinsky (26), Hilsenrath *et al.* (28), Hilsenrath and Klin (29), Keenan and Kaye (36), Landolt-Börnstein (41), Little and Neel (47), Matheson (51), Mellor (54), McBride *et al.* (57), Neel and Lewis (58, 59), Pascal (63), Sitting (73), Smith (74), Spiers (76), Strobridge (78), Vasserman *et al.* (88, 89, 91), and Wood *et al.* (99).

References

1 American Society of Heating, Refrigerating, and Air-Conditioning Eng., Inc., *ASHRAE Thermodynamic Properties of Refrigerants*, New York, 1969, pp. 303–8.

2 Andrussow, L., *Diffusion, Viscosity, and Conductivity of Gases*, Progr. Intern. Res. Thermodyn. Transport Properties, 2nd Symp., Princeton, Am. Soc. Mech. Eng., New York, 1962, pp. 279–87.

3 Armstrong, G. T., *J. Res. Natl. Bur. Std.*, 1954, **53** (4), 263–6.

4 Ayres, H. D., *Phys. Rev.*, 1913, **2** (2), 173.

5 Baly, E. C. C., Donnan, F. G., *J. Chem. Soc.*, 1902, **81**, 907–18.

6 Borovik, E., *J. Phys. USSR*, 1947, **11** (2), 149–55.

7 Brahinsky, H. S., Neel, C. A., *Tables of Equilibrium Thermodynamic Properties of Nitrogen. Vol. II. Constant Pressure*, U.S. Dept. of Commerce, Natl. Bur. Std., Arnol Eng. Dev. Center Techn. Rept. 69–126 (AD 692712), Washington, Aug. 1969, 270 pp.

8 Burton, J. T. A., Ziebland, H., *The Thermal Conductivity of Nitrogen and Argon in the Liquid and Gaseous States*, Ministry of Technology, Explosives Research and Development Establishment, Waltham Abbey, Rept. No. 11/R/57, 1957, 13 pp.

9 Canjar, L. N., Manning, F. S., Thermodynamic Properties of Nitrogen, in *Thermodynamic Properties and Reduced Correlations for Gases*, Gulf Publ. Co., Houston, Texas, 1967, pp. 149–58.

10 Chierici, G. L., Paratella, A., *AIChE Journal*, 1969, **15** (5), 786–90.

11 Clusius, K., *Z. phys. Chem. (Leipzig)*, 1929, **3B**, 41–79.

12 Curtiss, C. F., Hirschfelder, J. O., CM-472, (ATI 35673), 1948, 12 pp.

13 Data Book and Buyers Guide, Engineering Data Section, *Cryogenic and Ind. Gases*, 1971, **6** (3), 41–80.

14 Din, F., Ed., *Thermodynamic Functions of Gases*, Vol. 3, CH_4, N_2, C_2H_6, Butterworths, London, 1961, 218 pp.

15 Edwards, D. G., *The Vapour Pressure of 30 Inorganic Liquids between One Atmosphere and the Critical Point*, Univ. of California, Lawrence Rad. Lab. UCRL 7167, Livermore, California, June 13, 1963, 148 pp.

16 Engineering Science Data Units, *Specific Heats and Their Ratio as Functions of Temperature for Several Common Gases*, Royal Aeronautical Society, Data Sheet 00.01.08, London, 1965, 4 pp.

17 Engineering Science Data Units, *Approximate Data on the Viscosity of some Common Liquids*, Item No. 66024, London, July 1966, 26 pp.

18 Förster, S., *Cryogenics*, 1963, **3**, 176.

19 Friedman, A. S., *Pressure–Volume–Temperature Relationships of Gaseous Hydrogen, Nitrogen, and a Hydrogen-Nitrogen Mixture*, Ohio State Univ., Ph.D. thesis, 1950.

20 Friedman, A. S., White, D., *J. Am. Chem. Soc.*, 1950, **72** (9), 3931–2.

21 Giauque, W. F., Clayton, J. O., *J. Am. Chem. Soc.*, 1933, **55**, 4875–89.

22 *Gmelins Handbuch der anorganischen Chemie*, Syst. No. 4, Stickstoff, Verlag Chemie GmbH, Berlin, 1936, pp. 89–236.

23 Golubev, I. F., *Viscosity of Gases and Gas Mixtures*, Gosudarstvennoe Izdatel'stvo Fiziko-Matematicheskoi Literatury, Moskva, 1959, Table 4.

24 Golubev, I. F., Kalsinn, M. V., *Gaz. Prom.*, 1964, **9** (8), 41–3.

25 Gordon, J. S., WADC-TR-57-33 (AD 110735), 1957, 172 pp.

26 Grabau, M., Brahinsky, H. S., *Thermodynamic Properties of Nitrogen from 300 to 5000 K. and from 1 to 1000 Amagats*, Natl. Bur. Std., AEDC-TR-66-69 (AD 636730), Washington, Aug. 1966.

27 Gracki, J. A., *J. Chem. Phys.*, 1969, **51** (9), 3856–63.

28 Hilsenrath, J., Beckett, C. W., Benedict, W. S., Fano, L., Hoge, H. J., Masi, J. F., Nuttal, R. L., Touloukian, Y. S., Woolley, H. W., *Tables of Thermal Properties of Gases*, Natl. Bur. Std. Circular 564, Washington, Nov. 1, 1955, 488 pp.

29 Hilsenrath, J., Klein, M., *Table of Thermodynamic Properties and Chemical Composition of Nitrogen in Chemical Equilibrium Including Second Virial Correction from 1600 K to 15 000 K*, AEDC-TR-66-65 (AD 630461), April 1966.

30 Hilsenrath, J., Touloukian, Y. S., *Trans. Am. Soc. Mech. Eng.*, 1954, **76**, 967–85.

31 Himmelblau, D. M., *J. Chem. Eng. Data*, 1960, **5** (1), 10–15.

32 Honig, R. E., Hook, H. O., Vapour Pressure Data for Some Common Gases, *RCA Rev.* 1960, **21**, 360–8.

33 IUPAC, Atomic Weights of the Elements 1971, *Pure & Appl. Chem.*, 1972, **30** (3–4), 638–49.

34 Ibrahim, S. H., Kuloor, N. R., *Brit. Chem. Eng.*, 1961, **6** (12), 862–3.

35 Johannin, P., Thermal Conductivity of Nitrogen between 75° and 7000°C and 1000 to 1600 Atmospheres, *J. Res. Centre, Natl. Res. Sci. Lab., Paris*, 1953, **43**, 116–55.

36 Keenan, J. H., Kaye, J., *Gas Tables. Thermodynamic Properties of Air Products of Combustion and Component Gases Compressible Flow Functions*, Wiley, New York, 1948, 238 pp.

37 Kestin, J., Leidenfrost, W., *The Effect of Moderate Pressure on the Viscosity of Five Gases*, Progr. Internat. Res. Thermodyn. Transport Properties, 2nd Symp. Thermophys. Properties (J. F. Masi, D. H. Tsai, Eds.), Am. Soc. Mech. Eng., Academic Press, New York, 1962, pp. 321–38.

38 Keyes, F. G., *The Heat Conductivity, Viscosity, Specific Heat and Prandtl Number for Thirteen Gases*, Office Naval Res., Techn. Rept. 37, Project Squid (ATI 16173), April 1, 1952, 33 pp.

39 Keyes, F. G., Vines, R. G., *Trans. Am. Soc. Mech. Eng., J. Heat Transfer*, 1965, **87** (2), 177–83.
40 Klots, C. E., Benson, B. B., *J. Marine Res.*, 1963, **21** (1), 48–57.
41 *Landolt-Börnstein, Zahlenwerte und Funktionen aus Physik, Chemie, Astronomie, Geophysik, und Technik*, 7th ed., IV Band, Technik, 4 Teil, Bandteil a, Springer-Verlag, Berlin, 1967, pp. 242–52.
42 Lange, N. A., Ed., *Handbook of Chemistry*, 10th ed., McGraw-Hill, New York, 1967.
43 Langen, A., *Z. Elektrochem.*, 1919, **25**, 25–45.
44 Lax, E., Synowietz, C., Eds., *Taschenbuch für Chemiker und Physiker*, 3rd ed., Springer-Verlag, Berlin, 1967.
45 Leduc, A., *Chem. Rev.*, 1929, **6**, 1–16.
46 Liley, P. E., *The Thermal Conductivity of 46 Gases at Atmospheric Pressure*, Proc. 4th Symp. Thermophys. Prop. (Moszynski, T. R., Ed.), Am. Soc. Mech. Eng., New York, 1968, pp. 323–49.
47 Little, W. J., Neel, C. A., *Tables of the Thermodynamic Properties of Nitrogen from 100 to 1500 K*, Natl. Bur. Std., AEDC-TDR-62-170 (AD 283441), Washington, Sept. 1962.
48 Loeb, M. B., *Chem. Eng.*, 1971 (Oct. 4), **78** (22), 106.
49 Maitland, G. C., Smith, E. B., *J. Chem. Eng. Data*, 1972, **17** (2), 150–6.
50 Maryott, A. A., Smith, E. R., *Tables of Dielectric Constants of Pure Liquids*, Natl. Bur. Std. Circular 514, Washington, Aug. 10, 1951, 44 pp.
51 Matheson Gas Data Book, 4th ed., *Nitrogen*, The Matheson Company, Inc., East Rutherford, N.J., 1966.
52 Mathews, J. F., *Chem. Rev.*, 1972, **72** (1), 71–100.
53 Mathies, E., Crommelin, C. A., Kammerlingh Onnes, H., *Comm. Kammerlingh Onnes Lab., Leiden*, No. 145 c, 1914.
54 *Mellor's Comprehensive Treatise on Inorganic and Theoretical Chemistry*, Vol. VIII, Suppl. I, Nitrogen, Part I, Longmans, London, 1964, pp. 27–149.
55 Misic, D., Thodos, G., *AIChE Journal*, 1965, **11** (4), 650–6.
56 Murray, C. N., Riley, J. P., Wilson, T. R. S., *Deep-Sea Res.*, 1969, **16**, 297–310.
57 McBride, B. J., Heimel, S., Ehlers, J. G., Gordon, S., *Thermodynamic Properties to 6000 K for 210 Substances Involving the First 18 Elements*, Natl. Aeron. Admin. NASA SP-3001, Washington, 1963.
58 Neel, C. A., Lewis, C. H., *Interpolation of Imperfect Nitrogen Thermodynamic Data. I. At Constant Enthalpy*, Natl. Bur. Std., AEDC-TDR-64-212 (AD 449409), Washington, Oct. 1964.
59 Neel, C. A., Lewis, C. H., *Interpolation of Imperfect Nitrogen Thermodynamic Data. II. At Constant Pressure*, Natl. Bur. Std., AEDC-TDR-64-213 (AD 449381), Washington, Oct. 1964.
60 Nuttal, R. L., Ginnings, D. C., Thermal Conductivity of Nitrogen from 50° to 500°C and 1 to 100 Atmospheres, *J. Res. Natl. Bur. Std.*, 1957, **58** (5), 271–8.
61 O'Connell, J. P., Prausnitz, J. M., *Ind. Eng. Chem.*, 1968, **60** (1), 36–65.
62 O'Sullivan, T. D., Smith, N. O., *J. Phys. Chem.*, 1970, **74** (7), 1460–6.
63 Pascal, P., Ed., *Nouveau Traité de Chimie Minérale*, Vol. X, Azote, Phosphore, Masson et Cie, Éditeurs, Paris, 1956, pp. 13–43.
64 Powers, R. W., Johnston, H. L., Mattox, R. W., Proc. 8th Internat. Congr. Refrig., London, 1951, pp. 186–94.
65 Powers, R. W., Mattox, R. W., Johnston, H. L., *J. Am. Chem. Soc.*, 1954, **76**, 5968–71.
66 Reynes, E. G., Thodos, G., *J. Chem. Eng. Data*, 1966, **11** (2), 137–40.
67 Rossini, F. D., Wagman, D. D., Evans, W. H., Levine, S., Jaffe, I., *Selected Values of Chemical Thermodynamic Properties*, Natl. Bur. Std. Circular 500, Washington, Feb. 1, 1952.
68 Rudenko, N. S., *Zh. Expt. Theor. Fiz.*, 1939, **9**, 1078–80.
69 Rudenko, N. S., *Zh. Tech. Fiz.*, 1948, **18**, 1123–6.
70 Rudenko, N. S., Shubnikov, L. V., *Phys. Zh. Soviet*, 1934, **6**, 470–7.
71 Ruheman, M., *Z. phys. Chem.*, 1932, **76**, 380.
72 Schaefer, C. A., Thodos, G., *AIChE Journal*, 1959, **5** (3), 367–72.
73 Sitting, M., *Nitrogen in Industry*, Van Nostrand Co., Inc., Princeton, 1965, 278 pp.
74 Smith, C. E., *Thermodynamic Properties of Nitrogen*, Lockheed TR-6-90-62-111, Dec. 1962.
75 Spangenberg, C. D., *Chem. Eng.*, 1951, **58** (9), 170–1.
76 Spiers, H. M., Ed., *Technical Data on Fuel*, 6th ed., The British National Committee, World Power Conferences, London, 1961, 360 pp.
77 Stansfield, D., *Proc. Phys. Soc.*, 1958, **72**, 854–66.
78 Strobridge, T. R., *The Thermodynamic Properties of Nitrogen from 114 to 540°R. Between 1 and 3000 PSIA. Supplement A*, Natl. Bur. Std. Techn. Note 129A, Washington, Feb. 1963.
79 Stull, D. R., Dir., *JANAF Thermochemical Tables*, Dow Chemical Co., Midland, Mich., 1965–8.
80 Svehla, R. A., *Estimated Viscosities and Thermal Conductivities of Gases at High Temperatures*, Lewis Res. Center, Techn. Rept. R-132, Cleveland, Ohio, 1962, 140 pp.
81 Tsiklis, D. S., Polyakov, E. V., *Zh. Fiz. Khim.*, 1967, **41** (13), 3145.
82 Tsiklis, D. S., Polyakov, E. V., *Dokl. Akad. Nauk, SSSR*, 1967, **176** (2), 308.
83 Uhlir, A., *J. Chem. Phys.*, 1952, **20**, 463–72.
84 Van Itterbeek, A., *Physica*, 1953, **19**, 1158.
85 Van Itterbeek, A., Verbeke, O., Staes, K., *Cryogenics*, 1962, **32**, 489.
86 Vargaftik, N. B., Filippov, L. P., Tarzimanov, A. A., Yurchak, R. P., *Thermal Conductivity of Gases and Liquids*, Teploprovodnost' Gazov i Zhidkostey., Moskva, 1970, 185 pp.
87 Vargaftik, N. B., Zbaina, N. Kh., *Teplofiz. Vysok. Temp.*, 1964, **2** (6), 869–78.
88 Vasserman, A. A., Kazavchinskii, Ya. Z., Rabinovich, V. A., *Thermophysical Properties of Air and Air Components*, Izdatel'stvo NAUK, Moskva, 1966.
89 Vasserman, A. A., Rabinovich, V. A., Equation of State for Liquid Nitrogen, *Inzhenerno-Fizicheskii Zh.*, 1967, **12** (3), 342–8.
90 Vasserman, A. A., Rabinovich, V. A., Viscosity and Thermal Conductivity of Air and Its Components in a Wide Range of Temperatures and Pressures, *Teplo Massoperenos*, 1968, **7**, 47–57.
91 Vasserman, A. A., Rabinovich, V. A., *Thermophysical Properties of Liquid Air and Its Components*, Izdatel'stvo Komiteta Standartov, Mer i Izmeritel'nykh Priborov pri Sovete Ministrov SSSR, Moskva, 1968.
92 Vasserman, A. A., Selevanyuk, V. I., *Akust. Zh.*, 1967, **13** (1), 131.
93 Wagman, D. D., Kilpatrick, J. E., Taylor, W. J., Pitzer, K. S., Rossini, F. D., *J. Res. Natl. Bur. Std.*, 1945, **34**, 143–61.
94 Wagner, W., *Cryogenics*, 1972, **12** (3), 214–24.
95 Walden, P., *Z. phys. Chem.*, 1909, **65**, 129–225.
96 Washburn, E. W., Ed., *International Critical Tables of Numerical Data, Physics, Chemistry, and Technology*, Vol. VI, McGraw-Hill, New York, 1929, pp. 461–7.
97 Wiebe, R., Brevoort, M. J., *J. Am. Chem. Soc.*, 1930, **52**, 622–33.

98 Willis, W. L., *Cryogenics*, 1966, **6** (5), 279–84.
99 Wood, R. E., Baer, F. W., Boone, W. J., *Thermal Conductivities and Prandtl Numbers of Nitrogen from 133 to 740 K between 1 and 240 Atmospheres*, U.S. Dept of Interior, Bureau of Mines, Rept. 7541, Washington, July 1971, 34 pp.
100 Woolley, H. W., Natl. Bur. Std. Rept. 2287, Washington, 1953, 39 pp.
101 Yosim, S. J., *J. Chem. Phys.*, 1964, **40** (10), 3069–75.
102 Ziebland, H., Burton, J. T. A., *Brit. J. Appl. Phys.*, 1958, **9**, 52–9.

Table 31.1 Properties of Nitrogen

		Ref.
Chemical formula	N_2	(44)
Molecular weight	28.0134	(33)
Critical temperature	−147.0 °C	(52)
Critical pressure	33.94×10^5 N m^{-2}	(52)
Critical density	313.0 kg m^{-3}	(52)
Triple-point temperature	−210.0 °C	(44)
Triple-point pressure	0.1253×10^5 N m^{-2}	(44)
Boiling point at 1.0133×10^5 N m^{-2}	−195.79 °C	(78)
Melting point	−209.85 °C	(22, 63)
Transition temperature	−237.54 °C	(67)
Density of solid (beta-form) at −234.15 °C	995.0 kg m^{-3}	(71)
Compressibility of liquid at −208.15 °C	80.53×10^{-12} m^2 N^{-1}	(61)
Coefficient of thermal expansion of liquid at boiling point	5.56×10^{-3} °C^{-1}	(101)
Heat of formation of gas at 25 °C	0.0 J kg-mol^{-1}	(79)
Free energy of formation of gas at 25 °C	0.0 J kg-mol^{-1}	(79)
Heat of sublimation at −237.65 °C	7.2620×10^6 J kg-mol^{-1}	(43)
Heat of fusion	7.2010×10^5 J kg-mol^{-1}	(67)
Heat of transition	2.3030×10^5 J kg-mol^{-1}	(67)
Entropy of gas at 25 °C	1.9163×10^5 J kg-mol^{-1} °C^{-1}	(79)
Heat capacity ratio of gas at 15 °C and 1.0133×10^5 N m^{-2}	1.404	(45)
Velocity of sound at 0 °C and 1.0133×10^5 N m^{-2}	337.7 m s^{-1}	(96)
Diffusion in air at 0 °C and 1.0133×10^5 N m^{-2}	1.344×10^{-5} m^2 s^{-1}	(34)
Dielectric constant of liquid at −203.0 °C	1.454	(50)
Electrical conductivity of liquid at −198.0 °C	5.40×10^{-18} ohm^{-1} m^{-1}	(98)
Refractive index of liquid at −190.0 °C	1.2053	(4)

31.1 Vapour Pressure of Nitrogen (3,15,20,32,78)

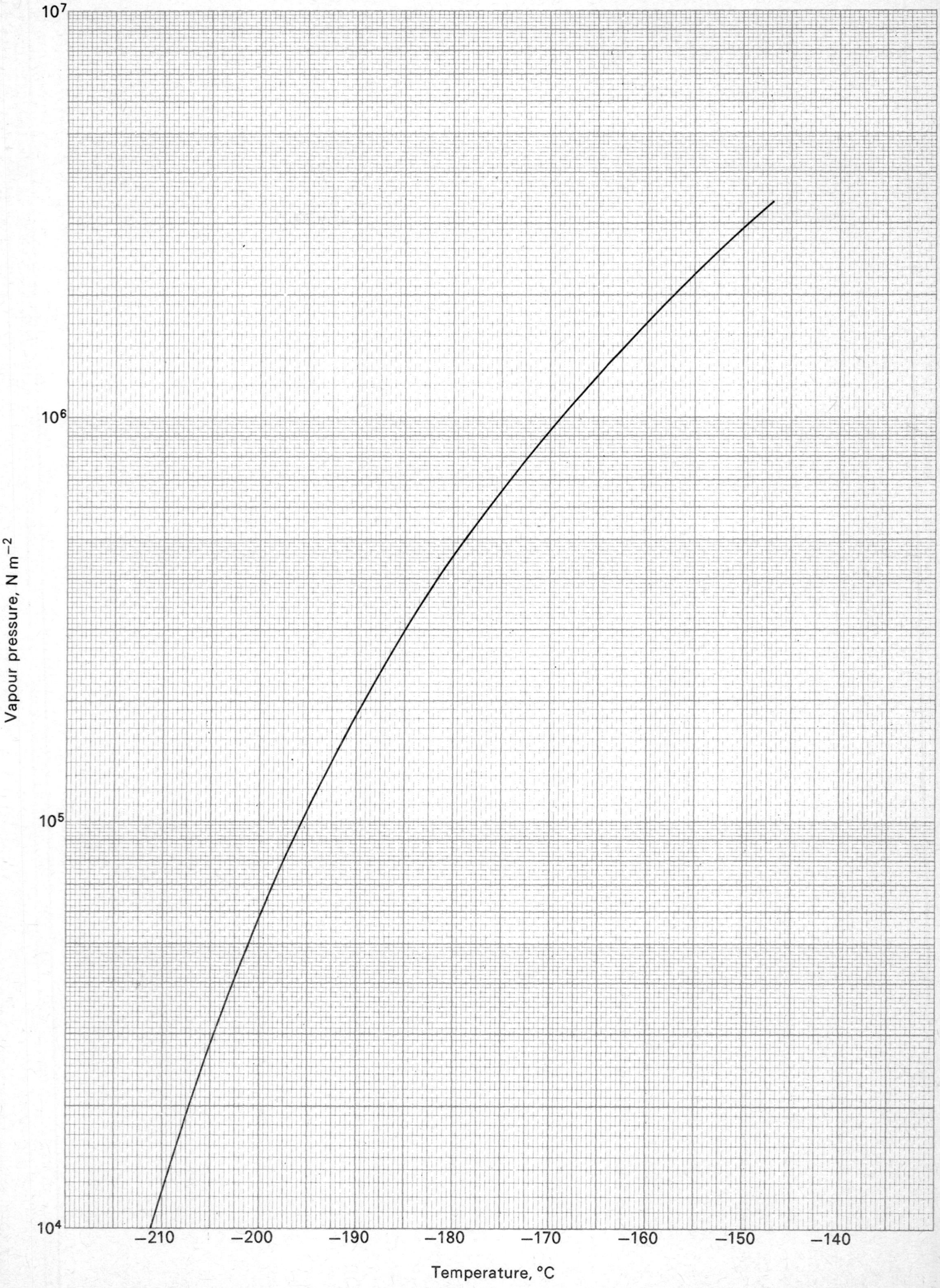

31.2 Orthobaric Density of Nitrogen (53,78,88,91,94)

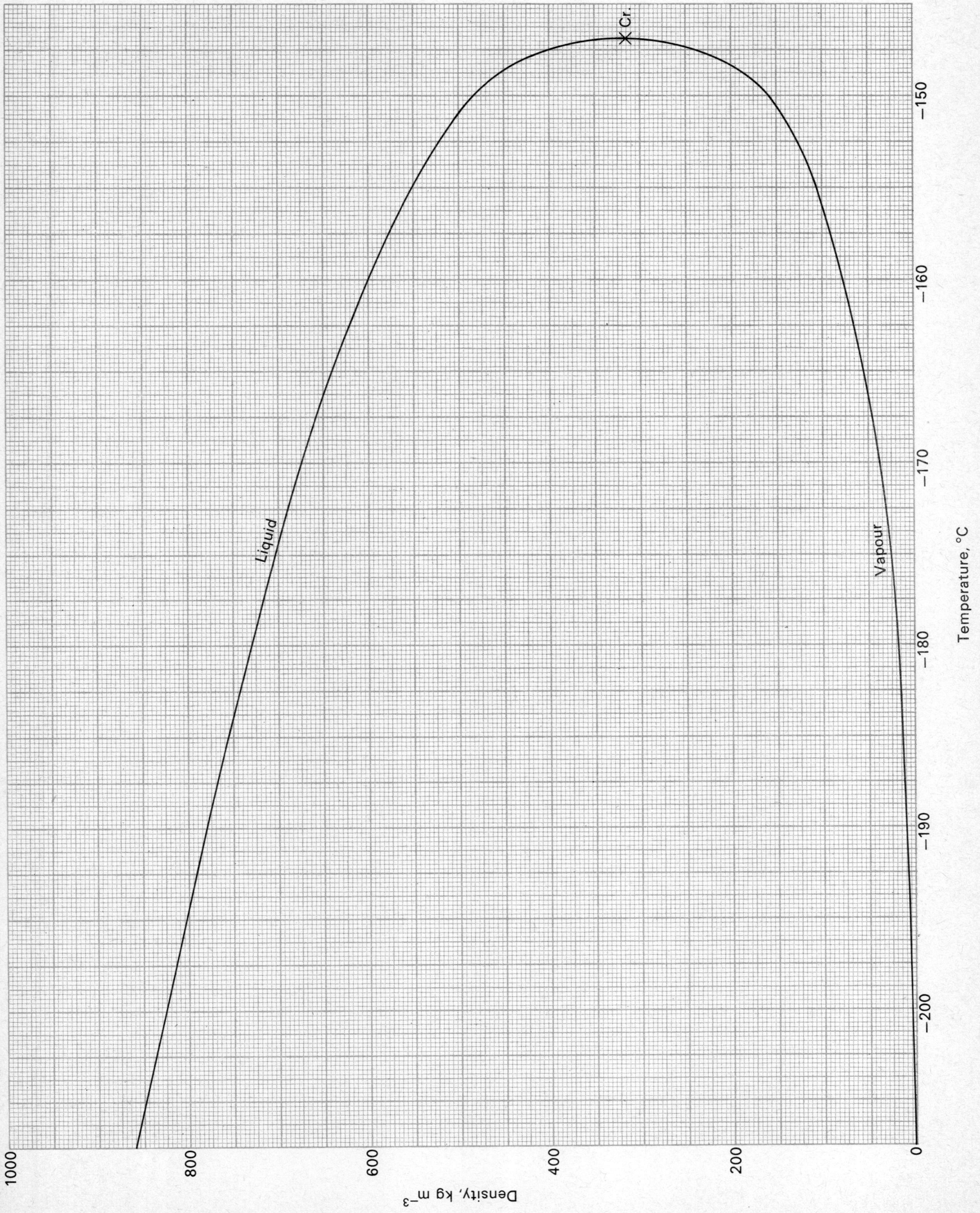

31.3 P-V-T Data for Nitrogen (19,81,82,88,89,91,94)

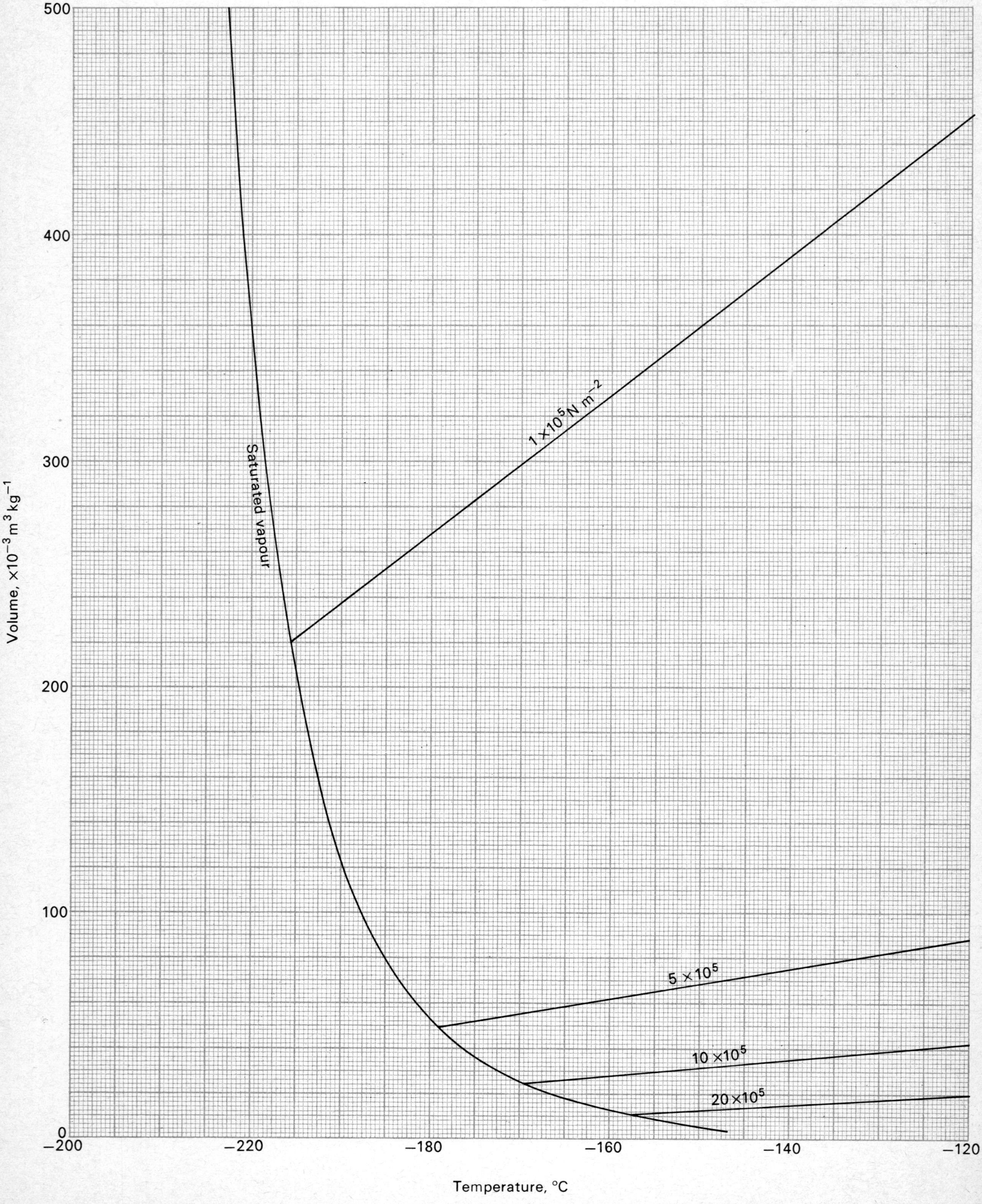

31.4 Latent Heat of Vaporisation of Nitrogen (88,91,94)

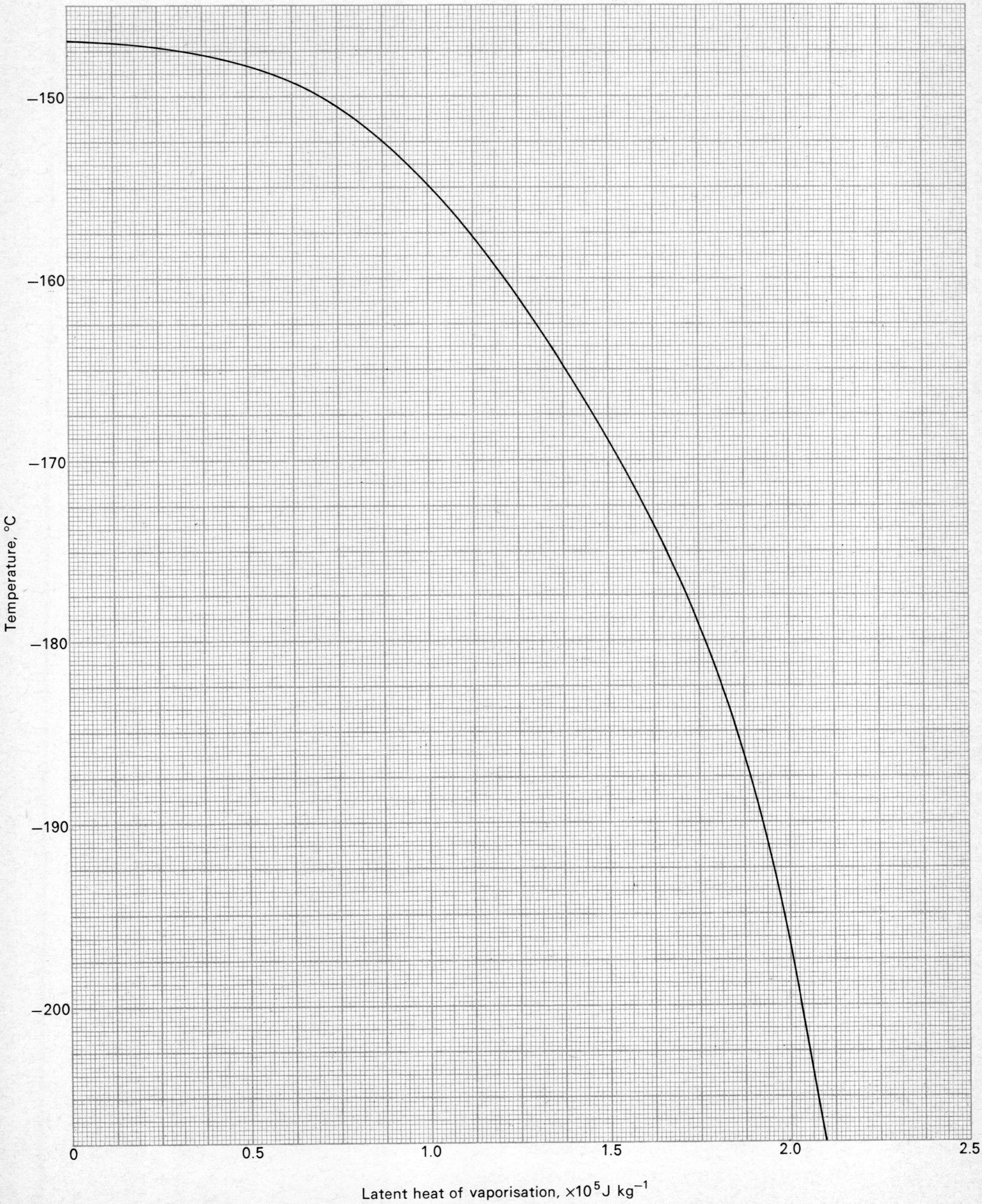

31.5 Heat Capacity of Liquid Nitrogen (11,21,78,97)

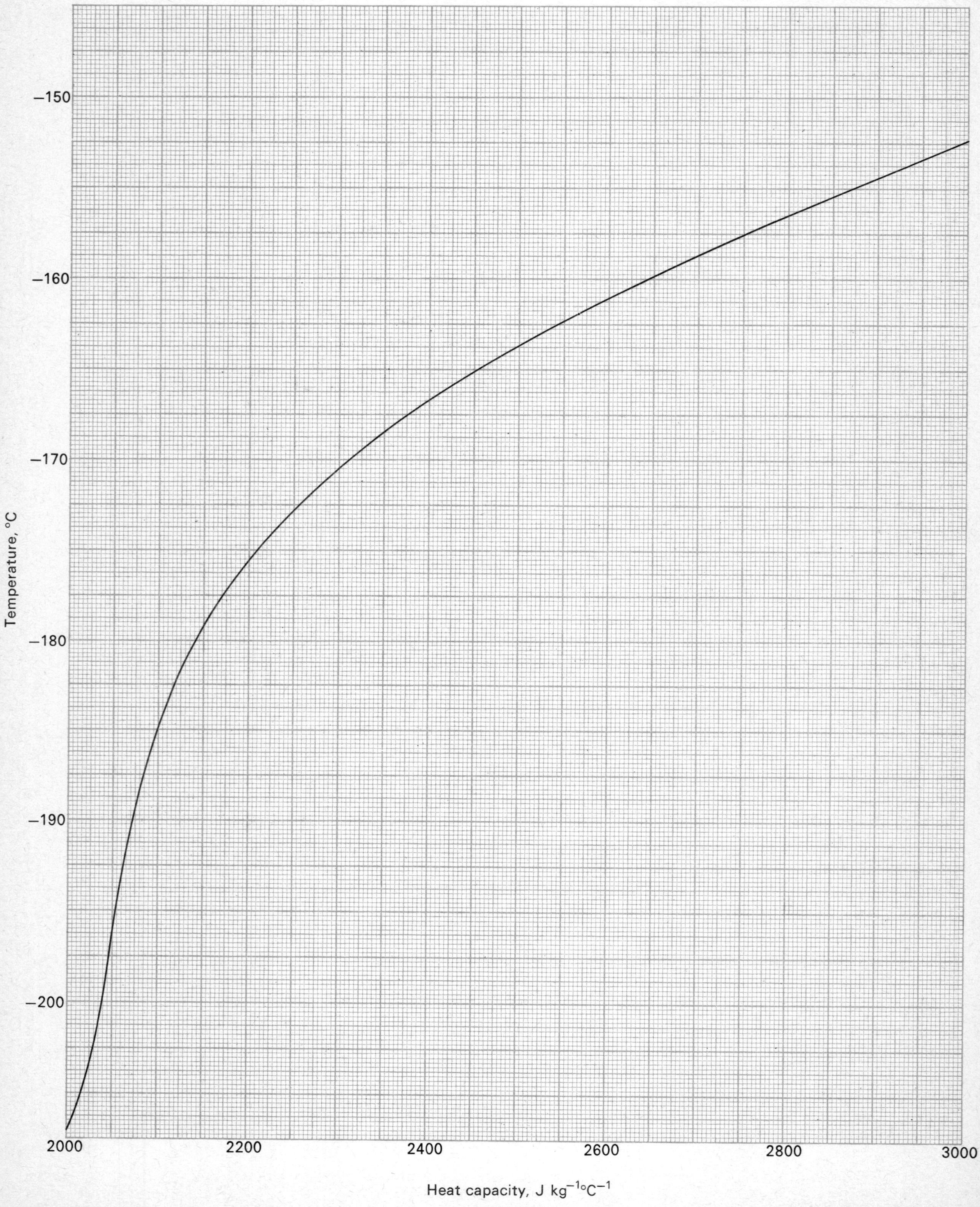

31.6 Heat Capacity of Gaseous Nitrogen at 1.0133×10^5 N m^{-2}(1,12,25,28,78,79,93,100)

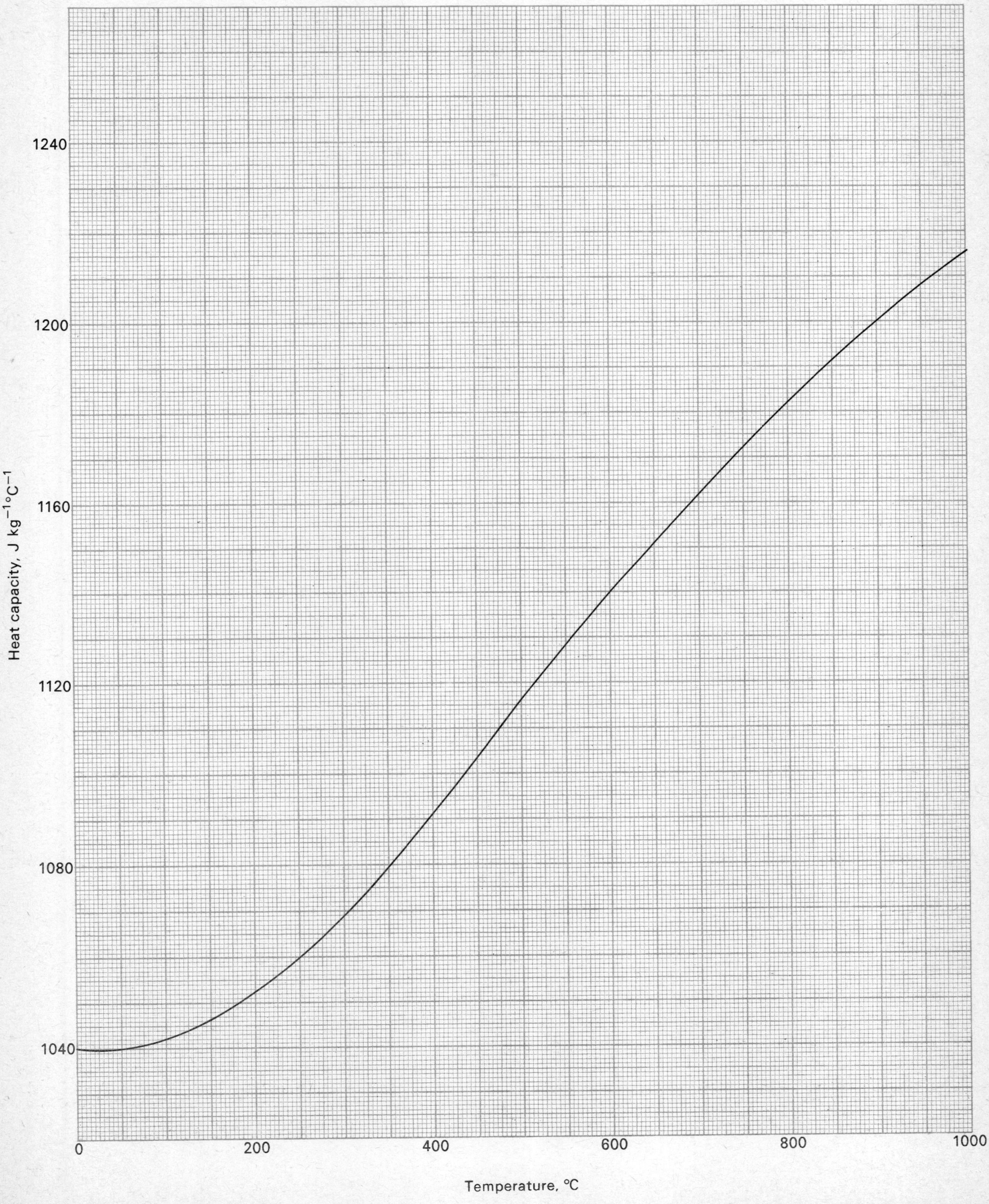

31.7 Heat Capacity Ratio (C_p/C_v) of Gaseous Nitrogen at 1.0133×10^5 N m^{-2} (16,48)

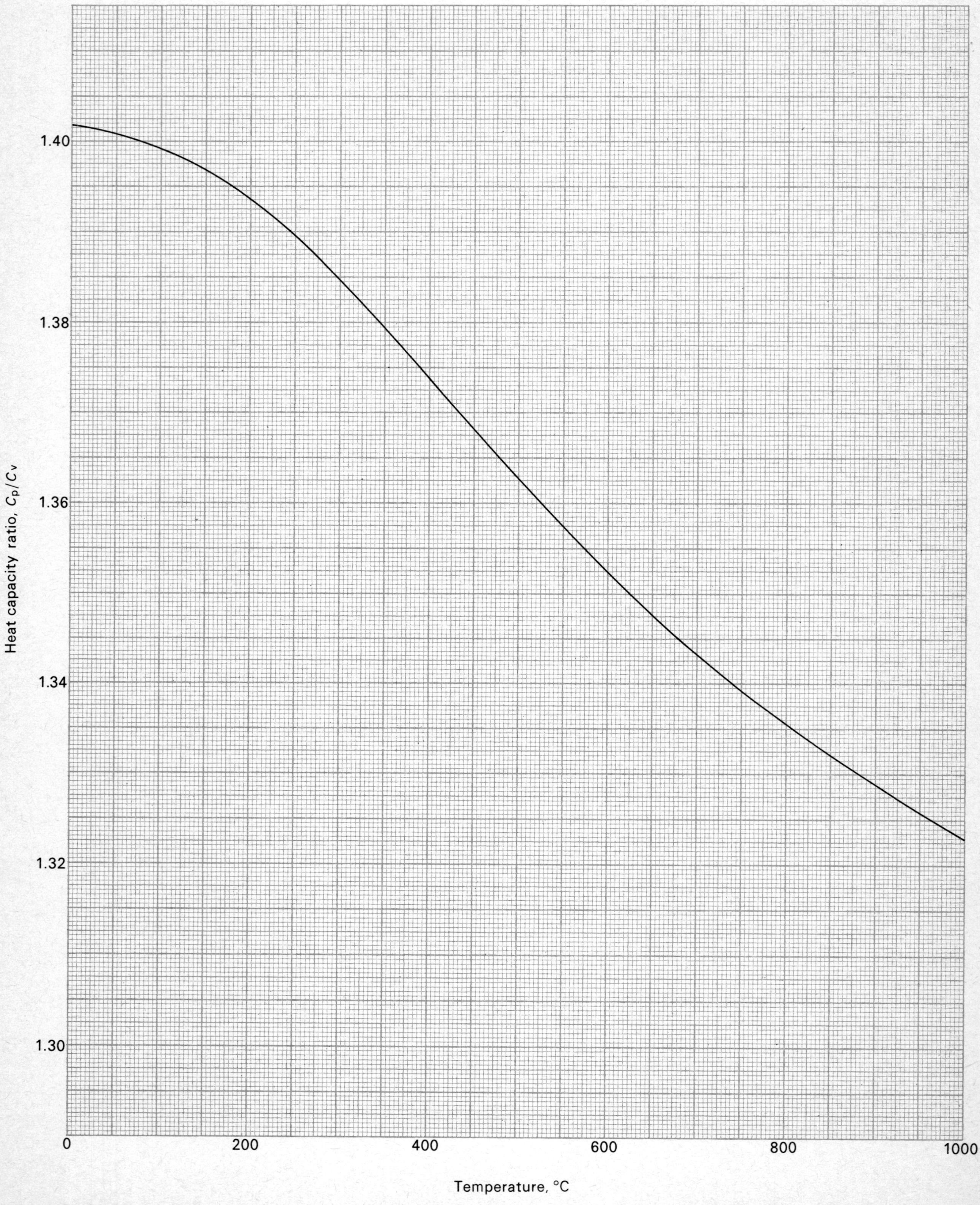

31.8 Viscosity of Liquid Nitrogen (17,18,68,69,70,85)

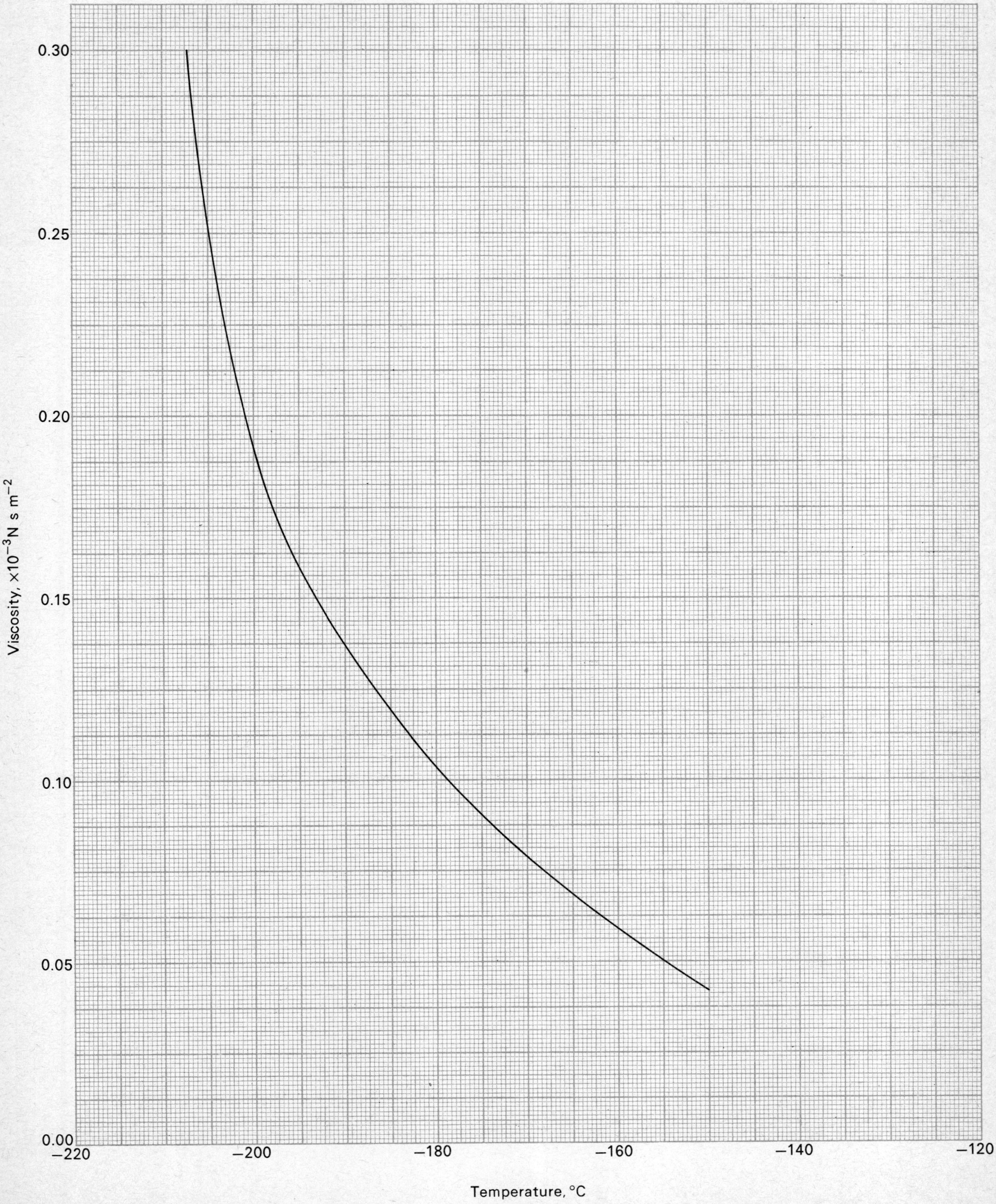

31.9 Viscosity of Gaseous Nitrogen at 1.0133×10^5 N m^{-2} (2,10,23,27,30,37,38,49,66,80,90)

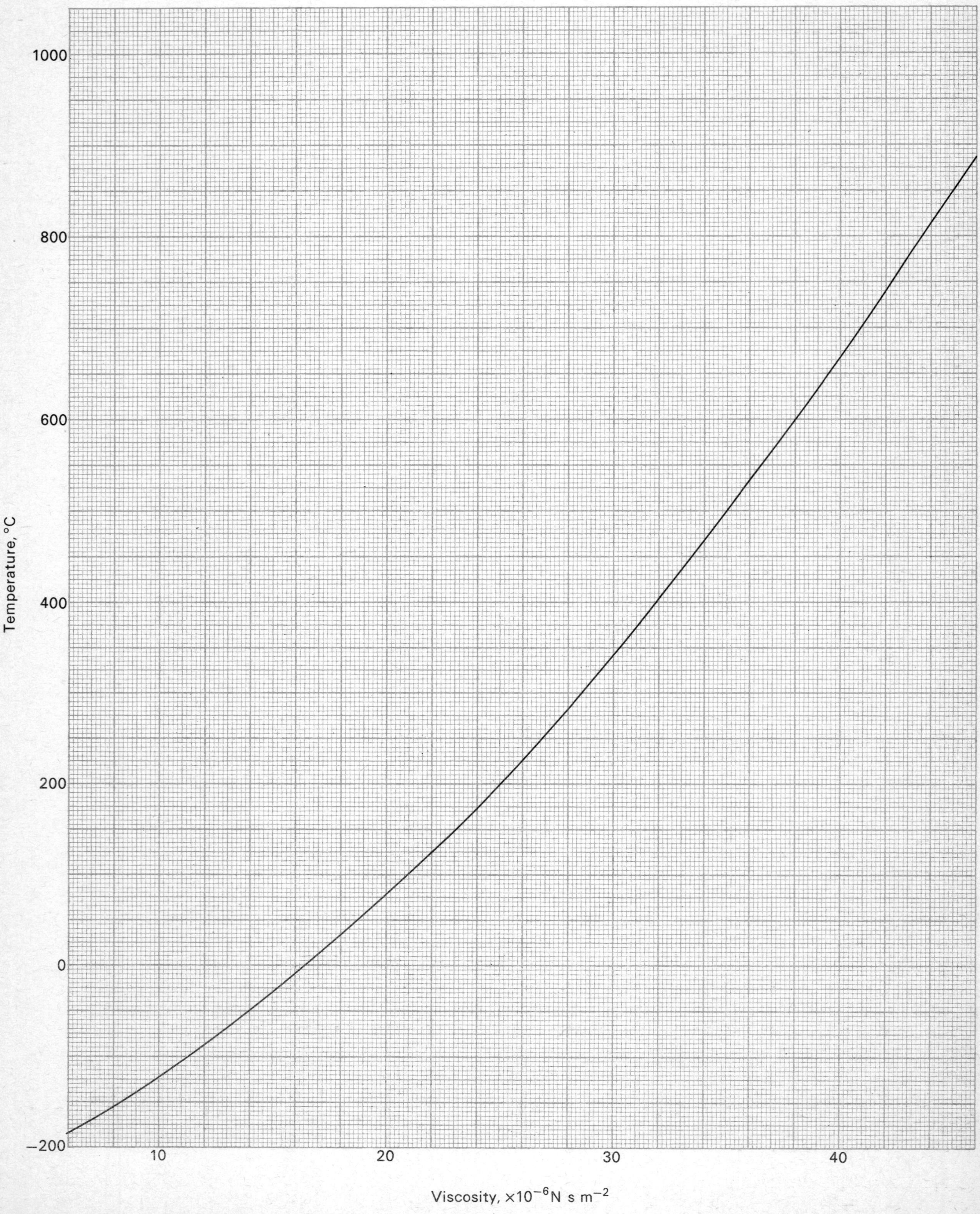

31.10 Thermal Conductivity of Liquid Nitrogen (8,64,65,83,86,102)

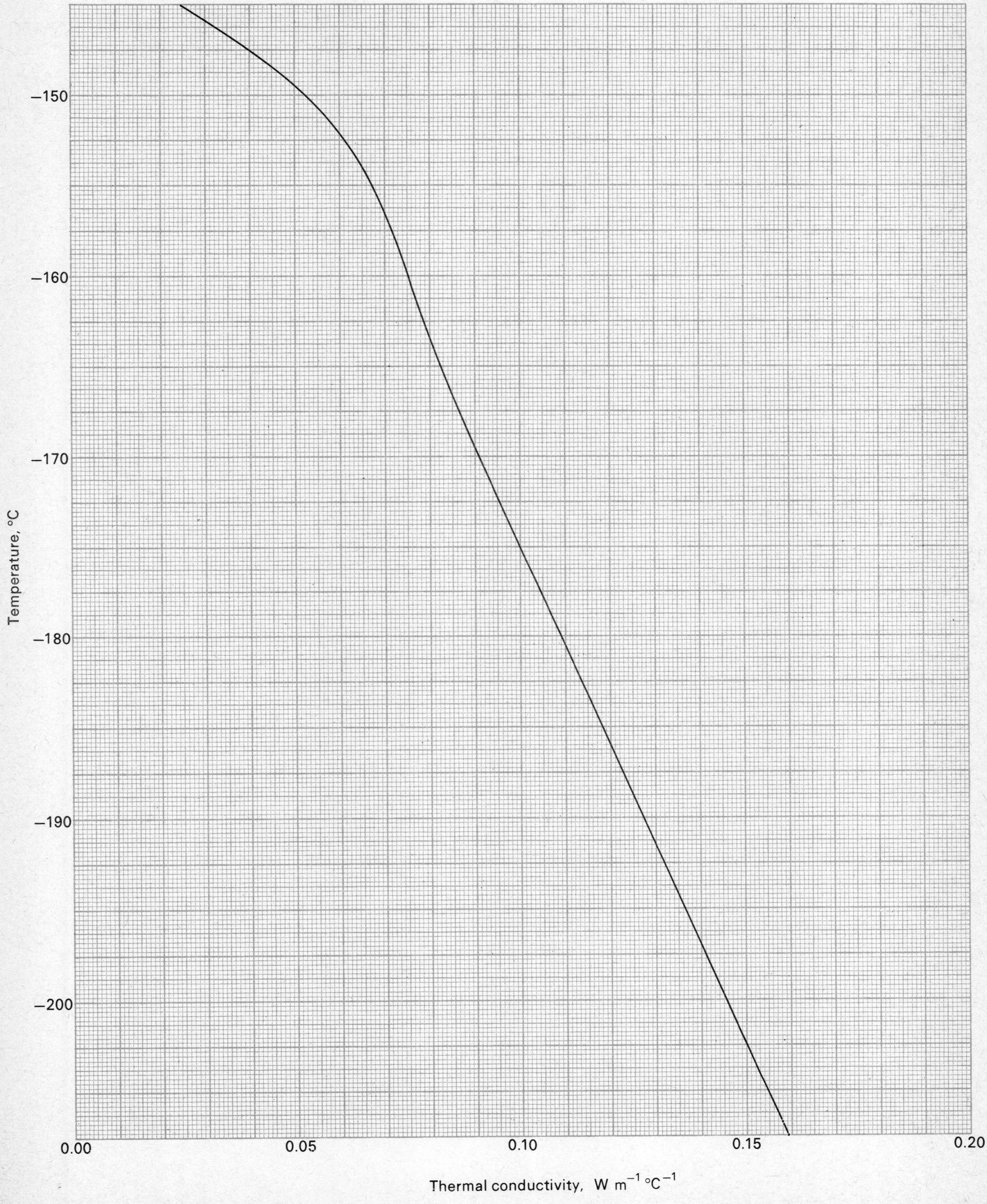

31.11 Thermal Conductivity of Nitrogen (2,6,8,24,30,35,37,38,39,46,55,60,72,86,87,90,99,102)

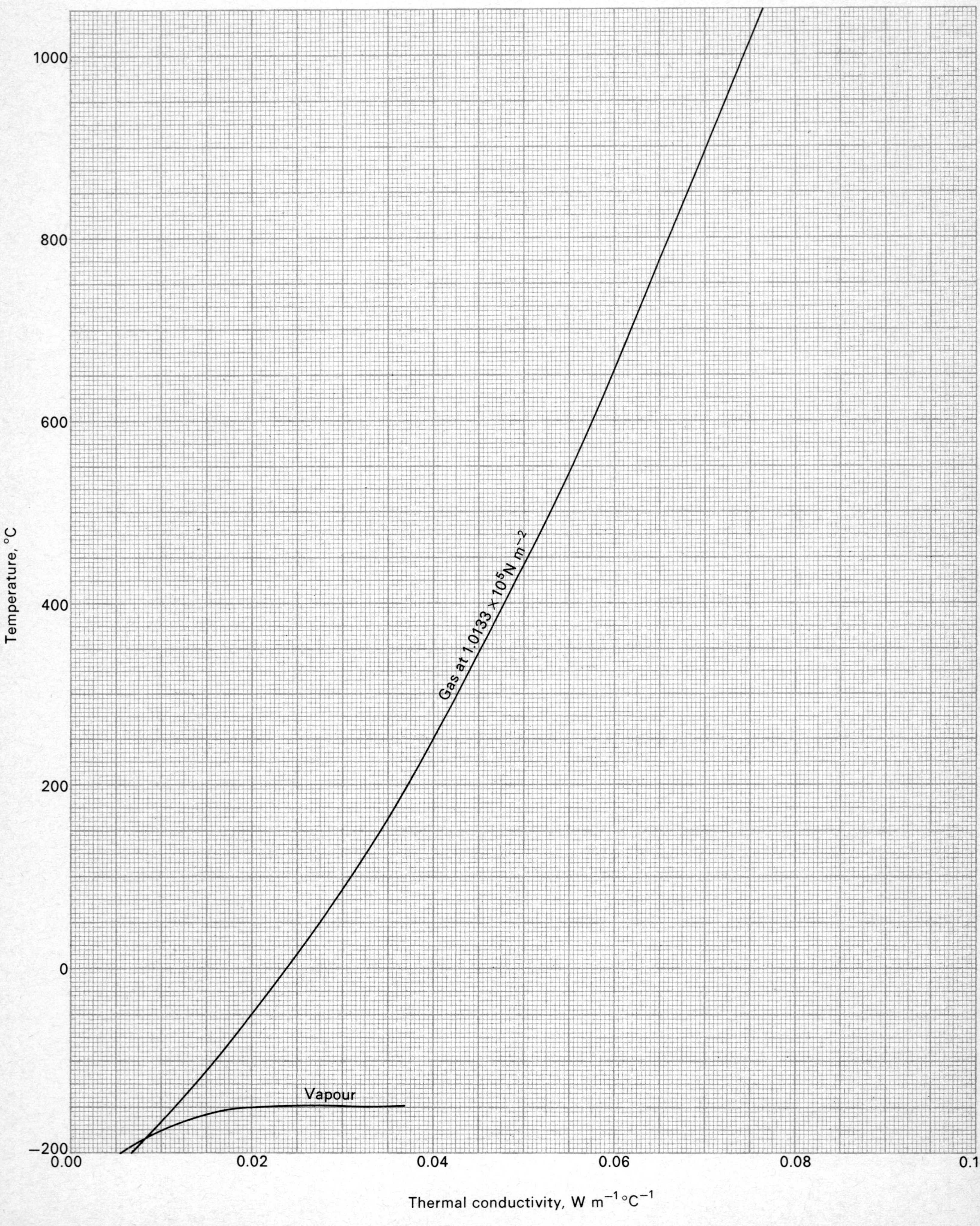

31.12 Surface Tension of Nitrogen (5,77,95)

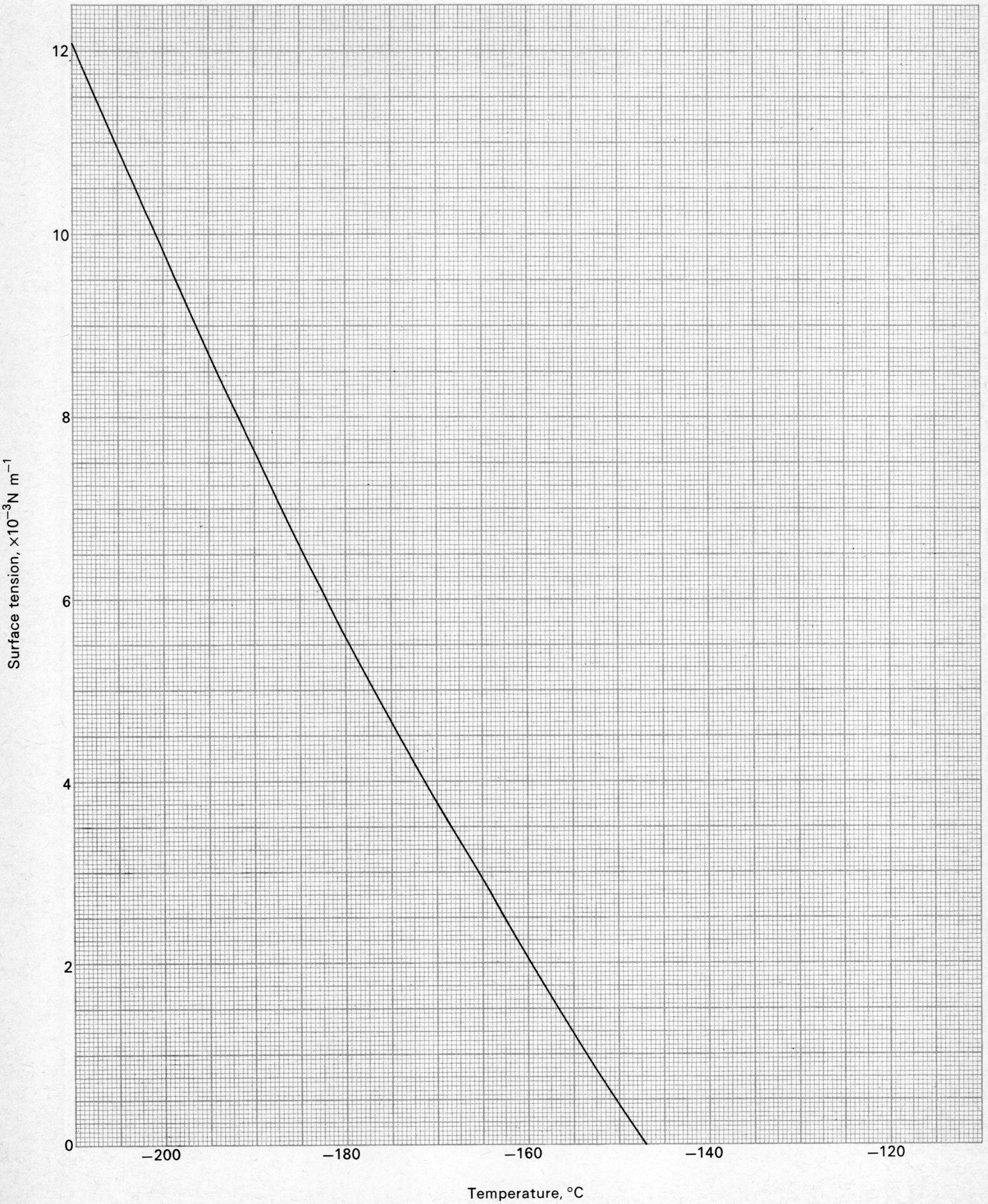

31.13 Solubility of Nitrogen in Water at 1.0133×10^5 N m^{-2} Total Pressure (31,40,42,56,62)

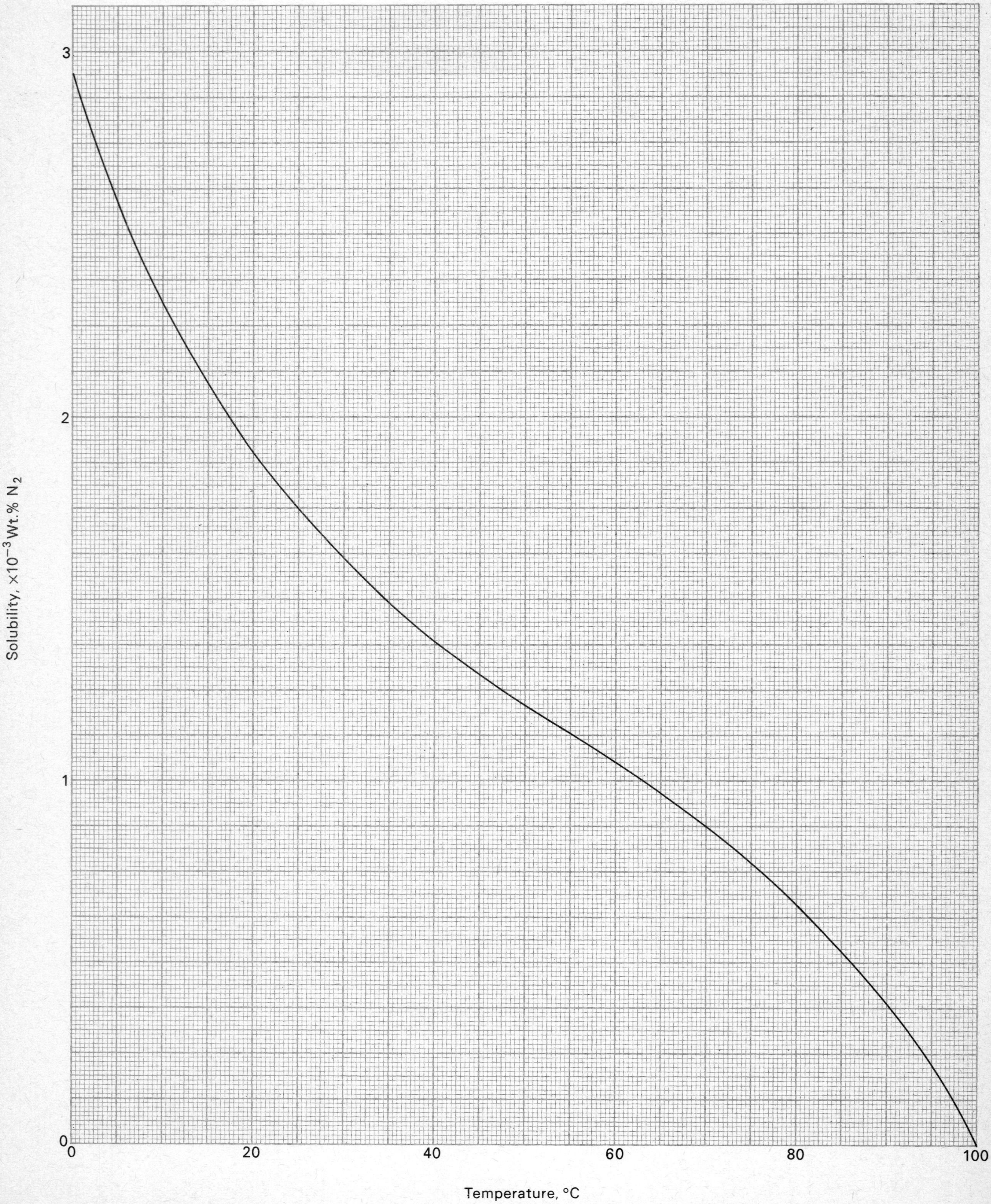

APPENDIX

Tables presented in the Appendix have been compiled from the following sources:

1 American Institute of Chemical Engineers, *Conversion Tables*, AIChE Journal, 1971, **17** (2), 511–2.

2 British Chemical Engineering, *SI Unit Conversion Factors for Chemical Engineers*, Br. Chem. Eng., 1971, **16** (9), 829–32.

3 British Standard Institution, *International System (SI) Units*, Br. Std. Inst., B.S. 3763, London, 1964, 45 pp.

4 British Standard Institution, *The Use of SI Units*, Br. Std. Inst., P.D. 5686, London, January 1969, 26 pp.

5 Canham, W. G., *The International Metric System*, Chem. Eng. Progr., 1972, **68** (7), 90–4.

6 Horvath, A. L., *SI Units—Absolute Units*, Chem. Ind., 1969 (Aug. 9), No. 32, 1083.

7 Horvath, A. L., *Specific Resistance*, Electronics & Power, 1970, **16** (4), 141.

8 Hughes, M. N., James, A. M., Silvester, N. R., *SI Units and Conversion Tables*, Machinery Publ. Co. Ltd., London, 1970, 23 pp.

9 Lees, F. P., *An SI Unit Conversion Table for Chemical Engineering*, Chem. Eng. (London), 1968 (Oct.), No. 222, 341–4.

10 Mullin, J. W., *SI Units in Chemical Engineering*, Chem. Eng. (London), 1967 (Sept.), No. 211, 176–7.

11 Mullin, J. W., *Recent Developments in the Change-Over to the International System of Units (SI)*, Chem. Eng. (London), 1971 (Oct.), No. 254, 352–6.

12 Mullin, J. W., *SI Units in Chemical Engineering*, AIChE Journal, 1972, **18** (1), 222–4.

13 McGlashan, M. L., *Manual of Symbols and Terminology for Physicochemical Quantities and Units*, Pure & Appl. Chem., 1970, **21** (1), 1–44.

14 National Bureau of Standards, *Policy for NBS Usage of SI Units* J. Chem. Educ., 1971, **48** (9), 569–72.

15 Norris, A. C., *SI Units in Physico-Chemical Calculations*, J. Chem. Educ., 1971, **48** (12), 797–800.

16 Page, C. H., Vigoureux, P., Eds., *The International System of Units (SI)*, Natl. Bur. Std. Spec. Publ. 330, Washington, D. C., January 1971, 42 pp.

17 Rossini, F. D., *Report on International Practical Temperature Scale of 1968*, J. Chem. Thermodyn., 1970, **2**, 447–59.

18 Taylor, B. N., Parker, W. H., Langenberg, D. N., *Determination of e/h, Using Macroscopic Quantum Phase Coherence in Superconductors: Implications for Quantum Electrodynamics and the Fundamental Physical Constants*, Rev. Mod. Phys., 1969, **41** (3), 375–496.

19 United Kingdom Atomic Energy Authority, *Metric Guide Nos. 1–4. SI Units for Everyone. SI Units in Everyday Use. Units for Measuring Nuclear Quantities. SI Units in Calculations*, U.K.A.E.A., London, 1969–70.

Fundamental Physical Constants

Physical quantity	*Value*	*SI unit*
Atomic mass unit	$1.660\ 43 \times 10^{-27}$	kg
Avogadro number	$6.022\ 52 \times 10^{26}$	atoms kg-mol^{-1}
Bohr magneton	9.2732×10^{-24}	m^2 A
Bohr radius	$5.291\ 771\ 5 \times 10^{-11}$	m
Boltzmann constant	$1.380\ 54 \times 10^{-23}$	J K^{-1}
Charge of proton	$1.602\ 10 \times 10^{-19}$	C
Charge to mass ratio for electron	$1.758\ 796 \times 10^{11}$	C kg^{-1}
Classical electron radius	$2.817\ 939 \times 10^{-15}$	m
Constant relating wavenumber and energy	$1.196\ 255 \times 10^{2}$	J m kg-mol^{-1}
Curie	37.0×10^{9}	disintegrations s^{-1}
Einstein constant	$8.987\ 554 \times 10^{16}$	J kg^{-1}
Electron magnetic moment	$9.284\ 851 \times 10^{-24}$	J T^{-1}
Electron rest mass	$9.109\ 558 \times 10^{-31}$	kg
Electron volt	1.602×10^{-19}	J
Electronic charge	$1.602\ 10 \times 10^{-19}$	C
Faraday constant	$9.648\ 70 \times 10^{7}$	C kg-mol^{-1}
Fine structure constant	$7.297\ 351 \times 10^{-3}$	
First radiation constant	$4.992\ 579 \times 10^{-24}$	J m
Gas constant	$8.314\ 33 \times 10^{3}$	J K^{-1} kg-mol^{-1}
Gravitational acceleration	9.807	m s^{-2}
Gravitational constant	6.6732×10^{-11}	N m^2 kg^{-2}
Gyromagnetic ratio of proton, corrected for diamagnetism in H_2O	$2.675\ 196 \times 10^{8}$	rad s^{-1} T^{-1}
Ice-point temperature	273.150	K
Magnetic flux quantum	$2.067\ 853\ 8 \times 10^{-15}$	T m^2
Molar volume of ideal gas at s.t.p.	22.4136	m^3 kg-mol^{-1}
Neutron rest mass	$1.674\ 920 \times 10^{-27}$	kg
Permeability of vacuum	$4\pi \times 10^{-7}$	J s^2 C^{-2} m^{-1}
Permittivity of vacuum	$8.854\ 185 \times 10^{-12}$	J^{-1} C^2 m^{-1}
Planck constant	6.6256×10^{-34}	J s
Proton magnetic moment	$1.410\ 620\ 3 \times 10^{-26}$	J T^{-1}
Proton rest mass	$1.672\ 614 \times 10^{-27}$	kg
Rydberg constant	$1.097\ 373 \times 10^{7}$	m^{-1}
Second radiation constant	$1.438\ 79 \times 10^{-2}$	m K
Speed of light in vacuum	$2.997\ 925 \times 10^{8}$	m s^{-1}
Standard pressure	$1.013\ 25 \times 10^{5}$	N m^{-2}
Standard temperature	273.15	K
Stefan–Boltzmann constant	5.6697×10^{-8}	W m^{-2} K^{-4}
Triple point of water	273.16	K
Wien's radiation constant	2.8978×10^{-3}	m K

Conversion Tables

Density

	kg m^{-3}	g cm^{-3}	lb ft^{-3}	lb in^{-3}	lb (UK $gal)^{-1}$	ton yd^{-3}	slug ft^{-3}
1 kg m^{-3}	1	10^{-3}	62.428×10^{-3}	36.127×10^{-6}	10.022×10^{-3}	75.248×10^{-5}	19.403×10^{-4}
1 g cm^{-3}	10^3	1	62.428	36.127×10^{-3}	10.0224	0.752 48	19.403×10^{-1}
1 lb ft^{-3}	16.0185	16.0185×10^{-3}	1	57.870×10^{-5}	0.160 544	$1.205\,36 \times 10^{-2}$	31.081×10^{-3}
1 lb in^{-3}	27.6799×10^3	27.6799	1.728×10^3	1	$2.774\,19 \times 10^2$	20.8286	53.707 97
1 lb (UK $gal)^{-1}$	99.776	99.776×10^{-3}	6.228 84	3.6046×10^{-3}	1	7.5080×10^{-2}	0.193 60
1 ton yd^{-3}	13.2894×10^2	1.328 93	82.9630	4.8011×10^{-2}	13.3192	1	2.578 571
1 slug ft^{-3}	5.1538×10^2	0.515 38	32.174	$1.861\,92 \times 10^{-2}$	5.1654	0.387 811	1

1 Amagat unit of density = 4.5162×10^{-8} kg-mol cm^{-3}

Pressure

	N m^{-2}	bar	atm	mmHg	in H_2O	in Hg	lbf in^{-2}	lbf ft^{-2}	dyn cm^{-2}	kgf cm^{-2}
1 N m^{-2}	1	10^{-5}	9.8692×10^{-6}	7.5006×10^{-3}	4.0146×10^{-3}	2.9530×10^{-4}	1.4504×10^{-4}	2.0885×10^{-2}	10	1.0197×10^{-5}
1 bar	10^5	1	9.8692×10^{-1}	7.5006×10^2	4.0146×10^2	29.530	14.504	2.0885×10^3	10^6	1.0197
1 atm	$1.013\,25 \times 10^5$	1.013 25	1	7.6000×10^2	406.8	29.92	14.6959	$2.116\,22 \times 10^3$	$1.013\,25 \times 10^6$	1.033 23
1 mm Hg	133.322	$1.333\,22 \times 10^{-3}$	$1.315\,79 \times 10^{-3}$	1	0.535 24	3.9370×10^{-2}	$1.933\,68 \times 10^{-2}$	2.784 50	$1.333\,22 \times 10^3$	$1.359\,51 \times 10^{-3}$
1 in H_2O	249.09	2.4909×10^{-3}	2.458×10^{-3}	1.8683	1	7.3556×10^{-2}	3.6127×10^{-2}	5.2023	2490.9	2.540×10^{-3}
1 in Hg	3.3864×10^3	3.3864×10^{-2}	3.3421×10^{-2}	25.40	13.595	1	$4.911\,55 \times 10^{-1}$	70.7263	3.3864×10^4	3.4532×10^{-2}
1 lbf in^{-2}	$6.894\,76 \times 10^3$	$6.894\,76 \times 10^{-2}$	$6.804\,60 \times 10^{-2}$	51.7149	27.680	2.0360	1	144.000	$6.894\,76 \times 10^4$	$7.030\,70 \times 10^{-2}$
1 lbf ft^{-2}	47.8803	$4.788\,03 \times 10^{-4}$	$4.725\,41 \times 10^{-4}$	$3.591\,31 \times 10^{-1}$	0.192 22	1.4139×10^{-2}	$6.944\,44 \times 10^{-3}$	1	$4.788\,03 \times 10^2$	$4.882\,43 \times 10^{-4}$
1 dyn cm^{-2}	10^{-1}	10^{-6}	$9.869\,23 \times 10^{-7}$	$7.500\,62 \times 10^{-4}$	4.0150×10^{-4}	2.9530×10^{-5}	$1.450\,38 \times 10^{-5}$	$2.088\,54 \times 10^{-3}$	1	$1.019\,72 \times 10^{-6}$
1 kgf cm^{-2}	$9.806\,65 \times 10^4$	0.980 665	0.967 841	$7.355\,59 \times 10^2$	393.70	28.959	14.2233	$2.048\,16 \times 10^3$	$9.806\,65 \times 10^5$	1

1 torr = 1 mmHg
1 pascal = 1 N m^{-2}

Specific Energy

	$J\ kg^{-1}$	$ft\ lbf\ lb^{-1}$	$kgf\ m\ kg^{-1}$	$Btu\ lb^{-1}$	$kcal\ kg^{-1}$	$cal\ g^{-1}$
1 $J\ kg^{-1}$	1	0.334 55	0.101 97	4.299×10^{-4}	2.388×10^{-4}	2.388×10^{-4}
1 $ft\ lbf\ lb^{-1}$	2.989	1	0.3048	1.2851×10^{-3}	7.139×10^{-4}	7.139×10^{-4}
1 $kgf\ m\ kg^{-1}$	9.807	3.281	1	4.216×10^{-3}	2.342×10^{-3}	2.342×10^{-3}
1 $Btu\ lb^{-1}$	2.3260×10^{3}	7.782×10^{2}	2.3719×10^{2}	1	0.555 556	0.555 556
1 $kcal\ kg^{-1}$	4.1868×10^{3}	1.4007×10^{3}	4.269×10^{2}	1.800 000	1	1
1 $cal\ g^{-1}$	4.1868×10^{3}	1.4007×10^{3}	4.269×10^{2}	1.800 000	1	1

1 $Chu\ lb^{-1}$ = 1 $kcal\ kg^{-1}$ = 1 $cal\ g^{-1}$

Heat Capacity

	$J\ kg^{-1}\ °C^{-1}$	$cal\ g^{-1}\ °C^{-1}$	$Btu\ lb^{-1}\ °F^{-1}$	$ft\ lbf\ slug^{-1}\ °C^{-1}$	$ft\ pdl\ lb^{-1}\ °F^{-1}$	$ft\ lbf\ slug^{-1}\ °F^{-1}$	$ft\ pdl\ lb^{-1}\ °C^{-1}$	$ft\ lbf\ lb^{-1}\ °F^{-1}$
1 $J\ kg^{-1}\ °C^{-1}$	1	2.3885×10^{-4}	2.3885×10^{-4}	10.764	5.980 00	5.980 00	10.764	0.18586
1 $cal\ g^{-1}\ °C^{-1}$	4.1868×10^{3}	1	1	4.5066×10^{4}	2.5037×10^{4}	2.5037×10^{4}	4.5066×10^{4}	7.7817×10^{2}
1 $Btu\ lb^{-1}\ °F^{-1}$	4.1868×10^{3}	1	1	4.5066×10^{4}	2.5037×10^{4}	2.5037×10^{4}	4.5066×10^{4}	7.7817×10^{2}
1 $ft\ lbf\ slug^{-1}\ °C^{-1}$	9.2903×10^{-2}	2.2190×10^{-5}	2.2190×10^{-5}	1	0.555 556	0.555 556	1	1.7267×10^{-2}
1 $ft\ pdl\ lb^{-1}\ °F^{-1}$	0.167 23	3.9941×10^{-5}	3.9941×10^{-5}	1.800 00	1	1	1.8000	3.108×10^{-2}
1 $ft\ lbf\ slug^{-1}\ °F^{-1}$	0.167 23	3.9941×10^{-5}	3.9941×10^{-5}	1.800 00	1	1	1.8000	3.1081×10^{-2}
1 $ft\ pdl\ lb^{-1}\ °C^{-1}$	9.2903×10^{-2}	2.2190×10^{-5}	2.2190×10^{-5}	1	0.555 556	0.555 556	1	1.7267×10^{-2}
1 $ft\ lbf\ lb^{-1}\ °F^{-1}$	5.3803	1.2851×10^{-3}	1.2851×10^{-3}	57.913	32.174	32.174	57.913	1

1 $Chu\ lb^{-1}\ °C^{-1}$ = 1 $Btu\ lb^{-1}\ °F^{-1}$

Viscosity (Absolute or Dynamic)

	N s m^{-2}	P	cP	kg m^{-1} h^{-1}	lb ft^{-1} s^{-1}	kgf s m^{-2}	lbf s ft^{-2}	lb ft^{-1} h^{-1}	lbf h ft^{-2}
1 N s m^{-2}	1	10	1000	3600	0.671 97	0.101 972	0.020 885 4	2419.09	$5.801\,38 \times 10^{-6}$
1 P	0.1	1	100	360	0.067 197	0.010 197 2	$2.088\,54 \times 10^{-3}$	241.909	$5.801\,38 \times 10^{-7}$
1 cP	10^{-3}	10^{-2}	1	3.6000	6.7197×10^{-4}	$1.019\,72 \times 10^{-4}$	$2.088\,54 \times 10^{-5}$	2.419 09	$5.801\,38 \times 10^{-9}$
1 kg m^{-1} h^{-1}	$2.777\,778 \times 10^{-4}$	$2.777\,778 \times 10^{-3}$	$2.777\,778 \times 10^{-1}$	1	$1.866\,58 \times 10^{-4}$	$2.832\,55 \times 10^{-5}$	$5.801\,51 \times 10^{-6}$	0.671 969	$1.611\,528 \times 10^{-9}$
1 lb ft^{-1} s^{-1}	1.488 16	14.8816	$1.488\,16 \times 10^{3}$	$5.357\,39 \times 10^{3}$	1	0.151 750	$3.108\,10 \times 10^{-2}$	3600.000 00	$8.633\,611 \times 10^{-6}$
1 kgf s m^{-2}	9.806 65	98.0665	$9.806\,65 \times 10^{3}$	$3.530\,39 \times 10^{4}$	6.589 76	1	0.204 816	$2.372\,31 \times 10^{4}$	$5.689\,333 \times 10^{-5}$
1 lbf s ft^{-2}	47.8803	$4.788\,03 \times 10^{2}$	$4.788\,03 \times 10^{4}$	$1.723\,69 \times 10^{5}$	32.1740	4.882 43	1	$1.158\,27 \times 10^{5}$	$2.777\,778 \times 10^{-4}$
1 lb ft^{-1} h^{-1}	$4.133\,79 \times 10^{-4}$	$4.133\,79 \times 10^{-3}$	$4.133\,79 \times 10^{-1}$	1.488 16	$2.777\,778 \times 10^{-4}$	$4.215\,29 \times 10^{-5}$	$8.633\,60 \times 10^{-6}$	1	$2.398\,222 \times 10^{-9}$
1 lbf h ft^{-2}	$1.723\,69 \times 10^{5}$	$1.723\,69 \times 10^{6}$	$1.723\,69 \times 10^{8}$	6.205×10^{8}	$1.158\,27 \times 10^{5}$	1.758×10^{4}	3600.0000	4.170×10^{8}	1

1 P (Poise) = 1 g cm^{-1} s^{-1} = 1 dyn s cm^{-2}
1 slug ft^{-1} s^{-1} = 1 lbf s ft^{-2}
1 reyn = 1 lbf s in^{-2} = 144 lbf s ft^{-2}

Thermal Conductivity

	W m^{-1} °C^{-1}	W cm^{-1} °C^{-1}	cal cm^{-1} s^{-1} °C^{-1}	kcal m^{-1} h^{-1} °C^{-1}	Btu ft^{-1} s^{-1} °F^{-1}	Btu ft^{-1} h^{-1} °F^{-1}	Btu in^{-1} h^{-1} °F^{-1}	Btu in ft^{-1} h^{-1} °F^{-1}
1 W m^{-1} °C^{-1}	1	10^{-2}	$2.388\,46 \times 10^{-3}$	$8.598\,45 \times 10^{-1}$	$1.604\,97 \times 10^{-4}$	0.577 789	$4.814\,91 \times 10^{-2}$	6.933 47
1 W cm^{-1} °C^{-1}	10^{2}	1	0.238 846	85.9845	$1.604\,97 \times 10^{-2}$	57.7789	4.814 91	$6.933\,47 \times 10^{2}$
1 cal cm^{-1} s^{-1} °C^{-1}	$4.186\,80 \times 10^{2}$	4.186 80	1	$3.600\,000 \times 10^{2}$	$6.719\,69 \times 10^{-2}$	$2.419\,09 \times 10^{2}$	20.1591	$2.902\,91 \times 10^{3}$
1 kcal m^{-1} h^{-1} °C^{-1}	1.163 00	$1.163\,00 \times 10^{-2}$	$2.777\,778 \times 10^{-3}$	1	$1.866\,58 \times 10^{-4}$	0.671 969	$5.599\,74 \times 10^{-2}$	8.063 63
1 Btu ft^{-1} s^{-1} °F^{-1}	$6.230\,64 \times 10^{3}$	62.3064	14.8816	$5.357\,39 \times 10^{3}$	1	$3.600\,000 \times 10^{3}$	$3.000\,000 \times 10^{2}$	$4.320\,000 \times 10^{4}$
1 Btu ft^{-1} h^{-1} °F^{-1}	1.730 73	$1.730\,73 \times 10^{-2}$	$4.133\,79 \times 10^{-3}$	1.488 16	$2.777\,777\,8 \times 10^{-4}$	1	$8.333\,333 \times 10^{-2}$	12.000 000 0
1 Btu in^{-1} h^{-1} °F^{-1}	20.7688	0.207 688	$4.960\,54 \times 10^{-2}$	17.8580	$3.333\,333\,3 \times 10^{-3}$	12.000 000 0	1	$1.440\,000 \times 10^{2}$
1 Btu in ft^{-1} h^{-1} °F^{-1}	0.144 228	$1.442\,28 \times 10^{-3}$	$3.444\,82 \times 10^{-4}$	0.124 014	$2.314\,81 \times 10^{-5}$	$8.333\,333 \times 10^{-2}$	$6.944\,444\,4 \times 10^{-3}$	1

1 Chu ft^{-1} h^{-1} °C^{-1} = 1 Btu ft^{-1} h^{-1} °F^{-1}
1 Chu in ft^{-2} h^{-1} °C^{-1} = 1 Btu in ft^{-2} h^{-1} °F^{-1}

Electrical Resistivity*

	ohm m	ohm mm^2 m^{-1}	ohm cm	ohm in	ohm (100 ft)$^{-1}$	ohm km^{-1}	ohm circ. mil ft^{-1}
1 ohm m	1	10^6	10^2	39.370 08	3.2808×10^{-2}	10^{-3}	6.0153×10^8
1 ohm mm^2 m^{-1}	10^{-6}	1	10^{-4}	3.9370×10^{-5}	3.2808×10^{-8}	10^{-9}	6.0153×10^2
1 ohm cm	10^{-2}	10^4	1	0.393 700 8	3.2808×10^{-4}	10^{-5}	6.0153×10^6
1 ohm in	2.5400×10^{-2}	2.5400×10^4	2.540 00	1	$8.333\ 33 \times 10^{-4}$	$2.540\ 00 \times 10^{-5}$	1.5279×10^7
1 ohm (100 ft)$^{-1}$	3.0480×10^1	3.0480×10^7	3.0480×10^3	$1.200\ 00 \times 10^3$	1	3.0480×10^{-2}	1.8334×10^{10}
1 ohm km^{-1}	10^3	10^9	10^5	3.9370×10^4	32.8083	1	6.0152×10^{11}
1 ohm circ. mil ft^{-1}	1.6624×10^{-9}	1.6624×10^{-3}	1.6624×10^{-7}	6.5450×10^{-8}	5.4542×10^{-11}	1.6624×10^{-12}	1

All values are in absolute units.
1 ohm m = 1 ohm−m = 1 ohm/m = 1 ohm/m^3 (all four terms have the same meaning.)
*Horvath, A.L., *Electronics & Power*, 1970, **16** (4), 141.

ACKNOWLEDGEMENTS

The contents of this book have been taken from numerous publications of others, from journals, books, bulletins, or manufacturers catalogues, etc. Because of the high frequency of citations on each page, it was not possible to express gratitude to the quoted sources on the particular page, but only through a general acknowledgement.

Herewith, the author wishes to acknowledge with thanks permissions from the following publishers, copyright owners, and manufacturers, who kindly furnished material for this book or gave permission to extract data from their publications, as follows:

1 American Chemical Society, for the reproduction of critical properties from: Mathews, J. F., *Chem. Rev.*, 1972, **72** (1), 71–100, and some data from ACS's publications, see references.

2 American Conference of Governmental Industrial Hygienists, Cincinnati, Ohio, for the reprinting of TLV data from the *Documentation of Threshold Limit Values*, revised edition, 1966.

3 American Society of Mechanical Engineers, for the reproduction of data from: Waterman, T. E., *Thermal Conductivity of Liquid Ozone*, 1st. Symp. on Thermophys. Properties, 1959, pp. 301–3; Chen, L. H., *Thermodynamic and Transport Properties of Gaseous Carbon Dioxide*, Ibid., pp. 358–69; Franck, E. U., Brose, M., Mangold, K., *Supercritical Hydrogen Chloride, Specific Heat up to 300°C and 300 atm. PVT-Data up to 400°C and 200 atm*, 2nd. Symp. on Thermophys. Properties, 1962, pp. 159–65; Thomas, W., *Volumetric Behaviour of Hydrogen Chloride*, Ibid., pp. 166–7; Sutton, J. R., *A Method of Calculating the Viscosities of Polar Gases*, Ibid., pp. 266–70; Andrussow, L., *Diffusion, Viscosity and Conductivity of Gases*, Ibid., pp. 279–87; Kestin, J., Leidenfrost, W., *The Effect of Moderate Pressures on the Viscosity of Five Gases*, Ibid., pp. 321–38; Eyring, H., Henderson, D., Ree, T., *Thermodynamic and Transport Properties of Liquids*, Ibid., pp. 340–51; Liley, P. E., *The Thermal Conductivity of 46 Gases at Atmospheric Pressure*, 4th. Symp. on Thermophys. Properties, 1968, pp. 323–49; Ziebland, H., Needham, D. P., *Recent Measurements of the Thermal Conductivity of Liquid Anhydrous Hydrogen Chloride*, Ibid., pp. 296–303.

4 British Oxygen Company Limited, Brentford, Middlesex, for the information on nitrous oxide.

5 Collet's Holdings Limited, Wellingborough, Northamptonshire, for the usage of data from: Vukalovich, M. P., Altunin, V. V., *Thermophysical Properties of Carbon Dioxide*, 1968.

6 Edward Arnold (Publishers) Limited, London, for the extraction of data from: Greene, E. F., Toennies, J. P., *Chemical Reactions in Shock Waves*, 1964, pp. 333–4, and Electrical Research Association, *1967 Steam Tables*, 1967.

7 Fisons Limited, Harston, Cambridge, for the information on the properties of anhydrous hydrazine.

8 Gosudarstvennaya Sluzhba Standartnykh i Spravochnykh Dannykh, Moscow, for the reproduction of data from numerous publications from 1959 to 1972.

9 Longman Group Limited, Harlow, Essex, for the used data from: Mellor, J. E., Ed., *A Comprehensive Treatise on Inorganic and Theoretical Chemistry*, 1922–67, 16 Vols + Suppls., see references for page numbers and data sources.

10 Masson et Cie, Éditeurs, Paris, for the extraction of data from: Pascal, P., Ed., *Nouveau Traité de Chimie Minerale*, 1956–64, 20 Vols, see references for further details.

11 Matheson Gas Products, Division of Will Ross Inc., East Rutherford, N.J., for the reproduction of data from: Matheson Gas Data Book, 4th ed., 1966, and further information on the liquid viscosity of sulphur hexafluoride.

12 McGraw-Hill Book Company, New York, to reprint material from: Lange, N. A., Ed., *Handbook of Chemistry*, 10th ed., 1967; pages are specified under the references, and Washburn, E. W., Ed., *International Critical Tables of Numerical Data, Physics, Chemistry, and Technology* 1926–33, 7 Vols plus Index.

13 National Academy of Sciences, Washington, D. C., for reproduction of data from: Washburn, E. W., Ed., *International Critical Tables of Numerical Data, Physics, Chemistry, and Technology, Vol. VI*, International Research Council, National Academy of Sciences—National Research Council, 1929.

14 Plenum Publishing Corporation, New York, Advances in Cryogenic Engineering, Vol. 16, for the reproduction of data by: G. C. Straty and R. Prydz, *The Vapour Pressure of Liquid Fluorine*, and *Calculation of Thermofunctions of Fluorine*, pp. 36–41 and 64–77 respectively.

15 Springer-Verlag, Heidelberg, for the extraction of data from: Lax, E., Synowietz, C., Eds., *Taschenbuch für Chemiker und Physiker*, 3 vols, 1964–70.

16 Verlag Chemie, GmbH., Weinheim, for the reproduction of data from: *Gmelins Handbuch der Anorganischen Chemie*, published by Gmelin-Institut für anorganischen Chemie in the Max-Planck-Gesellschaft zur Förderung der Wissenschaften e.V.; for full details regarding vols and page numbers, see references of each compound.

17 Wiley & Sons, Inc., New York, for the reproduction of data from: *Kirk-Othmer Encyclopedia of Chemical Technology*, 2nd ed., 22 Vols + Suppl., further details on the reproduced data are given under each compound.

Finally, I am especially indebted to my dear wife for her painstaking work on the preparation of the manuscript and reading of proofs.

1974

A. L. Horvath
Thelwall
Warrington

INDEX